PAPERS PRESENTED AT THE
FIFTH EUROPEAN CONFERENCE ON

MIXING

HELD AT
WURZBURG, GERMANY
JUNE 10-12, 1985

Conference sponsored and organised by:

**DVCV· Deutsche Vereinigung
für Chemie- und Verfahrenstechnik
VDI-Gesellschaft Verfahrenstechnik
und Chemieingenieurwesen**

Editor: Jane Stanbury

EDITORIAL NOTES

The Organisers are not responsible for statements or opinions made in the papers.

The papers have been reproduced by offset printing from the authors' original typescripts to minimise delay.

When citing papers from this volume the following reference should be used:-

Title, Author(s), Paper No., Pages, 5th European Conference on Mixing, Würzburg, Germany. Organised by VDI-Gesellschaft Verfahrenstechnik und Chemieingenieurwesen (GVC)

Printed and published by
BHRA, The Fluid Engineering Centre
Cranfield, Bedford MK43 0AJ, England.

© BHRA 1985

ISBN 0 947711 04 X

ACKNOWLEDGEMENTS

The valuable assistance of the Scientific Committee is gratefully
acknowledged.

SCIENTIFIC COMMITTEE

Prof. Mersmann, München/D, chairman
Prof. Couderc, Toulouse/F
Dr. Fort, Praha/CS
Dr. Kipke, Schopfheim/D
Prof. Nienow, Birmingham/GB
Dr. Pawlowski, Leverkusen/D
Prof. Staudinger, Graz/A

5th European Conference on
Mixing
Würzburg, Germany, 10-12 June, 1985

Contents

The Organisers regret that papers missing from the volume of papers were not received in time for printing.

STUDIES ON THREE-PHASE MIXING:
A REVIEW AND RECENT RESULTS

A.W. Nienow, M. Konno* and W. Bujalski

Department of Chemical Engineering
University of Birmingham
P.O. Box 363
Birmingham. B15 2TT

Summary

Firstly, solid suspension and solid-liquid mass transfer, gas dispersion and gas-liquid mass transfer are briefly discussed for two phase systems agitated by Rushton disc turbines. The discussion is then extended for this geometry to three phase systems. It is concluded that, though not the most energy efficient, disc turbines are suitable for three phase agitation and design equations are available for a recommended configuration of D/T = 1/2 with a clearance above the base of 1/4.

Secondly, recently published work on angled blade agitators is extended to include current research data. This work shows that though downward pumping agitators are inherently unstable when gas is sparged into them, this instability is minimised if the agitator has 6 blades, if D/T > ∿0.4 and if ring spargers are used. The power required for ungassed solid suspension is lower than with the other agitators considered. The speed and power required for suspension falls with the introduction of sparging but with increasing sparge rates both rise fairly rapidly. In the upward mode, again a large D/T ratio and 6 blades are most stable and gas dispersion is achieved at the lowest power input when using a large ring sparger (≃D). For this configuration, the speed and power required for solid suspension is almost independent of gassing rate.

*Department of Chemical Engineering, Tohoku University, Aoba, Aramaki, Sendai 980, Japan.

Held at Wurzburg, 10-12 June, 1985.

Organised by DVCV· Deutsche Vereinigung für Chemie- und Verfahrenstechnik
(German Association of Chemical and Process Engineering).

Organisation: GVC·VDI-Gesellschaft Verfahrenstechnik und Chemieingenieurwesen.  GVC

NOMENCLATURE

a = area of gas bubbles/unit volume of dispersion (m^{-1})

C = impeller clearance above the base (m)

D = impeller diameter (m)

$\mathcal{D}$ = diffusivity (m^2/s)

d_p = particle size (m)

Fl_G = gas flow number = Q_G/ND^3 (dimensionless)

Fr = Froude number = N^2D/g (dimensionless)

g = gravitational constant (9.81 m s^{-2})

H = liquid height (m)

k = solid liquid mass transfer coefficient (m s^{-1})

k_La = specific mass transfer coefficient (s^{-1})

N = impeller speed (rev s^{-1})

ΔN_{JS} = $N_{JSg} - N_{JS}$ (rev s^{-1})

N_{JSg} = the agitation speed required to just completely suspend all the particles under gassed conditions (rev s^{-1})

P = power drawn by impeller (W)

Po = power number (dimensionless)

Q_G = gas flow rate (m^3 s^{-1})

Q_{VG} = specific gas flow rate (vvm)

Re = impeller Reynolds number, ND^2/ν (dimensionless)

Re_p = particle Reynolds number, $u_s\,d_p/\nu$ (see Eqn. 2) (dimensionless)

S = suspension parameter in Zwietering equation (1) (dimensionless)

Sc = Schmidt number, $\nu/\mathcal{D}$ (dimensionless)

T = vessel diameter (m)

u_s = particle-fluid slip velocity (m/s)

vvm = (volumetric flow rate of gas/minute)/(volume of liquid)

X = concentration of solid - mass per mass of liquid x 100 (dimensionless)

ε_T = mean energy dissipation rate (W kg^{-1})

$(\varepsilon_T)_{JSg}$ = the mean energy dissipation rate required to just suspend all the particles under gassed conditions (W kg^{-1})

μ = dynamic viscosity (Pa s)

ν = kinematic viscosity ($m^2\ s^{-1}$)

ρ_L = density of liquid (kg m^{-3})

ρ_S = density of solid (kg m^{-3})

$\Delta\rho$ = $\rho_S - \rho_L$ (kg m^{-3})

Subscripts

g = under gassed conditions

CD = condition at which impeller just dispersed gas throughout the vessel

F = condition at which impeller is flooded

JS = at the speed at which solids do not spend more than 1 to 2 seconds on the bottom when observed

1. INTRODUCTION

Both gas-liquid and particle-liquid dispersions in mechanically agitated vessels
have been much researched. However, the suspension of solid particles whilst simultaneously dispersing a gas in a liquid has received less attention, despite numerous
applications in the process industries. These applications include various hydrogenations and oxidations, fermentations, evaporative crystallisation and froth flotation,
etc.

The study of these three phase systems poses special difficulties. There are
still a number of unresolved problems in solid-liquid mixing and even for the simplest
of hydrodynamic parameter, the minimum speed for just complete particle suspension,
N_{JS}, and the associated power per unit mass, $(\varepsilon_T)_{JS}$, there is no universally agreed
correlation. The same situation exists for gas dispersion processes and for the
prediction of N_F, the minimum impeller speed to prevent flooding at a particular
gassing rate, i.e., the transition from the two phase flow being predominantly controlled
by the sparged gas to the condition where the flow is controlled by the stirrer; and
N_{CD}, the minimum impeller speed to disperse gas to all parts of the vessel. These
uncertainties apply even to the most commonly used geometries for gas dispersion and
solid suspension, i.e., Rushton disc turbines and axial downward flow, high pumping
capacity agitators respectively. Thus, three-phase systems are having to be tackled
before the simpler two-phase mixing problems are adequately resolved.

Recent studies of three phase systems accentuate these difficulties. Agitators,
not considered previously for either of the two-phase systems, appear to have excellent
possibilities in three phase systems, e.g., upward-pumping, angled-blade agitators.
Thus both two phase systems have to be investigated with these agitators. Because
three-phase studies have shown that different agitators may offer advantages, there has
been an increased interest in other geometric variations, e.g., sparger types. These
points have been compounded by the recent introduction of agitators with special hydrodynamic features for two-phase flows, e.g., the aerofoil types for solid suspension and
disc turbines with more than six blades for gas dispersion.

To account for the above difficulties, therefore, this paper is divided into two
parts. Firstly, work on the well-tested Rushton turbine is reviewed and design
relationships are presented for both two and three phase systems. Secondly, a limited
amount of new data is introduced for two other geometries which shows that they are
superior for three phase systems under certain conditions. The review is not intended
to be fully comprehensive but is built up from a coherent set of papers over about 16
years. These papers give fuller details of the experimental techniques used.

2. A SYSTEM FOR WHICH DESIGN EQUATIONS ARE AVAILABLE: THE DISC TURBINE

2.1 Two Phase: Particle Liquid.

2.1.1 Particle suspension. Increases in agitator speed significantly enhance solid-
liquid rate processes up to the point where all the solid surface is available to
participate. This speed is N_{JS}. The Zwietering correlation (Ref. 1) is generally
accepted as the best for determining it. At N_{JS}, the homogeneity of the suspension may
be very good (for fine, low density particles) or very poor (large, high density
particles) (Ref. 2). Also at N_{JS}, the desired solids residence time distribution may not
be obtained (Ref. 2). The correlation is

$$N_{JS} = S \nu^{0.1} \, d_p^{0.2} \, (g\Delta\rho/\rho_L)^{0.45} \, X^{0.13} \, D^{-0.85} \qquad \dots\dots(1)$$

Whilst the Zwietering equation is best for predicting N_{JS} for small scales of
operation (say for vessels up to about 0.7m diameter), it tends to underpredict the
power required on the larger scale. From Eqn. 1, it can be shown (Ref. 2) that the
specific energy dissipation rate $(\varepsilon_T)_{JS} \, \alpha \, D^{-0.55}$ on geometrically-similar scale up.
However, experimental work up to $\sim$2m diameter suggest the relationship (Ref. 3)
$(\varepsilon_T)_{JS} \, \alpha \, D^{-0.28}$. Eqn. 1 can be applied for systems with a size distribution if the mass
mean is used (Ref. 3).

S is a geometric parameter very dependent on system geometry. Values are
available for disc-turbines (Ref. 3), paddles (Ref. 2) and angled-blade agitators
(Ref. 3). Good agreement between   different workers (Ref. 3) has been found for
disc turbines of different size and different clearance above the base for a square-
cylinder geometry with 4 baffles. Large D/T disc turbines of low clearance above the

base (about T/5) are most energy efficient for this type. However, disc turbines
require relatively high power inputs compared to axial flow types.

2.1.2 Solid-liquid mass transfer. Two models have mainly been used; the terminal
velocity-slip velocity theory and Kolmogoroff's theory of isotropic turbulence. It has
been argued in some detail that the slip velocity theory is preferable (Ref. 4). For
this case, the mass transfer coefficient, k_{JS}, at $N=N_{JS}$ can be calculated for any
agitator type from

$$Sh = 2 + 0.72 \ Re_p^{1/2} \ Sc^{1/3} \qquad \ldots\ldots (2)$$

where $Re_p = u_s \ d_p/\nu$. u_s is the slip velocity and, at N_{JS}, it can be estimated from the
still fluid terminal velocity (Ref. 4). The small increase in mass transfer coefficient,
k, at $N > N_{JS}$ is rarely worth the increased energy costs since $k \ \alpha \ N^{0.5}$ whilst $\varepsilon_T \ \alpha \ N^3$.

2.2   Two Phase: Gas-Liquid.

2.2.1 Gas-dispersion and power drawn. Fig. 1a to 1c shows successive flow regimes with
increases in agitator speed (Ref. 5), N: From N=O to $N=N_F$ (1a), a flow pattern
dominated by the gas-liquid flow up the middle, i.e., the impeller is flooded; from
$N_F < N < N_{CD}$ (1b), a flow pattern dominated by the agitator, horizontally out from the
agitator; for $N > N_{CD}$ (1c), a two-phase dispersion of gas to all parts. Increases in
gassing rate Q_G at constant N lead to the opposite sequence of events.

Alongside the changes in flow patterns indicated in Fig. 1, there are changes
in gas-filled cavity structure behind the agitator blades. Increases in Q_G from
$Q_G=O$ at constant N up to the transition from Fig. 1b to Fig. 1a generally lead to the
following changes of cavity type, Fig. 2:
2a) 6 vortex cavities (Ref. 6,7); 2b) 6 clinging cavities (Ref. 6,7); 2c) a "3-3"
combination of alternate large and small cavities (Ref. 8); 2d) when the transition
from Fig. 1b to Fig. 1a occurs, 6 equi-sized ragged cavities form (Ref. 9).

As the cavities change from Fig. 2a to 2c at constant N, the power drawn falls.
However, a step increase typically occurs at the point at which the transition to
6 ragged cavities takes place, Fig. 3. For point spargers and small ring spargers,
the gas flow rate which causes flooding is virtually independent of liquid height and
whether the liquid is coalescing or not (Ref. 10). No particular change in cavity
structure is seen at N_{CD}.

However, if the plot of Po_g versus Fl_G (= Q_G/ND^3) is made at constant Q_G and
increasing N, a minimum is found at N_{CD}, Fig. 4. In this case, the minimum is somewhat
sensitive to the degree of bubble coalescence and disappears at a high impeller
clearance off the base. It becomes shallower at higher gassing rates.

An impeller clearance, C, of 1/4 is recommended (Ref. 5) as this allows dispersion
into the lower part of the vessel at a reasonably low power input whilst still maintaining a satisfactory level of agitation in the upper part of the vessel. For this
configuration, N_F can be calculated (Ref. 10) from

$$(Fl_G)_F = 3O \ (D/T)^{3.5} \ (Fr)_F \qquad \ldots\ldots (3)$$

and N_{CD} from (Ref. 10)

$$(Fl_G)_{CD} = O.2(D/T)^{0.5} \ (Fr)_{CD}^{0.5} \qquad \ldots\ldots (4)$$

The power drawn by the agitator at different aeration rates can be estimated from
Fig. 5 based on data from a wide range of experimental conditions and vessel sizes
provided $N > N_{CD}$ (Ref. 5). However, on the large scale at typical gassing rates, N
tends towards N_F and $P_g \simeq O.5 \ P$, with a weak dependence on Q_G.

2.2.2 Gas-liquid mass transfer. Mass transfer coefficients expressed as k_La are system
specific. Traces of materials may considerably enhance k_La or reduce it. In general,
soluble substances reduce k_L from that of the pure liquid by collecting at the interface
but at the same time they repress bubble coalescence, thereby reducing the bubble size,
directly increasing "a" as a result of this and of the increased hold-up. In addition,
absolute values of k_La depend very critically on the method of measurement and on the

assumption made for the gas-phase mixing model when unsteady state techniques are used
(Ref. 11). However, on balance, it does appear that for a given gas and liquid and the
same measuring technique, $k_L a$ is independent of agitator type and only dependent on
gassing rate and power input (provided the gas is dispersed, i.e., $N > N_F$). Indeed,
it has been suggested (Ref. 12) that operation at $N = N_F$ is an optinum in that the
mass flux per unit of power input (both from gas sparging and the impeller) is a
maximum at this point.

2.3 Three-Phase Systems.

2.3.1 Gas-dispersion and power drawn.

Suspension in two-phase systemsis generally
improved by reducing the impeller clearance above the base. However, in three phase
systems, this may lead to instabilities (Ref. 13) as indicated in Figs. 6 and 7. The
change in flow pattern at a speed $N_C > N_{CD}$ from a double loop to a single loop is
accompanied by a drop in power number (or Po_g). However, at $C=T/4$, no instabilities
are found as shown in Fig. 7. Thus, this clearance is recommended, as it is also
reasonably good for solid suspension.

Solid suspension (Ref. 2) and gas dispersion (Ref. 5) are both achieved at
lower specific power inputs with large D/T ratio disc turbines. Therefore, it is to
these that this section mainly refers. The effect of solids on power number (Ref. 14),
Po_g, is shown in Fig. 8. It shows that, at 0.5 vvm,* concentration has only a small
effect on Po_g provided that the particles are suspended, i.e., for $N > N_{JSg}$. However,
at $N < N_{JSg}$, the concentration of particles in the system has a very significant effect
on the gassed power number. An explanation for this phenomenon is the false base
formed by the particles effectively reducing impeller clearance at low speeds. Fig. 8
also shows the position of the minimum in the Po_g versus Fl_G plot which marks N_{CD}.
For low concentrations (X < 15%), N_{CD} is independent of concentration. However, at
the higher concentrations (X > 15%), N_{CD} is achieved at lower speeds, i.e., higher
Fl_G. In these cases, at N_{CD}, the particle bed is so close to the impeller that gas is
easily dispersed and the overall liquid flow pattern tends to the single flow loop
normally found with low clearances and their associated lower power numbers, Fig. 6.

As particle bed heights diminished with increasing speed, so all the power
number versus flow number curves converge in spite of the fact that Po_g is based on
the liquid density, i.e., the effect of solids on Po_g (and Po) is considerably less
than might be expected from the increase in mean density. It can also be inferred
that the presence of solids has little effect on gas cavity formation at these
concentrations.

At very high solids concentration (>40% by weight or > 30% by volume glass
Ballotini), the power drawn is independent of the gassing rate (Ref. 15) as it is in
high viscosity viscoelastic fluids (Ref. 16). The remainder of this paper is only
applicable to solids concentration less than this. However, for these lower concentrations, Eqn. 3 and 4 and Fig. 5 still apply.

2.3.2 Gas-liquid mass transfer.

Solid particles may affect bubble size (and therefore
surface area and hold-up). They may also affect interfacial resistance to mass
transfer. Increases in effective viscosity should lead to increased bubble sizes and
reduced "a" and also possibly k_L; small particles may collect at bubble surfaces
reducing coalescence and enhancing "a" but increasing resistance and thus reduce k_L;
large particles may disrupt bubbles thereby enhancing coalescence (thus reducing "a")
whilst increasing k_L. The effect will depend on the wettability of the solids and
other characteristics, e.g., porosity. Recent works have reported large increases in
$k_L a$ (Ref. 17); no significant change in hold-up or $k_L a$ (Ref. 18); and reductions (up
to about 40%) in $k_L a$ (Ref. 19). There is much more work to be done. However, it is
clearly desirable that a method should be used for determining $k_L a$ which eliminates
the need to make assumptions about the gas phase mixing (Ref. 11).

2.3.3 Solid-suspension.

Experiments by Chapman et al. (Ref. 14) covered a wide range
of solids size, density and concentration. They showed that the minimum suspension
speed under gassed conditions, N_{JSg}, is related to them in a very similar way to that

*vvm is the specific gassing rate in (volumetric flow rate of gas/minute)/volume of
liquid in the vessel. It is expressed here in this way for two reasons. Firstly, it
compensates for the effect of scale of equipment. Secondly, values of 1/4 to about
2 vvm are commonly used particularly in fermentation and are easy to remember.

implied by Eqn. 1. However, density is slightly less important, i.e.,

$$N_{JS} \propto \Delta\rho^{0.22}$$

The introduction of gas into the impeller causes a decrease in the impeller pumping capacity and therefore in the liquid flow at all points in the vessel including the base. Thus, when this happens, some solids are deposited on the base and an increase in speed, ΔN_{JS} (= N_{JSg} - N_{JS}), is required to resuspend them. For particles of density from 1200 to 2900 kg/m^3 and of size from 80μm to 2800μm at concentrations up to 30% by weight in liquids from 1 to 5 m Pas. Fig. 9 shows how ΔN_{JS} is related to gassing rate in vessels from 0.29 to 1.8m diameter, i.e., N_{JSg} = 0.94 Q_{VG}, where N_{JSg} is in rev/s and Q_{VG} is in vvm (Ref. 14). For the solids noted above, and the gassing rate indicated in Fig. 9, $N_{JSg} > N_{CD}$.

2.3.4 Solid-liquid mass transfer. Nothing has been reported on this. The use of Eqn. 2, the terminal velocity and the enhancement factor suggested for solid-liquid systems is recommended to be applied at N = N_{JSg}. However, it may overestimate k_{JSg} since it has been shown (Ref. 20) that in solid-liquid systems highly aerated due to surface aeration, the mass-transfer coefficient is less than it is expected to be, probably due to sort of "vapour-blanketting".

3.   IMPELLER-VESSEL GEOMETRIES FOR WHICH DESIGNS CANNOT BE PRESENTLY FORMULATED

3.1 Solid-Liquid System.

There is a reasonable amount of information on solid suspension with 45° angled blade agitators pumping downwards. These are considerably more energy efficient than disc turbines. In baffled right cylindrical vessels, 6 blades are more efficient than 4. Contouring the base to reduce dead zones considerably reduces the speed and power required for particle suspension. Positioning axial flow agitators in draft tubes (if done without causing constriction to the flow) reduces the speed required to suspend particles. More measurements of power need to be made. Conical bases are very unsuitable (Ref. 2).

Our recent work (Ref. 23) on upward pumping angled blade agitators of D/T = 1/2 and with 6 blades shows that they can be more energy efficient for particle suspension than disc turbines (see Table 1).

3.2 Gas-Liquid Systems

Previously published work has shown that increasing the number of blades on the disc reduces the rate at which the gassed power falls with increasing gassing rate at a constant impeller speed and a higher gas rate is needed before flooding occurs. The same can be done by making the 6 blades semi-circular, concave-forwards in cross-section (Ref. 21). Sparger design is relatively unimportant unless of diameter greater than the impeller (Ref. 9, 22).

A downward pumping impeller at a clearance of T/4 produces a strong recirculation in ungassed systems from the impeller tip and back. However, there is also a weaker flow which is important for gas dispersion (Fig. 10). Clearly the interaction between the sparged gas and the flow pattern depends very much on whether sparging is by pipe sparger into the centre, i.e., into an upward flow; or into the outer region by ring sparger, i.e., into a downward flow. In general, the overall flow is downwards near the base and therefore there is a distinct chance for instabilities to occur.

Instabilities are most marked for central pipe sparging into small D/T ratios (1/3rd or less) with only 4 blades. Fig. 10 shows the flow patterns and Fig. 11, the change of gassed power with increasing agitator speed (Ref. 13). However, if a D/T of 0.4 to 0.5 with 6 blades and a ring sparger is used, then the dispersion is much more stable (Ref. 23), see Fig. 12. Nevertheless, at low agitator speeds or high sparging rates, the gas enters the impeller region directly (direct loading) whilst at high impeller speeds or low gassing rate, the gas is flushed away from the impeller and it only enters it after some recirculation (indirect loading) (Ref. 24). Thus these effects plus the natural opposition of impeller pumping direction to gas-liquid flow inherently leads to either gross instabilities of flow pattern which give large, low frequency torque oscillations; or alternatively, to smaller (but greater than with other agitators) high frequency torque fluctuations (see Table 2).

Upward pumping impellers are inherently more stable and again large D/T ratios with 6 blades are best. Fig. 13 gives an example (Ref. 23). The insensitivity to gassing rate should also be noted (Ref. 13, 23). Again the sparger choice makes a considerable difference and a sparger of the order of size of the impeller allows gas dispersion at the lowest power inputs (Ref. 23) (see Table 1) and with low torque fluctuations (see Table 2). It also leads to higher hold up at a particular gassing rate and power input.

The gas filled cavity structure goes through similar changes with angled blade agitator to those found with disc turbines. Fig. 14 shows examples that are equivalent to clinging/vortex cavities (depending on the relative sizes) and large cavities. However, in the latter case, it should be noted that all cavities are of the same size. Ragged cavities are not usually observed.

3.3 Three Phase Systems

Downward pumping agitators with six blades suspend solids at lower energy dissipation rates than other types at low gassing rates (see Table 1). In terms of power demands, both large and small are somewhat similar but the extreme sensitivity of small agitators (D/T $\leqslant \sim 1/3$) to gassing rates which leads to a catastrophic loss of solids suspension makes them unsuitable. However, large D/T ratios (0.4-0.5) do not have gross instabilities and particle suspension efficiency is good. The change in N_{JSg} and $(\varepsilon_T)_{JSg}$ is shown in Fig. 15. Surprisingly, small amounts of gas introduced by ring sparger lead to a reduction in N_{JSg}.

Upward pumping agitators, D/T = 1/2 are extremely stable in gas dispersion (see Tables 1 and 2). In addition both agitator speed and power drawn are for suspension, almost independent of gassing rate, Fig. 16.

It is hoped to produce design relationships for these agitators, Intermigs and other novel agitators in the present research programme.

4. CONCLUSIONS

Much work has been done on Rushton disc turbines and a set of reasonably accurate design relationship are available for two and three phase systems.

There are considerable advantages in using angled blade agitators for gas and three phase dispersions in that less power or greater stability may be achieved. However, the interactions with sparger type is much more complex than with disc turbines and more work is necessary before optimum geometries have been determined and design relationships are available.

No work is available on the effect of the presence of gas on solid-liquid mass transfer; the effect of solids on gas-liquid mass transfer has shown the problem to be complex and directly conflicting results have been reported. More work is required.

5. ACKNOWLEDGEMENTS

Thanks are due to the S.E.R.C. for support for this work over many years and especially for financial support for M.K. and W.B.

6.   REFERENCES

!)   Zwietering, T.N.: "Suspending of Solid Particles in Liquids by Agitators", Chemical Engineering Science, 8, 1958, 224-232.

2)   Nienow, A.W.: "Suspension of Solids in Liquids" in "Mixing in the Process Industries" (Editors, N. Harnby, M.F. Edwards, A.W. Nienow), London, Butterworths, Chapter 16, 1985 (to be published).

3)   Chapman, C.M., Nienow, A.W., Cooke, M. and Middleton, J.C.: "Particle-Gas-Liquid Mixing in Stirred Vessels: Part I - Particle-Liquid Mixing", Chemical Engineering Research and Design, 61, 1983, 71-81.

4)   Nienow, A.W.: "The Mixer as a Reactor - Solid-Liquid Systems", in "Mixing in the Process Industries" (Editors, N. Harnby, M.F. Edwards, A.W. Nienow), London, Butterworths, Chapter 18, 1985 (to be published).

5)   Nienow, A.W., Wisdom, D.J. and Middleton, J.C.: "The Effect of Scale and Geometry on Flooding, Recirculation and Power in Gassed, Stirred Vessels", Proc. of the 2nd European Conference on Mixing, Cambridge, April 1977, Cranfield, BHRA, 1978, pp. F1-1 to F1-16 and X52 to X55.

6)   Riet, K. van't and Smith, J.M.: "The Behaviour of Gas-Liquid Mixtures near Rushton Turbine Blades", Chemical Engineering Science, 28, 1973, 1031-1037.

7)   Nienow, A.W. and Wisdom, D.J.: "Flow over Disc Turbine Blades", Chemical Engineering Science, 29, 1974, 1994-1996.

8)   Warmoeskerken, M.M.C.G., Feijen, J. and Smith, J.M.: "Hydrodynamics and Power Consumption in Stirred Gas-Liquid Dispersions", Institution of Chemical Engineers Symposium Series, No.64, "Fluid Mixing", 1981, pp.J1-J14.

9)   Nienow, A.W., Allsford, K.V. and Kuboi, R.: "Using a De-Rotational Prism for Studying Gas Dispersion Processes", Paper to 9th Engineering Foundation Conference on Mixing, Henniker, New Hampshire, USA, 1983 (unpublished).

10) Nienow, A.W., Warmoeskerken, M.M.C.G., Smith, J.M. and Konno, M.: "On the Flooding/Loading Transition and the Complete Dispersal Condition in Aerated Vessels Agitated by a Rushton-Turbine", Proceedings of the 5th European Conference on Mixing, Wurtzburg, F.D.R., June, 1985; Cranfield, BHRA, 1985.

11) Chapman, C.M., Gibilaro, L.G. and Nienow, A.W.: "A Dynamic Response Technique for the Estimation of Gas-Liquid Mass Transfer Coefficients in a Stirred Vessel", Chemical Engineering Science, 37, 1982, 891-896.

12) Pollard, G.: "Flooding and Aeration Efficiency in Stirred Vessels", Proc. of International Conference on Mixing, Mons., Belgium, 1978; Faculty Polytechnique de Mons, March 1978, pp. C4-1 to C4-16.

13) Chapman, C.M., Nienow, A.W., Cooke, M. and Middleton, J.C.: "Particle-Gas-Liquid Mixing in Stirred Vessels: Part II - Gas-Liquid Mixing", Chemical Engineering Research and Design, 61, 1983, 82-95.

14) Chapman, C.M., Nienow, A.W., Cooke, M. and Middleton, J.C.: "Particle-Gas-Liquid Mixing in Stirred Vessels: Part III - Three Phase Mixing", Chemical Engineering Research and Design, 61, 1983, 167-181.

15) Greaves, M. and Loh, V.Y.: "Power Consumption Effects in Three Phase Mixing", Institution of Chemical Engineers Symposium Series, No.89, "Fluid Mixing II", 1984, pp. 69-96.

16) Nienow, A.W. and Ulbrecht, J.J.: "Gas-Liquid Mixing in High Viscosity Systems" in "Mixing of Liquids by Mechanical Agitation", (Editors, J.J. Ulbrecht and G.K. Patterson), New York, Gordons and Breach, Chapter 7, 1985 (to be published).

17) Brehm, A., Ledakowicz, S. and Kokunn, R.: "Effect of Suspended Inert Solid Particles on Gas-Liquid Mass Transfer in Stirred Vessels", Paper V.3.54 at CHISA Conference, Prague, September 1985 (unpublished),

18) Chapman, C.M., Nienow, A.W., Cooke, M. and Middleton, J.C.: "Particle-Gas-Liquid Mixing in Stirred Vessels* Part IV - Mass Transfer and Final Conclusions", Chemical Engineering Research and Design, 61, 1983, 182-185.

19) Frijlink, J.J., de Jong, P.G.T. and Smith, J.M.: "Properties of Gas-Liquid Dispersions with Suspended Solids", Paper V.3.53 at CHISA Conference, Prague, September, 1985 (unpublished).

20) Nienow, A.W.: "Dissolution Mass Transfer in a Turbine Agitated Baffled Vessel", Canadian Journal of Chemical Engineering, 47, 1969, 248.

21) Riet, K. van't, Boom, J.M. and Smith, J.M.: "Power Consumption, Impeller Coalescence, and Recirculation in Aerated Vessels", Transactions of the Institution of Chemical Engineers, 54, 1976, 124-131.

22) Nienow, A.W. and Allsford, K.V.: "The Effect of Sparger Type on Gas Dispersion by Rushton Turbines", Paper to Institution of Chemical Engineers "Multistream", Symposium, April 1985, to be published.

23) Konno, M., Bujalski, W. and Nienow, A.W.: "Gas Dispersion Using Angled Blade Agitators and a Variety of Spargers", to be published.

24) Warmoeskerken, M.M.C.G., Speur, J. and Smith, J.M.: "Gas-Liquid Dispersion with Pitched Blade Turbines", Chemical Engineering Communications", 25, 1984, 11-29.

Table 1 Date for Water-Air - 0.1% Lead Glass Ballotini (440-530μm)
(T = 0.45m; D = T/2; C = T/4)

Q_{GV}		Disc Turbine		6 Blade 45°(Down)		6 Blade 45°(Up)	
		SRS	LRS	SRS	LRS	SRS	LRS
0.0	N_{JS}	2.9	2.9	3.7	3.7	3.9	3.9
	$(\epsilon_T)_{JS}$	1.12	1.12	0.67	0.68	0.77	0.78
0.3	N_F	0.7	0.7	1.1	0.67	1.3	0.92
	$(\epsilon_T)_F$	0.014	0.013	0.02	0.005	0.03	0.011
	N_{CD}	1.8	1.8	1.8	1.7	2.3	1.7
	$(\epsilon_T)_{CD}$	0.23	0.25	0.066	0.058	0.16	0.078
	N_{JSg}	3.4	3.3	3.5	3.4	4.2	4.1
	$(\epsilon_T)_{JSg}$	1.6	1.5	0.53	0.47	0.93	0.87
1.0	N_F	1.0	1.0	1.3	0.92	1.6	1.1
	$(\epsilon_T)_F$	0.032	0.033	0.035	0.02	0.04	0.017
	N_{CD}	2.3	2.2	2.5	2.5	2.6	1.8
	$(\epsilon_T)_{CD}$	0.31	0.30	0.15	0.15	0.23	0.096
	N_{JSg}	3.9	3.9	4.7	4.4	4.4	4.3
	$(\epsilon_T)_{JSg}$	1.59	1.72	0.77	0.70	0.93	0.87
3.5	N_F	1.67	1.8	1.6	1.6	2.1	1.4
	$(\epsilon_T)_F$	0.12	0.14	0.091	0.095	0.10	0.03
	N_{CD}	2.5	2.5	4.9	4.6	2.6	2.2
	$(\epsilon_T)_{CD}$	0.32	0.31	0.63	0.55	.21	0.15
	N_{JSg}	5.1	5.1	7.3	6.9	4.8	4.7
	$(\epsilon_T)_{JSg}$	2.09	2.08	1.47	1.40	0.87	0.84

SRS - small ring sparger, 0.54D LRS - large ring sparger, 0.85D

Table 2 The percentage torque fluctuations (expressed as power) compared
to its time averaged value (T = 0.45m; D = T/2; C = T/4)

Gas flow rate (vvm)	Disc Turbine		6 Blade 45° (down)		6 Blade 45° (up)	
	0.2 W/kg	0.5 W/kg	0.2 W/kg	0.5 W/kg	0.2 W/kg	0.5 W/kg
0.28	9	8	20	16	14	9
1.0	12	11	35	51	17	12
3.5	9	10	22	20	15	10

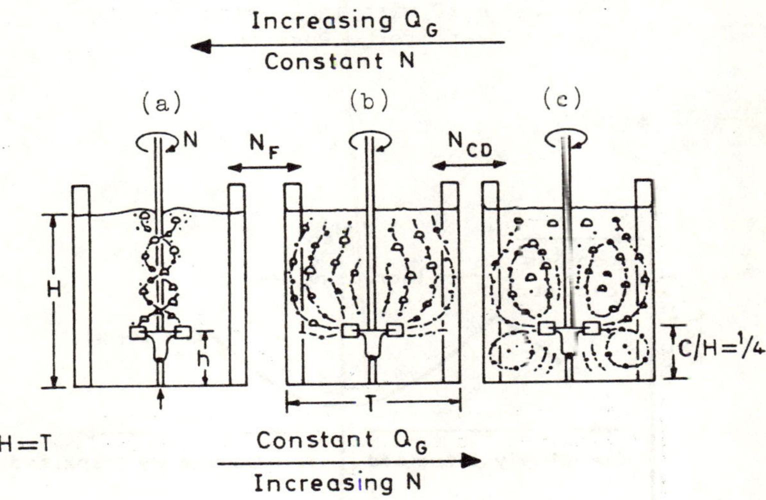

Fig. 1 The flooding-loading-complete
dispersion transitions for a
Rushton turbine

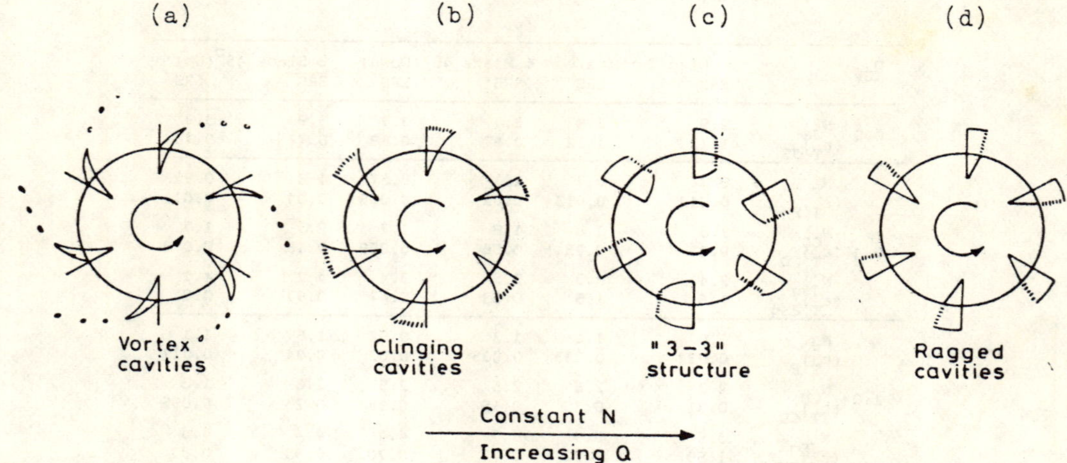

Fig. 2 Change in cavity structure at increasing gassing rate for a Rushton turbine

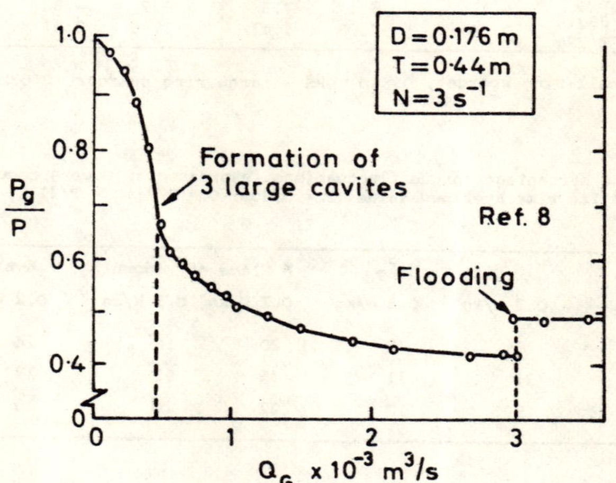

Fig. 3 P_g/P with increasing gassing rate for a Rushton turbine

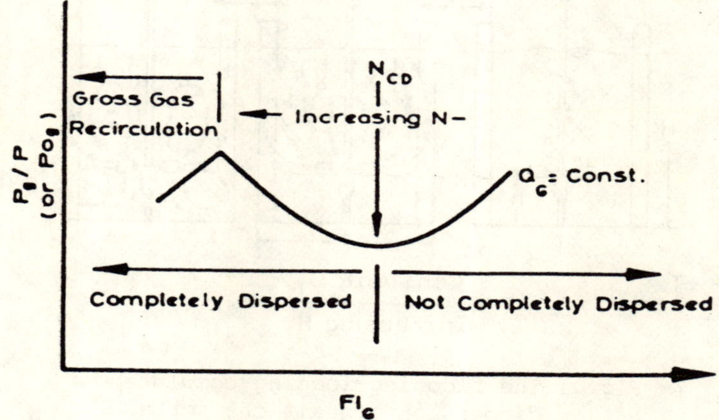

Fig. 4 P_g/P with increasing speed for a Rushton turbine (C/T=1/4)

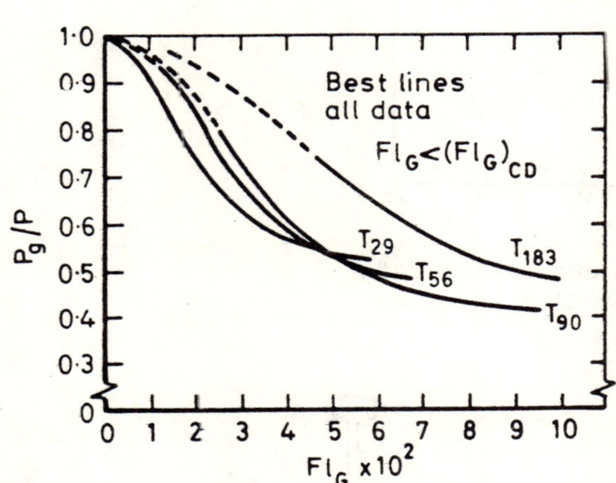

Fig. 5 Effect of scale on P_g/P for a Rushton turbine

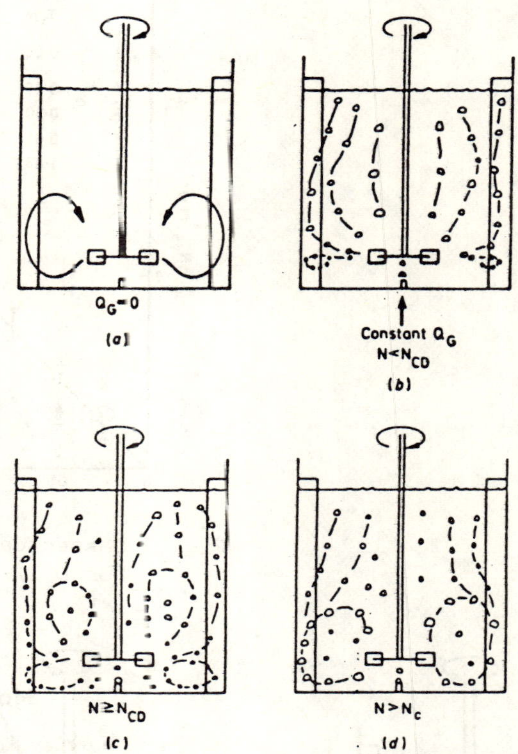

Fig. 6 Flow pattern changes for T/3 Rushton turbine at low clearance (C/T=1/6)

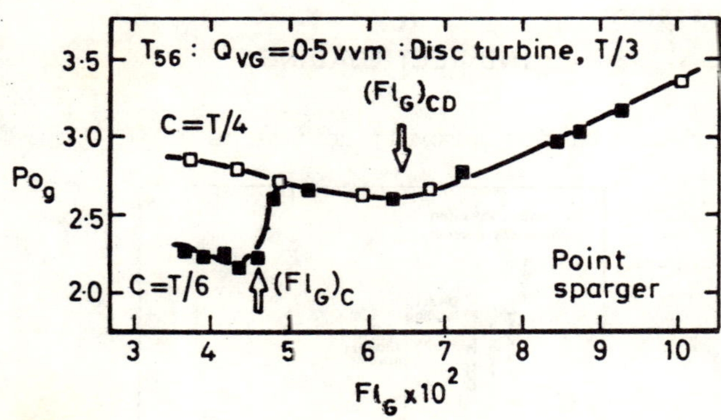

Fig. 7 Effect of clearance on gassed power number, Po_g

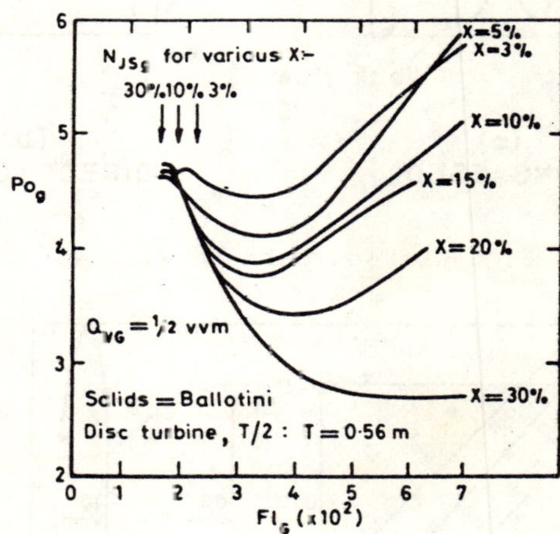

Fig. 8 Effect of particle concentration on gassed power number for a Rushton turbine

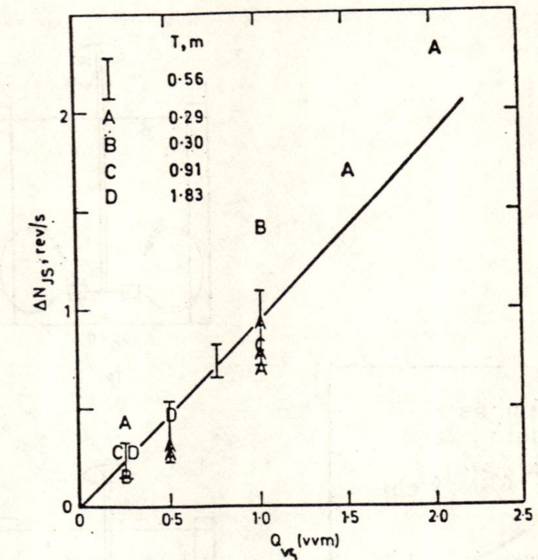

Fig. 9 Gassed particle suspension data
for D=T/2, C=T/4 disc turbines

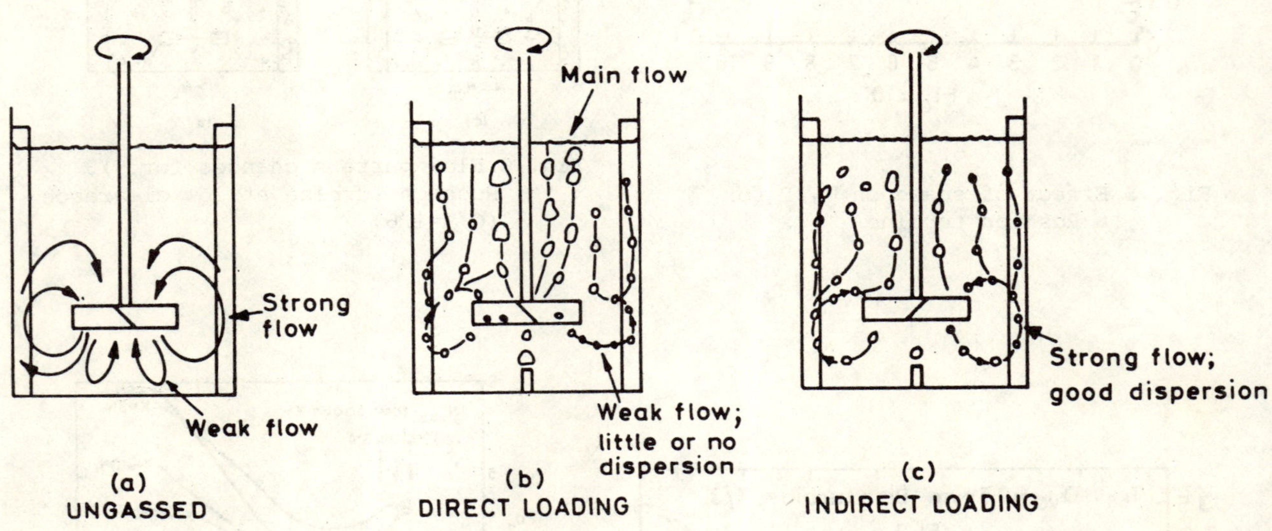

Fig. 10 Flow pattern with 45° pitch blade agitators pumping downwards

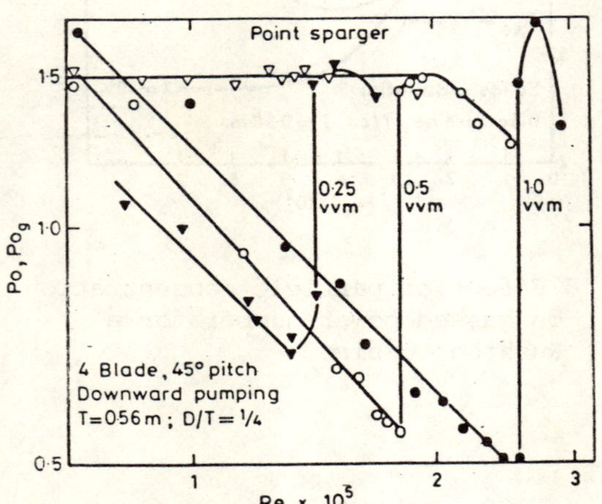

Fig. 11 Gross instabilities with T/4,
4 blade 45° pitched blade
agitators pumping downwards

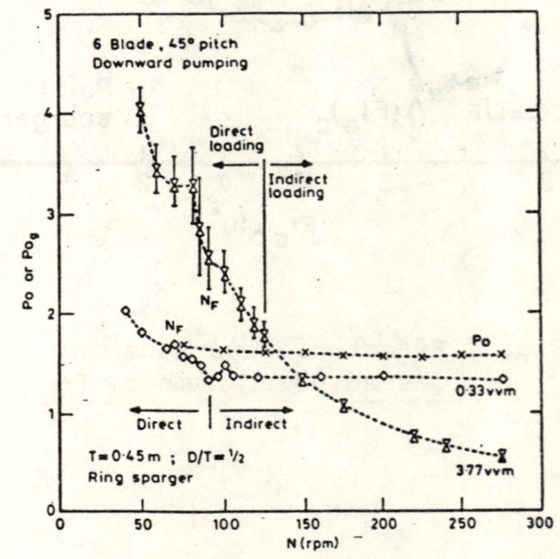

Fig. 12 Torque instabilities with T/2,
6 blade 45° pitched blade
agitators pumping downwards

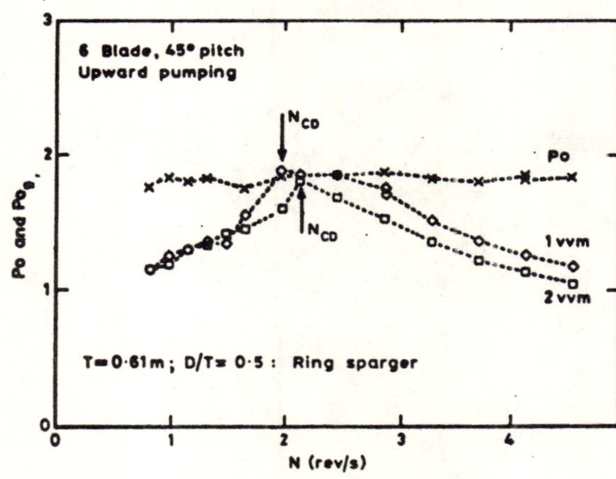

Fig. 13 Small effect of gassing rate on Po_g with T/2,
6 blade 45° agitators pumping upwards

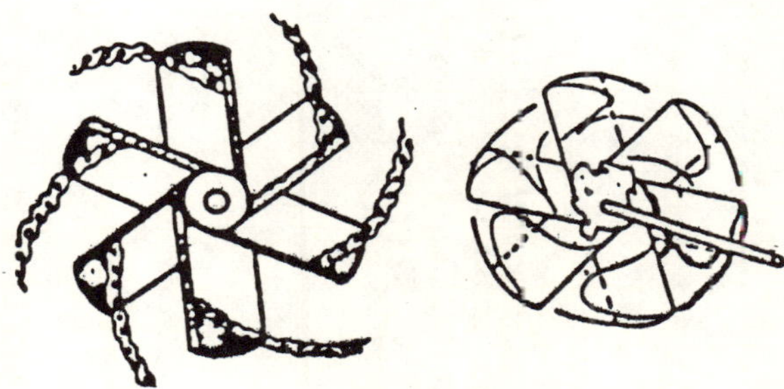

Fig. 14 Clinging and large cavities on a 45° pitch blade
agitator pumping upwards as seen from below

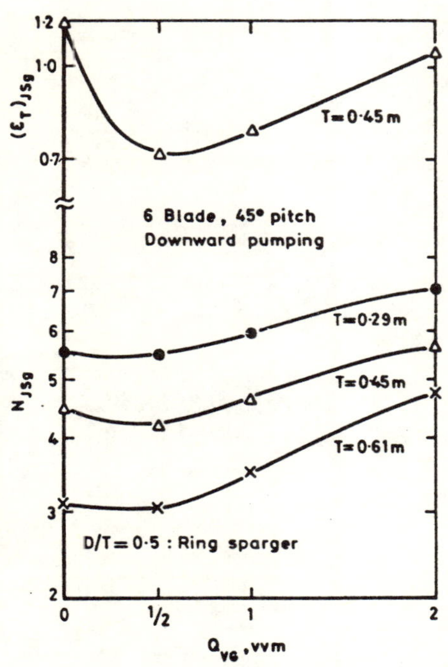

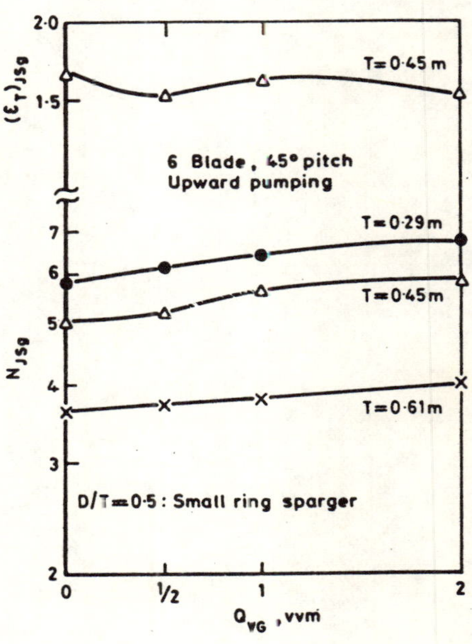

Fig. 15 Effect of gassing rate on N_{JSg} and
$(\varepsilon_T)_{JSg}$ with a downward pumping
agitator (1% 480 µm Ballotini)

Fig. 16 Effect of gassing rate on N_{JSg} and
$(\varepsilon_T)_{JSg}$ with an upward pumping
agitator (1% 480 µm Ballotini)

SCALING-UP RULES
FOR SOLIDS SUSPENSION
IN STIRRED VESSELS

C. Buurman, G. Resoort and A. Plaschkes

Koninklijke/Shell-Laboratorium, Amsterdam
(Shell Research B.V.)
Badhuisweg 3
1031 CM Amsterdam
The Netherlands

Summary

Although extensively investigated, the scaling-up of stirred vessels with
respect to solids suspension is still a difficult matter. In view of this we have
carried out a series of model suspension experiments in a large stirred vessel (4.3 m
in diameter). Complete and homogeneous suspensions have been studied and the tests
have been carefully repeated at the laboratory in a small-scale vessel under identical
conditions. From a comparison of the results it became clear that the scaling criterion for complete suspension is a constant specific power input. Homogeneity of the
suspension is better on the large scale than on the laboratory scale. However, absolute homogeneity is unattainable in practice. Therefore the position of the outlet in
a continuously-stirred tank reactor can have an important effect on the solids hold-up
in reactors.

The thickness of the stirrer blades has an important influence on the mixing performance, hence the blade thickness of a stirrer in model experiments must be taken into
account. Incorrect scaling of the blade thickness can easily cause too optimistic
figures for the power consumption on a commercial scale. The wide variety in scaling
rules for suspensions in the literature can be attributed to incorrect scaling of the
stirrer blade thickness. Moreover, this effect can be used advantageously for the
design of impeller blades. In our 80-l vessel a series of measurements was carried out
using stirrer blades with a modified profile. Simple skewing off appeared to be an
effective method to improve the performance of the stirrer or to maintain a certain
suspension quality at a lower power consumption. Savings of 17 % in power consumption
can easily be achieved in this way.

Held at Wurzburg, 10-12 June, 1985.

Organised by DVCV· Deutsche Vereinigung für Chemie- und Verfahrenstechnik
(German Association of Chemical and Process Engineering).

Organisation: GVC·VDI-Gesellschaft Verfahrenstechnik und Chemieingenieu-wesen. 

<u>NOMENCLATURE</u>

b	=	stirrer blade thickness,	m
c	=	solids concentration	
D	=	impeller diameter,	m
d_p	=	particle size,	m
$E(k,t)$	=	energy per wave number interval,	m^3/s^2
g	=	acceleration due to gravity,	m/s^2
H	=	liquid height,	m
k	=	wave number,	m^{-1}
n	=	stirrer speed,	s^{-1} unless indicated otherwise
n_c	=	stirrer speed for complete suspension,	s^{-1} unless indicated otherwise
T	=	vessel diameter,	m
v_e	=	fluctuating velocity component,	m/s
Z	=	height of homogeneous zone,	m
ε	=	specific power dissipation,	W/kg
ρ	=	liquid density,	kg/m^3
$\Delta\rho$	=	density difference,	kg/m^3

1. INTRODUCTION

In the process industry, a requirement frequently set is that suspensions of solid particles in agitated vessels should be complete and/or homogeneous. Many investigators have devoted considerable effort to finding reliable methods for predicting the stirring conditions required for given degrees of suspension. The equations which have been proposed in the literature still show a confusing variety, not only because of the numerous types of stirrers employed, but also because of the difference in the importance attached to the various physical and geometrical parameters (densities, viscosity, particle size, concentration, size of the stirrer and scale of the equipment.

Many of the uncertainties can be overcome by carrying out a small-scale experiment with representative materials in a geometrically scaled-down version of the mixing vessel. This approach reduces the problem of designing a large-scale installation to one of scaling rule. Even then, however, designers are faced with a variety of scaling rules in the literature which can easily result in, for example, a difference of a factor of two in the size of the motor selected for scaling-up from a 10-1 experiment to a mixing vessel of 50 m^3.

A few years ago Koninklijke/Shell-Laboratorium, Amsterdam (KSLA) was given the opportunity to carry out a limited series of experiments in a redundant mixing vessel of more than 50 m^3 in a chemical plant. The stirrer of this vessel was equipped with a speed variator (range 30 to 120 rev/min), which rendered the installation particularly suitable for suspension experiments. By carefully repeating such experiments on small-scale (80-1) laboratory equipment we hoped to find a reliable relation for the scaling rules for future design purposes.

2. EXPERIMENTS ON LARGE AND SMALL SCALE

2.1 General.

The large-scale reactor vessel was 4.26 m in diameter and 10 m high and had an elliptical bottom (see Fig. 1). It was equipped with four 0.426-m baffles. For our experiments we replaced the original impeller with a downward-pumping axial turbine impeller 1.72 m in diameter (0.4 T; see Fig. 2). For the concentration measurements three 1" sample tubes were installed at different heights, with flushing arrangements for tap water.

The properties of the solids had to be such that complete and homogeneous suspension could be expected well within the available speed range of the stirrer. Commercially available sand with a narrow particle size distribution and an average particle size of 157 μm, in combination with water, was found suitable for our purpose. The maximum solids concentration was limited to 15 %(v) since we could not take the risk of the impeller becoming buried in the settled sand and then not being able to start the stirrer again.

Further, the experiments were restricted to the one-stirrer configuration and the suspension height in the vessel was kept constant at 4.3 m (H = T).

The small-scale laboratory set-up was an accurately scaled-down version of the large installation, with a vessel diameter of 0.48 m. In the small installation we could easily incorporate a few extra sample points over the height of the vessel.

2.2 Detection of complete suspension.

In laboratory experiments transparent mixing vessels can be used and the assessment of whether the criterion for complete suspension (no particles longer than 1 s in rest at the bottom) has been met can be made visually. As our large-scale reactor was constructed from stainless steel an instrumental technique had to be found for this purpose. In the laboratory we tested an ultrasonic Doppler velocity meter for suitability. With this instrument we were able to observe a sharp transition between a stationary and a moving sand layer approximately 1 cm from the bottom of a vessel. In the large-scale mixing vessel the transition to complete suspension was distinct (see Fig. 3). In the laboratory model a less distinct transition was found, most probably because here the sand layers are proportionally thinner. At all events, comparison with the visual observations revealed that the transition to complete suspension occurred at the same apparent velocity as detected by the instrument.

2.3 Concentration measurements.

The concentrations at the various heights in the vessel were measured by taking
samples of about 7 litres (1 litre for the laboratory-scale tests) and determining
the sand concentration via drying and weighing.

3.   THEORY

In this section we will derive the relation between the motion of the particles in a
stirred vessel and turbulence in the liquid generated by the stirrer. Our first
assumption is that the turbulence in the fluid at a reasonable distance from the
stirrer is almost isotropic.

3.1 Complete suspension.

A solid particle is just lifted from the bottom of a stirred tank when the forces
exerted on that particle by the turbulent motion of the fluid become equal to the
weight of that particle. For the size of particles usually encountered in suspension
processes we may assume that the size of the eddies which lift the particles is in the
"inertial sub-range" of the turbulence spectrum (Ref. 1). The energy of these eddies
is:

$$E(k,t) \quad \propto \quad \varepsilon^{2/3} \; k^{-5/3} \tag{1}$$

The fluctuating velocity v_e in such eddies is found from:

$$E(k,t) \quad = \quad v_e^2 \; k^{-1} \tag{2}$$

For the condition where the eddy just lifts a particle from the tank bottom we can set
up the force balance:

$$\rho \; v_e^2 \; d_p^2 \quad \propto \quad g \; \Delta\rho \; d_p^3 \tag{3}$$

We now assume that the specific power of the eddies is proportional to the specific
stirring power:

$$\varepsilon \quad \propto \quad n^3 \; D^2 \tag{4}$$

and that the size of the eddies that will lift a particle is proportional to the
particle size itself:

$$k \quad \propto \quad d_p^{-1} \tag{5}$$

Combining equations (1) to (5) we find for the point of complete suspension the
following relationship:

$$n_c \; D^{2/3} \quad \propto \quad d_p^{1/6} \left(\frac{g \; \Delta\rho}{\rho} \right)^{1/2} \tag{6}$$

This equation can also be rearranged into the form of a modified Froude number:

$$\frac{\rho \; n_c^2 \; D^{4/3}}{g \; \Delta\rho \; d_p^{1/3}} \quad = \quad \text{constant} \tag{7}$$

Two parameters are not incorporated in these relations: the solids concentration c
and the viscosity of the liquid. The effect of solids concentration on the turbulence spectrum in a stirred vessel is an interesting subject but would be outside the
scope of this paper. The effect of viscosity simply does not come into this theoretical
approach. Probably it occurs in empirical suspension correlations only because the
scale of the experiments has been so small that the stirring conditions have been in
the transition region between turbulent and laminar mixing.

3.2 Influence of the stirrer blade thickness.

Van der Molen and Van Maanen (Ref. 2) have measured turbulence intensities in the impeller stream by laser-Doppler anemometry. They observed a distinct influence of the blade thickness of the stirrers, roughly corresponding to:

$$E(k,t) \propto (b/D)^{-1/2} \tag{8}$$

This effect can be incorporated in the suspension relation which we have derived above (Eq. 6). Thus we find:

$$n\, D^{2/3}\, (b/D)^{-1/4} \propto d_p^{1/6} \left(\frac{g\,\Delta\rho}{\rho} \right)^{1/2} \tag{9}$$

From this relation we can easily see what the effect is when the blade thickness is not correctly scaled down for small-scale laboratory experiments. When this is done correctly (i.e. when b/D is constant) the relation remains equal to the original one (Eq. 6), i.e.:

$$n_c \propto D^{-2/3}$$

However, if the scaling-down is done incorrectly, and for instance the blade thickness is kept constant irrespective of the scale (b = constant), relation (9) converts into:

$$n_c\, D^{11/12} \propto d_p^{1/6} \left(\frac{g\,\Delta\rho}{\rho} \right)^{1/2} \tag{10}$$

3.3 Homogeneous suspension.

To obtain a certain degree of homogeneity in a stirred vessel, not only must the solid particles be lifted from the bottom of the vessel; they must also be transported in sufficient quantities throughout the whole volume of the vessel. It is not only the eddies of the inertial sub-range that are responsible for this mechanism; the largest eddies (i.e. the circulation) also play a role. For these eddies we may simply say that the fluctuating velocity is proportional to the circulation velocity, i.e. to the stirrer tip speed:

$$v_e \propto n\, D \tag{11}$$

With respect to the size of these eddies we may assume that they are proportional to the stirrer size:

$$k \propto D^{-1} \tag{12}$$

With the force balance represented by Eq. 3 we then find for homogeneous suspension the simple Froude-type relationship:

$$\frac{\rho\, n^2\, D^2}{g\,\Delta\rho\, d_p} = \text{constant} \tag{13}$$

4. RESULTS

4.1 The effect of the scale on the completeness of suspension.

The stirrer speeds for complete suspension measured in the 4.26-m vessel and in the 0.48-m vessel show a relationship:

$$n_c \propto D^{-2/3} \tag{14}$$

(see Fig. 4). This is in agreement with theory (Section 3.1). However, in the literature (Refs. 3 to 6) and in other work from KSLA different exponents in this equation have often been measured.

4.2 Stirrer blade thickness.

Earlier measurements in 0.12-m, 0.24-m and 0.48-m vessels indicated values for the exponent in relation (14) in the range -0.9 to -0.8. In Fig. 5 the results of these measurements are summarized together with the present results and the literature data. It can be seen clearly that, in particular in the experiments in small-scale equipment, higher exponents in relation (14) are found. As explained above (Section 3.2) the stirrer blade thickness of the impellers used may have an influence on the apparent scaling rule. We therefore checked the thickness of the impeller blades of the stirrers which had been used in the present and earlier KSLA trials. In Fig. 6 it can be seen that, for the large scales, the blade thickness is proportional to the size of the equipment, whereas in the smaller equipment impeller blades of a constant thickness have been used. Most probably the blades for a set of small stirrers are often made out of the same piece of material in the workshop.

On the basis of equation (9) we calculated the apparent exponents for the scaling rule if suspension experiments were done in two similar vessels differing by a factor of two in size, assuming that the stirrer blade thicknesses are according to Fig. 6. The result is given in Fig. 5 as a set of dotted lines and shows good agreement with the experimental observations by ourselves and others. We conclude, therefore, that incorrect scaling of the stirrer blade thickness leads to erroneous scaling rules for suspension (and probably also for other mixing phenomena). This is the more serious because these errors lead to the prediction of two low stirrer speeds and motor powers for large-scale mixing equipment.

In a few additional experiments we tested the extent to which we could take advantage of this effect of the stirrer blade thickness. In fact, for reasons of strength, the stirrer blade thickness cannot be changed in most cases; however, the shape of the blade edges may also influence the turbulence spectrum and hence the suspension performance. We tested a number of blade-edge modifications (edges rounded off and skewed off in various combinations, Fig. 7) under identical conditions. The blade profile with the sharp edges, type C, gave the best improvement compared to the square-edged blades, type A. A reduction of more than 5 % in the stirrer speed for complete suspension was obtained, resulting in a saving of about 17 % in the power consumption. The power number of the stirrer was not noticeably altered by this modification.

For large mixers this effect can be attractive. Use can also be made of it in existing mixers where, for instance, an improvement in the suspension quality is necessitated by changed process conditions.

4.3 Homogeneity of suspensions.

For the conditions tested in the 4.26-m vessel we hardly found any differences in the solids concentrations at the three sample points. In general the concentrations were somewhat (i.e. about 85 %) below the average (Fig. 8). Increasing the stirrer speed did not seem to improve the homogeneity any further, so apparently the homogeneity was optimal. When these experiments were repeated in the 0.48-m laboratory vessel different behaviour was observed (Fig. 9) which, in fact, was more in line with that of all the earlier experiments with various solids and liquids, in the laboratory vessels of 0.24 and 0.48 m. Roughly, one can say that, in the lower part of the vessel, a zone of a relatively high solids concentration is formed, where moreover the concentration is approximately constant. We will call this the "homogeneous zone". On top of this zone the solids concentration decreases rapidly. The height of this homogeneous zone depends on the physical properties and the stirring conditions. Increasing the stirrer speed will increase the height of the homogeneous zone almost proportionally until a maximum value is attained of approximately 10 % of the vessel diameter below the liquid level (Fig. 10). It is striking that the average concentration has no noticeable influence on the height of the homogeneous zone, even for concentrations up to 40 %(v).

On the basis of these observations we correlated all our present and older data in terms of height of homogeneous zone with a modified Froude number:

$$\frac{Z}{T} = f\left(\frac{\rho \, n^2 \, D^2}{g \, \Delta\rho \, d_p} \cdot \left(\frac{d_p}{D}\right)^{0.45}\right) \tag{15}$$

The result is depicted in Fig. 11, which incorporates a correction for the deviating blade thicknesses of some of the stirrers. Unfortunately, no concentration measurements were made in the 4.26-m vessel at lower stirrer speeds, i.e. below-optimal homogeneity. Nevertheless, the measurements indicate clearly that $n^2 D^{1.55}$ is constant for homogeneous suspension.

On a large scale, therefore, homogeneity is attained at a lower specific power input than on a small scale. From Fig. 11 a relation can be derived for homogeneous suspension ($Z/T \sim 0.9\, T$; if $H/T = 1$):

$$\frac{\rho\, n^2\, D^2}{g\, \Delta\rho\, d_p} \cdot \left(\frac{d_p}{D}\right)^{0.45} \;\geqslant\; 20 \tag{16}$$

Assuming that the largest eddies of approximately the impeller size are responsible for the homogeneity we would have expected a relation like:

$$\frac{\rho\, n^2\, D^2}{g\, \Delta\rho\, d_p} \;\geqslant\; \text{constant} \qquad\qquad \text{(Section 3.3)}.$$

Relation (16) is in fact closer to the complete suspension relation (Eq. 7). Obviously the eddies of the inertial sub-range also make a contribution to the homogeneity of the suspension.

5.   CONCLUSIONS

1. Scaling rules for complete suspension in stirred vessels must contain the form:

   $$n_c \;\propto\; D^{-2/3}$$

   This is in agreement with the theory of isotropic turbulence and is supported by accurate measurements.
2. The broad variety of the scaling rules in the literature is due, at least partly, to incorrect scaling of the stirrer blade thickness for model experiments.
3. In homogeneous supension the largest eddies, i.e. the circulation, also play a role, as expected from the theory. This means that, on a large scale, homogeneous suspension is attained at a lower specific power input than on a small scale.

6.   REFERENCES

1.   Hinze, J.O.: "Turbulence". New York etc., McGraw-Hill, 2nd Edition, 1975.

2.   Van der Molen, K. and Van Maanen, M.R.E.: "Laser-Doppler measurements of the turbulent flow in stirred vessels to establish scaling rules". Chemical Engineering Science, 33, 9, Sept. 1978, pp. 1161-1168.

3.   Zwietering, Th.N.: "Suspending of solid particles in liquid by agitators". Chemical Engineering Science, 8, 3/4, 1958, pp. 244-253.

4.   Baldi, G., Conti, R. and Alaria, E.: "Complete suspension of particles in mechanically agitated vessels". Chemical Engineering Science, 33, 1, Jan. 1978, pp. 613-627.

5.   Kolár̆, V.: "Suspending solid particles in liquids by means of mechanical agitation". Collection of Czechoslovak Chemical Communications", 26, 1961, pp. 613-627.

6.   Nicolaus, F.:"Suspension of solids in liquids by stirring". (Suspendieren von Festkörpern in Flüssigkeiten durch Rühren.) Dissertation, University of Technology, Munich, 1961. (In German.)

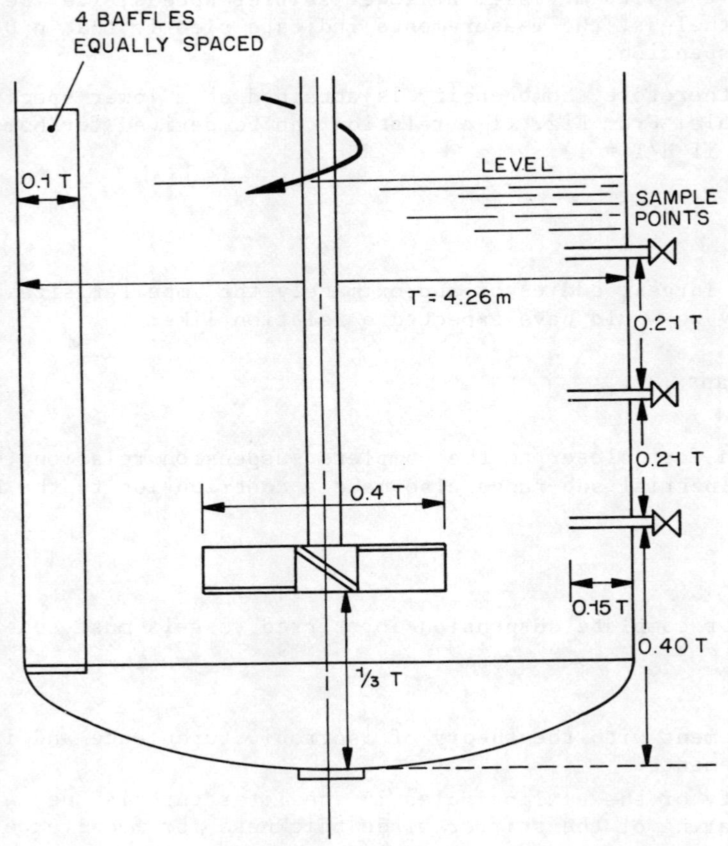

Fig. 1. Mixer geometry.

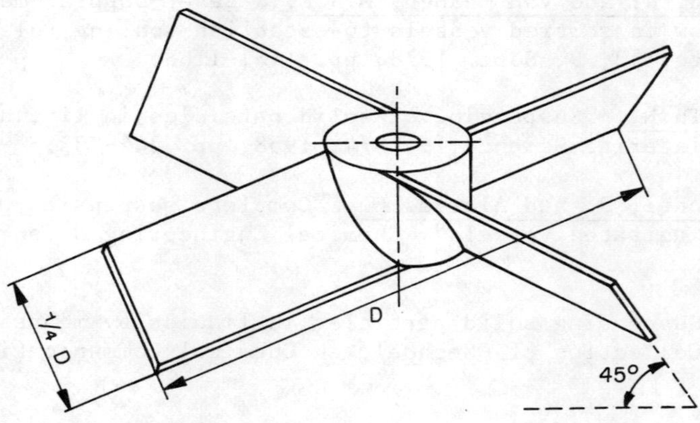

Fig. 2. Axial turbine impeller. D = 1.72 m.

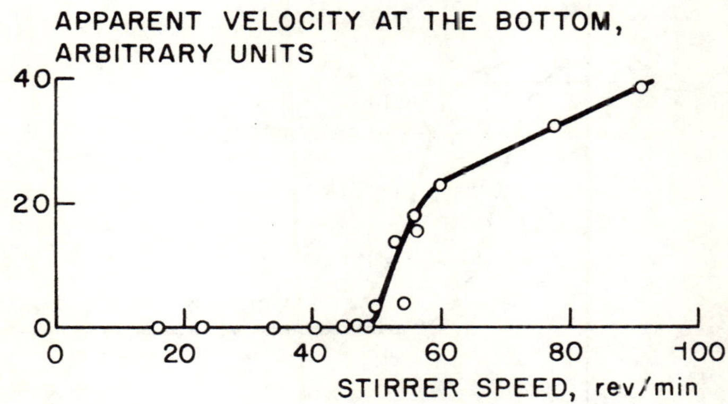

Fig. 3. Determination of stirrer speed for complete
suspension. 4.26-m vessel. 4.5 %(v) sand,
$d_p = 157$ µm.

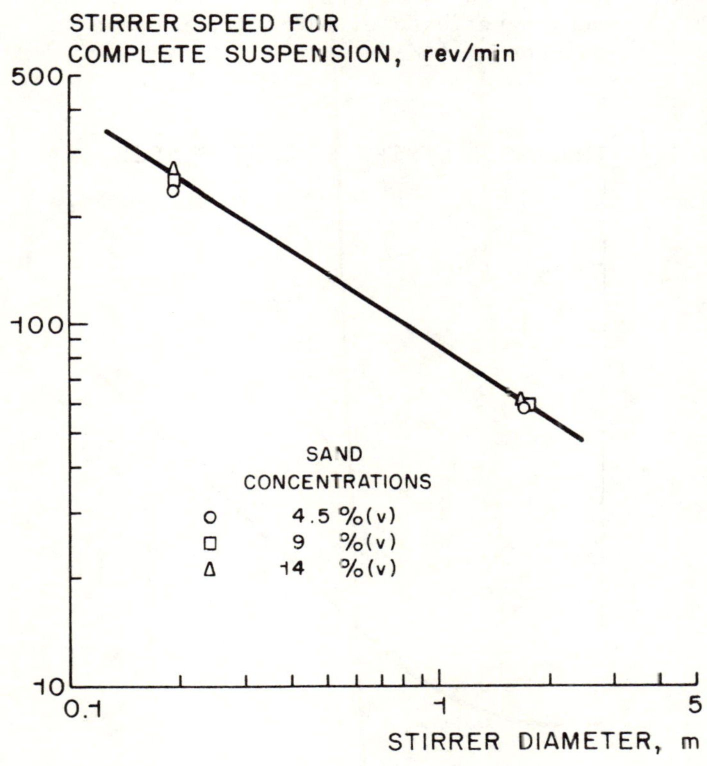

Fig. 4. Stirrer speeds for complete suspension
on two scales. Sand in water, $d_p = 157$ µm.

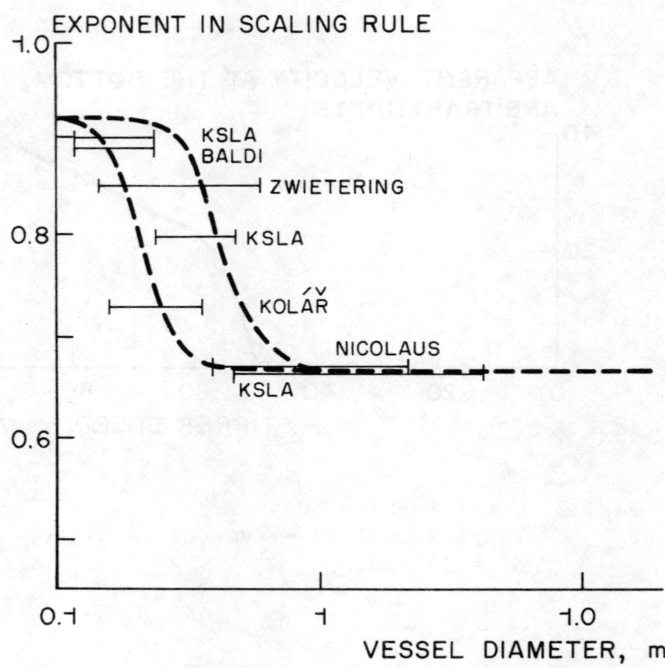

Fig. 5. Exponents in the scaling rule for complete
suspension from the literature and KSLA work.
The dotted lines are calculated for incorrect
scaling effects of the stirrer blade
thickness.

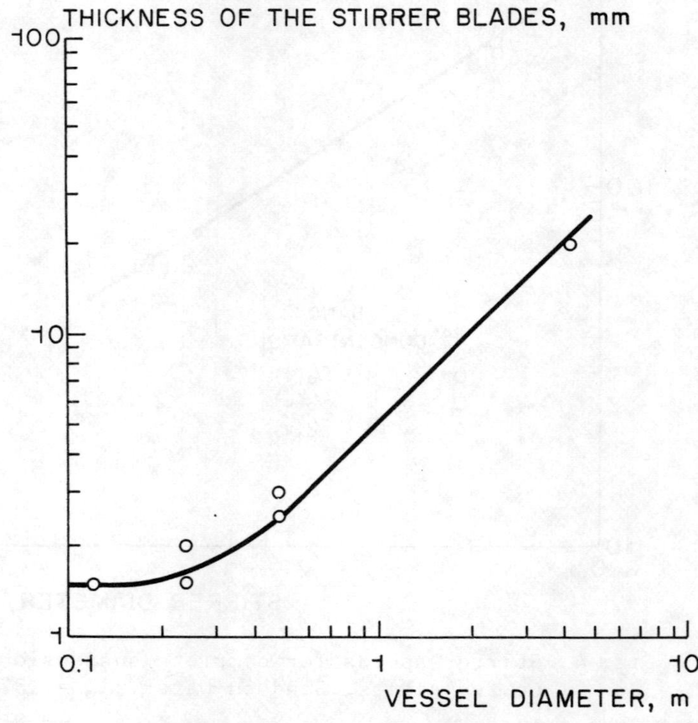

Fig. 6. Blade thicknesses of various stirrers in
the experimental set-up at KSLA.

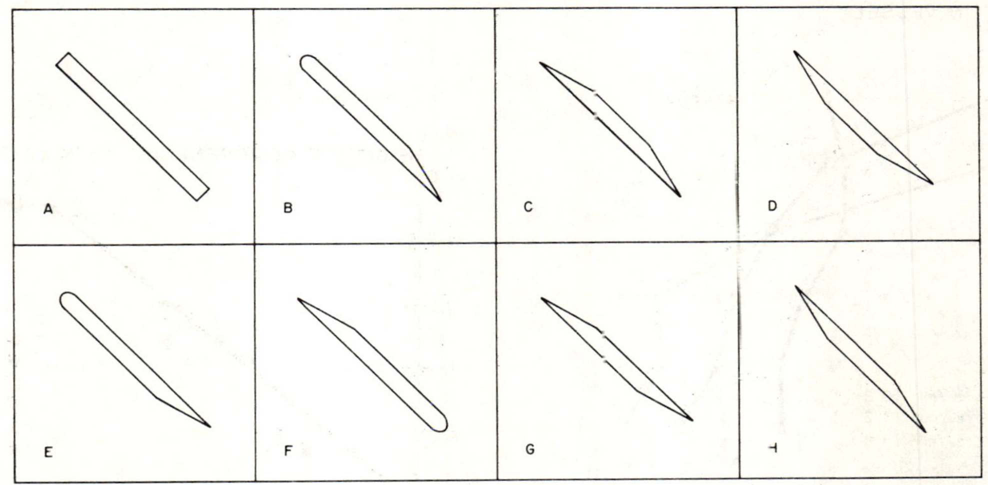

Fig. 7. Stirrer blade profiles. Stirrer: axial turbine, diameter 192 mm, blade width 48 mm, blade thickness 4 mm, skewing over 15 mm, edge thickness 0.5 mm.

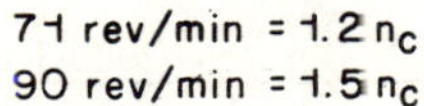

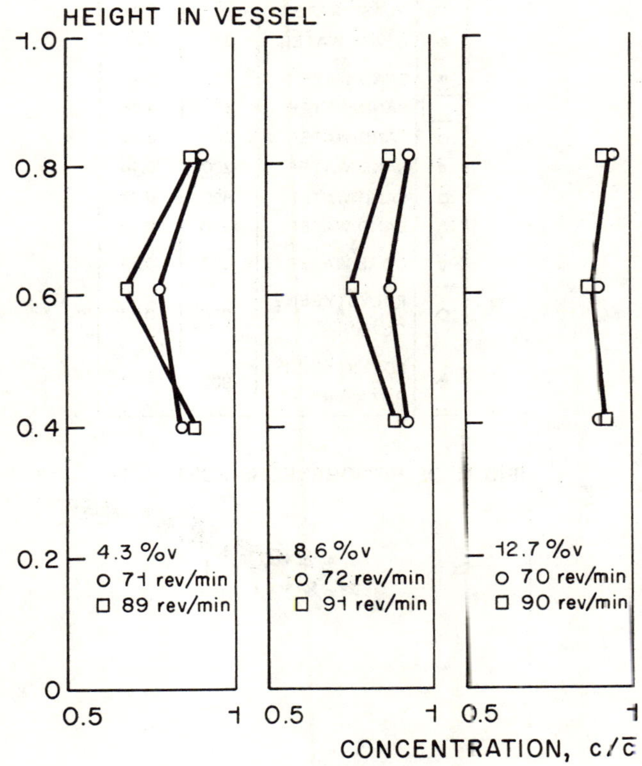

Fig. 8. Concentration profiles in 4.26-m vessel. Sand in water, d_p = 157 µm.

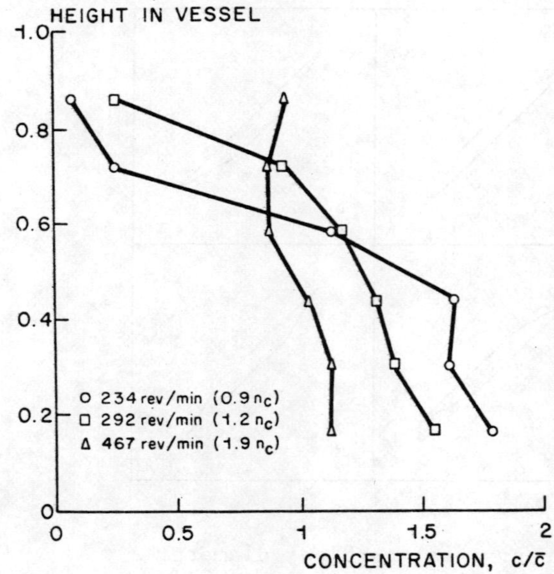

Fig. 9.
Concentration profiles in 0.48-m vessel.
Sand (9 %(v)) in water, d_p = 157 µm.

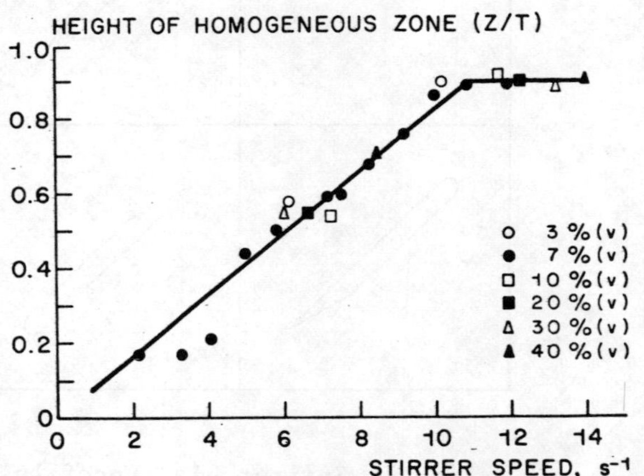

Fig. 10.
Effect of stirrer speed on homogeneity.
0.24-m vessel. Glass beads in water,
d_p = 200 µm.

	SUSPENSION	d_p, µm	T, m
◊	GLASS/WATER	200	0.24
◈	GLASS/WATER	800	0.24
▼	SAND/WATER	157	0.48
▭	SAND/WATER	157	4.26
○	SAND/WATER	200	0.24
●	SAND/WATER	200	0.48
□	SAND/WATER	800	0.24
△	SAND/WATER	2200	0.24
▽	COAL/WATER	800	0.24
◇	POLYSTYRENE/ OCTANE	800	0.24
◆	POLYSTYRENE/ BUTANOL	800	0.24

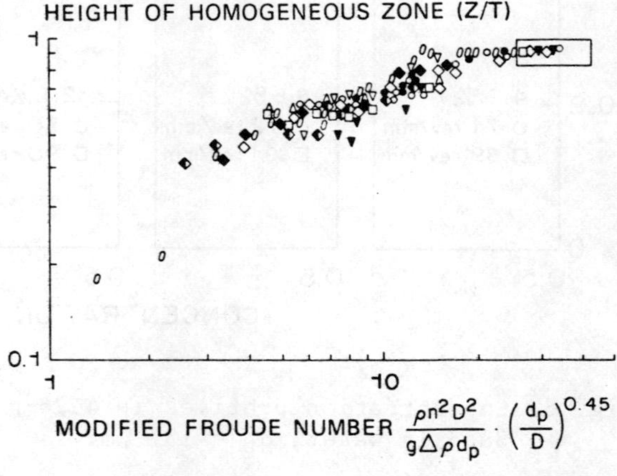

Fig. 11. Effect of stirring conditions on homogeneity.

THE APPLICATION OF TWO NOVEL TECHNIQUES
FOR MASS TRANSFER COEFFICIENT DETERMINATION
TO THE SCALE UP OF GAS SPARGED AGITATED VESSELS

S.N. Davies and L.G. Gibilaro

Department of Chemical and Biochemical Engineering
University College London, Torrington Place
London   WC1E 7JE
England

J.C. Middleton, M. Cooke and P.M. Lynch

I.C.I. New Science Group, Runcorn
Cheshire, England

Summary

Gas-liquid stirred vessels are widely used in industry as reactors and for other tasks, and reliable gas-liquid mass transfer design and scale up correlations are urgently required. The mass transfer coefficient, k_La, is usually determined in small scale research rigs by the dynamic response of dissolved gas concentration to a step change in inlet gas concentration, and the derivation of k_La from this reponse requires a gas phase mixing pattern to be assumed. Previous workers have assumed ideal gas mixing patterns with the resulting k_La values differing widely according to the assumptions made.

This paper discusses two techniques which avoid the necessity to make such assumptions: one by measuring both the liquid and outlet gas dynamic responses, and the other by studying the initial response of the dissolved gas concentration in the liquid. Results from the two methods are presented for a range of conditions in four scales of stirred vessel ranging from laboratory to production size.

Held at Wurzburg, 10-12 June, 1985.

Organised by DVCV· Deutsche Vereinigung für Chemie- und Verfahrenstechnik
(German Association of Chemical and Process Engineering).

Organisation: GVC·VDI-Gesellschaft Verfahrenstechnik und Chemieingenieurwesen. GVC

NOMENCLATURE

a_i, $i = 0,3$	polynomial coefficient	(s^{-i})
b_i, $i = 0,4$	polynomial coefficient	(s^{-i})
c_G^*	normalised gas concentration	$(-)$
c_L^*	normalised liquid concentration	$(-)$
c_M^*	normalised probe response	$(-)$
c_O^*	normalised outlet gas concentration	$(-)$
D_L	liquid phase diffusivity	(m/s^2)
K	Henry's Law constant	$(-)$
$k_L a$	mass transfer coefficient	(s^{-1})
P	power	(kW)
Q	gas flowrate	(m^3/s)
S	surface area of gas hold up	(m^2)
t	time	(s)
T	temperature	(K)
V_G	volumetric gas hold up	(m^3)
V_L	volumetric liquid hold up	(m^3)
v_s	gas superficial velocity	(m/s)
Y	normalised measured response of outlet gas	$(-)$

Greek symbols

μ	liquid viscosity	(Ns/m^2)
τ_G	probe time constant for gas phase	(s)
τ_L	probe time constant for liquid phase	(s)

1. INTRODUCTION

Gas-liquid mass transfer in an agitated vessel is an important, often rate controlling, step in many industrial processes: much effort has been devoted to obtaining scale up correlations, especially for the mass transfer coefficient, $k_L a$.

It has long been agreed that mass transfer data obtained using chemical methods are subject to many uncertainties and most workers now use one or other variation of the dynamic 'gassing out' method. These techniques, although superior to the chemical method in that they do not physically alter the system, suffer from the drawback that extensive knowledge of gas-liquid contacting patterns appears to be necessary.

Since Dunn and Einsele (Ref. 1) demonstrated the pitfalls in assuming that there is no depletion of the key component in the gas phase, various gas mixing models have been used, from the idealized well mixed (Ref. 1) and plug flow (Ref. 2) models, to their extensions, the tank-in-series (Ref. 3) and axial dispersion (Ref. 4) models. The degree to which these models can be considered acceptable depends upon the accuracy of the model in describing the gas mixing in the tank. Unfortunately, the actual measurement of the residence time distribution (RTD) of the gas phase is complicated because the effect of transfer of any tracer used, from the gas phase to the liquid phase, must be taken into account. These two interdependent processes would be extremely difficult to separate, and thus no independent measure of the degree of mixing in the gas phase is available. Mass transfer coefficients determined using the different gas phase mixing models often differ widely and gross errors can arise if the wrong model is chosen.

A method for measuring the mass transfer coefficient that circumvents the need to know the RTD of the gas phase has been presented previously (Ref. 5). It was shown that with knowledge of the liquid and outlet gas transients, following a step change in the inlet gas concentration, the mass transfer coefficient can be calculated without making any assumptions about gas mixing. As this technique involves measuring two transients, we shall refer to it as the *double response method*.

In the same paper, the authors suggested that in theory, at least, the mass transfer coefficient could be evaluated solely from the liquid phase transient or, to be precise, the initial second derivative of the liquid phase response. We shall refer to this as the *initial response method*; it is discussed in some detail elsewhere (Ref. 6).

In this paper we shall describe the development of these two techniques and discuss some of the experimental difficulties encountered. Results from a number of different vessels, covering a large range of scales, will be presented.

2. THEORY

The theoretical development of the double response method is essentially that given previously (Ref. 5). However, the original formulation assumed that the bubbles were of approximately equal size. We now do not impose this condition, but instead assume that the average concentration at the bubble interface is equal to the average concentration in the bulk gas. That is,

$$\frac{1}{S} \int_S C_G dS = \frac{1}{V_G} \int_{V_G} C_G dV_G \tag{1}$$

This condition can be shown to be satisfied if the gas phase is well mixed, in plug flow, or consists of a finite number of monosized bubbles. Thus the mass transfer coefficient, $k_L a$, can be related to the normalized liquid and outlet gas responses by,

$$\frac{dC_L^*}{dt} = k_L a \left[\frac{Q}{V_G} \int_0^t (1-C_O^*) \; dt - C_L^* \left(\frac{KV_L}{V_G} + 1 \right) \right] \tag{2}$$

This is a rearrangement of eqn (6) in (Ref. 5). The normalization is achieved by dividing the measured responses by their final equilibrium values. Thus

$$C_L^* (\infty) = C_O^* (\infty) = 1 \quad , \tag{3}$$

and no instrument calibration is necessary.

Differentiation of eqn (2) yields:

$$\frac{d^2 C_L^*}{dt^2} = k_L a \left[\frac{Q}{V_G} (1 - C_O^*) \, dt - \frac{dC_L^*}{dt} \left(\frac{K V_L}{V_G} + 1 \right) \right] \tag{4}$$

Evaluating eqn (4) at time zero and re-arranging gives,

$$k_L a = \frac{V_G}{Q} \left. \frac{d^2 C_L^*}{dt^2} \right|_{t=0} \tag{5}$$

since $C_O^* = 0$ and $dC_L^*/dt = 0$ at $t = 0$.

All the higher derivatives can be determined easily by repeated differentiation of eqn (2), and in principle these can provide additional information on the influence of the system parameters on the early dynamic response (Ref. 6) although in practice, measurement of higher derivatives would prove difficult.

In order to determine the second derivative of the liquid response at time zero, the initial portion of the transient is considered to be adequately described by the cubic equation,

$$C_L^* = a_0 + a_1 t + a_2 t^2 + a_3 t^3 \tag{6}$$

It turns out that both the coefficients a_0 and a_1 are zero because $C_L^* = dC_L^*/dt = 0$ at $t = 0$. Thus,

$$C_L^* = a_2 t^2 + a_3 t^3 \tag{7}$$

However, eqn (5) assumes there to be no measurement lag. Here, the measured response, C_{LM}^*, differs from the true liquid response , C_L^*, due to the lag imposed by the oxygen probe. This measurement lag can be considered to be first order (Ref. 5). Therefore the measured response can be related to the true response by the first order deconvolution relationship

$$C_L^* = \tau_L \frac{dC_{LM}^*}{dt} + C_{LM}^* \tag{8}$$

It can be easily shown by differentiation of eqn (8) that the initial second derivative of the actual liquid response is directly related to the initial third derivative of the measured response thus:

$$\left. \frac{d^2 C_L^*}{dt^2} \right|_{t=0} = \tau_L \left. \frac{d^3 C_{LM}^*}{dt^3} \right|_{t=0} \quad , \tag{9}$$

and the equivalent polynomial to fit to the measured response is now the quartic equation,

$$C_{LM}^* = b_3 t^3 + b_4 t^4 \quad . \tag{10}$$

Thus eqn (5) can be rewritten as,

$$k_L a = \frac{6 \tau_L \, V_G \, b_3}{Q} \tag{11}$$

and the mass transfer coefficient can be calculated once a numerical value for b_3 has been determined.

3. EXPERIMENTAL

The experiments were carried out in four geometrically similar tanks of diameters 0.31 m, 0.61 m, 0.91 m and 1.83 m, filled with water to a height equal to the vessel diameter. This gives an extreme volume ratio of 216:1. The tanks were cylindrical and of a flat base and open top construction. A six-bladed disc turbine of half tank diameter was used in each vessel at a constant clearance ratio of one quarter the static liquid height above the base of the tank.

Oxygen concentrations were measured using fast response dissolved oxygen probes (DO probes) which exhibited first order dynamic characteristics with time constants ranging from 0.5 to 1.25 secs. These were of the membrane covered, polarographic type, designed and built by ICI, New Science Group.

Shaft power measurements were obtained using shaft mounted strain gauges and telemetry systems on the two larger tanks, and bearings and dynanometers on the smaller tanks. Power measurements were reproducible to ± 5% over the ranges used.

Gas hold up measurements were taken by recording the level of dispersion above the static liquid height in one quadrant of the tank between two baffles using an ultrasonic sensor and monitoring the level over a period of 5 – 10 mins. These measurements were reproducible to within ± 10%.

The experimental procedure was basically the same as that described many times before. The stirrer speed and gas rate under investigation were set, and the liquid phase sparged using nitrogen gas. When the oxygen level in the tank has dropped to approximately zero, a fast acting three way valve was operated switching from the nitrogen supply to the air supply, and the subsequent uptake of oxygen in the liquid phase was followed by a DO probe until a new equilibrium was established.

Neither of the methods require assumptions concerning gas phase mixing. However, this advantage has to be paid for in the form of extra experimental measurements.

For the double response method, the average outlet gas concentration has to be measured. As in the original application (Ref. 5), this was done by sampling a representative area of the surface (approximately 10% of the total cross-sectional area of the tank) but using a much shallower funnel. Previously, the mixing time of the gas in the sampler was comparable to or even dominated the true outlet gas response. This was obviously not desirable. With the shallower sampler, very little mixing occured in the funnel, and the measurement lag could be modelled as a pure dead time plus the first order lag due to the dissolved oxygen probe. This assumption will be considered later.

To estimate the derivatives of the liquid response at time zero, it is obviously necessary to have a good indication as to the exact time at which oxygen enters the system. This was achieved by placing a fast responding DO probe at the entry point of the sparge pipe. This, therefore, measured the concentration of oxygen in the sparge gas. Using this method, we could mark the start time to within ± 0.05 seconds.

4. COMPUTATIONAL PROCEDURES

For the double response method, the outlet gas concentration was measured as well as the liquid concentration time history. This measured response was considered to contain a pure dead time, this being the time taken by the gas to travel through the sampler, hence the measured response, $Y(t)$ can be related to the actual response, C_O^*, by eqns (12) and (13)

$$C_{Om}^*(t) = Y(t + d) \tag{12}$$

$$C_O^*(t) = \tau_G \frac{dC_{Om}^*(t)}{dt} + C_{Om}^*(t) \tag{13}$$

The mass transfer coefficient is then found by simultaneously solving eqns (2), (12) and (13), and optimising the values of $k_L a$ and d by minimising the sum of the squares of the differences between the measured and calculated responses. This procedure was found to be very consistent and fairly insensitive to the value taken for the response time for the DO probe measuring the gas phase, τ_G, showing that the

assumption of no significant mixing in the funnel is not critical.

The initial response method requires the determination of the derivatives of the liquid response. Early results using smoothing splines proved very unreliable. Obviously a more stable method of evaluating the initial derivatives was required, and the method described in the theoretical section was adopted. Equation (10) was fitted to the initial portion of the liquid response by the method of least squares. Firstly, this portion is taken to be a long way into the response so that eqn (10) does not adequately describe the response. The length of time considered is then progressively reduced and the equation refitted until an acceptable fit is obtained. The acceptability of the fit was determined by a two-tailed chi-squared test on the fit relative to the noise level at steady state conditions. The 2.5% level was considered acceptable Effectively, this method determines how far into the response eqn (10) provides a satisfactory fit. A similar method is described in Ref. 6.

5. RESULTS

The results for the double response method for the 0.31 m and 0.91 m diameter tanks are shown in Fig. 1. A common form of correlations for $k_L a$ involves the specific power input and gas sparge rate. The present data were correlated in this form, giving,

$$k_L a = 0.17 \; \frac{P}{V_L}^{0.64} \; (v_s)^{0.22} \tag{14}$$

Although the experimental technique was much more involved than previously, the reproducibility of the results was very good, better than ± 10%.

Fig. 2 shows the preliminary results for the initial response technique for all four tanks. The reproducibility of these results was always within ± 20% and usually within ± 10%. Although the results for both methods agree well for low to medium power inputs, they diverge at higher power inputs, the dependency of $k_L a$ on power for the initial response method being higher. The correlation shown in Fig. 2 is,

$$k_L a = 0.49 \; \frac{P}{V_L}^{0.76} \; (v_s)^{0.45} \tag{15}$$

No temperature control was available on these tanks, so all runs were at ambient temperatures. The results quoted have been corrected to 20°C using the Stokes-Einstein equation,

$$\frac{D_L \, \mu}{T} = \text{const} \tag{16}$$

and assuming the liquid film mass transfer rate constant, k_L, was directly proportional to the liquid phase diffusivity, D_L. Previous work agrees well with this assumption (Ref. 7).

6. DISCUSSION

The double response method appears to be very reliable. It is self-consistent and gives results which are comparable with other methods in situations where it is reasonable to assume an ideal gas mixing model. The method of analysis is also insensitive to the parameters which need to be supplied for the model, that is, the volumetric gas flowrate, Q, the gas hold up, V_G, and the response time of the gas sampling system. Such insensitivity to poorly defined parameters is obviously desirable.

For the results obtained by the initial response method, reproducibility is also good. Each set of conditions was repeated five times, and these results always agreed to within ± 20%, and mostly much better than this. It would be expected that this method would be less reproducible than the double response method, using, as it does, only the initial portion of the curve, compared to the whole response. Also the dependency on the volumetric gas flowrate and hold up is greater for this method, as

can be seen from eqn (5).

The discrepancy between the two methods suggests that the assumption of a perfectly back mixed liquid phase is not as acceptable as previously thought. The effect on the double response method of this poor mixing would be smaller than that for the initial response method as it uses the whole response curve in the analysis. Unfortunately, no conclusive evidence can be taken from the data obtained so far. Experiments, where oxygen probes are placed in a number of positions around the tank and their responses simultaneously monitored are planned to provide direct experimental evidence to resolve this point. The effect of scale can been seen from Fig. 3 where the results from the double response technique are plotted on the correlation for the initial response method. For the 0.31 m diameter tank there is good agreement between the two methods, but on the larger scale there is a significant departure.

7. CONCLUSIONS

The goal of this work is the production of reliable data for the scale up of gas-liquid mass transfer in sparged stirred vessels. The methods described to evaluate the mass transfer coefficient, k_La, are believed to represent a marked improvement on previous techniques based on highly improbable simplifications of the gas mixing characteristics.

8. REFERENCES

1. Dunn, I.J. and Einsele, A.J.: "Oxygen transfer coefficients by the dynamic method". Journal of Applied Chemistry and Biotechnology, 25, 1975, pp. 707-720.

2. Shioya, S. and Dunn, I.J.: "Model comparisons for dynamic k_La measurements with incompletely mixed phases". Chemical Engineering Communications, 3, 1979, pp. 41-52.

3. Chandrasekhari, K. and Calderbark, P.H.: "Further observations on the scale up of aerated mixing vessels". Chemical Engineering Science, 36, 1981, pp. 819-823.

4. Shioya, S. and Dunn, I.J.: "A dynamic oxygen transfer coefficient measurement method for column reactors". Chemical Engineering Science, 33, 1978, pp. 1529-1534.

5. Chapman, C.M., Gibilaro, L.G. and Nienow, A.W.: "A dynamic response technique for the estimation of gas-liquid mass transfer coefficients in a stirred vessel". Chemical Engineering Science, 37, 1982, pp. 891-896.

6. Gibilaro, L.G., Davies, S.N., Middleton, J.C., Cooke, M. and Lynch, P.M.: "Initial response analysis of mass transfer in a gas sparged stirred vessel". Chemical Engineering Science, In Press.

7. Cooke, M.: Unpublished results.

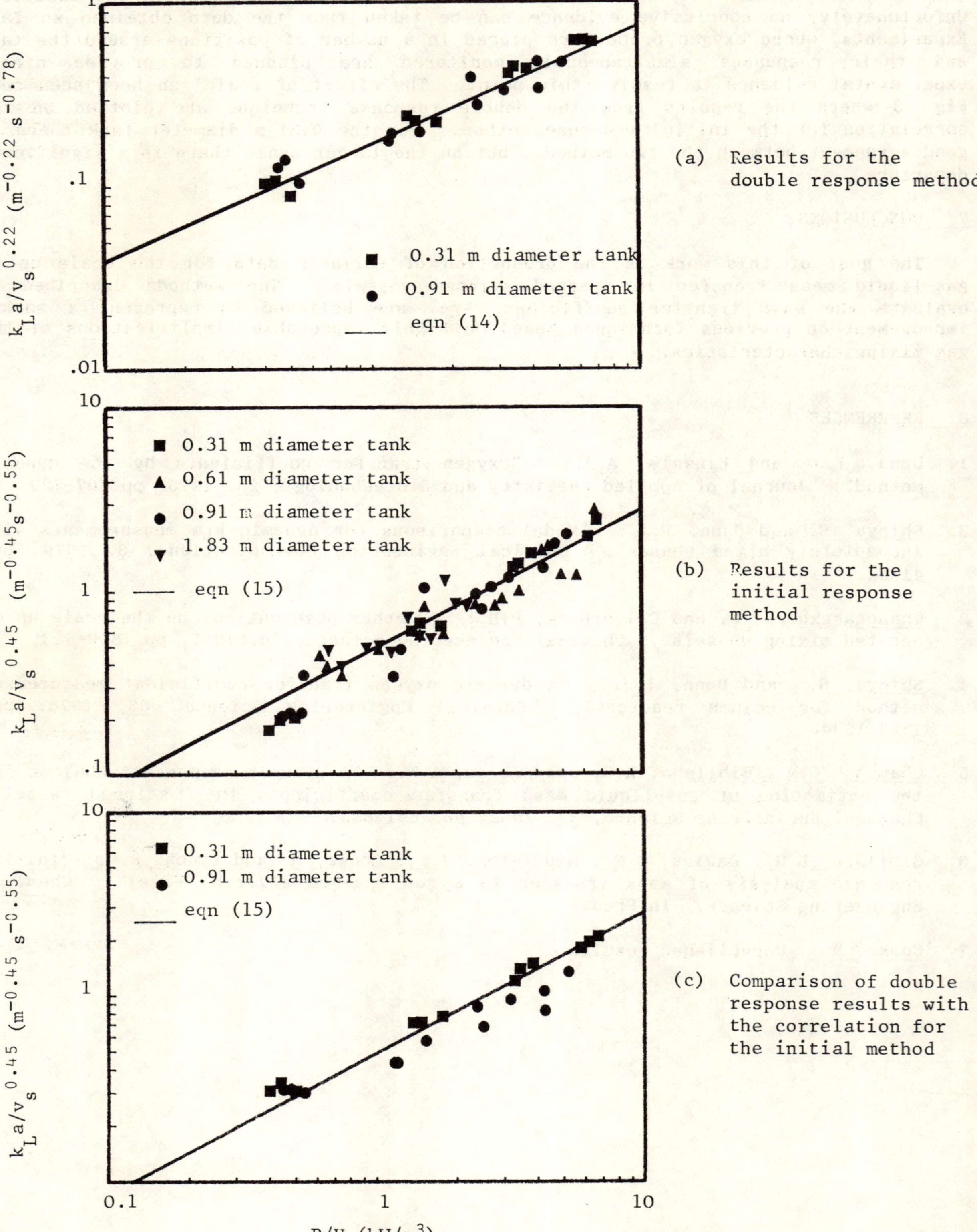

(a) Results for the double response method

(b) Results for the initial response method

(c) Comparison of double response results with the correlation for the initial method

SCALE-UP OF UNIQUE INDUSTRIAL FLUID MIXING PROCESSES

JAMES Y. OLDSHUE, Ph.D.
VICE PRESIDENT, MIXING TECHNOLOGY

MIXING EQUIPMENT CO., INC.
135 MT. READ BLVD.
ROCHESTER, NEW YORK 14611

SUMMARY

Many industrial processes involve several basic mixing steps. Each of these steps
may have an interaction with other parts of the process. Many times, scale-up practic-
alities of the various steps take different directions, so that the net result is that
a process in a large tank can have different controlling steps than it might have had
in the pilot plant.

Several basic differences exist between pilot plant tanks and full-scale industrial
tanks. The large tank will have a longer average circulation time, and may also behave in a manner similar to a series of small tanks within the large tank in terms of
the overall distribution of circulation times.

The large tank will also typically have a different spectrum of macro-scale shear
rates, since the tendency is for maximum impeller zone shear rates to go up, while
average impeller shear rates go down, and average tank zone shear rates go down compared to the pilot tank. In terms of micro-scale shear rates, the big tank can be
quite similar to the small tank.

The effect of some of these differences is that if one is concerned with detailed
shear rates, or droplet or bubble sizes in a large tank, they may be quite different
than the detailed spectrum of these parameters on the small-scale pilot tank. These
differences may have a major effect on the overall performance of the process.

On the other hand, if one is interested in the overall mass transfer rate, from gas
to liquid, liquid to liquid, or liquid to solid, it may not make much difference that
the bubble size or droplet size distribution in a large tank is different than a small
tank, since it is possible to get quite similar overall volumetric mass transfer coefficients, such as K_Gas or K_Las between small and large tanks.

In general, the Reynolds Number tends to increase on scale-up, so care must be exercised that the pilot tank is operating in the same characteristic fluid regime as a
full-scale tank will operate.

Held at Wurzburg, 10-12 June, 1985.

Organised by DVCV· Deutsche Vereinigung für Chemie- und Verfahrenstechnik
(German Association of Chemical and Process Engineering).

Organisation: GVC·VDI-Gesellschaft Verfahrenstechnik und Chemieingenieurwesen.  **GVC**

TABLE V

NOMENCLATURE

P   =   Power

Q   =   Impeller Flow Discharge

H   =   Impeller Velocity Head

K_L =   Mass Transfer Film Coefficient

a   =   Mass Transfer Surface Area – Square Feet/Cubic Feet

K_La =   Mass Transfer Coefficient

AFT =   Axial Flow Turbine

RFT =   Radial Flow Turbine

v   =   Individual Fluid Velocity At A Point

v^1 =   Fluid Velocity Fluctuation Away From Average At A Point

RMS =   Root Mean Square

θ =   Blend Time

N   =   Impeller Speed

D   =   Impeller Diameter

T   =   Tank Diameter

Z   =   Liquid Level

C   =   Impeller Distance Off-bottom

μ =   Viscosity

1. INTRODUCTION

Scale-up of fluid mixers in actual industrial processes almost always involves a combination of different phenomena and the role of mixing scale-up parameters must be examined in terms of the applicability for the particular combinations of processes involved.

A general principal that applies on scale-up is that if geometric similarity is used between the model and the prototype, the large mixing tank will have a much longer blend time, and much longer circulation time than the smaller pilot plant tank. In addition, it will usually have a higher maximum impeller zone shear rate than will the corresponding small-scale impeller. Also, it will have a lower average impeller zone shear rate and a lower average shear tank zone rate throughout the tank than does the small pilot tank.

One type of a mixing process is primarily influenced by the individual point velocities and point shear rates in different places in the tank. For example, the individual droplet size distribution in a liquid/liquid dispersion will normally be quite different in a large tank than in a small tank because the dispersion and recoalescence is quite different in a big system than a small. There could be a difference in the product from a polymerization reaction from different scale of systems. Another type of mixing process result depends only on the total integrated mass transfer rate and/or the average value of other parameters throughout the tank. It is quite possible to have the gas/liquid mass transfer coefficient, K_La, be quite similar in small and large tanks.

The **total** integrated average gas/liquid mass transfer coefficient in the entire tank can be quite similar even though individual bubble sizes are likely affected by scale; the total integrated average film coefficient K_L, and area per volume - a, turns out to be about the same in the different scales.

Another consideration involves determining whether the increase in blend time on scale-up is a serious factor in the overall process. Assume the blend time in small-scale is about two seconds. The blend time in full-scale could be 2 minutes or longer. It probably would not make much difference in a process that only shows deviation from optimum performance when blend times are 5 minutes or longer.

In another process in which significant deviations are found in the product when blend times exceed 30 seconds would very definitely show up to be a different situation in a big tank than in a small tank.

From this brief statement of some of the parameters, let's proceed to a more detailed look at individual situations.

It is preferred to use the term "scale-up" in talking about the physical relationships involved, and use the term "pilot planting" when discussing how to conduct the pilot plant experiment in order to get the proper data to apply the correct scale-up technique.

2. IMPELLER PUMPING CAPACITY

Pumping capacity refers to the overall flow pattern in the mixing tank, as shown in Figure 1. It is usually expressed as total cubic meters per second pumped normally through a defined surface area. In many respects, any fluid turbulence in this stream may not contribute to producing the maximum flow for a given level of power. Any turbulent eddies which decay into viscous shear forces transfers power from the mixer into heat at that point.

For the suspension of solids, it is normally thought that fluid velocities at various points in the tank and therefore total pumping capacity is the major variable in determining the performance of a mixer in this process. Any turbulence or any velocity gradience are not often thought to be of benefit to the process.

Thus, it becomes a worthy objective to try and design an impeller that has a minimum turbulent energy loss in the impeller zone itself, and thus try to obtain a given flow for a minimum horsepower. Fluid foil-type impellers (shown in Figure 2) are the result of optimization studies made by very elaborate fluid flow measurements, most conveniently done at the present time with laser velocity meters. Figure 3 shows a typical arrangement of the tank. Figure 4 shows a typical computer drafting machine result of the average velocity at various points across planes of interest.

A practical goal is to get a minimum drag coefficient, shown in Figure 5, in contrast
to a flow pattern around the impeller blade that would produce a higher value for the
drag coefficient shown in Figure 6.

The net result of experimental studies was that the angle of attack predicted from
wind tunnel tests on various airfoil sections did not prove out to be the correct
angle in the mixing tank. It turns out that the mixing tank flow patterns are so much
more random and are far from being a uniform pattern that a different blade angle was
determined experimentally than what might have been predicted to give the minimum
drag coefficient shown in Figure 5. Also it turns out that there is no advantage to a
large hub in a mixing tank. The impeller shown in Figure 2 is very typical of the impellers used in this type of service.

In general, the power required by the fluid foil impeller is about one-half that which
would be required by the 45° axial flow turbine (AFT), shown in Figure 7.

Another interesting characteristic of these fluid foil impellers is their marked tendency not to swirl and vortex like the 45° angle impeller. As a result, four baffles
(each one 12" diameter in width) are not required with the fluid foil impeller, and
normally three baffles of somewhat smaller width is normally adequate. Some tanks
cannot be baffled at all - in this case, these fluid foil impellers develop much less
swirl and vortexing than by other radial, axial flow impellers.

Most mixing operations require fluid motion while the tank is being drawn off and
often down to the very last volume contained in the vessel. Thus, impellers are
placed at the end of the shaft very far down in the liquid medium. Under these conditions the fluid foil-type impeller (Figure 6) requires about 30 - 40% less power for
the same overall pumping capacity of the tank than the AFT.

In a few applications where the impeller can be placed higher in the tank, then the
axial flow patterns from the fluid foil-type impeller behaves almost like a draft
tube. The power is about 30 - 50% as much as was in the AFT impeller as shown in
Figure 7.

In processes where there is a mass transfer requirement or a process where chemical
reactions take place, it cannot be assumed without some consideration; either experience, theoretical or experimental, that the reduction in micro-scale turbulence in
and around the impeller zone is the correct thing to do. As will be discussed later,
this micro-scale turbulence could be an important part of determining the mixer process requirement.

3.   MICRO-SCALE/MACRO-SCALE SHEAR RATES

In order to understand the various phenomenon in the mixing tank, a more detailed
look must be taken at velocity fluctuations at a given point in the system. If the
flow from an impeller is turbulent, a laser velocity meter will give a velocity versus
time curve on a chart schematically represented in Figure 8. From this data the average velocity can be calculated; either graphically or mathematically within a computer
circuit and one can also calculate the root/mean/square velocity fluctuation (RMS). In
a mixing tank, there is a basic velocity fluctuation of the flow pattern at a point due
to the number of blades passing by any given point, and can also be affected to some
degree by the number of baffles that are in the system.

It turns out that if the average velocity from a radial flow impeller, is plotted, a
curve shown in Figure 9, results. As shown in this figure, we can calculate shear
rates at any point on the curve. This shear rate is called the macro-scale shear rate,
and is determined by measuring the velocity gradient or a slope on the velocity profile to give a shear rate normally in reciprocal seconds. As one can see by going
around the profile, there is a maximum shear rate at the top and bottom of the impeller
and an average shear around the entire impeller. We can also calculate the average
tank zone shear rate as well as the minimum tank zone shear rate.

Most indications are that these shear rates operate on particles 500 microns and larger. Figure 10 shows that small particles are unaffected by these large, slow moving
shear rates.

If we look at the velocity fluctuation, it turns out that particles of 100 microns or
less respond only to these high frequency micro-scale shear rates, Figure 10.

The only way to go from power to heat is through viscous shear. In low viscosity, turbulent fluid situations, the fluid elements must be very small to set up viscous shear rates. Micro-scale shear rates are where most of the energy is dissipated. In the laminar or viscous flow region it turns out that there is viscous shear all the way from large-scale shear rates to micro-scale shear rates. Much of the energy is absorbed in the large-scale shear rates, and so the micro-scale shear energy is not a significant portion of the total energy input to the system.

Power per unit of volume is a pretty good criteria of micro-scale mixing in turbulent fluid regimes. Many processes involve both macro and micro-scale shear rates. This means that variables must be looked at from both standpoints and, of course, changes the optimization technique depending upon the growth of importance of the macro/micro-scale relationship.

Table I shows what can happen to some of the parameters on scale-up. The first column shows a list of parameters, such as power, power per unit volume, diameter of the impeller, speed, flow, flow per unit volume, tip speed, and Reynolds Number. These parameters in the second column are all given in numerical value of 1. That does not mean that they are all identical but that we are starting with a given value to see how these numbers change relative to each other on scale-up. The third column looks at the situation with geometric similarity and constant power per unit of volume. The new volume is 125 times the old volume and the linear dimension scale-up is 5:1. Shown in column three the speed drops, the total flow increases, but the power per unit of volume decreases, the impeller tip speed increases and the Reynolds Number increases.

Column three is further explained in Figure 12, in which we take the case of geometric similarity and equal power per unit of volume and look at what happens to the maximum impeller zone shear rate which is shown going up in response to tip speed. The average impeller zone shear rate goes down, it responds to the impeller speed. This brings up another point that geometric similarity in itself controls no mixing variable whatsoever. In fact, if we wish to control two or three parameters, we would normally go to non-geometric similarity between the pilot-plant and plant.

Looking further at Column 3, it is seen that the flow per volume is much less and thus we would expect the blend time to be greater in the big tank.

In Column 4 an attempt is made to maintain constant flow per unit of volume in the tank. When this is done, the power per unit of volume must go up with the square of the dimension ratio or, in this case, 25 times. This also requires the same speed in the plant as in the laboratory. While this might be possible on occasion, it is not very practical. Typically, maintaining constant flow per unit of volume is not something that can be done on scale-up.

In Column 5, it is attempted to maintain constant tip speed, which keeps the maximum shear rate the same around the impeller. Using geometric similarity causes the speed to drop drastically and the power per unit of volume to go down inversely with the linear dimension ratio; in this case, one-fifth. The flow per unit of volume also is much lower in value than is true in Column 3.

Constant tip speed coupled with geometric similarity does not normally work very well. It usually is too unconservative and usually causes unsatisfactory performance in full-scale. Constant tip speed coupled with appropriate geometry changes, however, can be useful on certain types of process scale-ups.

In all the scale-up calculations so far, Columns 3, 4 and 5, the Reynolds Number has increased. In Column 6 an attempt was made to maintain constant Reynolds Number on scale-up. This causes the total power to drop, which is generally an inpractical situation, and the net result is that the Reynolds Number almost always increases upon scale-up.

What sort of things can we learn from Table I and Figure 12. (1) We cannot control all the parameters using geometric similarity, (2) Although it has not been demonstrated yet in this paper, we can control, on occasion, several variables if we use non-geometric techniques to control mixing parameters of interest in certain cases. (3) In general, a pilot plant tank will be too good a blending device and have too low a maximum impeller zone macro-scale shear rate to be a good model of plant performance.

In general, we need to change the impeller by either using smaller D/T ratios or using more narrow blade widths to make the pilot tank more similar to full-scale. In many cases it will not be possible even with drastic changes in geometry to make the pilot tank completely comparable to the full-scale performance.

Why does the scale-up situation look so discouraging on this particular series of examples? The reason is we are trying to control individual velocities of fluids, specific shear rates in special locations, and concern ourselves with the total overall flow in the tank.

There is an entire group of processes that do not care about these particular problems. For example looking at Figure 12, if we plot the mass transfer coefficient (K_Ga) versus tank size at constant power per unit volume at significant gas velocity, it turns out that there is a very good correlation. The same thing holds true with liquid/liquid mass transfer. Apparently, the total integrated mass transfer rate of these two-phase systems does not really care much what the individual flow and fluid shear relationships are. The mass transfer result on the average throughout the small and large tanks is about the same. Thus, careful evaluation of geometry and other things in the tank for scale-up is not of great concern when mass transfer is the main criteria.

4. IMPELLERS

It is a general rule that axial flow impellers illustrated by Figure 1, 2 & 7 are inherently higher flow and lower shear rate and micro-scale turbulence losses than radial flow impellers, Figure 9. For processes requiring high flow and low shear rates, such as blending and solid suspension, these impellers are normally the choice.

Radial flow turbines clarified by Figure 9 typically give much higher shear rates and much lower pumping capacities for a given level of power. They are often used for gas/liquid mass transfer, liquid/liquid mass transfer and for carrying out various kinds of shear rate dispersions.

One of the things normally to do in determining how to run a pilot plant is to take a look at the kind of mixing application most likely to be required and then take a look at proposed full size tanks and normally chose an impeller which is a practical impeller for that particular combination.

The relationship for any given impeller type says that the flow to the fluid shear ratio is proportional to $(D)^{8/3}$.

$$Q/H : D^{8/3}$$

With any given impeller type, we can therefore change the pumping capacity and fluid shear relationship over a fairly wide limit by chosing different impeller and tank size ratios. However, it is seldom possible even with large diameter radial flow impellers to have as much flow and as little shear as you get with small diameter axial flow impellers. With the combination of impeller types (axial or radial flow), and impeller diameter it is possible to cover a wide variety of process requirements. It is possible to prepare quite reliable estimates of the macro-scale and micro-scale rates throughout the mixing vessel with any given impeller type, axial or radial flow.

The comparison between axial and radial flow impellers is difficult since the entire structure of the flow pattern is different and none of the various portions of the shear rate spectrum match each type impeller.

5. MICRO-SCALE VS MACRO-SCALE

One can look at a mixing tank as though it were a velocity field, and draw fluid motion vectors and calculate pumping capacities. In fact, a superficial liquid velocity can be calculated which is the total flow from the impeller divided by the cross-section area of the tank. This gives a very simplistic picture of the mixing system. Trying to characterize different applications by the proper "superficial liquid velocity" is a very difficult thing to use with any degree of perception as to the overall fluid mixing process.

Using this concept; if one doubles the settling velocity of a solid particle, to obtain the same degree of suspension, we would have to double the "superficial liquid velocity". This would require doubling, the speed and would require 8 times the horsepower. Experimentally, we observe that the power level goes up 2 or 3 times, not 8 times. This approach also gives no feel for the diffusion, turbulence and shear rate phenomenon going on in the mixing process.

Many mixing processes have several things occurring. This is illustrated by Table II
which lists 5 basic fluid areas; gas/liquid, liquid/liquid, liquid/solid, miscible
liquids and fluid motion. There are then 2 different major divisions (1) physical criteria, such as solid suspension and (2) chemical reaction or mass transfer criteria is
present such as gas/liquid mass transfer processes. Some processes are three-phase --
gas/liquid/solid, or possibly four-phase -- gas/liquid/liquid/solid.

6. HANDLING COMPLICATED PROCESSES

There are a large number of scale-up proposals in the literature. Figure 14 gives some
of the ones that have been proposed for just solid suspension. For example, for mass
transfer, or chemical reactions, constant power per unit of volume is often a suitable
candidate. For maintaining equal maximum shear rate, equal tip speed is prime factor,
but must be balanced with a change in geometric relationships to maintain flow and
power at reasonable levels. To maintain equal degrees of solid suspension, power often
decreases by an exponent of around .9. There are equations that relate droplet size to
dimensionless parameters. Heat transfer correlations use such dimensionless numbers
as the Nosselt Number, Reynolds Number, and Prandtl Number. Blend time has been correlated as a blend number, θN as a function of Reynolds Number.

All of these items are very useful, particularly in the relative effect that they de-
sire. In looking at pilot plant results, it is often possible to take several of these
relationships, and examine how they behave on scale-up in a relative sense, and then in
which direction the process will go in the bigger tank.

Many times one relationship has an affect on another. For example, it turns out that
the affect of gas dispersion on heat transfer coefficient is very minor, since the gas
bubble does not apparently blanket the liquid surface of a jacket or a coil.

Addition of solids to a tank enhances the heat transfer coefficient somewhat because
solids have inertia and tend to act to further increase turbulence in the fluid film.

7. GAS/LIQUID/SOLID PROCESSES

It turns out that unless the mixer energy is about 3 times higher in a radial flow impeller than in an expanding gas stream, the mixer will not control the flow pattern. At
a ratio of 1:1, the gas flow pattern controls, but the tank is still operable, and can
be designed for effective mass transfer.

With axial flow impellers the power, in the impeller must be about 10 times higher than
the expanding gas stream. If it is not, the gas flow tends to dominate and completely
destroy the axial flow pattern. It is common to use radial flow impellers with a combination gas/liquid/solid system, radial flow impellers for gas/liquid systems and
axial flow impellers for liquid/solid systems.

Normally what has to be done in a complicated mixer system is to look at all the potential mixer criteria, as referenced on Table II, and sub-divide and add refinements
where necessary.

For example, let's say there is a solid crystal produced in a gas/liquid/solid system.
Involved is mass transfer, nuecleation to form crystals, the shape and size of the
crystals as they are affected by the mixing variables in the tank. One set of correlations and regulations operates the gas/liquid mass transfer rate. A third set of mixer
relationships determines the affect of mixing on the size and shape of the crystal product. Scale-up will affect each of these three process parameters in a different fashion. An analysis must be made as to the probable affect of the change in flow, fluid
shear, mass transfer, turbulence, micro-scale/macro-scale on each of the important
steps of the process. Typically, a conservative design is used to estimate the mixer
requirement in the various steps and a prediction made of the probability of successful operation or the possibility of problems and what the nature of these problems
might be.

If it is quite apparent that increasing the power, speed or impeller diameter can do
nothing but good for the process and no harm could result, then a mixer with sufficient
design factor can improve the chance of a mixer operating successfully. On the other
hand, it may be that the things done to improve mass transfer may be detrimental to the
shear rate the mixer it produces, and thus to the size and shape of the crystal and the
product. A generous design factor may not be appropriate. Pilot plant studies are
used to evaluate such complexities.

REFERENCES

Oldshue, J.Y.: "Scale-Up, 'Mixing' 1 Chemical Engineering Review", J.J. Ulbrecht - Editor, Gorden & Breach, London 1985

Oldshue, J.Y.: "Agitation Scale-Up In The Chemical Process Industries", Bisio & Kabel, John Wiley, New York, 1985

Oldshue, J.Y.: "Perspectives On The Scale-Up of Impeller Mixers", Schmidtke & Smith, Butterworth, 1983

Oldshue, J.Y.: "Fluid Mixing Technology", McGraw-Hill, New York, 1983

<u>APPENDIX</u>

<u>Examples:</u>

1. Leaching & Extraction of Mineral Values from High Concentration of Solids - A uranium plant had 10 large slurry tanks for leaching and extraction (approximately 14 meters in diameter and 14 meters high). They had about 14,000 cubic meters capacity.

In a study designed to modify the leaching operation, it was desired to look at two different grind sizes of ore, one labeled fine grind and one labeled course grind. Also, the effect of various mixer designs and power levels on the extraction efficiency to arrive at the overall economic optimum was examined.

Figure 15 shows the result of a pilot study in which the impeller speed for a given impeller and tank geometry was measured for complete overall motion throughout the slurry for both the fine and course grind at various weight percent solids. As can be seen by that figure, the fine material required lower horsepower at low weight percent solids while the course grind required less horsepower up near the ultimate settled solids weight percentage.

The interpretation is that at lower percent solids, the viscosity of the fine grind aided suspension; while at higher percent solids, the higher viscosity of the fine material was detrimental to fluid mixing.

A mixing viscosimeter was used to measure the viscosity of the slurry. Figure 16 shows the viscosity of the fine and course slurries.

By combining the data from Figures 15 and 16 into Figure 17, it is seen that there is a correlation between the impeller speed required and the viscosity of the slurry regardless of whether the material was finely or coursely ground. This illustrates that viscosity is a key parameter in the process design for solid/liquid slurries.

The overall process economics examined the extraction rate as a function of power, residence time and grind size. The full-scale design possibilities were represented in the form on Table III, which were accompanied by other charts which gave different heights of suspension in the tank for the three different particle size fractions; fine, medium and course. These various combinations of power levels also gave various blending efficiencies and had different values of the effective residence time used in a system.

By calculating the residence time of the various solids in the tank and relating them to their corresponding extraction curves, the total uranium extraction for the entire train of mixers was estimated. The cost of the various mixer options, the production efficiency net result, and the cost of the installation and tank design could be combined to yield the economic optimum for the plant.

2. Fermentation - Figure 18 shows the results of fermentation yield for a typical batch process with 6 days total cycle time. Yield was a function of power input. As the power increases, the yield goes up until a point where the yield starts to decrease. One explanation is that this is a point where the fluid shear rate of the impeller becomes too great for the viability of the solid organism. Using an impeller with a larger D/T, it turns out that more power can be applied before this maximum shear rate is reached and the process result can be improved because the extra horsepower gives extra mass transfer which is desirable, and there is not a detrimental effect of shear rate on the viability of the organism.

In terms of overall plant design, a typical method is to use the chart shown in Table IV. The first column shows some of the important mixing variables and process variables involved in the pilot plant. Process variables are the liquid/solid mass transfer rate between the dissolved oxygen and the solid organism, and the gas/liquid mass transfer rate.

Referring now to the second column, which is the first column in the plant scale design, a D/T similar to the pilot plant is used and results in a shear rate 80% higher than there was in the pilot plant at the required conditions for liquid/solid and gas/ liquid mass transfer. By going to a larger D/T ratio. it is possible to cut this maximum shear rate ratio back to 1.4. By going to a still larger D/T ratio, 0.6, it is possible to reduce the shear rate in the plant unit to about the same level as it was in the pilot plant and still maintain the proper conditions for liquid/solid and gas/liquid mass transfer.

Since the speed required to drive impellers at large diameters is lower, than for
small diameters, it typically requires more torque from the mixer. The mixers shown
in columns 3 and 4 have an improved process result, but also have an increased capital
cost.

To explore the matter further, a new impeller was built for the pilot plant which increased the shear rate to a value 40% higher than it was in the original pilot plant
run. Using this impeller, as shown, the yield versus power curve works out pretty
well up to the design point and then a little bit beyond the point that higher shear
rate impeller causes a fall-off. This indicates that the shear rate proposed for
mixer C is o.k. What about the economies effected by using mixer B? Well, it turns
out that there is no way to build a practical impeller to increase the shear rate 80%
over what it was in the original pilot plant. If this were done, the blades would be
so narrow that there would be no relationship at all between the gas bubbles, liquid
particles and the other steps in the process.

It is not uncommon to have fewer data than we would like, so in looking at the chart
in Table IV, we can conclude that mixer D would certainly be safe and actually would
have improved blend times, heat transfer characteristics and might well be worth the
extra money in capital dollars it would cost to install.

Mixer C looks perfectly adequate, and from all that is known about the process, one
should have no difficulty in arriving at design production.

Mixer B looks risky even though this could not be tested in the pilot plant. There
would be a real risk taken in the extra shear rate that mixer B would provide as to
whether it would harm the viability of the organism and hurt the process results.

3. Effect of Multi-Size Particles - The effect of particle size in settling velocity
on power required for the uniform mixture of a given particle size is typically to an
exponent of 0.5. Some studies were made in which different degrees of suspension were
achieved where there was a mixture of particle sizes, and the settling velocity of the
particle size that was 60% suspended through the tank was curved. The curve in Figure
18 resulted showing the exponent on settling velocity to be 0.25. This illustrates
the fact that literature correlations based on visual studies using very carefully
screened and sized particles of a single uniform size do not give always the same
effect as this would happen where the particles were only one component of a multi-
size distribution of solids.

4. Homogeneous Chemical Reactions - A chemical reaction in the plant was observed to
give 70% yield. The chemicals were expensive and this was a continuous flow process,
and it was decided to go to a pilot plant to see whether improvements could be made
in the process.

A pilot plant was built geometrically similar to the plant. The retention time was
kept constant and when the yield was measured, it turned out to be 80%. This would
be a significant savings in chemicals and productivity so things were looked at in
the plant that might increase its yield also to 80%. Nothing that was done to the
plant that seem to help very much. On the other hand, studying variables in the
pilot plant did not change the results much from 80%.

An analysis of the pilot tank indicated that the pumping capacity per unit volume of
the pilot unit was 3 times higher than it was in the plant. This meant that the dilution ratio and the concentration of reagents around the impeller were different than
they were in the plant. Also the blend time throughout the reactor was different.

As a quick check, the flow through the pilot plant unit was increased to where it had
one-third the residence time, keeping the ratio of pumping capacity to feed rate about
the same. This reduced the yield into the 70% range.

The pilot plant was now "sensitized" to the plant. The variables that were studied
at the pilot plant will usually show up in similar fashion in plant-scale. Without
"sensitizing" the pilot plant it could have been operated extensively for years
without much useful data on relationships on the full-scale plant being obtained.

PROPERTIES OF A FLUID MIXER ON SCALE-UP

PROPERTY	PILOT SCALE, 80 LITERS	PLANT SCALE, 27,400 LITERS			
P	1.0	343	16,000	49	0.14
P/VOL	1.0	1.0	49	.14	0.0004
N	1.0	0.27	1.0	.14	0.02
D	1.0	7.0	7.0	7.0	7.0
Q	1.0	93	343	48	7.0
$Q/AREA$	1.0	1.9	7.0	1.0	0.14
Q/VOL	1.0	0.27	1.0	.14	.02
ND	1.0	1.9	7.0	1.0	0.14
$\dfrac{ND^2\rho}{\mu}$	1.0	13.2	49	7.0	1.0
$\dfrac{N^2D}{g}$	1.0	.4	7.0	1.0	.003
$\dfrac{N^2D^3\rho}{\sigma}$	1.0	8580	117,650	2300	47

MIXING   PROCESSES

PHYSICAL PROCESSING	APPLICATION CLASSES	CHEMICAL PROCESSING
SUSPENSION	LIQUID – SOLID	DISSOLVING
DISPERSIONS	LIQUID – GAS	ABSORPTION
EMULSIONS	IMMISCIBLE LIQUIDS	EXTRACTION
BLENDING	MISCIBLE LIQUIDS	REACTIONS
PUMPING	FLUID MOTION	HEAT TRANSFER

PROPOSAL FOR REVISED INSTALLATION

PROCESS FACTOR	RELATIVE TORQUE	MOTOR kW	D/T (DUAL)
1.2	1.0	300	.5
1.0	1.0	250	.5
0.8	.9	200	.5
0.7	1.0	150	.6

TABLE IV - Chart for scaling up fermentation process into three different full-scale mixer sizes, Column B, C & D.

| | PILOT | PLANT | | |
	A	B	C	D
T.	1.2	7.0	7.0	7.0
P.	1.0	340	340	340
N	1.0	.31	0.16	.075
D/T	.25	.25	0.38	0.6
NO. of IMP.	1	2	2	2
Z.	1.2	12.0	12.0	12.0
F.	F	10 F	10 F	10 F
L-S. MTR	1.0	1.0	1.0	1.0
G-L. MTR	1.0	1.0	1.0	1.0
MAX. I.Z. SHEAR RATE	1.0	1.8	1.4	1.0
TORQUE	---	1.0	2.0	4.1
COST ($)		10,000	18,000	35,000

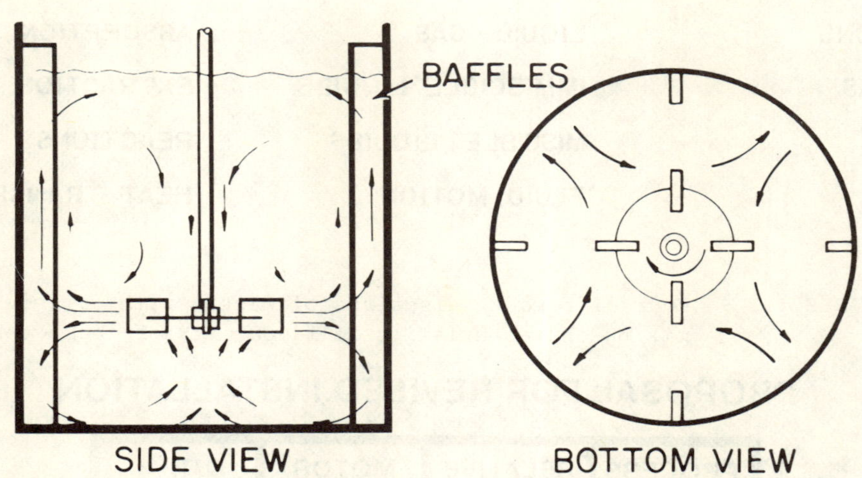

Figure 1 - Schematic of flow pattern with radial flow turbine.

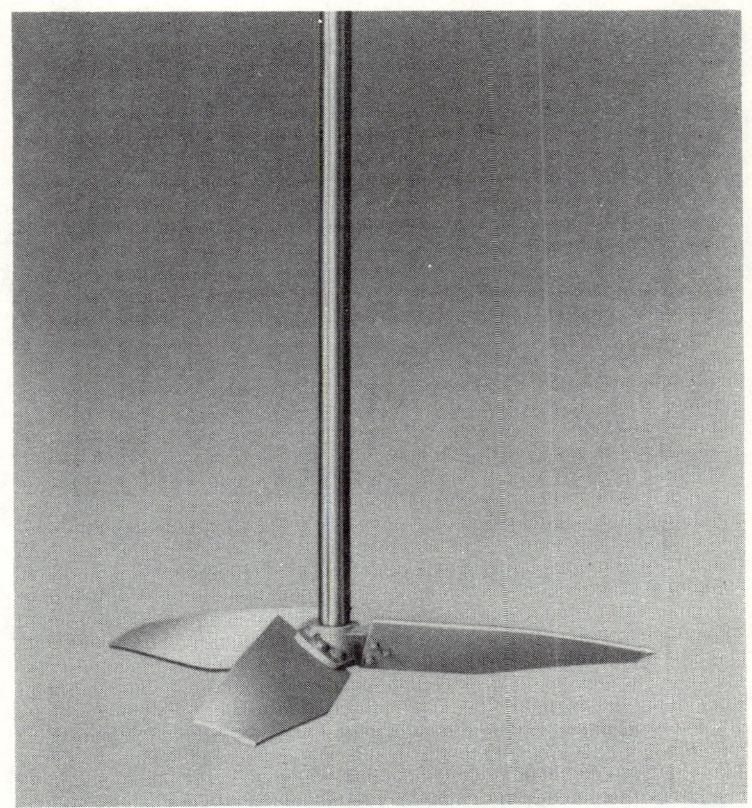

Figure 2 - Illustration of Fluid-Foil-type axial flow impeller.

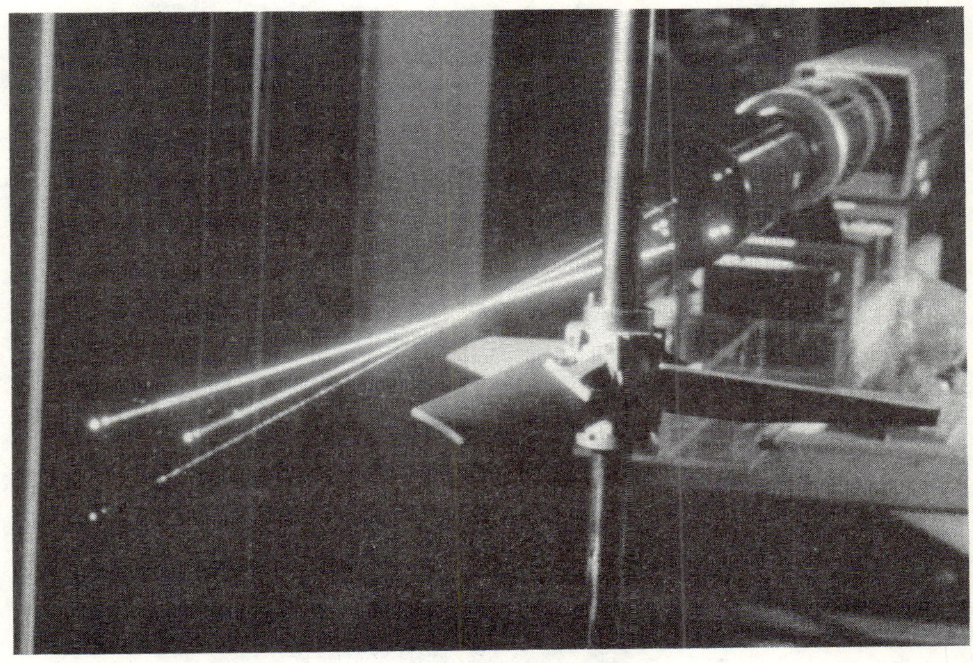

Figure 3 - Laser velocity meter for measuring fluid mixing parameters.

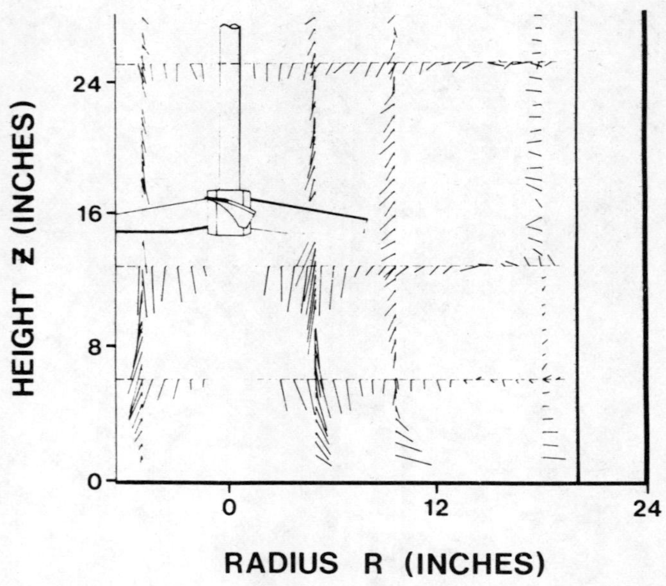

Figure 4 - Typical velocity profile of Fluid-Foil-type axial flow impeller.

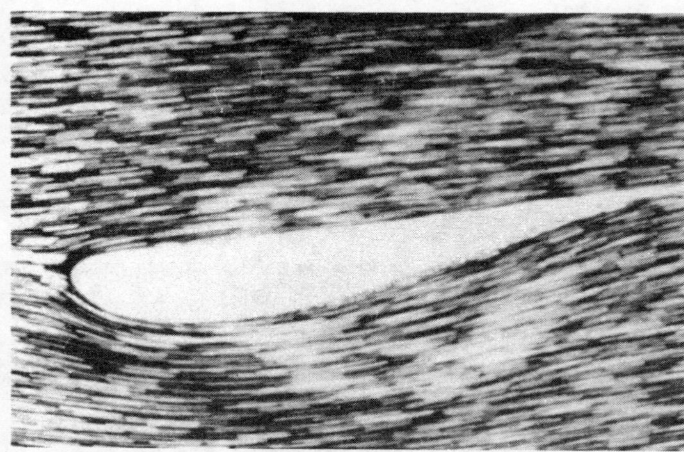

Figure 5 - Schematic of impeller orientation for minimum drag coefficient.

Figure 6 - Schematic of airfoil orientation with high drag coefficient corresponding to "stall".

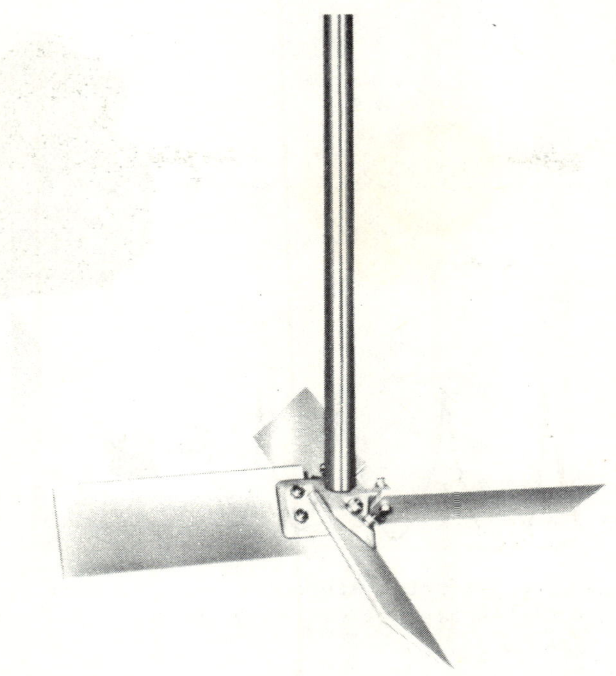

Figure 7 - Typical constant angle axial flow turbine.

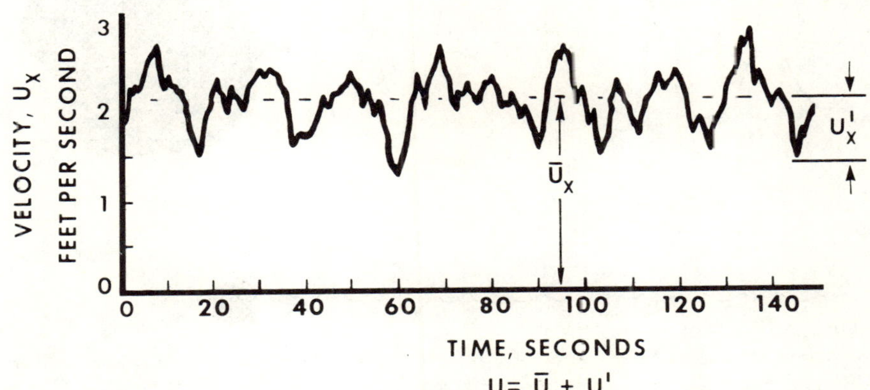

Figure 8 - Schematic of velocity probe measurement at a given point with time.

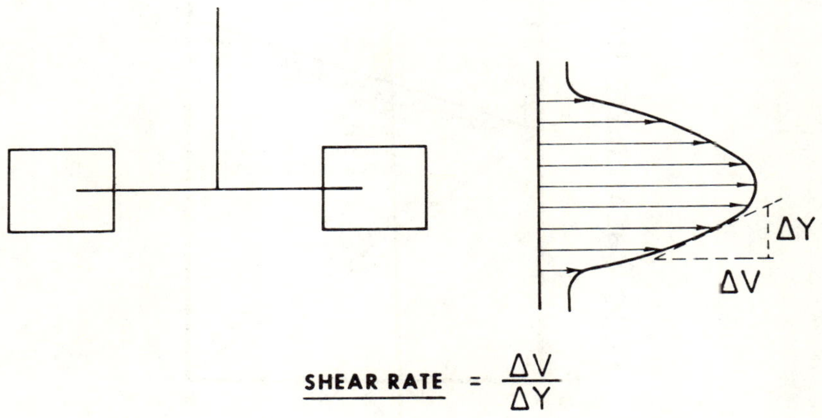

Figure 9 - Velocity profile from radial flow impeller illustrating
calculation of fluid shear rate.

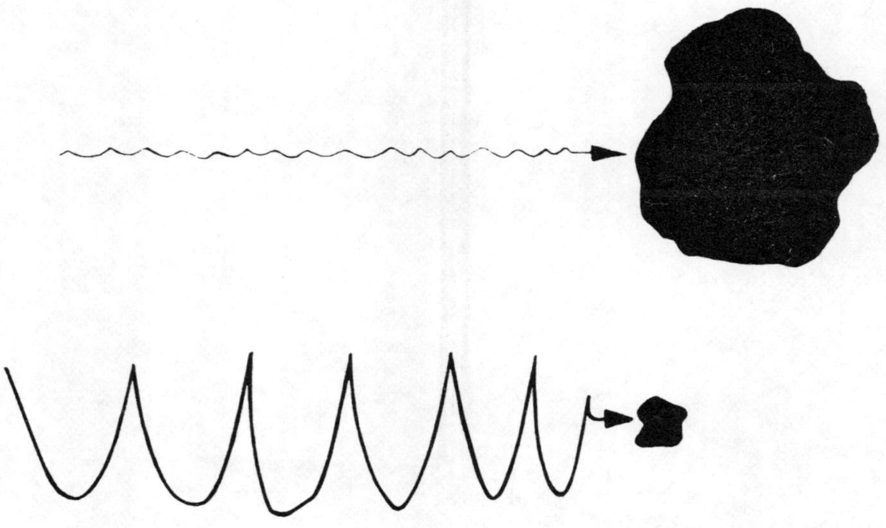

Figure 10 -   Schematic: small particles are not effected by macro-scale shear rates and large particles are not effected by micro-scale shear rates.

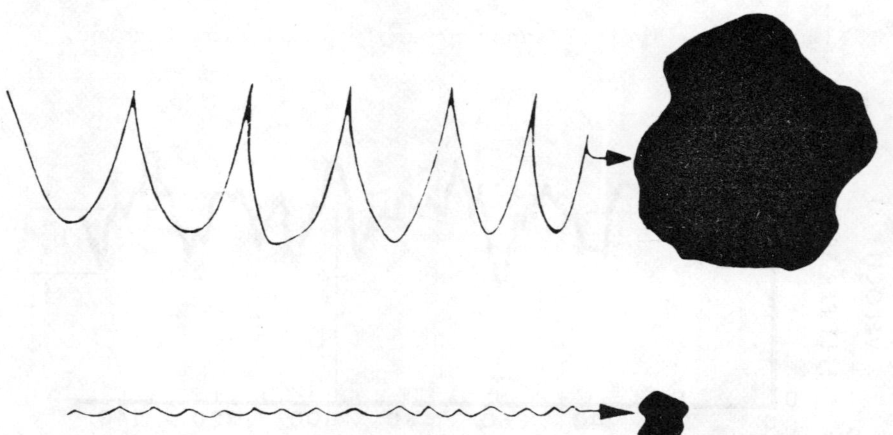

Figure 11 -   Small particles are effected by micro-scale shear rates and large particles are effected by macro-scale shear rates.

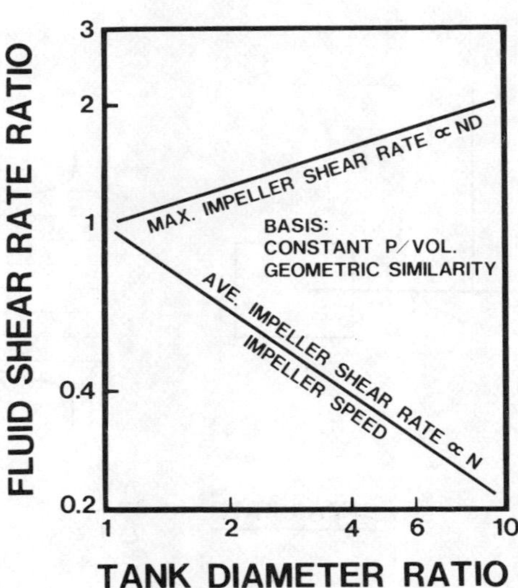

Figure 12 -   Illustration that maximum impeller zone macro-scale shear rate increases while average impeller zone macro-scale shear rate decreases upon scale-up.

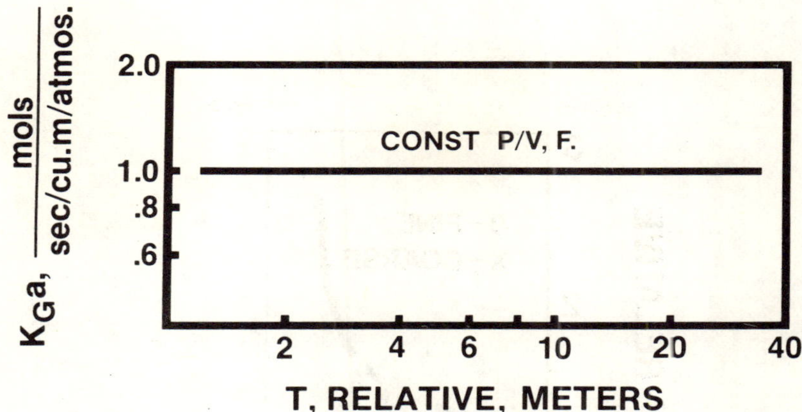

Figure 13 - Typical variation of K_Ga with tank size at constant power per unit of volume.

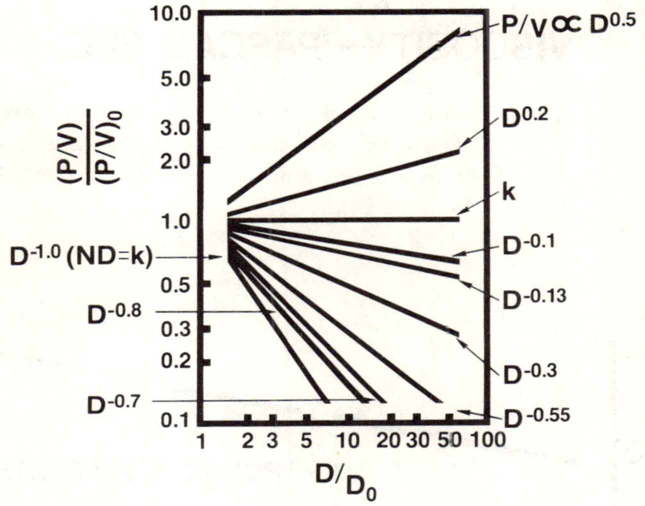

Figure 14 - A listing of the many different scale-up exponents determined experimentally by various investigators in the area of liquid/solid suspension.

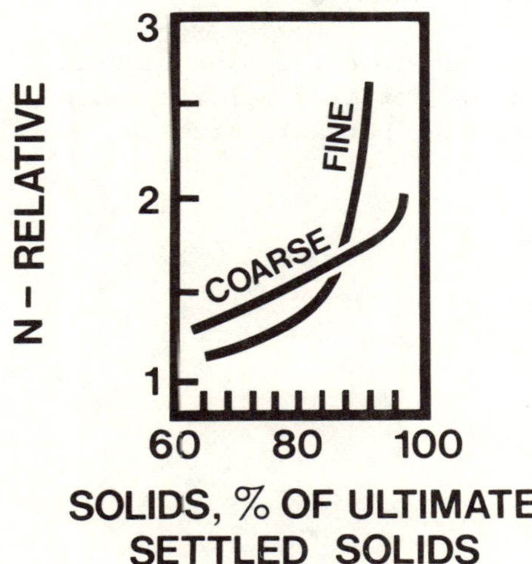

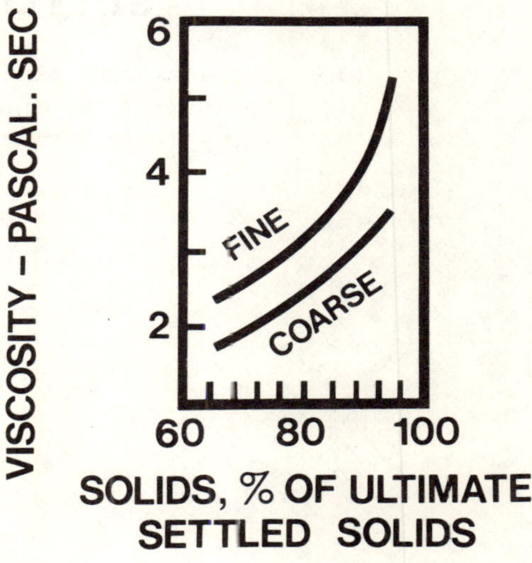

Figure 15 - Effect of percent solids on speed required for complete motion throughout the tank on two different grind sizes.

Figure 16 - Viscosity of fine grind and coarse grind at various weight percent solids.

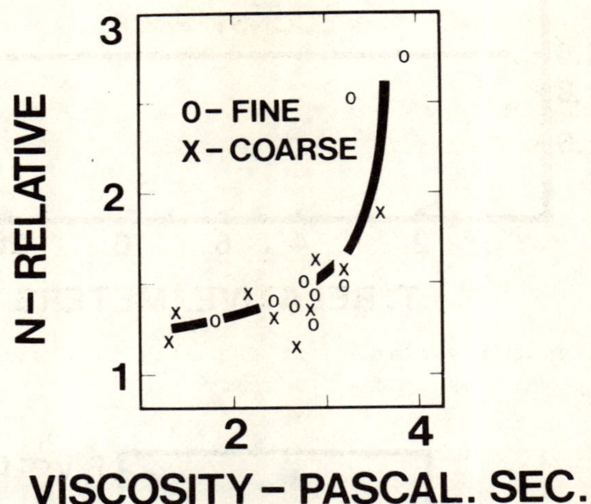

Figure 17 - Correlation of impeller speed versus viscosity of the pulp including both fine and course grind experimental data points.

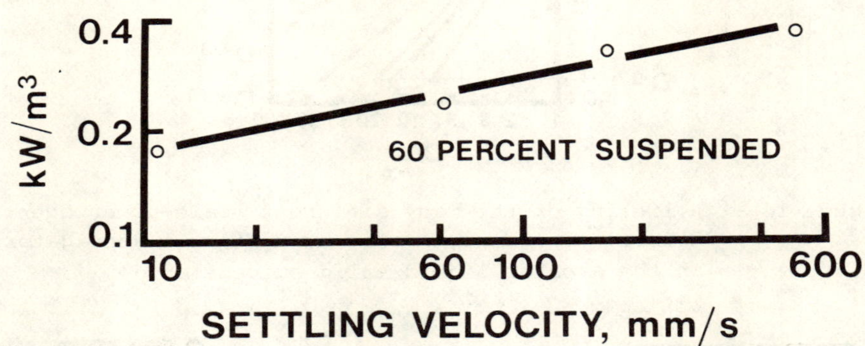

Figure 18 - Experimental summation of the increase in horsepower requried to suspend particles as a function of settling velocity when they are in a mixture of many different particle sizes.

SUSPENSION OF SOLID PARTICLES – RELATIVE
VELOCITY OF PARTICLES IN TURBULENT MIXING

Ditl,P., Rieger,F.

Czech Technical University, Department of Equipment Design
for Process Industries
Suchbátarova 4, 166 07 Prague 6
Czechoslovakia

Summary

In our previous papers [12],[5],the existence of different
particle-liquid hydrodynamic subregions depending on the existence of
intra-particle forces, particle diameter and the turbulence intensity
has been reported. We investigated separately suspension of "small"
and "large" particles with negligible influence of intra-particle forces
wherefrom we obtained different scaling-up criteria and different de-
pendences of impeller speed required for complete suspension vs. particle diameter. The dividing criterion between "small" and "large" particles has been proposed and experimentally verified as the ratio
$(d_p/\eta_T)_{CR}$ = 32. The equation for critical impeller speed was derived
theoretically in the form $Fr_m = C(T/D)^3$ which is in very good agreement with experiments. Overall correlation of suspension experiments
was statistically obtained in the form:

$$Re_f \;=\; C_{2,3}\; Ar^{0.45} \left(\frac{d_p}{T}\right)^{\beta} \left(\frac{T}{D}\right)^{-0.56}$$

with exponent $\beta = -1.42$ for $d_p/\eta_T \geq 32$ and $\beta = -1.25$ for
$(d_p/\eta_T) < 32$.

It was experimentally proved that the exponent of Archimedes
number has the constant value of 0.45 in the wide range of $Ar < 1 \div 10^6 >$
covering Stokes, intermediate and Newton particle-liquid hydrodynamic regions at settling in a still liquid. The obtained value 0.45 is
very close to the theoretical value of Ar - exponent 0.5 valid for the
turbulent particle settling.

Further, the relative particle-liquid velocity at the state
of suspension was determined indirectly from particle suspension
measurements

$$u_{Rf} \;=\; C_9\; \frac{\nu}{d_p}\; Ar^{0.45} \left(\frac{d_p}{T}\right)^{\alpha} \left(\frac{T}{D}\right)^{-0.23}$$

with exponent $\alpha = -0.09$ for $(d_p/\eta_T) \geq 32$ and $\alpha = 0.08$ for
$(d_p/\eta_T) < 32$.

Held at Wurzburg, 10-12 June, 1985.

Organised by DVCV· Deutsche Vereinigung für Chemie- und Verfahrenstechnik
(German Association of Chemical and Process Engineering).

Organisation: GVC·VDI-Gesellschaft Verfahrenstechnik und Chemieingenieurwesen **GVC**

©BHRA, The Fluid Engineering Centre, Cranfield, Bedford MK43 0AJ, England.

NOMENCLATURE

$B, C, C_1, C_2 \ldots C_{12}, C'$ = constants

c^v = volume concentration of solid phase

d_p = particle diameter, m

D = impeller diameter, m

g = acceleration due to gravity, m/s^2

H = liquid depth, m

H_2 = impeller clearance, m

l = characteristic eddy dimension, m

n_f = impeller speed of complete suspension, 1/s

P = impeller power input, W

T = vessel diameter, m

u_t = terminal settling velocity in a still liquid, m/s

u_T = terminal settling velocity in a turbulent field, m/s

u'_T = fluctuating component of u_T, m/s

u_{Rf} = vertical component of the relative velocity between a particle and fluid, m/s

V = bulk volume, m^3

$\mathcal{E}_v^{(e)}$ = kinetic energy of the flow per unit volume, W/m^3

$$\mathcal{E} = \frac{P}{\rho_{sus} \cdot V} = \frac{4\,Ne}{\pi} n^3 d^2 \left(\frac{d}{D}\right)^3 \; ; \; (H=D) = \text{specific power per unit mass, W/kg}$$

$\mathcal{E}_v$ = specific power per unit volume, W/m^3

μ = dynamic viscosity, Pas

$\nu = \mu/\rho$ = kinematic viscosity, m^2/s

η_T = characteristic eddie scale of the dissipating eddies (eq.(1))

ρ_{sus} = suspension density, kg/m^3

ρ = liquid density, kg/m^3

ρ_s = particle density, kg/m^3

$\Delta\rho$ = density difference, $\Delta\rho = \rho_s - \rho$, kg/m^3

φ_p, ϕ_p = functions given by eqs. (12a, b)

Ar = Archimedes number, $Ar = d_p^3 \Delta\rho\, g\, \rho/\mu^2$

Fr_m = modified Froude number, $Fr_m = \rho\, n_f^2 T / \Delta\rho g$

Ne = impeller power number, $Ne = P/\rho \cdot n^3 d^5$

Re_f = impeller Reynolds number at suspension, $Re_f = n_f D^2 \rho / \mu$

Re_{pf} = particle Reynolds number at suspension $Re_p = u_{Rf}\, d_p\, \rho/\mu$

1.   INTRODUCTION

The review and critical discussion of papers dealing with complete suspension, is given elsewhere [2],[1] ,[9],[4]. The results obtained by various authors and the proposed scaling-up criteria differ mutually.

In our previous paper [12] we reported the existence of different suspension mechanismus depending on the existence of intra particle forces, particle diameter and the intensity of turbulence in which the experiments haved to be correlated separately. The main aim of this paper is to determine the dividing criterion between the two mechanisms in suspensions of "small" and "large" particles with negligible influence of intra-particle forces, to compare equations obtained experimentally and theoretically for critical impeller speed necessary for complete suspension and to evaluate the results from the point of the relative particle velocity in a turbulent field.

2.   THEORETICAL

Impeller rotating in a baffled vessel generates the primary eddies, the scale of which is, for the given system geometry, proportional to the impeller diameter [10]. Such eddies having the relatively long age are the carriers of only the smaller part of kinetic energy ($\sim$ 20 %). The rest of energy is represented by the small and middle eddies which are dissipated due to viscous friction. As the particle size is comparable with the size of primary eddies, the suspension mechanism is governed by the motion of primary eddies. For smaller particles the suspension mechanism is determined by the motion of the middle and small eddies, having the characteristic scale:

$$\eta_T = \left(\frac{\nu^3}{\varepsilon} \right)^{1/4} \tag{1}$$

2.1   Dividing criterion

We postulated that the criterion between two above mentioned suspension mechanisms might be defined by the ratio d_p/η_T. Expressing the ratio d_p/η_T by means of equation (1) and subtituting for η_T one obtains:

$$\frac{d_p}{\eta_T} = \frac{d_p}{(\nu^3/\varepsilon)^{1/4}} = \left(\frac{4\,Ne}{\pi} \right)^{1/4} \frac{d_p}{D}\, Re^{3/4} \left(\frac{T}{D} \right)^{-3/4} \tag{2}$$

2.2 Theoretical models for suspension of "large particles" based on energy balances

For a solid particle suspended in a moving liquid, the liquid kinetic energy, being proportional to $\varrho u'^2$, is assumed be equal to the potential energy necessary to lift the particle by the eddy of scale l. Since this potential energy is proportional to $\Delta \varrho g l$ we have,

$$\varrho u'^2 \sim \Delta \varrho g l \tag{3}$$

The fluctating velocity u' can be expressed [10] as:

$$u' \sim \left(\frac{\mathcal{E}_v^{(e)} l}{\varrho} \right)^{1/3} \tag{4}$$

where $\mathcal{E}_v^{(e)}$ is the energy of primary eddies per unit volume. Comparing eqs. (3) and (4), one obtains the basic equation in the form:

$$\varrho \left(\frac{\mathcal{E}_v^{(e)} l}{\varrho} \right)^{2/3} \sim \Delta \varrho g l \tag{5}$$

The scale of primary eddies is in geometrically similar systems proportional to the impeller diameter $l \sim D$, so that one can write:

$$\varrho \left(\frac{\mathcal{E}_v^{(e)} D}{\varrho} \right)^{2/3} = C' \Delta \varrho g D , \tag{6}$$

where C' is a constant of the system.

Specific energy of the primary eddies $\mathcal{E}_v^{(e)}$ is, for a given system, particle concentration and hydrodynamic region, proportional to the specific impeller power input $\mathcal{E}_v$:

$$\mathcal{E}_v^{(e)} = C_4 \mathcal{E}_v = C_5 \ \varrho_{sus} n_f^3 d^2 \left(\frac{D}{T} \right)^3 \tag{7}$$

where $C_5 = 4 \, Ne \, C_4 / \pi$.

Substituting eq. (7) into eq. (6) and after rearrangements the final equation for critical impeller speed for suspension has been obtained in the following eguivalent forms:

$$n_f = C \left(\frac{\Delta \varrho g}{\varrho} \right)^{1/2} \left(\frac{T}{D} \right)^{3/2} D^{-1/2} \tag{8a}$$

$$\text{or} \quad Fr_m = \frac{\varrho n_f^2 T}{\Delta \varrho g} = C \left(\frac{T}{D} \right)^3 \tag{8b}$$

$$\text{or} \quad Re_f = C \, Ar^{0.5} \left(\frac{d_p}{T} \right)^{-3/2} \left(\frac{D}{T} \right)^{1/2} \tag{8c}$$

2.3 Model based on the terminal settling velocity in a turbulent field

The vertical component of the relative velocity between a particle and fluid $d_p \gg \eta_T$ at the state of complete suspension is [13] :

$$u_{Rf} = C_{11} (d_p \, \mathcal{E}_f)^{1/3} , \qquad (9)$$

where the review of other equations for u_{Rf} at different d_p / η_T can be found.

The same value of u_{Rf} can be expressed as:

$$u_{Rf} = u_T + u_T' , \qquad (10)$$

where u_T is a mean value of the terminal settling velocity in a turbulent field and u_T' is its fluctuating component. The ratio of u_{Rf} and the terminal particle velocity u_t in a still liquid can be expressed as:

$$\frac{u_{Rf}}{u_t} = \frac{u_T}{u_t} + \frac{u_T'}{u_t} = \frac{u_T}{u_t} \left(1 + \frac{u_T'}{u_T} \right) = \phi_p \, \varphi_p , \qquad (11)$$

where $\phi_p = u_T / u_t$ and $\varphi_p = 1 + u_T' / u_T$ (12a,b)

One can express u_{Rf} in terms of u_t as:

$$u_{Rf} = \phi_p \cdot \varphi_p \cdot u_t \qquad (13)$$

Terminal velocity u_t of a particle in a still liquid can be expressed as:

$$u_t = B^{1/3} \, \frac{\nu}{d_p} \, Ar^{\frac{\xi+1}{3}} , \qquad (14)$$

where the constant B and exponent ξ depend on Ar as shown in Table I:

Table I

Range of Ar	B	ξ	$(\xi+1)/3$
$Ar < 18$	1.715×10^{-4}	2	1
$36 < Ar < 0.25 \times 10^4$	1.045×10^{-2}	1.114	0.7
$8.25 \times 10^4 < Ar < 1.32 \times 10^{10}$	5.27	0.5	0.5

Combining eqs. (14), (13) and (9) one obtains the equation for the critical impeller speed for complete suspension in the form:

$$Re_f = C_8 \, Re_{pf} \left(\frac{d_p}{T} \right)^{-4/3} \left(\frac{T}{D} \right)^{-1/3} = C_8 \cdot B^{1/2} \, \varphi_p \, \phi_p \, Ar^{\frac{\xi+1}{3}} \left(\frac{d_p}{T} \right)^{-4/3} \left(\frac{T}{D} \right)^{-1/3} \qquad (15)$$

where $C_8 = C_{11}^{-1} (4 \, Ne / \pi)^{-1/3}$ and

$$Re_{pf} = B^{1/3} \phi_p \cdot \varphi_p Ar^{\frac{\delta+1}{3}} \qquad (16)$$

In most cases   is $u_T \gg u_T'$, so $\varphi_p = 1$.

3.   EXPERIMENTAL

The experiments were carried out with a standard mixing apparatus. A six-blade disc turbine, flat six-blade turbine, pitched six and three-blade turbine (Fig. 1a,b,c,d) were employed in flat - bottomed, baffled cylindrical vessels 0.15, 0.20, 0.30, 0.40 m in the inner diameter. The relative liquid depth was H/T = 1 in all cases investigated. The ratio T/D was varied in the range from 2.5 to 4. Impellers were tested in positions H_2 = 1/3, 2/3, 1 and 1.5 D. The Newtonian liquid viscosity was varied in the range from 10^{-3} to 0,47 Pas. Based on visual observation the impeller speed for complete suspension was determined as the average value of the independent observations. The glas spheres (ϱ_s = 2600 kg/m^3) of 0.085; 0.15; 0.25; 0.31; 0.64; 0.8; 1; 2; 3 and 4 mm in diameter and ion exchange particles (ϱ_s = 1259 kg/m^3) of 0.85 mm in diameter were used as a model solid phase. The volumetric concentration of solid phase was 0, 1, 2.5 and 5 %.

4.   RESULTS AND DISCUSSION

4.1   Dividing criterion

Critical impeller speed for suspension n_f increases with an increase of particle size d_p up to a certain value n_{fCR}. Further increase of d_p is not accompanied with an increase of n_f as it is seen in Fig. 4. Critical value of d_{pCR} was determined graphically as the break-point on the logarithmic curve $n_f = f(d_p)$. In such way 38 experimental sets (380 experiments) had been evaluated to obtain 38 values of d_{pCR}. Three impeller types had been tested in the positions H_2/D = 1/3, 2/3, 1 with relative impeller diameter T/D = 2.6, 3 and 3.9. Vessel diameter was 0.4 and 0.2 and the volumetric solid concentration was 1 %. Viscosity was 1 and 4.8 m Pas. Particles of previously specified diameters had been used in these experiments. From the obtained values of d_{pCR} the corresponding values of $(d_p/D)_{CR}$, $(d_p/T)_{CR}$, Ar_{CR} and $(d_p/\eta_T)_{CR}$ had been calculated. Whereas the values of $(d_p/D)_{CR}$, $(d_p/T)_{CR}$, Ar_{CR} had been varied in the wide ranges: $(d_p/D)_{CR} < 6.67 \times 10^{-3} \div 1.05 \times 10^{-2} >$, $(d_p/T)_{CR} < 4.05 \times 10^{-3} \div 1.7 \times 10^{-2} >$, $Ar_{CR} < 8362 \div 22300 >$, the values of $(d_p/\eta_T)_{CR}$ were found reasonably constant:

$$\left(\frac{d_p}{\eta_T}\right)_{CR} = \left(\frac{4\,Ne}{\pi}\right)^{1/4} \frac{d_{pCR}}{D}\ Re^{3/4} \left(\frac{T}{D}\right)^{-3/4} = 32 \pm 5 \qquad (17)$$

4.2 Critical impeller speed for complete suspension

The value $(d_p/\eta_T)_{CR} = 32$ divided our experiments into two groups wich had been statistically treated separately. The characterizing criteria varied in the following ranges in our experiments:
$Ar \langle 1.3 \div 10^6 \rangle$, $d_p/D \langle 0.8 \times 10^{-3} \div 10 \times 10^{-2} \rangle$, $T/D \langle 2 \div 4.4 \rangle$
$d_p/\eta_T \langle 0.3 \div 100 \rangle$.
On the basis of a detailed statistical analysis the final equation was obtained in the form:

$$Re_f = C_{2,3}\ Ar^{0.45} \left(\frac{d_p}{T}\right)^{\beta} \left(\frac{T}{D}\right)^{-0.56} \qquad (18)$$

with exponent $\beta = -1.42$ for $(d_p/\eta_T) \geq 32$ - "large particles" and $\beta = -1.25$ for $(d_p/\eta_T) < 32$ - "small particles". The values of constants C_2 and $C_3 = f(H_2/D\ ;\ c^V\ ;\ \text{impeller type})$ are given in Table II.

Table II: The values of constants C_2, C_3

C_2 , C_3	c^V [vol.%]	H_2/D		
		1/3	2/3	1
pitched	0	0.692 , 2.08	0.755 , 2.31	0.852 , 2.81
six-blade	1	0.770 , 1.73	0.831 , 1.99	0.957 , 2.28
turbine	2.5	0.880 , 2.31	– , 2.56	– , 3.08
	5	0.985 , 2.17	1.02 , 2.34	1.20 , 2.94
pitched	0	0.816 , 2.31	0.871 , 3.02	1.03 , 3.11
three-blade	1	0.910 , 1.99	0.953 , 2.24	1.04 , 2.57
turbine	2.5	1.09 , 2.56	1.13 , 3.04	1.19 , 3.64
	5	1.14 , 2.34	– , 3.19	– , –
flat	0	0.752 , 2.09	0.982 , 2.49	– , –
six-blade	1	0.797 , 1.75	1.03 , 2.31	1.08 , 2.81
turbine	2.5	0.885 , 2.27	– , 2.70	– , 3.21
	5	0.985 , 2.20	1.35 , –	– , –
Standard	0	0.857 , 2.35	1.11 , 2.89	– , –
disc-turbine	1	0.930 , 1.91	– , –	– , –
	2.5	1.01 , 2.51	– , –	– , –

Rewriting eq. (18) into dimensional form, one obtains:

$$n_f \sim d_p^{\beta_1} \; T^{\beta_2} \; \nu^{0.1} \; (\Delta \varrho \, g)^{0.45} \; (T/D)^{1.44} \tag{19}$$

with exponents $\beta_1 = -0.07$ and $\beta_2 = 0.58$ for "large" particles and $\beta_1 = 0.1$ and $\beta_2 = 0.75$ for "small" particles.

The condition $(d_p/\eta_T) \geqq 32$ was satisfied in almost all experiments treated without the knowledge of the dividing criterion in our previous paper [5] with result:

$$Fr_m = C_1 \, (T/D)^{3.1} \tag{20}$$

4.3 Relative velocity between a particle and fluid at the state of complete suspension

Comparing eq. (15) with equation (18), one obtains

$$Re_{pf} = C_2/C_8 \; Ar^{0.45}(d_p/T)^{\alpha} \; (T/D)^{-0.23} \tag{21}$$

where $\alpha = -0.09$ for $(d_p/\eta_T) \geqq 32$ and $\alpha = 0.08$ for $(d_p/\eta_T) < 32$.
Rewriting again eq. (23) into dimensional form one obtains:

$$u_{Rf} \sim d_p^{\beta} \; \nu^{0.1} \; (\Delta \varrho \, g)^{0.45} \; T^{\omega} \; (T/D)^{-0.23} \tag{22}$$

with exponents $\beta = 0.26$ and $\omega = -0.09$ for $(d_p/\eta_T) \geq 32$ and $\beta = 0.43$ and $\omega = 0.08$ for $(d_p/\eta_T) < 32$. The most important finding with respect to u_{Rf} is the constant value of exponent of Archimedes number (0.45) regardless of the particle - liquid subregion. This fact is clearly seen in Fig. 3 where $Re_{pf} \, (d_p/T)^{4/3} \, (T/D)^{1/3}$ vs. Ar is plotted for pitched six-blade turbine, $H_2/D = 1/3$, $c^v \sim 0$ as an example. The exponent value 0.45 is very close to the value 0.5 (Table I) which is valid for turbulent particle - liquid subregion, only. This means that any attempts to correlate u_{Rf} in turbulently mixed liquid by means of u_t other than for turbulent particle-liquid subregion is rather irrelevant.

REFERENCES

1   Baldi,G., Conti,R., Alaria,F.: "Complete suspension of particles in mechanically agitated vessels". Chemical Engineering Science 33, (1978), pp. 21 - 25.

2   Chapman,C.M., Nienow,A.W., Cooke,M., Middleton, J.C.: "Particle-Gas-Liquid Mixing in a Stirred Vessels". Chem. Eng. Res. Des. 61, 71 - 81 (1983).

3   Gates,L.E., Morton,J.R., Fondy,P.L.: "Selecting agitator systems to suspend solids in liquids". Chemical Engineering 83 (1976).

4   Kneule,F., Weinspach,P.M.: "Suspension of solid particles in agitated vessels". Verfahrenstechnik, 1 (1976),pp.531 - 540 (in German).

5   Rieger,F., Ditl,P.: "Suspension of solid particles in agitated vessels". Proceeding of 4th European Conference on Mixing, BHRA Fluid Eng. 263 - 269 (1982).

6   Weisman,J., Efferding,L.E.: "Suspension of slurries by mechanical mixers". AIChE Journal, 6 (1960) pp. 419 - 426.

7   Zwietering,T.N.: "Suspending of solid particles in liquid by agitators". Chem. Engng. Sci., 8 (1958), pp. 244 - 253.

8   Einenkel,W.D., Mersmann,A.: Verfahrenstechnik, 11, 90 (1977).

9   Einenkel,W.D.: Chem. Ing. Techn., 51, 7, 697 (1979).

10   Davies,J.T.: Turbulence phenomena, Academic Press, 1975.

11   Levins,D.M., Glastonbury,J.R.: "Particle-liquid hydrodynamics and mass transfer in a stirred vessel. Part I". Trans. Instn. Chem. Engrs. 50, 1 (1972), pp. 32 - 41.

12   Ditl,P., Rieger,F.: "Suspension of solid particles in agitated vessels". CHISA Conference, Paper No. V 3.61, Prague, August 1984.

13   Middleman,S.: "Mass transfer from particles in agitated systems: Application of the Kolmogoroff theory". AICHE Journal, 11,4 (1965) pp. 750 - 754.

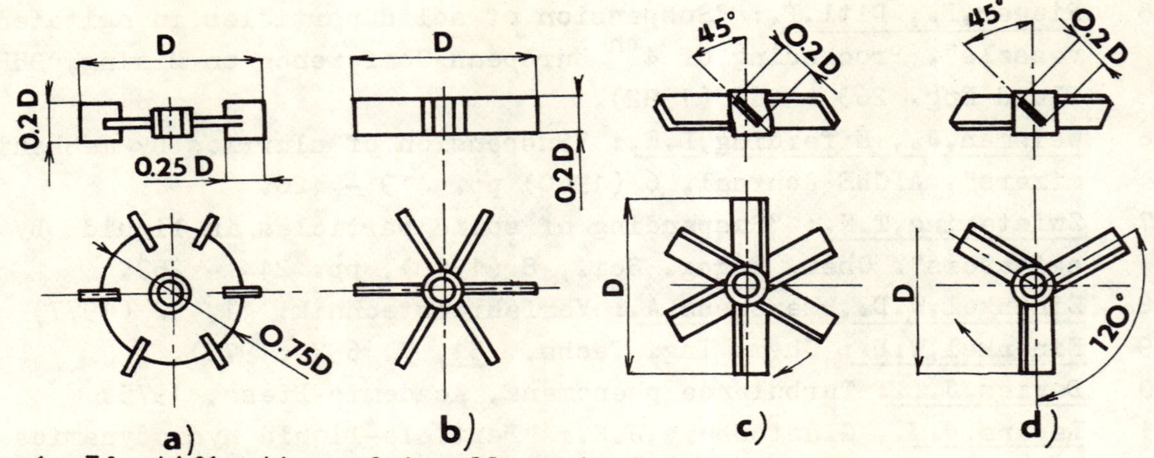

Fig.1: Identification of impellers tested.

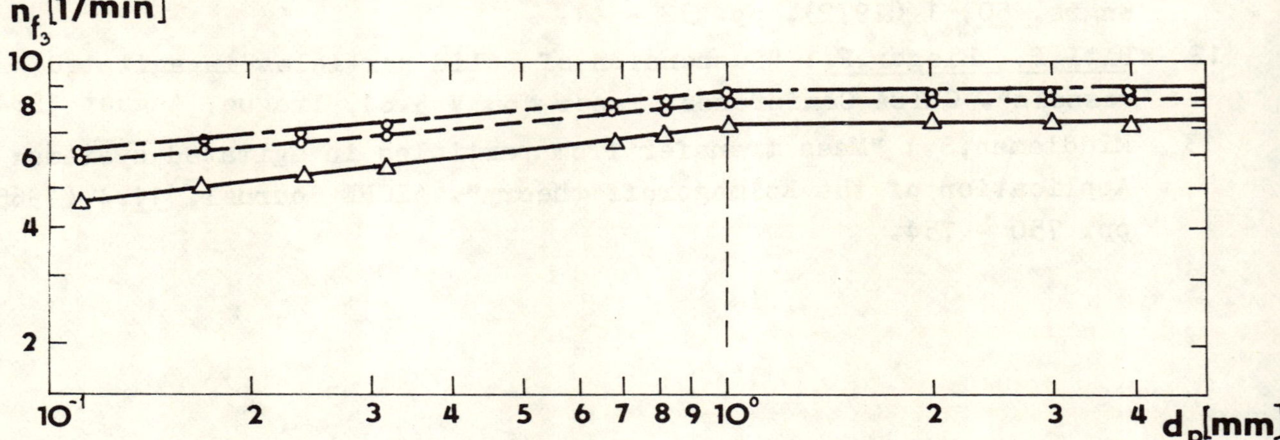

Fig.2: Typical logarithmic plot of $n_f = f(d_p)$. T = 0,39 m, T/D = 2,6,
H_2/D = 2/3. c^V = 1%, – impeller c) – impeller d) – impelller d).

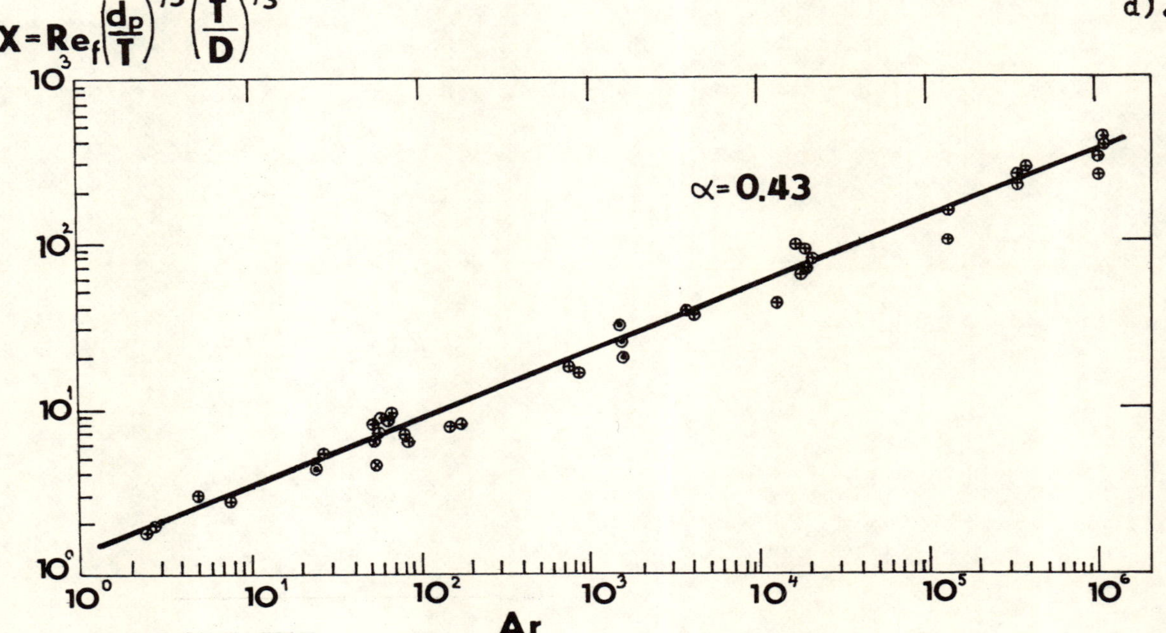

Fig.3: Typical graph of $Re_f(d_p/T)^{4/3}(T/D)^{1/3}$ vs. Ar. Pitched six-blade
turbine, H_2/D = 1/3; $c^V \doteq 0$.

MIXER CONFIGURATION, DESIGN GUIDELINE AND
SCALE-UP CRITERIA FOR EFFECTIVE MIXING OF IMMISCIBLE LIQUIDS

R. R. Hemrajani

Exxon Research & Engineering Company, USA

Summary

This presentation discusses a laboratory development of an energy efficient mixing system for mixing of immiscible liquids. This mixing system, consisting of a single axial flow turbine and four narrow baffles, has been shown to use only 10 to 50% of the energy of other mixing systems using conventional baffles. It also has been established that this partial baffles mixing (PBM) system can be scaled to commercial size tanks by maintaining the same power per unit volume.

Chemical processes involving the mixing of immiscible liquids often require effective and low energy mixers for minimizing batch times and for reducing jacket cooling load in exothermic reactions. This is particularly important when the process involves side reactions and the desired product is heat sensitive.

The approach taken in this study was to carry out laboratory experiments comparing several mixing systems on the basis of power required for good mixing. Colored water and Isopar M were used to simulate most common industrial systems. Various impellers used in this study include axial flow and radial flow turbines, retreat curve impeller, and Brumagin impeller. Liquid height, impeller diameter, number of baffles and baffle width were also varied to determine the optimum mixer configuration.

The PBM system thus developed has been commercially implemented thereby confirming the scale-up criteria. The mixing power requirements, which can be different for different processes, must be determined in small scale experiments for the specific applications under consideration.

Held at Wurzburg, 10-12 June, 1985.

Organised by DVCV· Deutsche Vereinigung für Chemie- und Verfahrenstechnik
(German Association of Chemical and Process Engineering).

Organisation: GVC·VD·Gesellschaft Verfahrenstechnik und Chemieingenieurwesen. 

©BHRA, The Fluid Engineering Centre, Cranfield, Bedford MK43 0AJ, England.

NOMENCLATURE

D   Impeller Diameter

FBT   Flat Blade Turbine

H_ℓ Liquid Height

N   Agitator Speed

N_p Power Number

P   Power

PBM   Partial Baffles Mixing

PBT   Pitched Blade Turbine

T   Tank Diameter

V   Batch Volume

W_b Baffle Width

Greek Letters

ρ Liquid density

1. <u>INTRODUCTION</u>

When mutually insoluble liquids are mixed in an agitated
tank, a dispersion of one phase is produced in the continuous phase
thereby increasing the interfacial area many folds between the
liquids. Mixing energy spent in maximizing this area is generally
cheaper than the resultant enhancement in the mass transfer rate.
However for process optimization, an effective low energy mixing system
must be designed for each immiscible liquids system.

The mixing of immiscible liquids is among the most important
chemical engineering operation in chemical, petroleum, pharmaceutical,
food and mining industries. Several reacting and non-reacting systems
include liquid-liquid extraction, alkylation, suspension polymerization, and emulsifications. Mixer configuration and energy input
controls the distribution of drops which is reflected in either mass
transfer rate and/or product quality. In addition, mixing energy
dissipated as heat can significantly increase the reactor cooling load
in highly exothermic reacting systems. Good mixing should, therefore,
be achieved at minimum power consumption for minimizing batch times,
reducing jacket cooling duty, and obtaining narrow drop size
distribution.

The approach taken in this study was to carry out laboratory
experiments comparing several mixing systems on the basis of power
required for good mixing. Colored water and Isopar M were used to
simulate most common industrial systems. Various impellers used in
this study include axial flow and radial flow turbines, retreat curve
impeller, and Brumagin impeller. Liquid height, impeller diameter,
number of baffles and baffle width were also varied to determine the
optimum mixer configuration. A reacting system of commercial
importance was also similarly studied to confirm the choice of mixer
configuration. Following the scale-up, commercial tests were carried
out to demonstrate validity of the mixer design guideline.

2. <u>THEORETICAL BACKGROUND</u>

The fluid dynamic mechanism responsible for formation of
liquid-liquid dispersions consists of simultaneous break-up and
coalescence of drops. In a mixing tank, the average size of dispersed
drops is determined from the mixing conditions when a dynamic balance
between break-up and coalescence is established (Reference 1). The
drop breakup occurs in regions of high shear, e.g., around the impeller
and where turbulent velocities and pressure variations are present.
Coalescence, on the other hand, can be expected in the low shear zones,
e.g., on tank floor or near liquid surface. Depending on the phase
concentrations, the turbulent flow may accelerate or slow down the
coalescence of droplets.

Several experimental studies reported in the literature
(References 1 to 5) have demonstrated that turbine agitators provide
the desired mixing conditions for mixing of immiscible liquids. Even
in high viscosity liquid emulsifications, turbines are more effective
than the agitators conventionally used for blending of viscous
liquids. This study, therefore, included evaluation of several
turbines commonly used in the industry.

Since in most common industrial systems the dispersed phase
is the lighter phase, the drops float in the continuous phase and tend
to rise up if agitation is reduced or stopped. These drops can,
therefore, be compared to buoyant particles which require draw-down
forces through surface vortices for producing the homogeneous slurry.
A mixing system (Reference 6), consisting of narrow baffles with a
pitched blade turbine and developed for buoyant particle slurries, was
also included in this study.

3. SELECTION AND DESIGN OF OPTIMUM MIXING SYSTEM CONFIGURATION

A wide variety of mixing systems are commonly used in the industry for mixing of immiscible liquids. Those include flat blade turbine (FBT) in alkylation reactors, Brumagin and Retreat Curve impellers in PVC manufacture, and pitched blade turbines (PBT) in several emulsifications. Brumagin impeller is conventionally used without wall baffles, while mixing with the Retreat curve impeller is aided by an 'h' shaped baffle. Baffles used with PBT's and FBT's are designed at width equal to 1/10th of the tank diameter. The above mentioned impeller types and baffle configurations were evaluated to determine the optimum mixing system for low energy mixing of immiscible liquids. Additional mixing parameters also studied include impeller/ tank diameter ratio, liquid height, and number of impellers.

3.1 Equipment and Procedure

Experiments were carried out in a 0.3 m diameter 0.7 m high plexiglas tank with a flat bottom shown in Figure 1. The tank was equipped with a top entering agitator driven by a 0.2 kW variable speed driver and was provided with a variety of inter-changeable impellers and wall baffles. The tank was supported on a hydraulic lift which could be raised or lowered to adjust the impeller height. A digital tachometer was used to measure the agitator speed. Various impellers used in the study (Figure 2) include pitched blade turbine (PBT), flat blade turbine (FBT), Brumagin impeller, retreat curve impeller.

The immiscible liquids system chosen in this study was water and Isopar-M which simulates common industrial systems. A typical experiment was carried out by pouring in equal volumes of colored water and Isopar M. The mixer was started at a low speed which was then raised in steps until complete homogeneity was achieved in the tank. At each agitator speed still pictures were taken for measurement of interface heights. The rise of interface was used to judge the degree of mixing. Figure 3 shows the progress of a typical experiment.

The mixer power required to achieve complete liquid homogeneity was used to compare various mixer configurations. Mixer power P at each mixing condition was estimated by:

$$P = N_p \rho N^3 D^5 \tag{1}$$

Where N_p is a power number, ρ is liquid density, N is agitator speed and D is impeller diameter. Values of power number N_p for each mixer was obtained from literature (Reference 7) and vendor technical information.

Following are additional mixing parameters used in this study:

- Baffle width of 2%, 4% and 8.3% of tank diameter

- Relative impeller diameter (D/T) in the range of 0.33 to 0.58

- Relative liquid height (H_ℓ/T) in the range of 1 to 2

- Two impeller system

3.2 Results

3.2.1 Comparison of Power Requirements with Various Impellers

Good mixing and complete homogeneity were achievable with all impellers used in this study, but at different mixer power levels. It was found that the minimum power for good mxing is 0.08 kW/m^3 when using a single PBT and 2% baffles. Other configurations required higher mixer power for the same mixing quality, as shown in Figure 4. Mixer power requirement increases in the following order of mixing systems.

PBT/2% Baffles < Retreat Curve < Brumagin < FBT/8.3% Baffles

Further comparison of the FBT and PBT using the same size baffles, of width equal to 1/50th (2%) of tank diameter, confirmed the superiority of the PBT. Power required for good mixing with FBT/2% Baffles is 0.2 kW/m^3 as shown in Figure 5.

3.2.2 Feasibility of Using Multiple Turbines in Tall Tanks

Certain immiscible liquids mixing tanks are required to be tall for maximizing jacket heat transfer area. Conventionally multiple turbines are used to provide good top to bottom mixing in such tall tanks. A two turbine system, with one PBT located centrally in each phase, was therefore studied to compare its effectiveness with that of a single turbine system. The power requirement for good mixing was found to be comparable for both mixing systems, as shown in Figure 6. However, at low agitator speeds with the two turbine system, water was dispersed in Isopar M while at higher speeds the reverse occurred. Isopar M was always the dispersed phase when using a single turbine.

This phase inversion depends on the surface tension of the liquids, therefore, it is not expected to occur in all liquid systems. In addition, if the mixer is designed well above the transition power, the lighter phase will always be dispersed. Laboratory experiments, however, should be carried out to identify the occurrence of phase inversion and the corresponding power level before designing the two-turbine mixer for a commercial system. The data from the two turbine and single turbine tests are shown in Figure 6. The power at which the phase inversion occurs is represented by the point where interface rise is extremely sensitive to mixer power.

3.2.3 Baffles Number and Width Optimization

The effect of baffle width was studied to determine if the partial baffles (2%) concept, previously developed for buoyant particle slurries mixing, can be extended to mixing of immiscible liquids. Experiments were carried out using a PBT and varying baffle widths from 2 to 8.3% of the tank diameter. The 8.3% baffle width represents conventionally used full baffles. Four equally spaced baffles were used in each case.

The results show that the power required for good
mixing increases from 0.08 to 0.88 kW/m^3 as the baffle
width is increased. Therefore, partial baffles are the
most energy efficient. These results, for a 127 mm
diameter impeller, are shown in Figure 7. The effect of
baffle width was somewhat reduced when a larger 178 mm
diameter impeller was used. This is illustrated in Figure 8
which is a plot of mixer power per unit volume against the
baffle width for two different impeller sizes.

A baffle configuration consisting of two 4%
baffles, which has been recommended by a vendor for PVC
reactors, was also studied. It was found to be less effective than the four 2% configuration as seen in Figure 9.

Effect of baffle width using the two-turbine was found to
be negligible as seen by the data plotted for four 2%, 4.2%
and 8.3% wide baffles in Figure 10.

3.2.4   Effect of Impeller/Tank Diameter Ratio and Batch Height

Several workers in the literature have reported
that the mixing system design for liquid-liquid mixing
should be scaled up on the basis of power per unit
volume. This is based on theoretical derivation of
dispersed phase drop size and experimental data. It is,
however, not known if the geometric dissimilarities require
deviation from this scale-up criterion. Therefore,
experiments were carried out by varying batch height and
impeller diameter to evaluate their effect on the mixer
power requirement for good mixing.

Mixing performance of PBT with 2% baffles was
found to be the same at constant power per unit volume,
0.08 kW/m^3, in wide ranges of liquid height and turbine
diameter. Relative liquid height (liquid height/tank
diameter) was varied from 1.0 to 1.42 and relative turbine
diameter (turbine diameter/tank diameter) was varied from
0.33 to 0.58 in these experiments. Results describing the
effects of relative liquid height and relative turbine
diameter are shown in Figures 11 and 12 respectively. When
using the PBT/2% baffles mixing system, the scale-up
criterion of power per unit volume is valid in a wide range
of liquid height and impeller sizes.

4. CONFIRMATION OF MIXER CHOICE AND SCALE-UP CRITERION IN A REACTING SYSTEM

The partial baffles mixing (PBM) system was tested for its
effectiveness to mix two immiscible liquids in a commercial reacting
system. A suitable system was chosen to facilitate the mixing quality
evaluation through measurement of reaction rate and product
selectivity. For the purpose of discussion, this reaction can be
generalized in the following form.

$$A_{(aqueous)} + B_{(organic)} \longrightarrow R_{(organic)} + S_{(aqueous)}$$

where component A of the aqueous phase reacts with a component B of the
organic phase to give a product R in the organic phase.

Prior to this study this batch process was carried out in a large sized mixing tank equipped with a flat blade turbine and full baffles. The mixing was inadequate and consequently a side product in the form of gummy solids would occasionally form resulting in yield loss, long batch times, and increased maintenance costs. Nitrogen sparging was occasionally used to augment the mixer to alleviate the fomation of the side product, but it created environmental problems caused by organic vapors entrained in the effluent nitrogen.

4.1 Equipment and Procedure

The mixing experiments were carried out in two different sized cylindrical glass tanks, 0.15 and 0.3 m in diameter. Each tank was equipped with a top entering variable speed agitator drive and was provided with a variety of interchangeable impellers and wall baffles. The reaction was carried out at a constant temperature which was maintained by a heater and a temperature controller. A typical experiment was carried out by heating the aqueous phase containing reactant A to the reaction temperature in the mixing tank and the organic phase containing reactant B to the same temperature in a separate container. The mixer was started and its speed adjusted to a predetermined level. The organic liquid was poured into the mixing tank in the minimum possible time and the reaction was started. Samples were withdrawn at various time intervals for analysis of the product R in the organic phase. All the experiments were carried out without nitrogen sparging.

A pitched blade downpumping turbine was used in all the experiments. Baffle widths of 1/50 and 1/12 of tank diameter were tested and the effect of D/T ratio on mixing was also studied. Performance of various mixer configurations was indexed by comparing the rise in concentration of R as a function of time. This concentration rose to a maximum value when the reaction was complete. The time to reach 90% conversior was also used to compare mixing quality of various mixing systems at different power levels. The power consumption at each mixing condition was estimated by Equation (1). The power number N_p for the range of Reynolds numbers employed was equal to 1.3 with a pitched blade turbine and 1/12 T baffles (Reference 7). The value of N_p for the same impeller and 2% baffles was equal to 0.78 in these experiments.

4.2 Results

4.2.1 Laboratory Evaluation of Mixer Configurations

The first step in the larboratory developmet was to compare the performance of partial baffles with that of full baffles. A faster increase in the concentration of R at the same mixer power would be evidence of better dispersion. Data were obtained in the 0.15 and 0.3 m diameter mixing tanks using D/T = 0.42 and varying mixer power in the range of 0.02 to 0.15 kW/m^3. The results, shown in Figure 13 for the 0.3 m tank as plots of concentration of R versus time, indicate faster reaction rate with the use of partial baffles. As expected, the reaction rate increases as the mixer power was increased with both baffle configurations.

Partial baffles provided the same mixing quality at 0.07 kW/m^3 as that obtained with full baffles at 0.15 kW/m^3. These results also demonstrate that about 0.07 kW/m^3 mixer power is adequate with partial baffles as mixing performance did not improve significantly at higher power. With full baffles at least 0.15 kW/m^3 would be required for equivalent mixing. The term optimum mixing will be used to describe just reaching the condition where further power increases do not significantly improve reactor yield. A point of inflection, observed in the plot for 0.02 kW/m^3 with partial baffles, was caused by formation of a solid side product which ultimately dissolved in the organic phase. These data were further interpreted as time required for 90% conversion at various mixer power levels. Figure 14 shows a rapid decrease in the reaction time for 90% conversion up to 0.07 kW/m^3 mixer power with partial baffles followed by insensitivity of the reaction time to the power. A similar behavior was obtained with full baffles excepting that the plateau is reached at about 0.2 kW/m^3. Therefore partial baffles provided more effective mixing than the full baffles.

The effect of impeller diameter to tank diameter ratio (D/T) was also studied. Experiments were carried out using three different sized impellers and the power required for equal mixing then compared. The data obtained in the 0.3 m reactor with D/T = 0.33 and 0.5 are plotted in Figure 15. Power required for optimum mixing was found not to depend on the impeller size. Formation of undesirable side product was prevented with the larger impeller as seen by continuous increase in the concentration of product R with D/T = 0.5 at all power levels studied. With the smaller impeller, D/T = 0.33, solids were formed which temporarily accumulated on the surface at power levels of 0.035 kW/m^3 or less. Therefore, if the mixer is designed at power levels adequate for optimum mixing, any impeller size between D/T - 0.33 to 0.5 would provide similar mixing quality with the use of the PBM system.

4.2.2 <u>Confirmation of P/V as Scale-up Criterion</u>

The criterion for scaling this PBM system to commercial sizes was determined by comparing the data obtained in the two reactor sizes, 0.15 and 0.3 m in diameter. Figure 16 shows the plots of concentration of R vs. reaction time at two different power levels for each reactor size. The product concentrations in the 0.15 m were adjusted to match the final concentrations in the 0.3 m reactor samples. The reaction rates in the two reactors appeared to be the same, within the experimental error, at the same mixer power per unit volume. Therefore, it was concluded that the PBM system for immiscible liquids mixing could be scaled up on the basis of power per unit volume. For this particular system 0.07 kW/m^3 appeared adequate.

This power per unit volume criterion of liquids can be contrasted with the tip speed criterion for solids. With liquids the dispersion into fine droplets, which must precede incorporation, is the limiting step. Since dispersion is power per unit volume dependent (Reference 7), this represents a reasonable criterion for scaling.

On the basis of these results, a PBM system was scaled to a commercial unit which has a 3.7 m tank diameter. Tests were carried out to evaluate the mixing at various agitator speeds and compared with the laboratory scale data. A direct comparison of reaction times to judge the mixing quality was not possible because filling the

tank with the organic phase was time consuming in the plant
but negligible in the laboratory. In addition, the fill
pumping rates were different for each test batch. To
account for the resultant differences in the batch reaction
times, the times for 90% conversion were normalized using
the pumping time. These normalized times are plotted
against mixer power in Figure 17. The results were nearly
the same as the laboratory data with partial baffles shown
in Figure 14, i.e., 0.07 kW/m^3 is required for optimum
mixing in this reaction system. Thus the power per unit
volume scale-up criterion is confirmed by this agreement in
results between these units of significantly different
size.

5. CONCLUSIONS

The laboratory tests for homogenization of immiscible liquids
indicated that a mixing system consisting of a single pitched blade
turbine and partial baffles provides the desired mixing at the lowest
energy consumption. This mixer can be scaled to commercial units on
the basis of same power per unit volume. The same were confirmed with
a reaction system in laboratory and commercial scale tests. For
different mixing needs, e.g., mass transfer with reaction or extraction, the power requirement should be established through laboratory
experiments.

REFERENCES

1. R. Shinnar, J. Fluid Mech., 10, 259 (1961).

2. T. Vermuelen, G. M. Williams and G. E. Langlois, Chem. Eng. Progr.,
    51 (2), 85F (1955).

3. F. B. Sprow, Chem. Eng. Sci., 22, 435 (1967).

4. F. B. Sprow, AIChE Jl., 13 (5), 995 (1967).

5. A Mersmann and H. Grossman, International Chem. Eng., 22 (4), 581
    (1982).

6. D. L. Smith, R. R. Hemrajani, R. M. Koros and B. L. Tarmy, "Mixing
    Technology for Homogeneous Suspension of Buoyant Solids and Liquid
    Drops," AIChE Annual Meeting, New Orleans, LA, November 8-12, 1981.

7. V. W. Uhl and J. B. Gray, "Mixing Theory and Practice", Vol. I,
    Academic Press, NY, 1966.

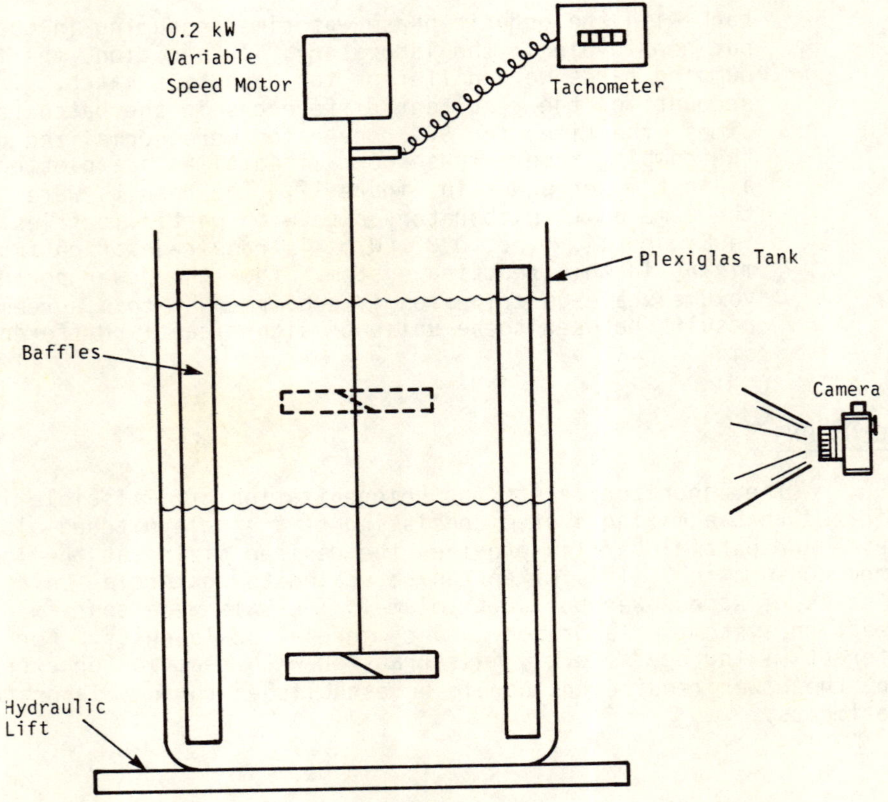

Figure 1: 12" DIAMETER LABORATORY MIXING TANK

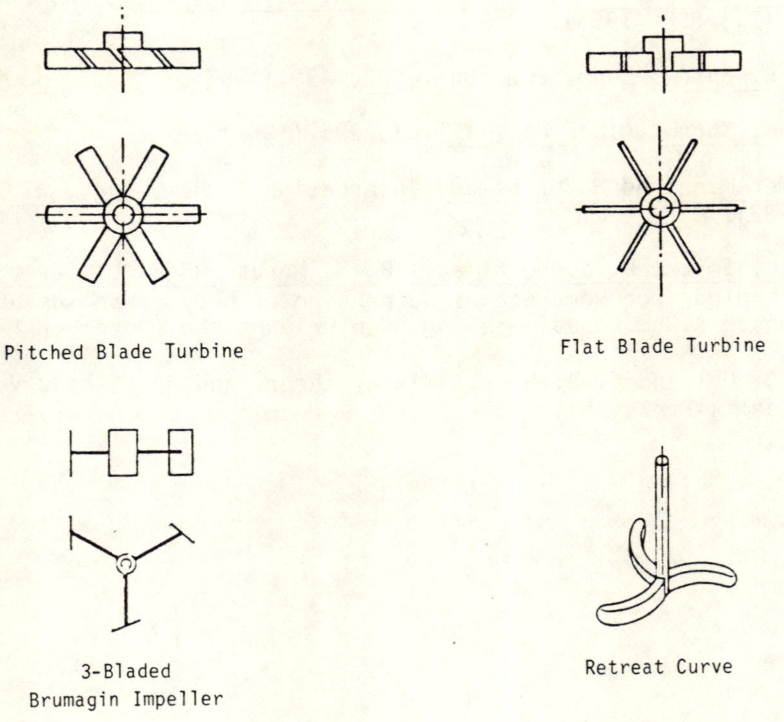

Figure 2: VARIOUS IMPELLERS USED FOR MIXING
OF IMMISCIBLE LIQUIDS

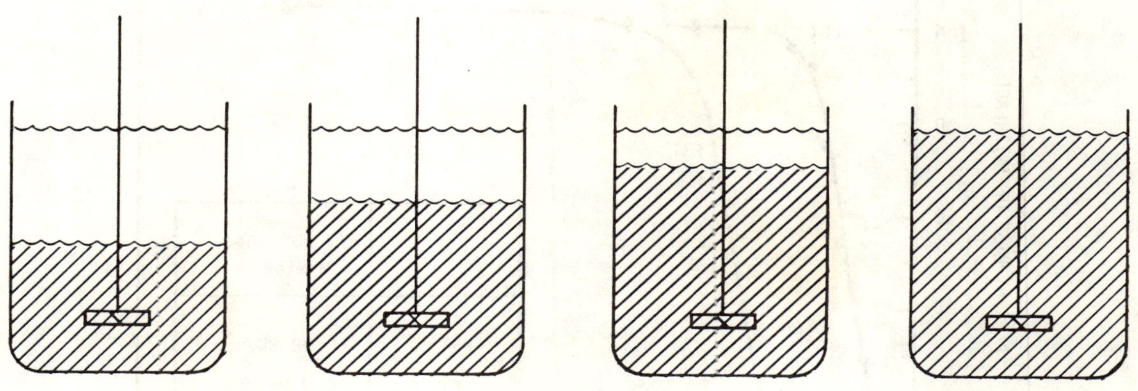

Figure 3: PROGRESS OF A TYPICAL WATER/ISOPAR M MIXING EXPERIMENT

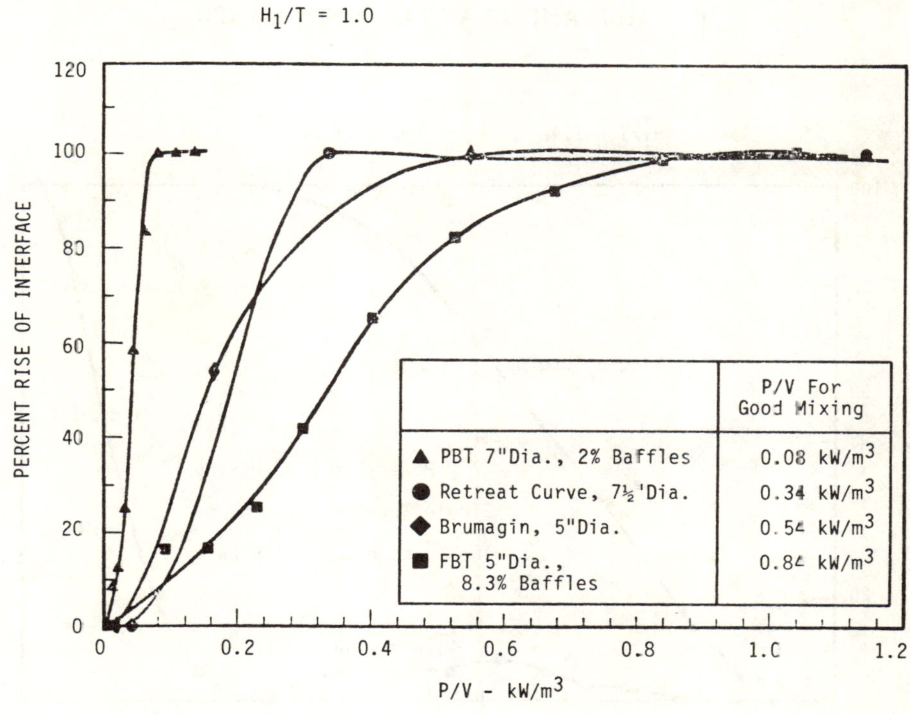

Figure 4: POWER REQUIREMENTS FOR GOOD MIXING
USING VARIOUS MIXING SYSTEMS

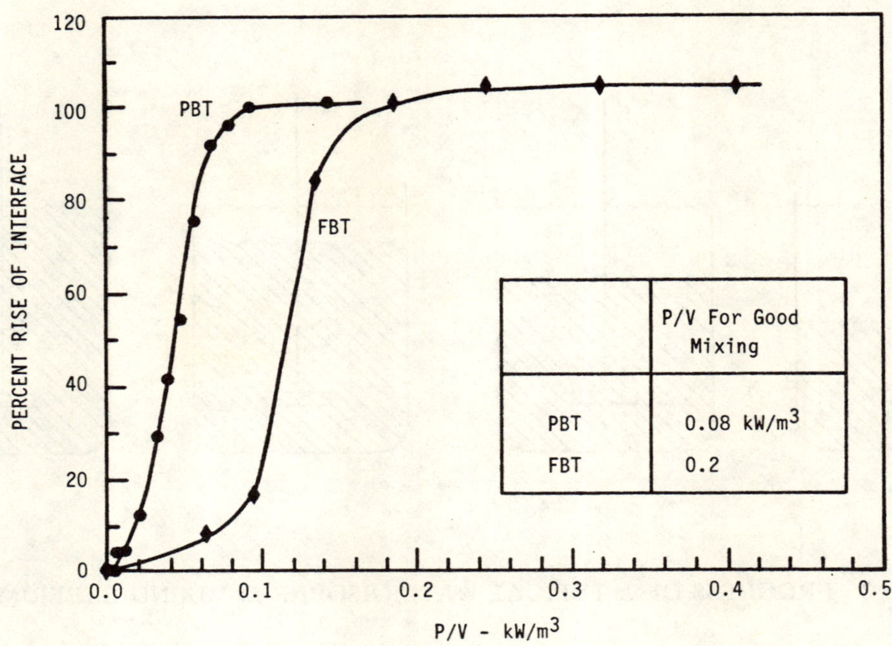

Figure 5: POWER REQUIREMENTS FOR GOOD MIXING USING PITCHED AND FLAT BLADE TURBINES

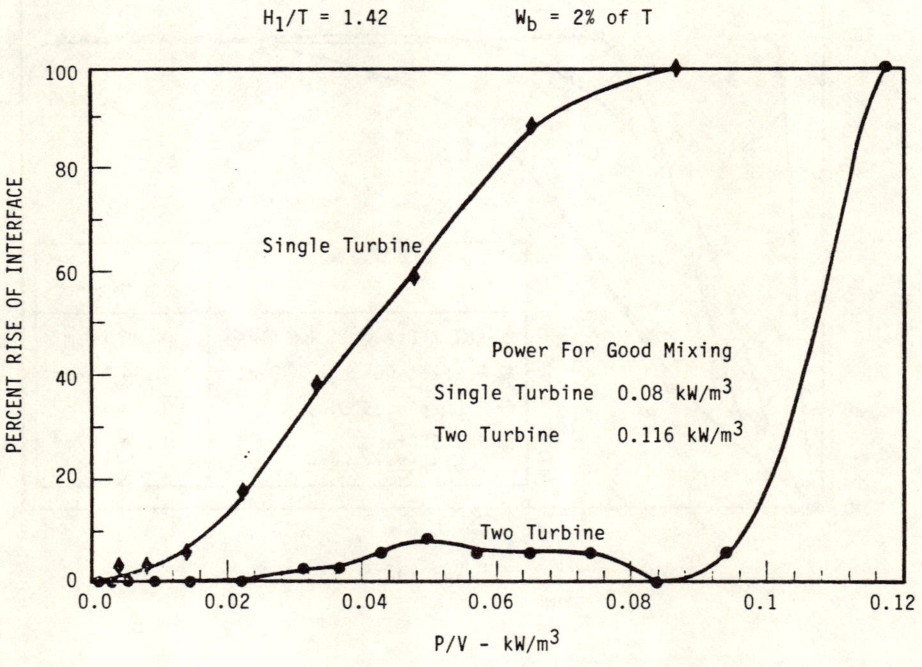

Figure 6: POWER REQUIREMENTS FOR GOOD MIXING USING SINGLE TURBINE AND TWO TURBINE SYSTEMS

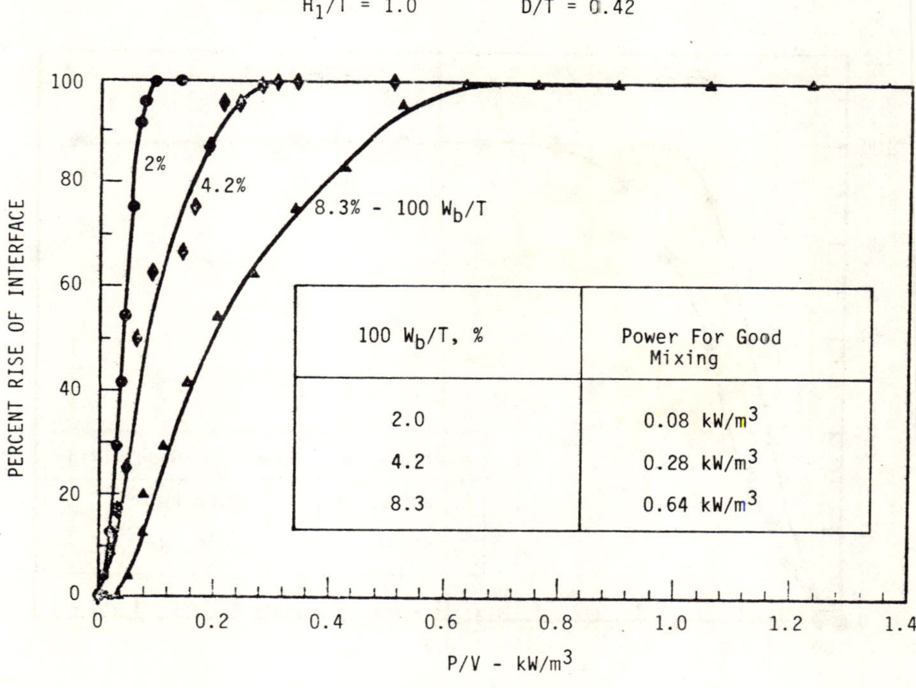

Figure 7: EFFECT OF BAFFLE WIDTH ON MIXER POWER REQUIREMENT FOR SINGLE PITCHED BLADE TURBINE

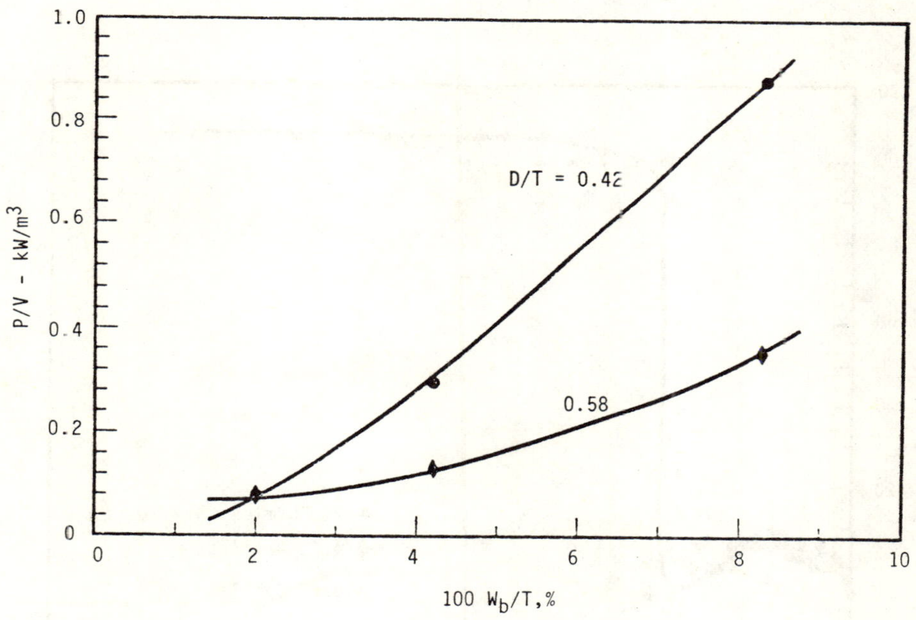

Figure 8: EFFECT OF BAFFLE WIDTH ON POWER REQUIREMENTS WITH DIFFERENT IMPELLER DIAMETERS

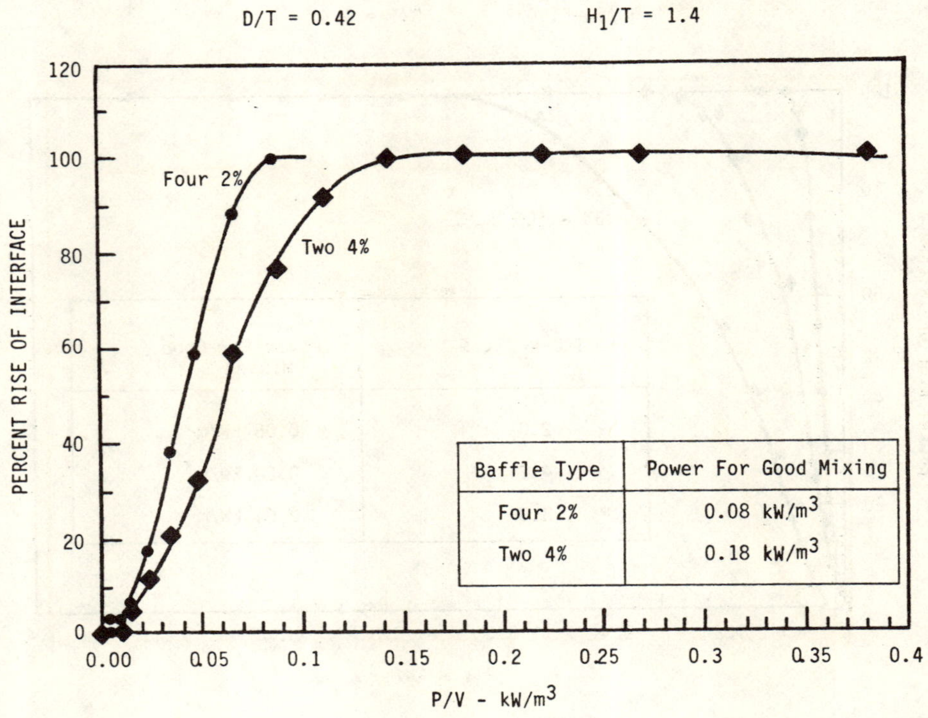

Figure 9: POWER REQUIREMENTS OF FOUR 2% AND TWO 4%
BAFFLES USING SINGLE PITCHED BLADE TURBINE

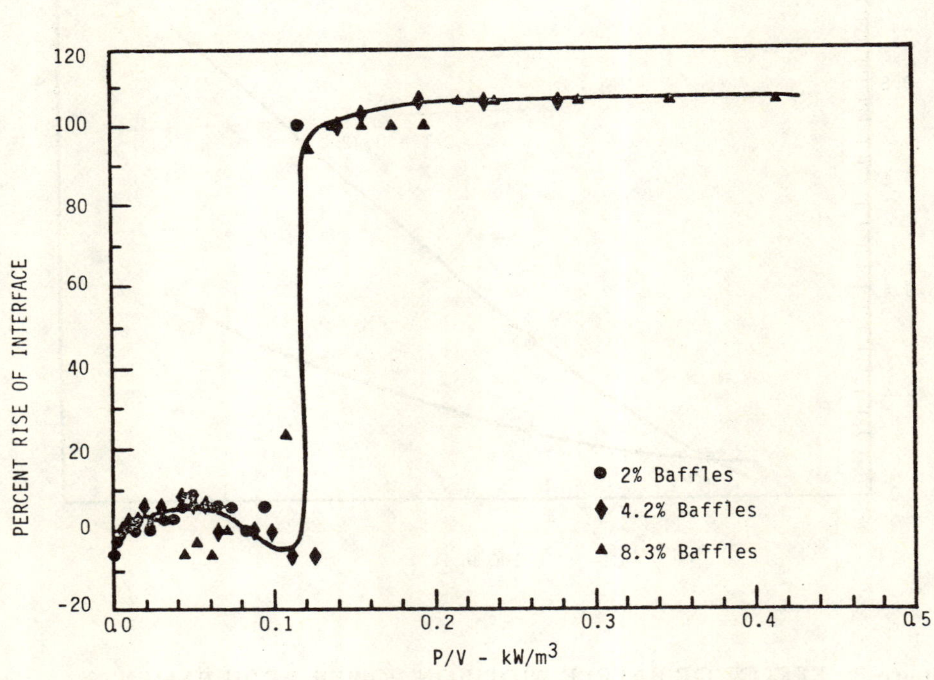

Figure 10: EFFECT OF BAFFLE WIDTH ON POWER REQUIREMENTS
USING TWO PITCHED BLADE TURBINES

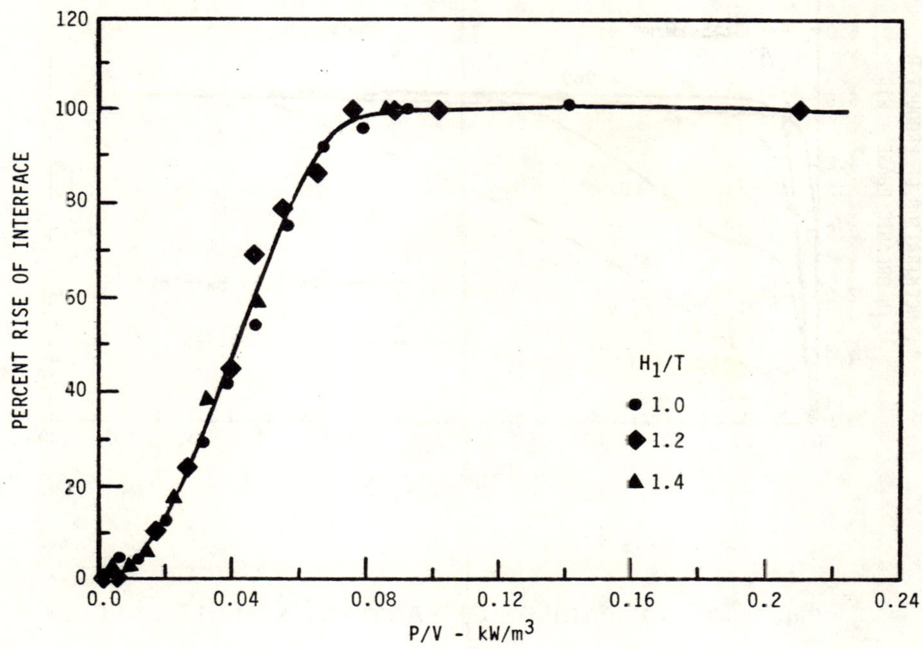

Figure 11: EFFECT OF LIQUID LEVEL ON POWER REQUIREMENTS

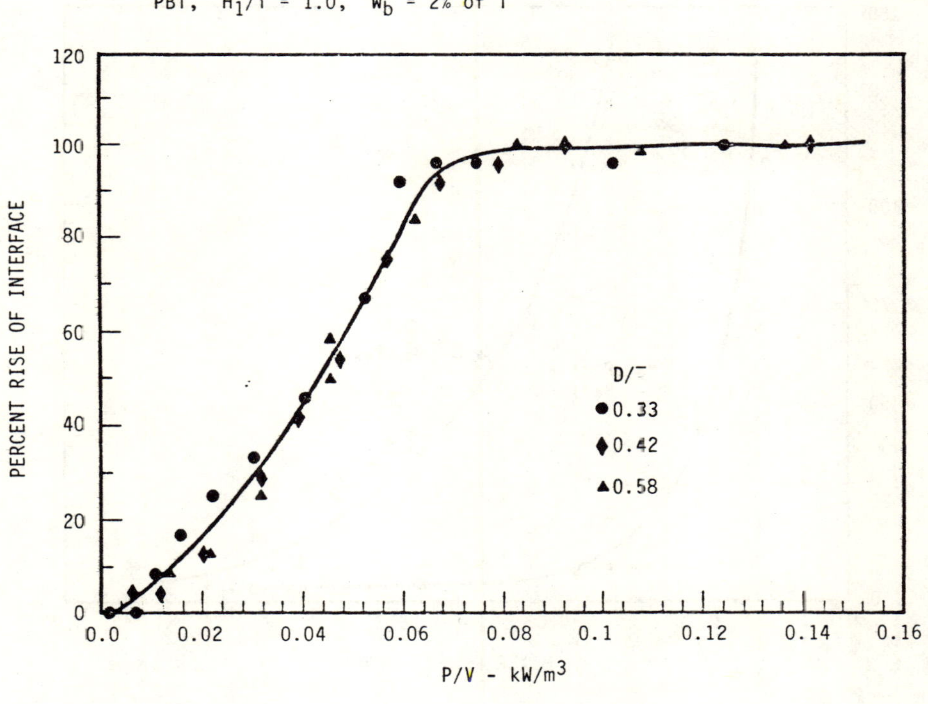

Figure 12: EFFECT OF RELATIVE IMPELLER DIAMETER ON
POWER REQUIREMENT USING PBT

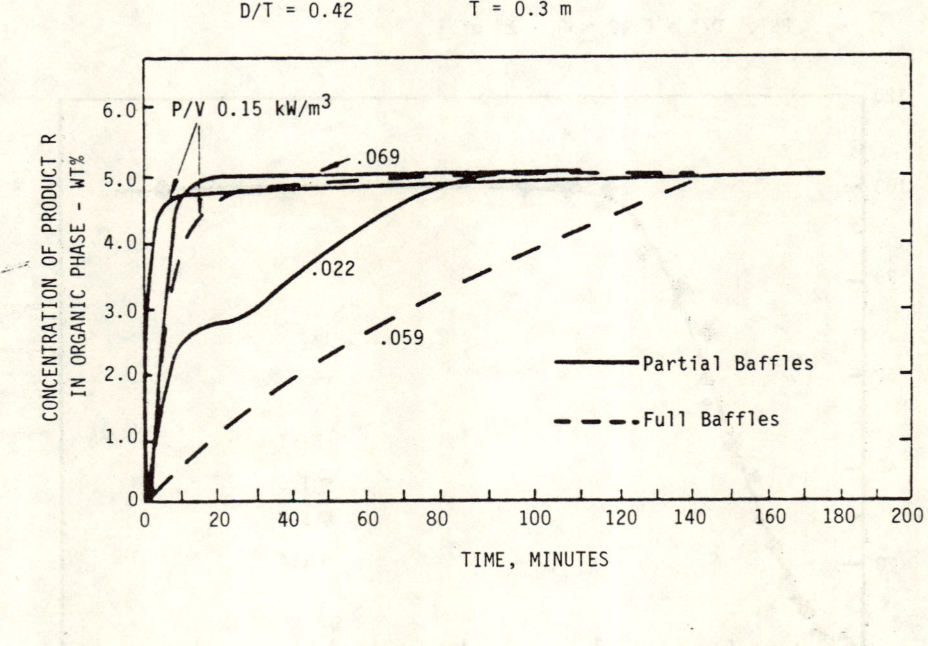

Figure 13: COMPARISON OF PARTIAL AND FULL BAFFLES

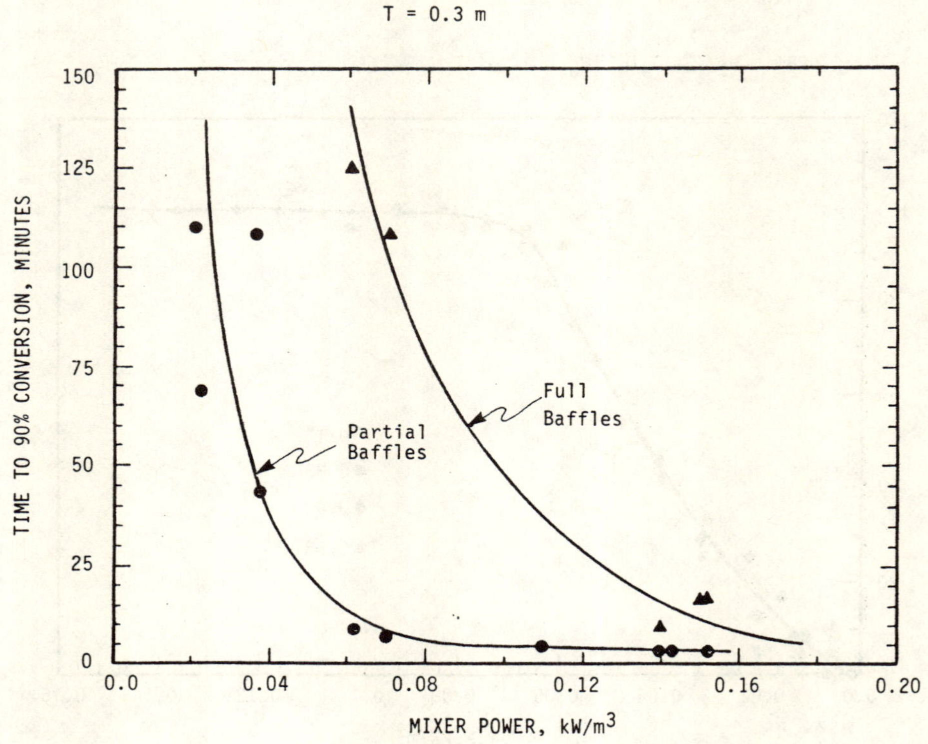

Figure 14: 90% CONVERSION TIMES WITH PARTIAL
AND FULL BAFFLES

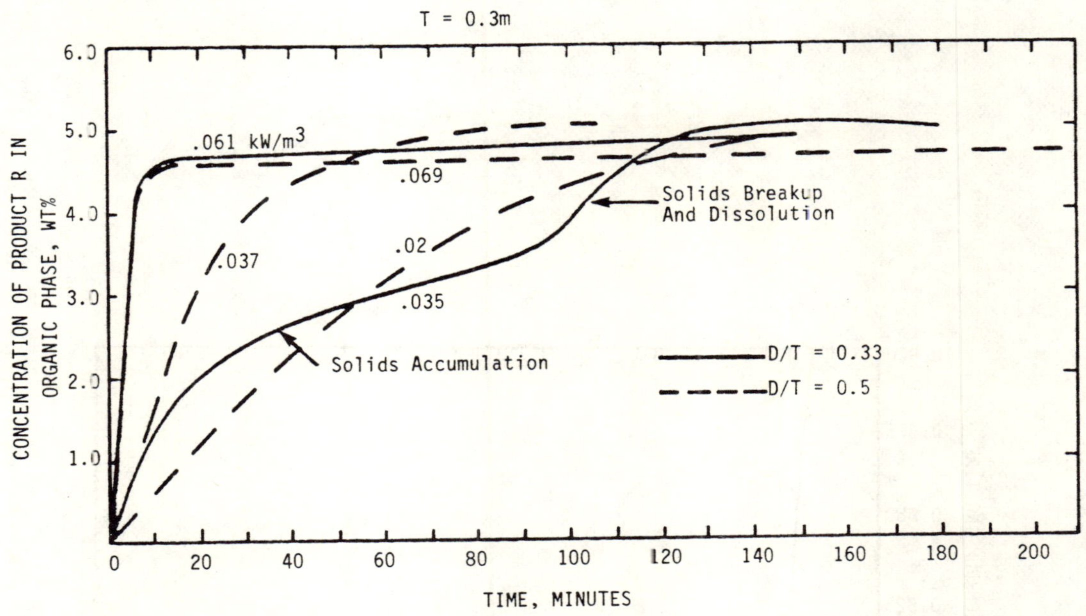

Figure 15: EFFECT OF D/T RATIO ON MIXING OF IMMISCIBLE
LIQUIDS USING PBM TECHNOLOGY

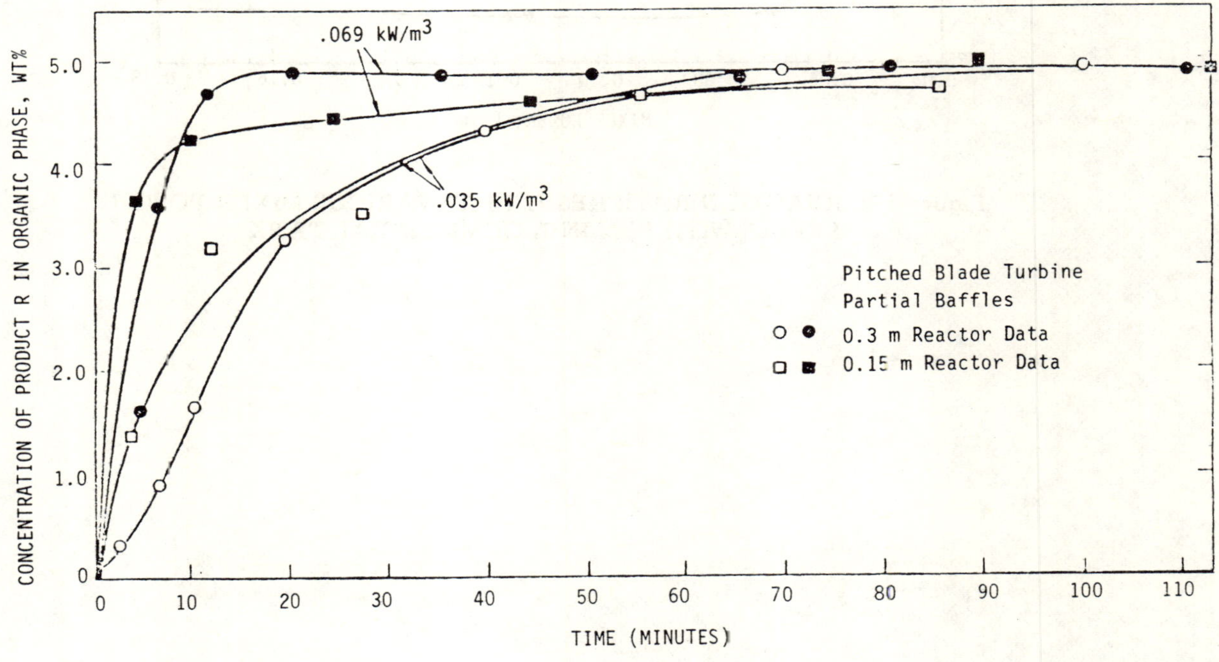

Figure 16: COMPARISON OF REACTION RATE DATA FROM TWO REACTOR
SIZES FOR SCALE-UP CRITERION DETERMINATION

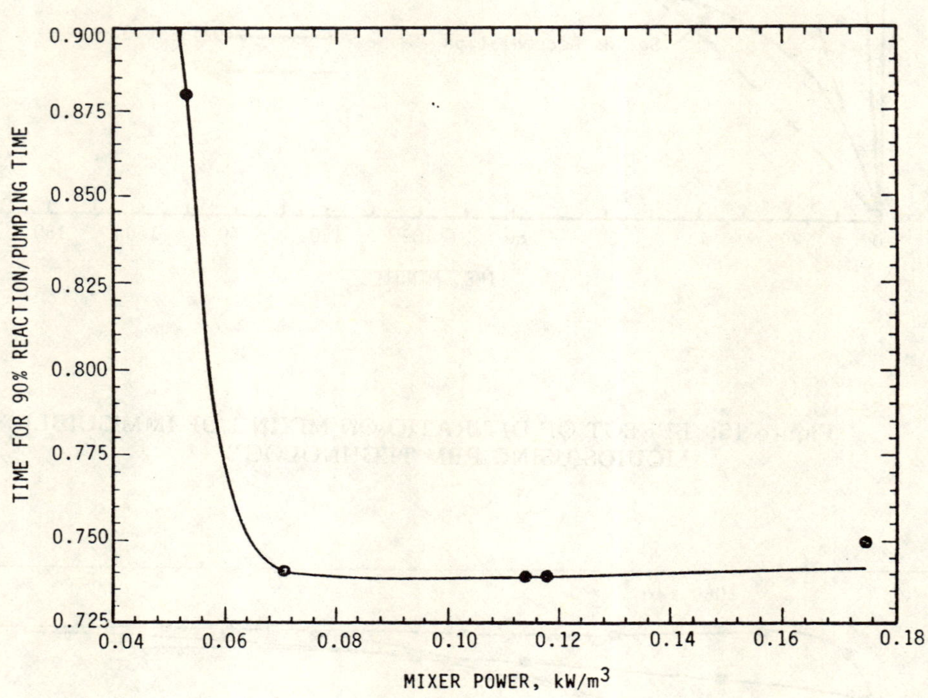

Figure 17: REACTION RATE RESULTS AT VARIOUS MIXER POWER LEVELS WITH PBM IN A COMMERCIAL TANK

FAST CHEMICAL REACTION IN HIGH INTENSITY YSTRAL DYNAMIC MIXER

J.R.Bourne and J.Garcia-Rosas

Technisch-chemisches Laboratorium ETH, CH-8092 Zurich, Switzerland

Summary

The Ystral Company, West Germany, manufactures a dynamic mixer, in which a rotor (speeds up to 20'000 rpm) turns inside a close fitting stator. The centrifugal force induces flow into the inside of the rotor, whence the liquid flows radially outwards through the rotor and the slits in the stator, and is subjected to a shearing action. This device can be fitted inside a chamber, so that it functions as an in-line mixer and simultaneously as a pump. It can, however, also be mounted in a tank, where it functions approximately like a conventional mixing impeller.

Rapid micromixing in a Ystral mixer has been followed using the diazo coupling reactions between 1-naphthol and diazotised sulphanilic acid in alkaline, aqueous solutions at room temperature. Variables studied included rotor speed, reagent concentrations, rates of flow of reagents, ratio of such flows and the stoichiometric ratio, as well as semi-batch and continuous operating modes. The influence of these variables on product distribution has been quantitatively related to a model of fast reactions. The high energy dissipation rates, which can be achieved by this mixer, make it suitable for carrying out processes, where mixing and reaction are on time scales of order 1 ms. These high energy dissipation rates are highly localised and care has to be taken to feed reagents into the high intensity mixing zone, which is limited to the mixing head and possibly also its discharge.

Held at Wurzburg, 10-12 June, 1985.

Organised by DVCV · Deutsche Vereinigung für Chemie- und Verfahrenstechnik
(German Association of Chemical and Process Engineering).

Organisation: GVC·VDI-Gesellschaft Verfahrenstechnik und Chemieingenieurwesen. 

<u>NOMENCLATURE</u>

A	Reagent in eq.(1), 1-naphthol
B	Reagent in eq.(1) and (2), diazotised sulphanilic acid
A_m/B_m	Molar initial stoichiometric ratio of reagents A and B
D	Diffusivity ($m^2 s^{-1}$)
k	Second-order rate constant ($m^3 mol^{-1} s^{-1}$)
M	Mixing modulus, defined in eq.(6)
N	Rotor speed (rpm)
R	Monoazo dye
S	Bisazo dye
V	Volume (m^3)
$\dot{V}$	Volumetric flow rate ($m^3 s^{-1}$)
X	Product distribution, defined in eq.(3)
Sc	Schmidt number (ν/D)
δ_o	Initial half-thickness of lamina (m)
ε	Rate of energy dissipation in reaction zone ($W kg^{-1}$)
ν	Kinematic viscosity ($m^2 s^{-1}$)

1. INTRODUCTION

Many single-phase chemical reactions are sufficiently fast that the rate of mixing of
the initially separate reagent streams influences the distribution of reaction products (Ref.1). This effect on selectivity disappears when the rate of energy dissipation is increased sufficiently, but in conventional stirred tank reactors the shortest
time scales for mixing are still of order 5 ms (Ref.2). Similar results have been obtained with a conventional centrifugal pump (Ref.3). The product distributions of certain fast reactions could be improved by using a mixer with a higher local rate of
energy dissipation. It is the object of this paper to report encouraging results in
this direction obtained with a commercial dynamic mixer of the rotor/stator type.

2. EXPERIMENTAL

A type X-20 Ystral mixer (Ystral GmbH, D-Ballrechten-Dottingen) with maximum power
260 W and speed 25'000 rpm, and a type 21-G mixing head were used. This head consists
of a rotor (a two-legged impeller) and a cage-type stator (a hollow cylinder with ob-
lique slits in its wall). A gasket seals the drive shaft from the rotor-stator (Fig.1A).
Fluid is drawn axially into the open stator and discharged through the rotor and subsequently through the slits in the stator. Intense fluid deformation occurs during
this discharge, the clearance between rotor and stator being only a few tenths of a mm.
Rotor speeds in the range 1000-13'000 rpm were employed.

The diazo coupling reactions between 1-naphthol (A) and diazotised sulphanilic acid (B)
at pH = 10.0 and room temperature conform to eq.(1) and (2). The concentrations of the
soluble dyes (R and S) formed in dilute aqueous solutions were determined spectrophoto-
metrically (Ref.4). The product distribution X, defined in eq.(3), represents the
fraction of the limiting reagent (B) which has reacted to form bisazo dye (S):

$$A + B \xrightarrow{k_1} R \qquad (1) \qquad\qquad R + B \xrightarrow{k_2} S \qquad (2)$$

$$X = \frac{2\{S\}}{\{R\} + 2\{S\}} \qquad (3)$$

The couplings were so rapid that they ran to completion. A mass balance check on the
diazonium ion was carried out for each run using eq.(4):

$$\{R\} + 2\{S\} = \{B_o\} \qquad (4)$$

The analytical errors in X were in the range $\pm$ 0.005 to $\pm$ 0.01. Eq.(4) was satisfied to
within 1-2 %.

The following conditions were used in most of the experiments. Temperature was 297 $\pm$ 1
K. The initial stoichiometric ratio of naphthol to diazonium ion A_m/B_m was 1.05. The
initial naphthol concentrations were 1.39, 0.69, 0.23 and 0.14 mol m^{-3}, corresponding
to volumetric (V_A/V_B) and volumetric flow rate ($\dot{V}_A/\dot{V}_B$) ratios of 10, 20, 50 and 100
respectively. (Other initial concentrations and stoichiometric ratios are indicated
later on the figures showing experimental results).

Semi-batch reactions were conducted in a 2 L beaker (diameter and height 0.13 m and
0.18 m respectively), containing initially 1.5 L of 1-naphthol solution (buffered to
pH = 10.0). The Ystral mixing head was mounted as shown in Fig.1B. The feed tube for
the diazotised sulphanilic acid solution was usually located 0.005 m below the open end
of the stator and tube bores of 0.001 to 0.004 m were used. Rates of feeding this solution varied between 0.003 to 0.01 L min^{-1} for rotor speeds below 3000 rpm and 0.03 to
0.18 L min^{-1} at speeds above 3000 rpm.

Continuous reactions were conducted when the Ystral mixing head was mounted inside a
housing (type 22-Z), so that it functioned as a mixer and a pump (Fig.1C). Flow rates
varied in the ranges $\dot{V}_A$ = 1.8 to 4.2 L min^{-1} and $\dot{V}_B$ = 0.02 to 0.54 L min^{-1}, which
corresponded to mean residence times in the housing of 0.5 to 1.3 s.

3. RESULTS

Fig.2-4 refer to continuous, in-line operation (Fig.1C). The ordinate X, defined in eq.
(3), increases with increasing segregation on the molecular scale (Ref.1,4) and thus

indicates slower micromixing with decreasing rotor speed in all three figures. The
initial stoichiometric ratio A_m/B_m was 1.05 in all runs, although as Fig.2 shows, product distribution also depended upon the ratio of the flow rates - a factor which is
irrelevant for slow reactions (Ref.5). Here the half-lifetimes of the first reaction,
eq.(1) with $k_1 = 7 \times 10^3$ m^3 mol^{-1} s^{-1}, ranged from 0.1 to 1 ms as $\dot{V}_A/\dot{V}_B$ increased from
10 to 100. Thus the increase in secondary product formation X in Fig.2 resulted from
higher concentration gradients on the molecular scale as $\dot{V}_A/\dot{V}_B$ fell. With increasing
rotor speed X becomes less dependent upon the flow rate ratio, as would be expected as
mixing accelerates and the chemical regime is approached. The asymptotic value of X in
this slow reaction regime depends only upon k_1/k_2, A_m/B_m and reactor type (Ref.5). In
this case X = 0.0006, assuming plug-flow through the housing, or X = 0.005, assuming
complete backmixing in the flow chamber. Such small values cannot be accurately measured and it cannot be decided from Fig.2 which asymptote corresponds to the measurements.

Fig.3 shows that the stoichiometric ratio also influences product distribution, as well
as the asymptote at higher rotor speeds. Similar results have been obtained in stirred
tank reactors (Ref.4) and are predictable for the slow regime (Ref.5). Fig.4 reveals
no significant effect of total flow rate (and therefore mean residence time) on product
distribution. It may thus be inferred that diffusion and reaction run to completion in
a time which is short compared to the mean residence time (of order 1 s).

Fig.5 and 6 present results from semi-batch reactions (Fig.1B). Fig.5 is analogous to
Fig.2 and so is its interpretation. No effect of the rate of feeding the B-solution was
noted, analogous to Fig.4, provided that a local stoichiometric excess of B in the
mixing head was avoided. {This situation, which was analysed elsewhere (Ref.3), leads
to S-formation and masks the effect of segregation}. The recirculation capacity of the
Ystral mixer seemed to fall substantially at rotor speeds below 3000 rpm and the reduced rate of feeding B, noted earlier, was then essential to avoid stoichiometric imbalance in the reaction zone.

When the vessel size was changed from 2 L to 1 L as well as to 3 L, no effect on X was
observed. Obviously reaction is so rapid that it is localised in or near the mixing
head. This was also shown by moving the addition point of B-solution to the alternative
in Fig.1B. X then rose from 1.3 to 14 %, reflecting a much lower rate of energy dissipation away from the mixing head.

Fig.6 shows that with increasing reagent concentrations more secondary product S was
formed. This result would be impossible in the slow reaction regime, where only the
stoichiometric ratio would be important, and together with the effect of rotor speed
reveals segregation and insufficiently fast micromixing. The reaction rate is a quadra-
tic function of concentration, whilst the rate of diffusion is linear in concentration:
segregation increases therefore with rising reagent concentrations.

4. APPLICATION OF MICROMIXING MODEL

A mathematical model of micromixing (Ref.6) will be applied, whose basis is to follow
diffusion and reaction in shrinking laminated structures within energy dissipating vortices. If the diffusivities of A, B and R are similar:

$$X = f(k_1/k_2, A_m/B_m, V_A/V_B \text{ or } \dot{V}_A/\dot{V}_B, Sc, M, \text{ reactor type}) \tag{5}$$

where the mixing modulus M is given by:

$$M = \frac{k_2 B_o \delta_o^2}{D} \tag{6}$$

and δ_o is set equal to half of the Kolmogoroff velocity microscale:

$$\delta_o = 0.5 (\nu^3/\varepsilon)^{1/4} \tag{7}$$

The function f has been computed for semi-batch as well as for completely backmixed
reactors (Ref.6). When the rate of energy dissipation in the reaction zone ε is known,
eq.(5) permits a first principles prediction of product distribution. Because the flow

field in the Ystral mixer has not been studied (at least in the literature), ε was unknown. Therefore one measured product distribution was fitted to the model (semi-batch, $V_A/V_B = 10$, $k_1/k_2 = 3800$, $A_m/B_m = 1.05$, $Sc = 1170$, $M = 1$ when $N = 5000$ rpm and $B_0 = 13.21$ mol m^{-3}, k_2 and D from Ref.6) and, assuming $\varepsilon \sim N^3$ under fully turbulent conditions, all other measurements could be plotted. Fig.7 and 8 compare these with the curves, predicted by the micromixing model (Ref.6).

Fig.8 for semi-batch reaction shows that the theory accounts fairly well for the effects of rotor speed, volumetric ratio and reagent concentrations (with the possible exception of the highest level). The agreement between model and experiments could have been improved by optimising the fit (Fig.8 refers to an arbitrary fit when $M = 1$), but this was not done.

Fig.7 refers to the continuous, in-line mixer and the solid curves were computed assuming complete backmixing of the fluid. No residence time distributions seem to have been measured and this assumption might well be inadequate e.g. the degree of backmixing could have depended on the rotor speed. That backmixing existed, was shown, however, by the appearance of coloured reaction products (dyes) at the plane P shown in Fig.1C. Moreover, under otherwise identical operating conditions, X was always higher for continuous in-line operation (Fig.7 and 8). It was unknown, however, whether this backmixing was complete. The agreement shown in Fig.7 is fairly satisfactory.

5. CONCLUSION

Fig.7 and 8 allow the rate of energy dissipation in the reactior zone of the Ystral mixer ε (W kg^{-1}) to be expressed as a function of rotor speed N (rpm):

$$\varepsilon = 3.93 \times 10^{-10} N^3 \tag{8}$$

e.g. N = 5000 rpm, ε = 49 W kg^{-1} and N = 10'000 rpm, ε = 393 W kg^{-1}. Such energy dissipations exceed the levels usual in tanks and pumps and are suitable for fast reactions on a time scale of order 1 ms, where intensive, local micromixing is required.

6. REFERENCES

1. Bourne J.R., Chem.Eng.Comm. 16 (1982) 79.
2. Angst W., Bourne J.R. and Dell'Ava P., Chem.Eng.Sci. 39 (1984), 335.
3. Bolzern O. and Bourne J.R., Proc.8th Int.Symp.Chem.React.Eng., Symp.Ser.No. 87 (1984) 543. Inst.Chem.Eng.
4. Bourne J.R., Kozicki F. and Rys P., Chem.Eng.Sci. 36 (1981) 643.
5. Levenspiel O. "Chemical Reaction Engineering", Chapter 7. Wiley, New York (1972).
6. Baldyga J. and Bourne J.R., Chem.Eng.Comm. 28 (1984) 243 and 259.

7. APPENDIX

The half-life for diffusion in deforming laminae t_D is given by (Ref.6):

$$t_D \simeq 2 \, (\nu/\varepsilon)^{1/2} \sinh^{-1} (0.05 \, Sc) \tag{9}$$

For typical aqueous solutions at 293 K, $\nu = 10^{-6}$ m^2 s^{-1}, $D = 10^{-9}$ m^2 s^{-1} and for the Ystral mixer ε is in the range 100-1000 W kg^{-1}. Thus, from eq.(9), t_D would range from 0.9 to 0.3 ms respectively, and the mixer would thus be suitable for reactions whose time scale was of order 1 ms.

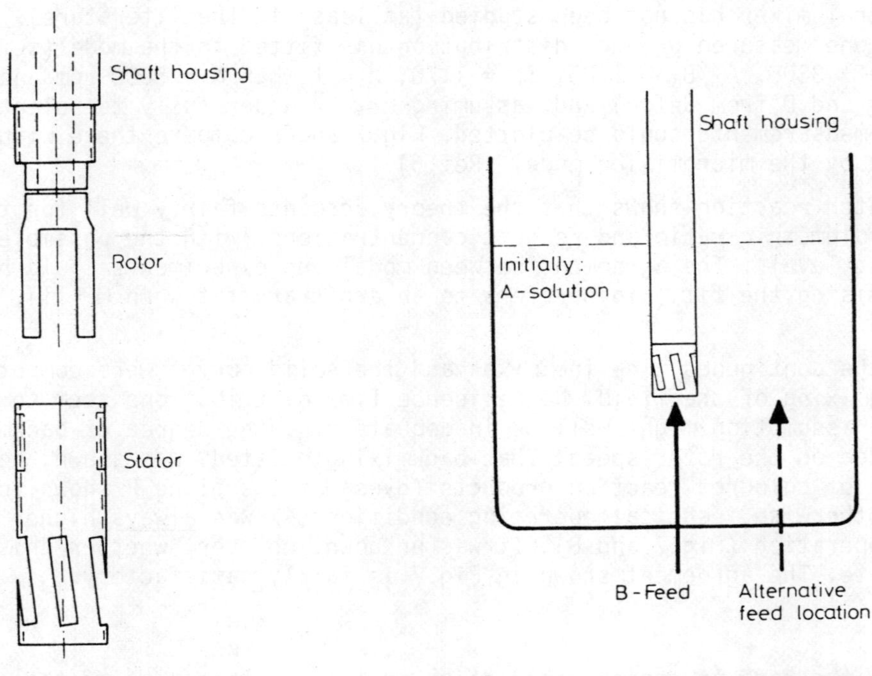

Fig. 1B   Semi-batch configuration.

Fig. 1A   Ystral mixing head (approx. to scale).

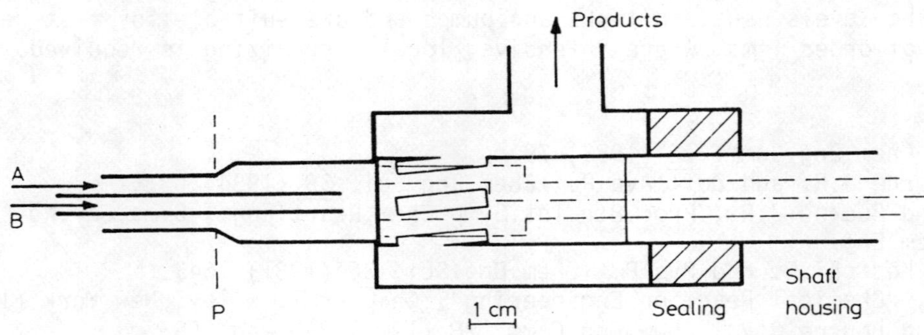

Fig. 1C   Flow chamber for in-line continuous operation of a Ystral mixer (approx. to scale).

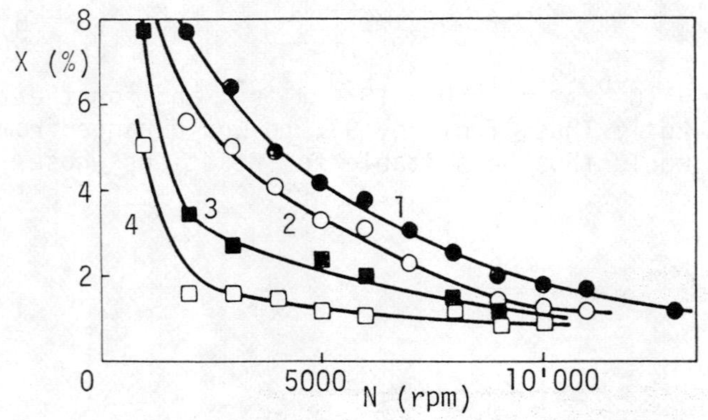

Fig. 2   Product distribution at different $\dot{V}_A/\dot{V}_B$ ratios (continuous operation).

$A_m/B_m = 1.055 \pm 0.025$

$B_o = 13.21$ mol m^{-3}

$\dot{V}_A/\dot{V}_B$:   1 = 10, 2 = 20, 3 = 50, 4 = 100

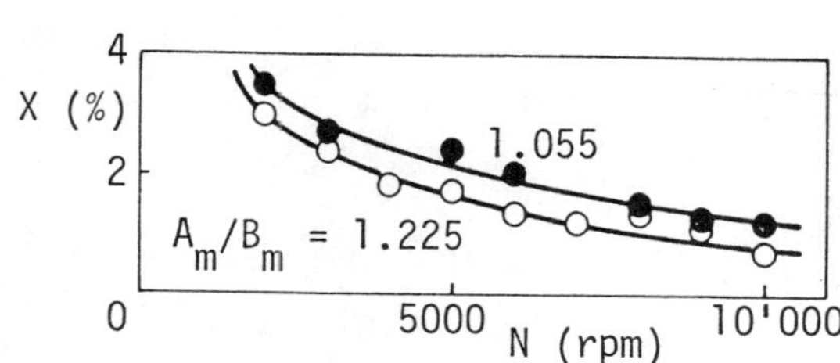

Fig. 3   Product distribution at different A_m/B_m ratios (continuous operation)
$B_o = 13.21$ mol m^{-3}, $\dot{V}_A/\dot{V}_B = 50$

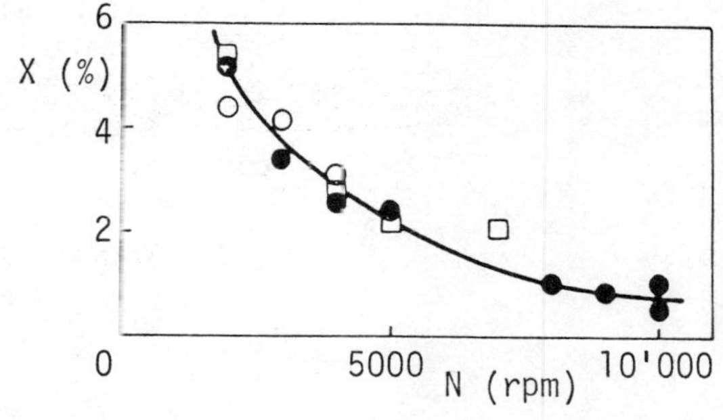

Fig. 4   Product distribution at different flow rates.

$A_m/B_m = 1.055$, $B_o = 6.6$ mol m^{-3}, $\dot{V}_A/\dot{V}_B = 5$
$\dot{V}_A + \dot{V}_B$ (L min^{-1}): ○ 2.1   □ 2.7   ● 3.3

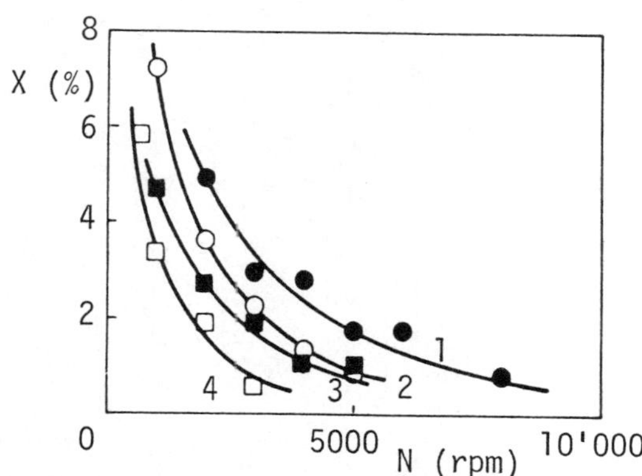

Fig. 5   Product distribution at different V_A/V_B ratios (semi-batch operation)

$A_m/B_m = 1.05$, $B_o = 13.21$ mol m^{-3}
V_A/V_B: 1 = 10, 2 = 20, 3 = 50, 4 = 100

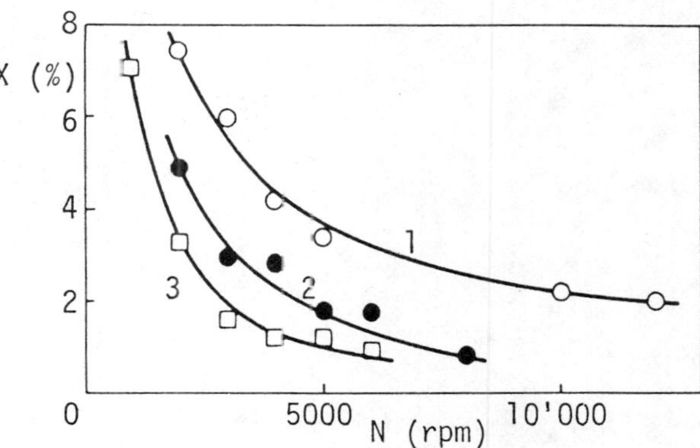

Fig. 6   Product distribution at different concentrations of the B-feed in semi-batch operation.

$A_m/B_m = 1.05$, $V_A/V_B = 10$
B_o (mol m^{-3}): 1 = 17.62, 2 = 13.21, 3 = 6.61

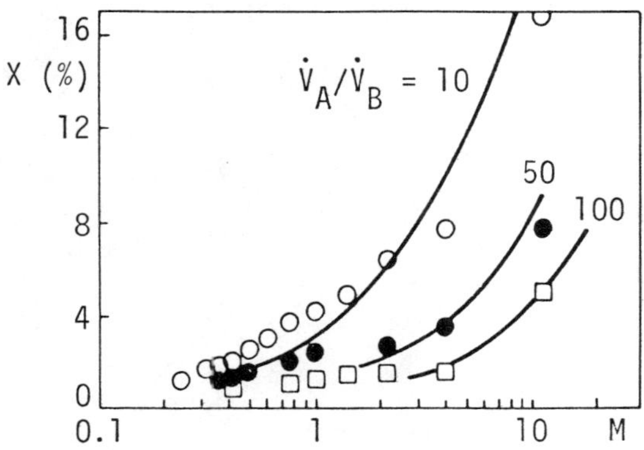

Fig. 7   Product distribution as a function of the mixing modulus M in continuous operation.

————   Predicted curves
Experimental results at $A_m/B_m = 1.055$,
$B_o = 13.21$ mol m^{-3}

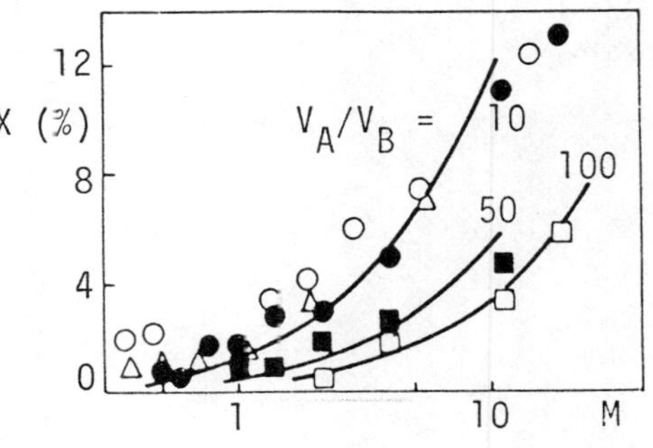

Fig. 8   Product distribution as a function of the mixing modulus M in semi-batch operation.

————   Predicted curves
Experimental results at $A_m/B_m = 1.055$,
B_o (mol m^{-3}): ○ 17.62   ●□■ 13.21   △ 6.61

COMPARATIVE MIXING TIMES FOR STIRRED TANK AGITATORS

S W RUSZKOWSKI AND M J MUSKETT

Fluid Mixing Processes (FMP)
BHRA   The Fluid Engineering Centre
Cranfield
Bedford
MK43 OAJ

SUMMARY

A conductimetric method for measuring mixing times has been modified to obtain a function which approximates to RMS concentration fluctuation or concentration variance. The method has been used for comparative testing of impellers in a 0.61 m diameter vessel using inert and reacting tracers. The method is described and some sample results are included.

The measured mixing times were found to be inversely proportional to impeller speed, and the decay rate constant of the RMS concentration fluctuation was proportional to specific power input raised to the one-third power. The mixing times using a reacting tracer were found to be shorter than those using an inert tracer.

Held at Wurzburg, 10-12 June, 1985.

Organised by DVCV· Deutsche Vereinigung für Chemie- und Verfahrenstechnik
(German Association of Chemical and Process Engineering).

Organisation: GVC·VDI-Gesellschaft Verfahrenstechnik und Chemieingenieurwesen. 

NOMENCLATURE

a	exponent	–
C_j	concentration of j'th data point	$mol\ dm^{-3}$
C_m	mean concentration in mixing vessel	$mol\ dm^{-3}$
C_T	concentration of tracer	$mol\ dm^{-3}$
$C(t)$	local concentration at time t	$mol\ dm^{-3}$
$C(0)$	concentration at time 0	$mol\ dm^{-3}$
$C(\infty)$	concentration when mixing is complete	$mol\ dm^{-3}$
C'_j	RMS concentration of j'th data point for all probes	$mol\ dm^{-3}$
C'_o	RMS concentration fluctuation at time t = 0	$mol\ dm^{-3}$
$C'(t)$	concentration fluctuation at time t	$mol\ dm^{-3}$
C'_{RMS}	root mean square concentration fluctuation	$mol\ dm^{-3}$
$C'_{RMS}(n)$	n'th estimate of RMS concentration fluctuation	$mol\ dm^{-3}$
j	subscript for j'th data point	–
k	number of points between calculation sections	–
m	number of points between calculation sections	–
M	mixing index	–
M_o	value of M at time t = 0	–
n	number of estimate of RMS concentration fluctuation time	–
N	impeller rotational speed	s^{-1}
t	time	s
T	time	s
V	vessel volume	m^3
V_T	tracer volume	m^3
β	mixing rate constant	–
$\bar{\epsilon}$	specific power input	Wkg^{-1}
ϵ'	turbulent power dissipation	Wkg^{-1}
ϵ_L	local power dissipation	Wkg^{-1}
σ_c^2	mean square concentration fluctuation	$mol^2\ dm^{-6}$
θ	mixing time	s

<u>INTRODUCTION</u>

There is a lack of consistent data in the literature on the comparative mixing performance of different agitators. Such data is of use to the engineer to predict the rates at which blending processes may be carried out, and to select an efficient vessel arrangement for any particular process. A great deal of effort has been made to understand the theoretical aspects of mixing. Mathematical models have been proposed which describe the way in which concentration fluctuations decay in a mixing vessel (Refs. 2, 3, 4) and which describe the effects of mixing on chemical reaction (Ref. 11).

Experimental work in the field has, however, been more limited. The manner in which data has been collected is somewhat haphazard. A wide variety of experimental methods has been used, with both inert and reacting tracers. The analysis of data has been very simple, and no systematic attempt has been made to fit the experimental data to physical models.

This paper describes a method which has been developed in an attempt to rationalise the measurement of mixer performance. The method described is intended to be a practical means of obtaining industrially relevant data which will be of benefit to engineers. It is by no means intended as an exhaustive analysis of practical or theoretical aspects of mixing. The references are intended only to illustrate points being made.

<u>EXPERIMENTAL METHOD</u>

A conductimetric method was used for measuring mixing times. A pulse of electrolyte was injected into the mixing vessel. The positions of the injector and probes, and the vessel dimensions are shown in figures 1 to 3.

The probes used are after a design described by Khang and Fitzgerald (Ref. 6). The probes are designed to be as small as possible, to minimise the effect of their intrusion into the vessel, and to minimise averaging effects due to their finite measuring volume. The probes are mounted at different heights within the vessel, at different radial distances from the mixer shaft and on different sides of the vessel. Thus any deficiencies in vertical or tangential convection within the vessel will be exposed.

The injector is situated in a well mixed region of the vessel so that local mixing effects due to injection will be small in comparison with the local mixing produced by the agitator. Two different electrolyte solutions are used as tracers: 2.4M nitric acid and 2.4M sodium hydroxide. In the 0.61m diameter vessel 35ml of tracer were injected for each experiment, in a pulse lasting 0.5 seconds. All experiments are carried out at a pH between 2.8 and 3.5, and acid and base injections are alternated. Stringent precautions are taken to ensure that the vessel and internals are inert to the tracers being used.

Strong acids and strong bases both have a high conductivity in relation to the salts formed when the acid and base neutralise each other. When an acid is injected into an acidic liquid, there will be a rise in conductivity proportional to the increase in acid concentration.

When a base is added to the same fluid, there will be an initial rise in conductivity, since the total electrolyte concentration is increased. The rise will be followed by a drop in conductivity as the base is micro-mixed with the acid and neutralised to form a salt. If only macro-mixing were to take place, even to a very small scale, there would be only minimal reaction between acid and base and hence only a rise in conductivity would be observed. To observe the drop in conductivity micro-mixing must take place.

For each experiment a continuous time-history of conductivity at each probe is obtained. The recording begins at the time the tracer pulse enters the vessel and is finished when the conductivity reaches a steady value. The conductivity data is stored in digitised form on a floppy disc.

DATA REDUCTION

The concentrations of tracer within the vessel are calculated by multiplying the conductivity measured at each probe by an appropriate calibration factor. This is done since the cell constant, and hence the response, varies between probes. The total step change in concentration is normalised between experiments to allow for small changes in the volume of tracer injected.

The time varying concentration at any point in the vessel may be regarded as being represented by a steady concentration, equal to the mean concentration in the vessel, and a time-varying component:

$$C(t) = C_m + C'(t) \tag{1}$$

where $C(t)$ is the local concentration at time t

C_m is the mean concentration in the vessel

$C'(t)$ is the local fluctuating component of concentration at time t.

The RMS concentration fluctuation over a time T is given by:

$$\sigma_c = \left(\frac{1}{T} \int_o^T (C(t) - C_m)^2 \, dt\right)^{\frac{1}{2}} = \left(\frac{1}{T} \int_o^T (C'(t))^2 \, dt\right)^{\frac{1}{2}} \tag{2}$$

This approach is commonly used for dealing with problems involving turbulent fluctuating quantities, and the RMS is generally used to characterise the fluctuating quantity. The situation in a mixing vessel is somewhat complicated, since the concentration fluctuations are transient. Not only C, but C'_{RMS} is a function of T, the time over which integration takes place. This problem may be overcome by calculating C'_{RMS} for small sections along the entire length of the conductivity time history.

Further, since the probes used have a finite measuring volume, the measured values of $C(t)$ will be averages and not point values. This will have the effect of reducing the apparent values of C'_{RMS}, depending on the probe measuring volume and the scale of the concentration fluctuations (Ref. 10). The fluid will appear fully-mixed when the scale of concentration fluctuations falls below the probe measuring volume.

The digitised trace obtained from the experiment will consist of a series of points corresponding to discrete values of conductivity.

A series of estimates of RMS concentration fluctuation may be obtained:

$$C'_{RMS}(n) = \left(\frac{1}{m} \sum_{kn}^{kn + (m-1)} \left[C_j - C(\infty)\right]^2\right)^{\frac{1}{2}} \tag{3}$$

where $C'_{RMS}(n)$ is the nth estimate of RMS concentration fluctuation

C_j is the concentration of the jth data point in the digitised time history

$C(\infty)$ is the concentration measured when mixing is complete, and is equal to C_m

m is the number of data points over which the RMS is calculated

k is the number of data-points by which the "calculation section" is moved along before calculating a new value of $C'_{RMS}(n)$

The RMS concentration fluctuation is calculated for a series of sections of the concentration time-history. There are m data points in each section, and the start of each section is k data-points after the start of the previous section.

A typical semilog plot of $(C'_{RMS})^2$ is shown in figure 6. The initial decay is very close to a straight line: this is the case for all the experiments.

The RMS concentration fluctuation has been used by several workers in theoretical studies of mixing in stirred vessels (Refs. 3 and 4). This work predicts that the mean-square fluctuations should decay exponentially with time:

$$\frac{d\sigma_c^2}{dt} = -\beta\sigma_c^2 \qquad (4)$$

where σ_c represents the RMS concentration fluctuation over the whole vessel, and β is a constant which varies with the mixer geometry, impeller speed and fluid used.

It should be noted that the calculated values of $C'_{RMS}(n)$ are only estimates of the true RMS concentration fluctuation within the vessel. Since the probes have a finite measuring volume, the values of C_j used are not point values, but mean values over the measuring volume. Also the measured values of concentration will be biased toward local conditions at the probe, rather than being a true RMS of the whole vessel. While the conductivity is being measured, fluid from different parts of the vessel flows past the probe. Hence an approximate measurement is made of concentration variance in the whole vessel, upon which a decay is superimposed. If the time over which the RMS is calculated is so short as to make the effect of the decay negligible, then the measured concentration will tend to be an instantaneous value of concentration at the probe, and will be unaffected by the concentrations in other parts of the vessel. If the RMS is calculated over a sufficiently long period of time that a meaningful average is taken over the whole vessel, then the effects of the decay of concentration fluctuations will become significant.

A better estimate of concentration variance throughout the whole vessel may be obtained by using the RMS concentration of all three probes for each data point:

$$C'_j = \left[\frac{C_{j1}^2 + C_{j2}^2 + C_{j3}^2}{3} \right]^{\frac{1}{2}} \qquad (5)$$

where C'_j is the RMS concentration of all three probes

C_{j1}, C_{j2}, C_{j3} are the concentrations of the jth data points for probes 1, 2 and 3

The values of C'_j are then used in expression (3) to calculate time-averaged RMS concentration fluctuations. This approach will remove the tendency for the RMS concentration fluctuation to be biased to conditions local to one probe. A typical semi-log plot obtained using this procedure, together with a test straight line fitted by the method of least squares, is shown in figure 7. The number of data points over which the RMS is calculated has been reduced from 32 in the case of a single probe (figure 6) to 4 in the case of the RMS of all three probes (figure 7). This has been done without appreciably affecting the linearity of the decay.

This approach differs considerably from those used in the past by workers measuring mixing times in stirred tanks (Refs. 1, 5, 7, 8 and 9). Previous methods have always used the maximum amplitude of concentration fluctuations within the vessel as a criterion of homogeneity. That is, the most extreme conditions found in the tank are used as a measure of mixer performance, rather than conditions which are representative of the whole contents. It is not possible to correlate the maximum amplitude of concentration fluctuations using any physical models. Individual measured values will tend to be biased toward the probe position used, although several workers have found little difference in mean mixing time between different probe positions (Refs. 1 and 9). Where several probes have been used (Refs. 5, 7 and 8) these have always been set up so that a mean value of concentration fluctuations is obtained, rather than an RMS, thus tending to average out the concentration fluctuations within the vessel.

The RMS concentration fluctuation may be used to obtain more detailed information about conditions within the vessel, for example estimates of the maximum concentration fluctuation, or the percentage of concentration fluctuations exceeding any given value. A probability density function (PDF) for the concentration fluctuations may be obtained from the original digitised data. The PDF describes the distribution of concentration fluctuations throughout the vessel and is itself a function of time, starting as a bimodal distribution at the instant of tracer injection, and finishing as a delta function at C_m after an infinite time. If the behaviour of the PDF is known, this may be used together with the RMS concentration fluctuation at any given time to calculate the distribution of concentration fluctuations in the vessel.

A mixing index may be defined by:

$$M = \frac{(C(\infty) - C(0)) - C'_{RMS}}{C(\infty) - C(0)} \tag{6}$$

where   C(0) is the concentration in the vessel before the injection of tracer has taken place

M expresses the RMS concentration fluctuation as a fraction of the total step change in concentration caused by injection of tracer.

A comparison between impellers may be made by comparing the values of β obtained for each impeller, or by measuring the time taken for M to exceed a given value. The second method of comparison is equivalent to measuring the time taken for the amplitude of concentration fluctuations to drop below a given value.

The two methods can be related. Expression (6) may be written:

$$M = 1 - \frac{C'_{RMS}}{\Delta C} \tag{7}$$

where $\Delta C = C(\infty) - C(0)$

Noting that C'_{RMS} is an estimate of σ_c, substitution into equation (4) gives:

$$\frac{d(1 - M)^2}{dt} = -\beta(1 - M)^2$$

Integration gives:

$$\ln (1 - M) = \frac{-\beta t}{2} + D$$

where D is a constant of integration, if $M = M_o$ at $t = 0$ then;

$$\frac{1 - M}{1 - M_o} = \exp \left(\frac{-\beta t}{2}\right) \qquad (8)$$

At time $t = 0$, if the volume of tracer which has just been injected is neglected, then $C = C(0)$ throughout the vessel

$$\therefore \quad C'_{RMS} = \left(\frac{1}{V} (C(0) - C(\infty))^2 . V\right)^{\frac{1}{2}}$$

$$= C(0) - C(\infty) = \Delta C$$

where V is the volume of the vessel

Substitution into equation (7) gives:

$$M_o = 0$$

Therefore equation (8) reduces to:

$$M = 1 - \exp \left(\frac{-\beta t}{2}\right) \qquad (9)$$

In practice the volume of tracer injected may not be negligible, and M_o will then become a function both of the volume, V_T, and concentration, C_T, of the tracer pulse:

$$(C'_o)^2 = \frac{1}{V} \left(\left[C(0) - C(\infty) \right]^2 (V - V_T) + \left[C_T - C(\infty) \right]^2 . V_T \right) \qquad (10)$$

where C'_o is the root mean square concentration fluctuation at time $t = 0$

V_T is the volume of tracer injected

C_T is the concentration of the tracer

A mass balance may be used to relate C_T and V_T at time $t = 0$:

$$V_T \, C_T + (V - V_T) \, C(0) = VC(\infty) \qquad\qquad (11)$$

Substitution of equation (11) into equation (10) gives:

$$(C'_o)^2 = \frac{1}{VV_T} \Big(\; [C(0) - C(\infty)]^2 \, (V - V_T).V_T + \qquad (12)$$

$$([\; C(\infty) - C(0)] . V + [C(0) - C(\infty)] . V_T)^2 \Big)$$

For small values of V_T, C'_o becomes very large and tends to infinity as V_T tends to zero. For the experimental conditions used log (ΔC^2) is equal to 6.62, and log $(C'_o)^2$ calculated using equation (12) is equal to 10.3. Examination of figures 6 and 7 shows that the intercepts, in fact, lie close to 6.62. In all the experiments, the intercept was found to be close to, but always greater than 6.62. Since the tracer injection is not instantaneous, and the pulse is diluted during injection due to jet mixing, C'_o will, in general be smaller than predicted by equation (12). However C'_o only approaches ΔC when the tracer volume, V_T, is close to half the vessel volume V. Clearly there is some discrepancy here, and to make use of equation (8) values of M_o must be measured for different impellers.

RESULTS

Different impellers have been compared at equal values of $\bar\varepsilon$, the power input per unit mass of fluid. Results for two impellers are given: a T/3 diameter square pitch marine propeller (axial-flow case), and a T/3 diameter 45° pitched-blade turbine (mixed flow case). Three different values of $\bar\varepsilon$ were used. Values of β, and the times taken for M to reach 0.95 are plotted in figures 8 to 11. Each point plotted represents the mean of nine experiments. The slope of the best straight line through the points has also been included in the graphs.

DISCUSSION

Typical time histories for an acid tracer and a base tracer are shown in figures 4 and 5. It is evident that they do not take the form of an exponentially-decaying sine wave which has been found by some other workers (Refs. 5 and 7). Workers who obtained an exponentially-decaying sinusoid used an axially symmetrical arrangement of conductivity probes around the impeller. This has the effect of averaging out the concentration fluctuations in the horizontal plane containing the probes. The vessel is effectively transformed into a two dimensional mixer, and the apparent mixing times are shortened. Other workers who used different probe arrangements (Refs. 1 and 8) have not obtained exponentially-decaying sine waves.

The data presented in this paper has been obtained by taking the root-mean-square of the concentration fluctuations, rather than the mean:

$$C'_j = \left(\frac{C_{j1}^2 + C_{j2}^2 + C_{j3}^2}{3} \right)^{\frac{1}{2}} \qquad\qquad (5)$$

The values of C'_j thus obtained are then used to calculate time-averaged RMS concentration fluctuations.

The points plotted in figures 8 to 11 all lie very close to straight lines. The quantity of data obtained for each impeller is rather limited and more work is required to confirm these trends.

The mixing times were found to be related to specific power input by:

$$\theta \ \alpha \ (\bar{\varepsilon})^{a}$$

a was found to be close to $-1/3$. Since $\bar{\varepsilon} \ \alpha \ N^3$, this result is equivalent to $N\theta$ being constant. The mixing rate constant, β, was found to be proportional to the one third power of $\bar{\varepsilon}$. This is in agreement with the value of the exponent which has been predicted on theoretical grounds (Refs. 3 and 4).

The measured mixing rates were found to be higher for experiments using a base tracer. The concentration fluctuations decayed more rapidly when the tracer used underwent an "instantaneous" chemical reaction at the same time as it was being dispersed by the turbulent flows in the vessel. This may, at first, seem surprising since when acid tracer is used only macromixing affects the concentration fluctuations, whereas for the case of a base injection the concentration fluctuations will not decay until micromixing has taken place. In fact, for the case involving chemical reaction two mechanisms operate to reduce the concentration fluctuations in the vessel: a decay due to turbulent dispersion and a decay to a removal of a tracer by chemical reaction. The combined effect of these two mechanisms is to increase the rate of decay of concentration fluctuations over that observed when a non-reactive tracer is used. This effect has been described by Toor (Ref. 11).

The analysis of the results using a base tracer is made difficult by the fact that the acid and base have different molar conductivities. It is only possible to calculate an average concentration fluctuation using an average molar conductivity. This will cause inaccuracies in the values obtained for RMS concentration fluctuation. In the case of a reacting tracer, the mean concentrations of the reacting species are also changing with time, and errors will arise in calculating the decay of mean concentrations.

The measurements using a base tracer are not as accurate a representation of conditions in the mixing vessel as the measurements using an acid tracer. The results using base tracer do not lie as close to a straight line as the results using an acid tracer, and further work is required to confirm whether the relationship between log β and log $\bar{\varepsilon}$ is linear.

The values of β obtained for the inert tracer (acid) may be compared with those predicted by an expression derived by Corrsin (Ref. 3):

$$\frac{1}{\beta} = (\tfrac{1}{2})\left[3 \ (5/\pi)^{2/3} \ (L_s^{2}/\varepsilon')^{1/3} \ + \ (\nu/\varepsilon')^{1/2} \ \ln Sc \right] \tag{14}$$

L_s is the integral length scale of concentration fluctuations

ε' is the turbulent power dissipation per unit mass

L_s may be replaced by $\frac{\pi}{5} \cdot \frac{2}{L}$ (Ref. 2)

where L is a characteristic length of the mixing system. For a predominantly axial pumping impeller this may be taken to be the impeller diameter. $\bar{\varepsilon}$ is given by $\varepsilon' = \eta\,\bar{\varepsilon}$ (Ref. 2) where η is an efficiency of conversion of power input to turbulent energy, η is approximately $\frac{1}{2}$ (Ref. 3). The following expression is thus obtained:

$$\frac{1}{\beta} = (^{1}/_{2})\left[3\,(D^2/2\bar{\varepsilon})^{1/3} + (2\nu/\bar{\varepsilon})^{1/2}\,\ln Sc\right] \tag{15}$$

Substitution of $D = 0.203$ m, $\nu = 10^{-6}$ m^{2}s^{-1}, $Sc = 10^3$ into equation (15) gives the following values of β for the three values of $\bar{\varepsilon}$ at which experiments were carried out:

$$\beta = 0.41,\ 0.84,\ 1.5 \qquad \text{s}^{-1}$$

$$\text{for }\ \bar{\varepsilon} = 0.0053,\ 0.043,\ 0.23 \qquad \text{Wkg}^{-1}$$

The predicted values are approximately twice the measured values (fig. 9). In deriving expression (14) Corrsin assumed stationary, homogeneous, isotropic turbulence, although, as was pointed out by him, these assumptions do not hold for a stirred vessel. The suggested values of L_s and η are also approximations. In view of this it would be surprising if there were exact agreement between the measured and the predicted values of β. The agreement might be improved by estimating local energy dissipation rates, ε_L, from measured values of turbulence intensity, and using the mean of $(\varepsilon_L)^{1/3}$ in place of $(\bar{\varepsilon})^{1/3}$ in equation (14).

CONCLUSIONS

A conductimetric method has been used to compare the mixing efficiencies of two impellers. By siting conductivity probes at appropriate points within the vessel, and calculating RMS concentration fluctuations, a function may be obtained which approximates to the intensity of segregation and concentration variance used in the past by several workers (Refs. 2, 3 and 4). An RMS concentration fluctuation gives a reliable indication of conditions within a mixing vessel. The measurements are dependent on the conditions in all the parts of the vessel where probes have been sited, and the concentration fluctuations are not averaged out by taking the mean values of several probes.

The values of RMS concentration fluctuation may be used with a PDF to describe fully the concentration fluctuations in the vessel. The results obtained so far indicate that mixing times measured using this method are inversely proportional to impeller speed for any one impeller. The decay rate constant for the concentration fluctuations is proportional to specific power input to the one-third power.

The method may be adapted for a chemically-reacting system. The experiments so far indicate that in a chemically-reacting system, mixing times will be shorter than those measured using an inert tracer.

ACKNOWLEDGEMENT

The authors wish to thank the members of FMP (Fluid Mixing Processes) for their financial and technical support, and their valuable advice: APV Osborne Craig Ltd, Chemineer, Ekato Mixing Ltd, Greaves Mixers Ltd, Lightnin Mixers, LPP Ltd, Pfaudler-Balfour Ltd, Plenty Mixers, Transkem Plant, Allied Colloids, Amoco plc, BNFL Ltd, Dow Chemicals plc, Du Pont plc, Exxon, ICI plc, Kodak plc, Merck Sharpe and Dohme plc, Norsk Hydro plc, Rohm and Haas, Unilever, Department of Trade and Industry. The authors are grateful for permission to publish general trends of work carried out for FMP.

REFERENCES

1.   Biggs R.D.
      "Mixing rates in stirred tanks".
      AIChEJ   Vol. 9, pp 636-640, 1963.

2.   Brodkey R.S.
      "Mixing in turbulent fields" in "Turbulence in mixing operations - theory and
      application to mixing and reaction"
      ed. R.S. Brodkey   Academic Press Inc. 1975.

3.   Corrsin S.
      "The Isotropic turbulent mixer: Part II. Arbitrary Schmidt Number".
      AIChEJ   Vol. 10, pp 870-877, 1964.

4.   Evangelista J.J., Katz S. and Shinnar R.
      "Scale-up criteria for stirred tank reactors"
      AIChEJ   Vol. 15, pp 843-853, 1969.

5.   Holmes D.B., Voncken R.M. and Dekker J.A.
      "Fluid flow in turbine-stirred, baffled tanks"
      Chem. Eng. Sci., Vol. 19, pp 201-213, 1964.

6.   Khang S.J. and Fitzgerald T.J.
      "A new probe and circuit for measuring electrolyte conductivity".
      Ind. Eng. Chem. Fundam., Vol. 14, pp 208-213, 1975.

7.   Khang S.J. and Levenspiel O.
      "New scale-up and design method for stirrer agitated batch mixing vessels".
      Chem. Eng. Sci., Vol. 31, pp 569-577, 1976.

8.   Kramers H., Baars G.M. and Knoll W.H.
      "A comparative study of the rate of mixing in stirred tanks".
      Chem. Eng. Sci., Vol. 2, pp 35-42, 1953.

9.   Prochazka J and Landau J.
      "Studies on mixing XII. Homogenation of miscible liquids in the turbulent
      region".
      Coll. Czech. Chem. Comm., Vol. 26, pp 2961-2973, 1961.

10.   Thyn J., Novak V., and Pock P.
      "Effect of the measured volume size on the homogenisation time".
      Chem. Eng. J., Vol. 12, pp 211-217, 1976.

11.   Toor A.L.
      "The non-premixed reaction: A+B → products". in "Turbulence in mixing
      operations - theory and application to mixing and reaction".
      ed. R.S Brodkey Academic Press Inc. 1975.

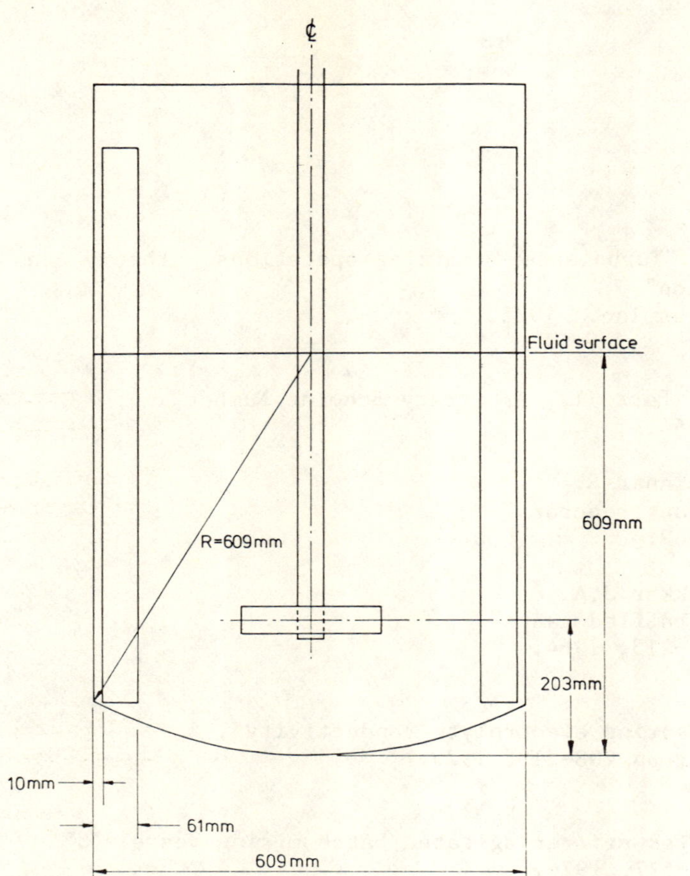

FIG. 1   VESSEL DIMENSIONS

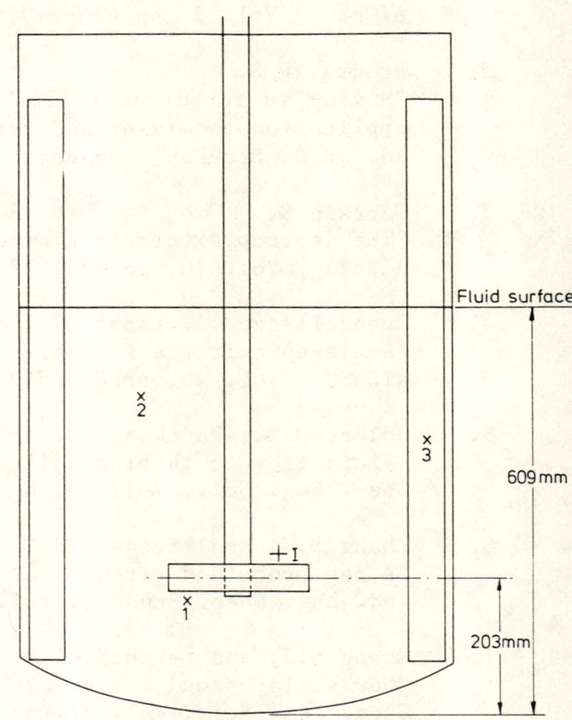

FIG. 2   PROBE AND INJECTOR POSITIONS
(Refer also to Fig.3)

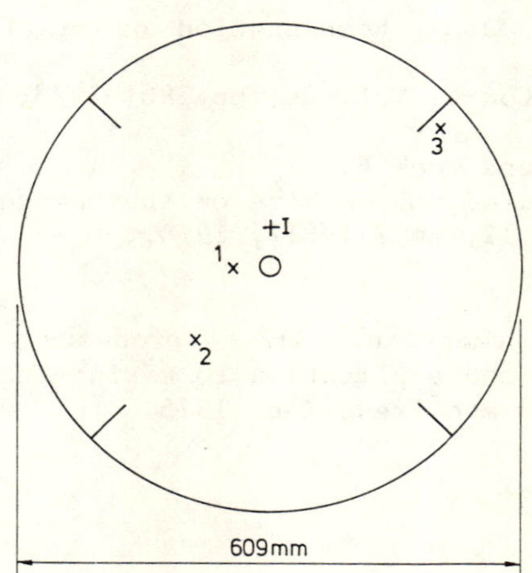

Injector	25mm above top of impeller. 75mm from shaft axis.
Probe 1	12mm below bottom of impeller. 75mm from shaft axis.
Probe 2	135mm below fluid surface. 130mm from shaft axis.
Probe 3	203mm below fluid surface. 274mm from shaft axis, adjacent to baffle ℄.

FIG. 3   PROBE AND INJECTOR POSITIONS

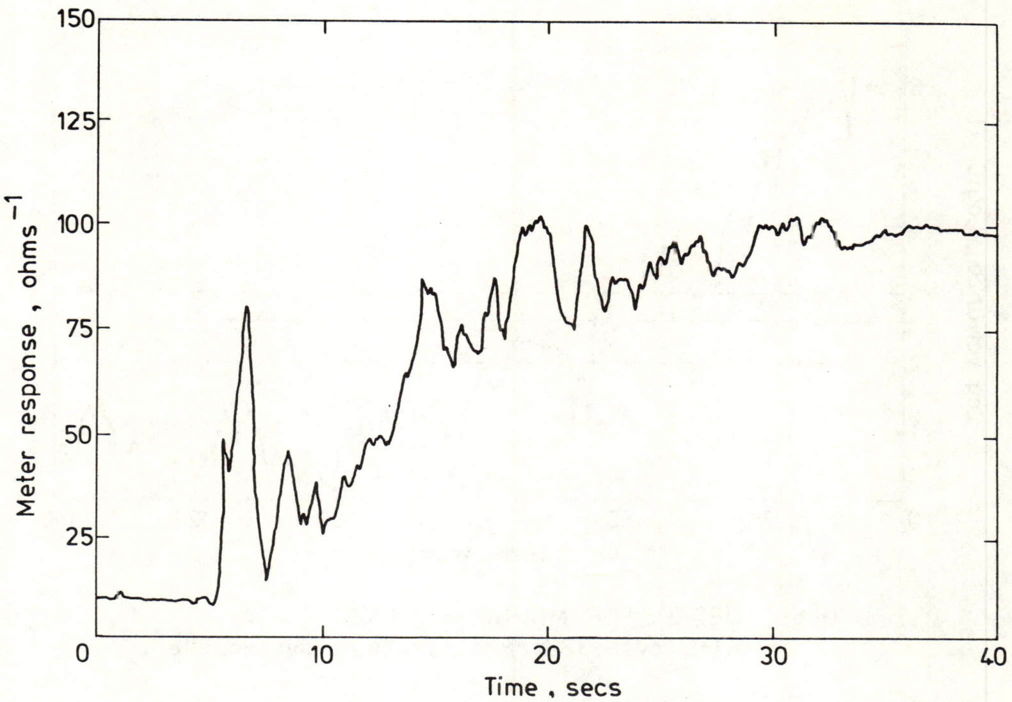

FIG. 4   CONDUCTIVITY METER RESPONSE vs TIME
Probe under impeller, acid injection.

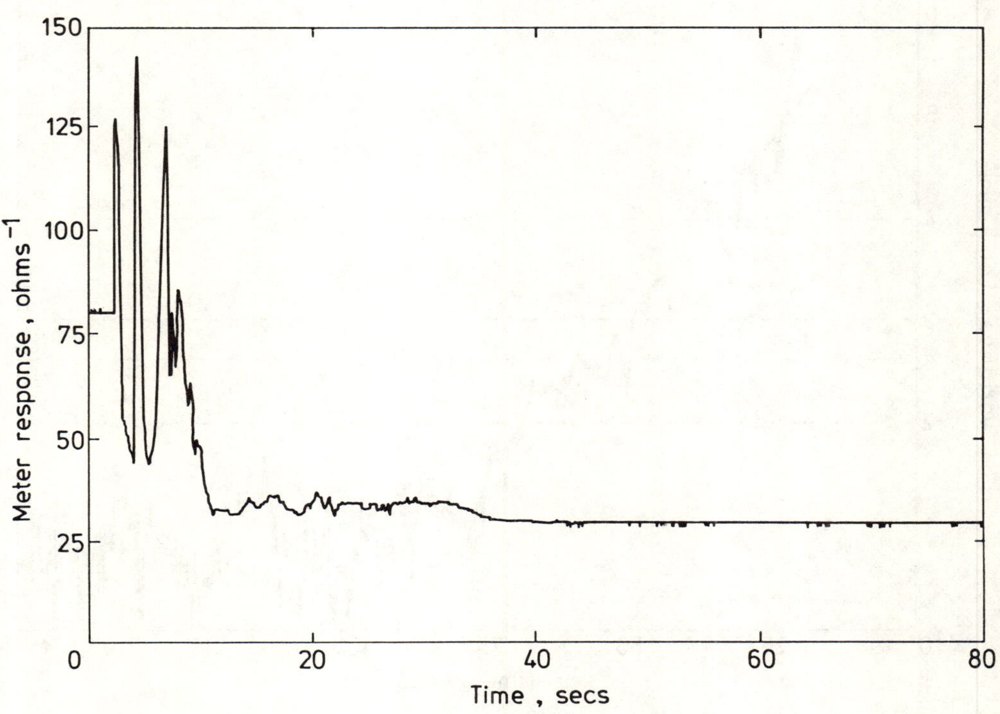

FIG. 5   CONDUCTIVITY METER RESPONSE vs TIME
Probe under impeller, base injection.

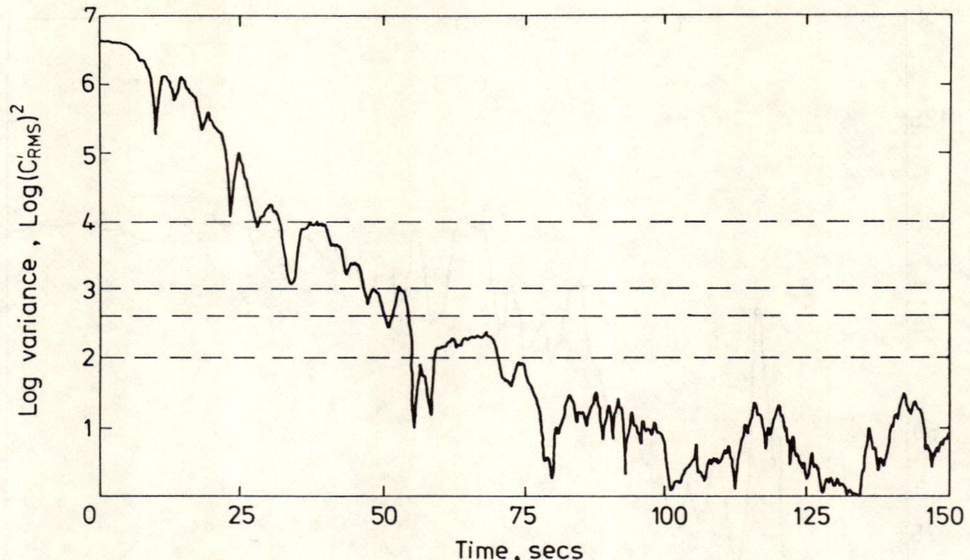

FIG. 6   DEGREE OF MIXING vs TIME
RMS over 32 points , probe under impeller .

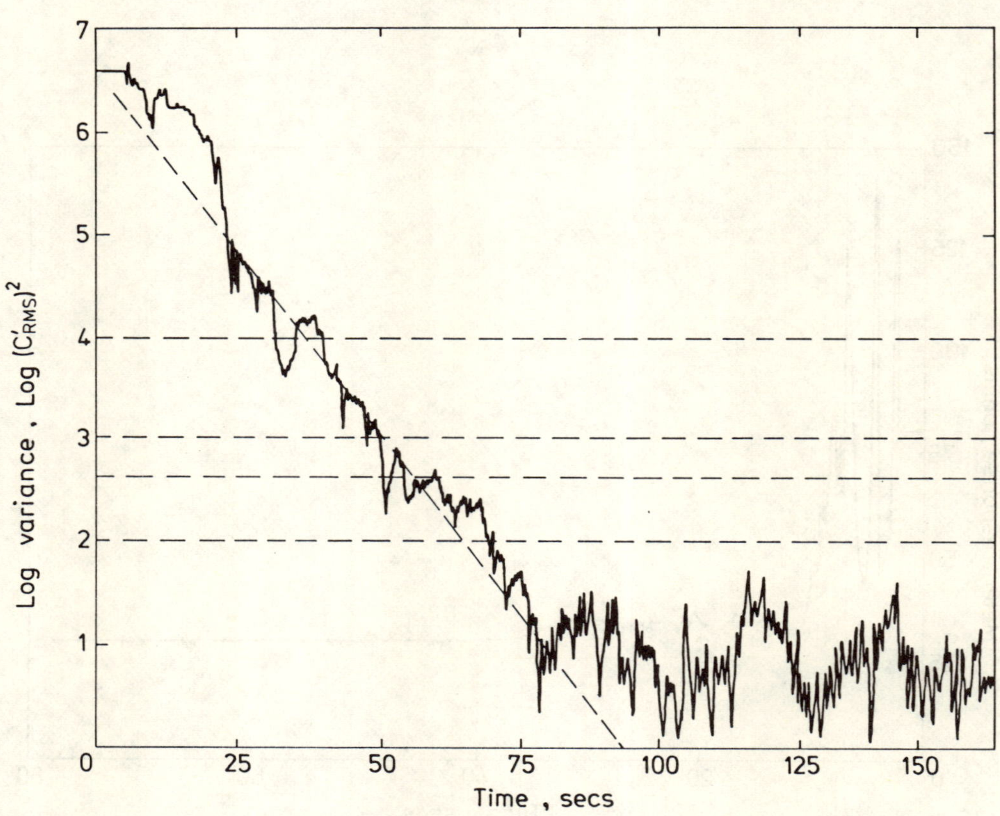

FIG. 7   DEGREE OF MIXING vs TIME
RMS over 4 points . RMS of all probes .

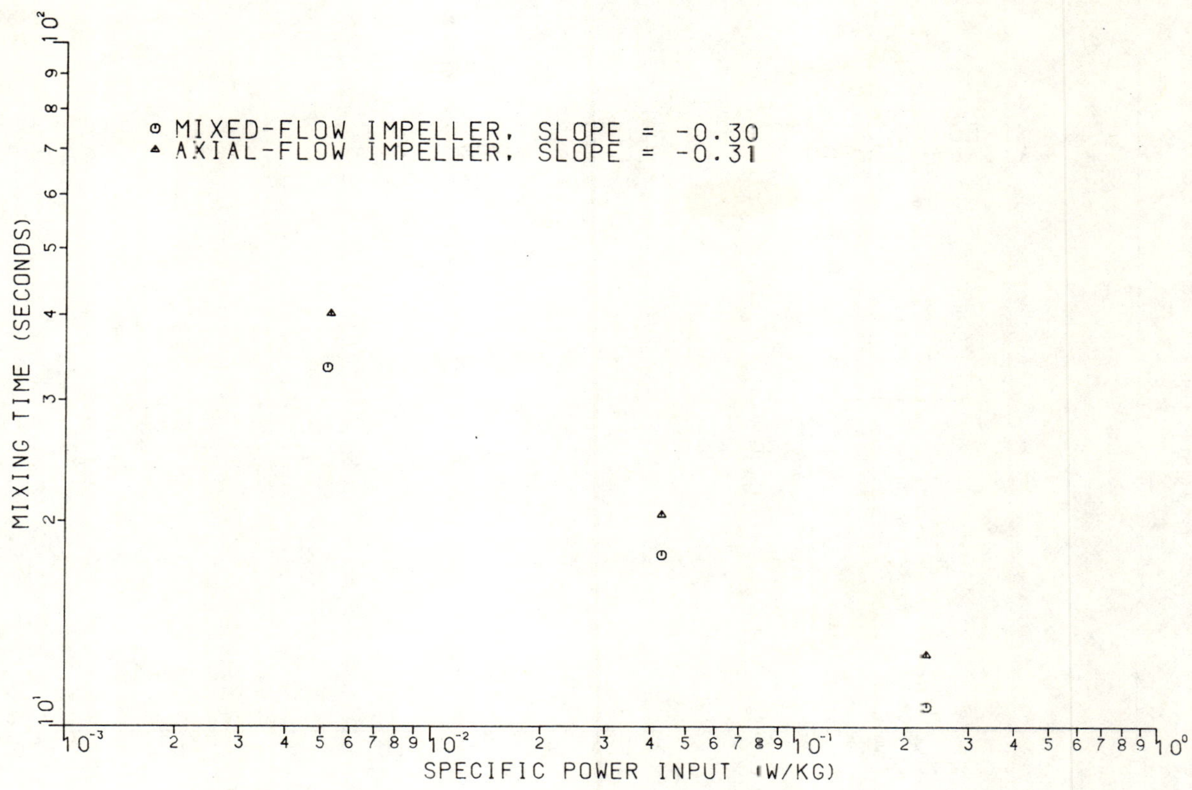

FIG. 8   MIXING TIME VS SPECIFIC POWER INPUT
TIME FOR M TO REACH 0.95 - ACID INJECTION

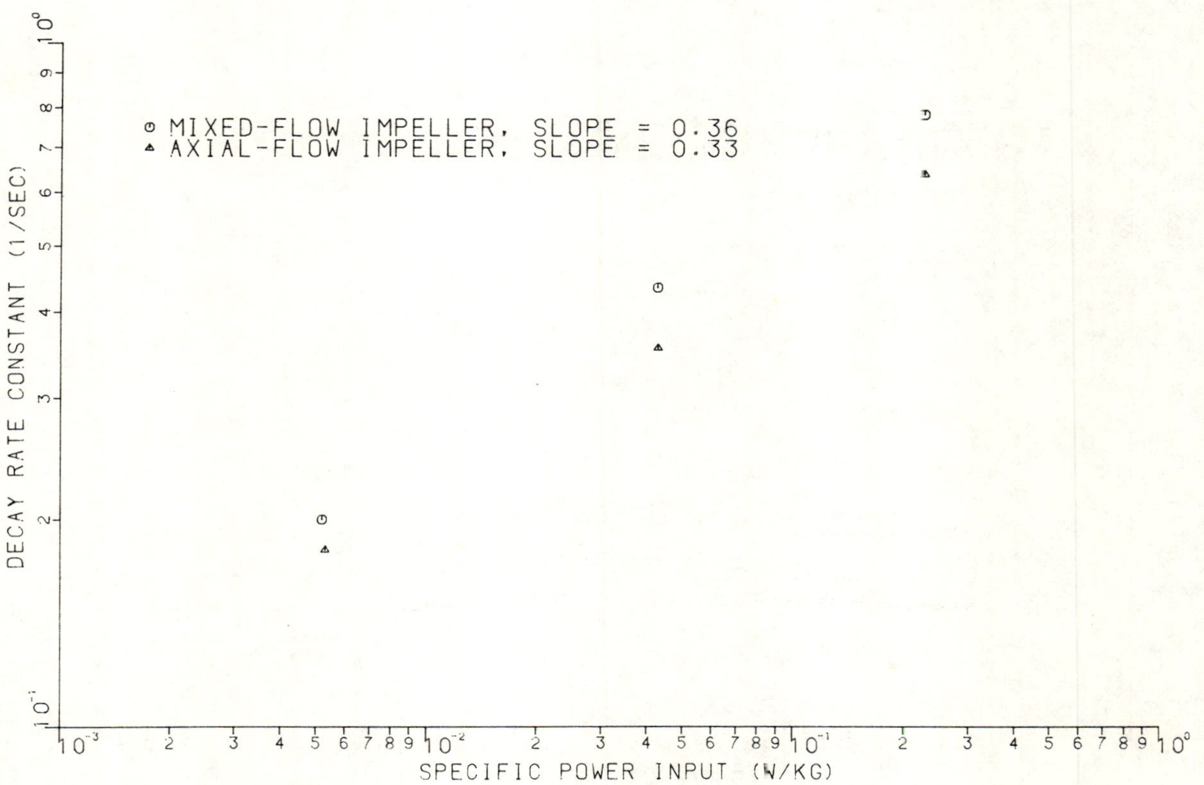

FIG. 9   DECAY RATE CONSTANT VS SPECIFIC POWER INPUT
ACID INJECTION

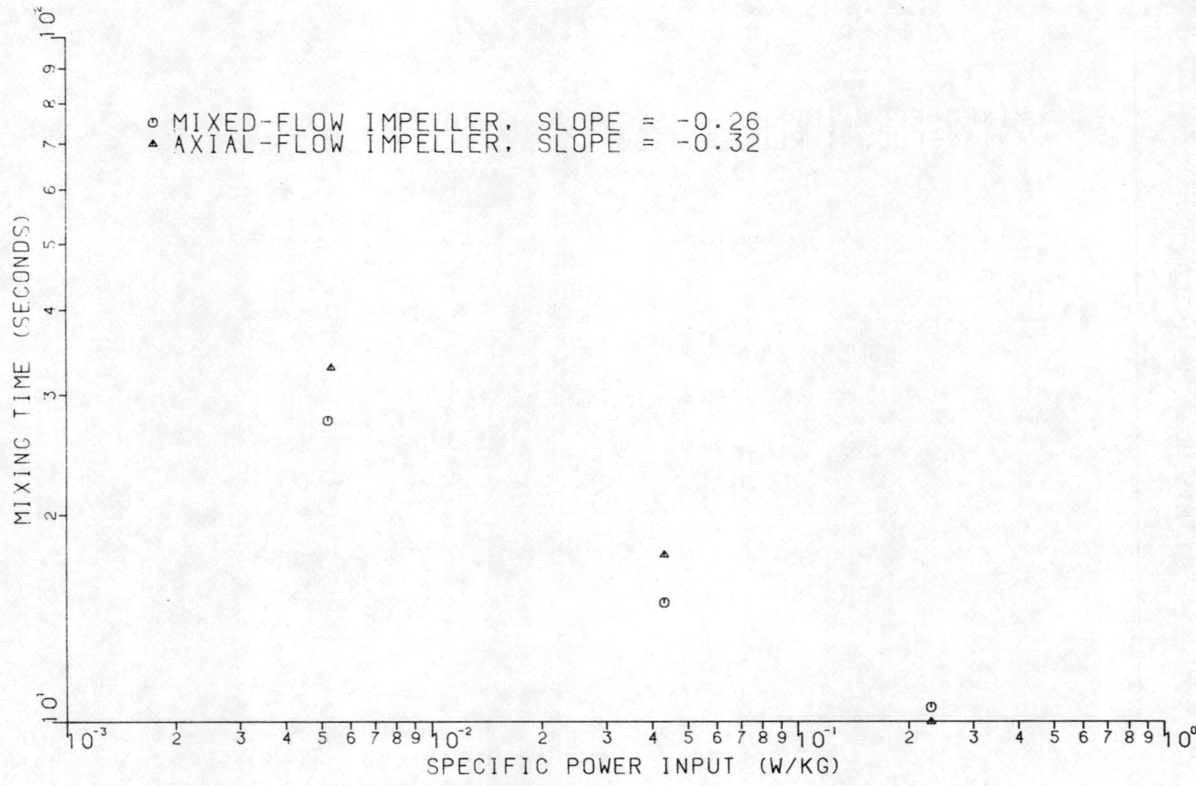

FIG. 10   MIXING TIME VS SPECIFIC POWER INPUT
TIME FOR M TO REACH 0.95 - BASE INJECTION

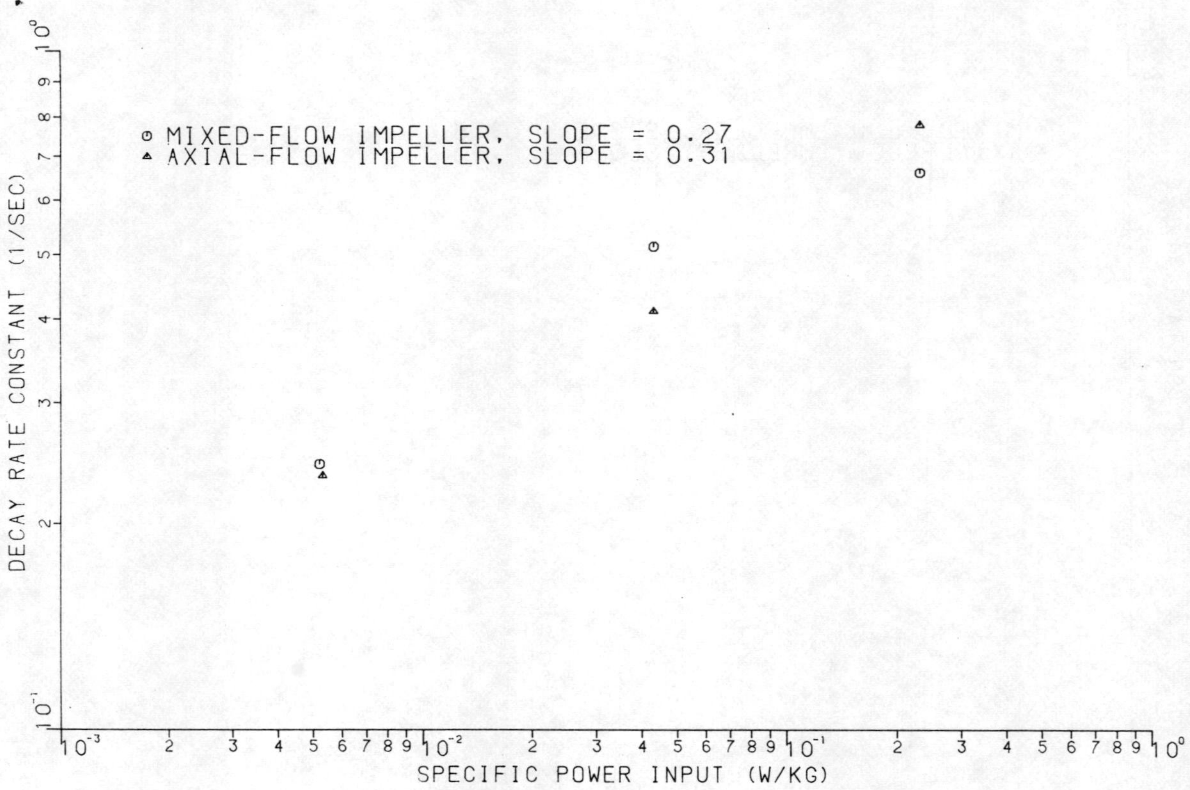

FIG. 11   DECAY RATE CONSTANT VS SPECIFIC POWER INPUT
BASE INJECTION

THE EFFECT OF MICROMIXING ON THE COURSE
OF PRECIPITATION IN AN UNPREMIXED FEED
CONTINUOUS TANK CRYSTALLIZER

R. Pohorecki, Ph.D. Sc.D., and

J. Bałdyga, Ph.D.

Department of Chemical Engineering,
Warsaw Technical University,
Warsaw, Poland

Summary

The precipitation of the solid product produced in a fast
chemical reaction between liquid substrates is described using a model
of micromixing in an isotropic, homogeneous turbulent field.

The model gives the most important characteristics of the
precipitation process: the size distribution of the solid product, the
degree of supersaturation in the crystallizer, etc.

The results calculated from the model are compared with the
experimental results obtained for the continuous precipitation of
$BaSO_4$ in a stirred tank crystallizer.

The most interesting conclusion, drawn form the calculations and
confirmed experimentally, is that the increase of mixing intensity in
the case of continuous precipitation brings about the increase of the
size of precipitating particles, contrarily to the batch case.

Held at Wurzburg, 10-12 June, 1985.

Organised by DVCV· Deutsche Vereinigung für Chemie- und Verfahrenstechnik
(German Association of Chemical and Process Engineering).

Organisation: GVC·VDI-Gesellschaft Verfahrenstechnik und Chemieingenieurwesen. 

©BHRA, The Fluid Engineering Centre, Cranfield, Bedford MK43 0AJ, England.

<u>NOMENCLATURE</u>

A_g = specific surface area — dm^{-1}
$\bar{a}$ = mean particle area — dm^2
b = baffle width — m
c_i = concentration of the component "i" — mol/dm^3
D = turbine diameter — m
d_s = particle diameter — $dm, \mu m$
$\bar{d}_{32}$ = particle diameter according to Sauter — $dm, \mu m$
G = rate of growth — dm/s
g = order of the growth process
H = liquid level height — m
K_g = $3 \cdot 10^3 \cdot \rho_s/M = 5.78 \cdot 10^4$ — $mol/(dm^2 s)$
K_s = nucleation rate constant — $dm^{12}/(s\,mol^4)$
k_{md} = micromixing parameter — s^{-1}
M = molecular mass — kg/mol
Ne = power number
n = agitator speed — s^{-1}
n_s = number of solid particles per unit volume — dm^{-3}
Re = Reynolds number
R_i = rate of reaction — $mol/(dm^3 s)$
Sc = Schmidt number
T = tank diameter — m
t = time, age — s
$\bar{t}_R$ = mean residence time — s
$\bar{t}_{Rm}$ = mean time of molecular mixing — s
t_{ms} = time constant for complete segregation — s
t_{md} = time constant for molecular mixing — s
V = volume — dm^3
V_{se} = volume of completely segregated fluid — dm^3
$\dot{V}$ = flow rate — dm^3/s
α = $4 \cdot$ (true surface area of particle) $/(\Pi d_s^2)$
ρ_L = liquid density — kg/dm^3
ρ_s = particle density — kg/dm^3
γ = kinematic viscosity — m^2/s

Subscripts

α = initial value   for any function F
A,B,C = reactants $F' = F - \bar{F}$
L = liquid $\bar{F}$ = mean value
s = solid
se = segregated

1. INTRODUCTION

In our previous publications $\left(\text{Ref.1, 2}\right)$ we have suggested that
the mixing intensity is one of the major factors affecting
precipitation process.

The precipitation of a solid product, resulting from a chemical
reaction between two liquid ionic solutions, can only occur as the
fluid phase becomes supersaturated with respect to the solid product
of reaction.

The process of precipitation can be considered, therefore, as
consisting of three succesive stages:
- mixing of the substrates on molecular scale, making possible
  the homogeneous chemical reaction;
- homogeneous chemical reaction producing supersaturation;
- crystallization (i.e. nucleation and crystal growth $\big)$
  controlled by the degree of supersaturation.

The intensity of mixing influences both the first stage of the above
sequence, and the mass exchange of the product to be crystallized
between fluid elements of different age (local redistribution of
supersaturation) . It affects therefore the overall rate of the
process, as well as the number and the size distribution of the solid
particles.

To account for these effects, the crystallizer will be considered
as an unpremixed - feed - reactor, and described by means of a
suitable micromixing model .

2. MODEL OF MICROMIXING

A detailed description of the model used in this paper to
characterise the micromixing process in the crystallizer, is presented
elsewhere $\left(\text{Ref.3, 4}\right)$; thus only the main points, important when
continuous precipitation is considered, are presented here.

The total volume of a crystallizer is considered to consist of
two zones: the zone of complete segregation and the zone of molecular
dissipation of concentration fluctuations.

The completely segregated fluid does not exchange mass with its
environment, thus in an unpremixed - feed - reactor this fluid forms
a dead zone. In consequence the existance of the completely segregated
zone decreases the effective volume and effective time of reaction.
The influence of the intensity of mixing on the rate of decay of the
completely segregated zone is given by Eq.1.

$$- \frac{dV_{se}}{dt} = \frac{V_{se}}{t_{ms}} \qquad /1/$$

Eq.1 was derived from Rosensweig's theory of turbulent mixer $\left(\text{Ref.5}\right)$.
The complete segregation time constant t_{ms} for the Rushton type mixer
used in our experiments, equals:

$$t_{ms} = \frac{2.07 \left(\frac{H}{D} - \frac{h}{D} - \frac{w}{2D}\right)^2 \cdot \left(\frac{T}{D}\right)^2}{n \cdot Ne^{1/3} \left(\frac{T}{D} - 1\right)^{1/3} \left(\frac{H}{D}\right)^{1/3}} \qquad /2/$$

Using Eq.1 one can estimate the completely segregated volume in the
system as follows:
- in a well mixed on macroscale CSTR

$$V_{se} = \dot{V} \, \bar{t}_R \cdot \left[1/\left(1 + \bar{t}_R/t_{ms}\right)\right] \qquad /3/$$

- in a semibatch system

$$V_{se} = \dot{V} \, t_{ms} \qquad /4/$$

For the zone of molecular dessipation of concentration fluctuations the rate of mixing can be obtained using equation analogous to
this given by Corrsin $\left(\text{Ref.6}\right)$ but valid only for the zone of molecular

mixing:

$$- \frac{d\overline{\left(c_i'\right)^2}}{dt} = \frac{\overline{\left(c_i'\right)^2}}{t_{md}} \qquad\qquad /5/$$

where t_{md}, the time constant for diffusional mixing is for a Rushton type mixer given by:

$$t_{md} = 2.73 \frac{\left(\frac{T}{D}\right)\left(\frac{H}{D}\right)^{1/2}\left[\ln\,(Sc) - 1.27\right]}{Ne^{1/2}\,n\,Re} \qquad\qquad /6/$$

In order to describe the course of a chemical reaction in the molecular dissipation zone an additional equation is necessary, namely the equation of micromixing with chemical reaction. The micromixing model proposed by Costa and Trevissoi Ref.7 , is sufficient for our purpose:

$$- \frac{dc_i}{dt} = k_{md}\left(c_i - \bar{c}_i\right) + R_i \qquad\qquad /7/$$

where the micromixing parameter k_{md} is:

$$k_{md} = \frac{1}{2\,t_{md}} = 0.183 \frac{Ne^{1/2}\,n\,Re}{\left(\frac{T}{D}\right)\left(\frac{H}{D}\right)^{1/2}\left[\ln\,(Sc) - 1.27\right]} \qquad\qquad /8/$$

3. DESCRIPTION OF PRECIPITATION OF A SOLID PRODUCT $\left(BaSO_4\right)$ IN A CONTINUOUS STIRRED TANK CRYSTALLIZER.

Let us consider an instantaneous, ionic reaction with a solid product $A + B \longrightarrow C$ proceeding in a Continuous Stirred Tank Crystallizer $(CSTC)$, as shown in Fig.1. The system is in a steady state. The mixture of reagents is stoichiometric and the flow rates are equal

$$V_A = V_B \;\; ; \;\; c_{A\alpha} = c_{B\alpha} \qquad\qquad /9/,\ /10/$$

3.1. The completely segregated zones.

The completely segregated volumes of A and B solutions are equal, because of equal flow rates (Eq.9) and constant t_{ms} in the CSTC. In an ideal well mixed stirred tank, these volumes would be given by Eq.3. However, in reality if A and B inlet points are placed far away from the outlet from the crystallizer, and the circulation time is longer than the segregation time constant t_{ms} (like in our experiments) a better estimation of these volumes is

$$V_{Ase} = V_{Bse} = V_A \cdot t_{ms} \qquad\qquad /11/$$

As it was already mentioned, reactant A cannot practically penetrate into the completely segregated B zone, and vice versa. The small particles of precipitate will follow the turbulent fluctuations of the fluid (and thus not penetrate to the completely segregated volume if the fluid does not penetrate there), if the inertia parameter $\psi \lll 1$ (Ref. 8).
It can be shown that for the stirred tank

$$\psi = 0.17\,Ne \cdot n \cdot \frac{D}{T^{2/3}\,H^{1/3}} \cdot \left(1 + \frac{2\,\rho_s}{\rho_L}\right)\frac{d_s^2}{\nu} \qquad\qquad /12/$$

For the $BaSO_4$ precipitation process in the Rushton type CSTC, for: $t = 25^{\circ}C$, $n \leqslant 10\ s^{-1}$ and $d_s \leqslant 10\ \mu m$, the parameter $\psi \leqslant 0.004$ and the $BaSO_4$ particles cannot penetrate into the completely segregated region.

3.2. The molecular mixing zone.

For any small fluid element called usually "point" of age "t" travelling through the molecular mixing zone, the micromixing equation are:

$$- \frac{dc_A(t)}{dt} = k_{md}\left[c_A(t) - \bar{c}_A\right] + R_A \qquad\qquad /13/$$

$$- \frac{dc_B(t)}{dt} = k_{md}\left[c_B(t) - \bar{c}_B\right] + R_B \qquad\qquad /14/$$

$$-\frac{dc_C(t)}{dt} = k_{md}\left[c_C(t) - \bar{c}_C\right] + R_C + K_g A_g \left[c_C(t) - c_C^*\right]^g \qquad /15/$$

where

$$\bar{c}_i = \int_0^\infty \frac{1}{\bar{t}_{Rm}} \cdot e^{-t/\bar{t}_{Rm}} \cdot c_i(t)\, dt \qquad /16/$$

The initial conditions for points containing initially only reactant A are: $c_A(0) = c_{A\alpha}$, $c_B(0) = 0$, $c_C(0) = 0$

and for points containing only B: $c_A(0) = 0$, $c_B(0) = c_{B\alpha}$, $c_C(0) = 0$
The reaction is instantaneous and irreversible with stoichometrical ratio equal to one; thus Eq.13, 14 and 16 can be solved independently of Eq.15, giving:

$$c_A = c_B = c_{A\alpha}\Big/\left[2\left(1 + k_{md}\bar{t}_{Rm}\right)\right] \qquad /17/$$

and

$$R_A = R_B = R_C = k_{md}\, c_{A\alpha}\Big/\left[2\left(1 + k_{md}\bar{t}_{Rm}\right)\right] \qquad /18/$$

After simple transformation of Eq.15 we have:

$$\frac{\bar{c}_C}{\bar{t}_{Rm}} = \frac{c_{A\alpha}\, k_{md}}{2\cdot\left(1 + k_{md}\,\bar{t}_{Rm}\right)} - K_g \cdot A_g \cdot \left(c_C - c_C^*\right)^g \qquad /19/$$

To find the $A_g\left(c_C - c_C^*\right)^g$ term let us consider equations describing growth and nucleation kinetics of $BaSO_4$.
To this end we have used the equations developed by Gunn and Murthy, (Ref.9) somewhat modified in our previous paper (Ref.1) . For nucleation and growth rates, we have respectively:

$$\frac{dn_s}{dt} = K_s \left(c_C - c_C^*\right)^4 \cdot n_s \qquad /20/$$

and

$$G = \frac{1}{2}\frac{dd_s}{dt} = 3\cdot10^3\left(c_C - c_C^*\right)^3 \qquad /21/$$

The dominating mechanism of nucleation in the CSTC is the secondary nucleation. Introducing Eq.21 into Eq. 19 and making the balance of particles in the mixing volume we have:

$$\frac{\bar{c}_C}{\bar{t}_{Rm}} = \frac{c_{A\alpha}\, k_{md}}{2\left(1 + k_{md}\,\bar{t}_{Rm}\right)} - K_g \cdot A_g \cdot \left(c_C - c_C^*\right)^3 \qquad /22/$$

and

$$\frac{\bar{n}_s}{\bar{t}_{Rm}} = K_s \left(c_C - c_C^*\right)^4 n_s \qquad /23/$$

As the average nucleation rate is constant in the crystallizer, an exponential distribution of crystal size can be assumed.
Thus, the average crystal size is:

$$\bar{d}_{32} = 6\,\bar{G}\,\bar{t}_{Rm} = 1.8\cdot10^4\cdot\left(c_C - c_C^*\right)^3\cdot\bar{t}_{Rm} \qquad /24/$$

and, for the shape factor $\alpha = 2.6$

$$\bar{a} = 0.454\,\bar{d}_{32} \qquad /25/$$

To solve the set of equations 22 ÷ 29 additional assumptions are necessary. These are as follows:
The destributions of concentration fluctuations c_C' surface area fluctuations A_g' and number of crystals per unit volume fluctuactions n_s' are normal and correlated with each other.
Introducing $n_s = n_s' + \bar{n}_s$, $c_C = c_C' + \bar{c}_C$ and $A_g = A_g' + \bar{A}_g$ into Eq.22 and 23, and substituting:

$$\overline{A_g' c_C'} = \overline{\left(c_C'\right)^2}\cdot\frac{\bar{A}_g}{\bar{c}_C} \;;\qquad \overline{A_g'\left(c_C'\right)^3} = \overline{\left(c_C'\right)^4}\cdot\frac{\bar{A}_g}{\bar{c}_C} \;; \qquad /26//27/$$

$$\overline{n_s' c_C'} = \overline{\left(c_C'\right)^2}\cdot\frac{\bar{n}_s}{\bar{c}_C} \;;\qquad \overline{n_s'\left(c_C'\right)^3} = \overline{\left(c_C'\right)^4}\cdot\frac{\bar{n}_s}{\bar{c}_C} \qquad /28/\ /29/$$

as well as neglecting c_C^* which is much smaller than $\bar{c}_C$ and expressing the higher order correlations by $\bar{c}_C$ and $\overline{\left(c_C'\right)^2}$ (according to the properties of the normal distribution) we have obtained:

$$\frac{\bar{c}_C}{\bar{t}_{Rm}} = \frac{c_{A\alpha} \cdot k_{md}}{2\left(1 + k_{md}\bar{t}_{Rm}\right)} - K_g \cdot \bar{A}_g \cdot \left\{ \bar{c}_C^{\,3} + 6\,\overline{(c_C')^2} \cdot \bar{c}_C + \frac{3\left[\overline{(c_C')^2}\right]^2}{\bar{c}_C} \right\} \qquad /30/$$

$$\frac{\bar{n}_s}{\bar{t}_{Rm}} = K_s \left\{ \bar{c}_C^{\,4} + 10\,\bar{c}_C^{\,2} \cdot \overline{(c_C')^2} + 15\left[\overline{(c_C')^2}\right]^2 \right\}\bar{n}_s \qquad /31/$$

and $\quad \bar{d}_{32} = \dfrac{1.8}{(c_C')^2} \cdot 10^4 \cdot \left[3\,(c_C')^2 \cdot \bar{c}_C + \bar{c}_C^{\,3} \right] \cdot \bar{t}_{Rm}$ $\qquad /32/$

To find the $(c_C')^2$ value, once again Eq.15 must be used.
Multiplying both sides of Eq.15 by c_C, averaging, and using again the
previous procedure for correlations, one gets:

$$\frac{\bar{c}_C^{\,2} - \overline{(c_C')^2}}{\bar{t}_{Rm}} = \frac{\overline{(c_C')^2}}{t_{md}} + 8\,K_g\bar{A}_g \left\{ 3\left[\overline{(c_C')^2}\right]^2 + \bar{c}_C^{\,2} \cdot \overline{(c_C')^2} \right\} \qquad /33/$$

From Eqs 31 ÷ 33 the values of $\bar{c}_C$, $\overline{(c_C')^2}$, $\bar{d}_{32}$ and $\bar{A}_g$ can be calculated for any set of conditions in CSTC. The number of crystals can be
found from the ratio of $\bar{A}_g$ and $\bar{a}$.
The $\bar{t}_{Rm}$ in Eqs 31 ÷ 33 is the real mean residence time in the
molecular mixing zone and, according to the considerations in section
3.1 (Eq.11) in our system $\bar{t}_{Rm} = \bar{t}_R - t_{ms}$.

4. EXPERIMENTS

Precipitation experiments were carried out in a Rushton type stirred
tank crystalliser of diameter T = 0.096 m, agitator (flat blade
turbine) of diameter D = 0.032 m, width of the agitator blades
w = 0.0096 m (there were 4 baffles in tank). The crystallizer was
operated in a continuous manner.
The aqueous solutions of $BaCl_2$ and Na_2SO_4 were fed into the vessel
through 2 pipes of diameter 0.008 m, placed at the bottom of the
tank, on its opposite sides. The measurements were carried out in the
range of impeller speed 1 rps $\leq$ n $\leq$ 6 rps, initial concentrations $c_{A\alpha}$ =
= $c_{B\alpha}$ equal to: 0.0095, 0.019, 0.0285 and 0.038 mol/dm_o^3, and residence time 28s $\leq \bar{t}_R \leq$ 124s, in constant temperature t = 25°C.
The samples of the precipitate were taken from the overflow stream,
and the density of the crystal distribution was measured by means of
a sedimentation balance.

5. RESULTS AND DISCUSSION

The experimental results were interpreted using the set of equations
developed in Section 3.
At first, the value of constant K_s was evaluated, using the whole set
of experimental results. It has been found that K_s is indeed a constant, independent of the experimental conditions , and equals K_s =
= 1.11 · 10^{12} /smol^4.
Some typical results are presented in Table 1.
In further calculations the above set of equations was solved using
the average value of K_s = 1.11 · 10^{12}, to find the effects of mixing
and of the mean residence time on the mean size of particles. The comparison of the experimental results with those calculated theoretically
is presented in Figs 3 and 4.
The increase of the intensity of mixing impeller speed increases the
size of the solid particles; this effect was observed in experiments
and predicted theoretically (Fig.3) . It is interesting, to note that
for a batch crystallizer, the opposite effect was observed and calculated (Ref.1) . In the case of CSTC it was also observed and predicted
that the increase of the mean residence time increases the mean size of
crystals (Fig.4) . In Fig.2a,b the measured size distributions are compared with those calculated theoretically using the mean, empirical
K_s value.
It should be pointed out, that the particle size is almost independent
of the initial concentration of the reactants.
This is only true in unpremixed-feed - systems; in premixed-feed

- systems larger influence of initial concentration and much higher
influence of mixing intensity are to be expected, (in unpremixed -
feed systems similar effects can be observed when the reactant inlets
are situated too close to each other).

6. CONCLUSIONS

A new model of micromixing in chemical reactors, assuming the existence of two different zones: the zone of complete segregation and the
zone of molecular dissipation of the concentration fluctuations, has
been used to describe the process of precipitation of the solid product in a continuous stirred tank crystallizer. The influence of the
intensity of mixing and mean residence time on the rate of precipitation and the crystal size of the product have been investigated. The
experimental results for $BaSO_4$ precipitation, although rather scattered, are in agreement with the theoretical predictions obtained from
the model.

REFERENCES
1 . Pohorecki, R. and Bałdyga, J.: "The use of a new model of micro-
mixing for determination of crystal size in precipitation".
Chemical Engineering Science, 38, 1, 1983, pp. 79 ÷ 83.

2 . Pohorecki, R. and Bałdyga, J.: "The influence of intensity of
mixing on the rate of precipitation". In: Proc. 7th Symposium on
Industrial Crystallization, Warsaw, Poland: 25 ÷ 27 Sep. 1978 ,
Industrial Crystallization 78, E.J. de Jong, S.J.Jancić (editors)
North-Holland Publishing Company, 1979, pp. 249 ÷ 258.

3 . Pohorecki, R. and Bałdyga, J.: "New model of micromixing in chemical reactors. 1. General development and application to a tubular reactor". Ind.Eng.Chem.Fundam., 22, 4, 1983, pp. 392 ÷ 397.

4 . Pohorecki, R. and Bałdyga, J.: "New model of micromixing in chemical reactors. 2. Application to a stirred tank reactor". Ind.
Eng.Chem.Fundam., 22, 4, 1983, pp. 398 ÷ 405.

5 . Rosensweig, R.E.: "Idealized theory for turbulent mixing in
vessels". A.I.Ch.E. Journal, 10, 1, 1964, pp. 91 ÷ 97.

6 . Corrsin, S.: "The isotropic turbulent mixer: part II. Arbitrary
Schmidt Number". A.E.Ch.E. Journal 10, 6, 1964, pp. 870 ÷ 877.

7 . Costa, P. and Trevissoi, C.: "Reactions with non-linear kinetics
in partially segregated fluids". Chemical Engineering Science,
22, 1972, pp. 2041 ÷ 2054.

8 . Gudmundsson, J.S. and Bott, T.R.: "Particle diffusivity in
turbulent pipe flow". J.Aerosol Sci. 8, 1977, pp. 317 ÷ 319.

9 . Gunn, D.J. and Murthy, M.S.: "Kinetics and mechanisms of precipi-
tations". Chemical Engineering Science, 27, 6, 1972, pp.1293 ÷
÷ 1313.

Table 1. Typical results.

Run No.	$c_{A\alpha}$ $\frac{mol}{dm^3}$	$\bar{t}_R$ s	n s^{-1}	k_{md} s^{-1}	t_{ms} s	$\bar{d}_{32} \cdot 10^5$ dm	$\bar{c}_C \cdot 10^4$ $\frac{mol}{dm^3}^2$	$\overline{(c')^2} \cdot 10^{10}$ $\frac{mol}{dm^3}$	$\bar{A}_g$ $\frac{dm^2}{dm^3}$	$\bar{n}_s$ $\frac{1}{dm^3}$	$K_s \cdot 10^{-12}$ $\frac{dm^{12}}{s\ mol^4}$
1	0.0095	124	6.21	8.40	3.3	5.33	2.904	0.394	25.98	2.015^{10}	1.159
4	0.0095	28	6.2	8.40	3.3	3.93	4.434	3.956	33.73	4.812	1.019
6	0.0095	124	1.2	0.75	16.4	4.73	2.891	2.942	28.67	2.820	1.285
14	0.0095	28	2.1	1.64	9.7	3.39	4.555	1.552	37.30	7.152	1.109
18	0.0095	57	4.1	4.48	5.0	5.07	3.774	2.545	26.58	2.278	0.929
20	0.019	57	2.1	1.64	9.7	4.83	3.829	4.267	57.48	5.429	0.954
24	0.0285	57	2.1	1.65	9.6	4.19	3.651	2.918	101.16	12.70	1.159
28	0.038	57	6.1	7.10	3.3	4.66	3.634	1.032	123.90	12.57	1.057

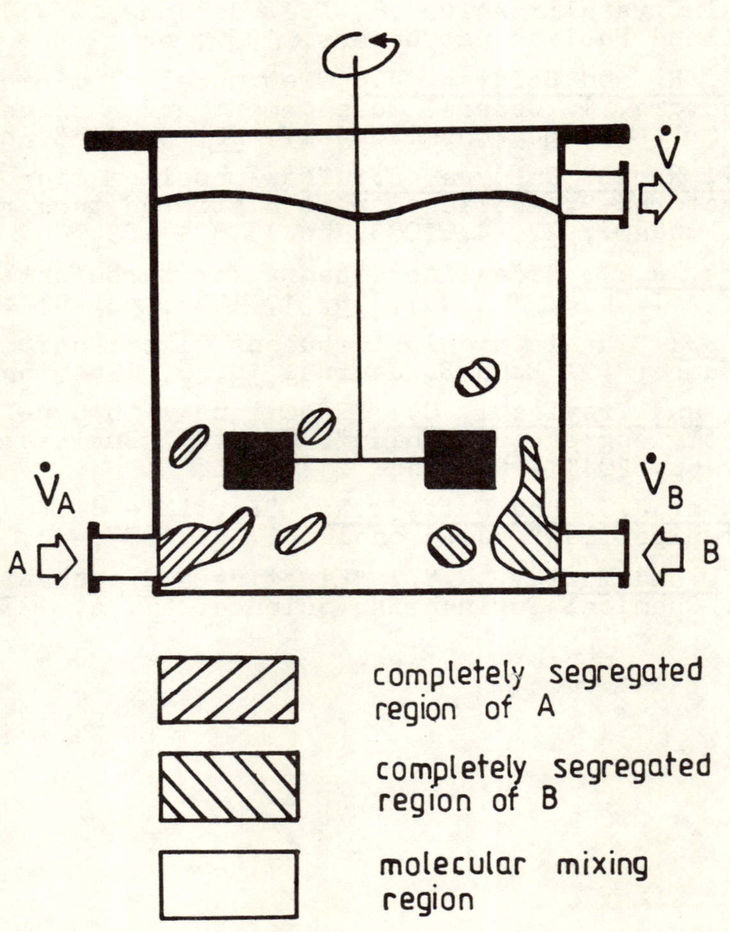

Fig.1. Scheme of the precipitation process in a
Continuous Stirred Tank Crystallizer.

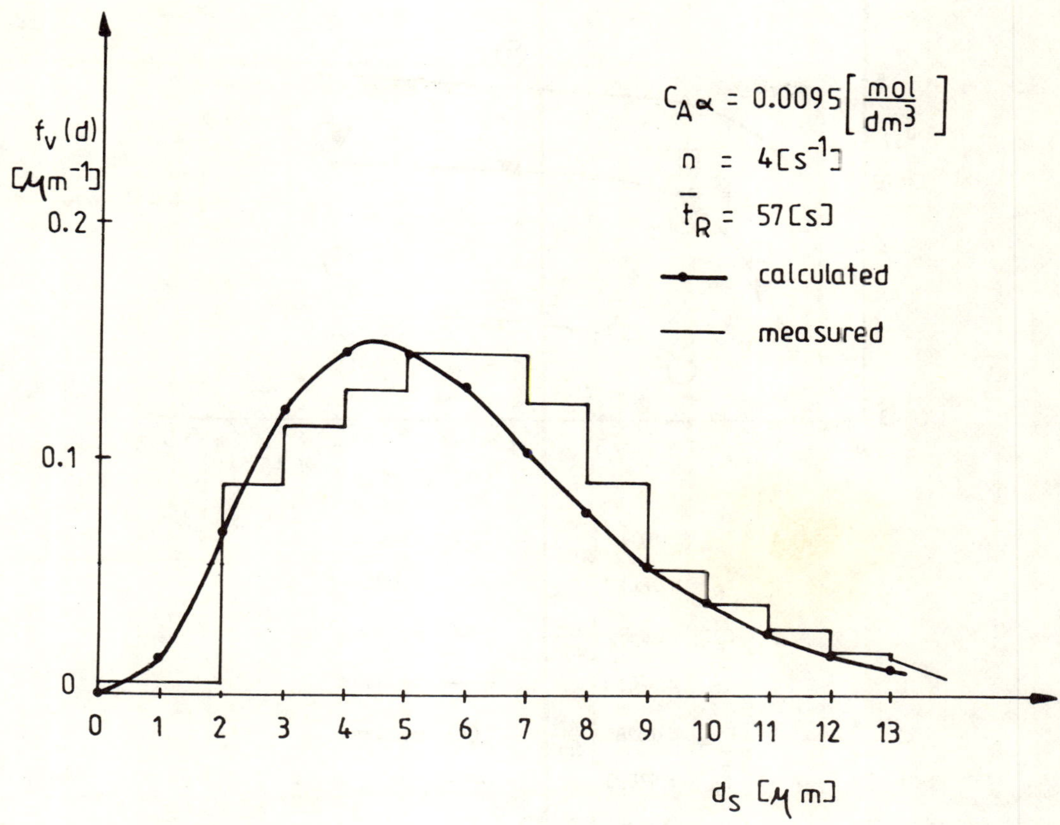

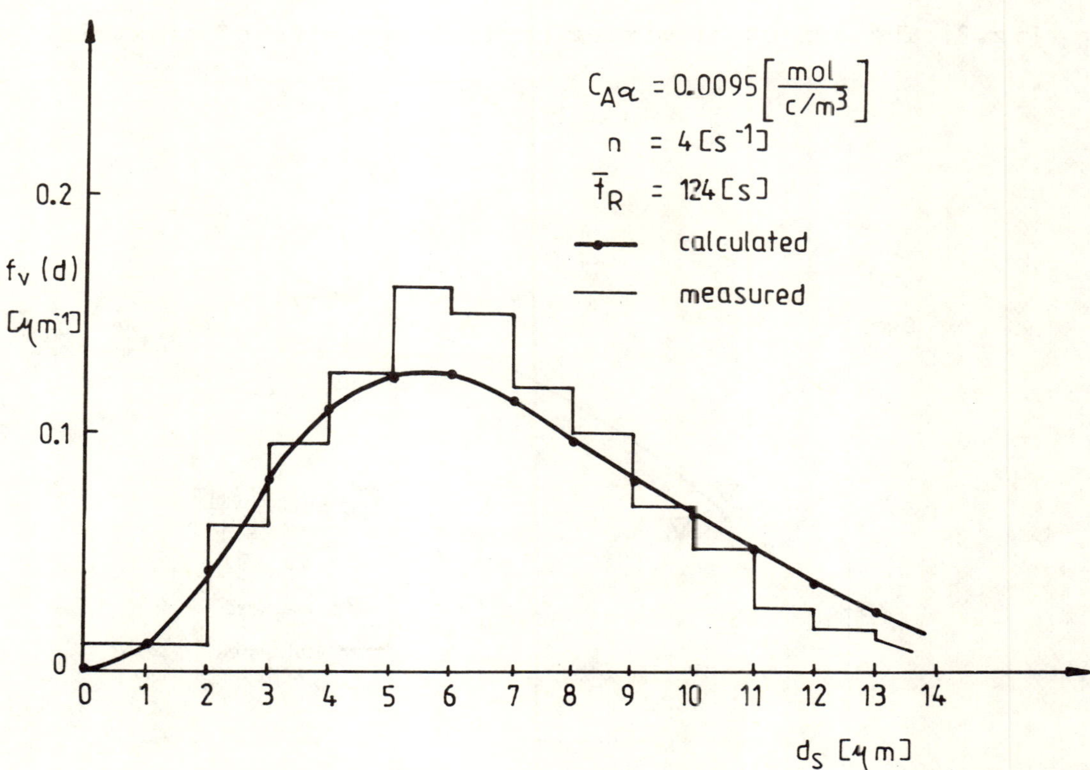

Fig.2a,b. Size distributions of crystals.

113

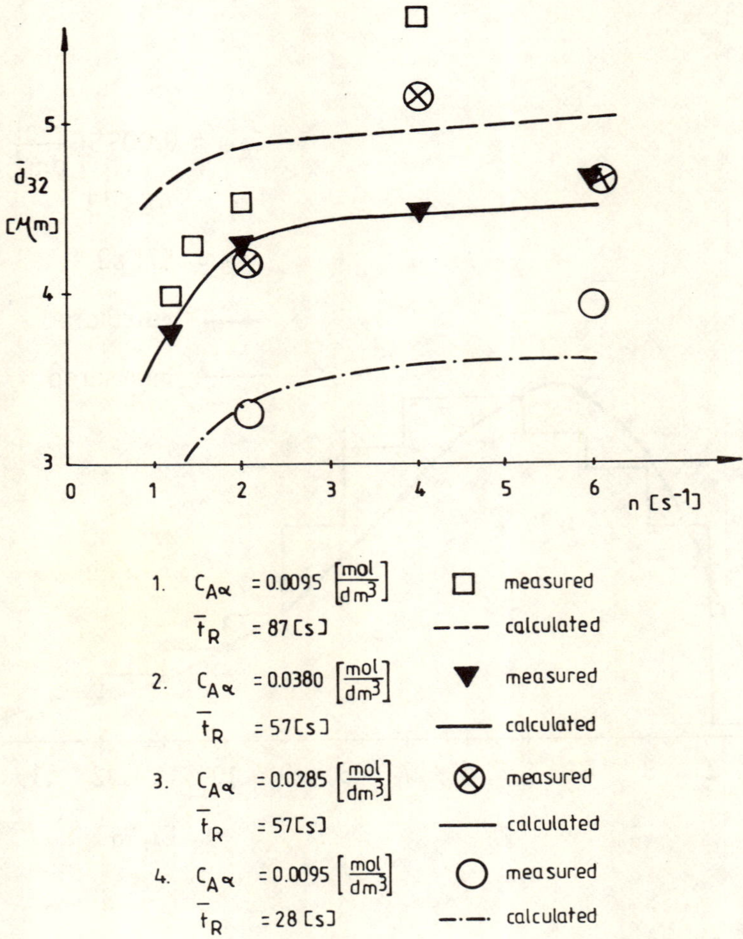

Fig.3. The effect of mixing on the mean size of crystals.

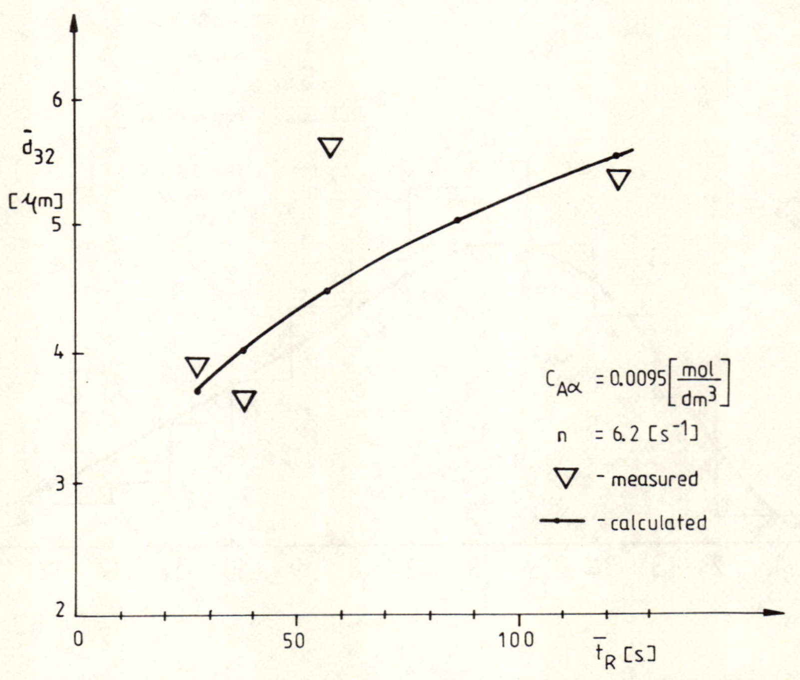

Fig.4. The effect of mean residence time on the mean
size of crystals.

THE DISPERSION OF GASES IN LIQUIDS WITH TURBINES

John M. Smith and M.M.C.G. Warmoeskerken.

Laboratory for Physical Technology,
Delft University of Technology,
Prins Bernhardlaan 6,
2628 BW DELFT, The Netherlands

Summary

This paper considers some recent work on the two phase flow hydrodynamics of disc turbine impellers operating in low viscosity systems. Similar flow regimes are developed with impellers of scales from 0.17m to 0.48m diameter in vessels from 0.30 to 1.20m diameter. Scale independent correlations, which include the effects of the ratio of the diameters of the tank and impeller, have been developed to allow prediction of operating conditions.

Held at Wurzburg, 10-12 June, 1985.

Organised by DVCV· Deutsche Vereinigung für Chemie- und Verfahrenstechnik
(German Association of Chemical and Process Engineering).

Organisation: GVC·VDI-Gesellschaft Verfahrenstechnik und Chemieingenieurwesen. 

©BHRA, The Fluid Engineering Centre, Cranfield, Bedford MK43 0AJ, England.

Nomenclature.

a	Specific contact area	(m^2/m^3)
D	Impeller diameter	(m)
Fl	Gas Flow Number (Q_g/ND^3)	$(-)$
Fr	Impeller Froude Number (N^2D/g)	$(-)$
g	Gravitational acceleration	(m/s^2)
k_l	Mass transfer coefficient	(m^2/s)
N	Impeller rotational speed	$(1/s)$
N_r	Impeller speed for 'recirculation' (eq.4)	$(1/s)$
Q_g	Gas supply rate	(m^3/s)
Re	Impeller Reynolds Number $(ND^2 rho/mu)$	$(-)$
T	Tank diameter	(m)
epsilon	Gas volume fraction in tank	$(-)$
mu	Liquid dynamic viscosity	$(Pa.s)$
rho	Liquid density	(kg/m^3)

Introduction.

        Knowledge and understanding of the operation of turbine impellers has
developed swiftly following the pioneering work published by Henry Rushton and co-
workers, (Rushton et al., 1959). The crucial observation was that the wide variety of
impellers used in practice brings few benefits and that there are great advantages in
the standardisation of equipment. It so happens that disc turbines were selected as
the basic design, and this pattern, now often referred to as a Rushton turbine, has
become a generally accepted reference standard for many mixing operations. The
dispersion of gases in low viscosity liquids, especially in aqueous solutions, has
been an area where this increase in knowledge has been particularly notable. Until a
few years ago the conventional view was that gas dispersion results from bubbles
breaking up in the intense turbulence of the liquid discharged from the impeller.
Modification of the flow around the impeller blades as a result of the presence of
the gas was scarcely considered relevant.

        Following the work of Biesecker (1972), Nienow, (1974), van't Riet (1975)
and Judat (1976), it is now generally recognised that the two phase hydrodynamics of
the impeller play a central role. The flow regime around the impeller has a direct
influence on the power demand and closely related aspects of performance such as the
mass transfer between dispersed gases and the liquid, Warmoekerken and Smith (1981)
and (1982).

        One of the most significant aspects of the two phase flow around the
agitator is the development of ventilated cavities behind the impeller blades. These
cavities have a streamlining action and thereby lower the drag coefficient and thus
the power demand of the impeller.

        Detailed study of the flow field near the impeller has led to the defini-
tion of alternative stable cavity configurations each with characteristic influences
on the nature of the dispersion process. In recent times considerable effort has been
devoted to mapping the operating conditions that lead to the development of the
various stable flow regimes.

The Stirred Tank.

        The six blade disc turbine we are considering is illustrated in Fig. 1. The
construction of flat blades slotted into a disc is simple to fabricate, strong and
easily cleaned. Although many detailed differences are found in practical industrial
designs, the flat bottomed tank with a liquid depth equal to the tank diameter is
arbitrarily adopted as a reference configuration. Usually four vertical symmetrical
wall baffles are fitted, each with a width 0.1 times the tank diameter. The ratio
between the tank and impeller diameters is commonly in the range between two and
four. The phenomena which will be quantitatively described in this review of recent
work have been observed while dispersing air in tap water in tanks of four different
diameters, 0.30, 0.44, 0.65 and 1.20 m, generally using a T/D ratio of 2.5.

Ventilated Cavities.

It is generally recognised that in an ungassed system roll vortices develop behind
the impeller blades. These trailing vortices were studied by van't Riet (1975). More
recently, detailed laser-doppler measurements of the velocity field within the vortices have become available from Popiolek et al. (1984). The rotation of each of
these vortices leads to a relative underpressure along the axis. This draws gas in
the vicinity inwards towards the vortex core and leads to the formation of ventilated
cavities. This phenomenon should not be confused with true cavitation in which vapour
develops as the local pressure falls below the saturation pressure of the liquid at
the ambient temperature, though obviously such cavitation can easily occur if the
liquid phase is close to its boiling point.

Three distinct types of cavity are formed in low viscosity liquids. In given equipment the cavity form is determined by the stirrer speed, (N) and the gas input rate, (Q_g). At low gas rates the dominant action is that of the vortices drawing the gas inwards which has been described above. The ventilated vortex cavities first attract gas in the vicinity of the impeller and subsequently loose the gas from the elongating downstream tip of the cavity, (Fig. 2). The rotational structure is still clearly present. If the gas rate is increased the cavity grows in diameter until it interferes with the rotation of the liquid in the region between the cavity and the blade. As a result of the reduced rotation there is a distinct change in the form of the gas cavity. It begins to cling to the rear of the blade, though the flow is still primarily outwards with gas loss from the cavity along the turbulent outer surface, (Fig. 3.). A further increase in the gas rate leads to the development of large cavities, (Fig. 4), which are characterised by clear smooth surfaces and from which gas is lost as a result of much more gentle breakaway from the rear surfaces. When large cavities have been formed scarcely any remnant of the original vortex motion remains.

The ´3-3´ Structure.

If the two phase flow around a disc turbine impeller is considered as a whole, a remarkable three-symmetric structure is observed, (Warmoeskerken et al., 1984a). Although both vortex and clinging cavities form the expected six-symmetric structure, (Figs. 5a and 5b), when large cavities develop these are formed perferentially behind alternate blades so that three clinging cavities and three large cavities coexist, (Fig. 5c). At higher gas flow rates six large cavities develop, but these are of two different sizes and distribute themselves in an alternating large-small 3-3 configuration, (Fig. 5d).

These 3-3 structures are remarkably stable. The cavities remain associated with the given blades and do not change their relative configurations, though which of the two alternative 3-3 patterns develops initially is a matter of chance. Cavities formed with various agitators are drawn in Fig. 6. The 2-2 formation of the four blade impeller (Fig. 6b) is perhaps to be expected in the light of the finding with the six blade disc turbine.

The differing cavity forms on alternate blades lead to unstable hunting when impellers with an odd number of blades are used. The location of the largest cavities on the five and three blade impellers shown in Figs. 6a and 6c changes continuously and will lead to an unpredictable periodic non axial loading on the impeller shaft. Although the frequency is much lower than that of the shaft rotation it is strongly dependent on the gas rate and almost independent of the impeller speed. A non-axial loading will also result with the strange 1-1 cavity distribution formed with the two blade impeller of Fig. 6d, but in this case the structure is stable.

At extreme gas rates the impeller becomes overloaded. The stable 3-3 structure is then replaced by one with six large but irregular and rather unstable vibrating ragged edged cavities, (Nienow et al., 1985).

Hub mounted turbines, including pitched blade designs, do not develop this 3-3 and related configurations; all the cavities are alike, (Warmoeskerken et al. 1984b). It can be concluded that the interactions between sucessive blades are primarily a result of the distribution of the gas by the disc, though the mechanism of this has not yet been elucidated.

Power Demand of Aerated Turbines.

Typical power demand curves for an aerated six blade disc impeller are shown in Fig. 7. It is usual to set out the relative power demand, P_g/P_u as a function of the dimensionless gas flow number Q/ND^3, even though it is clear from this figure that this does not accommodate the independent variables Q and N adequately. The curves have two distinct regions. There is a slight reduction in power at first while at larger gas rates much less energy is needed to maintain the impeller speed. It appears that at in this latter region the drag coefficient of the impeller approaches a relatively constant value which extends until the gas handling capacity of the impeller is exceeded and the impeller is said to become flooded; a point where a sudden increase in the shaft power has to be supplied if speed is to be maintained.

The Cavity Formation Line.

As well as this transition to the stable 3-3 cavity structures directly affecting the power demand of the impeller, there is an associated influence on such other process variables as the average retained gas fraction and gas-liquid mass transfer. (Warmoeskerken and Smith, 1979 and Warmoeskerken and Smith, 1984). It is therefore useful to have some means of predicting when this transition will occur.

It has been found that the transition is rapid. It can be precisely determined with a very simple technique. A flexible vane is mounted in the impeller outflow so that the liquid causes it to vibrate. The liquid discharge between two blades is affected directly by the presence and nature of any ventilated cavity in the gap between the two succeeding blades. Any changes in the character of the cavities will be reflected in an alteration in the vibration of the vane. This is easily detected by analysing the output from strain gauges mounted on the vane with a frequency analyser. In an ungassed flow, or with six similar cavities, the frequency spectrum shows a clear peak at the blade passage frequency (6N) as shown in Fig. 8a. As soon as the 3-3 structure is formed this output is modified to a bimodal distribution with peaks at 6N and 3N, Fig. 8b. It is possible to generalise the results of various scales and T/D ratios in a single line, Fig. 9. To the left of and above the line six clinging cavities are present while to the right of and below the line large cavities have been formed and assume the 3-3 configuration. The transition from clinging to the 3-3 large cavity structure is found to occur at a gas flow number given by

$$Fl = 3.8*10^{-3}(Re^2/Fr)^{0.067}(T/D)^{0.5} \qquad (1)$$

In interpreting Fig. 9, Q_g/D^2 can be regarded as a superficial gas velocity based on the impeller cross sectional area while the ND term represents the tip velocity. In this way the slope of the line, which is the constant of Eq. (1), corresponds to the Gas Flow Number Q_g/ND^3. The Re^2/Fr term compensates for scale, (i.e. D), in terms of the physical properties of the system, but it must be admitted that this has not yet been checked for systems other than air-water. Although equation (1) is written in terms of Qg the effects of gas recirculation are implicitly included, partly embodied in the scale factor just mentioned and partly as a result of the much less efficient capture of gas by large cavities which do not have the strong rotational vortex motion.

Flooding.

Although in recent years much research effort has been devoted to conditions of moderate gas loading, (typically with Flow Numbers up to about 0.01), industrial applications usually involve much larger gas rates.

If an impeller is driven at a constant speed while the gas rate is steadily increased, various successive flow regimes can be distinguished. Small volumes of gas will be dispersed effectively by the impeller and the liquid circulation carries bubbles to all regions of the tank, Fig. 10a. Somewhat larger gas supply rates apparently lead to a reduction in the liquid pumping action as the cavities develop, with the result that the liquid velocities are no longer sufficient to force much of the gas to circulate below the impeller level, Fig. 10b. At very large gas rates the impeller ceases to be able to pump the two phase mixture at all, the buoyant pumping action of the bubble column becomes dominant, Fig. 10c. The impeller is said to be flooded. There is much confusion as to the precise criteria which should be applied in order to define flooding. For our present purposes it is convenient to regard the cessation of radial pumping as defining this point.

A semi-theoretical analysis of the circulation forces (Warmoeskerken and Smith, 1984), leads to the conclusion that the transition to flooding must be governed by a linear relationship of the form

$$Fl = Const.\ Fr \qquad\qquad (2)$$

The constant in this equation has to be determined experimentally. It will depend on the details of the geometry, such as whether the tank has a rounded or flat bottom, the sparger diameter, the clearances between the sparger and impeller and the impeller and the tank bottom, and the T/D ratio. The radial pumping action of the impeller can be detected easily if a micro-propellor is mounted in the disc plane. Fig. 11 shows the results of experiments made while the impeller speed has been varied both with and without gas throughput. In the absence of gas there is a steady linear increase in the rotation rate of the micropropellor with impeller speed whereas in the gassed condition a critical minimum agitator speed must be exceeded before the radial flow can be detected. The transition is well defined and forms a convenient criterion for the flooding transition. The results of a very large number of determinations of this transition are presented in Fig. 12. The line corresponds to a value of 1.2 for the constant of Equation (2). Above and to the left of the line the impeller is said to be loaded and operating satisfactorily while below and to the right of the line the agitator is flooded. More recent work (Nienow et al., 1985) has shown that the T/D ratio can be satisfactorily included in an equation of this type and leads to the relationship

$$Fl = 30\ Fr\ (T/D)^{-3.5} \qquad\qquad (3)$$

High gas loadings are often desired in biotechnological applications, frequently in fermentors or other reactors which are much deeper than the H/T ratio of 1.0 which has hitherto been used for most investigations. Deep vessels usually have multiple impellers mounted on a single shaft, normally separated by distances of the orderof the tank radius. Apart from the added complication in the geometric variables, it is a matter of observation that in equipment in which the gas is introduced below the lowest impeller, it is that one which overloads first, (Kurpiers et al. 1985). It seems likely that similar Froude Number - Gas Flow Number relationships could be developed for these multi impeller plant designs, though further work is needed to produce the quantitative justification for this opinion.

It has been remarked above that very high gas rates lead to the breakdown of the stable 3-3 structure and the development of six ragged vibrating cavities. Because of this reversion to six similar cavities the vane technique can again be used to detect the transition. Fig. 13 shows how the results of such experiments in a 0.44m diameter tank confirm that this further change in cavity form occurs at the same time as the transition to flooding.

Regime Map.

The sucessive flow fields established in a mixing vessel with a six blade disc impeller dispersing gases in low viscosity liquid can be summarised as follows:

a1. A stable regime of six similar vortex cavities. The power demand in this region differs little from that of the ungassed system.

a2. A stable regime of six similar clinging cavities. The power demand in this region differs at most by 10% from that of the ungassed system.

b1. A stable regime in which three large and three clinging cavities are distributed behind alternate blades of the impeller. The power demand in this region may fall to about 40% of the original ungassed value.

b2. A stable regime in which six large cavities are present. However, three of these are larger than the others. These are again distributed in a 3-3 configuration around the impeller. The power demand remains low.

c. An unstable regime in which the impeller is oversupplied with gas and there are six vibrating ragged cavities behind the impeller blades. When this transition occurs there is an increase of about 25% in the absolute power demand of the impeller.

The transition from regime a2 to b1 is given by equation (1) and that from regime b2 to c by equation (3).

Nienow et al. (1977) identified another transition, which they termed a recirculation, on the basis of a peak in the gassed power demand at constant gas rate, and this can also be expected to affect reactor performance. They correlated the transition with the equation

$$N_r D^2 / Q_g^{0.2} T = 1.5 \ (m^{0.3} s^{-0.5}) \qquad\qquad (4)$$

which, like the relationships found for cavity structure changes, has a scale dependent term.

It is not possible to depict all the transitions in a single figure valid for all scales of equipment and all values of the D/T ratio. However for a given vessel a single diagram can be prepared. As an example Fig. 13 is presented as a flow regime chart applicable to a 1.2 m diameter flat bottomed tank with a 0.48m diameter impeller (T/D ratio 2.5). From this figure it can be seen how the hydrodynamic operating regime for any desired process conditions can be determined.

General observations.

The objective of this review has been to consider present knowledge of the two phase flow hydrodynamics of disc turbine impellers with flat blades. It should be possible to relate this to the performance of these impellers in practice. Almost invariably the primary aim of mixing vessels working in a two phase regime is mass transfer, and such factors as $k_l a$ and retained gas fraction are of great importance. Some progress has been made in systematising data on the basis of the impeller hydrodynamics. It can be expected that the various flow regimes will have quite different performance characteristics. This has been confirmed for gas fraction and mass transfer factor predictions where the scatter of the usual correlations can be considerably reduced by separating into lower from higher gas loading regimes. Gas holdup can be accurately predicted by different relations depending on whether the 3-3 regime has been established or not. In a vessel of 0.44m diameter with a 0.176m impeller the gas fraction is given within $\pm$ 15% by two distinct (but unfortunately dimensional) equations:

before large cavities are formed

$$epsilon = 0.62 \ Q_g^{0.45} . N^{1.6} \qquad\qquad (5)$$

and after large cavities have been formed

$$epsilon = 1.67 \ Q_g^{0.75} . N^{0.7} \qquad\qquad (6)$$

The mass transfer factor for physical absorption (solution of oxygen from air) can likewise be much better represented with alternative expressions:

before large cavities are formed

$$k_1 a/N = 1.1*10^{-7} Fl^{0.6} Re^{1.1} \qquad\qquad (7)$$

while after large cavities have been formed

$$k_1 a/N = 1.6*10^{-7} Fl^{0.42} Re^{1.02} \qquad\qquad (8)$$

Another reflection of the importance of cavity formation in plant operation can be found in a study of solids suspension in a gassed vessel with a dished bottom. (Warmoeskerken et al., 1984(b)). In this work it was reported that the critical impeller speed needed to achieve the ´just suspended´ condition is at first reduced by small quantities of gas, and only as the gas flow is increased does the impeller have to be speeded up again to maintain the suspension, reaching the original value at about the time large cavities are formed. It follows from this that the power required to maintain the suspension is actually reduced by about twenty percent as a result of the introduction of a small amount of gas into the system. Although cavity formation and structure behind pitched blades follows a somewhat similar course as with disc turbines, (Warmoeskerken et al. 1984(a)), simultaneous gas dispersion does not lead to easier suspension of solids with these more axial flow designs. (Frijlink et al., 1984).

There is increasing interest in the development of ´hydrodynamically profiled´ impellers, and this is evident from the number of manufacturers offering these designs for sale. After years of neglect this appears to justify a more fundamental approach to the study of the induced flows. However, the operation of mixing impellers in the multi-phase environment is far from straightforward, and it is only by building up our understanding of the complex processes involved that worthwhile improvements will be achieved.

References

Biesecker, B. O.
Begasen von Flüssigkeiten mit Rühren,
V.D.I.-Forschungsheft, 554, (1972).

Frijlink, J.J., Kolijn, M. and Smith,J.M.
The suspension of solids withaerated pitched blade turbines.
in Fluid Mixing II, Inst. of Chem. Engrs. Sympos. Ser. 89, (1984), 49 - 58.

Judat, H.
Zum Dispergieren von Gasen
Doctoral Thesis, Dortmund, (1976).

Kurpiers, P., Steiff, A., and Wienspach, P-M.
Zum Überflutungsverhalten ein- und zweistufiger Rührbehälter.
Chem. Ing. Tech. 57, (1985), 62-63. (MS 1307/85)

Nienow, A.W. and Wisdom, D.J.,
Flow over Disc Turbine Blades
Chem. Engg. Sci., 29, (1974), 1994 - 1997.

Nienow, A.W., Wisdom, D.J., and Middleton, J.C.
Effect of scale and geometry on flooding, recirculation and power in gassed stirred vessels.
Second European Conference on Mixing, (1977), Fl 1 - 16 (BHRA, UK.)

Nienow, A.W., Warmoeskerken, M.M.C.G., Smith, J.M. and Konno, M.
The Flooding-Loading Transition in Aerated Vessels
Submitted to Fifth European Conference on Mixing, Würzburg, (1985). (BHRA,UK.)

Popiolek, Z., Whitelaw, J.H., Yianneskis, M.
Unsteady Flow over Disc Turbine Blades,
Proc. Second International Conf. on Applications of Laser Doppler Anemometry to Fluid
Mechanics, Lisbon, Portugal 2-5 July 1984. paper 17.1, 1-6.

Riet, K. van´t,
Turbine Agitator Hydrodynamics and Dispersion Performance.
Doctoral Thesis, Delft (1975).

Rushton, J.H., Costich, E.W., and Everett, H.J.,
Power Characteristics of Mixing Impellers.
Chem. Engg. Progress., 46, (1959), 395 - 404 and 467 - 476 .

Warmoeskerken, M.M.C.G., Feijen, J. and Smith, J.M.
Hydrodynamics and Power Consumption in Stirred Gas-liquid Dispersions.
in Fluid Mixing, Inst. of Chem. Engrs. Sympos. Ser. 64, (1981), J1 - 14.   (a).

Warmoeskerken, M.M.C.G. and Smith, J.M.,
Hydrodynamics and Mass Transfer in Agitated Systems
Proc. CHISA Conference, Prague, (1981), Paper b 3.4   (b).

Warmoeskerken, M.M.C.G.and Smith, J.M.
Description of the Power Curves of Turbine Stirred Dispersions
Fourth European Conference on Mixing, (1982), 237-246. (BHRA,UK).

Warmoeskerken, M.M.C.G., Speur, J. and Smith, J.M.,
Gas-liquid Dispersion with Pitched Blade Turbines,
Chem. Engg. Commun., 25, (1984), 11 (a)

Warmoeskerken, M.M.C.G., Speur, J. and Smith, J.M.,
The Flooding Transition with Gassed Rushton Turbines,
in Fluid Mixing II, Inst. of Chem. Engrs. Sympos. Ser. 89, (1984), 59 - 67. (b).

Warmoeskerken, M.M.C.G., van Houwelingen, M.C., Frijlink, J.J. and Smith, J.M.
Role of cavity formation in stirred gas-liquid-solid reactors.
Chem.Eng.Res.Des.62, (1984), 197-200.   (c)

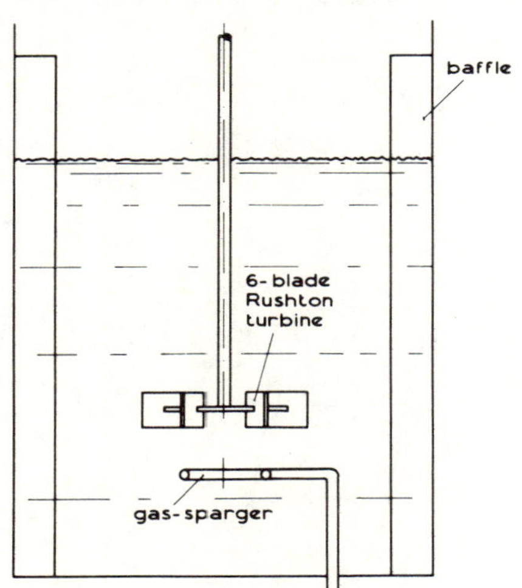

Fig. 1. Standard tank geometry.

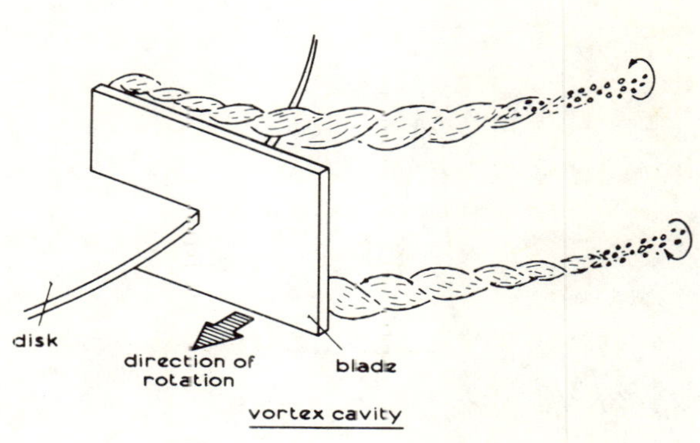

Fig. 2. A pair of vortex cavities.

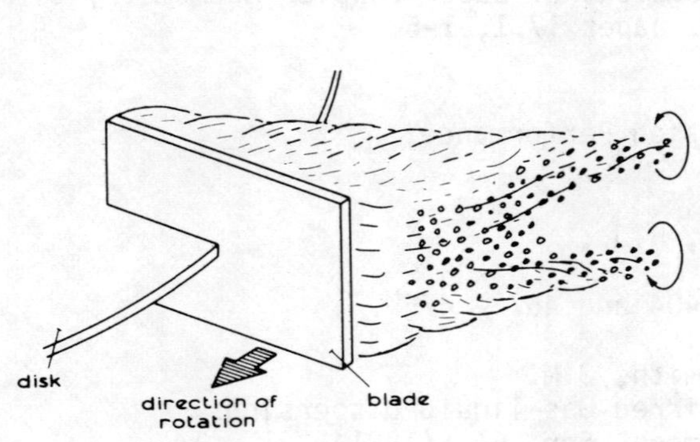

Fig.3. A clinging cavity.

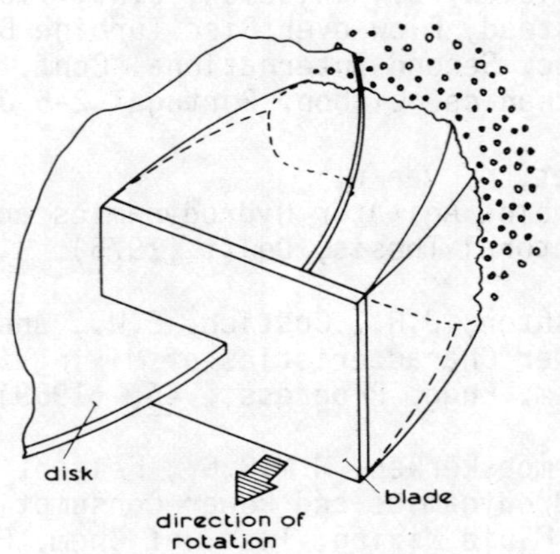

Fig. 4. A large cavity

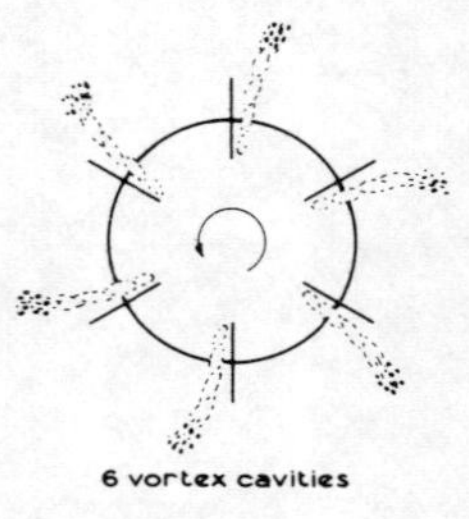

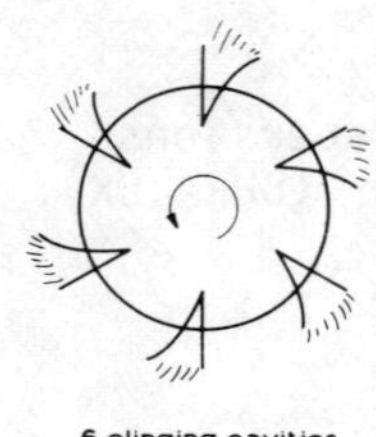

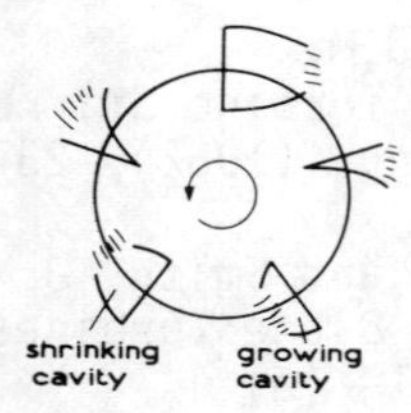

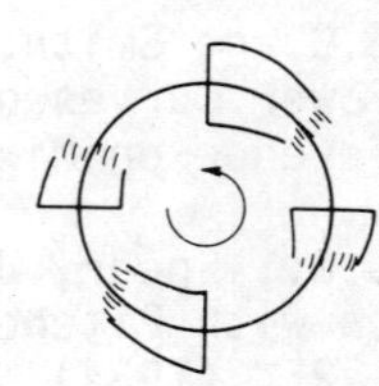

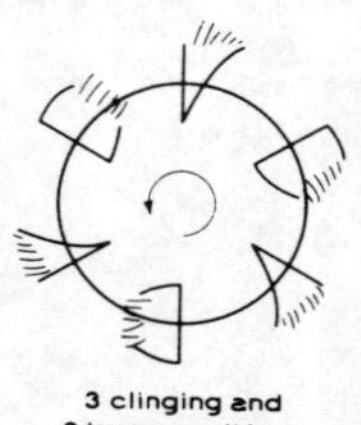

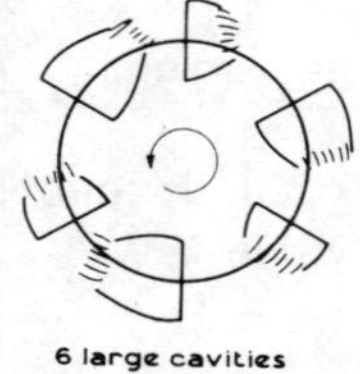

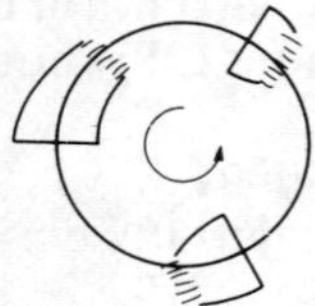

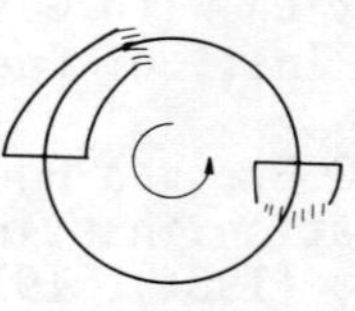

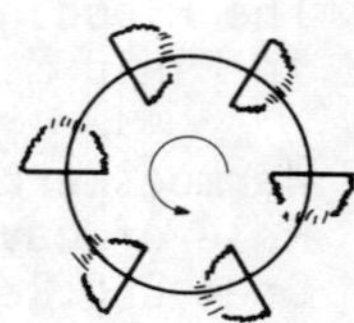

Fig.5. Successive stages in cavity distribution

Fig.6 Cavities with five, four, three and two blade impellers

Fig.7. Symmetrical ragged cavities.

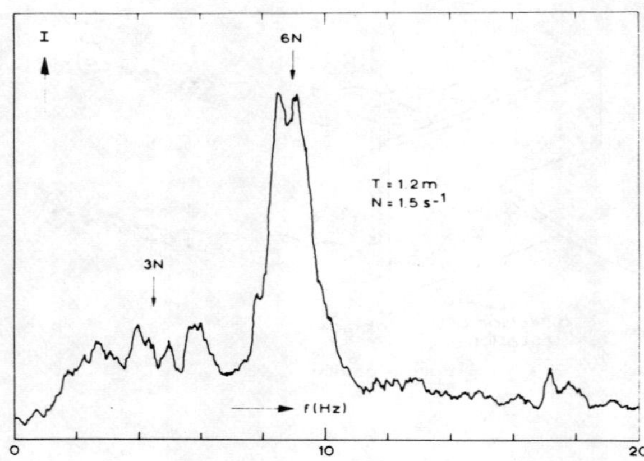

Fig.8a. Frequency spectrum of vane signal, vortex cavities.

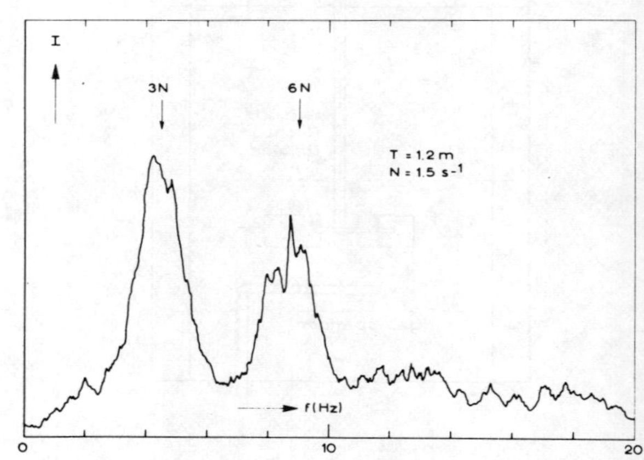

Fig. 8b. As 8a but with large cavities present

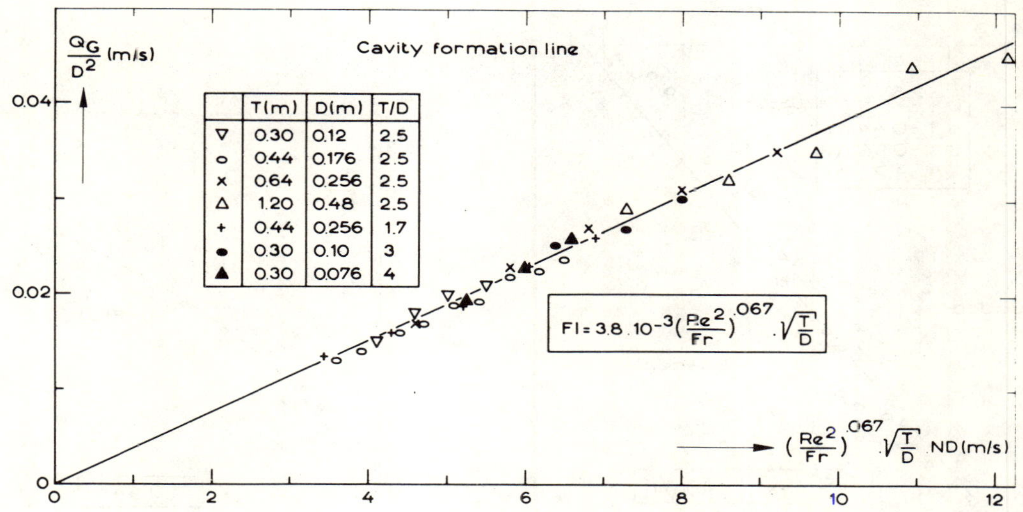

Fig.9. The cavity formation line

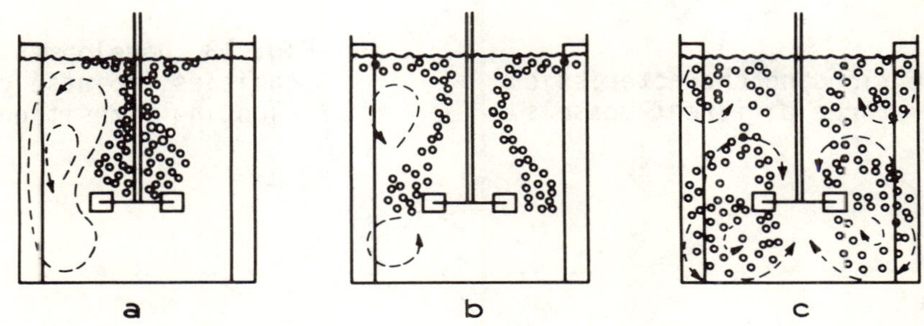

Fig. 10. Successive flow regimes
with increasing stirrer speed.

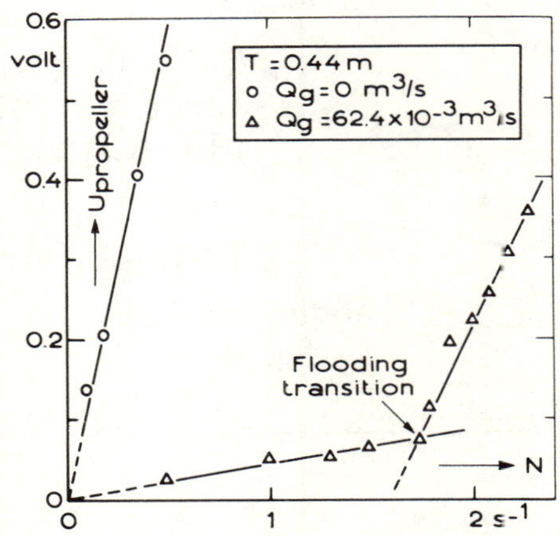

Fig. 11. Measurements with micro-
propeller to determine the flooding
transition.

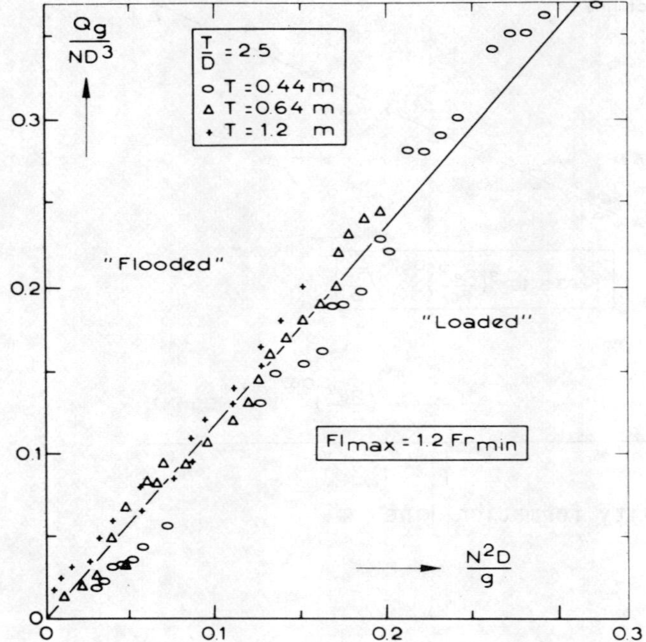

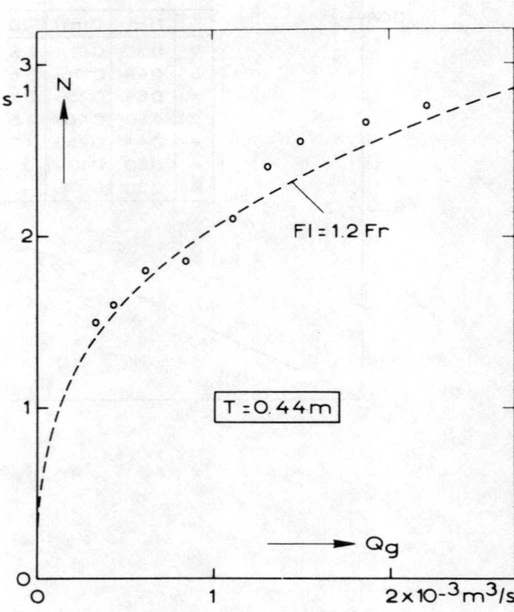

Fig. 12. The flooding characteristic
for three different vessels.

Fig. 13. Development of vibrating
cavities (points) compared to the
flooding transition line

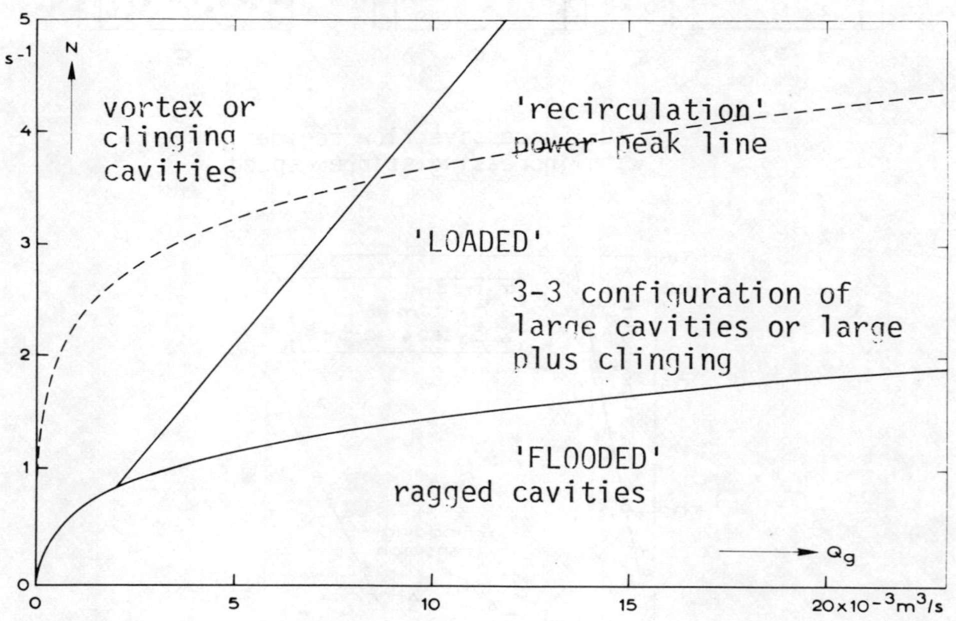

Fig. 14. General flow map for a 1.2m diameter tank with a 0.48m impeller.
Lines are Eq.(1) for the cavity formation, Eq. (3) for the
flooding transition and Eq. (4) for the ´recirculation´ power peak.

POWER CONSUMED BY RUSHTON TURBINES IN NON
STANDARD VESSELS UNDER GASSED CONDITIONS

M. Roustan

Institut National des Sciences Appliquées
Toulouse, France

Summary

Very little data is found concerning the power consumed for air dispersion in non standard vessels. So the measurement of the power dissipated has been studied in two non standard vessels, with two or three Rushton turbines. By comparison with classical results obtained in standard vessels, the configuration with 2 or 3 impellers gives results slightly different, concerning the shape of the curve $K = P_g/P_o$ versus Na at n = constant. The lower impeller is still flooded while the upper one disperses the gas phase correctly. The importance of the determination of the working conditions (flooding and loading) is shown. The presence of three impellers, instead of two, doesn't increase the gas hold-up but only the power consumed.

Held at Wurzburg, 10-12 June, 1985.

Organised by DVCV· Deutsche Vereinigung für Chemie- und Verfahrenstechnik
(German Association of Chemical and Process Engineering).

Organisation: GVC·VDI-Gesellschaft Verfahrenstechnik und Chemieingenieurwesen. 

©BHRA, The Fluid Engineering Centre, Cranfield, Bedford MK43 0AJ, England.

NOMENCLATURE

a	=	specific interfacial area	m^{-1}
b	=	baffle width	m
D	=	agitator diameter	m
Fr	=	Froude number $n^2 D/g$	
Fr'	=	modified Froude number $V_s/\sqrt{g\,T}$	
g	=	gravitational acceleration	m/s^2
G	=	gas flowrate	m^3/s
h	=	agitator distance form bottom of the vessel	m
H	=	liquid level height	m
k_L	=	film liquid mass transfert	m/s
K	=	ratio of gassed power/ungassed power	
n	=	agitator speed	s^{-1}
n_L	=	agitator speed corresponding to loading	s^{-1}
n_f	=	agitator speed corresponding to flooding	s^{-1}
Na	=	aeration number G/nD^3	
Ne	=	power number $P/\varrho\, n^3 D^5$	
Po	=	ungassed power consumption	W
Pg	=	gassed power consumption	W
R	=	ratio of gassed radial velocity/ungassed radial velocity	
T	=	tank diameter	m
V_G	=	radial velocity of the liquid	m/s
V_L	=	radial velocity of the liquid without air	m/s
V_s	=	gas flowrate per unit of cross section of the vessel $4G/\pi T^2$	m/s
φ	=	gas hold-up	%

1. INTRODUCTION

Gas-liquid reactors with mechanical agitation are often used in chemical and biochemical industries. Many papers have been published during the last twenty years concerning mass transfer, gas hold-up, power dissipated, hydrodynamics in these reactors. However, in spite of the research carried out, the power consumption in the presence of air is still a parameter difficult to predict. The difficulty is due to the complexity of the hydrodynamic phenomena taking place in a vessel in the presence of air.

The aim of our study is to give information on the power dissipated in non standard vessels H = 2 T with two or three impellers, in comparison with phenomena observed in standard vessels. The gas hold-up will serve also as an element of comparison.

2. MATERIALS AND EXPERIMENTAL METHODS

The experiments have been performed in two non standard vessels (H = 2 T) of diameter T 0.43 m and 0.78 m with hemispherical bottoms, and in a standard vessel (H = T) of 1 m in diameter with a flat bottom. The characteristic dimensions of vessels and impellers are given in Table 1 and Fig. 1. Two or three six blade Rushton turbines of variable diameter were mounted into the non standard vessel, with 4 baffles of width 0.1 T. The liquid was water and the gas (air) was dispersed into the vessel by a perforated ring with holes of 2 mm in diameter, below the lower turbine (for H = 2 T vessel). For H = T vessel, air was dispersed through a perforated plate, situated at the bottom of the vessel.

3. GENERALITIES ON THE POWER DISSIPATED IN AN AERATED MEDIUM

In the presence of air, the power dissipated Pg decreases. This cannot be explained simply by a decrease in the mass per unit volume of the gas-liquid dispersion. In fact, this decrease is above all due to the decrease in the power number of the impeller which represents its drag coefficient.

The power Pg will therefore by conditionned by the hydrodynamic behaviour of the gas at the impeller blades. Van't Riet (Ref. 1) has shown the formation of different types of cavities behind the impeller blades : vortex, clinging and large cavities. Warmoeskerken (Ref. 2 and 3) has described these phenomena in detail and has stated the conditions under which they appear. These cavities are both function of the gas flow G and the rotational speed of impeller n.

In general, one uses the ratio $K = Pg/Po$ where Pg is the power dissipated with air and Po the power dissipated without air, at the same rotational speed n. The evolution of the ratio K as a function of the aeration number Na ($Na = G/ND^3$) at constant speed gives the classical graph of the Fig. 2. One can distinguish a convex and a concave portion and a point (or rather a zone) of inflexion which corresponds to K values of the order of 0.6 to 0.7. Each portion corresponds to a different type of cavity.

Another phenomenon which characterises the hydrodynamics of an agitated and aerated vessel is the condition of flooding and loading. Roustan, Bruxelmane (Réf. 4), for standard vessels, have given a precise definition of the flooding and loading points (see Fig. 3) and have described a method for their determination. In practice a knowledge of these 2 points is important. For an optimum impeller performance, it is indispensable that the impeller functions beyond its flooding point. A gas is introduced with the aim of causing mass transfer from the gas phase towards the liquid phase (aerobic fermentation, oxidation...). This task must be accomplished whilst consuming the least amount of energy possible. One criterion used is the quantity of mass transfered per unit of energy consumed (ex. $Kg\ O_2/kWh$). Pollard (Ref. 5) has shown that for oxygen transfer from air, this value is an optimum for operating conditions close to the loading point.

4. RESULTS FROM THE STANDARD VESSEL H = T = 1 m

Before presenting the results for non standard vessels, we would like to present some results obtained with standard vessel (H = T = 1 m and D = 0.333 m).

The graph of $K = f\ (Na)$ for n = constant is given in Fig. 4. For each rotational speed a separate curve is obtained which evidently renders difficult extrapolation of these results, since the aeration number Na is not an adequate dimensionless number, capable of representing the whole mass of results. The point (or zone) of inflexion corresponds to K values comprised between 0.6 and 0.7.

The curve Pg = f (n) for G = constant shows some irregularities as shown by Pharamond (Ref. 6). Pg is not proportional to n^3 all along the curve (as is the case without air) except, either for low gas flowrate or for high gas flowrate and rotational speeds higher than a critical value (see Fig. 5).

The graph of K = Pg/Po as a function of n for G = constant confirms the previous observations. As shown in Fig. 6, beyond a certain critical rotational speed, one can consider that K is a constant for all values of speed n. This critical speed takes a value very close to that for loading. Nienow (Ref. 7) considers that the minimum of the graph K = f (Na) for G = constant corresponds to a value very close to that for complete dispersion of the gas.

Recently we reviewed certain experimental results concerning the evolution of the radial velocity of the liquid being pumped by the impeller (Ref. 4). We have used the ratio R (at fixed n) which is the ratio of the radial velocity of liquid in the presence of air, to the radial velocity of the liquid in the absence of air. The Fig. 7 shows the evolution of R as a function of n at G = constant. For low gas flowrates, and for velocities superior to that for loading, R takes a value of 1. These values of gas flowrates correspond to the concave portion of the graph K = f (Na) at n = constant (Fig. 4) (clinging and vortex cavities). For high gas flowrates (convex portion of the graph, large cavities), the value R = 1 is never reached, and R gets smaller as the flowrate G increases.

One sees therefore that the type of cavity influences the pumping flowrate of the impeller and thus the circulation flow engendered in the vessel. From a practical point of view a good circulation flow in the vessel (which conditions for example solids suspension, gas hold-up...) requires that the impeller not only operates beyond the flooding point but also close to (and even beyond) the loading point. The correct determination of these 2 points is therefore extremely important.

5. POWER DISSIPATED IN NON STANDARD VESSELS H = 2 T (2 impellers)

Very little data concerning these types of vessels is to be found in the litterature. Kuboï and Nienow (Réf. 7) have conducted studies with 2 Rushton type impellers in standard vessels H = T, and have determined each of their contributions to the total power. They have shown that the value of the ratio K = Pg/Po is greater for 2 impellers than for a single impeller under the same operating conditions, and that the lower impeller was flooded whilst the upper one achieved a good dispersion of the gas. For a good dispersion of the gas by both impellers the ratio of the speeds must be comprised between 1.4 and 1.7.

As for the total power dissipated in the absence of air by 2 impellers mounted in non standard vessels, one can consider that, in our case, the power number Ne)2 is equal to twice the power number Ne for a single impeller in a standard vessel as is shown by Table 2.

In the presence of a gas, the behaviour of the 2 impellers is different. At constant gas flowrate, if the velocity is increased one observes that the uppermost impeller already achieves a good dispersion of the gas whilst the lower one is still flooded. A global measurement of the power dissipated will take into account both impeller contribution without estimating the contribution of each of them. A detailed analysis of the phenomenon would require that one defines for each impeller a velocity characterising flooding and loading. However from a practical point of view, it is the operating conditions of the lowest impeller that must be particulary taken into account.

5.1. Results obtained with D/T = 0.33

Table 3, gives the values of the rotational speed corresponding to the loading point for the lowest impeller in the vessel H = 2T, T = 0.43 m and D/T = 0.33. Visual observations made concerning the velocity of the uppermost impeller at its loading point allow us to state that n_{L_1} is equal to 1.5 n_{L_2} . The value 1.5 is very close to that given by Kuboï and Nienow (Réf.7).

Gas hold-up as a function of rotational speed n can give very precise information, but only if one works with a very large range of values. The general shape of these curves is shown in Fig. 8. At low speeds, the gas hold-up increases slowly upto $n = n_{L_2}$; it then increases rapidly upto $n = n_{L_1}$. Beyond this point, there is a regular increase in the gas hold-up. Though the determination of the loading point n_{L_1} does not involve much difficulty (see Table 3) one cannot say the same for $n = n_{L_1}$ which is sometimes difficult to locate on the graph. This method thus requires very precise determination of the gas hold-up.

From an analysis of Table 4 giving the evolution of the ratio $K = Pg/Po$ as a function of Na for different gas flowrates G, we can make the following conclusions :
- the graphs of $K = f(Na)$ for n = constant (see Fig. 9) show a difference in behaviour between the non standard and standard vessels. One does not observe (or at least it is very small) the concave portion of the curve. One observes only a convex portion. The point of inflexion, which is seen for standard vessels at the moment of formation of the 3 large cavities, is not marked for this configuration. This does not mean that the cavity formation does not occur here, but that in the presence of 2 impellers, the sudden decrease in the ratio K is compensated by the contribution of the uppermost impeller, which does not behave in the same manner.

- the data of Table 4, giving $K = f(Na)$ at G = constant, show that for rotational speed n superior to the loading point, K can be considered as being constant (Ref. 9) and thus independant of Na and thus of n. All the data for the 2 vessels having a ratio $D/T = 0.33$ can be represented on a graph by plotting K versus $V_s/\sqrt{g} T$ (see Fig. 10) which is a dimensionless number (Froude number) where V_s represents the gas flowrate per unit cross section of the vessel. It is evident that this graph must be verified using vessels of much larger diameter. The advantage of this graph is that it gives one rapid information about the value of K if vessel diameter and gas flowrate are known. (condition $n > n_{L_1}$)

5.2. Influence of the ratio D/T

We have studied the influence of the ratio D/T on the power dissipated, and thus on the ratio $K = Pg/Po$. For the values of $D/T = 0.30$ and 0.37, the results obtained are similar to those for $D/T = 0.33$. The curves $K = f(Na)$ at n = constant have the same shape and if $n > n_{L_1}$, we can consider that the values of K are constant for a given gas flowrate G (see Table 5).

On the other hand, for ratios $D/T = 0.42$ and 0.46, we cannot consider K = constant, except for low gas flowrates (G = 0.27 10^{-3} m^3/s and D/T = 0.46, see Fig. 11) where the average value of K is identical to that found with the other ratio D/T. This variation of $K = f(n)$ is more important with $D/T = 0.46$ than with $D/T = 0.42$. For the same conditions (G = constant), the values of K are always lower with these two ratios of D/T than with for example $D/T = 0.33$. This same result has been found with standard vessels (Ref. 10).

One characteristic of the curves of Fig. 11 is that the rotational speeds are always greater than speeds corresponding to the loading point of the lower impeller.

One explanation for the decreasing K could be the following : with these ratios of D/T (0.42 or 0.46) in comparison with $D/T = 0.33$, the liquid circulation into the vessel is more important. Thus, this recirculation under the impeller of an important quantity of gas would be equivalent to a supplementary gas flowrate.

An increase of the ratio D/T has for result a decrease in $K = Pg/Po$, but not in great proportions (see Fig. 11 and Table 4). Another result of the increase of D/T, is the decrease of the gassed power Pg, such that the lower turbine is loaded (at G = constant).

The final choice of the ratio D/T must therefore take into consideration all the information concerning the mixing process.

<u>6. POWER DISSIPATED IN NON STANDARD VESSELS H = 2 T (3 impellers)</u>

The results have been obtained in the vessel T = 0.43 m using 3 impellers of diameter 0.143 m (see Table 1 and Fig. 1). In non gassed conditions the measured power number Ne)3 is 14.2 (i.e. Ne = 4.7 for each impeller). Comparison with the value for 2 impellers, Ne)2 = 10.4 (i.E. Ne = 5.2 for each impeller) shows that the power dissipated with 3 impellers is not entirely proportionnal to the number of impellers used.

Under gassed conditions, the values of K = Pg/Po are given in Table 6 and are compared with those obtained with 2 impellers for various speeds of rotation n and for 3 gas flowrates G. One notices that globally the difference between the 2 configurations is very small except maybe for the higher rotational speeds where K is much larger ; but it is difficult to conclude on this hypothesis.

Given the identical values of K, the power dissipated Pg, for n and G constant, would thus be more important with 3 impellers than with 2. However if one compares the values of gas hold-up, one sees that they are identical (as shown in Table 6).

This means that an increase in the power dissipated by the addition of a third impeller does not involve an increase in the gas hold-up, and thus in the specific surface area (if the bubble diameter does not vary very much). The gas hold-up depends obviously on the degree of turbulence in the vessel (fractionnation of the gas into bubbles) but it is above all related to the circulation engendered in the vessel. The 3 impeller configuration increases the energy supplied to the fluid without improving the circulation. However in the case of gas liquid transfer, the quantity transfered depends on the film transfer coefficient k_L and the specific surface area a. Though the 3 impeller configuration does not increase a, it can improve the value of k_L which, as a rule, depends on the degree of turbulence. Unfortunately, we do not have data on gas liquid transfer with this latter configuration to verify the above.

<u>7. CONCLUSIONS</u>

The results presented and their interpretation are only the 1st step in solving the problem of power dissipated in aerated non standard vessels. The complex hydrodynamics of the system make it difficult to predict the power dissipated for the whole range of rotational speeds and gas flowrates. The problem of scale-up has not been completely solved either. We think that two important parameters to be further studied and understood are :

- the conditions of flooding and loading of the two and three impeller systems. From a practical point of view one must be capable of predicting the values of these two operating conditions.

- the types of cavities which form behind the impellers. It has been demonstrated that the correlations giving either gas hold-up or mass transfer are different depending on the types of cavities (Ref. 2 and 8).

Finally we think that the laboratory studies must be carried out in vessels having a minimum size of T = 0.5 m in order that the results obtained are representative of real systems.

<u>8. REFERENCES</u>

1. <u>Riet, K. Van't</u> : Ph. D. Thesis, Techn. Univ. of Delft, The Netherlands, 1975.

2. <u>Warmoeskerken, M.M.C.G. and Smith, J.M.</u> : "Hydrodynamics and mass transfer in agitated systems : recent results". 7th CHISA, Prague, 1981, Paper no.B3.4.

3. <u>Warmoeskerken, M.M.C.G. and Smith, J.M.</u> : "Description of the power curves of turbine stirred gas-liquid dispersions". In : Proc. 4th European Conference on Mixing, Noordwijkerhout, Netherlands, BHRA Fluid Engineering, 1982, Paper G1, 237 pp.

4. <u>Roustan, M. and Bruxelmane, M.</u> : "Loading and flooding points in turbine agitated gas-liquid dispersions". In : 7th CHISA Congress, Prague, 1981, Paper no B3.3.

5. Pollard, G.J. : " Flooding and aeration efficiency in standard stirred vessels". In : Proc. International Symposium on Mixing, Faculté Polytechnique de Mons, Belgique, 1978, Paper no C4.1.

6. Pharamond, J.C. : "Contribution à l'étude de l'agitation en milieu aéré". Thèse Docteur-Ingénieur, INSA Toulouse, France, 1973.

7. Kuboï, R. and Nienow, A.W. : "The power drawn by dual impeller systems under gassed and ungassed conditions". In : Proc. 4th European Conference on Mixing, Noordwijkerhout, Netherlands, BHRA Fluid Engineering, 1982, Paper G2, 247 pp.

8. Roustan, M. : "Hydrodynamic and mass transfer in agitated gas-liquid reactors with different geometrics and air liquid systems". Réunion interne du groupe de travail "Mixing" de la Fédération Européenne de Génie Chimique. Cologne, Allemagne, GVC-VDI, 1980.

9. Charles, J.M. : "Contribution à l'étude de l'extrapolation des réacteurs agités mécaniquement". Thèse Docteur-Ingénieur, INSA Toulouse, France, 1978.

10. Roustan, M. : "Contribution à l'étude des phénomènes d'agitation et de transfert de matière dans les réacteurs gaz-liquide'. Thèse Doctorat ès-Sciences, INSA Toulouse, France, 1978.

Table 1. Geometrical characteristics of the
agitated system used.

T	0.43	0.43	0.78	1
H/T	2	2	2	1
number of impellers	2	3	2	1
D	0.128 0.143 0.160 0.179 0.200	0.143	0.230 0.260 0.290 0.327	0.333
10^3 G	0.277 - 2.21	0.55 - 1.66	1.4 - 11.1	2.8 - 11.1
Gvvm	0.13 - 1	0.27 - 0.81	0.11 - 0.91	0.21 - 0.84
n	3 - 8	3 - 8	1.6 - 4	1.5 - 4

Table 2. Values of power number.

D/T	T = 0.43 Ne)2	T = 0.43 Ne)3	T = 0.78 Ne)2	T = 1 Ne)1
0.30	10.35		10.9	
0.33	10.42	14.2	10.8	5.1
0.37	10.15		11.1	
0.42	10.83		11.4	

Table 3. Values of rotational speed n_{L_1} corresponding to the loading of the lower impeller.

	10^3 G	n_{L1}	n_{L1}
T = 0.43 m	0.55	3	2.3
	0.83	3.7	2.9
	1.11	4.3	3.3
	1.39	5	3.5
	1.66	5.5	3.8
	1.94	5.8	4.2
T = 0.78 m	2.78	2.3	
	4.1	2.8	2.1
	5.56	3.3	2.4
	6.9	3.6	2.7
	8.3	3.8	2.8
		D/T = 0.33	D.T = 0.37

Table 4a. Experimental data K = f(Na) for different gas flowrates G (T = 0.43 m D/T = 0.33).

Fr	n s^{-1}	G = 0.277 10^{-3} m^3/s		G = 0.55 10^{-3} m^3/s		G = 0.83 10^{-3} m^3/s		G = 1.11 10^{-3} m^3/s		G = 1.39 10^{-3} m^3/s		G = 1.66 10^{-3} m^3/s		G = 1.94 10^{-3} m^3/s		G = 2.21 10^{-3} m^3/s	
		K	100 Na	K	100 Na	K	100 Na	K	100 Na	K	100 Na	K	100 Na	K	100 Na	K	100 Na
0,13	3,0	0,84	3,15	0,78	6,3	0,66	9,45	0,56	12,6	0,56	15,7	0,56	18,5	0,53	22,0	0,51	25,2
0,20	3,66	0,84	2,58	0,77	5,16	0,64	7,74	0,53	10,3	0,55	12,9	0,55	15,5	0,52	18	0,52	20,6
0,28	4,41	0,85	2,14	0,81	4,28	0,66	6,42	0,61	8,6	0,59	10,7	0,52	12,8	0,45	15	0,47	17,2
0,39	5,16	0,90	1,83	0,8	3,66	0,68	5,49	0,6	7,3	0,57	9,1	0,51	11	0,48	12,8	0,45	14,6
0,5	5,83	0,91	1,62	0,8	3,24	0,67	4,86	0,6	6,5	0,57	8,1	0,53	9,7	0,45	11,3	0,44	13
0,63	6,6	0,82	1,43	0,79	2,86	0,65	4,29	0,59	5,7	0,56	7,1	0,52	8,6	0,44	10	0,44	11,4
0,78	7,33	0,9	1,29	0,79	2,58	0,65	3,87	0,58	5,2	0,55	6,5	0,51	7,8	0,45	9	0,42	10,4
0,94	8,03	0,87	1,18	0,76	2,26	0,62	3,54	0,57	4,7	0,51	5,9	0,48	7,1	0,46	8,3	0,41	9,4

Table 4b. Experimental data K = f(Na) for different gas flowrates G (T = 0.78 m D/T = 0.33).

Fr	n s^{-1}	G = 1,39 10^{-3} m^3/s		G = 2,78 10^{-3} m^3/s		G = 4,1 10^{-3} m^3/s		G = 5,56 10^{-3} m^3/s		G = 6,9 10^{-3} m^3/s		G = 8,3 10^{-3} m^3/s		G = 9,7 10^{-3} m^3/s		G = 11,1 10^{-3} m^3/s	
		K	100 Na	K	100 Na	K	100 Na	K	100 Na	K	100 Na	K	100 Na	K	100 Na	K	100 Na
0,10	1,97	0,82	4	0,63	8	0,60	12	0,60	16	0,56	20	0,55	24	0,55	26	0,54	30
0,136	2,27	0,83	3,48	0,66	7	0,57	10,4	0,56	13,9	0,54	17,4	0,53	20,8	0,52	24,4	0,51	27,8
0,17	2,53	0,84	3,12	0,68	6,2	0,59	9,4	0,56	12,5	0,51	15,6	0,5	18,7	0,5	22,8	0,49	25
0,205	2,78	0,83	2,84	0,70	5,7	0,62	8,5	0,54	11,4	0,51	14,2	0,49	17	0,48	19,9	0,47	22,7
0,241	3	0,82	2,63	0,70	5,3	0,62	7,9	0,53	10,5	0,50	13,1	0,48	15,8	0,46	18,4	0,44	21
0,29	3,3	0,83	2,4	0,62	4,8	0,62	7,2	0,54	9,6	0,49	12	0,46	14,4	0,45	16,8	0,43	19,2
0,34	3,6	0,83	2,20	0,69	4,4	0,61	6,6	0,54	8,8	0,48	11	0,46	13,2	0,44	15,4	0,42	17,6
0,39	3,83	0,84	2,06	0,70	4,1	0,61	6,2	0,54	8,2	0,49	10,3	0,47	12,3	0,43	14,4	0,42	16,5

Table 5. Data of K for different ratios D/T.

$1000 \dfrac{V_s}{\sqrt{gT}}$		D/T = 0.30 $\overline{K}$	D/T = 0.33 $\overline{K}$	D/T = 0.37 $\overline{K}$
T = 0.43 m	0.93	–	0.83	0.83
	1.86	0.79	0.79	0.77
	2.79	0.68	0.66	0.65
	3.72	0.57	0.58	0.57
	4.65	0.53	0.55	0.52
	5.58	–	0.51	0.50
	6.52	–	0.45	0.45
	7.45	–	0.43	
T = 0.78 m	1.05	–	0.83	
	2.1	0.66	0.69	0.71
	3.15	0.58	0.61	0.61
	4.20	0.50	0.54	0.55
	5.25	0.45	0.49	0.50
	6.30	–	0.47	0.48
	7.35	–	0.45	
	8.40	–	0.42	

Table 6. Comparison of two and three impellers configuration T = 0.43 m D/T = 0.33.

n s^{-1}	G = 0,55 10^{-3} m³/s				G = 1,11 10^{-3} m³/s				G = 1,66 10^{-3} m³/s			
	K (2)	K (3)	φ % (2)	φ % (3)	K (2)	K (3)	φ % (2)	φ % (3)	K (2)	K (3)	φ % (2)	φ % (3)
3	0,78	0,72	1,8	1,9	0,60	0,58	2,6	2,9	0,60	0,53	2,7	3,4
3,3		0,78		2,2		0,58		3,1		0,57		3,8
3,8	0,77	0,78	2,2	2,5	0,53	0,56	3,5	3,6	0,55	0,53	4,1	4,5
4,1		0,84		2,7		0,58		4		0,52		5,1
4,4	0,81	0,82	2,7	2,9	0,59	0,60	4,5	4,3	0,52	0,51	5,6	5,6
4,8		0,83		3,1		0,6		4,7		0,5		6
5,2	0,8	0,84	2,9	3,3	0,61	0,59	4,9	5	0,51	0,47	6,7	6,7
5,6		0,84		3,4		0,6		5,4		0,49		7
5,9	0,8	0,83	3,5	3,5	0,59	0,62	5,7	5,9	0,53	0,49	7,6	7,5
6,2		0,83		3,7		0,63		6,3		0,51		7,8
6,6	0,79	0,84	3,6	3,9	0,59	0,64	6,2	6,6	0,52	0,51	8,1	8,2

K (2) φ (2) value of K and φ for 2 turbines

K (3) φ (3) value of K and φ for 3 turbines

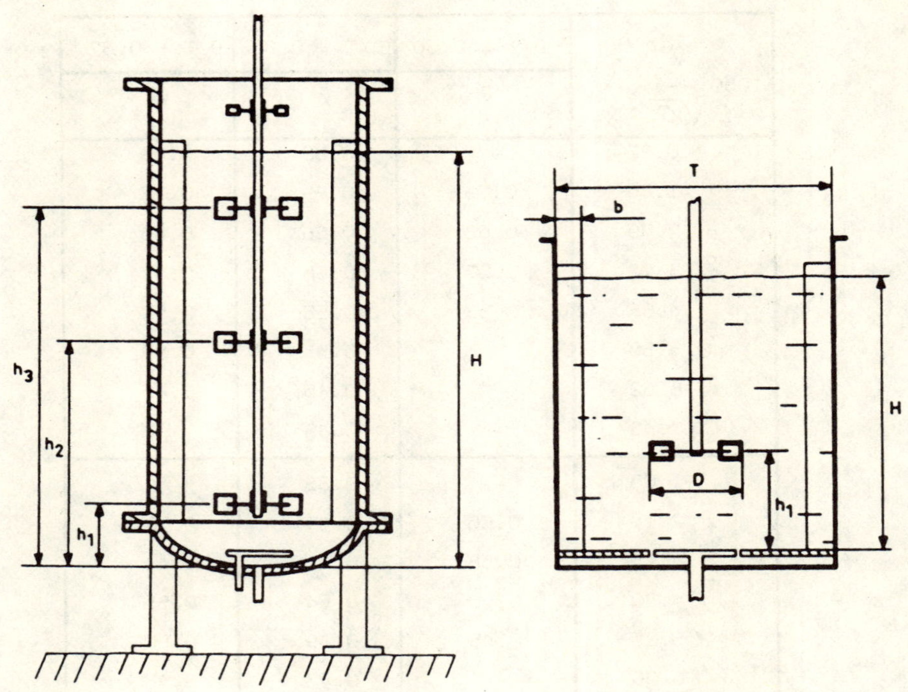

Fig. 1. Schematic diagram of equipment

$h_1 = 0.143$	$h_1 = 0.143$	$h_1 = 0.33$
$h_2 = 0.573$	$h_2 = 0.378$	
	$h_3 = 0.613$	
2 impellers	3 impellers	1 impeller

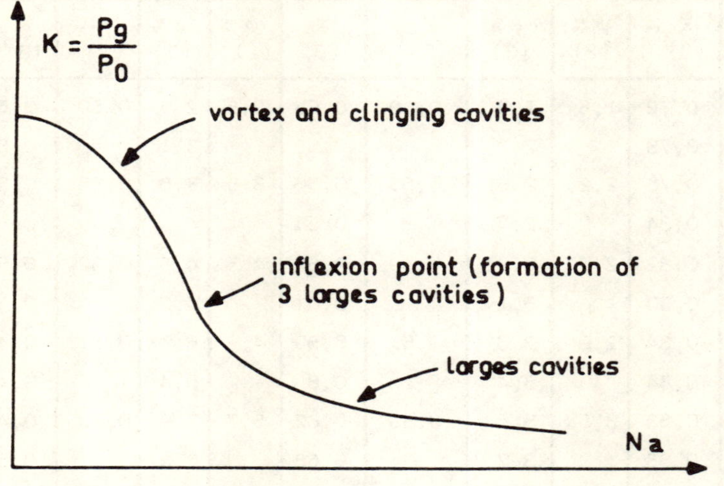

Fig. 2. Evolution of K as a function of
aeration number.

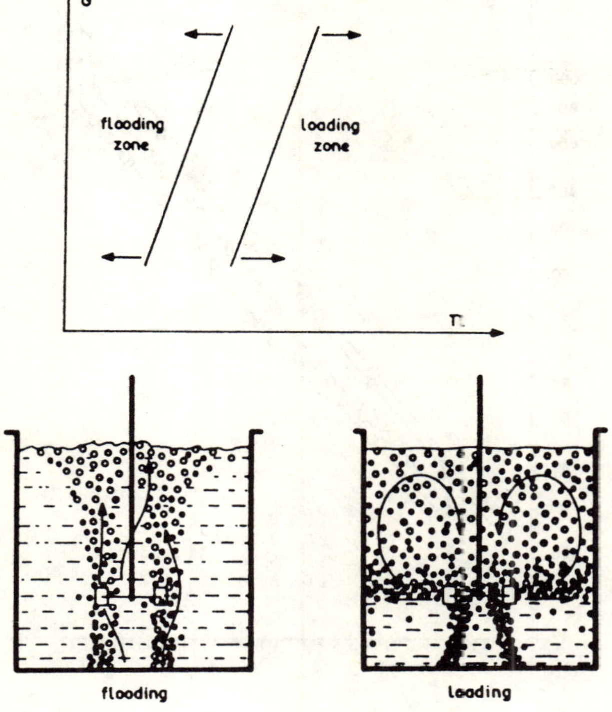

Fig. 3. Definition of flooding and
loading for standard vessels.

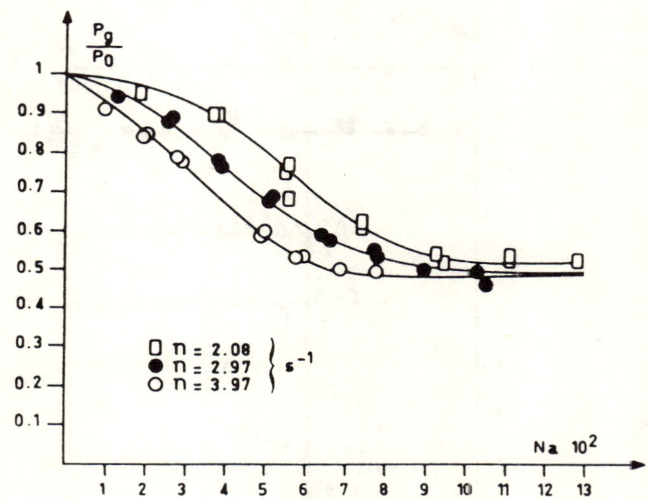

Fig. 4. K = f(Na) at constant n for
standard vessel H = T = 1 m.

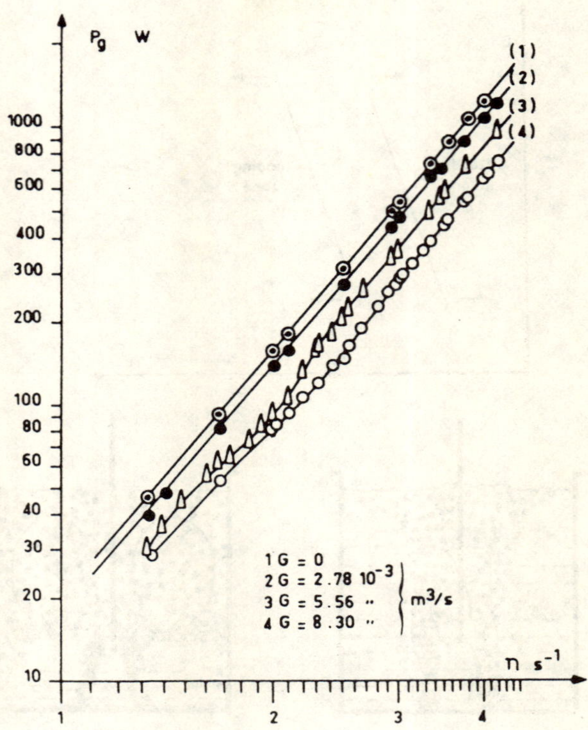

Fig. 5. Pg = f(n) at constant gas
flowrate for standard
vessel H = T = 1 m.

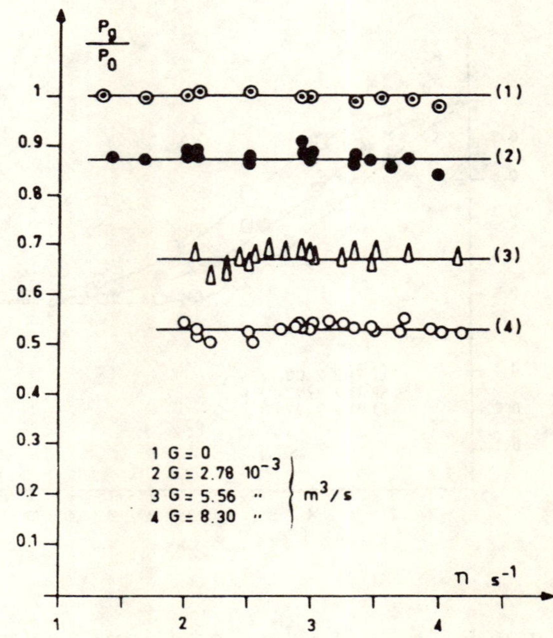

Fig. 6. K = f(n) at constant gas
flowrate for H = T = 1 m.

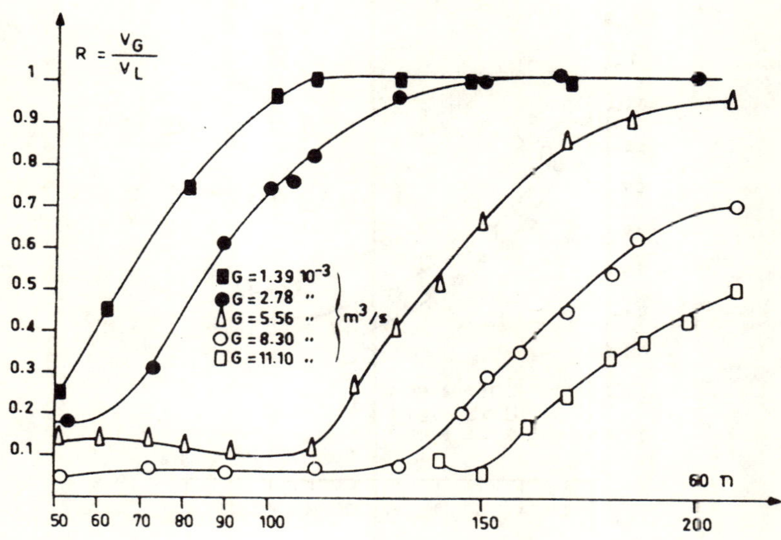

Fig. 7. R = f(n) at different gas flowrates
for standard vessel.

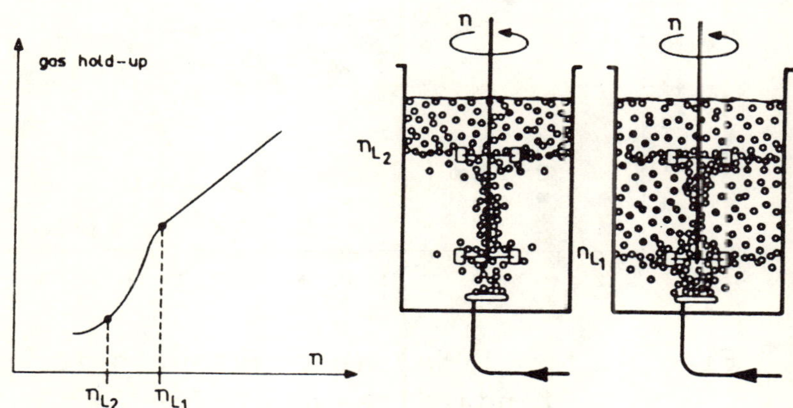

Fig. 8. Evolution of gas hold-up in non standard
vessel.

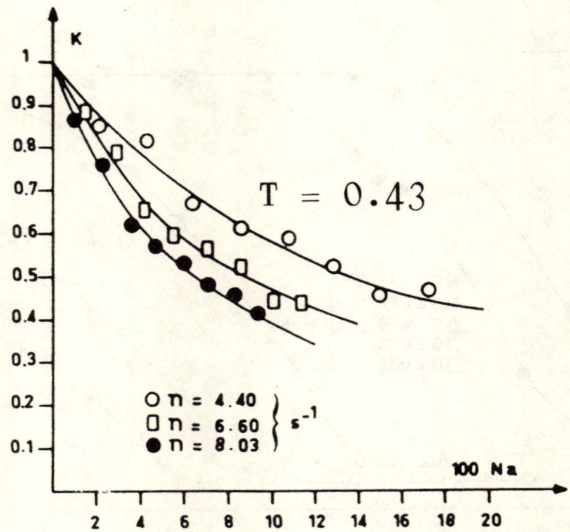

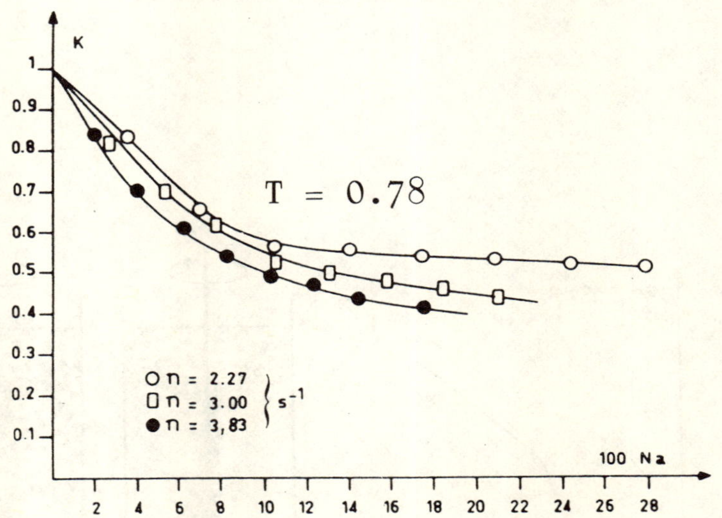

Fig. 9. K = f(Na) at constant n for non standard vessel D/T = 0.33.

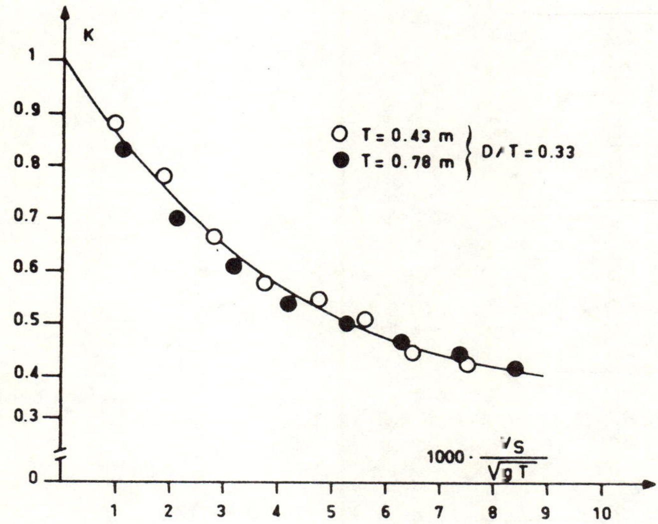

Fig. 10. $K = f(V_S/\sqrt{gT})$ for non standard vessel $D/T = 0.33$.

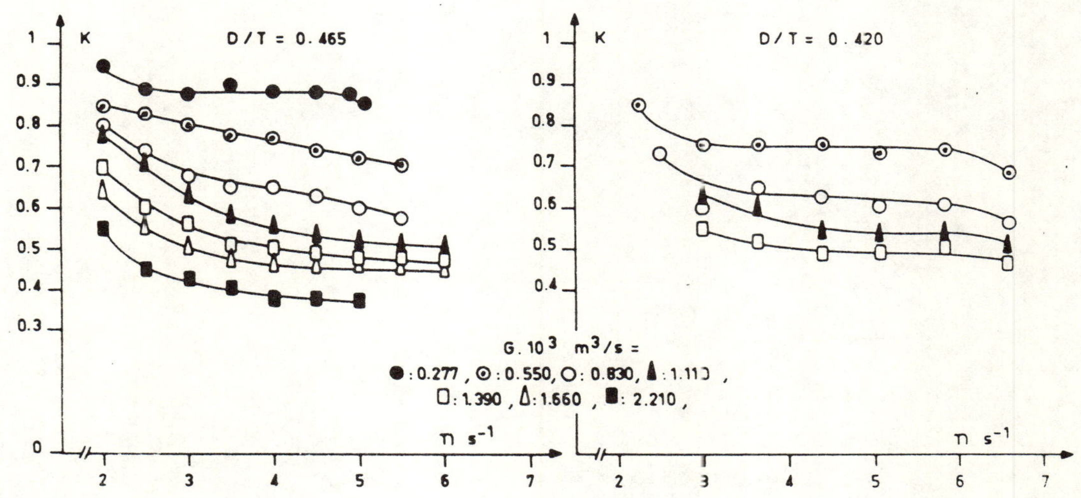

Fig. 11. $K = f(n)$ at constant gas flowrate for different ratios D/T (T = 0.43 m).

ON THE FLOODING/LOADING TRANSITION AND THE COMPLETE DISPERSAL CONDITION IN AERATED VESSELS AGITATED BY A RUSHTON-TURBINE

A.W. Nienow, M.M.C.G. Warmoeskerken*, J.M. Smith* and M. Konno

Department of Chemical Engineering
University of Birmingham
P.O. Box 363
Edgbaston
Birmingham B15 2TT

SUMMARY

Two main series of experiments are reported using air and tap water. In Delft, experiments have been conducted in vessels of 0.44, 0.64 and 1.2m diameter of liquid depth equal to tank diameter with an impeller-to-tank diameter of ratio of 0.4 and the same off-bottom clearance. In Birmingham, experiments have been conducted in vessels of 0.29, 0.45, 0.56 and 0.61m diameter with D/T ratios of 0.33 and 0.5 and an off-bottom clearance of 0.25. In addition, the following variables have been briefly studied: other D/T ratios. sparger types and position, impeller off-bottom clearance, non-coalescing electrolyte solutions and liquid depth.

In each laboratory, the basic measurements made were the variation of impeller torque as a function of agitator speed and aeration rate especially in the region where the flooding/loading transition occurred. In each case, the bulk-flow transition was viewed simultaneously with the flow in the impeller region by means of a de-rotational flow visualisation technique.

At realistic impeller speeds, an increase in gassing rate which causes the transition from loading to flooding is accompanied by a change from 3 large plus 3 clinging cavities to 6 ragged cavities with D/T ratios of $\leq$ 0.4. At the same time, there is a step increase in the power drawn. With a D/T ratio of 0.5, the transition is from clinging cavities to ragged cavities. In this case, there is a step decrease in the power drawn. These results are the same for all the other variables investigated.

For the main experiments, the equation found to fit the data well is:

$$(Fl_G)_F = 30 \; (D/T)^{3.5} \; Fr_F$$

where $(Fl_G)_F$ and $(Fr)_F$ are the gas flow number and the Froude number at the flooding/loading transition. This equation is similar to earlier ones in the literature for this transition. Using this equation and a literature equation for predicting the agitator speed, N_{CD}, for gas dispersion throughout the vessel, maps are given indicating the bulk flow as a function of gassing rate and agitator speed for different scales of operation.

It is also shown that a) the specific power required to disperse gas on scale-up increases when maintaining the gas residence time constant, b) large D/T ratios are less susceptible to flooding than small ones and require less power to disperse a given aeration rate.

*Laboratory for Physical Technology, Delft University of Technology, Prins Bernhardlaan 6, 2628 BW Delft, The Netherlands.

Held at Wurzburg, 10-12 June, 1985.

Organised by DVCV· Deutsche Vereinigung für Chemie- und Verfahrenstechnik
(German Association of Chemical and Process Engineering).

Organisation: GVC·VDI-Gesellschaft Verfahrenstechnik und Chemieingenieurwesen. 

<u>NOMENCLATURE</u>

A   constant in Eqn. 1 (dimensionless)

C   impeller clearance above the base (m)

D   impeller diameter (m)

Fl_G gas flow number $= Q_G/ND^3$ (dimensionless)

Fr Froude number $= N^2D/g$ (dimensionless)

g   gravitational constant (9.81 m s^{-2})

H   liquid height (m)

N   impeller speed (rev s^{-1})

P   power drawn by impeller (W)

Po   power number (dimensionless)

Q_G gas flow rate (m^3 s^{-1})

Q_{VG} specific gas flow rate (vvm)

T   vessel diameter (m)

vvm (volumetric flow rate of gas/minute)/(volume of liquid)

ε_T mean energy dissipation rate (W kg^{-1})

Subscripts

g   under gassed conditions

CD   condition at which impeller just disperses gas throughout the vessel

F   condition at which impeller is flooded

1. INTRODUCTION

In studies of gas dispersion, it is desirable to observe the two phase bulk flow
and the flow in the region of the agitator. The bulk flows when using Rushton turbines
at steadily increasing speeds, starting from zero, to disperse a particular gassing
rate go through three well-defined regimes, (Fig. 1): From $N=0$ to $N=N_F$, a flow
dominated by the gas flow, consisting of a two-phase gas-liquid flow up the middle and
a liquid flow down at the walls, i.e., the impeller is flooded, Fig. 1a; From
$N_F < N < N_{CD}$, a two-phase pattern dominated by the agitator and driven out horizontally
from it, Fig. 1b. For $N > N_F$, the agitator is loaded; For $N > N_{CD}$, a two-phase
dispersion of gas throughout the vessel. In the region $N \gg N_{CD}$, large amounts of gas
recirculate (Ref. 1).

Increases in Q_G lead to the opposite sequence of events provided the speed is high
enough to disperse even a very low gassing rate. The N_F transition is a sharp and
obvious one. N_{CD} is more subjective. Because these transitions are easy to observe
and are of significance in gas-liquid reactor design, many attempts have been made to
obtain predictive equations for them.

Recently, Warmoeskerken and Smith (Ref. 2) reviewed the correlations for predicting
the N_F correlation. They also proposed a theory leading to the equation

$$(Fl_G)_F = A\, Fr_F \tag{1}$$

where $(Fl_G)_F$ is the gas flow number $(Q_G/N_F D^3)$ and Fr_F is the Froude number $(N_F^2 D/g)$ at
the flooding-loading transition. Eqn. 1 is similar (but of a simpler form) to equations
proposed by other workers. They found that the dimensionless constant A was 1.2 for
dispersing air in water using Rushton turbines of size $D/T = 0.4$ and clearance $C/H = 0.4$
in vessels of 0.44, 0.64 and 1.2m diameter ($H = T$).

Earlier, Nienow et al. (Ref. 1) gave a correlation for N_{CD} for dispersing air in
electrolyte solutions using a range of vessels from 0.3 to 1.8m diameter.

$$(Fl_G)_{CD} = 0.2\, (D/T)^{0.5}\, (Fr)_{CD}^{0.5} \tag{2}$$

Eqn. 2 is similar to Eqn. 1 and it might be expected therefore that the constant A
would be a function of D/T ratio. Nienow et al. showed that N_{CD} was linked to the
impeller clearance C. Eqn. 2 applied when $C/H = 1/4$, a value chosen as a good
compromise. Lower clearances lead the upper part of the vessel to function like a
bubble column; greater clearances rapidly increase the speed and power required to
disperse the gas into the lower part of the vessel.

This paper reports additional recent data relating to N_F from work carried out in
Delft and in Birmingham. The N_F values are linked to changes in gas-filled cavity
structure as viewed by derotational techniques (Ref. 3,4) and the relationship between
impeller torque and power drawn, agitator speed and gassing rate. Water and electrolyte
solutions have been used and D/T ratios from 0.22 and 0.58. Sparger type and position
and impeller off-bottom clearance have been varied too. Except where otherwise stated
the liquid height was maintained equal to the vessel diameter. Vessels with flat-
bottoms, $H = T$ and 4, 10% strip baffles, up to 1.2m in diameter have been used. The
equipment has been described previously (Ref. 2,4).

With this extra data, a general correlation for N_F is proposed. From it and the
N_{CD} correlation, maps of operating conditions are drawn up and comments are made about
scale-up and D/T ratio.

2. THE EFFECT OF GASSING RATE ON POWER DRAWN, GAS-FILLED CAVITY STRUCTURE AND FLOODING

2.1 Previous Work and Confirmation within this Study

For the disc turbine, increases in gassing rate at constant speed have been
reported generally to lead to the following changes of cavity type, Fig. 2:
a) 6 vortex cavities; b) 6 clinging cavities; c) 3-3 combination of alternate large and
clinging cavities.

Vortex and clinging cavities were first named by Van't Riet and Smith (Ref. 3)
and noted by them and Nienow and Wisdom (Ref. 5). The extreme stability of the 3-3
combination, missed by the earlier workers, was reported by Warmoeskerken et al.
(Ref. 6) in 1981. All these changes have been confirmed in this work with water and
electrolyte solutions as the working fluids.

A plot of gassed-to-ungassed power against flow rate Q_G shows typically a step
increase at the point at which flooding occurs (Ref. 2). In 1983, Nienow et al.

(Ref. 7) reported that this step coincided with a change to six ragged cavities behind each agitator blade (Fig. 3). These cavities oscillate violently and often only cover the upper half of a blade. Thus, the pressure difference between the front and back is greater than with the "3-3" structure and the power drawn increases. In plan view, the appearance of ragged cavities is like that of clinging cavities except that they are foreshortened and the gas breaks away at their rear end rather than at the edge of the cavity where it trails off behind the impeller blade, (compare Fig. 2 and 3).

Fig. 4 shows a plot of P_g/P versus Fl_G confirming the step increase at N_F for a T/3 impeller in a 0.61m diameter vessel using a sparger of diameter less than the size of the impeller disc. Similar results were found for this transition in all vessels and all gassing rates investigated up to 3.5 vvm* for D/T ratios of 0.33 and 0.22. Only with D/T = 0.22 was significant hysteresis found.

Warmoeskerken and Smith (Ref. 2) reported the same interactions for all conditions studied at a D/T ratio of 0.40 in their vessels. However, they also suggested that the large cavities found in the "3-3" structure could not form at Froude numbers less than about 0.045. Thus the "3-3"-ragged transition should not occur at such Froude numbers. Tests were carried out in this work to examine this anomaly.

2.2 The Clinging Cavity-Ragged Cavity Transition

Agitator speeds were selected to give Foude numbers of 0.040 (1.5 rev/s) and 0.045 (1.6 rev/s) in the 0.44m vessel with D/T = 0.4 at C/H = 0.4. Fig. 5 shows a graph of P_g/P versus Fl_G for these two Froude numbers. At the lowest gassing rates, vortex cavities form and the power barely falls. However, a rapid fall occurs as the clinging cavities develop. The shape of the curve is similar to that found at higher speeds (one at 2.5 rev/s is included for comparison) in spite of the fact that the "3-3" structure never forms with increases in gassing rate. Flooding as observed by the change in bulk flow from Fig. 1b to Fig. 1a is accompanied by a change in cavity structure from clinging-to-ragged and is accompanied by a step decrease in power consumption. Once flooded, a significant reduction in gassing rate is required to return to loading and a sizeable hysteresis is indicated in Fig. 5.

The reasons for the step down in power consumption as flooding occurs can be postulated to be as follows: Prior to flooding, the impeller with clinging cavities behind its blades is still able to pump efficiently. With the onset of flooding, the flow is dominated by the sparged gas and the pumping capacity falls dramatically. Thus, the power drawn falls suddenly too.

It has been reported (Ref. 1) that large D/T disc turbines are able to completely disperse a given volumetric flow rate of gas at lower power inputs than small ones. This work has shown that to be so for the flooding-loading transition too. In the 0.29, 0.45 and 0.61m vessels at Birmingham using 1/2 D/T impellers up to about 3.5 vvm, the Froude numbers obtained at the gas flow rate that caused flooding were always less than about 0.05. Thus in all cases, a step down was noticed at the flooding point, coin- ciding with a clinging-ragged cavity transition. Fig. 6 gives an example of the typical P_g/P versus Q_G plot for such geometries for the 0.61m vessel. The high value of P_g/P at the flooding point should be noted.

2.3 Effect of Sparger Type

Warmoeskerken and Smith (Ref. 2) used a ring sparger and a point sparger. The ring sparger was of a diameter less than that of the turbine disc. A range of impeller- sparger clearances were used. Essentially, the flooding loading transition always occurred at conditions within ±10% for each case for the D/T = C/H = 0.4 configurations used. The same results were found in this work wherever different sparger types were tested provided the ring sparger was smaller in diameter than the disc. However, if the sparger diameter is greater than the disc, considerable differences are found (Ref. 8,9) and the cavity structure transitions are reported elsewhere (Ref. 9).

2.4 Effect of Electrolytes (Coalescing/Non-Coalescing Systems)

Sodium carbonate, deca-hydrate was added in sufficient quantities to produce a solution with an ionic strength of about 0.4 ($\sim$0.13 moles/litre). This ionic strength severely represses coalescence. Fig. 7 shows data for two agitator speeds in the 0.44m

*vvm ((volume of gas/minute)/ volume of liquid) is commonly used in fermentation technology with operating values in the range 0.5 to 2 vvm. It is approximately inversely proportional to the gas phase residence time, normally quoted in seconds.

vessel giving a clinging-ragged and a "3-3"-ragged transition at flooding. The results are essentially identical for water and the non-coalescing electrolyte solution and this was so (<±10%) in all flooding-loading transitions measured as has previously been found for clinging-"3-3" cavity transitions (Ref. 6,8).

2.5 Effect of Impeller Clearance above the Base

Tests on the 0.44m vessel (D/T = 0.4) with C/H = 0.25 and 0.4 showed no effect on the flooding-loading transition.

2.6 Effect of Liquid Height

Tests on the 0.44m vessel (D/T = 0.4, C/T = 0.4 and 0.25) with water and electrolyte solutions from 0.3m to 0.6m deep (0.6 < H/T < 1.36) showed no effect on the flooding-loading transition. Fig. 8 illustrates this independence. This insensitivity to liquid depth (or of height of impeller relative to the base or the sparger) is surprising because the theory of Warmoeskerken and Smith implies that it should be important. More work is planned over wider range of H/T values.

3. CORRELATION FOR THE FLOODING-LOADING TRANSITION

Warmoeskerken and Smith (Ref. 2) found that for D/T = C/H = 0.40 in vessels from 0.44 to 1.2m diameter, Eqn. 1 with A = 1.2 applied to a wide range of conditions, the bulk flow transition coinciding with the "3-3"-ragged cavity transition. This work has confirmed this correlation for all the conditions discussed within Section 2 even though both clinging-ragged and "3-3"-ragged transitions were noted at the flooding-loading point. Only with this geometry were both cavity transitions observed.

Similar correlations were found to apply (D/T = 0.33 and 0.5) at Birmingham in 0.29, 0.45, 0.56 and 0.61m vessels at C/H = 0.25. Fig. 9 shows the data for D/T = 1/3. The values for the smallest vessel lie somewhat above the best line, perhaps because the relative size of the shaft, hub and disc slightly reduce the relative pumping capacity in spite of care being taken to maintain geometric similarity. For D/T = 1/3, the cavity transition was "3-3"-ragged at the flooding-loading point and for D/T = 1/2, all transitions were clinging-ragged with minimal hysteresis,(see Fig. 10).

Finally, two other D/T ratios were used in the 0.44m vessel at Delft. For D/T = 0.22, the data is shown in Fig. 10 along with the lines of the correlating equations for the two other D/T ratios. For this small D/T ratio, considerable hysteresis was noted and all transitions were "3-3"-ragged ones. The constants of proportionality of Fig. 10 are plotted against D/T ratio in Fig. 11 giving an overall correlation.

$$(Fl_G)_F = 30 \ (D/T)^{3.5} \ (Fr)_F \tag{3}$$

which is similar to one proposed earlier by Zwietering (Ref. 10) and Mikulkova et al. (Ref. 11).

4. DISCUSSION OF RESULTS AND CONCLUSIONS

4.1 Implication for Scale-Up

Though the power input from an impeller falls when it is gassed, the practical range of P_g/P is usually around 0.6 to 0.5. To a first approximation, therefore, the ungassed power-speed relationship holds for the fully turbulent region, i.e., $P_g \dot{\alpha} P \alpha N^3 D^5$ and for geometrically-similar systems, the power/unit volume or specific energy dissipation rate is linked to speed and scale by $(\varepsilon_T)_g \dot{\alpha} (\varepsilon_T) \alpha N^3 D^2$. Also for reasons of stoichiometry and reaction kinetics, the volume of the vessel on the larger scale is made such that the mean residence time of the gas is kept constant. Again to a first approximation this means that for geometrically similar systems, Q_G/D^3 = constant. In terms of the commonly used fermentation concept of vvm, scale-up with geometric similarity and constant Q_G/D^3 is equivalent to constant vvm.

Using these simple relationships, then the equation for the flooding-loading transition, $(Fl_G)_F = A \ (Fr)_F$, implies that for geometrically similar scale-up at constant vvm,

$$1/N_F \ \alpha \ N_F^2 D \tag{4}$$

so that

$$(\varepsilon_T)_F \ \dot{\alpha} \ (D^{-1/3})^3 \ D^2 \ \dot{\alpha} \ D \tag{5}$$

From the complete-dispersion relationship, $(Fl_G)_{CD} \ \alpha \ (Fr)_{CD}^{1/2}$, similar reasoning shows that

$$(\varepsilon_T)_{CD} \ \dot{\alpha} \ D^{1.25} \tag{6}$$

Thus, on the larger scale, more power/unit volume is required to prevent flooding or achieve complete dispersion when scale-up is done at constant vvm. Alternatively, it suggests that the reactor is likely to be operated closer to the flooding-loading transition on the larger scale as often in practice the power/unit volume is held constant or even reduced on scale-up.

4.2 Implications for Choice of D/T Ratio

From Eqn. 3, for constant gassing rate and vessel size,

$$1/N_F D^3 \ \alpha \ (D)^{3.5} \ N_F^2 D \tag{7}$$

so that

$$N_F \ \alpha \ D^{-2.5} \tag{8}$$

For constant pumping rate, $N \ \dot\alpha \ D^{-3}$. Thus, this relationship might be taken to imply that, at a fixed scale of operation, constant pumping rate is required to disperse a given flow rate of gas. The $D^{-2.5}$ rather than D^{-3} relationship might arise because the flow number is a weak function of D/T. From Eqn. 8,

$$P_F \ \dot\alpha \ N^3 D^5 \ \alpha \ D^{-2.5} \tag{9}$$

i.e., less power is required to disperse a given gassing rate with large D/T ratios at a particular scale of operation. Carrying out the same algebraic manipulations, Eqn. 2 gives $N_{CD} \ \alpha \ D^{-2}$ so that

$$P_{CD} \ \alpha \ D^{-1} \tag{10}$$

The flooding-loading transition is more sensitive to gassing rates at small D/T. Rearranging Eqn. 3 gives

$$N_F \ \alpha \ \left[(T/D)^{3.5} \ (1/D^4) \right]^{1/3} \ Q_G^{1/3} \tag{11}$$

so that at constant Q_G and T

$$dN_F/dQ_G \alpha \ D^{-2.5} \tag{12}$$

The form of this relationship is shown by Fig. 10. The flooding-loading transition was also studied at D/T=0.58 in the 0.44m vessel. Both the bulk flow and cavity transition were hard to observe. Either, the impeller speed was so low that gas dispersion did not occur even at a very low gassing rate or the speed was so high that the impeller would not flood at $\sim$2vvm. Both of these observations fit in qualitatively with the trends predicted by Eqn. 8 and 12, i.e., the speed required to disperse the gas is extremely low (too low to form cavities at all?) and virtually independent of gassing rate.

4.3 Flow Regime Maps

Warmoeskerken and Smith (Ref. 2) developed flow regime maps for 0.44m and 1.2m vessels for D/T = 0.4. From these, given an agitator speed and flow rate, the regime can be determined. Figs. 12 and 13 are adapted from their work (Ref. 2).

In this case, the regimes correspond to the bulk flows indicated by Fig. 1. In addition, lines of Fr = 0.045, of constant vvm and of constant unaerated power/unit volume (assuming a power number of 5) have been marked on them. These lines indicate likely operating values of Q and N at typical specific power inputs and gassing rates. Comparing Figs. 12 and 13 confirms that higher specific powers are required on scale-up in order to achieve a similar flow regime at the same specific gassing rate. Alternatively, it can be inferred that on the larger scale it is more likely that agitators will be operating in the region around N_{CD} and N_F where the power drawn is rather insensitive to gassing rate and typically about 1/2 the ungassed power.

5. REFERENCES

1. Nienow, A.W., Wisdom, D.J. and Middleton, J.C.: "The Effect of Scale and Geometry on Flooding, Recirculation and Power in Stirred Vessels", Proc. of the 2nd European Conference on Mixing, Cambridge, April 1977; Cranfield, BHRA, 1978, pp. F1-1 to F1-16 and X52 to X55.

2. Warmoeskerken, M.M.C.G. and Smith, J.M.: "Flooding of Disc Turbines in Gas-Liquid Dispersion", Chemical Engineering Science, 1985, to be published.

3. Riet, K. van't and Smith, J.M.: "The Behaviour of Gas-Liquid Mixtures near Rushton Turbine Blades", Chemical Engineering Science, 28, 1973, 1031-1037.

4. Kuboi, R, Nienow, A.W. Allsford, K.V.: "A Multipurpose Stirred Tank Facility for Flow Visualisation and Dual Impeller Power Measurements", Chemical Engineering Communications, 22, 1983, 29-39.

5. Nienow, A.W. and Wisdom, D.J.: "Flow over Disc Turbine Blades", Chemical Engineering Science, 29, 1974, 1994-1996.

6. Warmoeskerken, M.M.C.G., Feijen, J. and Smith J.M.: "Hydrodynamics and Power Consumption in Stirred Gas-Liquid Dispersions", Institution of Chemical Engineers Symposium Series, No.64, "Fluid Mixing", 1981, pp.J1-J14.

7. Nienow, A.W., Allsford, K.V. and Kuboi, R.: "Using a De-Rotational Prism for Studying Gas Dispersion Processes" Paper to 9th Engineering Foundation Conference on Mixing, Henniker, New Hampshire, USA, 1983 (unpublished).

8. Riet, K. van't: "Turbine Agitator Hydrodynamics and Dispersion Performance", Ph.D. Thesis, University of Delft, 1975.

9. Nienow, A.W. and Allsford, K.V.: "Effect of Sparger Type on Gas Dispersion by Rushton Turbines", Institution of Chemical Engineers Multistream Symposium, April 1985.

10. Zwietering, Th.N.: "Untitled contribution to the discussion (in Dutch)", De Ingenieur, 75, 1963, ch.60-ch.61.

11. Mikulcova, E., Kudrna, V. and Vlcek, J.: "Conditions for Impeller Flooding (in Czech)", Scientific Papers of the Institute of Chemical Technology, Prague, K1, 1967, 167-183.

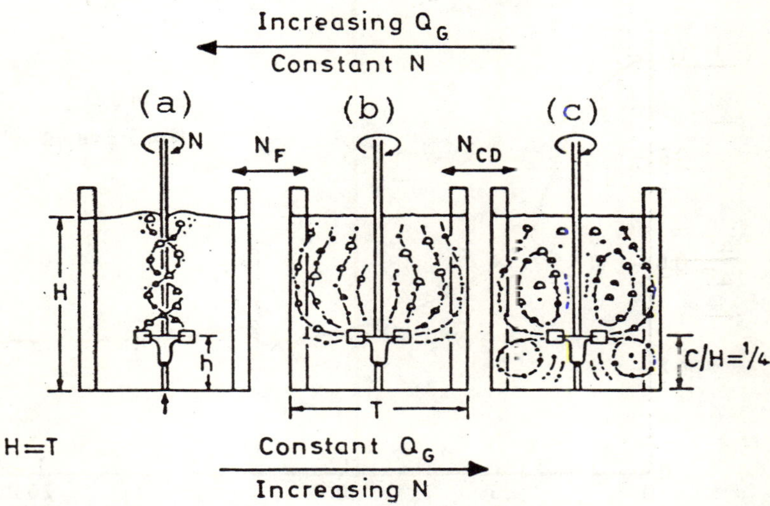

Fig. 1 Bulk flow patterns

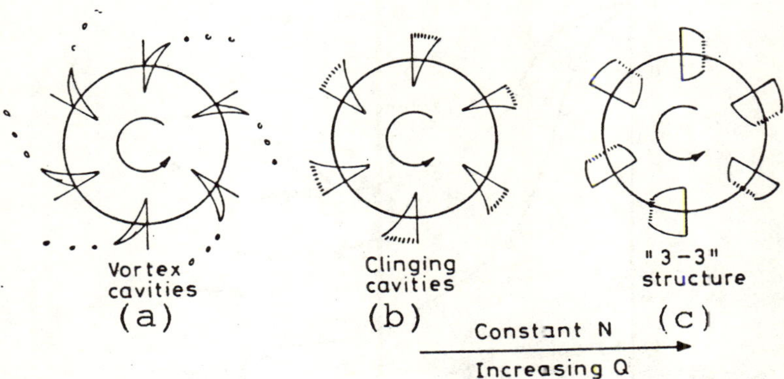

Fig. 2 Cavity changes in the loading
regime i.e. N > N_F

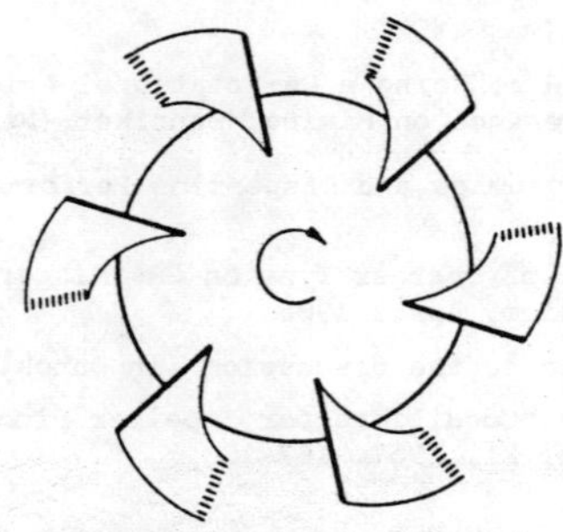

Fig. 3   Typical ragged cavities found
under flooding conditions
(T=0.61m, D/T=1/3:
Q_{GV}=0.50 vvm; N < N_F)

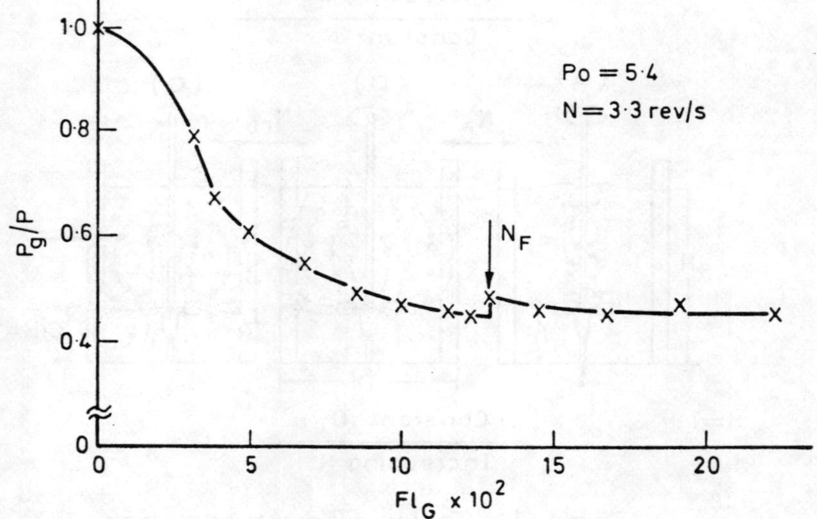

Fig. 4   Confirmation of the step
increase at N_F (D/T=0.33,
T=0.61, C/H=0.25)

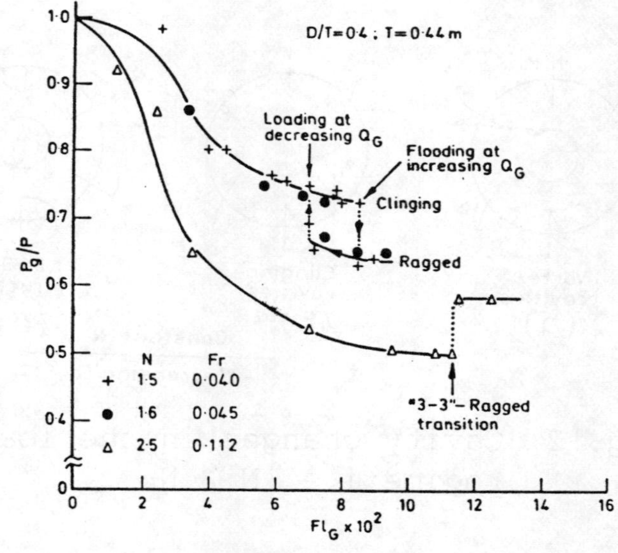

Fig. 5   The effect on P_g/P of a
clinging-ragged cavity transition

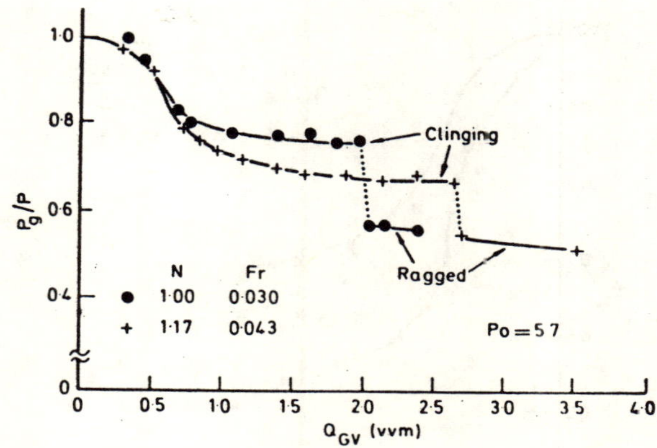

Fig. 6   The change in P_g/P at the flooding-loading transition for D/T=0.5 agitator in 0.61m vessel

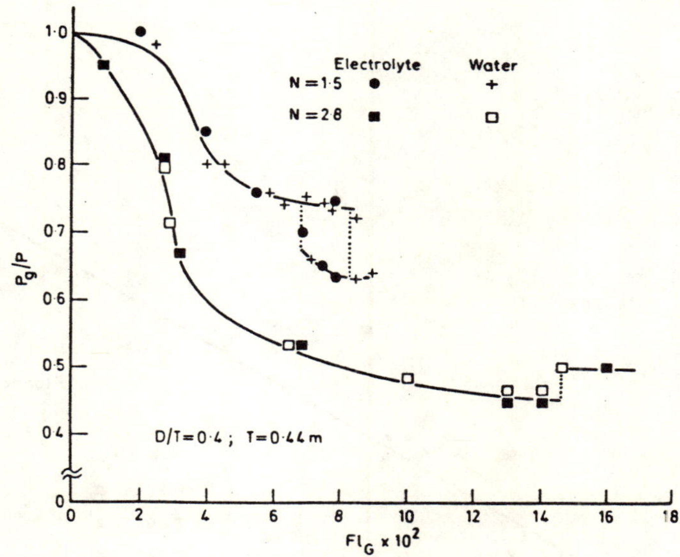

Fig. 7   Results for electrolyte solution and water

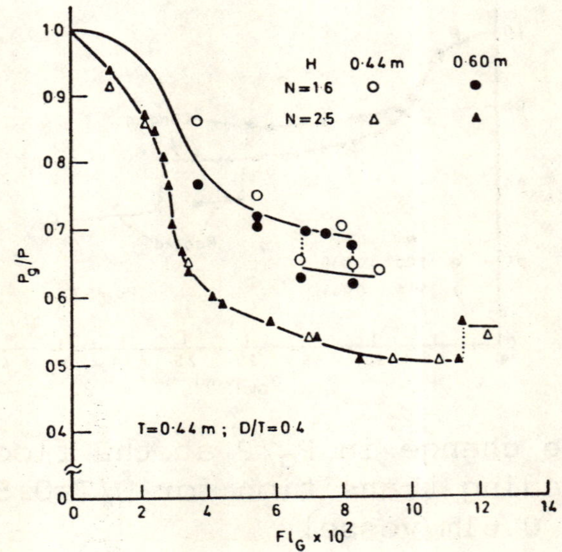

Fig. 8 Effect of height on
flooding-loading
transition

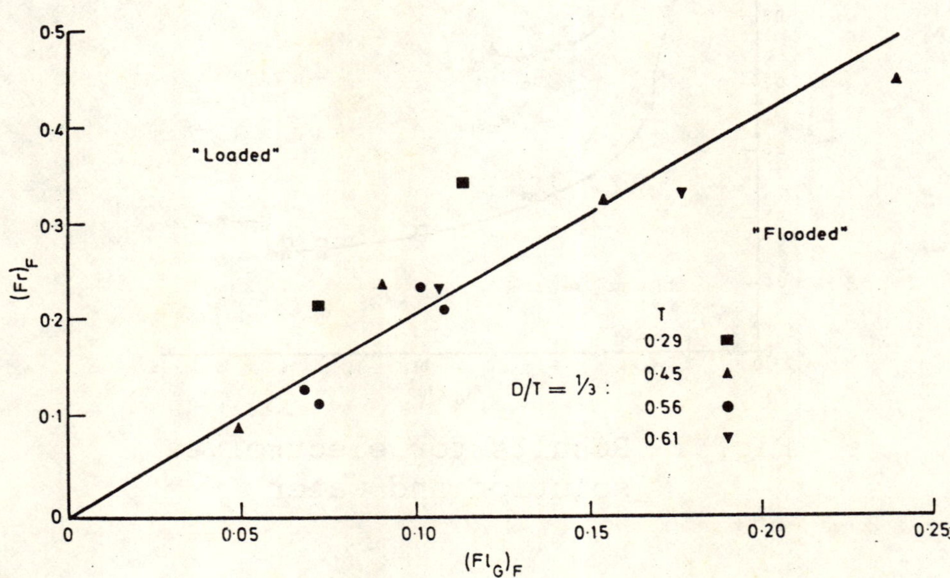

Fig. 9 Flooding-loading relationship
for D/T=1/3

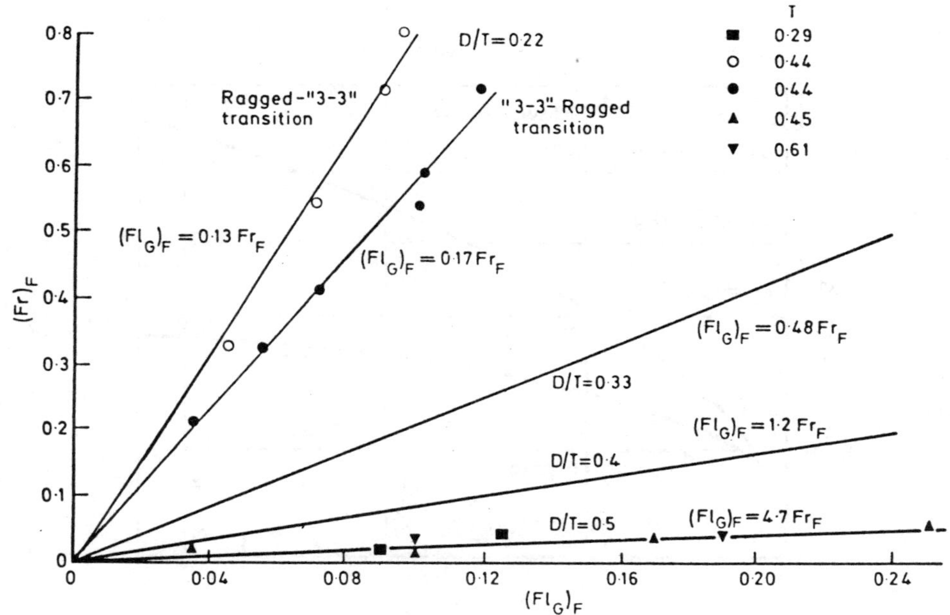

Fig. 10 Flooding-loading relationships for different D/T ratios

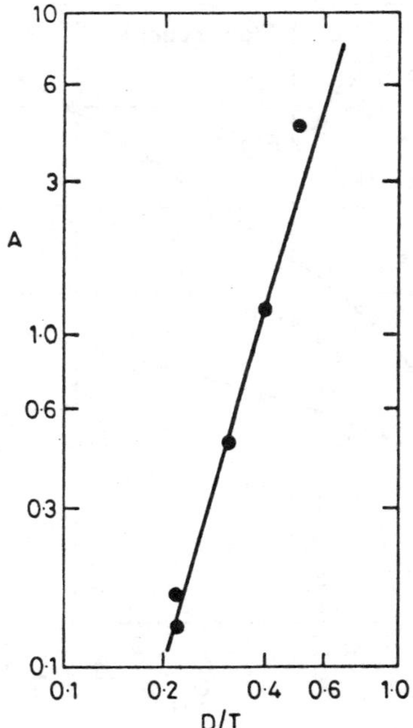

Fig. 11 Proportionality constant versus D/T ratio for flooding-loading

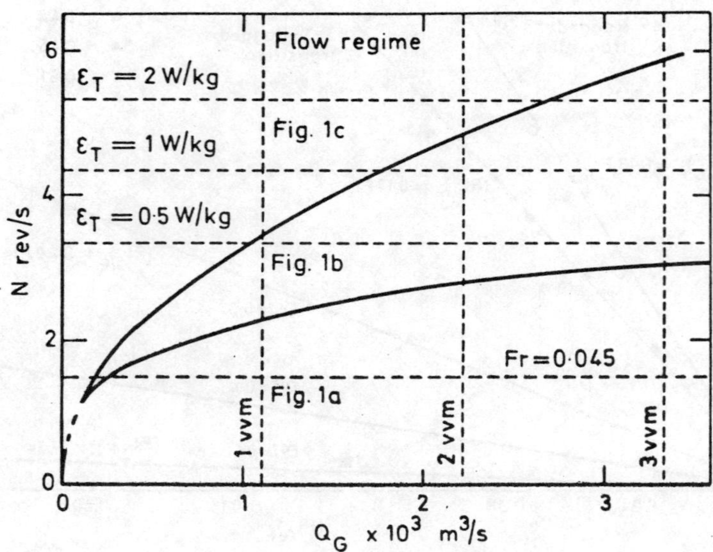

Fig. 12   Regime map for 0.44m vessel,
D/T=0.4

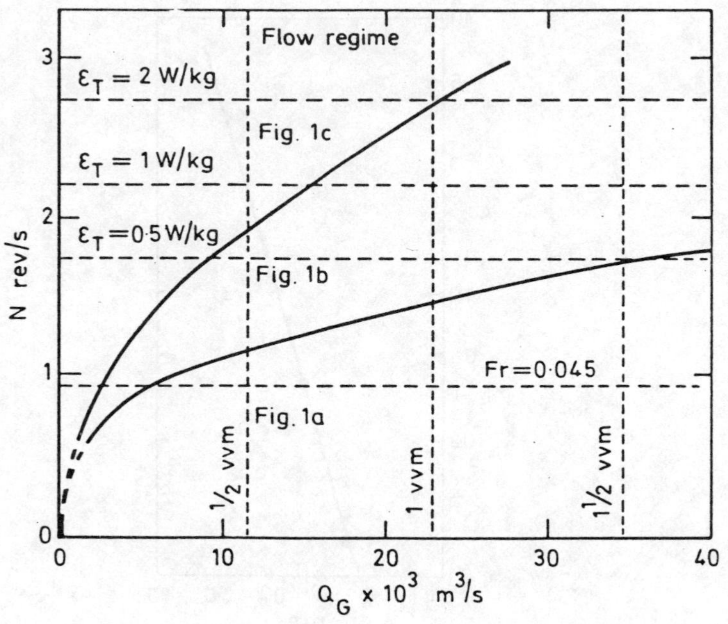

Fig. 13   Regime map for 1.2m vessel,
D/T=0.4

DUAL IMPELLER SYSTEMS FOR AERATION OF LIQUIDS: AN EXPERIMENTAL STUDY

V. Machoň, J. Vlček, J. Skřivánek

The Prague Institute of Chemistry and Technology
Czechoslovakia

Summary

The hydrodynamic characteristics of a gas-liquid two-phase system were studied. The experimental setup is a model of a type frequently used in industrial fermentors. It consists of a tall cylindrical vessel equipped with baffles. The stirrer itself is designed as two mixers on the same shaft. The lower mixer is a standard Rushton turbine, whereas the upper mixer is interchangeable so that either a Rushton turbine or an axial mixer having inclined blades and so oriented as to provide upward or downward flow can be used. Power imput was measured for different mixer types, changing stirrer frequency and gas flow. The results are compared with those for single stirrer setups under similar conditions. Also the loading and flooding conditions were established for this system and compared with data obtained previously for single stirrer systems. The advantages and drawbacks of dual impeller systems compared with standard designs are discussed.

Held at Wurzburg, 10-12 June, 1985.

Organised by DVCV· Deutsche Vereinigung für Chemie- und Verfahrenstechnik
(German Association of Chemical and Process Engineering).

Organisation: GVC·VDI-Gesellschaft Verfahrenstechnik und Chemieingenieurwesen.  **GVC**

©BHRA, The Fluid Engineering Centre, Cranfield, Bedford MK43 0AJ, England.

Nomenclature

A	constant in Eq.5	-
b	width of the baffle	m
D	impeller diameter	m
g	gravitational acceleration	ms^{-1}
H_o	height of the liquid	m
H_2	distance of turbine from bottom of the vessel	m
K	constant in Eq.1	-
K´	constant in Eq.5	-
$k_L a$	volumetric mass transfer coefficient	s^{-1}
l	distance between the stirrers	m
N	frequency of impeller rotational speed	s^{-1}
N_p	power input number $[P/(\rho N^3 D^5)]$	-
N_{Re}	Reynolds number $(ND^2 \rho / \eta)$	-
N_Q	gas flow number $[Q/(ND^3)]$	-
N_F	Froude number $(N^2 D/g)$	-
P	power input	W
Po	power input (without aeration)	W
Pg	power input (with aeration)	W
Q	volumetric gas flow rate	$m^3 s^{-1}$
q	constant in Eq.1	-
r	correlation coefficient	-
T	tank diameter	m
η	liquid viscosity	Pas
ρ	liquid density	kgm^{-3}

1. Introduction

In a number of areas and in particular in industrial fermentation systems, aeration of gases takes place in fermentors having several impellers on the same shaft. While such devices are fairly common in practice, their optimization sofar has to rely on only a very modest theoretical background. Information on power input, gas hold-up, the problem of flooding, and the mass transfer problems for aerated systems involving a single impeller is plentiful enough to allow for generalization, whereas no sufficient description of all these aspects is available sofar for the systems with multiple impellers where the hydrodynamic conditions are much more complicated. When attempting to optimize the apparatus setup and scale-up, the impeller power input in the aerated charge is the primary information required; however, it would not do here to simply add together the inputs of the different impellers carried on the same shaft. Further, it is very important to define the conditions under which the phenomenon known as flooding is provoked into existence. These are the major problems dealt with in this paper.

2. Experimental

The power input was measured in a cylindrical tank made of an organic glass tube having an interior diameter T = 0.29 m and a height of 0.7 m (Fig.1). The tank was fitted with four radial baffles 0.7 m high and 0.03 m wide. The tank bottom was fixed to a cylindrical vessel of 0.41 m in diameter. This vessel, supported in two oil bearings, was free to rotate about its vertical axis. The shaft with stirrers was located in the tank axis and was coupled at its top end to the driving system composed of a motor, gear, and stirrer frequency controller. This system, including the shaft with impellers, was vertically adjustable. Pairs of sparging rings located beneath each stirrer served to feed gas into the vessel. Each of the sparging rings, of 0.067 m in diameter, had six orifices of 0.002 m in diameter uniformly spaced atop the ring. The torque was measured by balancing the rotational movement of the vessel, as described in detail by Prochazka and Landau (Ref.1). Power input was measured in an air-water dispersion. Two impeller types were used: a six-blade Rushton disc turbine (type A) and a turbine mixer with six inclined blades pumping either downward (B) or upward (B′). The diameter of all impellers was 0.1 m, i.e., one third of the tank diameter. The height of liquid was equal to the

tank diameter in experiments with one impeller, and was twice this
diameter in experiments where two impellers were used. The lower impeller was always located at a distance of $H_2 = T/3 = 0.1$ m above the
bottom, and the distance between impellers in experiments involving
two impellers was equal to the tank diameter $1 = T = 0.29$ m. A summary of the experimental configurations used is given in Fig.2. The
stirrer frequencies were in the 250-500 rpm range and the gas flow rates were varied from 2 to 60 liter per minute. The experiments were
performed at room temperature and the physical properties of water were taken at 19°C.

3. Measurements and results

3.1. Impeller power input without aeration – comparison of
single-impeller and dual-impeller configurations

It is a question of importance to practical designing whether or not
the power inputs to individual single impellers can be meaningful for
the calculation of the total power input to a multi-impeller configuration. The total power input is known to depend on the impeller spacing. With impellers spaced at a distance approaching the stirrer diameter, the total power input is less than the sum of the power inputs
to the individual stirrers. For the arrangement described above, i.e.,
with the impeller spacing equal to the vessel diameter, the total input and the inputs to the individual single impellers were compared in
terms of the power number N_p. The results are shown in Table 1 indicating the power numbers obtained at three stirrer frequencies; it is
clear from the N_{Re} values that the experiments were run in the fully
turbulent region and that the reduction in N_p due to an increase in
stirrer frequency was slight. From a comparison of the power numbers
obtained with the dual impeller and those calculated as the sum of the
power numbers for the two single impellers used it follows that, at
this impeller spacing, the resultant power number (power input) equals
the sum of the power numbers (power inputs) of the two impellers, even
if, as it were, the impellers should be of different types. The absence of any interaction between the impellers is also documented by the
fact that the impeller with inclined blades yields the same values regardless of whether it pumps upwards or downwards.

3.2. Effect of aeration on power input and the flooding conditions

The power input will drop when the charge is aerated. Quite a number
of papers (Ref.2,3) deal with how much it will drop and how this drop
can be estimated, for single-impeller systems. As a rule, the data are
interpreted using the Pg/Po ratio as function of the gas flow number
N_Q. Data is rather scarce (Ref.4) however for systems with multiple
impellers, in spite of their fairly wide application in practice. The
data processing procedures derived for single-impeller systems shall
now be adopted to interpreting our experimental data. We shall use the
term loading (Ref.2,5) to denote the desirable situation of disperga-
ting a gas by a stirrer, which occurs at a suitable ratio of gas flow
rate to stirrer frequency. In turn, the term flooding (Ref.2,4,5)
shall be applied to conditions poorly suited for operation, i.e., a
situation where the gas fed into the tank is poorly dispergated in the
bulk of liquid. Under such conditions the impeller is incapable of di-
spergating the gas properly, which means that the stirrer frequency is
too low to handle the quantity of gas being brough in, or that the gas
flow rate is too high for the given stirrer frequency to cope with.
The transition from the loading to the flooding condition is thought
to occur in some relation to the tearing away of large cavities which
form behind the impeller blades. Various authors (Ref.2,5) have forwarded somewhat differing notions of the loading-to-flooding transition depending on the behaviour of the dispersion in the agitated vessel. The question is, namely, whether the transition occurs when the
bubbles cease reaching the vessel wall due to a radial effect or, alternatively, when they cease reaching the vessel bottom. Having bypas-
sed the examination of which of the two was the case with our configurations, we have based our estimates of what are the conditions under
which flooding occurs on the notion (Ref.2) that it ought to be at
such conditions which are reflected by a minimum on the Pg/Po versus
N_Q curves at a constant Q. The reason is that given a constant gas
flow rate the rising branch of these curves to the right of the minimum represents in fact an increase in power input hand in hand with
a reduction in stirrer frequency. This suggests a significant change
in the flow conditions of both the liquid and the gas in the stirrer-
-adjacent region, accounted for by the formation of various kinds of
cavities behind the blades and/or their tearing away in the flooded
condition (Ref.6). Figure 3 is a plot of our data for Setup 1, i.e.,
with a single turbine impeller. The probable curves to be obtained at
constant gas flow rates are shown, with well-defined minima. The

points corresponding to the region of flooding are marked, and as unsuitable are also indicated the points for $N_Q < 1.5 \cdot 10^{-2}$, i.e., the
region of high stirrer frequencies and low gas flow rates which is regarded by Kudrna (Ref.7) as being out of bounds of technical interest,
owing to imperfect aeration of the charge. It corresponds with the region of the six vortex cavities (Ref.8).

Data for Setup 2 employing two turbines is plotted in Fig.4. Here the
gas flow rate was the same (max.60 liters per minute) as with the sin-
gle-turbine setup. The diagram gives a similar impression as that for
the single-impeller arrangement, including the minima on the curves
plotted for constant gas flow rate. In this case too the points to the
right of the minima as well as the points for $N_Q < 1.5 \cdot 10^{-2}$ are marked
and will be struck out from further data processing. The experimental
data for Setups 3 and 4, i.e., systems employing a dual impeller with
a turbine as the lower impeller and a stirrer with inclined blades giving either a downward or an upward pumping effect as the upper impeller, if plotted as functions of Pg/Po on N_Q, give the same characteristics once again, with minima on the curves for constant gas flow rate. Again, the points corresponding to the flooding region and to
$N_Q < 1.5 \cdot 10^{-2}$ were excluded from further data treatment.

3.3 Relative power input as function of the gas flow number

(Pg/Po versus N_Q)

To process the power input data, an equation proposed by Kudrna
(Ref.7) will be used, which can be fitted rather easily onto the experimental data. The equation is of the form

$$1/\left[1-Pg/Po\right] = K + q(Pg/Po)^{0.5}/N_Q \tag{1}$$

where K and q are constants for given geometry and stirrer type. The
constants K and q of the Eq.1 were evaluated by linear regression for
each of the setups used. The results are listed in Table 2 and the
correlation coefficients are also shown. Plots corresponding to the
equations obtained for individual configurations are reproduced in
Fig.5. Both the Table 2 and the Fig.5 indicate that the Eq.1 can be
adopted in evaluating either the single-impeller or the multi-impeller
systems, as long as only the conditions of good dispersion of gas
(i.e., in the loading region) are considered. The goodness of fit of
the calculated curves is rather high, as is documented by the values
of the correlation coefficients (even though these were computed for

different variates, viz., $1/[1-Pg/P_0]$ and $(Pg/P_0)^{0.5}/N_Q$, (cf.Eq.1).
Correlations higher than 0.9 suggest that the equation proposed is
well suited for this kind of modelling. Also the values of the constants are rather similar for all the configurations which means that
there is little difference between the plots. In fact, this is surprising, taking into account the principial differences between the configurations. Obviously the absolute power input to charge is much lower for the dual-impeller system employing an upper impeller with inclined blades (cf.Table 1); while this is desirable from the viewpoint
of power savings, the decision on whether to use impeller with inclined blades can only be taken after aeration efficiency of this configuration will have been established by measuring the volumetric mass
transfer coefficient $k_L a$.

3.4. Prediction of flooding conditions

The flooding conditions were examined by a number of authors
(Ref.2,5,6). The treatment mostly results in some description of the
relationship between the Froud number N_F and the gas flow rate number
N_Q where actual values are substituted for the gas flow rate and the
stirrer frequency under the transient conditions between loading and
flooding. Formally these equations as proposed by different authors
are identical but their constants differ. For instance, if such
equations are expressed in the same form we arrive at different expressions dependingon whether the relation (for the same geometry) was
proposed by Mikulcová (Ref.9)

$$N_Q N_F^{-1} = \text{const.} \tag{2}$$

by Zlokarnik (Ref.10)

$$N_Q N_F^{-0.75} = \text{const.} \tag{3}$$

or by Nienow (Ref.2)

$$N_Q^{-0.5} N_F^{0.25} = \text{const.} \tag{4}$$

On substitution, the expressions for stirrer frequency N in relation
to gas flow rate Q at a constant stirrer diameter reduce to the following:

Mikulcová $\qquad N^{3.0} Q^{-1} = \text{const.} \tag{2a}$

Zlokarnik $\qquad N^{2.5} Q^{-1} = \text{const.} \tag{3a}$

Nienow $\qquad N^{2.0}Q^{-1} = \text{const.}$ $\hfill (4a)$

Our set of loading/flooding transient conditions as ascertained by experiments, i.e., paired values of flow rates and stirrer frequencies
coinciding with the minima on the Pg/Po vs. N_Q curves at constant flow
rates, were correlated using this type of equation. The constants A
and K' of the equation

$$N^A Q^{-1} = K' \qquad\qquad (5)$$

were calculated by linear regression using the logarithmic form of the
equation, viz.,

$$A \ln N = \ln K' + \ln Q \qquad\qquad (5a)$$

The constants thus obtained for the Setups 1,2, and 4 are listed in
Table 3 together with the respective correlation coefficients. It is
clear from the table that Setup 1 with a single turbine where the exponent A = 3.01 gives a very good agreement with the value by Mikulcová derived from theory, and that the experimental data characterizing
the flooding condition are scattered within a very narrow band about
the calculated values, as is documented by the correlation coefficient
attained. For the dual-impeller configuration with two turbines, Setup 2, the exponent has risen to 3.43 indicating that a lesser incre-
ment of stirrer frequency is required to dispergate gas entering at
higher flow rates than is the case with the single-impeller configuration. At given gas flow rate, the stirrer frequency values are lower
by ca. 20% for this configuration than for the single turbine system.
An even higher value of A is obtained for the setup with one turbine
and one impeller with inclined blades giving an upward pumping effect,
Setup 4; however, the absolute stirrer frequency values required to
dispergate a given quantity of air were higher than for the two-turbine configuration. As suggested by our data, this appears to be a useful and economical configuration, inasmuch as the dispergation of gas
is good and the absolute power drawn is less. It would be desirable,
of course, to have yet another proof of the usefulness of this configuration; this could be provided by measuring the gas held-up and the
mass transfer coefficient.

It is worth noting that a considerable inaccuracy creeps in when taking the readings of the minima of the curves, cf. Figs.1 and 2. It
can nevertheless be concluded that the exponent A of the Eq.5 would
probably exceed the value of 3 for the multi-impeller systems.

The aforementioned procedure used in treating the data proved unsuitable in the case of those dual-impeller systems where the upper stirrer exerted a downward pumping effect. The plot of experimental data shown in Fig.6 indicates that if the points corresponding to constant gas flow rates are interconnected, two extrema occur, of which one is a maximum and the other is a less pronounced minimum. The left-hand part of these plots (i.e., the points having the lowest Pg/Po ratios) can probably be regarded as corresponding to so-called recirculation conditions or even to gas suction by the free liquid surface. The desirable aeration mode described as loading sets in a region denoted by the arrow 1 in Fig.6, whereas at the point denoted by the numeral 2 the relative power input starts growing higher as N_Q increases. This should give rise to the flooding condition but the readings taken at this point indicated that the stirrer frequency was nearly independent of the gas flow rate; hence, the relationship as described by the Eq.5 cannot be applied. Here the conclusion is that the setup involving an upper stirrer with inclined blades giving a downward pumping effect is not very suitable, owing to a relatively narrow region of prevalence of the loading condition, i.e., of adequate aeration. This finding coincides with the results of the power input measurements undertaken in the case of aeration by a single stirrer with inclined blades giving a downward pumping effect (Ref.10).

4. Conclusions

1. It has been found that, with an experimental configuration such that the spacing of the two impellers used equals the vessel diameter, the power input (the power number) is given by the sum of the power inputs (power numbers) of the two impellers. This applies even to the case where a turbine impeller is used in combination with an impeller having inclined blades.

2. The power drawn by dual-impeller systems was reduced in case of charge aeration in a manner similar to that encountered with single-impeller systems. On eliminating those data points which corresponded to the flooding condition, the remaining data for both single-impeller and dual-impeller systems lent themselves easily to correlation by the Eq.1.

3. The flooding condition for the turbine impeller may be expressed by the relation N^3Q = const. The exponent of N is higher than 3 for two-impeller systems.

4. Gas recirculation took place when an impeller with inclined blades
   and a downward pumping effect was used as the upper stirrer. On
   the basis of our data, this configuration must be regarded as undesirable in applications involving gas dispergation in agitated liquids.

5. References

1. Procházka J., Landau J.: "Accurate measurement of rotational stirrer power input". Collection of Czechoslovak Chem.
   Commun., 28, 1103 (1963).

2. Nienow A.W., Wisdom D.J., Middleton J.C.: "The effect of scale and
   geometry on flooding, recirculation and power in gassed stirrer vessels". Proceedings of Second European
   Conference on Mixing, Cambridge, England, p.F1,
   April 1977.

3. Warmoeskerken M.M.C.G., Smith J.M.: "Rushton Turbines in Gas-Liquid
   Systems: The Real Variables". Presented at the 8th
   Conference on Mixing, Henniker, New Hampshire, USA,
   9.-14. August 1981.

4. Nienow A.W., Lilly M.D.: "Power Drawn by Multiple Impellers in
   Sparged Agitated Vessels". Biotechnol. and Bioengng,
   XXI, 2341 (1979).

5. Warmoeskerken M.M.C.G., Smith J.M.: "The Flooding Transition with
   gassed Rushton Turbines". The Institution of Chemical
   Engineers, Symposium Series No.89, Fluid Mixing II.
   p.59, 1984.

6. Warmoeskerken M.M.C.G., Smith J.M.: "Description of the Power Curves of Turbine Stirred Gas-Liquid Dispersions". Proceedings of Fourth European Conference on Mixing,
   Leeuwenhorst, The Netherlands, p.237 April 1982.

7. Kudrna V.: "Power input of rotational stirrer in aerated liquids".
   PhD Thesis, Prague Inst. of Chem.Technol., 1965.

8. Riet, k. van't: PhD Thesis, Techn.Univ. of Delft, The Netherlands,
   1975.

9. Mikulcová E., Kudrna V., Vlček J.: "Estimation of Flooding Conditions of Aerated Turbine". Scientific Papers of Pra-
   gues Inst. of Chem.Technol., k1, p.167, 1967.

10. Machoň. V., Vlček J.: "Aeration of Liquids in a Vessel Equipped
with Multistage Impellers". Collection of Czechoslovak Chem. Commun., in press.

Table 1   Values of Power numbers for all configurations

stirrer frequency, min^{-1}

	300		400		500	
Configuration	dual stirrer	sum of single stirrer	dual stirrer	sum of single stirrer	dual stirrer	sum of single stirrer
1	4.81		4.78		4.63	
2	9.50	9.62	9.42	9.56	9.37	9.26
stirrer B	1.44		1.27		1.26	
3	6.16	6.25	6.07	6.05	5.90	5.89
stirrer B'	1.44		1.27		1.31	
4	6.39	6.25	6.08	6.05	5.97	5.94
N_{Re}	48 302		64 408		80 503	

Table 2   Values of constats K and q, Eq.1

Configuration	K	q	r
1	1.43	0.029	0.979
2	1.44	0.066	0.988
3	1.62	0.055	0.989
4	1.46	0.042	0.978

Table 3   Values of constants A, K', Eq.5

Configuration	A	$\dfrac{K'}{10^{-2}}$	r
1	3.01	1.34	0.948
2	3.43	1.47	0.962
4	4.05	0.24	0.992

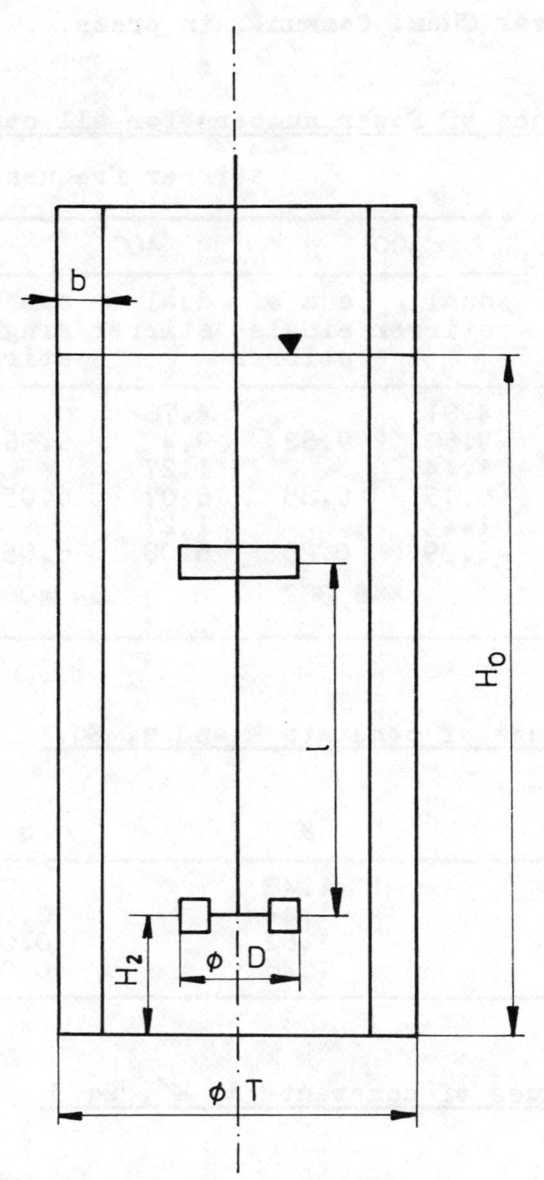

Fig.1   Sketch of stirred vessel

b = 0.03, T = 0.296, D = 0.1, H_0 = 0.58 (=0.29 when one turbine was used)
H_2 = 0.1, l = 0.29
all dimensions are in meters.

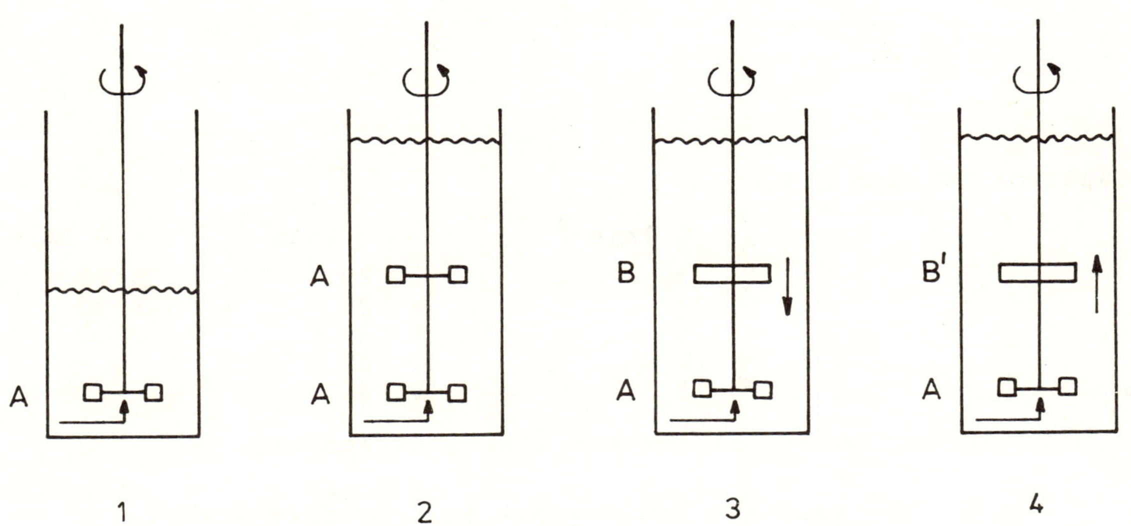

Fig.2   Sketch of configurations
         1 – one turbine, 2 – two turbines, 3 – upper stirrer
         with inclined blades and downward pumping effect
         4 – upper stirrer with inclined blades and upward pumping
         effect.

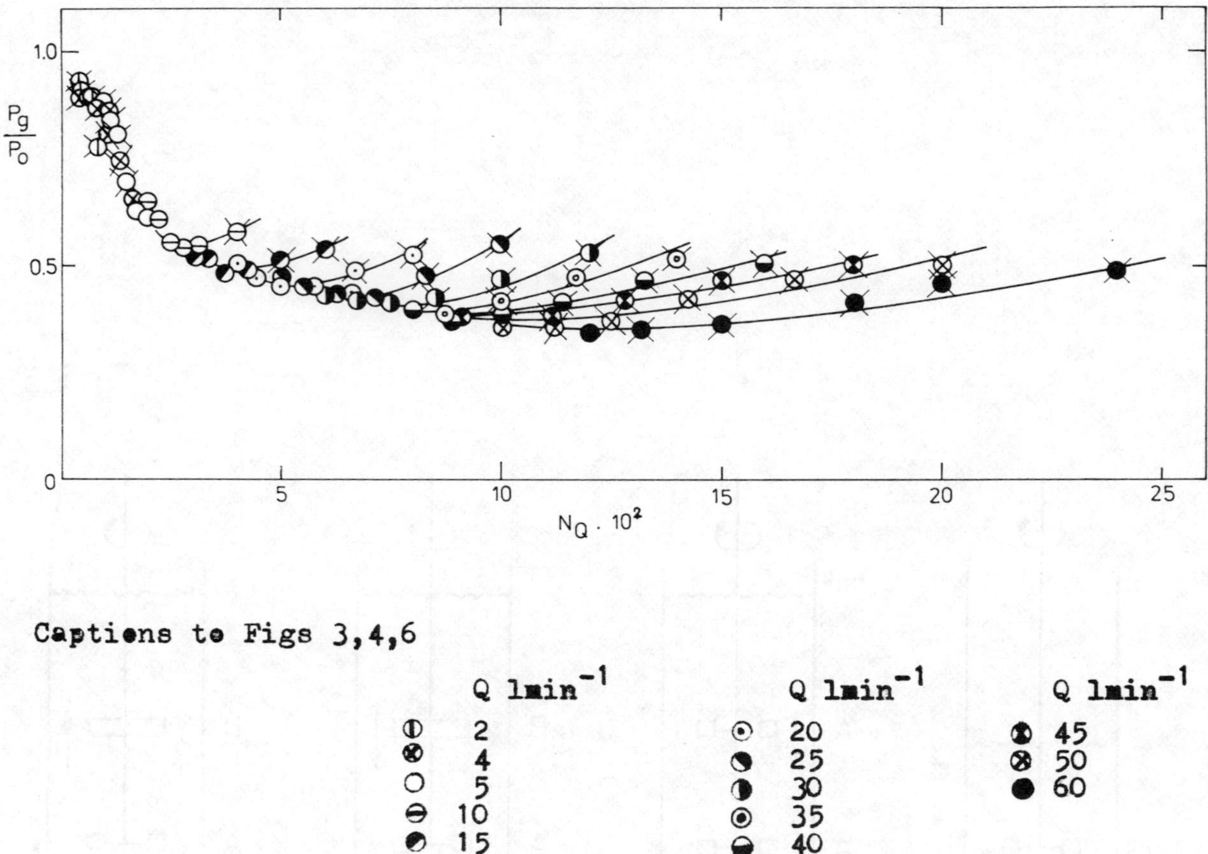

Captions to Figs 3,4,6

	Q lmin^{-1}		Q lmin^{-1}		Q lmin^{-1}
	2		20		45
	4		25		50
	5		30		60
	10		35		
	15		40		

Marked points in further treatment ommited.

Fig.3   The relative power input vs. gas flow number.
         Configuration 1. All data.

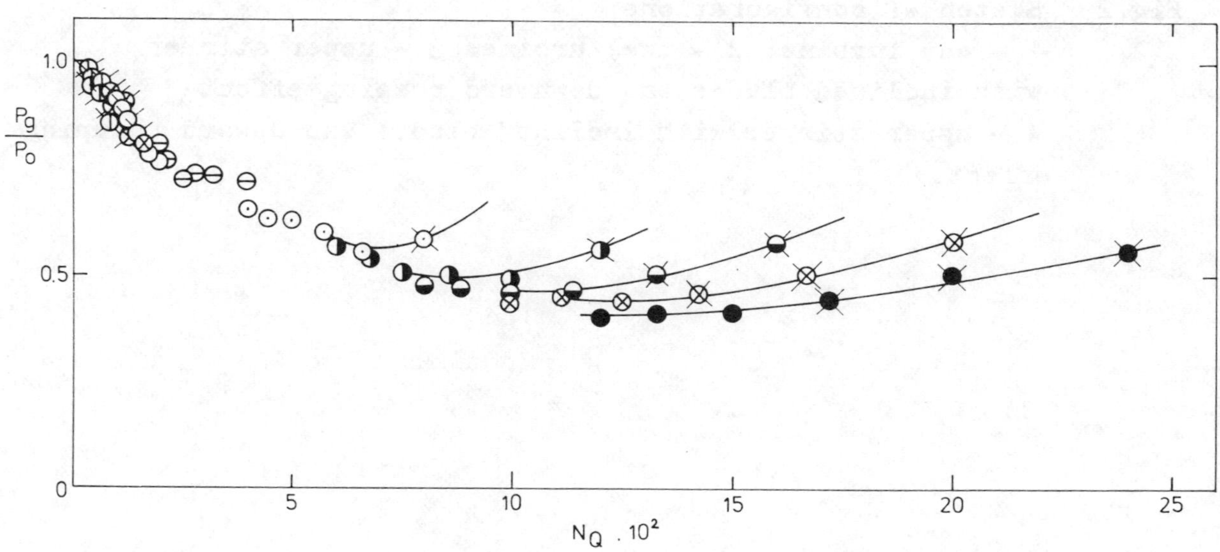

Fig.4   The relative power input vs. gas flow number.
         Configuration 2. All data.

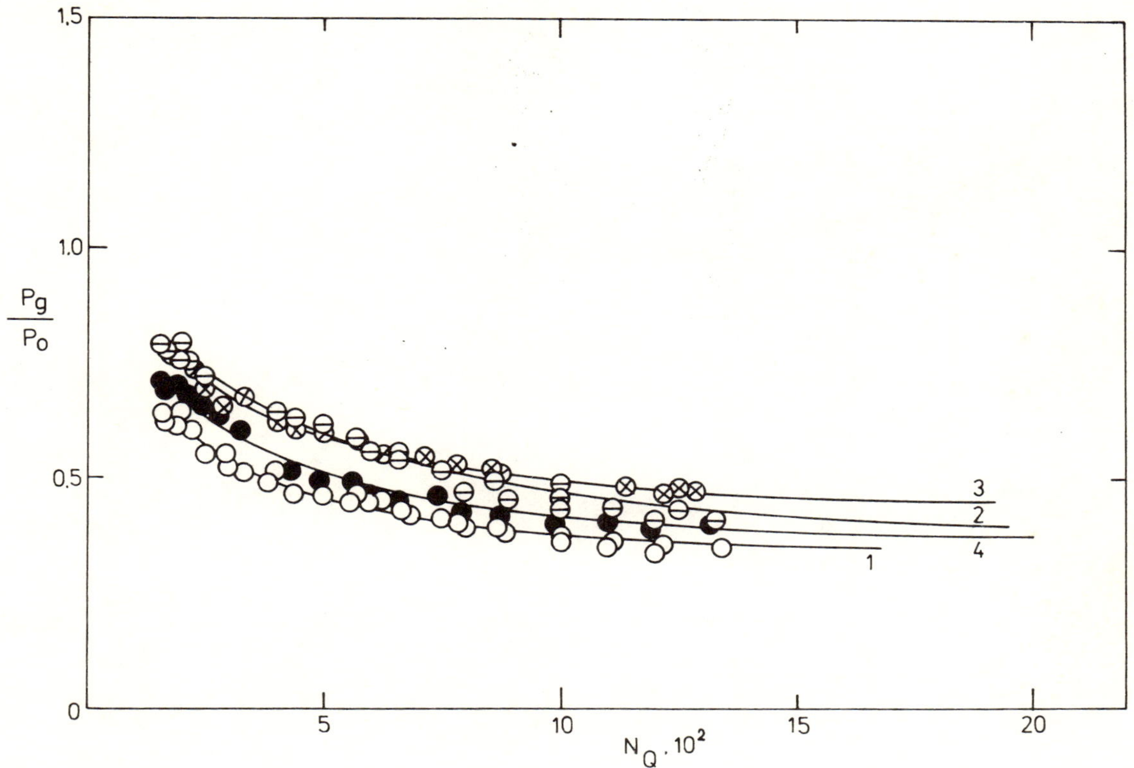

Fig.5   Comparison of our data (loading regime) with
        Eq.1 using constants in Table 2.
        ◯ – configuration 1,   ◓ – configuration 2 ,
        ⊗ – configuration 3,   ● – configuration 4 .

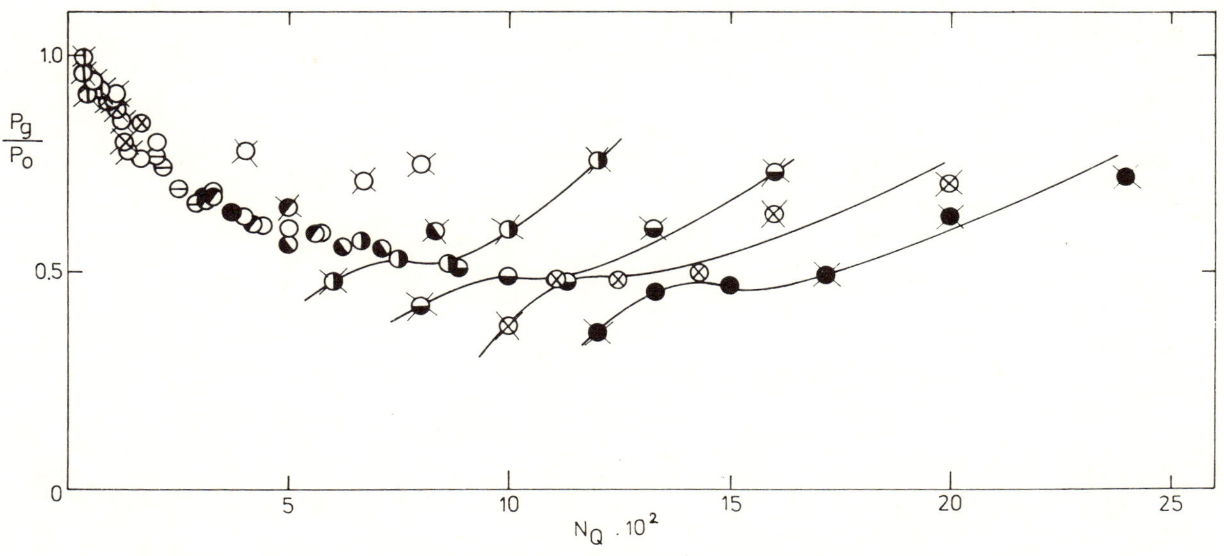

Fig.6   The relative power input vs. gas flow number.
        Configuration 4. All data.

MIXING AND GAS/LIQUID MASSTRANSFER
IN THE HORIZONTAL STIRRED TANK

A. Bauer and A. Moser

Institute of Biotechnology, Microbiology
and Waste Technology, T.U. Graz,
A-8010 GRAZ/AUSTRIA

Summary

A new design of G/L-reactor, the socalled horizontal stirred tank reactor as part of the multipurpose-bioreactor is characterized here. The main aim of this new reactor is to serve as a model reactor in bench scale for purposes of process kinetic analyses especially in the area of bioprocessing. The reactor consists of three different modes of construction and operation. The first is a conventional vertical stirred tank. The reactor vessel can be turned from vertical to horizontal position and can be equiped with rotating drums of constant area of liquid exposed to the G-phase with a certain contact time depending on rotational speed. The third mode of construction and operation is a cylinder with perforations and baffles operated in horizontal position.

Recent results on macro- and micromixing are shown together with data on G/L-masstransfer in case of O_2.

Mixing times in dependence of degree of mixing are referred, which were measured with a conventional conductivity method. Results are demonstrated in the form of mixing time vs. power consumption.

Residence time distribution measurements are also presented, leading to a new evaluation method for Bodensteinnumber resp. number of equivalent stages based on the value of the ratio between time of maximum value and mean residence time. This method seems to be advantageous in use especially in the range of transition of flow behaviour of tubular to stirred tank reactors.

Finally data on O_2-transfer of the three constructions are presented which were measured according to a simplified momentum method. The volumetric mass transfer coefficient is shown as a function of power consumption and gas velocity. Comparison to other horizontal contactors is made to some extent.

The main conclusion from this work is that the horizontal tank with special mixing and aeration devices offers extra ordinary advantages in realizing higher transfer rates at lower values of power consumption, thus presenting an effective alternative to the conventional vertical stirred tank.

The work presented in this paper was supported by an austrian research grant (FFWF, project no. 4496)

Held at Wurzburg, 10-12 June, 1985.

Organised by DVCV· Deutsche Vereinigung für Chemie- und Verfahrenstechnik (German Association of Chemical and Process Engineering).

Organisation: GVC·VDI-Gesellschaft Verfahrenstechnik und Chemieingenieurwesen. 

©BHRA, The Fluid Engineering Centre, Cranfield, Bedford MK43 0AJ, England.

NOMENCLATURE

a, b, c, d, e, f		coefficients in empirical correlations
a	m^{-1}	specific area
Bo	–	Bodenstein-number
c_L	$kg.m^{-3}$	liquid concentration
c_L^*	$kg.m^{-3}$	saturation concentration in liquid phase
f (t)	–	pulse function of RTD
k_{L1}	$m.s^{-1}$	G/L-masstransfer coefficient
M	J (Nm)	torque (restraining)
m	–	degree of mixing
n	s^{-1}	rotational speed (rps)
N_{eq}	–	ideally stirred tank
P	$W (Nm.s^{-1})$	power
t_{MAX}	s	time of maximum value of RTD-function
t	s	time
t_e	s	characteristic time of electrode response to reach 63,2 % of end-value
t_E	s	electrode response time
t_G	s	mean gas residence time
$t_{M,90}$	s	mixing time at m = 90 %
$\bar{t}_R$	s	mean residence time
$V. (V_L)$	m^3	volume
$\dot{V}_G$	$m^3.h^{-1}$	gas flow rate
$v_{S,G}$	$m.s^{-1}$	superficial gas velocity

ABBREVIATIONS

HSTR	horizontal stirred tank reactor
OTR	oxygen transfer rate $[kg.m^{-3}.s^{-1}]$
RTD	residence time distribution
STR	stirred tank reactor (vertical)
ThLR	thin-layer reactor
G	gas
L	liquid
E	electrode
rps, rpm	rotation per second, per minute
Δ_1, Δ_2 (%)	mean relative error in eq. 5 resp. eq. 6

1. INTRODUCTION

Usually several types of reactors are applied in gas/liquid-processing all of them taking the shape of vertical vessels, e.g. stirred tanks, bubble columns. However, this design has its drawbacks which become only significant on a very large scale: a significant part of power consumed is required for compression of the feed air or gas to overcome the high static liquid pressure. Therefore, horizontal vessels would avoid this disadvantage. Furthermore, horizontal reactors could be operated with highest gas hold up, which case is hard to be realized in vertical position.

Horizontal shaped reactor constructions are described in literature e.g. the tubular loop reactor, the horizontal rotary fermenter, the horizontally rotating cylinder with baffles, the paddle wheel reactor and tubular reactors with gas spargers. They were recently summarized (ref. 1, 2). Generally these constructions suffer from the fact that only relatively low mass transfer rates can be achieved.
This paper describes a horizontal contactor which is able to achieve highest values of G/L-mass transfer at relatively low power consumption. Supplementary the new reactor is characterized by showing results on mixing time and residence time distribution. Finally some empirical correlations for mixing time, oxygen mass transfer and power consumption are presented.

2. MATERIAL / METHODS

2.1 Reactor

A horizontal vessel of conventional stirred tank shape is equiped with a rotating special mixing device as shown in Fig. 1. The rotating cylinder is constructed in such a form, that the surface is full of perforations and baffles to the inner and outer side. This horizontal stirred tank reactor (HSTR) is part of the multi-purpose-reactor system recently designed (ref. 3) and constructed by MBR/Zürich. The reactor vessel can be turned from the vertical position, operating as a conventional STR, to the horizontal position. Here, different drums as mixing resp. aeration devices can be installed as shown in Fig. 1 or operating as thinlayer-film fermenter (ThLR) when using a simple drum with smooth surface (ref. 4). The dimension of the vessel with a working volume of V_L = 42 l corresponds to Dechema-standards. The diameter of rotating drum was 0,166 m.

2.2 Measuring methods

Mixing times (t_M) resp. RTD ($f(t)$) were measured by the traditional conductivity method at different values of degree of mixing m. A pulse of a concentrated solution of potassium nitrate (200 g/l) was injected at the liquid surface. Conductivity was measured by a simple, non-standardised electrode consisting of two steel bars cast in PVC. The signal from the electrode was displayed by a conductivity meter (WTW, LF 39T) and a plotter as well.
Mass transfer-coefficients of O_2 (k_La) were measured in pure water with different viscosity employing the gassing-in-gassing-out method. The oxygen was stripped by a stream of nitrogen, which was, heaving reached a sufficiently low oxygen concentration, immediately replaced by an air stream. This manner was employed to prevent gas hold-up to collapse which would have influenced the accuracy of the measurements negatively.
Concentration of dissolved oxygen was measured by a fast oxygen electrode of the polarographic type (WTW, EO 16B). Time dependency was recorded by a plotter again. Specific power input (P/V) was determined automatically by the system itself. The torque at the stirring shaft was evaluated from the total power uptake of the engine   minus bearing friction. Power input per unit volume was than evaluated from torque and stirrer speed.

2.3 Evaluating methods

Mixing time was defined as the time taken for the conductivity i.e. for the salt concentration to finally reach a value at a given degree of mixing m (e.g. 90 %). Curves from the plotter were sometimes very blurred and thus hard to evaluate. It can be expected that a modified conductivity method (using electrodes made of several platinated wire ends of smallest diameter) or alternative methods based on temperature or pill flow followers will overcome this problem (see ref. 5). At the same time evaluation methods based on the first order approach from Khang and Levenspiel (ref. 6) are to be preferred.

Concerning the evaluation of RTD-measurements, a new method was developed based on the
well-known equation using the equal-size tanks-in- series (N_{eq}) model (e.g. ref.6):

$$f(t) = \frac{N_{eq}^{N_{eq}} \cdot t^{N_{eq}-1}}{(N_{eq}-1)! \; \bar{t}_R^{N_{eq}}} \cdot e^{-N_{eq} \cdot t / \bar{t}_R} \tag{1}$$

The time of the maximum value of the function (t_{max}) is thought to be significant
in the range of $1 \leq N_{eq} \leq 5$. Thus the first derivation of eq. 1 was set to zero,
leading to the following relation between t_{max} and N_{eq} (with $t_{max} = 0$ for $N_{eq} = 1$
and $t_{max} = \bar{t}_R$ for $N_{eq} \gg 5$):

$$N_{eq} = \frac{1}{1 - t_{max}/\bar{t}_R} \tag{2}$$

Fig. 2 represents a graphical plot of eq. 2. It can be used for the determination
of N_{eq} from evaluated values of t_{max} especially in case of reactors exhibiting of
flow behaviour in the transition range $1 \leq N_{eq} \leq 5$.

$K_{L1}a$-coefficients of mass transfer (OTR) were evaluated from oxygen concentration
vs. time plots. The time which elapsed from the moment of changing the nitrogen
to oxygen to the moment when dissolved oxygen concentration reached 63.2 % of
saturation was defined as characteristic (t_e). Time constants of the electrodes
(t_E) and residence time of the oxygen in the gaseous phase (t_G) were assumed to be
resistances in series to oxygen transfer along with the transfer coefficient itself. $K_{L1}a$ was avaluated from the measured characteristic time and the time
constant according to a concept of Ruchti et al. (ref. 7) presented as

$$k_{L1}a = 1/(1/t_e + 1/t_E + 1/t_G) \tag{3}$$

Power consumption was calculated based on measured torque M and rotational speed n
(rps) using the following equation:

$$P = M \cdot n \cdot 2\pi \tag{4}$$

3. RESULTS

3.1 Mixing Behaviour (t_M and N_{eq}):

Fig. 3 gives a comparison of mixing times $t_{M,90}$ at 90 % degree of mixing in dependence on specific power consumption P/V as a function of gas flow rates $\dot{V}_G$ in case
of vertical and horizontal STR.
Mixing time in the conventional vertical STR with turbine impellers and Dechema-
standard dimensions showed virtually no dependency on gas flow rate ($0 \leq \dot{V}_G \leq 15$
$1.min^{-1}$), also stirrer speed is not significant in this case when $\dot{V}_G > 0$.
It must be mentioned that the measured mixing times refer to axial mixing rather
than to radial mixing because of the experimental arrangement. In case of the
HSTR with the perforated and baffled drum again the influence of gas flow on
mixing time is disregardable, which vary within only ten percent at a given power
input. On the other hand, mixing time decreased sharply with increased stirrer speed,
as can be seen in Fig. 3. Mixing took up to 4 minutes at a stirrer speed of 30 rpm,
with respect to a homogeneity of 95 percent. There was no such influence of the
gas flow as with the turbine impeller, since both gas flow and stirrer mixed
radially. This is the reason why axial mixing is very poor with this stirring
device. However, this drawback was thought to be neclectible in operating practice
because oxygen is gassed in on the total reactor length.
Concerning the macromixing characterization from RTD-measurements it can be mentioned here, that the resulting value of the equivalent number of ideal stirred
tanks was in the range $1,05 \leq N_{eq} \leq 1,2$ in case of the HSTR at $30 \leq n \leq 190$ rpm
at $0 \leq \dot{V}_G \leq 15 \; 1 \cdot min^{-1}$.

3.2 O_2-transfer coefficients at G/L-interface ($k_{L1}a$)

Fig. 4 compiles all data on OTR in the multipurpose bioreactor including the
conventional vertical STR, the ThLR and the HSTR.
In case of the STR with turbine impeller, the resulting $k_{L1}a$-coefficients were
increasing with both stirrer speed and gas flow, i.e. exponentially with the
further and decreasingly with the latter. Highest measured $k_{L1}a$-coefficients were

around 200 (h^{-1}) at 500 rpm and 30 ($l.min^{-1}$) gas flow. In a double-logarithmic diagramm $k_{L1}a$-coefficients showed linear dependency on power input.
In case of the ThLR with the horizontal smooth drum the range of measured values can be devided into two parts. In the first part oxygen transfer obviously takes place only through the well defined, but relative small gas-exposed liquid surface on the drum. Here $k_{L1}a$-coefficients were very low compared to the other stirring devices but are in accordance with surface renewal theory (ref. 4).
In the second part the revolutional speed of the drum becomes that high that liquid drops are cast from the drum. $k_{L1}a$-coefficients thus increase along with interfacial area and are comparable to OTR in STR's, at higher values of power consumption however. At very low rotational speeds values are still lower than in part one due to the collapse of the film.
In the case of the HSTR $k_{L1}a$-coefficients were increasing exponentially with stirrer speed, whereas the influence of gas flow was not significant. Highest measured values (at P/V $\sim$ 1 W/l) lay around 1000 (h^{-1}) with a stirrer speed of the perforated and baffled drum of 210 rpm, but are affected by very high errors due to the evaluation method (deviations due to t_G and t_E in eq. 3). Plotting $k_{L1}a$ versus power input on a double-logarithmic scale yielded straight lines throughout a wide range of the measured values as shown in Fig. 4.

4. CONCLUSION / DISCUSSION

For process engineering purposes (quantification for better comparability) data of $k_{L1}a$-coefficients and power input were fitted (using linear regression for double logarithmic plots) to empirical correlations of the following type:

$$k_{L1}a = a \; (P/V)^b \cdot (v_{S,G})^c \qquad (5)$$

and

$$P/V = d \; (10^{-3}.n)^e \cdot (v_{S,G})^f \qquad (6)$$

The resulting values of coefficients a, b, c, d, e and f are compiled in Table 1 together with the corresponding mean relative error Δ_1 and Δ_2.

Table 1: Values of empirical coefficients a to f in eq. 5 and 6
for the investigated reactor constellations of the
vertical STR and the horizontal ThLR and HSTR:

reactor	a	b	c	Δ_1 (%)	d	e	f	Δ_2 (%)
STR	27	0.50	0.70	8	22	3.00	–0.46	10
HSTR	490	1.20	0.05	12	100	2.30	–0.12	6
ThLR V_L = 5 (l)	50	0.8	0.00	6	28	1.70	0.00	28
10.3	28	0.5	0.00	10	33	2.00	0.00	11
14.2	28	0.5	0.00	11	29	2.40	0.00	25

It should be noted that there was the difficulty in measuring power input, that (1) there was a slow shift in torque-readings and (2) the horizontal devices (HSTR and ThLR) include an increased zero-value due to the fact that in this position two seals were necessary.
In comparison to vertical G/L-contactors, thus, it can be concluded that empirical correlations in horizontal vessels exhibit significant differences (see power numbers b and c). Data from literature reflect a similar tendency (e.g. ref,8,9,10). Less power is required in the HSTR for higher transfer rates compared to vertical stirred and sparged tanks.
As a consequence of this fact, the HSTR looks very promising for research purposes and technical scale processing in case of highly viscous media e.g. solid substrates and similar fermentations. Recent investigations with bioprocesses will contribute to the final characterization of this new reactor type (e.g. ref. 11).

5. REFERENCES

(1) <u>Moser, A.</u>: "Tubular Bioprocessing - a review", in Proc. 33rd Canad. Chem. Engng.
Conf. 1983, Toronto vol. 2, 417-423 (1983)

(2) <u>Moser, A.</u>: "Imperfectly mixed bioreactor systems", in Comprehensive Biotechnology,
(Moo-Young M., ed.-in-chief) Pergamon Press, Oxford, (1985) vol. 2, chap. 4

(3) <u>Moser, A.</u>: "Multi-purpose bioreactor for process kinetic analysis" in Proc. Adv.
in Ferm. 83, Suppl. Process Biochemistry 1983, 202-211 (1983)

(4) <u>Moser, A.</u>: "Dünnschichtreaktoren in der Biotechnologie" Chem.Ing.Techn. <u>49</u>,
<u>612-625</u> (1977)

(5) <u>Moser, A.</u>: "Mixing in Bioreactors - Quantification and Modelling", working party
bioreactor performance of the Europ. Fed. Biotechnol. Dechema/Frankfurt, 1986,
in preparation

(6) <u>Khang, S.J. and Levenspiel, O.</u>: "New scale-up and design method for stirrer
agitated batch mixing vessels" Chem. Eng. Sci., <u>31</u>, 569-577 (1979)

(7) <u>Ruchti, G., Dunn, I.J. and Bourne, J.R.</u>: "Comparison of dynamic oxygen electrode
methods for the measurement of $k_L a$", Biotechnol. Bioengng. <u>23</u>, 277-290 (1981)

(8) <u>Ziegler, H., Dunn, I.J. and Bourne, J.R.</u>: "Oxygen Transfer and Mycelial Growth
in a tubular loop fermentor", Biotechnol. Bioengng., <u>22</u>, 1613-1635

(9) <u>Joshi, J.B. and Sharma, M.M.</u>: Canad. J. Chem. Engng. <u>54</u> (1976) 56Off.

(10) <u>Herzog, P., Gschwend K., Widmer, F., Fiechter, A.</u>: "Physikalische und biologische
Charakterisierung des Torusreaktors", Chem. Ing. Techn. <u>55</u>, 566-567

(11) <u>Küng, W. and Moser, A.</u>: "Bioprocess Engineering characteristics of Multi-purpose
bioreactor" in preparation (1985)

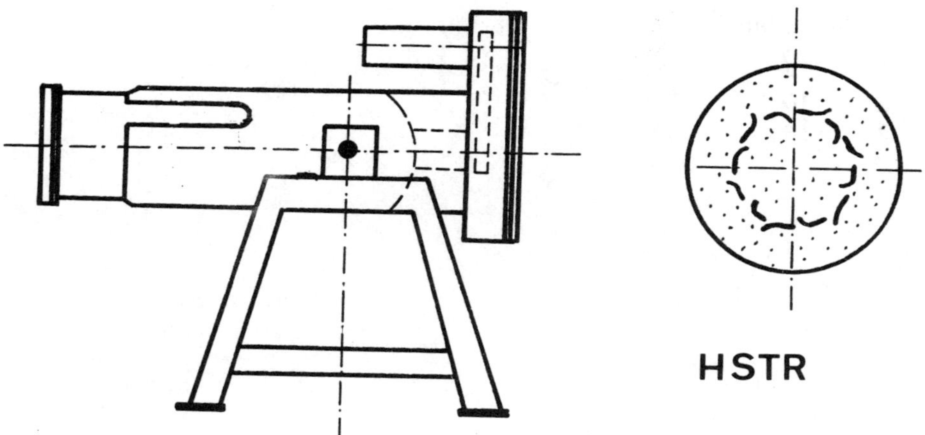

Fig. 1: The horizontal stirred tank reactor (HSTR)
        equiped with a rotating perforated and
        baffled cylinder

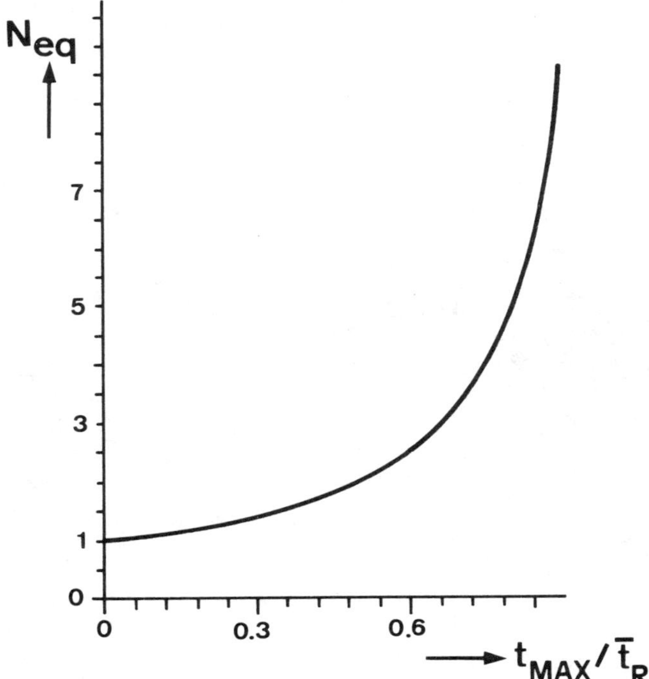

Fig. 2: Graphical representation of the estimation
        method for the number of equivalent stages
        N_{eq} from the value of time t_{max} where the
        pulse residence time distribution function
        reaches a maximum

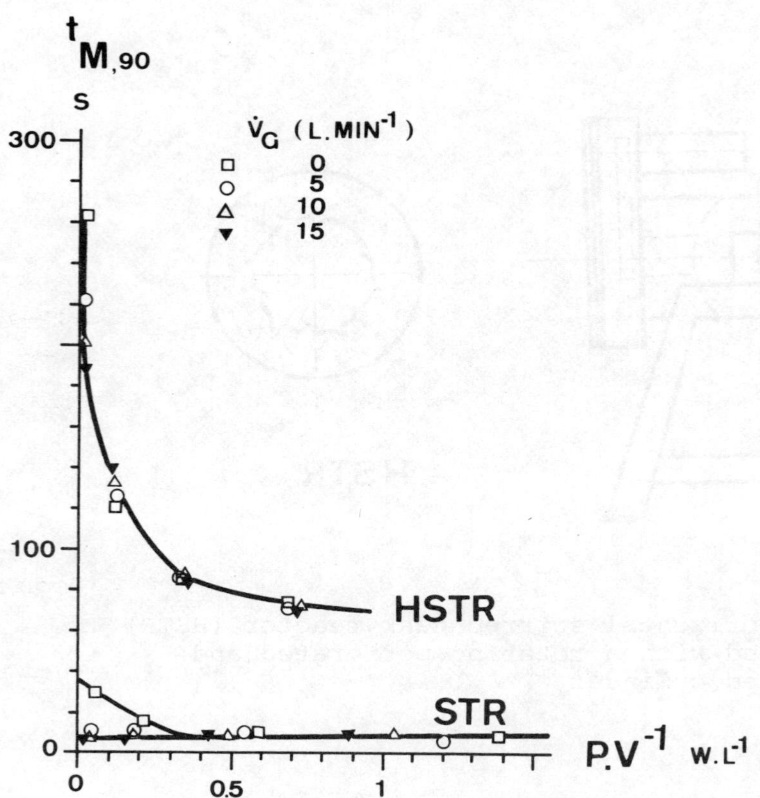

Fig. 3: Mixing time at 90 % degree of mixing $t_{M,90}$ in dependence on specific power consumption $P.V^{-1}$ as a function of gas flow rate $\dot{V}_G$ in the case of the vertical stirred tank (STR) and the horizontal reactor HSTR

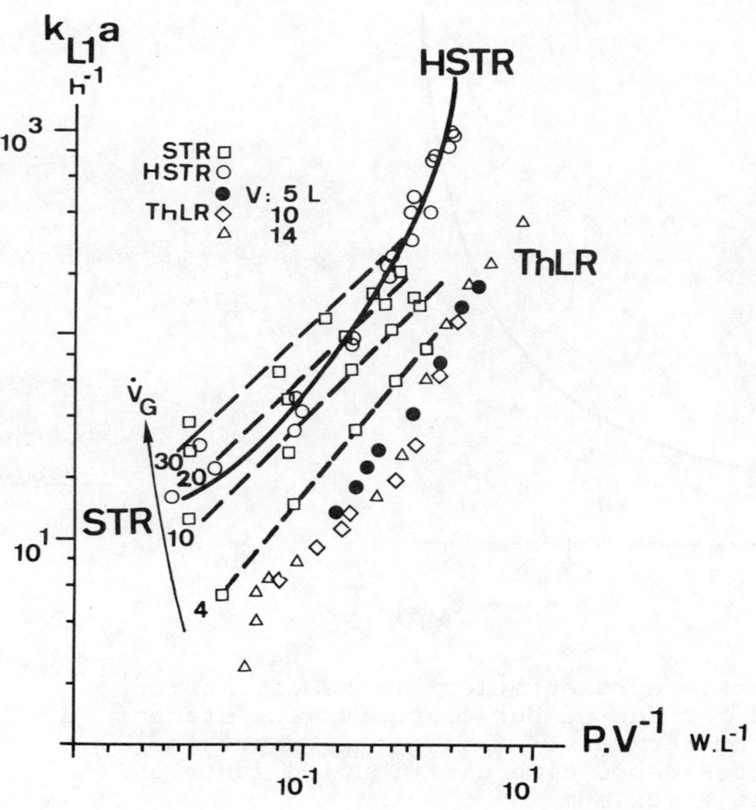

Fig. 4: Volumetric mass transfer coefficient at gas-liquid-interface $k_{L1}a$ in dependence on specific power consumption $P.V^{-1}$ in case of the horizontal stirred tank reactor (HSTR), the thin-layer reactor (ThLR) and the vertical tank (STR) as a function of gas flow rate $\dot{V}_G$ [l.min^{-1}]

HYDROMECHANICS AND MASS TRANSFER
IN AGITATED SUSPENSIONS

P. Ditl

Czech Technical University - Department
of Equipment Design for Process Industries
Suchbátarova 4, 166 07 Prague 6
Czechoslovakia

Summary

The present paper summarizes results obtained recently by the author and his co-workers in the theoretical and experimental research dealing with hydrodynamical and mass-transfer aspects of mixing in suspensions.

On the basis of dimensional analysis of the governing equations for system - dynamics and on the basis of an extensive experimental program, equations for the estimation of the most important parameters have been proposed, particularly for: suspension of solid particles, vortex depth and aeration in unbaffled vessels and also for mass transfer coefficient between a particle and liquid. Numerous experiments have been carried out to obtain a practical information about concentration profiles in mixed suspensions, mixing of concentrated suspensions of fine particles and dispersion of powders in liquids. The main attention has been focused to the scaling - up and to the effect of system geometry on the equipment performance and "quality" of the product. A mathematical model of the physical process has been formulated and solved for the dissolution of a polydisperse mixture of solid particles in a mixed fluid under conditions of variable driving force. The models for nonisothermal and isothermal adsorption have been formulated and solved. The results published can be used both for further research in solid - fluid systems and for engineering calculations and design of mixing equipments used for solid particle - liquid processing in chemical, biochemical, food, nuclear and hydrometallurgical industries.

Held at Wurzburg, 10-12 June, 1985.

Organised by DVCV· Deutsche Vereinigung für Chemie- und Verfahrenstechnik
(German Association of Chemical and Process Engineering).

Organisation: GVC·VDI-Gesellschaft Verfahrenstechnik und Chemieingenieurwesen. 

<u>NOMENCLATURE</u>

c^v	= volumetric concentration of solid phase, 1
$\bar{c}^v$	= mean value of c^v
C_1, C_2, C_3	= constants
d_p	= particle size, m
D	= impeller diameter, m
D_f	= diffusion coefficient, m^2/s
D_{ef}	= effective diffusivity for intra-particle transport, m^2/s
e	= energy consumption per unit of the product volume, J/m^3
g	= acceleration due to gravity, m/s^2
H	= height of static liquid surface above the vessel bottom, m
H_2	= impeller clearance; the shortest distance between the impeller blade and vessel bottom, m
n	= agitator spead, 1/s
n_f	= agitator speed at which the particles become completely suspended, 1/s
$p_o(d_p)$	= size frequency distribution, 1/m
P	= mixing power, W
T	= inside diameter of stirred tank, m
t	= time, s
v	= variation coefficient defined in eq. (7)
V	= volume of the tank, m^3
α, δ	= exponents
β	= mass-transfer coefficient, m/s
ε	= specific power input per unit mass, W/kg
ε_V	= specific mixing power per unit volume, W/m^3
η_T	= eddy-size characteristic, m
μ	= viscosity, Pas
$\nu = \mu/\rho$	= kinematic viscosity, m^2/s
ρ	= liquid density, kg/m^3
ρ_s	= particle density, kg/m^3
$\Delta\rho$	= density difference, kg/m^3, $\Delta\rho = \rho_s - \rho$
Ar	= Archimedes number, $Ar = d_p^3 \Delta\rho\, g\, \rho/\mu^2$
Bi	= Biot number, $Bi = \beta d_p/D_{ef}$
Fr	= Froude number, $Fr = n^2D/g$
Fr_f	= modified Froude number, $Fr_f = n^2T\rho/(\Delta\rho\, g)$
Ga	= Galileo number, $Ga = Re^2/Fr = gD^3/\nu^2$
Ne	= power number, $Ne = P/\rho n^3D^5$
Re	= Reynolds number, $Re = nD^2\rho/\mu$
Re_f	= Reynolds number at complete suspension, $Re_f = n_fD^2\rho/\mu$
Sc	= Schmidt number, $Sc = \mu/\rho D_f$
Sh	= Sherwood number, $Sh = \beta d_p/D_f$

1. INTRODUCTION

Operations occuring in agitated suspensions can be divided into two groups:

1. The total mass of solid particles is kept constant during the operation. These cases are characterized by the absence of mass transfer between both phases involved, such as:

   a) Preparation or homogenization of suspensions or slurries (e.g. feedstocks for atomizers, flotators, etc. or chemical reactions with solid catalyst).

   b) Wetting of fine particles in liquids (e.g. pigments or atomized products, etc.).

   c) Dispersion or predispersion (e.g. production of paints, various pastes, food products etc.).

2. Mass of solid particles is changed during the operation due to the mass-transfer between the solid and liquid phase, namely:

   a) Dissolution which is accompanied by a decrease of the particle size.

   b) Leaching which is accompanied by changes in physical properties and by changes of their structure.

   c) Crystallization which is accompanied by the crystal growth.

   d) Adsorption and ion exchange e.g. in removal of contaminants from solution and fractionation of liquid mixtures.

   e) Chemical reactions in which the solid particles are one of the reactants.

Each of the above mentioned operations has some specific features and requirements on mixing intensity and geometry. However, the design and calculations for a given case is a combination of some of the following general steps:

1. Complete definition of the mixing requirements based on "product quality" and mechanical and material limitations.

2. Design of mixer capable to secure the required uniformity of particles suspended in liquid or sufficient particle motion on the vessel bottom. This includes:

   a) An estimation of the critical impeller speed for complete suspension.

   b) The knowledge of concentration profiles in suspension.

   c) Requirements connected with resuspendation of solid particles.

   d) Limitation of the operating agitator speed range by unwanted aeration.

3. The design of mixing equipment to achieve:

   a) Dispersion of pigments or the others fine particles in a liquid,

b) Wetting of solid materials.

4. Start-up and steady-state impeller power imput.

5. The design of agitated equipments for s-l mass transfer operations further includes:
   a) Estimation of mass transfer coefficients in the agitated s-l system
   b) Solution of mass transfer kinetics which allows to predict the operation time and tank volume, consenquently.

6. The design of stirred tanks with heat transfer in suspension includes:
   a) Estimation of heat transfer coefficients in the s-l system
   b) Calculations of the heated or cooled surface, the heat transfer course or of the temperature profile.

7. Comparison of possible variants on the basis of energy consumption.

8. Mechanical design respecting material, operating ang manufacturing requirements.

9. Final economical evaluation of the design and comparison with the similar operating equipments.

In spite of the fact that mixing in suspension has been frequently studied in the past, there is still a lot of unsolved problems or uncertainties following from results obtained by various authors which differ mutually. The aim of this paper is not to review all the results obtained in this field or to pick up the problems unsolved but only to summarize recently obtained results by the present author dealing with mixing in s-l systems.

2. SUSPENSION OF SOLID PARTICLES

Different particle - liquid hydrodynamic subregions depending on the the existence of intra - particle forces, particle diameter and the intensity of turbulence has been observed in our experiments [2]. If the particle size is comparable with the size of primary eddies, the suspending mechanism is governed by the motion of these eddies. For smaller particles the suspending mechanism is determined by the motion of middle - and small eddies having the characteristic scale $\eta_T = (\nu^3/\varepsilon)^{1/4}$. Suspension mechanism is then controlled by the flow around the sediment - liquid interface, the geometrical properties of particles and their polydispersity, intra - particle forces and by concentration of suspension. It seems to be reasonable to distinquish three cases:

2.1 Suspension of "large particles" ($d_p/\eta_T > 32$)

Theoretical model leading to the derivation of the final

equation which allows to estimate the critical speed for complete suspension is presented in our other paper in the conference proceedings or in [1] . A very good agreement between the theoretically obtained equation and our experimental results has been observed. For suspension of "large particles" the equation published by Rieger and Ditl [2] can be recommended:

$$Fr_f = \frac{Re_f^2}{Ar} \left(\frac{T}{D}\right)^4 \left(\frac{d_p}{T}\right)^3 = \frac{\varrho \, n_f^2 \, T}{\Delta\varrho \, g} = C_1 (T/D)^{3.1} \qquad (1)$$

which implies the scaling - up condition $nT^{0.5} = $ const which does´nt indicate any dependance of critical impeller speed for suspension on the particle diameter d_p. The values of the C_1 constant are given for four impeller types, four impeller clearances and four solid concentrations in [1] . Typical values of C_1 for pitched six and three-blade turbine, clearances $H_2/D = 1/3$, $2/3$ and concentrations ~ 0, 1 and 5 vol. % are given in Table I. There is a lack of published experimental results in this area representing suspension of particles larger than at least 1 mm.

The same experimental data have been statistically treated to obtain the function $Re_f = f(Ar, d_p/T, D/d)$ which follows from inspection analysis of the governing equations describing hydrodynamics of the system with the result:

$$Re_f = C_2 \, Ar^{0.45} \, (d_p/T)^{-1.42} \cdot (T/D)^{-0.56} \qquad (2)$$

which may be rewritten into the dimensional form:

$$n_f \sim d_p^{-0.07} \, T^{-0.58} \, \nu^{0.1} \, (\Delta\varrho g)^{0.45} \, (T/D)^{1.44} \qquad (3)$$

The examples of the constant C_2 are given in Table I.

Table I: Typical values of the constants C_1, C_2, C_3

C_1, C_2, C_3	H_2/D	c^V [vol %]		
		~ 0	1	5
Pitched six-blade turbine	1/3	0.056;0.692;2.08	0.068;0.77 ;1.73	0.112;0.985;2.17
	2/3	0.065;0.755;2.31	0.078;0.831;1.99	0.126;1.02; 2.34
Pitched three-blade turbine	1/3	0.075;0.816;2.60	0.104;0.91; 2.04	0.165;1.14; 3.35
	2/3	0.139;0.871;3.02	0.111;0.953;2.24	0.197; - ;3.19

2.2   Suspension of "small particles" without the influence of
       intra - particle forces ($d_p/\eta_T < 32$)

Most results published so far belong into this area where the particle size was smaller than about 1.5 mm. An excellent review of results published previously is given e.g. in [15], [13] . The exponent α in scaling - up criterion $n_f T^{\alpha} = $ const varied in the range from

0.67 (constant ε_V) to 0.85 (clasical Zwietering's paper). An interesting approach to explain the variation of exponent α by Mersman, Zehner and Tebel [13] is under way. They derived theoretically that the value of α increases from the limiting value 0.67 with decreasing Galileo number (Ga = gT^3/ν^2).

We correlated our experiments in this area by the semiempirical equation (1):

$$Re_f = C_3\ Ar^{0.45}\ (d_p/T)^{-1.25}\quad (T/D)^{-0.56} \tag{4}$$

Typical values of the constant C_3 are given in Table I. Equation (4) can be rewritten into the dimensional form:

$$n_f \sim d_p^{0.1}\ T^{-0.75}\quad \nu^{0.1}\ (\Delta\varrho g)^{0.45}\ (T/D)^{1.44} \tag{5}$$

Our experiments from wich eqs. (2) and (4) have been obtained are good in the following ranges of dimensionless criteria: $Ar < 1.3 - 10^6 >$, $T/D < 2 - 4.4 >$, $d_p/T < 0.8 \times 10^{-3} - 10 \times 10^{-3} >$, $d_p/\eta_T < 0.3 - 100 >$.

It is worth to note that our experiments proved constant value of the exponent of Archimedes number 0.45 in the wide range of $Ar < 1.3 - 10^6 >$ under the turbulent impeller Reynolds number. This observation supports an idea that the relative particle-fluid velocity in the turbulent field does'nt depend on hydrodynamic subregions as is the case at terminal settling velocity in a still liquid.

2.3 Dividing criterion between large and small particles

This criterion was determined experimentally from the breakpoint on the curves $n_f \sim d_p^\alpha$. It was found that only the following criterion was reasonably constant while changing d_p, ν, T, T/D and impeller type used [1]:

$$\frac{d_p}{\eta_T} = \frac{d_p}{(\nu^3/\varepsilon)^{1/4}} = \left(\frac{4Ne}{\pi}\right)^{1/4}\left(\frac{d_p}{T}\right) Re^{3/4}\left(\frac{T}{D}\right)^{-3/4} \doteq 32 \tag{6}$$

2.4 Suspension of fine particles with significant intra-particle forces

Such suspensions are characterized by a very low value of the minimum settling velocity. It often takes hundreds of hours before the steady interface between the sediment and a clear solvent is established even in laboratory scale tank. It is very important for investigators and also for equipment purchasers to specify the time for ressuspendation because the energy required to lift up a slurry strongly depends on the "history" of previous sedimentation, besides this energy is different from the energy necessary to maintain a slurry. The choice of impeller type depends on physico - chemical properties of suspension and also on its concentration. Fig. 1 shows the comparison of suspension ability between pitched six-blade turbine and EKATO - MIG agitator. With an increasing solid concentration the multistage agitators of the relatively larger diameter (T/D = 2) become more efficient if compared with turbine impellers. Lower impeller speed of multistage agitators is also more convenient from the point of mechanical design, especially, in large volume tanks (long shaft). Further conclusions have been made during the testing of about 30 different geometrical configurations of multistage and screw-agitators for mixing of bauxite pulps. The mixing efficiency of agitators tested has been compared with standard EKATO - MIG impellers. Some results have been published in [3].

The scaling-up condition of EKATO-INTERMIG agitators for similar cases was estimated in [16],[17] as $n_f T^{\alpha}$ = const. with exponent α in the range from 0.86 to 0.95. We obtained value of the exponent α = 0.92. From this value it follows that the specific power imput per unit volume drastically decreases with an increase of vessel volume $\mathcal{E}_V \sim V^{-0.25}$.

3.   CONCENTRATION PROFILES IN AN AGITATED VESSEL

Concentration profiles have been experimentally investigated [4] using the γ-adsorption radiometric method. Suspension nonuniformity was quantified by the variation coefficient:

$$v = \frac{\left[\frac{1}{N} (c_{i\ local} - \bar{c}^V)^2 \right]^{1/2}}{\bar{c}^V} \cdot 100 , \qquad (7)$$

where N denotes the number of measuring stations.
Typical concentration profiles obtained for glas beads at impeller speed n_f = 9.67 1/s and n = 15.6 1/s using pitched six-blade turbine are shown in Fig. 2. The changes of concentration profiles with an increase of impeller speed are demonstrated in Fig. 4 for glas beads suspension and for the bauxite pulp agitated by 4-stage agitator (2 pitched blades 45° in each stage) in Fig. 3. The effect of the relative impeller diameter on suspension uniformity is demonstrated in Fig. 6 where the variation coefficient v is plotted vs. impeller speed n for T/D = 2.5, 2.88 and 3.17 as a parameter. The best uniformity of glas beads suspension was achieved by the pitched six-blade turbine at T/D = 2.88.

4.   MASS TRANSFER COEFFICIENT

The theoretical progress in estimation of β is closely related to the relative velocity between a particle and turbulent liquid which is also very important for particle suspension. Before this problem with be satisfactorily solved, it seemed to be usefull to create the "data bank" of own and other reliable published experimental results. From these data the equation for estimation of mass transfer coefficient in tanks agitated by properly designed, and in vessel situated, axial impellers was evaluated in the form:

$$Sh = 2 + 0.49 \left(\frac{d_p^{4/3} \, \varepsilon^{1/3}}{\nu} \right)^{0.60} \left(\frac{T}{D} \right)^{-0.25} Sc^{0.39} \qquad (8)$$

Equation (8) reflects about 2.500 experiments and it can be recommended in the following ranges of the dimensionless criteria ($d_p^{4/3} \varepsilon^{1/3}/\nu$)
$\langle 10^{-2} \div 10^3 \rangle$ Sh$\langle 4 \div 150 \rangle$, $d_p/T \langle 10^{-4} \div 5.5 \times 10^{-3} \rangle$, T/D$\langle 4 \div 2.5 \rangle$, Ar$\langle 10^{-1} \div 3 \times 10^4 \rangle$ Re$\langle 500 \div 5 \times 10^5 \rangle$.

5.   MASS TRANSFER KINETICS IN AGITATED VESSELS

The dimension of the tank volume depends on the operation time, which can be determined from the mass transfer kinetics describing the solute concentration changes with time.

5.1   External mass transfer is the controlling mechanism

For operations where the external mass or heat-transfer is a controlling mechanism, a very flexible model of mass or heat-transfer kinetics across the interface boundary between the fluid and polydisperse particle mixture has been proposed [8]. This model assumes a variable driving force of the process and can be applied regardless of whether the particle size increases or decreases during the operation. On the concept of this general model the dissolution of polydisperse mixtures of solid particles in agitated tanks has been described under the condition of variable driving force changing due to temperature changes or due to significant increase of the solute concentration.The problem solved is depicted qualitatively in Fig.5. An analytical solution was obtained for the special case of dissolution under the constant driving force in which the dissolving polydisperse particle mixture was described by the Rosin - Rammler - Sperling distribution [9].

5.2   Kinetics of batch agitated adsorbers and ion exchange devices

Adsorption kinetics was solved under the isothermal condition and under the assumption of significant solute concentration changes inside the mixed volume for nonlinear or linear adsorption isotherm [11]. The similar case of adsorption under the nonisothermal conditions has been solved and analyzed in [12]. When Bi $>$12 pore diffusion was found to be the controlling mechanism. An increase in mixing intensity influences adsorption kinetics only when Bi $<$ 12. No

significant intraparticle temperature profile was found for typical
cases of liquid adsorption up to heat of adsorption values of 400 kJ/mol.
However, an optimal temperature of adsorption exists resulting in the
shortest time to achieve the required concentration.

6. COMPARISON OF VARIANTS ON THE BASIS OF ENERGY CONSUMPTION

Energy consumption per unit of the product volume can be
expressed as:

$$e = \frac{P \cdot t_{op}}{V} = \mathcal{E}_v \cdot t_{op} \tag{9}$$

Knowing calculated or meassured dependence of the agitator power P
and the operation time t_{op} on the tested parameter (e.g.: impeller
speed, T/D, H_2/D, etc.) one can simply optimize the value of the parameter tested from the point of minimum energy consumption. As an
example of this procedure, the optimizing of dissolution tanks for
water works was reported in [10] .

7. POWDER DISPERSION AND WETTING

A quantitative model of powder dispersion has been proposed [5] . This model explains the diminishing of powder clusters on
the time and mixing conditions and leads to the scaling-up condition
$nT^{0.89}$ = const. under the assumption of a product of the same quality.
Numerous experiments with TiO_2 were focused on the influence of impeller type, vessel shape and geometry, relative impeller diameter, impeller speed, temperature and scaling-up.

Wetting of powders is connected with the creation of the
vortex in unbaffled vessels and our contributions to this group of
problems have been published in [6] , [7] .

8. REFERENCES

1.   Ditl,P., Rieger,F.: "Suspension of solid particles in agitated
vessels". CHISA Conference, Prague, August 1984, Paper No V3.61.

2.   Rieger,F., Ditl,P.: "Suspension of solid particles in agitated
vessels". 4th European Conference on Mixing Proceedings, Noord-
wijkerhaut, Netherlands, BHRA Fluid Engineering, April 1982,
pp. 263 - 273.

3.   Rieger,F., Ditl,P., Novák,V.: "Mixing of concentrated suspensions
of fine particles". Applied Rheology Proceedings, Velké Karlovice
1981, pp. 139 - 144 (in Czech).

4.   Ditl,P., Thýn,J., Rieger,F.: "Concentration profiles in agitated
suspensions". Chemický průmysl, 34,59, 1984, pp. 618 - 622.

5. Ditl,P., Novák,V., Rieger,F.: "Powder dispersion in liquids using high speed mixers". Chemický průmysl 31, 1981, pp. 3 - 10.

6. Rieger,F., Ditl,P., Novák,V.: "Vortex depth in mixed unbaffled vessels". Chem. Eng. Sci., 34, 1979, pp. 397 - 403.

7. Novák,V., Ditl,P., Rieger,F.: "Mixing in unbaffled vessels". 4th European Conference on Mixing Proceedings, Noordwijkerhout, Netherlands, BHRA Fluid Engineering, April 1982, pp. 57 - 70.

8. Ditl,P., Vaněček,V.: "Mass transfer kinetics of polydisperse mixtures with variable driving force". 29. National CHISA Conference, Mariánské Lázně, October 1982 (in Czech).

9. Ditl,P., Šesták,J., Partyk,K.: "Mass transfer kinetics in dissolving polydisperse solid materials, Int. J. Heat Mass Transfer, 19, 1976, pp. 635 - 641.

10. Ditl,P., Rieger,F., Ječmen,J., Vařeka,J.: "Design of dissolution tanks for preparation of chemical solutions". 11th OSCHI Conference Procedings, Madarasz, May 1983, pp. 127 - 132 (in Czech).

11. Ditl,P., Coughlin,R.W., Jerre,E.H.: "Mass transfer kinetics of adsorption on suspended particles". Journal of Coll. Interf. Sci 63, 3, 1978, pp. 410 - 420.

12. Roušar,I., Ditl,P.: "Kinetic characteristies of a batch adsorber or an ion exchange device operated under nonisothermal conditions". Chem. Eng. Commun. 19, 1983, pp. 341 - 353.

13. Chapman,C.M., Nienow,A.W., Cooke,M., Middleton,J.C.: "Particle - gas - liquid mixing in stirred vessels". Part I - Particle - liquid mixing. Chem. Eng. Res. Des. 61, 1983, pp. 71 - 81.

14. Zehner,P., Tebel,K.H.: "Suspension hydrodynamics in agitated vessels". Mischvorgänge, Freising - Weihenstephan, 1984.

15. Einenkel,W.D.: "Description of fluid mechanics by suspension in agitated vessel". PhD Thesis, TU Munich, 1979 (in German).

16. Kipke,R.: "Application of fluid-mechanical interference effects upon the development of the interference-flow-mixer". Verfahrenstechnik 11, 9, 1977 (in German).

17. Müller,W., Todtenhaupt,E.K.: "Large agitators in basic industries". Aufbereitungs-Technik 13, 1972, p. 32, (in German).

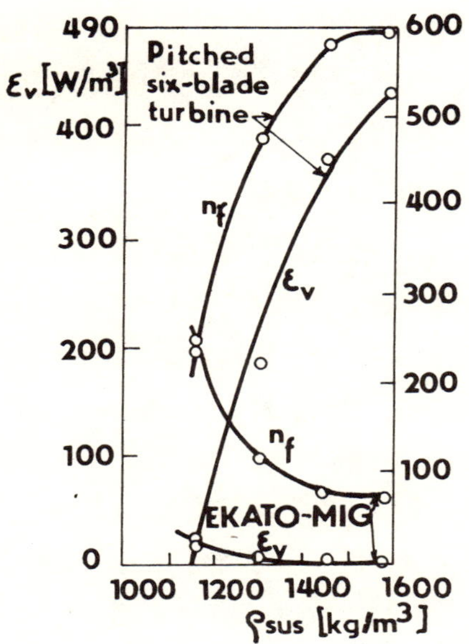

Fig.1: T = 0.2 m, bauxite pulp

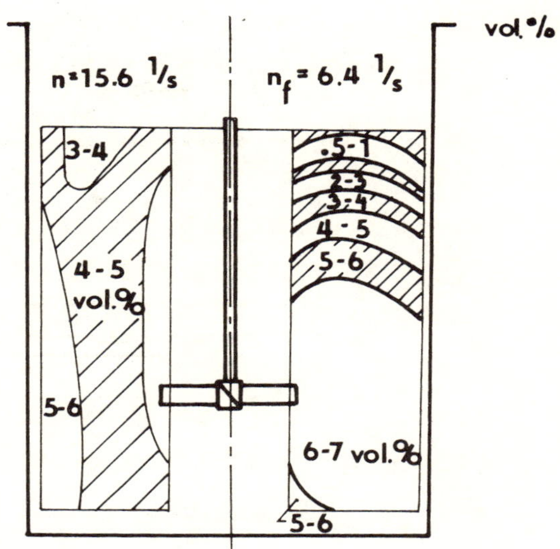

Fig.2: Pitched six-blade turbine glas balotine $d_p \langle 80-100 \rangle \mu$m, $\bar{c}^v = 4.85$

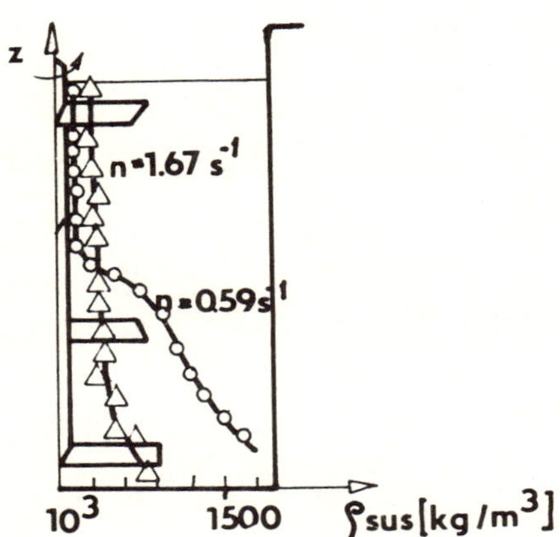

Fig.3: Bauxite pulp

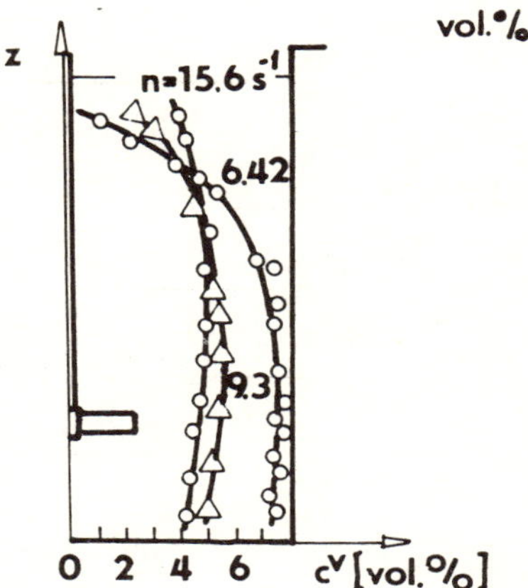

Fig.4: Pitched six-blade turbine, glas balotine $d_p \langle 80-100 \rangle \mu$m, $\bar{c}^v = 4.85$

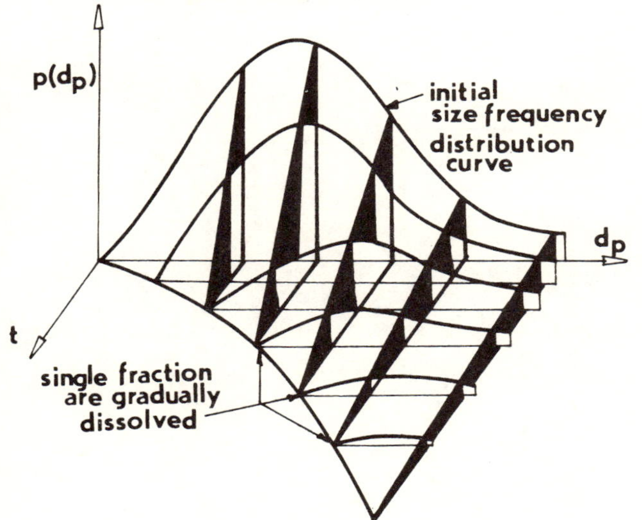

Fig.5: Dissolution of polydisperse solid materials

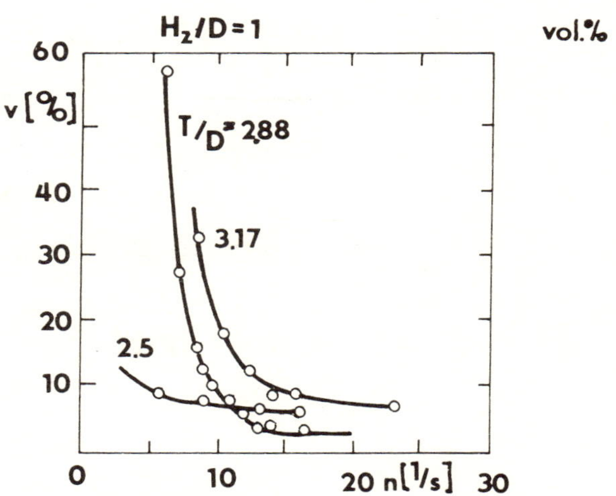

Fig.6: Pitched six-blade turbine glas balotine $d_p \langle 80-100 \rangle \mu$m, $\bar{c}^v = 4.85$

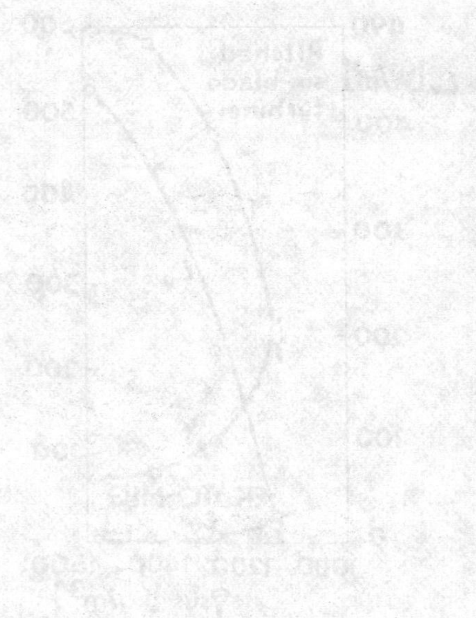

LIQUID-LIQUID MIXING:

INVESTIGATIONS ON THE COMPLETE DISPERSION

M. GAGGERO, E. ARATO, M. GIAMA°, B. CANEPA, P. COSTA, G. LODI

Istituto di Scienze e Tecnologie dell'Ingegneria Chimica,
Facoltà di Ingegneria dell'Università di Genova,
Via dell'Opera Pia 15 - 16145 Genova - ITALY

° Università di Mogadiscio

Summary

The minimum impeller speed for complete dispersion of a liquid into another and the
prevision of which phase is the continuous one have not received , up to now, sufficient attention. As a contribution to the study concerning the aforesaid questions,
in this paper are presented some experimental results on the influence of the viscosity of the phases as well as of their volume ratio in a wide range of these parameters.

Held at Wurzburg, 10-12 June, 1985.

Organised by DVCV· Deutsche Vereinigung für Chemie- und Verfahrenstechnik
(German Association of Chemical and Process Engineering).

Organisation: GVC·VDI-Gesellschaft Verfahrenstechnik und Chemieingenieurwesen. 

©BHRA, The Fluid Engineering Centre, Cranfield, Bedford MK43 0AJ, England.

<u>Nomenclature</u>

d   agitatore diameter, m

D   vessel diameter, m

g   acceleration due to gravity, m/s^2

K   constant

N_{min} minimum impeller speed, 1/min   (N_c: corrected N_{min})

α_1 empirical exponent

ρ mean density, kg/m^3

η_d dynamic viscosity of disperse phase, kg/m.s

η_c dynamic viscosity of continuous phase, kg/m.s

ν kinematic viscosity, m^2/s

φ_v volume fraction of the oils

σ interfacial tension, Kg/s^2

1. INTRODUCTION

In contacting two immiscible liquids the goal is often to obtain the heighest
interfacial area by an appropriate stirring which induces an intimate dispersion of one
liquid into the other, with droplets as small as possible. In such situations, which
are very common in the chemical industry, it is very important a good knowledge of the
correlations between geometrical, physical and operating parameters and the resulting
interfacial area. Moreover the design of a mechanically stirred apparatus should   be
based   also on correlations concerning the power consumption of the stirring   device,
the minimum stirring speed to obtain complete dispersion and finally on an   affidable
prevision of which phase is the continuous one.

Up to now these questions have not been sufficiently investigated (Ref. 1):
while much work has been done on the corresponding problems for solid-liquid systems,
only very few studies have concerned with liquid-liquid systems. For example, in the
field of batch agitation there are some uncertainties on the prevision of which of the
two fluids will form the continuous phase and which the dispersed one. On the   other
hand, the choice between these two types of dispersion can have an important effect on
the resulting interfacial area and on the sedimentation behaviour (Ref. 2).   Similar
uncertainties are present in the generalized prevision of the minimum stirrer   speed
which is necessary in order to reach complete dispersion of one liquid in the other.

The existing correlations between the minimum impeller speed for complete dispersion and some physical parameters can be considered valid only for particular systems and for small ranges of the physical properties of the phases. In fact for the
Nagata's correlation (Ref. 3)

$$N_{min} = K\,D^{-2/3} \left(\frac{\mu_c}{\rho_c}\right)^{1/9} \left(\frac{\Delta\rho}{\rho_c}\right)^{0.26}$$

no information is available on the range of validity. Similarly the Skelland and Seksaria's correlation (Ref. 4)

$$N_{min} = C_1\,(g/d)^{1/2}\,(D/d)^{\alpha_1} \left(\frac{\mu_c}{\mu_d}\right)^{1/9} \left(\frac{\sigma}{d^2\rho_c g}\right)^{0.30} \left(\frac{\Delta\rho}{\rho_c}\right)^{0.25}$$

is referred to a 0.5 volume fraction and a narrow range of viscosity.

This paper reports some results concerning stirring apparatus of standard geometry and operated with different fluids covering large ranges of viscosity and of volume fraction. The aim is to give a contribution to the understanding of the behaviour
of liquid-liquid dispersions in terms of minimum impeller speed and in terms of con -
ditions for the phase inversion.

2. EXPERIMENTAL APPARATUS AND PROCEDURE

Our experimental apparatus, which is schematized in figure 1, is baffled flat
bottomed tank made in plexiglass to permit visual observation. Table 1 shows the main

Table 1. Apparatus Dimensions

Internal diameter of vessel	0,21	m
Liquid height in vessel	0,21	m
Height of vessel	0,30	m
Diameter of shaft	0,10	m
Baffle lengh	0,25	m
Baffle width	0,021	m
Volume fraction of organic liquid	0,10-0,95	
Impeller diameter	0,07	m
Impeller height	0,07	m

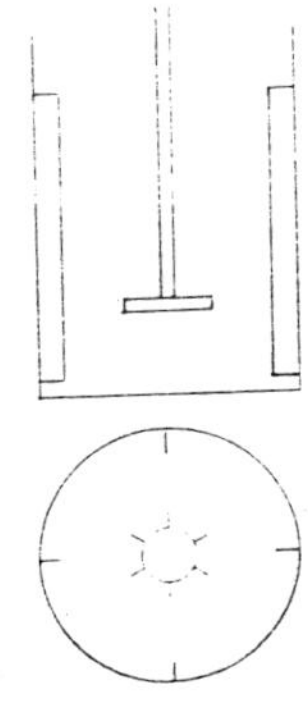

Figure 1

geometrical characteristics of the vessel. The impeller used is a six flat blade turbine drived by an electrical motor with controlled rotation velocity. The stirring speed
was measured by a stroboscope.

Up to now tests were done by using water as the heavier liquid and a silicone
oil "Tegiloxan" as the lighter one. These oils present low vapor pressures and relatively flat viscosity-temperature curves. Five oils were alternatively employed, with viscosities ranging from 3 to 1000 m^2/s . The volume fraction engaged by the heavier
phase was varied from 0.1 to 0.95 . All runs were made at 25° C : the fluids properties at this temperature are reported in table II.

Table II. Fluid properties

Liquid	density, Kg/m^3	viscosity, m^2/s	sup.tensio N/m
Water	1000	$1 . 10^{-6}$	72.10^{-3}
Oil 3	885	$3 . 10^{-6}$	$18,3 . 10^{-3}$
Oil 10	909	$10 . 10^{-6}$	$19,3 . 10^{-3}$
Oil 50	957	$50 . 10^{-6}$	$20,3 . 10^{-3}$
Oil350	968	$350 . 10^{-6}$	$20,3 . 10^{-3}$
Oil 10^3	970	$1000 . 10^{-6}$	$20,3 . 10^{-3}$

The minimum stirring speed was measured starting from a very gentle agitation and increasing slowly the velocity until the complete dispersion of one liquid in the other
was reached.
According to other researches(Ref 5) the criterion to establish the complete dispersion
is that at this point no clear liquid must be observed either at the top or at the bottom of the vessel. Now , in our observations, clear liquid pockets of approximately
$0,5 - 5 . 10^{-6}$ m^3 may persist near the baffles (at the top and the botton)   although the rest of the liquids is well dispersed. These small liquid pockets can be
suppressed only by increasing the speed up to 100% or more. Therefore a completly
dispersed state was assumed when only small , relatively non stationary liquid pockets
remain unmixed in the bulk dispersion. For this reason the absolute values of the minimum stirring speed could be affected by the "operator sensibility". Nevertheless this
fact seems to have no significant effect on the relative values and then on the general
trend. This conclusion was confirmed through the comparison of the results obtained by
different operators.

3. RESULTS AND DISCUSSION
First of all it is to be noted that in our tests which cover as said before, the complete range of volume ratios we were able to obtain the dispersion of the lighter phase in the heavier as well as the dispersione of the denser phase in the lighter. In
such a way the influence of the viscosity of both the continuous and the dispersed
phase can be investigated. The summary of our results is shown in Figure 2 where the
minimum impeller speed is plotted versus the volume ratio of the phases for the various systems investigated. The left side concerns with the dispersion of the light phase
in the heavier, while the right side represents the dispersion of the heavy phase in
the lighter one. A critical analysis of the results enables us to point out th following considerations:
-the minimum impeller speed necessaray for disperding the heavy phase in the lighter
is, in our case, lower than the one corresponding to the dispersion of the light phase
in the heavier one;
-the volume ratio of the two phases influences the minimum impeller speed, increasing
which increases in both directions in approaching the point in which the inversione
between continuous and dispersed phase occurs;

-the volume ratio at which the phase inversion occurs corresponds to a maximum difficulty in achieving a complete dispersion. Moreover this point seems to be indipendent from the viscosities of the phases. The effect of the impeller height will be investigated in our following work. However, some experimental points in figure 2 corresponds to a situation in which the volume fractions are not very different and the continous phase is not the one surrounding the agitator at rest. In other words the inversion of the dispersion does not corresponds to the change of the nature of the liquid which surrounds the agitatore at rest (Ref. 6). The viscosity of both the dispersed and the continuous phase has a remarkable effect on the minimum impeller speed for complete dispersion. Figure 3 shows the minimum impeller speed , corrected by accounting for the small density differences of the oils, versus the viscosity of the dispersed phase, for two different volume ratios. Similarly figure 4 shows the minimum impeller speed versus the viscosity of the continuous phase. In figure 3 it can be noted that the heigher is the viscosity of the dispersed phase the heigher is the minimum impeller speed for complete dispersion. This trend is in contrast with the corresponding one deduced by Skelland and Seksaria in a small viscosity range. More difficult is the interpretation of the results showed in figure 4 : in fact, while for viscosities below $50 . 10^{-6}$ m^2/s the effect of the viscosity of the continuous phase is in agreement with the results reported in the literature, increasing the viscosity of the continuous phase above $50 . 10^{-6}$ m^2/s the effect seems to change with an opposite trend. In this range of viscosity an increasing of the viscosity of the continuous phase is connected with a decreasing of the minimum impeller speed. This opposite trend induces to assume that for viscosities around $50 . 10^{-6}$ m^2/s a change in the mechanism of the dispersion occurs.

4. CONCLUSIONS

This paper presents some contribution to the understanding of the dispersion mechanism of a liquid phase into another. Although reference is made only to a particular geometrical configuration, the results here reported increase the knowledge with regard to wider viscosity and volume ratio ranges. In our opinion, some of the experimental findings presented in this paper can be underlined.

First of all the inversion of the dispersion occurs, for a given impeller height, at a volume ratio which corresponds to a maximum volume in the minimum impeller speed (figure 2). This critical volume ratio is not affected by the viscosities of the two phases. So, the nature of the continuous phase does not seem to be determined by either the impeller position or by the volume fraction, but rather by a combination of these two factors.

Another interesting effect who needs further investigation is the change in the trend of minimum impeller speed versus the viscosity of the continuous phase.

Literature cited

1.Short, D.G.R. and Etchells,A.W. "Unsolved problems of industriale mixing"In: 4th European Conference on Mixing (Noordwijkerhout,Netherlands:Sep.15-17, 82) BHRA F.E.
2.Mersman,A. and Grossman,H.:"Dispersion of immiscible liquids in agitated vessels" Int.Chem.Eng. 22 Oct.82 pp. 581 - 590
3.Nagata, S. "Mixing, Principles and Applications" John Wiley and Sons,New York-London (1975).
4.Skelland,A.H.P. and Seksaria, R."Minimum impeller speeds for liquid-liquid dispersion in baffled vessels"Eng. Chem.,Proc.Des.Devel.17 N° 1 pp 56 - 61 (1978)
5.Quin, J.A. and Sigloh, D.B., Cand. J.Chem.Eng. 41 15-18 (1963).
6.Uhl,W.V. and Gray, J.B. "Mixing" Academic Press, New York-London (1966).

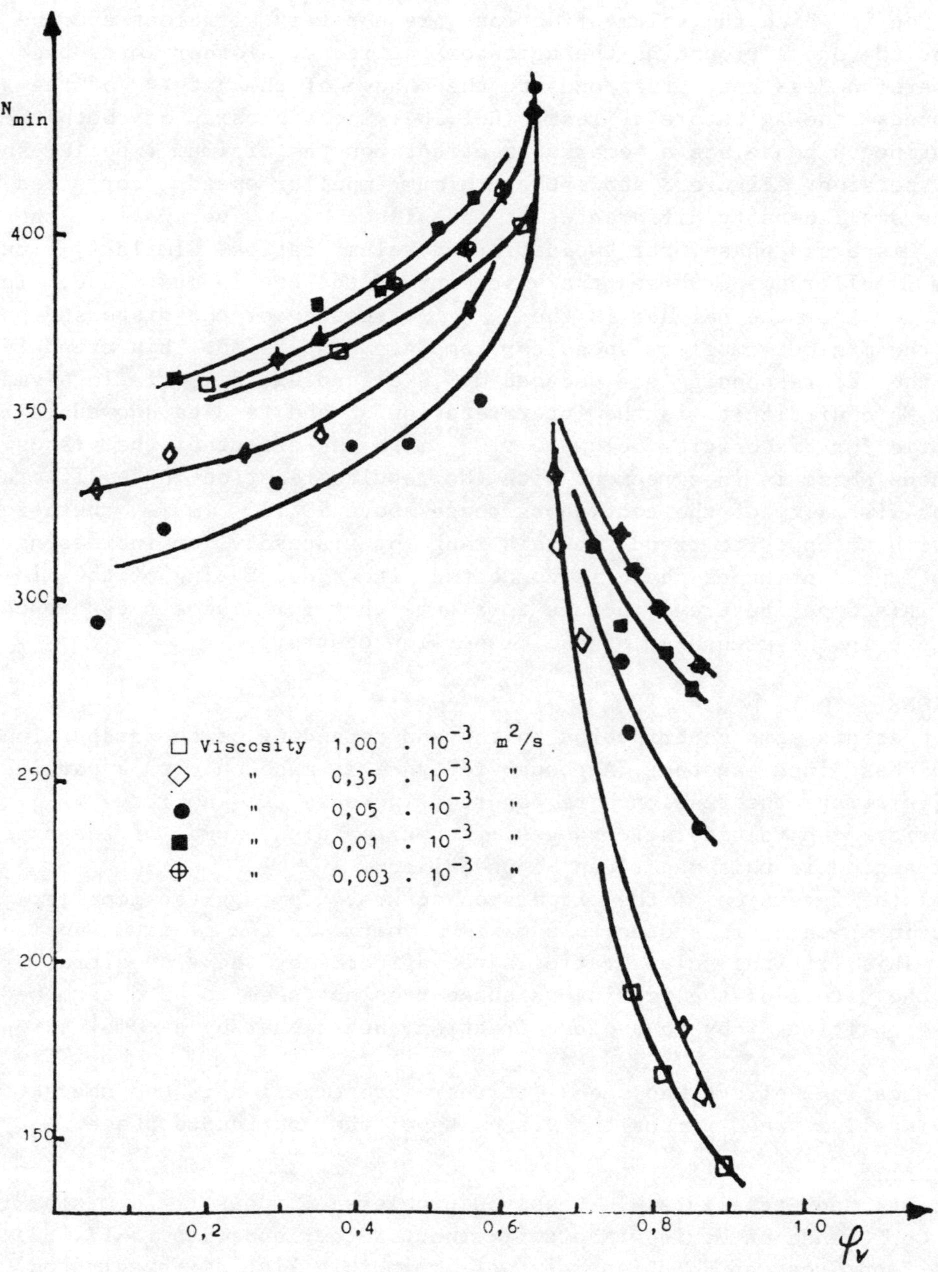

Figure 2

Minimum impeller speed for complete dispersion versus
volume fraction of the oils.

196

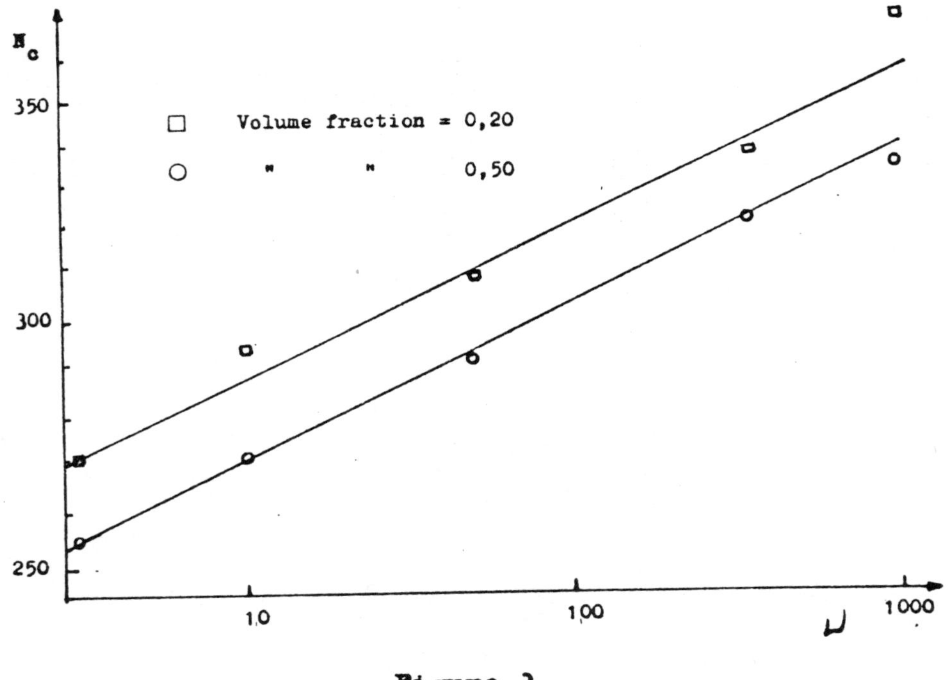

Figure 3

Corrected minimum impeller speed for complete dispersion versus   viscosity of the dispersed phase.

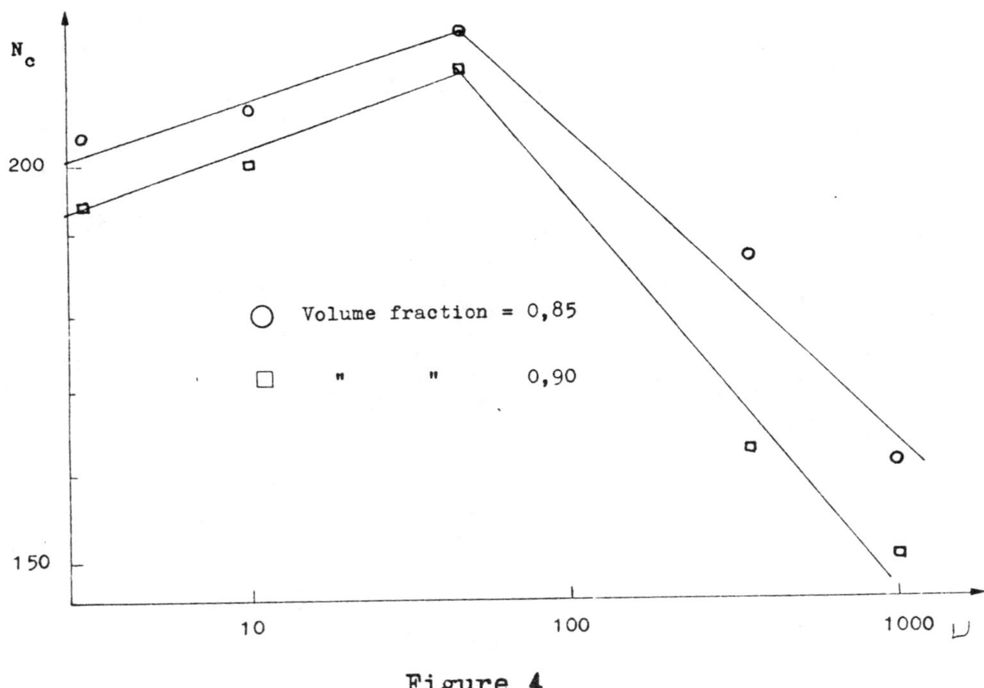

Figure 4

Corrected minimum impeller speed for complete dispersion versus   viscosity of the continuous phase.

LOCAL AND AVERAGE MASS AND HEAT TRANSFER DUE TO TURBINE IMPELLERS

J.R.Bourne, O.Dossenbach and T.Post

Technisch-chemisches Laboratorium ETH, CH-8092 Zurich, Switzerland

Summary

Local mass transfer coefficients between agitated fluids and a tank wall were accurately determined using the electrochemical method, whereby the limiting current for the ferri/ferro-cyanide system in alkaline solution was measured. By setting 172 nickel electrodes in the cylindrical tank wall as well as the dished base, several hundred locations were generated. The tank diameter was 0.158 m and the ratio of height to diameter was 2.2. 6-Bladed Rushton disc turbines and 4 vertical baffles were employed. A data acquisition system controlled experiments and collected and to some extent also evaluated the local mass transfer rate measurements.

Average mass transfer coefficients, calculated for the whole wall surface, were also evaluated and agreed to within 1 % with average heat transfer coefficients, accurately measured in a heat flow calorimeter (Ciba-Geigy). The analogy between heat and mass transfer was thus confirmed. The following variables, affecting average transfer rates, were investigated: contribution of tank base, effect of liquid height and effect of a second turbine in deep tanks on mass transfer and power consumption.

Tangential (horizontal) variations in transfer rate seldom exceed 2 % of their mean values. Axial (vertical) variations of up to nearly one order of magnitude in deep tanks were found. Such vertical profiles were measured for various liquid heights and with one and two turbines. Exponents on the Schmidt and Reynolds numbers varied in the ranges 0.24 to 0.46 and 0.49 to 0.93, respectively. Theoretical interpretations for these variations are offered, whereby for example the exponent on Re is related to the direction of the flow relative to the wall. Local flow patterns deduced from the Re-exponents were confirmed visually and explain, why simplified heat and mass transfer correlations, using constant exponents on Re, Sc and Pr, are not universally valid.

Held at Wurzburg, 10-12 June, 1985.

Organised by DVCV· Deutsche Vereinigung für Chemie- und Verfahrenstechnik
(German Association of Chemical and Process Engineering).

Organisation: GVC·VDI·Gesellschaft Verfahrenstechnik und Chemieingenieurwesen. 

<u>NOMENCLATURE</u>

A	Dimensionless constant in heat/mass transfer correlations
B,C	Exponents on Re and Sc in heat/mass transfer correlations
D	Diffusivity ($m^2 s^{-1}$)
d	Turbine diameter (m)
H	Liquid height in tank (m)
h	Height of turbine above tank base (m)
s	Standard error of fit of Nu and Sh to heat/mass transfer correlations (%)
T	Tank diameter (m)
α	Angle between fluid flow and heat/mass transfer surface
Nu	Nusselt number for process-side, film heat transfer coefficient
Pr	Prandtl number of process-side liquid
Re	Impeller Reynolds number ($N d^2/\nu$)
Sc	Schmidt number of process-side liquid
Sh	Sherwood number for process-side, film mass transfer coefficient ($k_L T/D$)

INTRODUCTION

The results of numerous determinations of the process-side (inside) film heat transfer coefficient in jacketed tanks fitted with a Rushton (straight bladed) turbine have been correlated by equations of the form (Ref.1):

$$Nu = A\ Re^{2/3}\ Pr^{1/3} \tag{1}$$

There are, however, substantial deviations and inconsistencies between some of these correlations (Ref.2,3), which might be explained by the almost limitless number of configurations of tank, impeller, impeller position etc. as well as by differences in measuring and evaluating the inside film coefficient. For instance application of the Wilson method to evaluate the film coefficient from the overall value assumes that the exponent on Re is 2/3 and this biases the exponent obtained when Nu is finally regressed on Re (Ref.3). The frequently used exponents 2/3 and 1/3 in eq.(1) are less well established than is commonly assumed (Ref.3).

Under appropriately chosen conditions, electrochemical mass transfer measurements lead to local film coefficients, which are more accurate than most thermal measurements and which, being local, provide more detailed information about the effects of geometric and operating parameters. By varying the temperature and especially the concentration of supporting electrolyte (NaOH), substantial variations in physical properties and therefore also in Re and Pr are readily obtained. Limiting currents are easily processed with a data acquisition system and rapid and efficient experimentation is possible. The advantages of electrochemical measurements are illustrated here.

EXPERIMENTAL

Fig.1 shows the cylindrical tank (T = 0.158 m, H/T = 2.2), having a dished base, which had four vertical wall baffles of width 0.1 T. A Rushton disc turbine, having 6 blades (d/T = 0.32), was installed at one of two heights above the tank base (h/T = 1/3 and 1/2). There were 144 nickel electrodes (typical area 3×10^{-4} m^2) in the cylindrical wall and 28 in the dished base. The row G^+ were the main monitor cathodes for the limiting current from the electrochemical reaction:

$$Fe''' (CN)_6^{-3} + e^- \rightarrow Fe'' (CN)_6^{-4} \tag{2}$$

The concentrations of potassium ferri- and ferro-cyanide were 10 and 20 mol m^{-3} respectively: the supporting electrolyte (NaOH) was used in concentrations from 500 to 3000 mol m^{-3}. Temperature was in the range 288-308 K and Sc of the ferri-cyanide ion varied between 873 and 6279. Densities, kinematic viscosities and diffusivities - measured with a rotating disc electrode - are available (Ref.4). The limiting current method to measure film mass transfer coefficients has been described elsewhere (Ref.5,6). Turbine power consumption was found by measuring the couple with the head of a Contraves "Rheomat 30" viscometer. Runs were controlled by a microprocessor, which sampled the current from a particular cathode at 6 Hz over 60 s. Further details are given in Ref.3.

AVERAGE HEAT AND MASS TRANSFER COEFFICIENTS

Measured, local film mass transfer coefficients were averaged vertically and horizontally, giving an average Sh number (Sh_{av}) for correlation against Re and Sc in the form $Sh = A\ Re^B\ Sc^C$. With h/T = 1/3

$$Sh_{av} = 0.563\ Re^{0.690}\ Sc^{0.305} \tag{3}$$

(1531 < Re < 25548; 843 < Sc < 6375; s = 2.3 %)

(95 % limits: A ± 18 %; B ± 1.8 %; C ± 3.6 %)

When the exponent on Sc was adjusted to 1/3, eq.(3) became:

$$Sh_{av} = 0.426\ Re^{0.694}\ Sc^{1/3} \tag{4}$$

Accurate thermal measurements in a Ciba-Geigy heat flow calorimeter (Ref.7) for the same configuration had been correlated by:

$$Nu_{av} = 0.420 \ Re^{0.694} \ Pr^{1/3} \tag{5}$$

$$(8 < Re < 46'000; \ 6 < Pr < 30'000; \ s = 3.2 \ \%)$$

The difference between eq.(4) and (5) is 1 %, which is not statistically significant, suggesting that mass and heat transfer follow analogous laws. With h/T = 1/2:

$$Sh_{av} = 0.656 \ Re^{0.698} \ Sc^{0.279} \tag{6}$$

$$(1222 < Re < 33'007; \ 833 < Sc < 6167; \ s = 4.5 \ \%)$$

$$(95 \ \% \ \text{limits}: \ A \pm 23 \ \%; \ B \pm 2.1 \ \%; \ C \pm 5.3 \ \%)$$

The correlations (3) and (6) indicate that the exponents on Re and Sc deviate significantly from the widely accepted values of 2/3 and 1/3.

Using the correction factor of Strek (Ref.8) that $Sh \sim (h/T)^{0.12}$ and adjusting the exponents of eq.(6) to those in eq.(4), eq.(6) becomes:

$$Sh_{av} = 0.422 \ Re^{0.694} \ Sc^{1/3} \tag{7}$$

which agrees so well with eq.(4) and (5) that Strek's correction may be accepted.

Eq.(3-7) apply to an active heat/mass transfer surface which extends from H/T = 1 over the cylindrical, but also over the dished parts of the tank wall (Fig.1). If only the cylindrical surface had been active, Sh_{av} would have been 8 % higher than the values in eq.(3-7), but the loss of the dished base would have reduced the surface area by 28 %. Thus the overall heat flux would have been 20 % lower with a flat, inactive base.

Liquid height was varied, using an unchanged turbine position (h/T= 1/3). The dished base was active. Sh_{av} was correlated by eq.(8):

$$Sh_{av} = A \ Re^{0.709} \ Sc^{1/3} \tag{8}$$

where A decreased from 0.387 (H/T = 1) through 0.316 (H/T = 1.5) to 0.284 (H/T = 2). The mass flux density fell with increasing liquid height, but the overall mass flux increased (Fig.2) e.g. the total surface area when H/T = 2 was 90 % more than when H/T = 1. The power consumption of the turbine remained unchanged. Thus, for example, when H/T = 2 the mass/heat flux would be 40 % higher than when H/T = 1 with no corresponding increase in power.

With increasing liquid height, the local mass transfer coefficient decreases towards the liquid surface when the turbine is located deep in the tank (details see next section). This may be partially offset by adding a second turbine, the spacing between the turbines being of order T. Results were correlated with eq.(9), when H/T = 2:

$$Sh_{av} = A \ Re^{0.698} \ Sc^{1/3} \tag{9}$$

For one turbine (h/T = 1/3) A = 0.314, whilst for two turbines A = 0.41 - 0.42, the value depending somewhat on the location and spacing of the turbines (Ref.3). The power numbers of one and two turbines were 5.3 and 9.5 - 10.0 respectively. Thus, with H/T = 2, adding a second turbine improves the mass/heat flux by some 30 %, but calls for some 85 % more power.

Further results on reducing the degree of baffling are reported elsewhere (Ref.3).

LOCAL MASS TRANSFER COEFFICIENTS

440 positions of the cathode were investigated (Ref.3), a few typical results being presented here. Tangential variation of Sh was of order ± 2 % of its mean value: only axial variations are considered. The maximum and minimum values of Sh were usually observed at the same level as the turbine and just below the liquid surface respectively. Correlations of local Sh (h/T = 1/3, H/T = 1 and 4 wall baffles) produced exponents on Re and Sc in the ranges:

$$0.49 < B < 0.93 \qquad \text{and:} \qquad 0.24 < C < 0.46$$

An interpretation of why B varies will be offered in the next section.

Fig.3 illustrates a typical vertical variation of Sh when H/T = 1. The ratio of maximum and minimum values is 2.2. Fig.4 extends these results to show the much greater variation with increasing liquid height: the ratios rose to 3.2 (H/T = 1.5) and 7.3 (H/T = 2). A second turbine was added to reduce this axial variation in Sh when H/T = 2, the spacing being T, with the result shown in Fig.5: here the ratio of maximum and minimum Sh is only 2.4, comparable to a tank with H/T = 1. Despite the unfavourable rise in power consumption relative to the modest improvement in Sh_{av} (noted in the last section), a second turbine greatly improves the uniformity of mass/heat transfer.

<u>LOCAL MASS TRANSFER AND LOCAL FLOWS</u>

In the correlation:

$$Sh \sim Re^B \tag{10}$$

Re is formed using a characteristic velocity and not the local value, relating to the area over which heat/mass transfer is measured. Fig.6 suggests that B might possibly correlate with the angle α between the flow and the surface. A tentative equation describing at least the two limiting cases in Fig.6, where α = 0 and 90°, is:

$$B = 0.3 (1 - \sin \alpha) + 0.5 \tag{11}$$

For the cases A, B and C in Fig.6, eq.(11) predicts:

A $\alpha = -90^o$ B = 1.1

B $0^o < \alpha < 90^o$ 0.5 < B < 0.8

C $-90^o < \alpha < 0^o$ 0.8 < B < 1.1

Considering mass and heat transfer measurements in the literature for simplified geometries (plate, cylinder, sphere) with various flow angles, reasonable agreement with eq.(11) was found for positive angles (Fig.6, case B) and for negative angles down to about -30°, where B was nearly 1. Flow separation occurred for smaller angles (Fig.6, case C with $\alpha < -30^o$) and B values scattered in the range 0.5 - 0.7 (Ref.3).

Eq.(11) was applied to the stirred tank of Fig.1 in the following way. Local Sh was correlated against Re to give B. The flow angle α was then calculated from eq.(11), giving the flow direction at the wall. The flow pattern was then completed from the principle of continuity. Fig.7 gives an example. This flow pattern was confirmed qualitatively by visualizing the motions of tracer particles in intense planes of light (Ref.3). Further confirmation was obtained from other types of impeller (Ref.3).

<u>CONCLUSIONS</u>

The electrochemical method permitted a detailed picture to be obtained showing the vertical and horizontal variations in local, film mass transfer coefficients for the inside wall of an agitated tank. Once the apparatus had been constructed, rapid, accurate and numerous data collection became possible under the control of a small microprocessor.

The average Sh numbers, obtained by vertical and horizontal averaging of the local measurements, were correlated against the Sc number and the impeller Re number. The exponents B and C on Re and Sc respectively were significantly higher and lower than the commonly used values of 2/3 and 1/3. Local B-values ranged from 0.49 to 0.93 and a hypothesis was advanced relating B to the angle between the flow and the surface. {Further insight into variations of B and C may be obtained by considering the turbulence intensity (Ref.3)}. Since flow patterns in stirred tanks are complex and diverse, there is no reason why a single set of exponents (B and C) should apply to all situations.

Accurately determined heat and mass transfer data agree within 1 %, confirming the analogy between these processes also in a stirred tank. The dished base offers a useful contribution to the total heat/mass flux and it is worthwhile to make the base active. By increasing the liquid depth to H/T = 2, the total heat/mass flux can be

increased by about 40 % with no rise in turbine power consumption. The additon of a
second turbine nearly doubles the power consumed, whilst raising the heat/mass flux by
only about 30 %. {Other examples of a less than proportionate increase in heat and
mass transfer rates relative to the increase in power consumption are available (Ref.9,
10)}. However, the local Sh becomes much more uniform, when using a second turbine in
a deep tank (H/T = 2).

REFERENCES

1. Edwards M.F. and Wilkinson W.L., The Chem.Engr. (1972) No. 264, 310.
2. Buerli M., Diss.No. 6479 (1979), ETH, Zurich.
3. Post T., Diss.No. 7249 (1983), ETH, Zurich.
4. Bourne J.R., Dossenbach O. and Post T., J.Chem.Eng.Data (1985)
5. Grassmann P., Ibl N. and Trueb J., Chem.Ing.Tech. 33 (1961) 529.
6. Selman J.R. and Tobias C.W., Adv.Chem.Eng. 10 (1978) 211.
7. Bourne J.R., Buerli M. and Regenass W., Chem.Eng.Sci. 36 (1981) 347.
8. Strek F., Int.Chem.Eng. 3 (1963) 533.
9. Bourne J.R., Dossenbach O. and Post T., Inst.Chem.Eng.Symp.Ser. 89 (1984) 177.
10. Ibl N., Electrochim.Acta 22 (1977) 465.

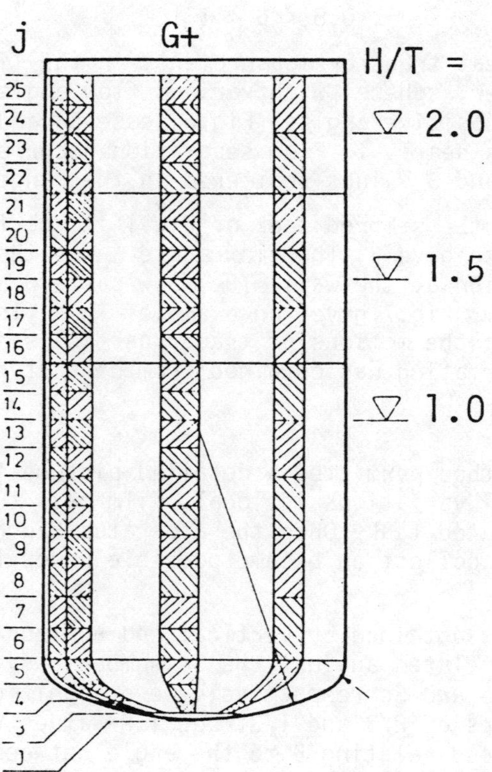

Fig. 1 Tank showing cathode configuration.

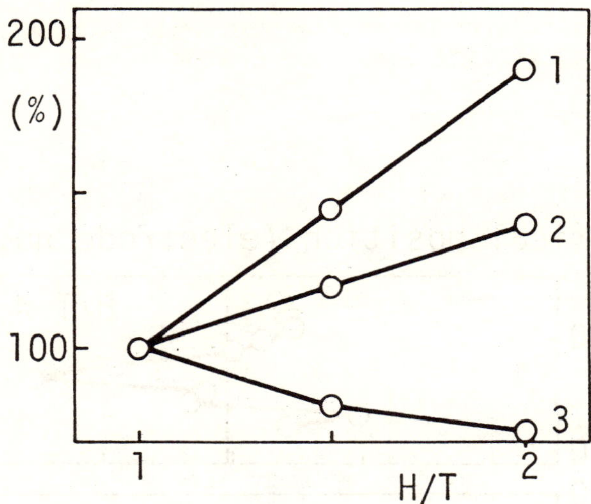

Fig. 2   Effect of liquid height on active
         wall area (1), overall mass flux
         (2) and coefficient A (3) in eq.
         (8), all normalised to H/T = 1.

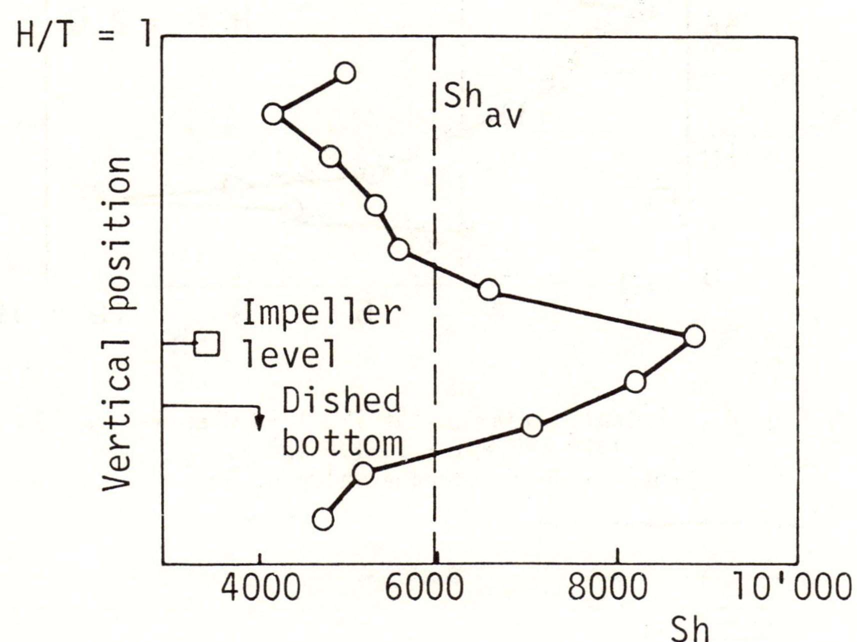

Fig. 3   Vertical variation of local Sh
         when Re = 27'000, Sc = 1470 and
         H/T = 1.

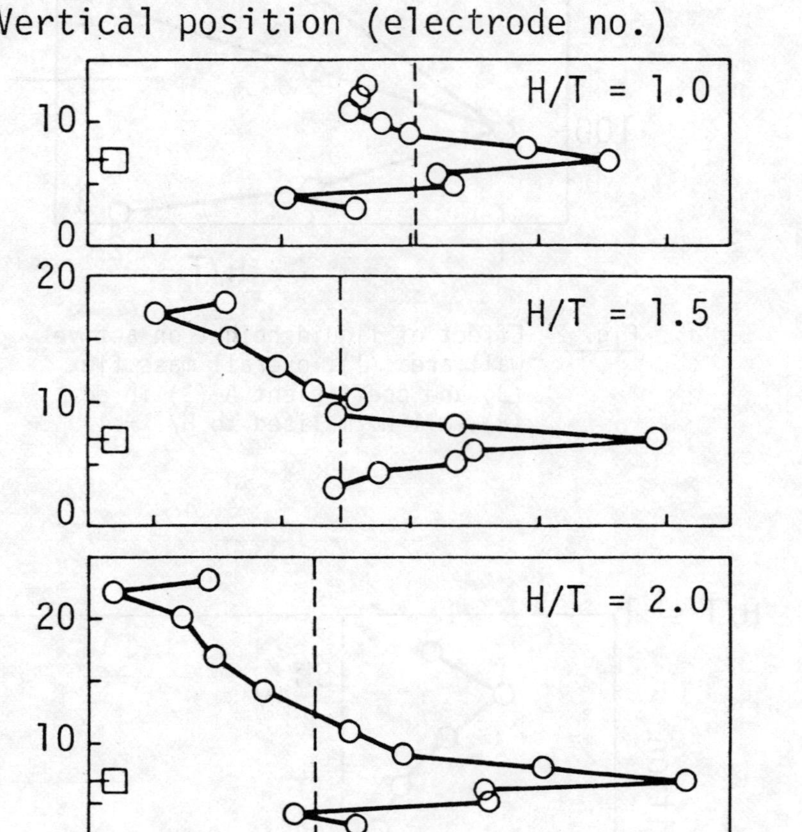

Fig. 4 Vertical variation of local Sh when Re = 27'000,
Sc = 1470 and H/T = 1, 1.5 and 2.
Broken line is average value of Sh.

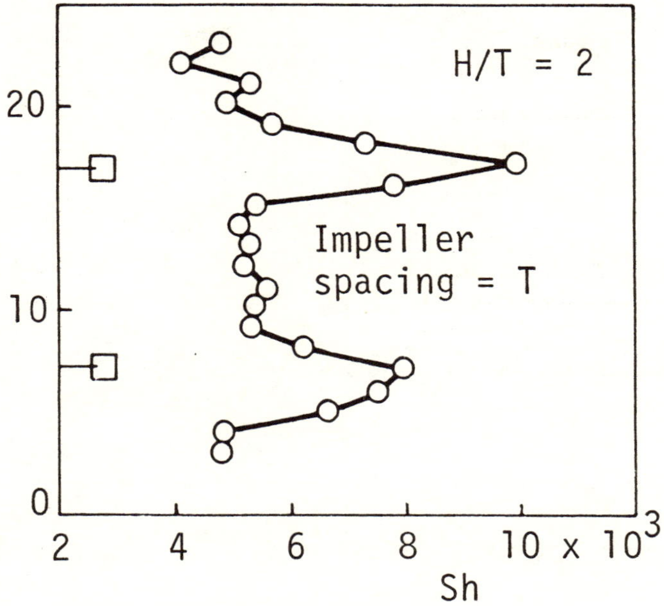

Fig. 5 Vertical variation of local Sh
when Re = 27'000, Sc = 1470,
H/T = 2 and two turbines.

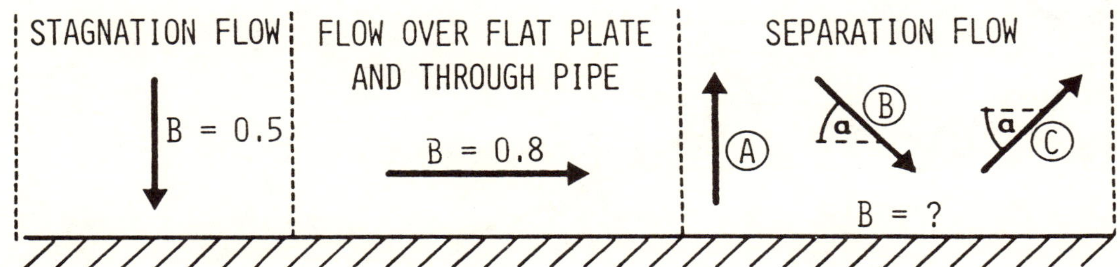

Fig. 6 Values of exponent B on Re for defined flow directions.

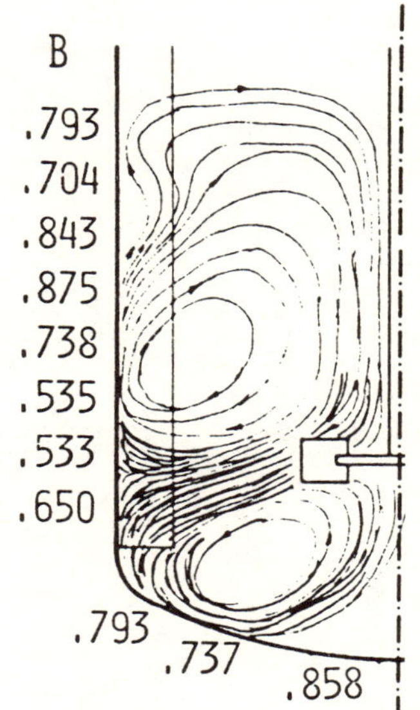

Fig. 7 Values of exponent B and flow pattern at
various positions in stirred tank.

POWER CONSUMPTION AND HEAT TRANSFER IN STIRRED GAS/LIQUID AND GAS/LIQUID/SOLID SYSTEM

A. Steiff

Department of Chemical Engineering,
University of Dortmund
D-4600 Dortmund 50, P.B.O. 500 500
(Germany)

Summary

This contribution deals with the aspects of power consumption, fluid dynamically limiting cases (suspension and flooding points) and heat transfer in stirred gas/liquid- and gas/liquid/solid-systems.

The flooding behaviour in three-phase systems was found resembling to that of two-phase systems. At the same specific power consumption, in large vessels, the maximum distributable gas flow per unit area of cross-section increases with increasing reactor diameter. The influence of particle density, particle diameter and solids mass fraction is small.

In aerated suspensions a higher suspension rotational speed to achieve the state of complete suspension is necessary compared to un-aerated systems. Suspension rotational speed and volume specific power input rise with increasing particle diameter, solids mass fraction and density difference.

Heat transfer in stirred gas/liquid- and gas/liquid/solid systems can be modelled by formally applying the concept of the penetration theory. The heat transfer coefficient is mainly dependent on the effective mass specific energy dissipation near the heat exchange area, the kinematic viscosity and the heat penetration coefficient of the liquid. An increase in density, specific heat capacity and heat conductivity results in an improvement and an increase in viscosity results in a deterioration of the heat transfer.

In stirred three-phase systems the solid phase influences the heat transfer very strongly. The deterioration compared to the two-phase gas/liquid system is caused by the dampening of the turbulence by the particles.

The two-stage stirrer with sufficient stirrer distance on the shaft generally leads to higher heat transfer coefficients at a given overall power input than the corresponding single-stage stirrer.

Held at Wurzburg, 10-12 June, 1985.

Organised by DVCV· Deutsche Vereinigung für Chemie- und Verfahrenstechnik (German Association of Chemical and Process Engineering).

Organisation: GVC·VDI-Gesellschaft Verfahrenstechnik und Chemieingenieurwesen. 

NOMENCLATURE

a = coefficient
b = heat penetration coefficient
c_p = specific heat capacity
D = stirrer diameter
d_p = particle diameter
e = bottom distance of the
    sparger ring
g = acceleration due to gravity
H = height
H_R = stirrer clearance
K = constant
m = mass
n = stirring speed
P = power requirement of the
    stirrer
P_G = power input of the gas
T = temperature
T = reactor diameter
t_s = time scale of Kolmogorov
V = volume
V* = volumetric flow rate
v_{Go} = superficial gas velocity
α = heat transfer coefficient
$\varepsilon_G \equiv \dfrac{V_{aer}-V}{V_{aer}}$ = gas hold-up
ε = mass specific overall
    energy dissipation
ε_B = mass specific power input
    of the gas
ε_{Rw} = effective mass specific power
    input of the stirrer
η = dynamic viscosity
λ = heat conductivity
ν = kinematic viscosity
ξ = energy distribution parameter
ρ = density
σ = surface tension
τ = contact time of the eddies
    at the wall
$\psi_S \equiv m_S/(m_L+m_S)$ = solids mass
    ratio
$\pi_h \equiv H_{Sch}/H$ = relative layer
    thickness

Subscripts

aer = aerated
G = Gas
L = Liquid
S = Solid
Sch = layer
Sp = at "complete" suspension
tot = total
ü = shortly before flooding
w = effective
w = wall
0 = un-aerated
1,2 = lower and upper

Dimensionless number

$$Fr \equiv \frac{n^2 \cdot D}{g} = \text{Froude number}$$

$$Ga \equiv \frac{D^3 \cdot g}{\nu^2_L} = \text{Galilei number}$$

$$Ne \equiv \frac{P}{\rho_L n^3 \cdot D^5} = \text{Newton number}$$

$$Nu \equiv \frac{\alpha \cdot T}{\lambda_L} = \text{Nusselt number}$$

$$M \equiv \frac{g \cdot \eta_L^4}{\sigma^3 \rho_L} = \text{Morton number}$$

$$Pr_L \equiv \frac{\eta_L \cdot c_{pL}}{\lambda_L} = \text{Prandtl number}$$

$$(P/V)^* \equiv \frac{P/V}{\rho_L (\nu_L g^4)^{1/2}} = \text{Specific power number}$$

$$Re \equiv \frac{n \cdot D^2}{\nu_L} = \text{Reynolds number}$$

$$Re_G \equiv \frac{v_{Go} \cdot T}{\nu_L} = \text{Reynolds number (gas)}$$

$$We \equiv \frac{n^2 \cdot D^3 \cdot \rho_L}{\sigma} = \text{Weber number}$$

$$\pi_G \equiv \frac{V_G^*}{n \cdot D^3} = \text{Gas through-put number}$$

$$\pi_k \equiv \left(\frac{P/m}{\nu_L^3 \cdot d_p^{-4}}\right)^{1/3} = \text{Kolmogorov number}$$

1. INTRODUCTION

In gas/liquid- and gas/liquid/solid systems the liquid forms
the continuous phase, whereas gas and solid exist in dispersed state.
The function of the stirrer in gas/liquid systems is to disperse the gas
needed for the reaction. In gas/liquid/solid system it additionally has
to suspend the solid phase.

2. FLUID DYNAMICS

2.1 Gas Dispersion

The gas can be fed into the reactor via forced aeration
(pressure aeration), suction aeration or surface aeration. In forced
aeration pressurized gas is introduced and at the same time predistri-
buted through static gas distributors. Suction aeration uses the underpressure behind the edges of a hollow stirrer. The gas flow rate suct
in through the hollow agitator shaft directly depends on the stirrer
speed.

Surface aeration occurs in systems without baffles when the
vortex reaches down to the stirrer at high stirrer speeds.

Fig. 1 showing the mean gas hold-up as a function of volume
specific power consumption gives a qualitative comparison of all three
kinds of aeration. The geometrical relations used here were found to be
optimal in our preinvestigations. The hollow agitator even at high
stirrer speeds only allows to reach a relatively small gas hold-up. The
complete aeration starts only above a characteristic agitator speed, at
which the suction overcomes the hydrostatic pressure and the pressure
loss due to friction in stirrer shaft.

By surface aeration using a pitched blade turbine a higher gas
hold-up can be achieved compared to the hollow stirrer. However a substantially higher power consumption is necessary. By forced aeration a
higher gas hold-up can generally be reached at same power input. The gas
hold-up in this case depends as well on stirrer speed as on gas velocity.

As the self-aerating stirrer systems are practically only interesting in special cases, only forced aerated systems are regarded in
the following. The stirrer is placed near the bottom to allow simultaneously dispersing the gas and suspending the solid particles.

2.2 Flooding

In stirred gas/liquid reactors as also in stirred three-phase
reactors, the gas through-put which can be distributed by the stirrer,
is limited. Increasing the gas flow, under otherwise constant operating
conditions, will eventually exceed the dispersing capacity of the
stirrer which will then become flooded. At this system state gas only
occurs above the stirrer forming a channel of large bubbles agglomerates.
Thus the effect of the stirrer on the circulation flow of the liquid is
largely reduced. In suspensions this leads to settling of the particles
on the bottom. For this reasons the state of flooding has to be prevented in both systems.

On introduction of gas into the stirred liquid the power consumption and thus the Newton number decreases (compare Fig. 2). At small
aeration rates the decrease is relatively weak. With increasing gas flow
rate and at high stirrer speed the influence of aeration on power consumption becomes more clear as the spreading of the curves shows. Before
reaching the state of flooding the power consumption of the stirrer gets
independent of gas flow rate. On reaching the point of flooding a jump

in power consumption is observed, whereas it slightly decreases again
on further increase of the gas flow rate.

Correlations describing the flooding characteristics in aerated
liquids and suspensions

$$\pi_{G\ddot{u}} = 0.973 \left(-0.15 + \sqrt[4]{M} + 0.05 \ln \left(\frac{We}{Fr} \right) \right) Fr^a + 0.2 \cdot Fr^{1.6} \tag{1}$$

with

$$a = 0.605 + 1 \, / \, \left(\ln \frac{We}{Fr} \right) .$$

and the power consumption at the point of flooding

$$(P/V)^* = 0.0489 \cdot Fr^{1.18} \cdot Ga^{0.176} \cdot (1 + \psi_s)^{0.36} . \tag{2}$$

are given by Wiedmann (Ref. 1).

The main influential parameters regarding the flooding at a
given agitator speed are the superficial gas velocity and the reactor
diameter. The presence of particles and thus also their properties, as
well as the liquid viscosity have only small influence. At given volume
specific power input a higher gas flow per unit cross-sectional area
is distributable in larger vessels than in smaller ones with the pre-
sumption of geometrical similarity. This yields for two- as well as for
three-phase systems.

2.3 Suspending

An essential purpose of stirring in suspensions is to whirl up
and to distribute the solid particles in the reactor. Suspending of
particles is only possible if the continuous phase is lighter than the
solid phase. The difference between gravity and buoyant force produces
a more or less strong tendency to phase separation and settling of the
solid phase. The following characteristic states of suspension can be
distinguished:

- "complete" suspension,
- layer thickness π_h,
- "homogeneous" suspension.

The state of complete suspension is considered as the state at
which just all particles are suspended and are set in motion after
resting on the bottom of the vessel for not longer than 1 s.

Independently of the fact whether all particles are suspended,
two distinct zones may occur. The lower one contains almost the whole
solid content, whereas the upper one is almost particle-free. Both zones
can be distinguished by a distinct, more or less fluctuating border
(layer thickness π_h). Homogeneous distribution of particles requires a
high power input into the stirred reactor. As both, the state of layer
thickness π_h and the state of homogeneous suspension are only applied
in special cases, further considerations regarding suspensions will be
restricted to the state of "complete" suspension, at which just the
whole particle surface is available for transport processes.

In aerated suspensions a higher stirring speed n_{Sp} to achieve
the state of complete suspension is necessary compared to un-aerated
systems (compare Fig. 3, the values of the latter are presented on the
left margin in the diagram). The measurements also reveal a distinct
increase of the suspension rotational speed with increasing superficial
gas velocity.

With higher rotational speed for complete suspension at the same time a higher volume specific power input is required. Both variables rise with increasing particle diameter, solids mass fraction and density difference $\Delta\rho = \rho_S - \rho_L$, with $\rho_S > \rho_L$. The specific power input decreases with increasing vessel diameter.

Wiedmann (Ref. 1) developed the following correlations for the Reynolds number required for complete suspension

$$Re_{Sp} = 5.58 \cdot (Re_G + 1)^{0.083} \cdot \psi_S^{0.12} \cdot Ga^{0.462} \cdot \left(\frac{d_p}{D}\right)^{0.26} \cdot \left(\frac{\Delta\rho}{\rho_L}\right)^{0.51} \qquad (3)$$

and for the corresponding dimensionless power input:

$$\pi_k = 1.99 \cdot Ga^{0.477} (Re_G + 1)^{0.035} \cdot \left(\frac{d_p}{D}\right)^{1.57} \cdot \left(\frac{\Delta\rho}{\rho_L}\right)^{0.53} \cdot \psi_S^{0.12} \qquad (4)$$

with

$$\pi_k = \left(\frac{P/m}{v_L^3 \cdot d_p^{-4}}\right)^{1/3} \quad .$$

3. HEAT TRANSFER

In order to obtain desired reaction temperatures and for temperature control in stirred vessels it is often necessary to supply or withdraw heat either through the vessel wall or through additional heat transfer installations.

Heat transfer "wall/fluid" in stirred vessels can be modelled by formally applying the concept of the penetration theory to heat transfer. In contrast to bubble columns (compare Ref. 3), in stirred vessels the locally very differing energy dissipation has to be considered.

Regarding the temperature function in eddies at contact with the wall and estimating the mean contact time using Kolmogorov theory yields the following expression for the mean heat transfer coefficient "wall/fluid":

$$\overline{\alpha}_w = \frac{2}{\sqrt{\pi}} \sqrt{\varrho_L \cdot c_{pL} \cdot \lambda_L} \cdot \sqrt{\frac{1}{\tau}} \qquad (5)$$

with heat penetration coefficient $b = \sqrt{\rho_L \cdot c_{pL} \cdot \lambda_L}$
and contact time $\tau = ct_S = c \left(\frac{v_L}{\overline{\varepsilon}}\right)^{1/2}$.

In the aerated stirred vessel the overall energy dissipation consists of two portions, a portion ε_{Rw} induced by the stirrer and a portion ε_B induced by the gas.

$$\overline{\varepsilon} = \varepsilon_{Rw} + \varepsilon_B \quad . \qquad (6)$$

Estimations for the effective power input of the stirrer and the energy dissipation by the gas (compare Ref. 4) lead to following expression for the heat transfer coefficient in stirred aerated systems:

$$\alpha_w = \text{konst.} \left[\xi \frac{P/V}{\rho_L} + v_{G0} \cdot g \left(1 - \frac{2}{g \cdot T(H_0 - e)} \cdot \sqrt{\frac{P \cdot n \cdot T}{\rho_L}}\right)\right]^{0.25} \left(\frac{\rho_L^3 \, c_{pL}^2 \, \lambda_L^2}{\eta_L}\right)^{0.25} \quad . \qquad (7)$$

The formal discussion of this equation shows that the effective specific energy dissipation near the heat exchange area is a main variable.

At a given rotational speed the power consumption of the stirrer (compare Fig. 4) initially decreases with increasing superficial gas velocity and than remains constant until the stirrer gets flooded. The specific power input by the gas thereby increases steadily

but gains importance only at high aeration intensity. The effective
volume specific power input is the sum of the gas power input and the
weighted stirrer power input. The effective power input initially
decreases slightly and than increases again with rising superficial gas
velocity.

The corresponding dependence of the heat transfer as a function of superficial gas velocity for different rotational speeds is
shown in fig. 5. The curves for the energy dissipation and the heat
transfer are quite similar.

The employment of stirrers in systems which are determined by
the heat transfer seems only to be valuable in case of small aeration
intensities. At high gas flow rates in a bubble column even a little
higher heat transfer coefficient can be achieved than in a stirred
vessel (Ref. 5).

Extensive measurements by Kurpiers (Ref. 4) in single- and
two-stage stirred gas/liquid systems confirmed the validity of the
model equation. Only the exponent of the Prandtl number had to be modified from the theoretical value 0,5 to 0,4 due to the measurements. In
dimensionless form the correlation finally yields:

$$N_{U_w} = K\left[\xi\left(\frac{T}{D}\right)\left(\frac{T^3}{V}\right)Ne\cdot Re^3 + Ga\cdot Re_G\left\{1 - 2\left(\frac{D}{T}\right)\left(\frac{D}{H_\delta e}\right)Ne^{0.5}\,Fr\,\right\}\right]^{0.25}\cdot Pr_L^{0.4}\cdot\left(\frac{\eta}{\eta_w}\right)_L^{0.23} \quad . \tag{8}$$

The single-stage stirred system with T/D = 3 could be well
described with the only empirical constants of the model K = 0.1 and
ξ = 0.2.

In stirred three-phase systems solid phase is often very unevenly distributed and influences at the same time the heat transfer
very strongly. A deterioration of the heat transfer compared to the
two-phase system gas/liquid is clearly visible. This is caused by the
dampening of the turbulence by the particles.

The disadvantageous distribution of the solid particles in two-
and three-phase systems can be positively influenced by employing multi-
stage stirrers. Kurpiers (Ref. 4) showed that the distance of two
stirrers on the shaft should be at least two stirrer diameters in order
to achieve optimal improvement of the heat transfer.

A clear evaluation of the single- and two-stage installation
regarding the stirring purpose "heat transfer" is possible when the
power consumption required for the same heat transfer in both systems
is known (compare fig. 7).

It is clearly visible that the two-stage stirrer generally
leads to higher heat transfer coefficients at a given overall power
input than the corresponding single-stage stirrer.

4. CONCLUSIONS

When the volumetric gas flow rate is too high for a particular
stirring speed, flooding occurs, the stirrer loses its effectiveness,
and the gas passes up through the reactor without being dispersed. In
three-phase systems the solid settles down on the bottom.

For the same stirring power per unit volume, a higher gas flow
rate per unit cross-sectional area of the vessel can be dispersed in a
larger reactor, depending on the superficial gas velocity, than in a
smaller reactor, for both two- and three-phase systems.

The influence of the particle density, the particle diameters and the solids mass fraction is insignificant. For reaching the state of complete suspension (1 s-criterion) in three-phase systems usually a higher power input is necessary than in two-phase systems.

The power consumption and the corresponding stirring speed to achieve the state of "complete" suspension rise with increasing particle density, increasing particle diameter and solids mass fraction. The heat transfer in stirred single- and multi-phase systems can be described with an uniform concept. By formally applying the concept of the penetration theory to the heat transfer "wall/fluid" also considering the local specific energy dissipation, equations were developed, which allow a precise description of the heat transfer in stirred liquids, gas/liquid- and gas/liquid/solid systems. These equations were verified by extensive measurements in stirred single-, two- and three-phase systems with single- und multi-stage stirring arrangements. - The heat transfer coefficient in the suspended state usually is smaller than the one in the non-aerated systems, but it rises with increasing stirring speed and a higher gas through-put.

5. ACKNOWLEDGEMENTS

This work was carried out under support provided by the Deutsche Forschungsgemeinschaft and the Arbeitsgemeinschaft der Großforschungseinrichtungen, Bonn.

The author is appreciate to P.-M. Weinspach, who motivated this investigations. The investigations on fluid dynamics and heat transfer were carried out by Mr. J.A. Wiedmann and Mr. P. Kurpiers.

6. REFERENCES

1.   Wiedmann, J.A.: "Zum fluiddynamischen Verhalten zwei- und dreiphasig betriebener Rührreaktoren".
     Thesis, University of Dortmund, 1982.

2.   Judat, H.: "Zum Dispergieren von Gasen".
     Thesis, University of Dortmund, 1976.

3.   Deckwer, W.-D.: "On the Mechanism of Heat Transfer in Bubble Column Reactors".
     Chem. Eng. Sci. 35, 1980, pp. 1341-1346.

4.   Kurpiers, P.: "Wärmeübergang in gerührten Ein- und Mehrphasenreaktoren beim Einsatz ein- und zweistufiger Rührer".
     Thesis, University of Dortmund, 1984.

5.   Nottenkämper, R.: "Indirekter Wärmeübergang in Blasensäulen".
     Diploma-thesis, University of Dortmund, 1979.

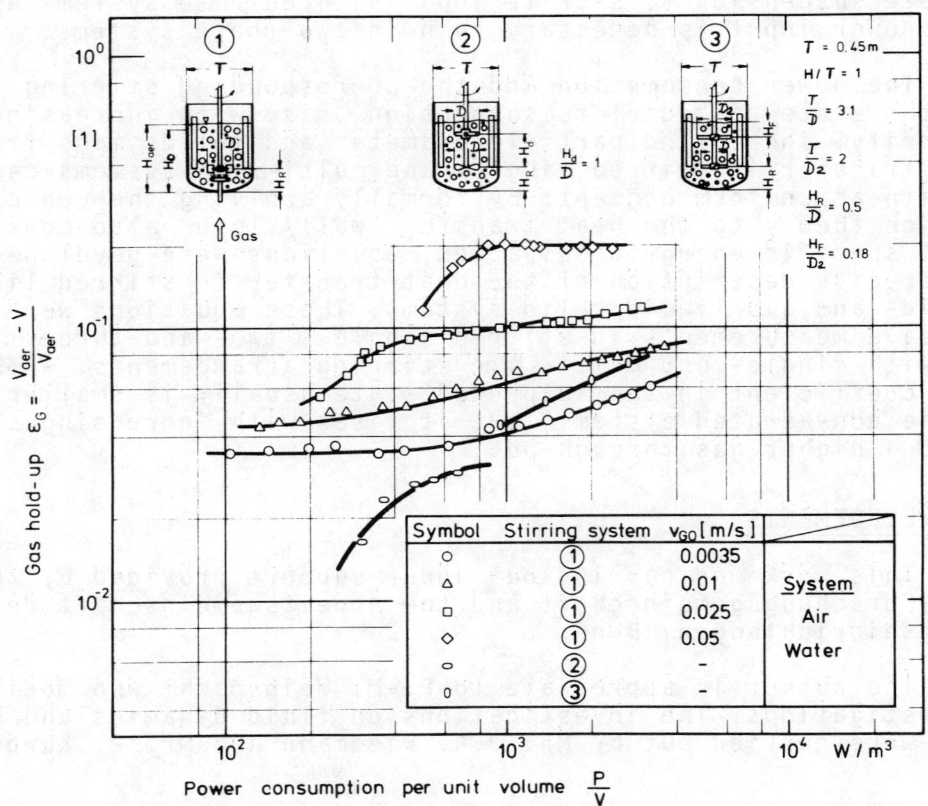

Gas hold-up $\varepsilon_G = \dfrac{V_{aer} - V}{V_{aer}}$

Power consumption per unit volume $\dfrac{P}{V}$

Figure 1   Gas hold-up as function of stirring power per unit volume in the gas/liquid systems

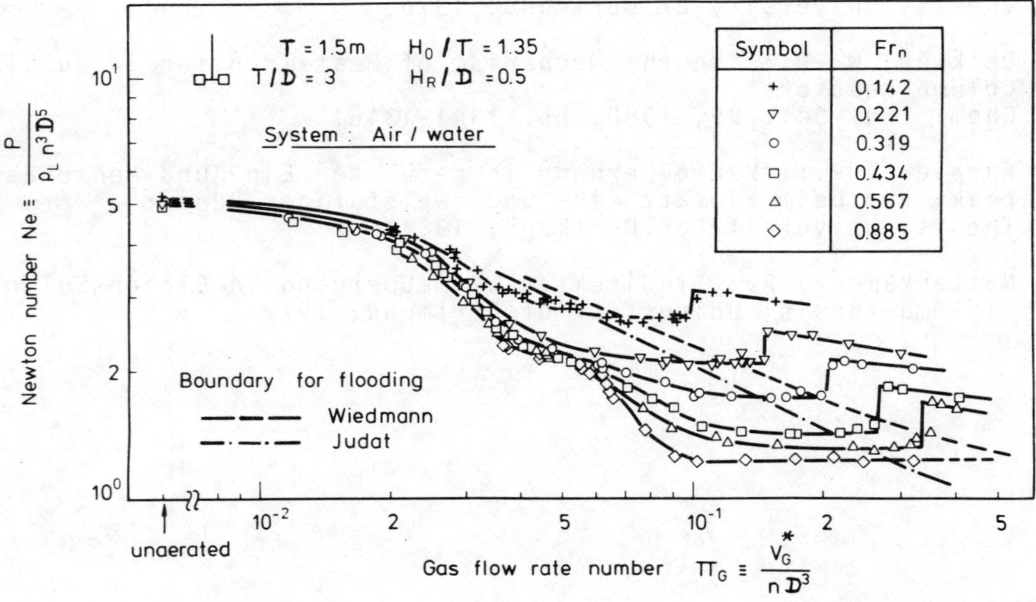

Newton number $Ne \equiv \dfrac{P}{\rho_L n^3 D^5}$

Gas flow rate number $\pi_G \equiv \dfrac{V_G^*}{n D^3}$

Figure 2   Ne = f $(\pi_G,\ Fr)$ in a vessel with turbine stirrer (T = 1,5 m); curves for the boundary for flooding (Ref. 1 and 2)

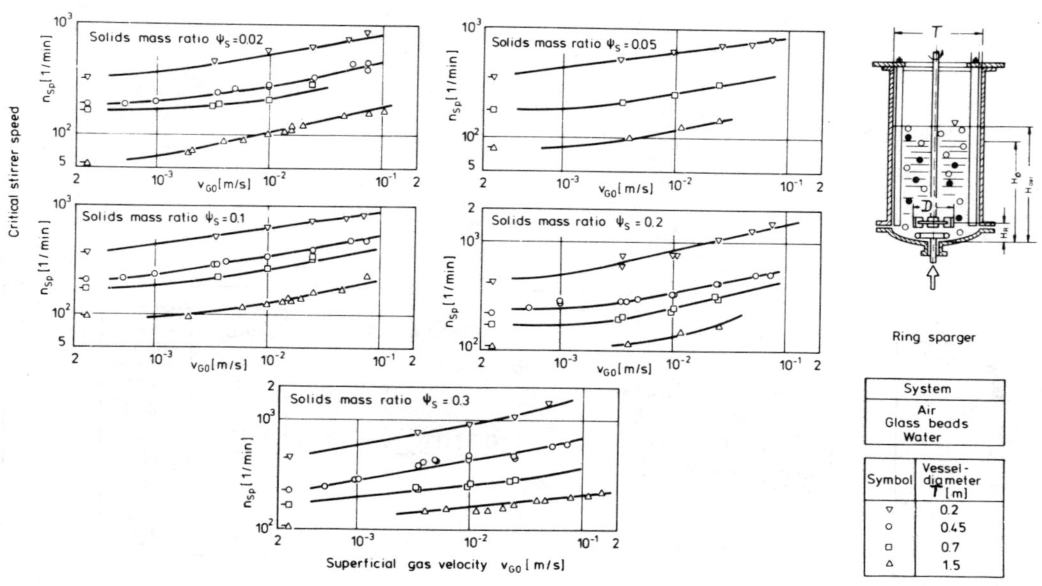

Figure 3   Critical stirring speeds as function of superficial gas velocity, vessel diameter and solids mass fraction

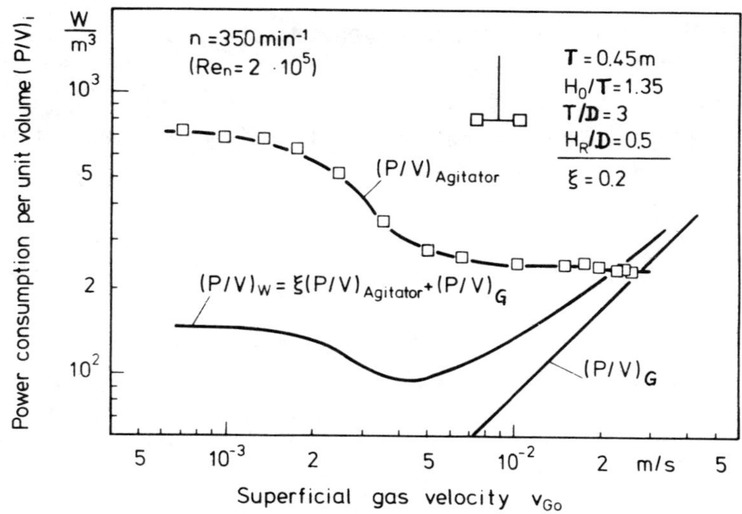

Figure 4   Determination of the effective specific energy dissipation near the heat exchange area

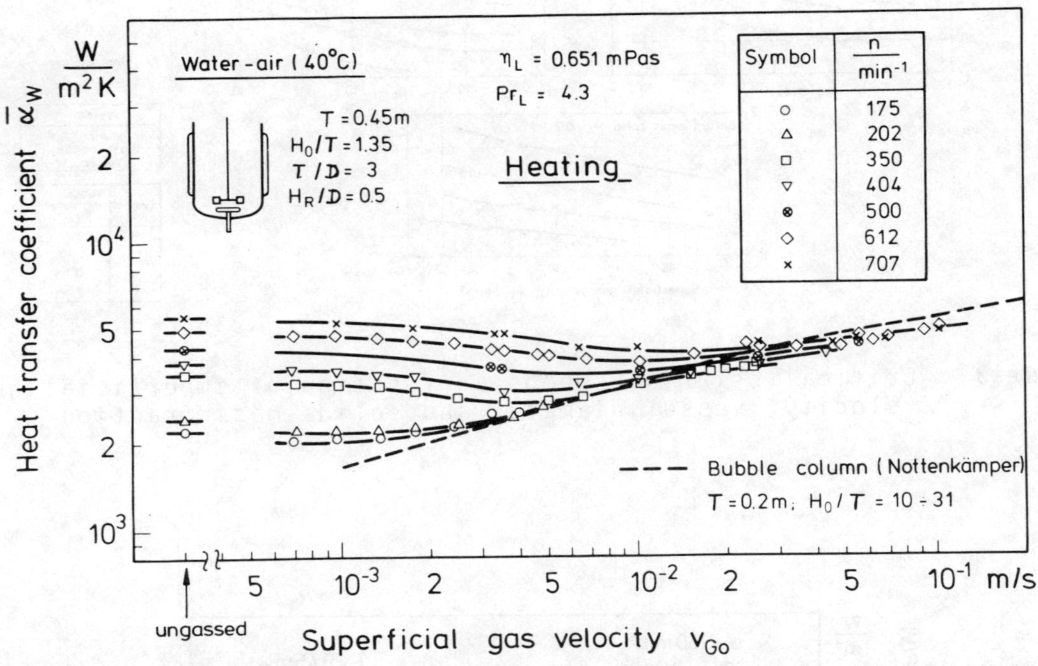

Figure 5   Influence of aeration intensity on heat transfer

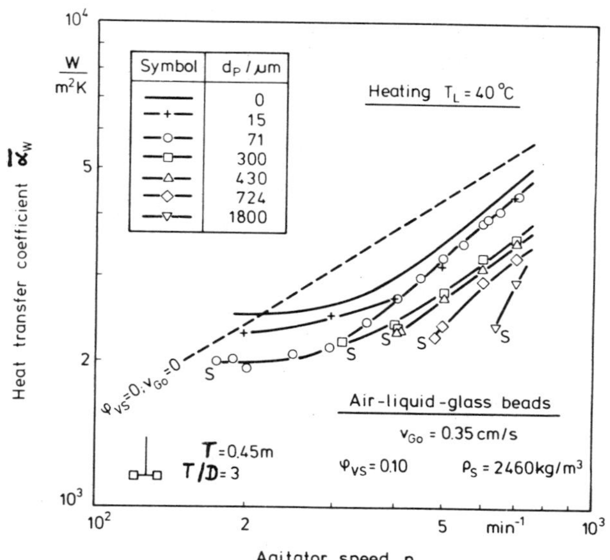

Figure 6   Influence of the particle size on heat transfer in stirred three-phase systems

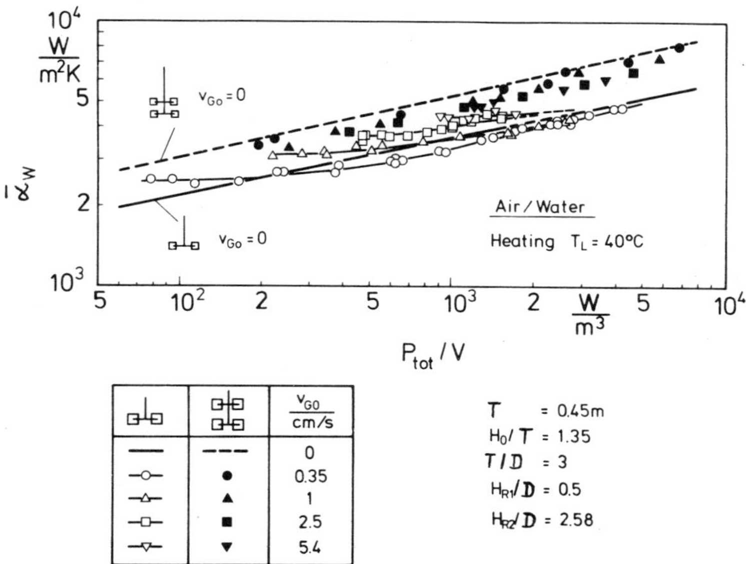

⊡	⊞	v_Go / cm/s
—	- - -	0
○	●	0.35
△	▲	1
□	■	2.5
▽	▼	5.4

Figure 7   Influence of the total power input per unit volume on heat transfer in a gas/liquid system

A STUDY OF LOCAL HEAT-TRANSFER COEFFICIENTS IN
MECHANICALLY AGITATED GAS-LIQUID VESSELS

by

K. L. Man

National Engineering Laboratory, East Kilbride, Glasgow, UK

SUMMARY

Local variations in mass-transfer coefficient to the vessel wall of a 0.3 m diameter aerated, agitated vessel have been determined using an electrochemical technique. Agitation was provided by a six-bladed disk turbine. Using an appropriate analogy, the mass-transfer results were converted to heat-transfer data.

The experimental data indicated that the profile of local heat-transfer coefficient at the vessel wall depends on the gas flowrate and the impeller speed. These observations could be qualitatively related to the flow pattern in gas-liquid vessels.

The average heat-transfer coefficients have been calculated from the local values and these results indicated that for the aeration conditions employed (the superficial velocity V = 0.0014-0.005 m/s) the heat-transfer rates are within ±15 per cent of the values obtained without gas addition.

Held at Wurzburg, 10-12 June, 1985.

Organised by DVCV· Deutsche Vereinigung für Chemie- und Verfahrenstechnik
(German Association of Chemical and Process Engineering).

Organisation: GVC·VDI-Gesellschaft Verfahrenstechnik und Chemieingenieurwesen. 

NOMENCLATURE

A	= Cathode area	m^2
B	= Baffle width	m
C_b	= Bulk concentration	$kg\ m^{-3}$
C_p	= Specific heat	$J\ kg^{-1}\ K^{-1}$
D	= Impeller diameter	m
E	= Constant in equation (1)	–
F	= Faraday constant	–
H	= Total liquid height	m
I	= Electric current	amp
n	= Impeller speed	s^{-1}
T	= Tank diameter	m
x	= Axial distance measured from centre of impeller plane	m
α	= Heat-transfer coefficient	$W\ m^{-2}\ K^{-1}$
η	= Liquid dynamic viscosity	$N\ s\ m^{-2}$
λ	= Liquid thermal conductivity	$W\ m^{-1}\ K^{-1}$
ρ	= Liquid density	$kg\ m^{-3}$
$\mathscr{D}$	= Molecular diffusivity	$m^2\ s^{-1}$

Dimensionless Groups

Nu	= Nusselt number for wall	$\alpha T/\lambda$
Pr	= Prandtl number	$\eta C_p/\lambda$
Re	= Reynolds number	$\rho n D^2/\eta$
Sc	= Schmidt number	$\eta/\rho \mathscr{D}$
Vi	= Viscosity ratio	η/η_w

1. INTRODUCTION

Mechanically agitated vessels are in common use throughout the chemical, food and
fermentation industries. In many instances heat removal or addition is of major
importance to maintain the optimum processing conditions. This can be achieved using
either a jacketted vessel or a helical coil immersed in the vessel. A large number of
experimental heat-transfer studies in single-phase liquid systems have been carried out
by research workers. Results of these studies are frequently reported in the
conventional, forced convection form ie

$$Nu_{j,c} = ERe^a Pr^b Vi^c \qquad (1)$$

where the constants, E, a, b and c are found by fitting equation (1) to experimental
data. A review of many such correlations has been presented by Edwards and Wilkinson
(Ref. 1), and more recently by Poggemann et al (Ref. 2).

Many chemical processes such as hydrogenation and chlorination require the processing
of two-phase gas-liquid mixtures in mechanically agitated vessels. Therefore, a knowledge of the effect of a second gas phase on the rate of heat transfer in the agitated
vessel is of considerable importance. In recent years several papers (Refs 3-7), have
been published pertaining to gas-liquid heat-transfer studies in agitated vessels.
However, no quantitative data are available at present on the study of local heat
transfer in agitated aerated systems. These data will provide a useful guide of the
transport processes occurring at the heat-transfer surfaces. The purpose of this
present paper is to report experimental results of local heat-transfer measurements in
aerated Newtonian liquids at the wall of an agitated vessel. Agitation is provided by
a turbine impeller. Local mass-transfer coefficients have been measured using an
electrochemical technique and corresponding heat-transfer coefficients determined from
the Chilton-Colburn analogy (Ref. 8). An attempt has also been made to explain these
local heat-transfer profiles by relating them to the observed gas-liquid flow pattern.

2. LITERATURE SURVEY

Several researchers have presented heat-transfer data in mechanically agitated aerated
systems. Rao and Murti (Ref. 3) measured heat-transfesr coefficients using Newtonian
liquids in two separate jacketted vessels having diameters of 0.2 and 0.7 m respectively. Each vessel was unbaffled and equipped with a helical coil. Agitation was
provided by a six-bladed turbine impeller. They were able to correlate their data
using an extended form of equation (1) with a modified Reynolds number to include the
superficial gas velocity term. They found that at all impeller speeds sparging of gas
improved the heat-transfer coefficients in a specific manner; the enhancement factor is
reduced with increased impeller speed. However, the results obtained in a 0.3 m^3
stirred tank by Pollard and Topiwala (Ref. 4) suggested that the relationship between
heat-transfer coefficient and impeller Reynolds number is not explicit. They proposed
that power input and gas hold-up should be the main parameters to correlate heat-
transfer data in gas-liquid systems.

Edney and Edwards (Ref. 5) carried out an experimental investigation on heat transfer
to a helical coil for aerated agitated Newtonian and non-Newtonian liquids in a 1.22 m
diameter baffled vessel. Their heat-transfer data indicated that for the aeration
conditions used ($\leqslant 0.02$ m/s) the heat-transfer rates remained essentially unchanged from
those observed with no aeration. This finding is somewhat different from those
reported by Steiff and Weinspach (Ref. 6) and De Maertelaire (Ref. 7). Steiff and
Weinspach (Ref. 6) carried out their work in three different sizes of vessel (T = 0.19,
0.45, 0.7 m) and with different liquid height-tank diameter ratio. They found that the
jacket heat-transfer coefficient increased with superficial gas velocity in the range
of less than 0.05 m/s at low impeller speeds. However, further increase in the gas
velocity did not improve the heat-transfer rate. At high impeller speeds, the gas
flowrates had no significant effect on the heat transfer. De Maertelaire (Ref. 7)
measured the rates of heat transfer to a coil in a baffled vessel of 0.18 m diameter.
He reported that below a critical impeller speed the coil heat-transfesr coefficient
increased with gas flowrates. For speed above the critical value, the α_c decreased
with gas flowrates.

It can be seen that there is serious disparity in these studies. Although part of the
reason for this is probably that different vessels as well as gas sparger geometries
were employed in these various investigations. In the study reported it is clear that
the situation is far from satisfactory. This paper set out to gain more understanding
of the process by concentrating on local heat-transfer studies and then relating these
results to the gas-liquid flow pattern in the agitated vessel.

3.   EXPERIMENTAL METHOD AND APPARATUS

The electrochemical system used in this study utilises the reduction of potassium
ferricyanide to potassium ferrocyanide

$$Fe(CN)_6^{3-} + e^- \rightarrow Fe(CN)_6^{4-} \text{ at cathode}$$

$$Fe(CN)_6^{4-} \rightarrow Fe(CN)_6^{3-} + e^- \text{ at anode.}$$

To accomplish this method nickel electrodes are placed upon the surface of interest (in
this case curved nickel electrodes were embedded in the vessel wall) and a potential is
applied across the electrodes. The electrochemical reactions take place at the surface
of the electrodes.

If the supply of electrons (current) is sufficiently large to remove all of the ferri-
cyanide ions in contact with the electrodes the reaction becomes diffusion controlled.
Under these circumstances the current travelling between the cathode and anode becomes
independent of the applied voltage and is a direct measure of the reaction/mass
diffusion rate. The concentration of ferricyanide at the cathode is virtually zero and
since the bulk concentration is known (by titre) the mass-transfer coefficient can be
quickly obtained from the measured current. Application of the Chilton-Colburn analogy
gives

$$\alpha = \frac{I\,\rho C_p}{FAC_b} \left(\frac{Sc}{Pr}\right)^{\frac{2}{3}}. \tag{2}$$

The application of the electrochemical technique with the ferri-ferrocyanide redox
couple has been discussed by Man et al (Ref. 9).

The experimental rig is illustrated in Fig. 1. The vessel had a diameter of 0.3 m and
was fabricated from Perspex. The height of the vessel was 0.4 m with the liquid
initially charged to a height of 0.3 m. The vessel was fitted with four equally spaced
Perspex baffles of 0.03 m width.

Agitation was provided by a flat-bladed turbine of 0.1 m diameter. The turbine had six
blades (0.02 m width and 0.025 m long) on a disk of 0.078 m.

The type of sparger used in this study was a cylindrical stainless steel tube of
0.0095 m with eight equispaced holes of 0.002 m diameter drilled on the upper surface
along a length of 0.08 m. The tube was located approximately 0.03 m beneath the
turbine impeller and placed such that the discharge holes were within the diameter of
the impeller. The sparge tube was rigidly fixed within the vessel and when aeration
was in progress the gas bubbles were dispersed through the agitated liquid with the
assistance of the rotating turbine blades. Since air degrades the electrolyte employed
in the tests nitrogen was used for sparging.

Ten equally sized (0.0762 × 0.0254 m) nickel electrodes were sealed into holes made at
regular intervals (0.0635 m separation) up the height of the vessel. These electrodes
were curved, smoothed to a fine finish and inserted in a way that sought to minimise
flow disturbances. 'Araldite' adhesive was used to seal the electrodes at the outer
vessel wall. Measurement of local transfer coefficients is often complicated by an
'entrance effect', which is caused by the developing boundary layer. The size of the
electrodes is considered sufficient (Ref. 10) to eliminate entrance effects.

Typical current-voltage curves are shown in Fig. 2. The plateau on the resultant plot
indicates the diffusion controlled region and gives the desired 'limiting current'.
The heat-transfer coefficient is directly calculable from equation (2). Physical

property data across the temperature range (20-60°C) have been collected from various
sources and reported by MacBeth and Neilson (Ref. 11).

A photographic technique has been used to view the flow characteristics around the
turbine impeller for both water and electrolyte solutions. When using this technique
to minimise optical distortion, the Perspex vessel was submerged in a square Perspex
tank filled with water. Both sides of the square tank were shielded with black card-
board except for a small vertical slit running parallel to the impeller shaft. The
slits allowed light from a 1000 W tungsten lamp to pass through the tank from each
side.

All the still photographs were taken with a Hasselblad 500 ELM camera using Kodak film
and a 150 mm lens. An exposure time of a quarter second and an aperture of f = 22 were
used.

4.   RESULTS AND DISCUSSION

4.1 Local Heat-transfer and Flow Pattern

Local and average heat-transfer coefficient data are presented below for the turbine
impeller used in this study. Observations of the gas-liquid flow pattern characteristics in agitated vessels are also presented.

The experiments conducted in this study were completed using wall electrodes in the
mid-point position between adjacent baffles. Five different impeller speeds with four
different gas flowrates for each speed were studied and the variation of heat transfer
with axial position is presented in Figs 3 and 4. Data for the single liquid phase
were taken from an earlier study by Man et al (Ref. 9).

A study of these figures indicates that the local heat-transfer coefficient at the
vessel wall depends on the gas flowrate and impeller speed. At low impeller speeds
$(1.67-3.33 \text{ s}^{-1})$ the profile variation with axial position in two-phase experiment is
very different from the single phase. Single-phase data indicate that the local heat-
transfer coefficient is largest in the plane of the impeller and falls away as one
moves above and below this plane. The shape of this profile can be explained by the
radial flow pattern created by a flat-bladed turbine. However, this trend is not
observed in gas-liquid systems.

In Fig. 3a, it is seen that in the gas-liquid system at n = 1.67 r/s the local heat-
transfer coefficient reduces markedly at the impelller plane and it increases slightly
as one moves up to an axial position of x/T = 0.36 above this plane. A large rate of
increase of local heat-transfer coefficient is noted between this axial position and
the liquid surface with the exception of the lowest gas flowrate condition. The local
heat-transfer coefficient is almost constant at the bottom of the vessel. The shape of
this heat-transfer profile may be explained by the gas-liquid flow pattern as shown in
Fig. 6a. From this figure, it is seen that the impeller has blanketted with gas and
the gas is not dispersed outward. This behaviour is similar to that of a partial
bubble column where the pumping motion of the liquid phase is hindered and the local
heat-transfer coefficient at the impeller plane reduces markedly from that obtained in
the single liquid phase. Above the impeller plane, the ascent of bubbles displaces the
liquid outwards producing a transverse motion which gives an increase in local heat-
transfer coefficient values. Extra turbulence is produced by the movement of the
bubble swarm and by the wavy motion near the liquid surface causing a larger increase
of heat-transfer coefficient in this region.

In the region below the impeller plane, values of the local heat-transfer coefficient
are lower than the single-phase data. This can be seen to be due to the flooding of
the impeller which causes a weaker liquid motion at this region.

At the higher impeller speed of 3.33 r/s, the shape of the heat-transfer profiles also
depends on the gas flowrate (see Fig. 3b). At low gas flowrates it is seen that the
heat-transfer coefficient distribution in the upper part of the vessel is similar to
that shown for the single-phase results. This observation indicates that the impeller

acts as a pump and disperses the gas-liquid phase radially resulting in a maximum value
of heat-transfer coefficient at the impeller plane. As the gas flowrate increases
(V = 3.8 ×10^{-3} m/s) the impeller is flooded with gas and this causes a reduction of
about 50 per cent in the heat-transfer coefficient at the impeller plane. This result
can be related to the flow pattern in the vessel as shown in Fig. 6b. It is seen that
the impeller cannot disperse the gas radially outwards. Under such conditions, the gas
flowrate dominates the liquid flow pattern giving a much lower heat-transfer coefficient than the single-phase data at the impeller plane and at the regions below the
impeller plane. In the upper part of the vessel, the rising gas bubbles impart extra
turbulence and cause an increase in the heat-transfer coefficients.

Further increase in impeller speeds (5.0-6.7 s^{-1}) give heat-transfer profiles which are
similar to the profiles obtained in single phase (Figs 4(a and b)). These can again be
related to the flow patterns as shown in Figs 6(c and d)). It is seen that the system
is thoroughly aerated, with air distributed throughout the upper part of the vessel
(Fig. 6c); and gas recirculation in both lower and upper loops as seen in Fig. 6d. In
the latter case the vertical flows near the tank wall are rapid enough to carry gas
below the impeller plane.

4.2 Average Heat Transfer

The average heat-transfer coefficients have been calculated by taking the arithmetic
mean values of the experimental local values for each speed and gas flowrate. Direct
comparison of this work with previous investigation is difficult because of the
different vessel geometries used as well as the different gas-supply devices and heat-
transfer systems employed. As a result, only qualitative comparison is made in this
paper and the mean heat-transfer coefficient has been plotted as a ratio of the single
phase to gas-liquid phase data versus gas flowrate in Fig. 5. It is observed that for
the aeration conditions employed the heat-transfer rates except that of the lowest
impeller speed are within ±15 per cent of the values obtained without gas addition. At
the lowest impeller (n = 1.67 s^{-1}), the heat-transfer rates increase with gas
flowrates. These results may be explained by the heat-transfer profiles and flow
patterns obtained in this study.

At low impeller speed, the gas is not dispersed outwards by the impeller and the vessel
behaves as a partial bubble column. As a result, the heat-transfer rates increase with
gas flowrates due to the axial flush of gas through the impeller plane up to free
liquid surface. As the rotation speed increases the rising bubbles are dispersed such
that extra turbulence is produced by the movement of the gas bubbles passing through
the agitated liquid which in turn will produce higher heat-transfer coefficients.
However, the presence of the gas near the heat-transfer surface will have the effect of
reducing the rate of heat transfer since gas has a much higher resistance to the
transfer of heat than that of liquid. Furthermore, the presence of the gas at the
impeller reduces the pumping capacity of the impeller resulting in decrease of heat-
transfer rates at the impeller plane and in the lower part of the vessel. The net
effect of these factors balancing each other, thus producing no significant increase or
decrease in heat-transfer rate. These findings agree qualitively with the results
obtained by Edney and Edwards (Ref. 6) and Steiff and Weinspach (Ref. 7). Edney and
Edwards (Ref. 6) reported that the heat-transfer rates to a coil remained essentially
unchanged from those observed with no aeration. Steiff and Weinspach (Ref. 7) found
that the jacket heat-transfer coefficient increased with superficial gas velocity at
low impeller speeds. At high impeller speeds, the gas flowrates had no significant
effect on the heat transfer.

5. CONCLUSIONS

The electrochemical technique has been used to measure local mass-transfer (and hence
heat-transfer) coefficients at the wall of an 0.3 m aerated, agitated vessel.
Agitation was provided by a six-bladed disk turbine.

Measurement revealed that the profile of local heat-transfer coefficient at the vessel
wall depended on the gas flowrate and the impeller speed. These observations could be
qualitatively related to the flow patterns obtained.

The average heat-transfer coefficients calculated from these local values indicated that, for the aeration conditions employed, the heat-transfer rates, except that of the lowest impeller speed, were within ±15 per cent of the values obtained without gas addition.

6.   ACKNOWLEDGEMENT

The contribution of Mr R Houston in photographic work is gratefully acknowledged. The work reported here was supported by the Mechanical and Electrical Engineering Requirements Board of the Department of Trade and Industry and was carried out under the bioreactor research programme at the National Engineering Laboratory. This paper is published by permission of the Director, National Engineering Laboratory, Department of Trade and Industry.

7.   REFERENCES

1.   Edwards, M.F. and Wilkinson, W.L.: "Heat transfer in agitated vessels". Chemical Engineers, 1972, Part I, No 264, pp. 310-318; Part II, No 265, pp. 328-334.

2.   Poggemann, R., Steiff, A. and Weinspach, P.M.: "Heat transfer in agitated vessels with single-phase liquids". German Chem. Engng, 1980, Vol. 3, No 3, pp. 163-174.

3.   Rao, K.B. and Murti, P.S.: "Heat transfer in mechanically agitated gas-liquid systems". Ind. Engng Chem. Process Des. Develop., 1973, Vol. 12, No 2, pp. 190-197.

4.   Pollard, R. and Topiwala, H.H.: "Heat-transfer coefficients and two-phase dispersion properties in a stirred-tank fermentor". Biotechnol. and Bioengineering, 1976, Vol. 18, Part 11, pp. 1517-1535.

5.   Edney, H.G.S. and Edwards, M.F.: "Heat transfer to non-Newtonian and aerated fluis in stirred tanks". Trans. Instn Chem Engrs, 1976, Vol. 54, pp. 160-166.

6.   Steiff, A. and Weinspach, P.M.: "Heat transfer in stirred and non-stirred gas-liquid reactors". German Chem. Engng., 1978, Vol. 1, Part 3, pp. 150-161.

7.   De Maertelaire, E.: "Heat transfer in turbine agitated gas-liquid dispersions". International Symposium on Mixing, Faculté Polytechnique de Mons, 1978, pp. XC7-XC35.

8.   Chilton, T.H. and Colburn, A.P.: "Mass transfer (absorption) coefficients". Ind. Engng Chem., 1934, Vol. 26, pp. 1183-1187.

9.   Man, K.L., Edwards, M.F. and Polley, G.T.: "A study of local heat-transfer coefficients in agitated vessels". Fluid Mixing, Institution of Chemical Engineers Symposium Series No 89, 1984, pp. 193-207.

10. Askew, W.S. and Beckman, R.B.: "Heat and mass transfer in an agitated vessel". I. & E. C. Process Design and Development, 1965, Vol. 4, No 3, pp. 311-318.

11. Macbeth, R.V. and Neilson, A.J.: "Heat transfer and pressure drop in shell-and-tube heat exchanger using orifice baffles". Heat Transfer and Flow Service. Research Symposium, 1977, Paper 169, AEEW-R104P.

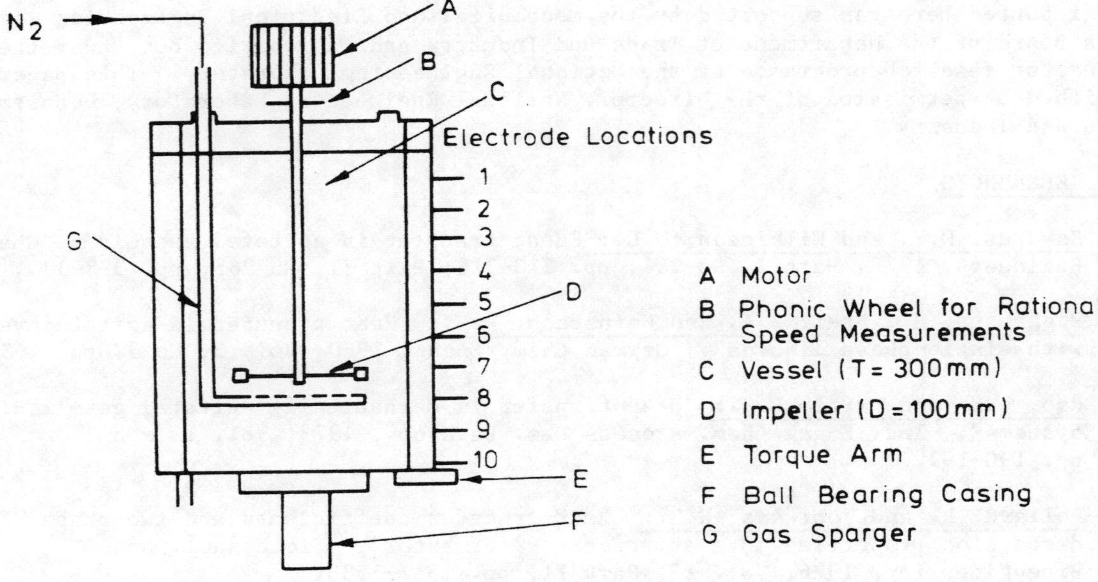

1   Test vessel

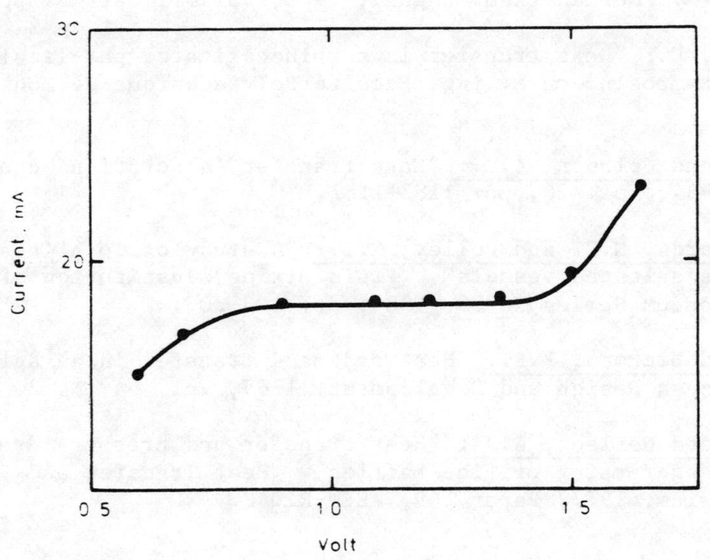

2   Current-voltage curve

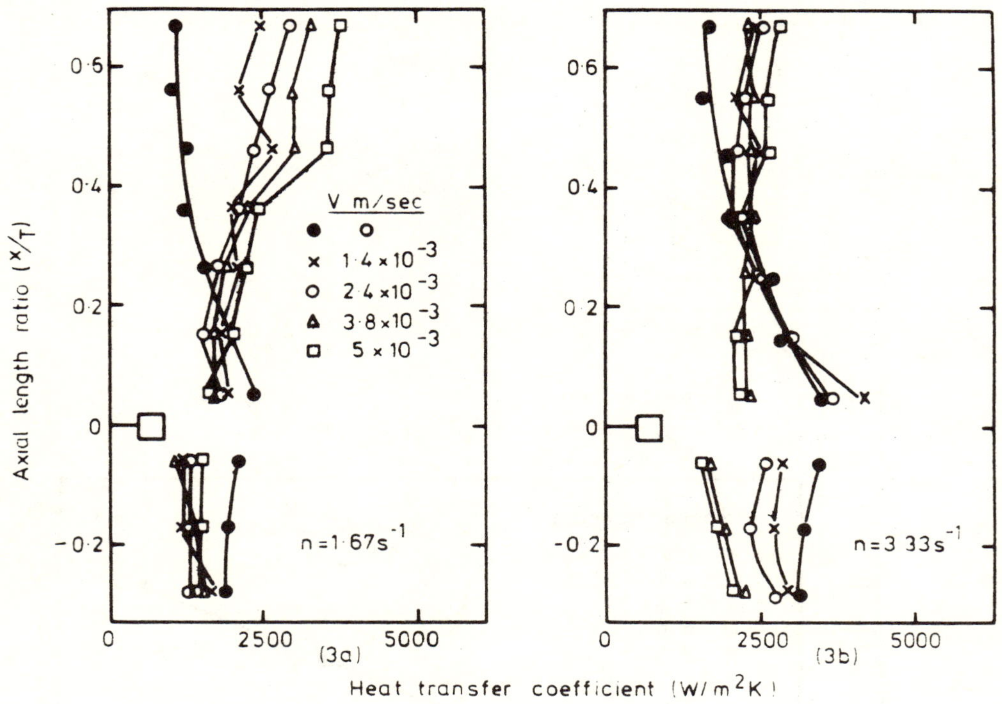

3 Variation of heat-transfer coefficient with axial length

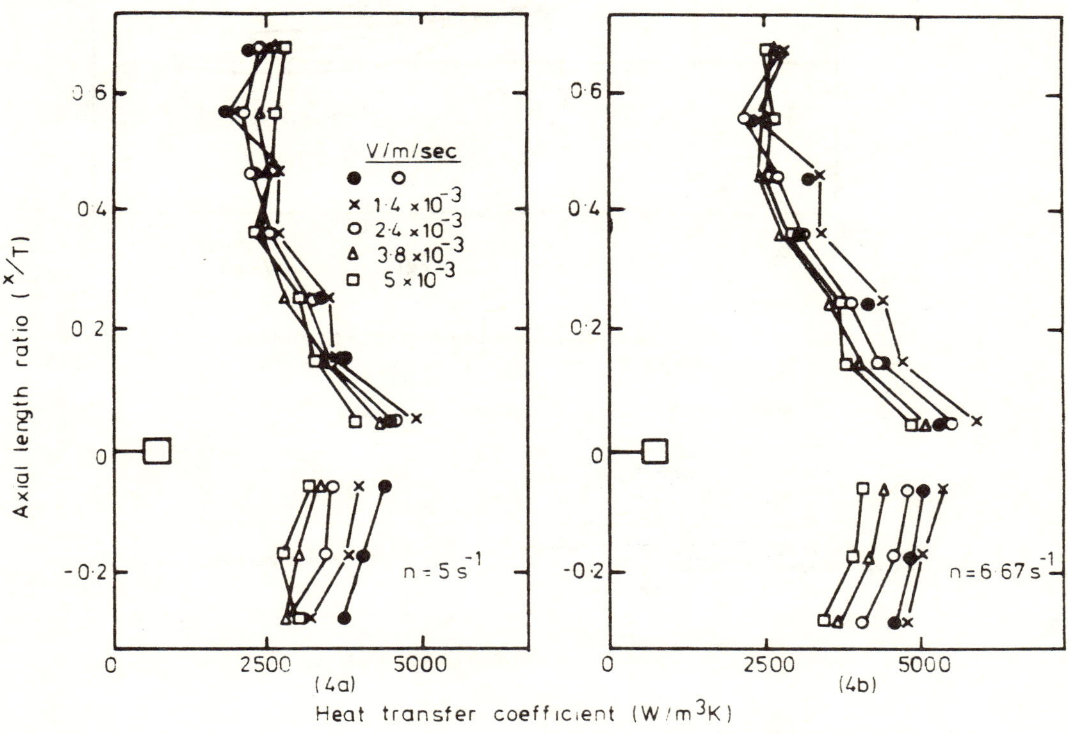

4 Variation of heat-transfer coefficient with axial length

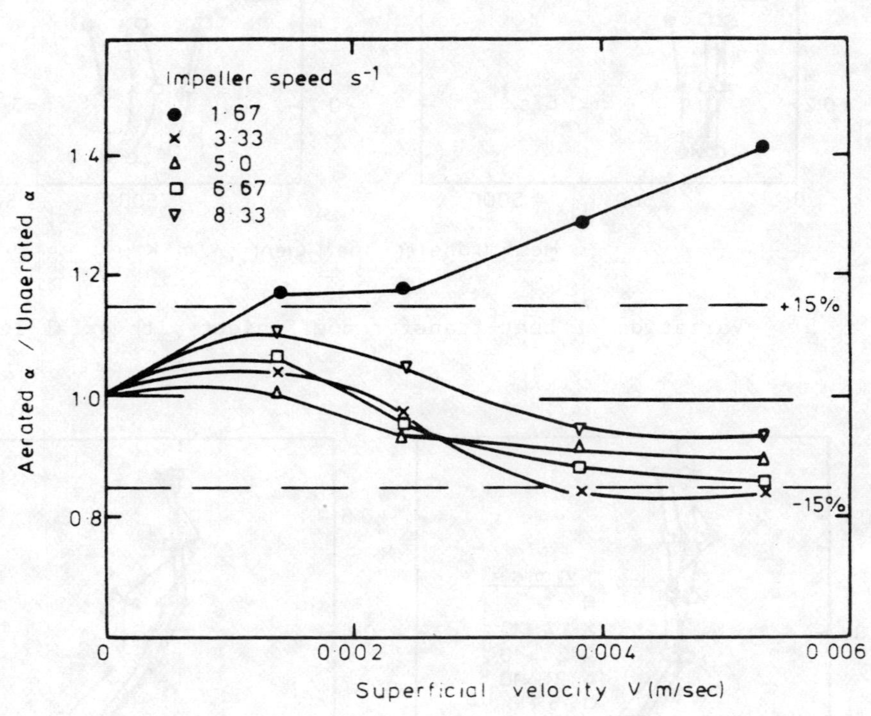

5   Ratio of two-phase to single-phase heat-transfer coefficient
to wall versus superficial gas velocity

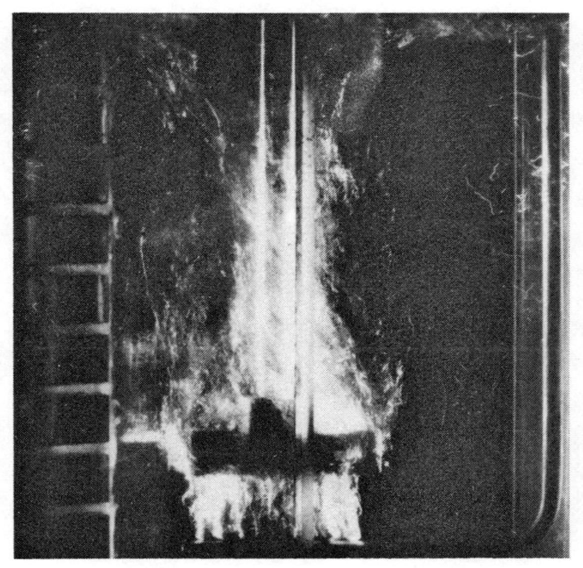

n = 1.67 s^{-1}
(6a)

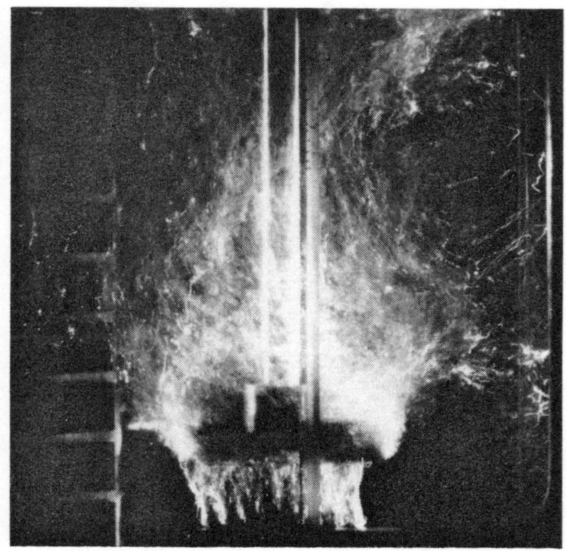

n = 3.33 s^{-1}
(6b)

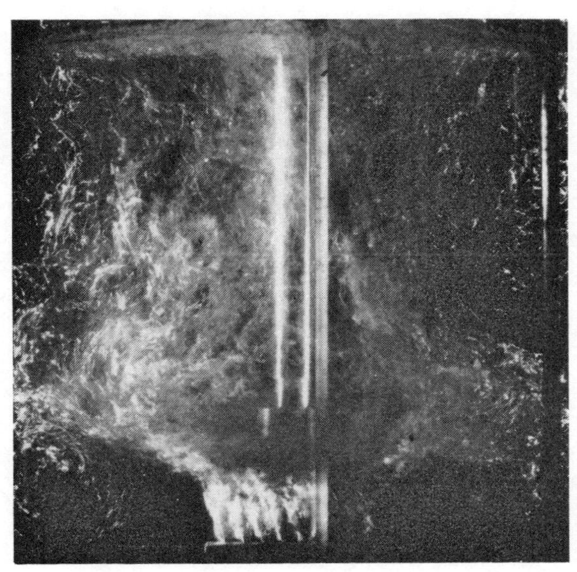

n = 5.0 s^{-1}
(6c)

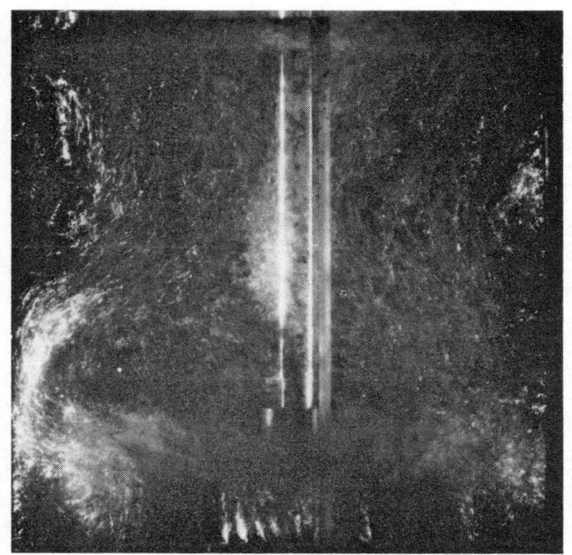

n = 6.67 s^{-1}
(6d)

6   Gas-liquid flow pattern characteristics in agitated vessels at constant superficial velocity of 5×10^{-3} m/s.

A FINE CHEMICALS VIEW OF THE MIXING WORLD

K.J.Carpenter PhD. C.Eng. M.I.Ch.E

I.C.I. Fine Chemicals Manufacturing Organisation
Blackley, Manchester M9 3DA
England

Summary

The Fine Chemicals side of the Chemical Industry is a major growth area. A special feature is the use of stirred vessels for an enormous range of duties, and mixing-related problems are common. The nature of many of the chemicals involved causes additional problems of mechanical design and has implications on the process performance. These problems serve to compound those caused by gaps and inadequacies in the published literature in several areas of major significance.

Held at Wurzburg, 10-12 June, 1985.

Organised by DVCV· Deutsche Vereinigung für Chemie- und Verfahrenstechnik
(German Association of Chemical and Process Engineering).

Organisation: GVC·VDI-Gesellschaft Verfahrenstechnik und Chemieingenieurwesen. 

©BHRA, The Fluid Engineering Centre, Cranfield, Bedford MK43 0AJ, England.

1. INTRODUCTION

Most of the current growth in the Chemical Industry, certainly in the U.K., is in what can be called "Fine Chemicals". I.C.I.'s stated policy of increasing the proportion of "Speciality Business" (1) is a clear example of this, and many other companies are making similar moves.

A special feature of the Fine Chemicals Business is the proliferation of mixing-related problems. Multi-purpose and multi-product batch plants predominate and the many stages of multiple reactions, isolations, solvent extractions and even supposedly simple blending or holding operations are all carried out in some form of mixing equipment - the agitated vessel being the norm. Temperatures and pressures are usually moderate but chemical conditions can be very severe and a single piece of equipment will often be required to perform several very different and difficult tasks.

Chemical conditions are not only severe because of the number and complexity of, often competing, reactions involved but also because of the aggressive nature of the chemicals themselves. It is quite normal for the shape of an impeller and internal fittings in a vessel, to be determined by what can be made from a suitable material rather than those which the literature may recommend. In addition, in much of the literature it is apparent that certain impellers have become academic favourites, with volumes of data on their performance, and many well-used systems totally ignored. The Rushton Disc Turbine is a clear example. It is by far the most widely tested agitator and yet in I.C.I. Fine Chemicals we have only a handful in over 1000 stirred vessels, and one of those is glass-coated and the vessel only poorly baffled.

As if to compound these problems, not only is relevant and reliable performance data scarce, but it is now needed more than ever before. Many of the chemicals involved are extremely expensive requiring very high reaction selectivity and efficient recovery and purification stages. Equipment has to be well designed and work well from the outset - we cannot permit inefficiency and evolutionary process development which may have previously been acceptable.

This, then, is the background against which this review is set. It touches on Power, Blending Time, Immiscible Liquids, Gas-Liquid Reactors, Solid-Liquid Duties and Reaction Selectivity. In each area points of special relevance to the Fine Chemicals Industry are brought out. Comments and criticisms are based on our own experiments and operating experience. Criticism is meant to be constructive - provoking discussion and hopefully arousing interest in Fine Chemical manufacturers' problems.

2. IMPELLER POWER

Impeller power is certainly well-researched, possibly the earliest reference being over 100 years ago (3). In practice there are few problems in this area. In the more 'normal' baffled systems our own measurements usually agree well with Bates et al (4). We differ slightly in multiple tier agitators. Our own measurements and those of BHRA (5), made on I.C.I. standard single, double and triple 60° turbines show that in the turbulent region:

$$Po \text{ (3 tiers)} \sim 1.5 \, Po \text{ (2 tiers)} \sim 1.5^2 Po \text{ (single tier)}$$

whereas for a 45° impeller Bates shows

$$Po \text{ (2 tiers)} \sim 1.73 * Po \text{ (single tier)}.$$

For multiple tiers with a mixture of impeller types, the combined Po approaches the sum of the individual values as the tiers differ more widely in flow pattern.

In unbaffled and partially baffled systems we employ our own data supported by that produced by BHRA's FMP service (6).

Problems can arise with impellers and baffles manufactured from laminated plastics, which for mechanical reasons may require a slightly unusual shape and thicker blades. Plant measurements indicate lower power numbers than the more 'normal' shaped metal counterpart. Medek (7) has briefly discussed the effect of blade thickness on power for 'open' type turbines.

Similarly glass-coated impellers can suffer from the same problem. Although
our own measurements generally support Pfaudler-Balfour published values of Po and
correction factors (8), it is apparent that flat or angled blade impellers fabricated
from "squashed tube", with rather fat blades and wide radius edges, perform rather
poorly. In our experience Tychon have been the most successful manufacturer in
supplying impellers of a more acceptable shape.

3.   MIXING TIME

       Most mixing time experiments as exemplified by Kang and Levenspiel (9),
consist of blending a tracer into a Newtonian liquid, usually to an arbitrary end
point. The resulting time is useful in comparing impeller and baffle arrangements,
but impossible to use to quantify practical applications, which are usually only the
prerequisite to, for example, a reaction or mass transfer operation. Taking a system
of competing reactions as an example; it is fairly clear that in a vessel with a
"mixing time" (to achieve reasonable tracer homogeneity) of minutes, by-product
reactions with characteristic times of days will cause little problem, whereas times
in seconds could be a disaster. However, it is not possible to use mixing time to
actually quantify such effects (see section 7.) It is therefore only the crudest
guide in such systems.

       An additional complication in many practical systems is the viscosity of
the reagents. One area which has received very little attention in the literature
but is the cause of many problems in the Fine Chemicals industry, is that of adding
viscous materials to a low viscosity medium in a vessel. Figure 1 shows the result
of a very simple mixing time experiment used to illustrate such problems. The
experiment measures the time from starting an impeller with two static layers one
acid and one alkaline, to reach the end point.   In one case the lower layer is
alkaline Glycerol and in the other alkaline brine.   There is clearly an effect which
can span orders of magnitude and is in serious need of further investigation.

4.   IMMISCIBLE LIQUIDS

       Immiscible liquid duties are extremely common, usually as organic-aqueous
reactions or solvent extractions. One of the major criticisms in this area concerns
the correlations published to predict the minimum impeller speed required to achieve
complete dispersion of both phases. Godfrey et al (14) have recently reviewed the
state-of-the-art and indicate that most of the correlations proposed to date are of
the form:

$$Re = K \, Su_i^x \, Ar^y$$

       However, from attempting to use many of these correlations, typified by
Godfrey (14), Van Heuven and Beek (15), Skelland and Seksaria (16) and Esch et al
(17), it is apparent that there are severe shortcomings in this approach for the
following reasons:

Dimensionless Groups

       The dimensionless groups employed are not based on a fundamental
understanding of the underlying hydrodynamic principles, rather as a convenient way of
grouping what are believed to be the relevant physical properties which are easily
known. Thus the Archimedes number (Ar) and modified Suratman number (Su_i) employed
include the _impeller_ diameter – relating performance to the properties of a drop or
particle the size of the impeller. Mersmann and Grossman (18) state that drop
diameter is a function of impeller diameter only for a small dispersed phase fraction
and similar liquid viscosities.

       Forcing experimental data to fit a dimensionless equation on these grounds
has in this case forced unreal interrelations between variables. For example
Godfrey (14) states that his data can be represented equally well by:

$$Re \propto Su_i^{.47}$$
$$\text{and} \quad Re \propto Su_i^{.14} \, Ar^{.33}$$

which when reduced to single variables becomes:

$$N \propto \sigma_i^{0.47} \rho_c^{-0.53} \mu_c^{0.06}$$
$$\text{and} \quad N \propto \sigma_i^{0.14} \Delta\rho^{0.33} \rho_c^{-0.53} \mu_c^{0.06}$$

Clearly using these equations to assess the effect on a working extraction vessel of changing the solvent (dispersed) phase would give very different results.

<u>Scale and Geometry Dependence</u>

It is well recognised (e.g. Godfrey (14)) and our own experience confirms that these correlations are not independent of scale, nor are they independent even of the details of impeller and baffle geometry.

<u>Limited Range</u>

Many of the correlations are based on a limited range of fluids. For example although Van Heuven and Beek (15) state:
$$N \propto (\rho_c - \varepsilon_d \Delta\rho)^{-0.53}$$
$$\text{and } N \propto \mu_c^{0.08}$$

They never actually varied ρ_c and μ_c in their experiments at all.

Similarly Skelland Seksaria (16) make no allowance for dispersed phase fraction (ε_d) which they did not change from 0.5. Others report no effect of ε_d through working in a limited range (e.g. 0.3 to 0.5) or including both light dispersed in heavy and heavy dispersed in light, which give opposing relationships when studied independently (I.C.I. data (19)).

These criticisms are based on an experimental programme undertaken within I.C.I. (19) covering angled blade turbines and Pfaudler impellers. For example our observations on Pfaudlers briefly include:

1. The light phase is dispersed up to $\varepsilon_d \sim 0.7$, coming up in speed from separate layers, after 0.7 the heavy phase becomes dispersed.

2. The best fit to the data is obtained by a direct regression on physical properties rather than employing the above dimensionless groups. A much clearer picture of the significance of individual physical properties results, allowing an accurate appreciation of, for example, the effect of changing solvent.

3. Light phase dispersed in heavy and heavy dispersed in light, behave very differently and should be represented by individual correlations.

4. Some heavy organic materials, e.g. Perchloroethylene, can behave in very strange ways, giving inverse dependence of speed on volume fraction between 0.3 and 0.6, whereas others, e.g. Chlorobenzene, give a monotonic positive dependence as do the light organics. Experiments with Perchloroethylene can generally show some very strange effects, for example an inverse vortex of clear organic beneath a dispersion.

5. For a light phase volume fraction above around .4 to .5 there exists an ambivalent region where which phase is dispersed depends upon how the particular operating conditions have been arrived at. For example starting with two static layers and increasing speed will cause the light phase to disperse (vol. fraction up to .7) whereas starting from high speed can cause the heavy phase to disperse. A similar phenomenon can be observed by charging one of the phases whilst the agitator is running.

Clarke and Sawitowski (20) and Mersmann and Grossmann (18) have studied this phenomenon. It is one which can cause serious problems in Fine Chemicals operations, particularly where phases 'invert', that is the dispersed phase switches to become continuous. Serious problems with subsequent separation can occur because of very fine droplets which can be produced when this happens.

In general therefore, despite their significance in industry, the formation of dispersions and particularly the occurrence of the ambivalent region are not well-researched areas, and lack fundamental understanding.

Further aspects of immiscible liquid systems worthy of further investigation are the effect of impeller type on drop-size distribution - important as it can have a serious effect on settling rate, as described by Forschner(21) and the change in drop size distribution as scale increases, which again can be significant where phase separations are difficult.

5. GAS-LIQUID REACTORS

Until recently almost all reported gas-liquid work concerned the Rushton Disc Turbine, probably the most significant development being the multi and curved blade versions investigated by van't Riet (10). However, for many reasons disc turbines are not common in Fine Chemicals plant and many gassing reactions are carried out with angled blade turbines.

More recently Chapman et al (11) and Warmeoskerken et al (12) have studied angled blade turbines. Table 1 indicates that our own plant measurements, although well outside Warmeoskerken's (11) experimental range, tend to confirm his prediction of gassed to ungassed power ratio. In addition it has become apparent, also reported by Warmeoskerken (11), that mass transfer performance tends to be a function of (P/V), independent of impeller type, at least in fully baffled systems.

The range of gas rates covered in the literature to date has been rather limited and we have undertaken our own work (Kenyon (13)) to extend this range. Whilst not wishing to discuss this work in detail here, Figures 2 and 3 are typical of the power response at high gas rates. Clearly there are some very strange effects. In addition Kenyon has demonstrated that the larger diameter imepeller (T/2) is much more stable, even at the same power, as compared to the smaller impeller (.3T), and is therefore our preferred design.

A criticism of the open literature is the lack of large scale data and, coupled with this, the use of the dimensionless gas flow number, Q/ND^3. It is clear that even at small scale Q/ND^3 cannot fully account for changes in N or in D, so there is no reason at all to suppose it will be independent of scale. On the contrary it is painfully apparent that it isn't.

In addition, there is almost no published data on the performance of partially baffled systems, as typified by glass-lined reactors, nor on such "unsuitable" impellers as Pfaudlers. Whilst it is true that they do not perform as well as other types of impeller, it is helpful to characterise their performance. Our own data (Kenyon (13)) indicates that a disc turbine with two "beavertail"-type baffles gives roughly one order of magnitude better gas utilisation than a Pfaudler in the same vessel at the same speed.

6. SOLID-LIQUID DUTIES

The traditional methods of Zwietering (22) and Chemineer (23) appear to work reasonably well for solids-suspension. The only real criticism in this area is the lack of reliable data for partially baffled vessels and Pfaudlers. The major criticism regarding solids duties from the Fine Chemicals point of view concern incorporating solids through surfaces and handling non-Newtonian pastes.

Firstly, incorporating solids through surfaces. There are as many solutions as there are problems, from the twin impeller, half-baffled design through the partial baffles, for example of Joosten et al (24) or Smith et al (25), to "Heath-Robinson" surface raking devices. Each design has its own place. The partial baffles apparently work quite well at a specific batch height, for powder-like materials which do not agglomerate to any significant degree. The twin impeller type of design has greater capability in breaking soft agglomerates caused by partial wetting - a simple Rushton turbine with sufficient power to generate good surface turbulence may be as effective. For solids which form extremely hard agglomerates the only solution may be to prevent agglomeration in the first place, hence the surface "rakes" with solids-scattering addition. This area is badly in need of bringing together, perhaps starting with even a crude way of classifying solids and ranking design proposals alongside the type of solid for which it is best suited.

Secondly dispersing solids, for example filter cakes, to form non-Newtonian pastes. In this area we are at the mercy of manufacturers who generally have very little performance data on their own machines. Many pastes can be 'damaged' by excess

work being done on them - work being used loosely to describe the shear rate and the
amount of time each particle sees that shear rate, thus a function of peak shear rate
and a mean 'circulation rate'. Yet manufacturers generally have no information on
shear and circulation in their machines and rely on crude tests to select one of their
off-the-shelf devices.

Figure 4 is a clear illustration of the damage - shown by high apparent
viscosity - which can be done to a paste in an inappropriate machine.

7.   REACTION YIELD AND SELECTIVITY

For extremely fast, millisecond, competing reactions, the work of Bourne
(see e.g. 26) needs little introduction. The major problems with this approach is
that it is really only valid to construct what are very complex mathematical models
when accurate data is available. Unfortunately our knowledge of eddy size and
diffusivity of complex organic molecules is sketchy to say the least. In this context
the simpler early 'billiard-ball' model of Nabholz et al (27), or its development by
Jenson (28) may be sufficient to give a crude estimate of selectivity. The insight
provided by Zollinger (31) is invaluable in tackling such problems "semi-empirically".

In practice it is our experience that a majority of selectivity problems are
of the timescale of bulk mixing effects (secs. to mins.) rather than milliseconds.
Until recently this area has remained unquantified, however the work of Middleton (29),
Mann and Mavros (30) and currently of Drain (2) gives great promise for the future.

8.   CONCLUSIONS

The general conclusion of this review is that despite the fact that mixing
is a rather mature area of research, and one in which many papers are published each
year, there are still many areas which are crying out for investigation. One of the
major criticisms must be the persistent use of rather inappropriate dimensionless
groups - sadly very many correlations simply cannot be applied to large scale reactor
design. The special problems of the Fine Chemicals Industry only serve to compound
the problems - if correlations do not work for "academic standard" impellers what
chance do they have for some of the strange things we are forced to employ?

9.   ACKNOWLEDGEMENTS

Thanks to Prof. A.W.Nienow and Dr.J.C.Middleton for suggesting this paper,
to M.G.Peacock, J.G.Kenyon, P.C.Lines and D.McKay for producing most of the quoted
I.C.I. results and to I.C.I. F.C.M.O. for kind permission to publish.

10.   REFERENCES

1.   Interview with J.Harvey-Jones. Chem.Week. 28, 25 July 1984.

2.   S.Drain. Currently studying for Ph.D., University of Birmingham.

3.   Unwin W.C.: Proc.Roy.Soc. (London) A31, 54, 1880.

4.   Bates R.L., Fondy P.L. and Corpstein R.W. I and E.C. Proc.Des.Dev., 2 (4)
      October 1963.

5.   King R., BHRA Project RPG01332, Report CR2086, Sept. 1983.

6.   Muskett M.J. and Ruszkowski S.W., FMP Report 007, Jan 1985 (Confidential to
      members of FMP).

7.   Medek J., Int. Chem.Eng. 20 (4) pp664, 1980.

8.   "Agitator Power Consumpton in Sybron/Balfour Glassed Steel Vessels". Bulletin
      from Pfaudler-Balfour, Leven, Fife, Scotland.

9.   Khang S.J. and Levenspiel O. Chem.Eng.Sci. 31 pp569-577, 1976.

10.   v'ant Riet K., Ph.D. Thesis, Delft University of Technology, 1975.

11.   Chapman C.M., Nienow A.W., Cooke M. and Middleton J.C. Chem.Eng.Res.Dev.,61,
      1983.

12.   Warmeoskerken M.M.C.G., Speur J. and Smith J.M., Chem.Eng. Commun, 25, pp11-
      29, 1984.

13. Kenyon G.J. Internal ICI Documents, 1985 (Confidential to I.C.I.).

14. Godfrey J.C. Reeve R.N., Grilc V. and Kardelj E., I.Chem.E.Symp.Series No.89 (Fluid Mixing II), 8, 107 (1984).

15. van Heuven J.W., Beek W.J., Proc. ISEC71, 70, 1971.

16. Skelland AHP. and Seksaria R., I. and E.C. Proc.Des.Dev., 17, 56, 1978.

17. Esch D.D., D'Angelo P.J., Pike R.W. Can.J.Chem.Eng. 49, 872, 1971.

18. Meismann A. and Grossman H., Int.Chem.Eng. 22(4), 581, October 1982.

19. Peacock M.G., Carpenter K.J. and Lines P.C., Internal ICI Documents, 1985 (Confidential to ICI).

20. Clarke S.I. and Sawitowski H., Trans.I.Chem.E. 56, 50, 1978.

21. Forschner P., Kipke K.D., "The Influence of Mixing Impellers on the Performance of Mixer-Settlers in Hydrometallurgy" from Ekato Reuhr-und Mischentechnik GmbH, Schopfheim, W.Germany.

22. Zwietering T.N., Chem.Eng.Sci., 8, 224, 1958.

23. Gates L.E., Morton J.R., Fondy P.L., Chemical Engineering, May 24 1976.

24. Joosten G.E.H., Schilder J.G.M. and Broere A.M., Trans.I.Chem.E. 55 1977.

25. Smith D.L., Hemrajani R.R., Koros R.M. and Tarmy B.L. A.I.Ch.E. Annual Meeting (New Orleans), November 8-12, 1981.

26. Baldyga J. and Bourne J.R., Chem.Eng.Commun., 28, pp231-241, 1984.

27. Nabholz F., Ott R.J. and Rys P., Proc 2nd European Conference Mixing, 1977, paper B2 p27, BHRA, Cranfield, England.

28. Jenson V.G., Chem.Eng.Sci., 38(8), pp 1151-1157, 1983.

29. Middleton J.C. Pierce F. and Lynch P.M., I.Chem.E.Symp.Series 87 (ISCRE8, Edinburgh, 1984), paper 5.1, p239.

30. Mann R. and Mavros P., Proc. 4th European Conference Mixing, 1982, paper B3 p35, BHRA, Cranfield, England.

31. Zollinger H., Chemtech, September 1979 p 578.

Table 1: Predicted and Measured Performance of a Gas-Liquid Reactor

	Vessel dia(m)	Speed (rps)	Q/ND^3	Pu/V	Pg/Pu	Pg/V
Predicted*	2.13	1.33	0.14	1.03	0.6	0.62
Measured	2.13	1.33	0.14	0.95	0.5	0.56

* predicted by extrapolating Warmeoskerken (12).

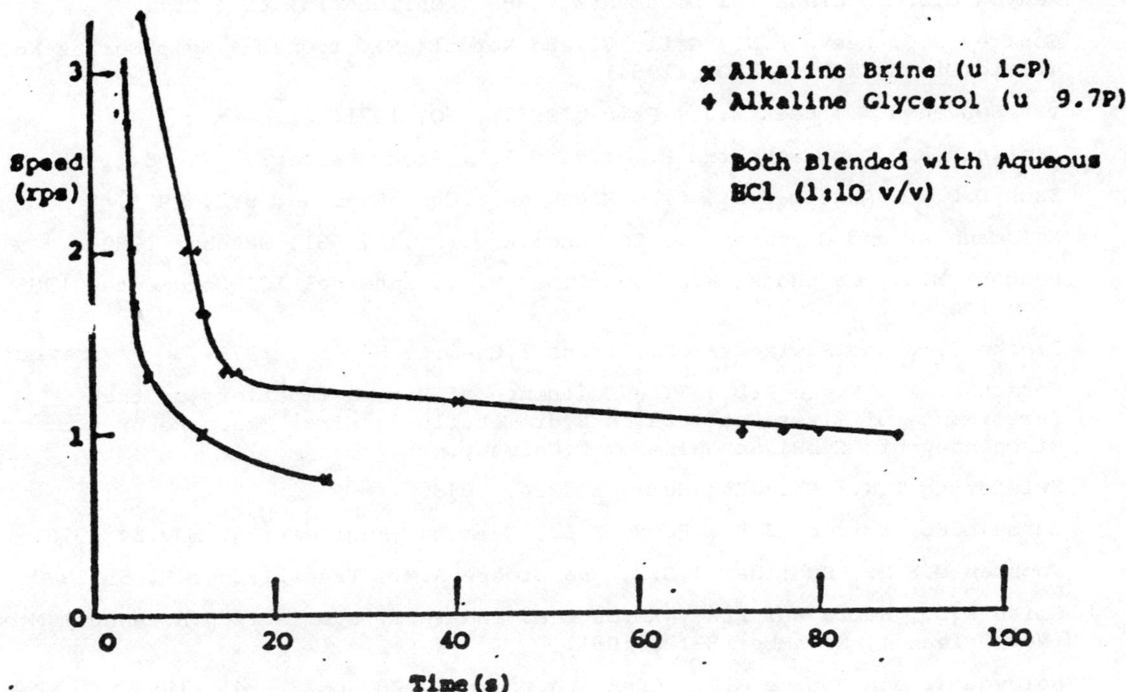

Figure 1: The Effect of Viscosity Ratio on Blending Time

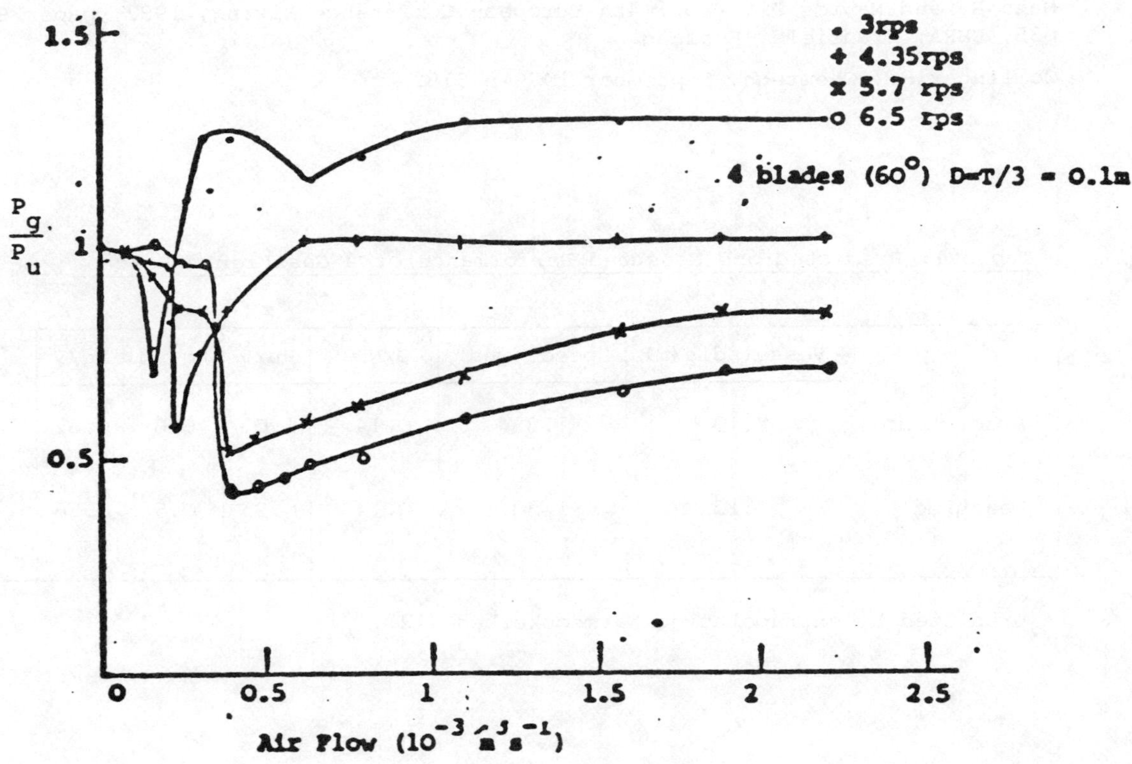

Figure 2: The Effect of Gas Rate on Power for Angled Blade Turbines.

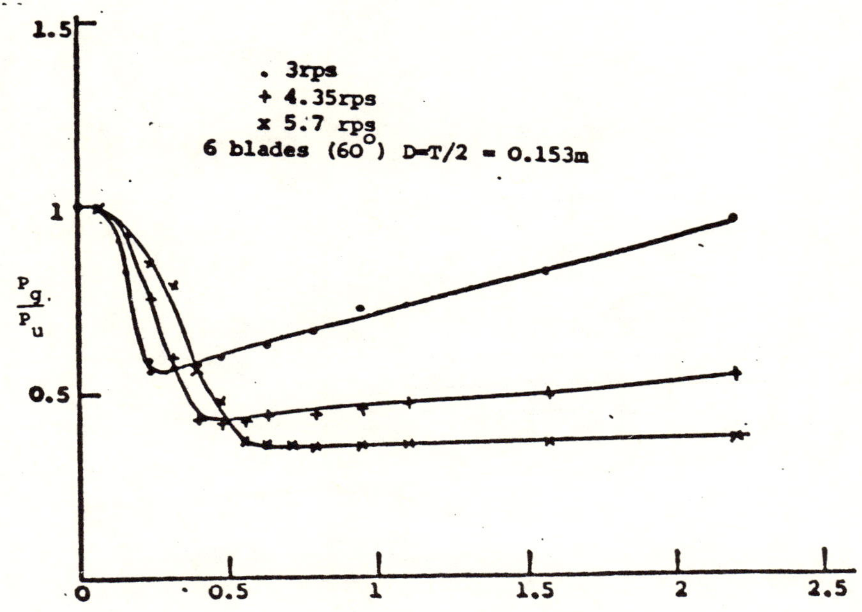

Figure 3:   The Effect of Gas Rate on Power for Angled Blade Turbines

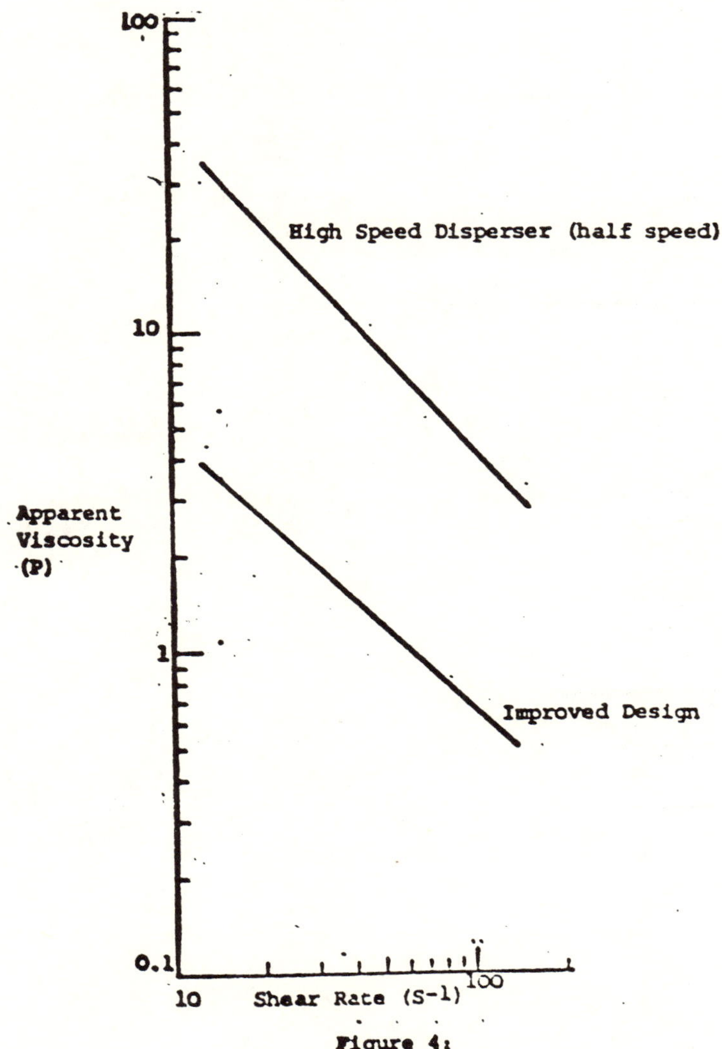

Figure 4:

Increased Viscosity in Standard Disperser
(even at half-speed)

241

DESIGN OF UNGASSED BAFFLED MIXING VESSELS TO MINIMISE
SURFACE AERATION OF LOW VISCOSITY LIQUIDS

N.I. Heywood, P. Madhvi, M. McDonagh

Materials Handling Division
Warren Spring Laboratory
Stevenage, Herts, SG1 2BX, UK

Summary

As part of a programme of mixing research, sponsored by the British
Government and nine industrial companies, a series of experiments has been carried out
to determine the agitator speed at which the onset of surface aeration occurs in low
viscosity liquids agitated in standard four-baffled circular tanks. The tanks ranged
in diameter from 0.128 m to 0.54 m. Five different agitator types were used (Rushton
turbine, 6-bladed pitched impeller, marine propeller and two types of Torrance
disperser) in liquids of different surface tensions and either with or without
electrolyte (dissolved potassium sulphate). In all experiments, the liquids were not
deliberately sparged. A range of impeller diameter (D) to tank diameter (T),
clearance (h_1) to T and height of liquid (H) to T ratios were used.

Three critical agitator speeds were defined and measured: the lower speed
defined the condition under which bubbles were first broken off surface vortices
appearing in the agitated liquid, the second speed defined the point at which bubbles
were first drawn down into the impeller region and the third speed defined the point
at which power number dropped with increasing Reynolds number in the turbulent mixing
regime. The first two speeds were correlated successfully with the geometric
variables, D, T, h_1 and H.

These correlations can be used to design mixing systems to reduce, or
virtually eliminate, surface aeration while still achieving the main process result,
e.g. solids suspension, heat transfer, mass transfer and blending. Two separate
situations arise: design of a new system and modification of an existing system.
Recommendations are given in the paper regarding appropriate design features or
changes to minimise surface aeration whilst satisfying the main process requirement.

Held at Wurzburg, 10-12 June, 1985.

Organised by DVCV· Deutsche Vereinigung für Chemie- und Verfahrenstechnik
(German Association of Chemical and Process Engineering).

Organisation: GVC·VDI-Gesellschaft Verfahrenstechnik und Chemieingenieurwesen. 

<u>NOMENCLATURE</u>

C_p = heat capacity

D = impeller diameter

e = impeller shaft eccentricity (distance of shaft from tank axis)

h_1 = impeller clearance

h_j = jacket heat transfer coefficient

H = height of liquid in tank under quiescent conditions

k = thermal conductivity

n_{SA1} = critical impeller speed for incipient surface aeration

n_{SA2} = critical impeller speed at which bubbles are first seen drawn into the impeller region

n_{SA3} = critical impeller speed at which a fall in power number is first observed to occur on the power curves in the transitional or turbulent mixing regimes

Ne = Newton number = $P/(\rho\, n^3\, D^5)$

P = power to agitator

p = air pressure above liquid surface

P_A = atmospheric pressure

Re = Reynolds number = $n\, D^2/\nu$

ν = kinematic viscosity

η_L = liquid viscosity

ρ_L = liquid density

1. INTRODUCTION

The drawing-in of air through the free surface of liquid or suspension agitated in a tank can often be detrimental to achieving the main process result. This phenomenon, generally referred to as surface aeration, occurs in baffled tanks when fluid agitation is sufficiently vigorous to create surface vortices (see Fig. 1). Local turbulence in the fluid surrounding these vortices enables bubbles to be broken off the base of these vortices, and the bubbles can then be transported downwards into the impeller region through fluid circulation in the tank. Once in the impeller region, these bubbles may be reduced in size as a result of the high shear stresses imparted to the continuous fluid by the impeller.

Often the presence of air bubbles in the impeller region will cause a significant reduction in the power required by the impeller, at a given rotational speed. This reduction may be undesirable because high power inputs may be required to achieve certain process results such as solids suspension or heat transfer. In addition, on cessation of the agitation of the fluid, while large bubbles may disengage quite readily and rise rapidly to the surface, finer, unwanted bubbles may remain suspended in higher viscosity liquids or suspensions.

Somewhat similar phenomena occur in unbaffled tanks but in this case a central, large vortex is created, as a result of the circulatory motion of the fluid instead of a number of smaller surface vortices. The depth of this central vortex increases progressively as agitator speed is increased until bubbles are created from the rupture of the vortex surface by the agitator motion either at or just before the bottom of the vortex reaches the impeller blades. Studies have been made of this and results published in the literature, including data and analysis from a separate part of the present study (Ref. 1) but the present article is confined to surface aeration in baffled tanks. Table 1 summarises work published previously on baffled unsparged tanks.

To investigate surface aeration effects an experimental programme of work was carried out at Warren Spring Laboratory within the Cooperative Project "Thick Liquids and Pastes: Processing and Flow". The work was funded by the British Government's Department of Trade and Industry and nine industrial companies.

2. CRITICAL IMPELLER SPEEDS

Experiments were designed to observe the impeller speeds at which a number of surface aeration criteria occurred for a range of conditions. In all, three critical impeller speeds were defined and measured whenever practical:

(i) n_{SA1} – This is the lower critical speed for incipient surface aeration and is defined as the impeller speed at which bubbles first become detached from surface vortices (Fig. 1);

(ii) n_{SA2} – This is the critical speed at which bubbles are first observed visually to be drawn into the impeller region;

(iii) n_{SA3} – This is the upper critical speed defined as the speed at which a fall in Newton number (Ne) is first observed to occur with increasing Reynolds number in the transitional or turbulent agitation regimes.

In general, $n_{SA1} < n_{SA2} \leqslant n_{SA3}$. Under most circumstances, n_{SA1} was the easiest to measure and n_{SA3} the least. The relative position of all critical speeds on the Ne/Re plot is depicted in Fig. 2. All three critical speeds depend on h_1, the impeller clearance, H, the liquid height, T, tank diameter, D, impeller diameter, e, shaft eccentricity, η_L, liquid viscosity, σ, surface tension, ρ_L, liquid density, and g, gravity. Thus, for example,

$$n_{SA1} = f (h_1, H, T, D, e, \eta_L, \sigma, \rho_L, g) \tag{1}$$

3. EXPERIMENTAL DETAILS

The following variables affecting critical agitator speeds for incipient surface aeration were investigated

(1) impeller type, shape and dimensions (see Table 2);

(2) impeller clearance (h_1) from the bottom of the tank;

(3) impeller shaft eccentricity (e);

(4) liquid height (H), this allowed an estimation of the importance of H when tanks are either being filled or emptied while the agitator is continuing to rotate at a fixed speed, since most industrial mixers operate at a fixed speed, and do not have a variable speed facility for use during emptying and filling operations;

(5) tank diameter (T), the use of larger tanks allowed an assessment to be made of scale-up problems of design methods obtained using small-scale equipment;

(6) liquid viscosity (see Table 3);

(7) surface tension (see Table 3);

(8) dissolved solids, creating what is often referred to as a 'non-coalescing' system, i.e. once bubbles are formed their rate of coalescence within the liquid is substantially retarded by the presence of dissolved electrolyte;

Table 4 contains the range of geometric variables used overall; however, in few cases were all variables varied over the full range for a given liquid.

Experiments were performed using four liquids. The work was undertaken using circular baffled tanks containing four equi-spaced vertical baffles of width 0.1 T. All the tanks were made of glass except the 240 mm and the 540 mm tanks. The former was made of perspex and the latter of steel. It was necessary to fit two flat perspex windows into the walls of the larger tank in order to observe visually the values of n_{SA1} and n_{SA2}. The use of a light diffuser and a light source behind one window allowed observations to be made of what was happening at and below the surface of the liquid.

Power consumption of the agitation systems was also measured to determine n_{SA3}. The torque-speed relationship was assessed by using three different sets of equipment and Newton number/Reynolds number plots derived.

4. RESULTS FROM EXPERIMENTAL WORK

The following observations were made from experimental data of impeller Reynolds numbers in excess of 10,000:

(1) n_{SA1} and n_{SA2} rise as either D or h_1 is reduced, all other variables remaining constant;

(2) n_{SA1} and n_{SA2} rise as H is increased, all other variables remaining constant; the dependence appears to be strongly related to the type of circulation created by the impeller. Impeller designs producing high axial flows give a much greater dependence of critical impeller speeds on H (i.e., larger exponent on H) than designs which create predominantly radial flows;

(3) The effect of tank diameter on n_{SA1} and n_{SA2} appears, from limited

evidence using distilled water only, to depend on the impeller design used. For high axial flow impellers (6-bladed, raked and marine), n_{SA1} and n_{SA2} <u>decrease</u> with increasing tank diameter, while for high radial flow impellers (Rushton, Torrance P2 and P7), n_{SA1} and n_{SA2} <u>increase</u> with increasing tank diameter, all other variables remaining constant;

(4) The effect of impeller shaft eccentricity is small over the range $0 \leqslant e/T \leqslant 0.229$;

(5) Visual estimates of n_{SA1} and n_{SA2} were repeatable to within ±15 to 20% accuracy. Within this order of repeatability, the effects of dissolved solids (potassium sulphate at 0.1 M and 0.5 M concentrations) on n_{SA1} and n_{SA2} compared with data obtained for distilled water appear to be insignificant as shown in Figs 3 and 4. Similarly the use of a liquid (isopropanol) with a low surface tension ($\sigma = 0.022$ N/m) appears to have little effect on critical impeller speeds compared with data obtained using distilled water ($\sigma = 0.073$ N/m), as shown in Fig. 5;

(6) Attempts were made to estimate n_{SA3} from sudden drops in power number during surface aeration in the turbulent region. Although previous workers (Refs 2,4,8) claim to have had some considerable success using this technique, it was found difficult to identify n_{SA3} from the Newton number/Reynolds number plots. Of the estimates for n_{SA3} that were obtained, most exceeded the corresponding n_{SA2} value, as is to be expected, although some fell below n_{SA2} values. The n_{SA2} data were insufficient for the purposes of both regression analysis and the development of correlations relating n_{SA3} to geometric factors, as has been done for most of the n_{SA1} and n_{SA2} data;

(7) All experiments were carried out on the 6-bladed, raked turbines and the marine propeller when they were pumping <u>downwards</u>. The effects on n_{SA1} and n_{SA2} values when these axial flow impellers are pumping <u>upwards</u> remains to be explored.

Many of the data for n_{SA1} and n_{SA2} have been summarised by applying regression analysis. For the case of a centrally-mounted impeller shaft ($e = 0$), two alternative regression analyses were performed on the same set of data. In the first case, the following model was assumed:

$$n_{SA1} \text{ (or } n_{SA2}) = A \, T^f \, D^a \, h_1^{\,b} \, H^c \tag{2}$$

$$\text{or} \quad \log n_{SA1} = \log A + f \log T + a \log D + b \log h_1 + c \log H \tag{3}$$

Thus $\log n_{SA1}$ was regressed linearly on the variables $\log T$, $\log D$, $\log H$ and $\log h_1$. With some data sets there was no variation in one or more of the variables and hence eqn (2) was correspondingly reduced to a model involving only the independent variables which were varied.

In the second case, estimates of regression coefficients were made using a non-dimensionalised form of eqn (2),

$$Re_{cr} = \left(\frac{D^2 \, n_{SA1} \, \rho_L}{\eta_L} \right) = A' \left(\frac{D}{T} \right)^{a'} \left(\frac{h_1}{T} \right)^{b'} \left(\frac{H}{T} \right)^{c'} \tag{4}$$

With this regression model, for any set of data, the number of regression coefficients was reduced by one, and it was observed that this had the effect of reducing the accuracy of estimates of n_{SA1} (or n_{SA2}) from eqn (4) compared with corresponding estimates made from eqn (2). Again, multiple linear regression was performed by writing eqn (4) in the form

$$\log \left(\frac{n^2 \, n_{SA1} \, \rho_L}{\eta_L} \right) = \log A' + a' \log \left(\frac{D}{T} \right) + b' \log \left(\frac{h_1}{T} \right) + c' \log \left(\frac{H}{T} \right) \qquad (5)$$

Table 5 summarises the regression analyses undertaken with the range of variables included and Figs 6 and 7 compare estimates of critical impeller speeds with measured values. In some instances, the number of variables investigated was reduced. For example, all potassium sulphate solution and isopropanol tests were carried out in the 240 mm diameter baffled tank, and hence the variable T is removed from the regression model. On another occasion, only one marine impeller diameter was used when the impeller shaft eccentricity effect was explored. Hence the variable D does not appear in the appropriate regression model.

All regression models employed are summarised in Table 6.

5. DESIGN OF AGITATED TANKS TO MINIMISE SURFACE AERATION WHILE ACHIEVING THE MAIN PROCESS RESULTS

Agitation is generally used to achieve a particular process result more quickly than would be possible without it. The main reason for agitation might be to enhance heat transfer from a jacket or to improve mass transfer. Agitated vessels are also used to suspend solids, blend two or more materials together, promote solids agglomeration, etc. Correlations usually exist which relate the critical parameters for design purposes and by comparing these with the correlations for surface aeration critical speeds it is possible to suggest design changes to satisfy the main process requirement whilst reducing the likelihood of surface aeration. This technique has been used to suggest the design changes summarised in Table 9 for heat transfer, solids suspension, mass transfer and blending systems when surface aeration is predicted to be a problem and is enlarged on below. In some cases the correlations include terms that were not varied in the experiments used to derive them but are included on a part-theoretical basis. Where the user has available other correlations preferred to the ones given or correlations for processes not included here then the same approach can be used to derive acceptable design changes. It should be noted that correlations with significantly different exponents on the variables from those used here might alter the design change recommendations and therefore only well-tested correlations should be used. When it is proposed to install a system of unusual geometry eg high H/T ratio, high h_1/H ratio, low D/T ratio, etc reference should be made to the geometry used in the original work in order to judge whether use of the correlations is still justified.

The surface aeration correlations used are those derived in this study and they are in a suitable form for comparison (see Tables 7 and 8). It should be remembered that these correlations were derived from results on small scale systems and should be used with caution for large tanks. The correlations for electrolyte solution or isopropanol could be substituted if considered more relevant to the system concerned. When non-Newtonian materials are to be agitated for which the correlations quoted below do not apply, it may be necessary to carry out experimental work preferably using a full scale system in order to obtain a design for which surface aeration problems are minimised.

The impeller types discussed below are not necessarily recommended for the particular duty but are the ones for which data exist. In addition to the design changes suggested there is also the possibility of changing the type of impeller used in which case the critical speed for surface aeration can be recalculated and compared with the probable new speed necessary to achieve the desired process result. Table 10 ranks some impellers in order of likelihood to result in surface aeration and can be used to suggest alternatives.

For the heat transfer systems an additional option is to increase the temperature difference between the jacket and the tank contents so that the impeller speed can be reduced without increasing the batch time. Other possibilities in some cases are simply to accept a longer batch cycle time by using a slower impeller speed, or to use an off-centred or inclined impeller. Novak et al (Ref. 10) have demonstrated for the off-centred impeller that there is a minimum critical speed at some distance from the centre and this position must be avoided.

It is recommended that the changes suggested in Table 9 are used as first pointers as to what to do if surface aeration is predicted to be, or found to be, a problem. Clearly it is necessary to alter the parameter or parameters until the agitator speed needed to satisfy the main process requirement is below the critical speed for surface aeration. In general n_{SA1} correlations have been used in the following comparisons but these can be converted to n_{SA3} if appropriate as described below. In many cases the system parameters of interest will fall outside the ranges used in the experiments to derive the various correlations. Hence the absolute values of n and n_{SA1} calculated should be treated with caution and if no practical experience of the system exists then changes should be made such that $n < 0.75\ n_{SA1}$. If possible small or full scale experiments should be carried out to verify the findings and the system should be designed flexibly such that, for example, impeller speed could be increased or reduced after the tank is installed. A variable speed agitator motor would be ideal and provision made to operate satisfactorily with a longer cycle time and slower agitator speed than anticipated.

An example is given below to show how a number of correlations for the various process results listed in Table 9 have been compared with the surface aeration correlations. Here we consider heat transfer on the jacketed side of an agitated tank. For the 6 blade disc turbine, the heat transfer correlation quoted by Oldshue (Ref. 11) can be used:

$$\frac{h_j T}{k} \propto \left(\frac{D^2 n \rho}{\eta}\right)^{0.66} \left(\frac{C_p \eta}{k}\right)^{0.33} \left(\frac{\eta}{\eta_w}\right)^{0.14} \left(\frac{H}{T}\right)^{-0.56} \left(\frac{D}{T}\right)^{0.13} \tag{6}$$

Assuming that h_j is specified and k, C_p, η, η_w and ρ cannot be varied for the particular system, then:

$$n \propto D^{-2.2}\ H^{0.85}\ T^{0.86} \tag{7}$$

The surface aeration correlation (B1) in Table 9 for the 6 blade disc turbine gives:

$$n_{SA1} \propto D^{-0.97}\ H^{0.59}\ T^{0.616}\ h_1^{-0.23} \tag{8}$$

$$\frac{n}{n_{SA1}} \propto D^{-1.23}\ H^{0.26}\ T^{0.244}\ h_1^{0.23} \tag{9}$$

For a system where surface aeration is predicted to be a problem ie $n > n_{SA1}$, it is desired to change the design such that $\dfrac{n}{n_{SA1}}$ decreases until $n < n_{SA1}$. The parameters with exponents furthest from zero will have the greatest effect. Possible design changes in order of decreasing effectiveness are as follows:

$$D\uparrow, H\downarrow, T\downarrow, (h_1\downarrow)$$

In this case increasing agitator diameter is probably the best option. Decreasing H or T reduces the batch volume and may not be an acceptable option.

Other correlations for various process requirements have been selected and treated in a similar way to the above analysis in order to develop Table 9. The process requirements together with the appropriate reference to the correlation involving impeller speed are listed below:

(1)   <u>Heat Transfer</u>

      A.   Jacketed Tanks : 6-blade pitched turbine (Ref. 12)
                             : propeller (Ref. 11)

	B. Helical Coil	: 6-blade disc turbine (Ref. 13)
		: 4-blade flat paddle (Ref. 14)
		: propeller (Ref. 11)
	C. Vertical Tubes	: 4-blade disc turbine (Ref. 15)

(2)   <u>Solids Suspension</u>   : Ref. 16

(3)   <u>Liquid-Solid Mass Transfer</u>

      6-blade disc turbine   : Ref. 17
      6-blade pitched turbine : Ref. 18

(4)   <u>Liquid-Liquid Mass Transfer</u> : Ref. 19

(5)   <u>Blending</u> : Ref. 20

6. RELATIONSHIP BETWEEN n_{SA1} and n_{SA3}

In most cases surface aeration critical speed correlations for a given system as shown in Table 8 exist in terms of n_{SA1} or n_{SA3} and not both. However in practice the requirement may be to minimise surface aeration in terms of n_{SA3} when a correlation only exists for n_{SA1}. It is necessary therefore to have some means of converting from one to the other.

Greaves and Kobbacy (Ref. 2) derived both n_{SA1} and n_{SA3} values for systems in which tap water, distilled water and 0.11M K_2SO_4 solution were agitated. However in their experiments H = T and h_1 was fixed at H/4 so the way in which the relationship between n_{SA1} and n_{SA3} varied with h_1/H ratios was not found. The following procedure therefore assumes that:

(i)   As H → h_1 then n_{SA1} → n_{SA3}

(ii)   As H → ∞ then $\dfrac{n_{SA1}}{n_{SA3}}$ → 0

There are insufficient data available to know the precise form of the equation relating n_{SA1} to n_{SA3} but the following equation satisfies the above assumptions and should be sufficiently accurate for its intended purpose:

$$\frac{n_{SA1}}{n_{SA3}} = e^{-X(H-h_1)} \tag{10}$$

where X is a constant for the system concerned.

The general procedure to find X and hence the relationship between n_{SA1} and n_{SA3} for <u>tap water</u> is as follows:

1.   Knowing h_1 for the system as designed, set $H = 4h_1$.

2.   Using Greaves and Kobbacy's data, ie $\dfrac{n_{SA1}}{n_{SA3}} = 0.54$ when $H = 4h_1$, substitute in equation (10) to find X.

3.   Substitute X, h_1, and the value of H to be used in practice, into equation (10) to give n_{SA1}/n_{SA3}.

For <u>distilled water</u> follow the same procedure but use $n_{SA1}/n_{SA3} = 0.58$ in step 2.

For systems where <u>surfactant solution</u> is to be agitated follow the same procedure but use $n_{SA1}/n_{SA3} = 0.605$ in step 2.

For systems involving <u>other liquids</u> no data are available so either carry out small scale experiments to find the relationship between n_{SA1} and n_{SA3} or use a general value of n_{SA1}/n_{SA3} of 0.57 and assume that the critical speed is 20% below that calculated.

7. CONCLUSIONS

This paper provides additional correlated data for critical impeller speeds for surface aeration to occur. In Table 10, the impellers investigated are compared for their tendency to cause surface aeration. The experimental data have been related to a number of principal processing requirements to show how geometric variables could be modified at the design stage to minimise or eliminate unwanted surface aeration and to meet the main processing requirement. However, existing agitation systems are often used for new liquid or suspension processing requirements and changes to existing systems are potentially far more expensive than those made to a new system at the design stage. Careful consideration should be given to selecting the best option when more than one choice exists and Table 11 indicates some of the possible problems.

8. ACKNOWLEDGEMENTS

The work described in this paper was jointly financed by the Department of Trade and Industry through the Process Plant Committee of a Research Requirements Board and the following companies: Akzo NV, Baker Perkins Ltd, BP Ltd, Donald Macpherson Ltd, Evode Ltd, ICI plc, Mars Ltd, Shell Research Ltd and Turner and Newall Ltd.

9. REFERENCES

1. Heywood, N.I., Madhvi, P. and McDonagh, M.: to be published.

2. Greaves, M. and Kobbacy, K.A.H.: I.Chem.E. Symposium Series No. 64, 1981, pp H1-H21.

3. Greaves, M.: In: Proc 3rd European Conference on Mixing, BHRA Fluid Engineering, Cranfield, Beds, Paper F5.

4. Clarke, M.W. and Vermeulen, T.: A.I.Ch.E.J., 10, 3, 1960, pp 420-422.

5. Nagata, S., Yamamoto, K., Hashimoto, K. and Naruse, Y.: Mem. Fac. Engng Kyoto, 21, 1951, pp 260-274.

6. Metzner, A.B. and Taylor, J.S.: A.I.Ch.E.J., 6, 1960, p 109.

7. Matsumura, M., Masunaga, H. and Kobayashi, J.: J. Ferment. Technol., 55, 4, 1977, pp 388-400.

8. Sverak, S. and Hruby, M.: Int. Chem. Eng., 21, 3, 1981, pp 519-526.

9. Dierendonck, J.M.H., Fortuin, J.M.H. and Venderbos, D.: In: Proc. 4th European Chemical Reaction Symposium, Brussels, pp 205-213.

10. Novak, V., Ditl, P. and Rieger, F.: In: Proc. 4th European Conference on Mixing, BHRA Fluid Engineering, Cranfield, Bedford, England, Paper C1.

11. Oldshue, J.Y.: "Fluid Mixing Technology". New York, McGraw-Hill Publications Co., 1983, p 283.

12. Nagata, S.: "Mixing Principles and Applications". John Wiley & Sons, 1975.

13. Chen, C-C.: Proc. Nat. Sci. Counc. Part 3 (Tanva), 7, 1974, pp 213-225.

14. De Maerteleire, E.: Chem. Eng. Sci., 33, 1978, pp 1107-1113.

15. Dunlap, J.R. and Rushton, J.H.: Chem. Eng. Prog. Symp. Series No. 5, 49, 1955, pp 137-151.

16. Zweitering, T.N.: Chem. Eng. Sci., 8, 1958, pp 244-253.

17. Barker, J.J. and Treybal, R.E.: A.I.Ch.E.J., 6, 1960, p 239.

18. Askew, W.S. and Beckmann, R.B.: Ind. Eng. Chem. Proc. Des. Dev., 4, 1965, pp 311-365.

19. Skelland, A.H.P. and Lee, J.M.: A.I.Ch.E.J., 27, 1, 1981, pp 99-111.

20. Khang, S.J. and Levenspiel, O.: Chem. Engng., 83, 21, 1976, pp 141-143.

TABLE 1. - Summary of Previous Work on Unsparged Liquids in Baffled Tanks

Workers	Liquids Used	Agitator Type (D, mm)	Tank Type and Size (T, mm)	h_1/H Values	H/T Values
Greaves and Kobbacy (1981) (Ref. 2)	filtered tap water distilled water 0.11 M potassium sulphate solution	disc blade turbine (76.2, 106, 135)	fully baffled, flat-bottomed tank (152, 203)	h_1 = H/4	
Greaves (1979) (Ref. 3)	filtered tap water	Rushton turbine (51, 76, 101.6)	flat-bottomed, cylindrical 4-baffled tank (152)	= 1/3	= 1.0
Clark and Vermeulen (1964) (Ref. 4)	water carbon tetrachloride	4-bladed flat paddle (127, 152.4)	fully-baffled tank (254)		= 1.0
Nagata et al (1959) (Ref. 5)	tap water	8-flat bladed paddle (300) 8 retreated blade paddle (300) 8-bladed arrow head turbine (300) 8-bladed Brumgin type impeller (300) 8-pitched blade paddle (300) 8-flat blade paddle (180)	cylindrical 8-baffled tank (585)		= 1.0 in all cases
Metzner and Taylor (1960) (Ref. 6)	Karo syrup 0.7% sodium carboxymethyl cellulose	6-flat bladed impeller (101.6, 152.4)	4-baffled tank	cylindrical h_1 = 101.6	
Matsumura et al (1977) (Ref. 7)	water ethyl alcohol benzyl alcohol ethylene glycol	3 6-bladed turbine impellers (87-133)	3-baffled flat-bottomed tank (218)		0.333 and 1.0
Sverak and Hruby (1981) (Ref. 8)	distilled water carbon tetrachloride ethyl iodide water with 'tenside' (a surfactant) 25%, 44%, 74%, 91% water solution of glycerine	2-vertical bladed agitator 6-vertical bladed turbine with a separating disc (for both agitators T/D = 3.0)	4-baffled tank	h_1 = 0	0.5, 0.75, 1.0
Dierendonck et al (1978) (Ref. 9)	water cyclohexane n-hexane	6-bladed turbine (60, 80, 100, 120, 160, 250, 800)	4-baffled tank (165, 290, 450, 2600)		= 1.0 in all
	10% aminocaprolactum solution in water	6-bladed turbine (140)	4-baffled tank (290)	T/D = 2.07	
	10% aminocaprolactum solution in water	6-bladed turbine (300)	4-baffled tank (450)		
		6-bladed turbine (140)	4-baffled tank (290)		

TABLE 2. – Details of the Impellers Used

Impeller Type	Impeller Diameter, mm
Rushton turbine	41.5, 71.5, 101, 125
3-bladed equi-pitched marine propeller	72.0, 120
6-bladed 45° raked open impeller	84.0, 128
Torrance P2 dispersers	103, 126
Torrance P7 dispersers	103, 122

TABLE 3. – Liquids Used in Surface Aeration Experiments

Liquid	Viscosity, Pa s	Density, $kg\ m^{-3}$	Surface Tension, $N\ m^{-1}$
Distilled water	0.001	1000	0.073
0.1 M potassium sulphate solution	0.001	1010	0.072
0.5 M potassium sulphate solution	0.001*	1000	0.073
Isopropanol analar	0.0037	782	0.0222

* Quoted by literature.

TABLE 4. – Range of Geometric Variables Used

Variable	Range
D/T (T)	0.22 – 0.938 (0.128 m – 0.54 m)
h_1/T	0.25 – 0.50
H/T	0.56 – 1.29
e/T	0 – 0.229

TABLE 5. – Summary of Regression Analyses

Fluid	Impeller Type	T varied? (mm)	D varied? (mm)	h_1/T varied?	H/T varied?	e/T varied?
Distilled Water	Rushton	216-540	42-126	0.50-1.29	0.25-0.50	No
	6-bladed raked	"	"	"	"	No
	marine	"	"	"	"	No
	P2	"	"	"	"	No
	P7	"	"	"	"	No
0.1 M K_2SO_4 Solution (T = 240 mm)	Rushton	No	42-125	No)	0.56-1.29	No
	6-bladed raked	No	84,128	No)	"	No
	marine	No	72,120	No)(0.38)	"	No
	P2	No	103,126	No)	"	No
	P7	No	103,122	No)	"	No
Isopropanol (T = 240 mm)	Rushton	No	101,126	0.25-0.5	0.56-1.29	0-0.229
	6-bladed raked	No	No(D=128)	"	"	No
	marine	No	No(D=72)	"	"	0-0.229
	P2	No	103,126	"	"	0-0.229
	P7	No	103,122	"	"	0-0.229

TABLE 6. – Summary of Regression Equations used and Tables
of Regression Coefficients

Fluid	Equations Used
__Distilled Water__	$n_{SA1} = AT^f D^a h_1^{\ b} H^c$
	$n_{SA2} = AT^f D^a h_1^{\ b} H^c$
	$\dfrac{D^2 n_{SA1}\ \rho_L}{\eta_L} = A' \left(\dfrac{D}{T}\right)^{a'} \left(\dfrac{h_1}{T}\right)^{b'} \left(\dfrac{H}{T}\right)^{c'}$
	$\dfrac{D^2 n_{SA2}\ \rho_L}{\eta_L} = A' \left(\dfrac{D}{T}\right)^{a'} \left(\dfrac{h_1}{T}\right)^{b'} \left(\dfrac{H}{T}\right)^{c'}$
__0.1 M K_2SO_4 Solution__	$n_{SA1} = AD^a H^b$
	$n_{SA2} = AD^a H^b$
__Isopropanol__	$n_{SA1} = AD^a h_1^{\ b} H^c \quad (e = 0)$
	$n_{SA2} = AD^a h_1^{\ b} h^c \quad (e = 0)$
	$n_{SA1} = AD^a h_1^{\ b} H^c \ (T - e)^d$
	$n_{SA2} = AD^a h_1^{\ b} H^c \ (T - e)^d$

or, when only one impeller diameter used (marine impeller)

$$n_{SA1} = A h_1^{\ b} H^c \ (T - e)^d$$

$$n_{SA2} = A h_1^{\ b} H^c \ (T - e)^d$$

TABLE 7. – Baffled Tanks for which Usable Critical Speed Correlations Exist

Workers	Batch vol m^3	Impeller type	Correlation No in Table 8
Present study	0.0024–0.16	6–blade disc turbine	B1,B2,B3,B4
		Marine propeller	B5,B6,B7
		6–bladed 45° turbine	B8,B9
		Torrance P2 disperser	B10,B11,B12,B13
		Torrance P7 disperser	B14,B15,B16,B17
Greaves and Kobbacy	0.0066	Disc blade turbine	B18,B19
Clark and Vermeulen	0.013	4 blade 90° paddle	B20
Dierendonck et al	0.003–13.8*	6 blade disc turbine	B21

* It is not clear from the reference whether the 13.8 m^3 tank was used in deriving the surface aeration correlation.

<u>TABLE 8. – Baffled Tank Correlations</u>

Correlation No	Correlation	Conditions
B1	$n_{SA1} = 1.04 \ T^{0.616} \ D^{-0.97} \ h_1^{-0.23} \ H^{0.59}$	Distilled water
B2	$n_{SA1} = 0.42 \ D^{-2.2} \ H^{1.92} \quad T=0.24\text{m}, \ h_1 = 0.09\text{m}$	0.1M K_2SO_4 soln
B3	$n_{SA1} = 0.0725 \ D^{-1.93} \ h_1^{-0.355} \ H^{0.968} \quad T \text{ constant}$	Isopropanol $\sigma = 0.022$ N/m
B4	$n_{SA1} = 0.132 \ D^{-1.94} \ h_1^{-0.35} \ H^{0.951} \ (T-e)^{0.448}$	Eccentric shaft $0.125 < \dfrac{e}{T} < 0.229$
B5	$n_{SA1} = 7.56 \ T^{-1.11} \ D^{-0.116} \ h_1^{-0.295} \ H^{1.65}$	Distilled water
B6	$n_{SA1} = 0.974 \ D^{0.097} \ H^{1.45} \quad T=0.24\text{m}, h_1=0.09\text{m}$	0.1M K_2SO_4 soln
B7	$n_{SA1} = 39.9 \ h_1^{-0.75} \ H^{1.20} \ (T-e)^{-0.123}$	Eccentric shaft $0.125 < \dfrac{e}{D} < 0.229$
B8	$n_{SA1} = 0.139 \ T^{-0.545} \ D^{-2.05} \ h_1^{-0.25} \ H^{1.64}$	Distilled water
B9	$n_{SA1} = 0.315 \ D^{-2.17} \ H^{1.52} \quad T=0.24\text{m}, h_1=0.09\text{m}$	0.1M K_2SO_4 soln
B10	$n_{SA1} = 0.134 \ T^{0.292} \ D^{-1.76} \ h_1^{-0.97} \ H^{0.93}$	Distilled water
B11	$n_{SA1} = 30.3 \ D^{-0.32} \ H^{1.19} \quad T=0.24\text{m}, h_1=0.09\text{m}$	0.1M K_2SO_4 soln
B12	$n_{SA1} = 1.48 \ D^{-1.07} \ h_1^{-0.37} \ H^{0.92} \quad T \text{ constant}$	Isopropanol $\sigma = 0.022$ N/m
B13	$n_{SA1} = 1.28 \ D^{-1.09} \ h_1^{-0.38} \ H^{0.957} \ (T-e)^{-0.111}$	Eccentric shaft $0.125 < \dfrac{e}{T} < 0.229$
B14	$n_{SA1} = 0.0014 \ T^{0.724} \ D^{-4.11} \ h_1^{-1.06} \ H^{0.94}$	Distilled water
B15	$n_{SA1} = 0.207 \ D^{-2.53} \ H^{0.91} \quad T=0.24\text{m}, h_1=0.09\text{m}$	0.1M K_2SO_4 soln
B16	$n_{SA1} = 0.06 \ D^{-2.64} \ h_1^{-0.49} \ H^{1.07} \quad T \text{ constant}$	Isopropanol $\sigma = 0.022$ N/m
B17	$n_{SA1} = 0.071 \ D^{-2.64} \ h_1^{-0.5} \ H^{1.14} \ (T-e)^{0.065}$	Eccentric shaft $0.125 < \dfrac{e}{T} < 0.229$
B18	$n_{SA1} = k_{1GK} \ \left(\dfrac{(T^2H^2)}{D^2}\right)^{1/3} \left(1 - \dfrac{h_1}{H}\right)^{1/3} \left(\dfrac{P_A}{P}\right)^{0.13}$ $k_{1GK} = 0.476 \quad$ distilled water $k_{1GK} = 0.429 \quad$ tap water $k_{1GK} = 0.375 \quad$ 0.11M K_2SO_4 soln	$Re > 10^4$ P_A = atmos pressure P = pressure above liquid surface
B19	$n_{SA3} = k_{2GK} \ \left(\dfrac{(T^2H^2)}{D^2}\right)^{1/3} \left(1 - \dfrac{h_1}{H}\right)^{1/3} \left(\dfrac{P_A}{P}\right)^{0.13}$ $k_{2GK} = 0.82 \quad$ distilled water $k_{2GK} = 0.80 \quad$ tap water $k_{2GK} = 0.62 \quad$ 0.11M K_2SO_4 soln	$Re > 10^4$
B20	$n_{SA3} = 0.221 \ \left(\dfrac{T^2H}{D^3W}\right)^{1/2} \left(\dfrac{H}{h_1}\right)^{1/3}$	T and H constant
B21	$n_{SA1} = 1.55 \ \dfrac{T}{D^2} \left[\dfrac{H-h_1 - \dfrac{W}{2}}{T}\right]^{1/2} \left[\dfrac{g\sigma}{\rho}\right]^{1/4}$	$0.1 < \dfrac{H-h_1 - \dfrac{W}{2}}{T} < 0.2 + 1.75\dfrac{D}{T}$ $1.4 < \dfrac{T}{D} < 7.5$

TABLE 9. - Design Changes to Minimise Surface Aeration in Baffled Tanks Whilst Still Achieving the Main Process Requirement

Main Process Requirement	Impeller type	Design changes (in order of decreasing effectiveness)	Comments
1. Heat transfer from jacket	6 blade disc turbine	$D\uparrow$, $H\downarrow$, $T\downarrow$	$D\uparrow$ probably the best option. $T\downarrow$ and H reduce the batch size and therefore may not be feasible. Could also consider reducing N such that $N < N_{SAl}$ and accepting slower heat transfer: or increasing temp difference between jacket and tank contents
2. Heat transfer from jacket	6 blade pitched turbine	$T\downarrow$, $D\downarrow$, $H\uparrow$	The batch size can be unchanged if T and H are altered appropriately. Note possible change of heat transfer area. Could also reduce N or increase temp difference (see 1 above)
3. Heat transfer from jacket	propeller	$D\uparrow$, $T\downarrow$, $H\uparrow$	$D\uparrow$ quite effective and probably a better alternative than $T\downarrow$ and $H\uparrow$. Could also reduce N or increase temp difference (see 1 above)
4. Heat transfer from immersed helical coils	6 blade disc turbine	$d_c\downarrow$, $D\uparrow$, $T\downarrow$	Could also reduce N or increase temp difference (see 1 above)
5. Heat transfer from immersed helical coils	4 blade flat paddle	$T\downarrow$, $D\uparrow$	(see 4 above) Could also change coil dimensions
6. Heat transfer from immersed helical coils	propeller	$D\uparrow$, $T\downarrow$, $d_c\downarrow$	(see 4 above)
7. Heat transfer from vertical tubes	6 blade disc turbine	$D\uparrow$	Tubes as baffles provide a reasonably, but not fully, baffled tank. Could also reduce N or increase temp difference (see 1 above)
8. Heat transfer from vertical tubes	6 blade pitched turbine	$T\downarrow$, $D\uparrow$	(see 7 above)
9. Heat transfer from vertical tubes	propeller	$D\uparrow$, $T\downarrow$	(see 7 above)
10. Heat transfer from vertical tubes	4 blade flat paddle	$D\uparrow$, $T\uparrow$	(see 7 above)
11. Just off bottom solids suspension	6 blade disc turbine	$D\uparrow$, $T\downarrow$, $H\uparrow$, $h\downarrow$	Although h_l appears to have a fairly small exponent, no account has been taken of h_l in the correlation for solids suspension. It would be sound practice to reduce h_l.
12. Just off bottom solids suspension	6 blade pitched turbine	$T\downarrow$, $H\uparrow$, $D\uparrow$, $h_1\downarrow$	(see 11 above)
13. Just off bottom solids suspension	propeller	$T\downarrow$, $D\uparrow$, $H\uparrow$, $h_1\downarrow$	(see 11 above)
14. Liquid-solid mass transfer	6 blade disc turbine	$D\uparrow$, $T\downarrow$	H and h_l not included in mass transfer correlation
15. Liquid-solid mass transfer	6 blade pitched turbine	$T\downarrow$, $h_1\downarrow$	H and h_l not included in mass transfer correlation though it is stated that $k_D\uparrow$ as $h_l\downarrow$; D does not have much effect overall
16. Liquid-solid mass transfer	propeller	$T\downarrow$, $D\uparrow$, $h_1\downarrow$	(see 15 above)
17. Liquid-solid mass transfer	4 blade flat paddle	$D\downarrow$	T, H and h_l not included in mass transfer correlation
18. Liquid-liquid mass transfer	4/6 blade flat paddle	$T\uparrow$	High and low surface tension systems. H and h_l not included in mass transfer correlation. Exponent on T is relatively small so this design change may not be very effective
19. Blending	propeller	$T\downarrow$, $D\uparrow$	Based on blending of Newtonian liquids of similar density and viscosity. $10^4 < Re < 10^6$

<u>TABLE 10. – Ranking of Impellers for Surface Aeration in Baffled Tanks</u>

H/T < 1	Impeller Type	Comments
Most likely to result in surface aeration	6 blade disc turbine	
n_{SA1} increasing	6 blade raked turbine	
	propeller	
Least likely to result in surface aeration	P2 and P7 Torrance dispersers	Similar results from the two Torrance dispersers used

H/T > 1		
Most likely to result in surface aeration	6 blade disc turbine	
n_{SA1} increasing	6 blade raked turbine P2 and P7 Torrance dispersers	Similar results from the two Torrance dispersers used
Least likely to result in surface aeration	propeller	

<u>TABLE 11. – Effects of Design Changes to an Existing System</u>

Parameter Changed	Comments
N	May be necessary to buy a new motor (preferably variable speed)
D	Relatively cheap option provided shaft design still acceptable
$H\uparrow$	If there is sufficient ullage to increase the liquid height without modifying the tank then this is a very cheap option. If tank modifications or a new tank are necessary it may become costly. If layout constraints require modifications to the building or moving of other vessels this option will probably become too expensive to be feasible. $H\uparrow$ alone results in a bigger batch size and possible resultant problems upstream and/or downstream.
$H\downarrow$	$H\downarrow$ alone results in a smaller batch size and hence a possible reduction in output.
T	Changes to T almost certainly require a new tank and if it is jacketed or a pressure vessel this option will be expensive. Changes to T affect the batch size unless H can also be changed to compensate.
$h_1\downarrow$	Fairly cheap option although it probably requires a new impeller shaft and the mechanical design should be checked in case a greater diameter shaft is necessary.
e (off centred impeller)	Cost of this option will vary according to the difficulty of remounting the agitator system in a new position. Modifications to the tank lid (if it has one) could also be expensive.

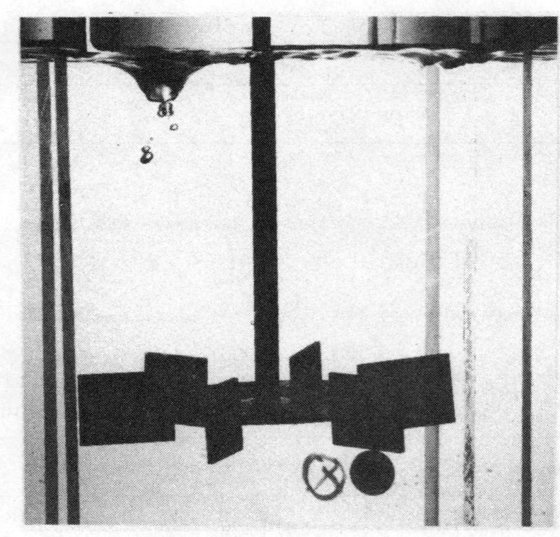

distilled water, $H = 200$ mm,
$h_1 = 74$ mm, $T = 240$ mm, $D = 125$ mm,
$n_{SA1} = 1 \cdot 68$ r.p.s.

FIG.1 THE FORMATION OF THE FIRST
BUBBLES FROM A SURFACE VORTEX

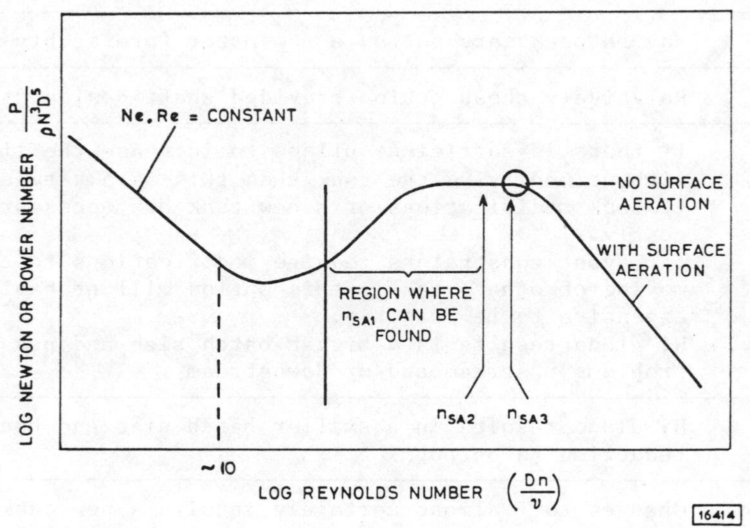

FIG.2 TYPICAL NEWTON NUMBER OR POWER CURVE FOR BAFFLED
SYSTEMS, SHOWING CRITICAL IMPELLER SPEED LOCATIONS

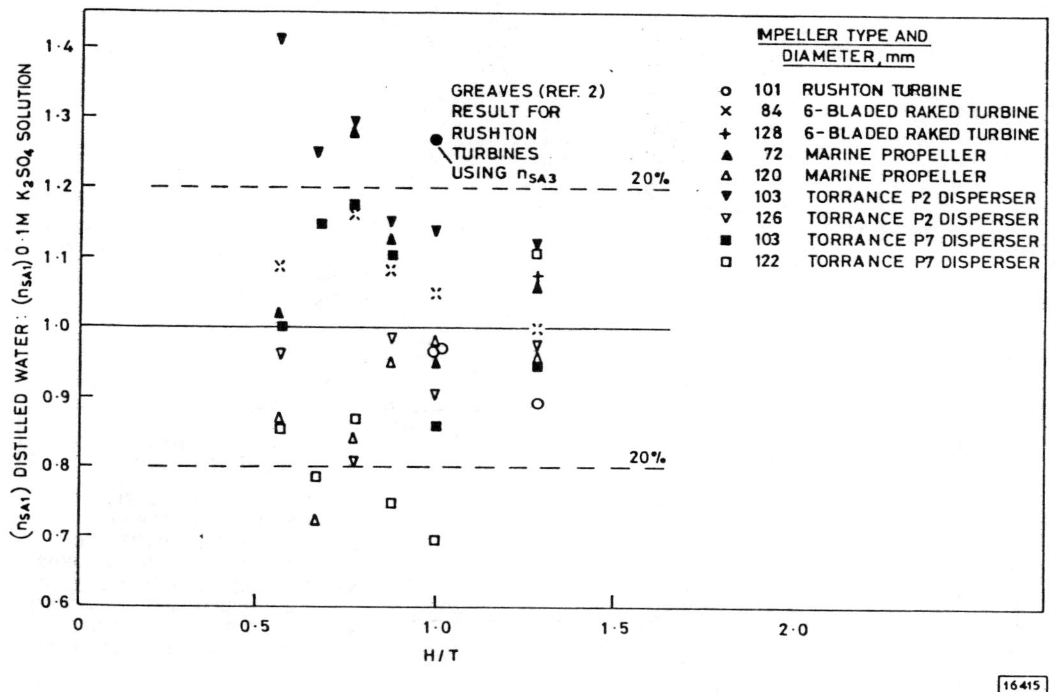

FIG. 3   COMPARISON BETWEEN n_{SA1} DATA FOR DISTILLED WATER AND FOR POTASSIUM SULPHATE SOLUTION (0·1M) (T = 240 mm , h_1 = 90 mm)

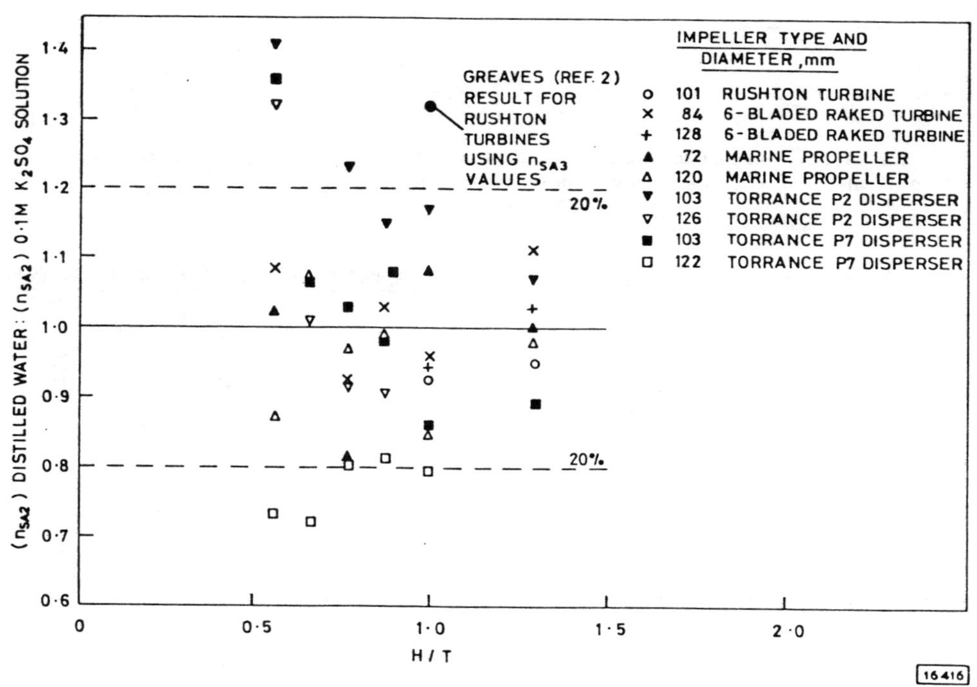

FIG. 4   COMPARISON BETWEEN n_{SA2} DATA FOR DISTILLED WATER AND FOR POTASSIUM SULPHATE SOLUTION (0·1M) (T = 240 mm , h_1 = 90 mm)

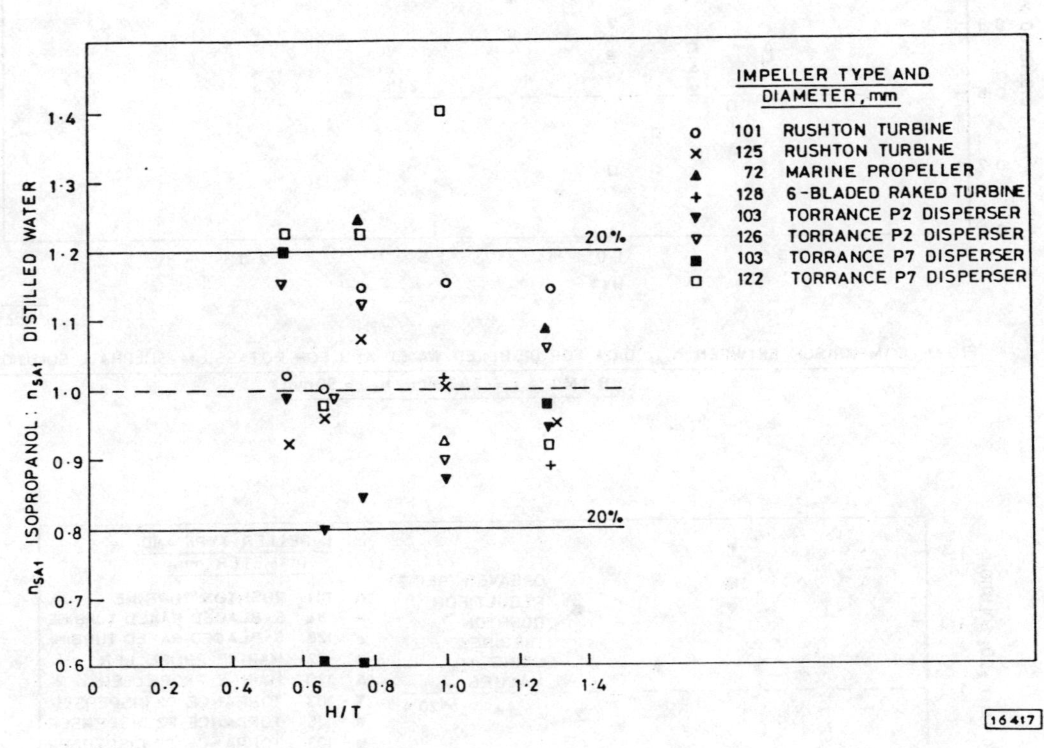

n_{SA1} ISOPROPANOL : n_{SA1} DISTILLED WATER

FIG. 5   COMPARISON OF n_{SA1} DATA FOR DISTILLED WATER AND ISOPROPANOL (T = 240mm, h_1 = 90mm)

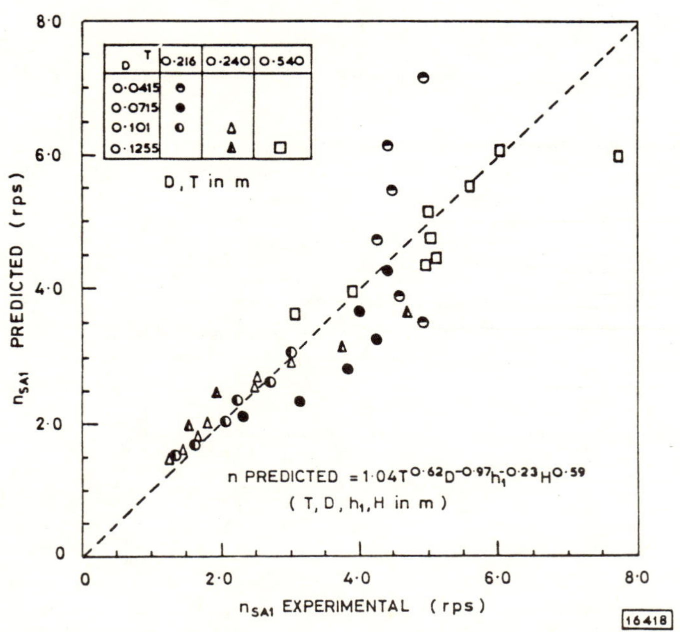

FIG. 6 COMPARISON BETWEEN PREDICTED AND EXPERIMENTAL
n_{SA1} VALUES FOR DISTILLED WATER USING RUSHTON TURBINES
(T = 240 mm)

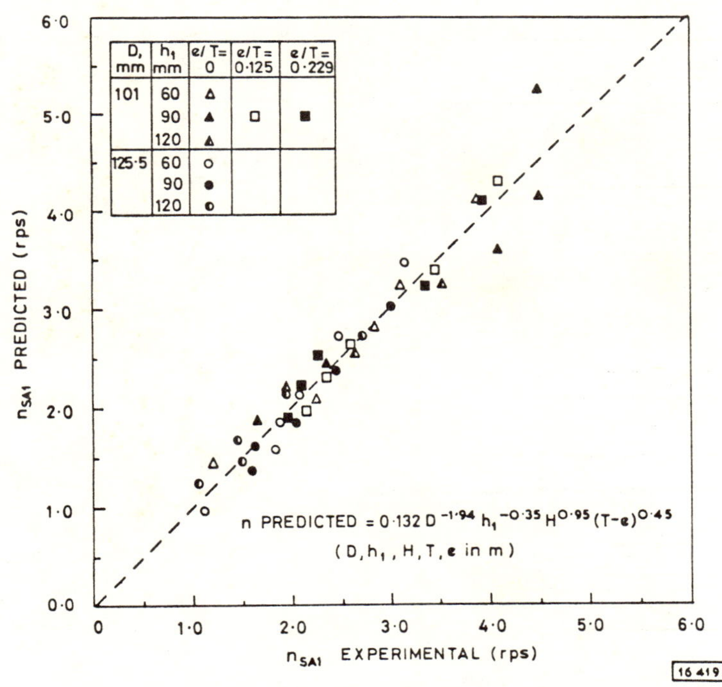

FIG. 7 COMPARISON BETWEEN PREDICTED AND EXPERIMENTAL
n_{SA1} VALUES FOR ISOPROPANOL USING RUSHTON TURBINES
(T = 240 mm)

MIXING IN VESSEL WITH ECCENTRICAL MIXER

<u>J. Medek</u> - Department of Chemical Engineering,
Technical University - Brno, Kraví hora II,
602 00 Brno, Czechoslovakia

<u>I. Fořt</u> - Department of Chemical Engineering,
Institut of Chemical technology - Prague,
Suchbátarova 1905, 166 28 Praha 6,
Czechoslovakia

Summary

Most problems connected with mixing are realized in cylindrical baffled vessels equipped with a mechanical impeller located in the vessel axis of symmetry. Such a geometrical arrangement is usually called the "standard one" (see Fig. 1). For the "standard arrangement of agitated system" there are sufficient data available in the literature, so that simple problems can be solved by computations only, without expensive pilot-plant experiments [1,2,3,4]. For the solution of complex problems the pilot-plant experiments are usually necessary, and then there is no reason to use the "standard arrangement" if another geometric shape of the system suits better to the aim of mixing. The case may concern both the mixing itself and a design of the equipment or technological operations.

Some possibilities of the non-standard geometrical arrangements of the agitated systems are described in this contribution. For all arrangements, only one high-speed eccentrically located impeller (D/d > 2) was used in the vessel equipped with different combinations of the baffles at the wall. From the experimental results it was possible to draw conclusion about energy consumption as well as about the circulation ability of mixers at the non-standard arrangements.

Held at Wurzburg, 10-12 June, 1985.

Organised by DVCV· Deutsche Vereinigung für Chemie- und Verfahrenstechnik
(German Association of Chemical and Process Engineering).

Organisation: GVC·VDI-Gesellschaft Verfahrenstechnik und Chemieingenieurwesen.

GVC

NOMENCLATURE

b   = baffle width /m/

D   = vessel diameter /m/

d   = impeller diameter /m/

e   = impeller eccentricity /m/

H   = height of liquid above bottom at rest /m/

H_2 = height of impeller lower edge above bottom /m/

h   = impeller blade width /m/

n   = impeller frequency of rotation /s^{-1}/

n_b = number of baffles /-/

P   = Power input of impeller /W/

$\dot{V}$ = impeller pumping capacity /m^3s^{-1}/

$\dot{V}_c$ = total pumping capacity /m^3s^{-1}/

v   = liquid velocity through the body of rotating impeller

α = angle of inclination of blades

ϱ = density /$kg\ m^{-3}$/

μ = dynamic viscosity /Pas/

Dimensionless Variables

E(Ec) = Impeller Energetic Efficiency /-/

Kp   = Flow rate number   /-/

Kp_c = Total flow rate number   /-/

Po   = Power number   /-/

Re_M = Reynolds number for mixing   /-/

1. THEORETICAL

In the text below the term "standard arrangement of the agitated system" will refer to a mixing vessel equipped with four baffles at the wall (b=0.1D) with a high-speed impeller located in the vessel axis of symmetry (see dimensionless simplexes in Tab. I)[1]. The location of the baffles in the vessel is necessary to avoid tangential vortex flow in the vessel (central vortex at the surface). Then according to the impeller used, either prevailing axial or radial circulation of the liquid will originate. The impeller eccentricity in the vessel having no baffles can have the same effect, at the impeller eccentricity e = D/3, no tangential vortex appears when an axial impeller is used, the impeller energy consumption being the same as with "standard arrngement"[1]. In certain cases, however, it is useful to place in the vessel a limited number of baffles (1, 2 or 3) in combination with the eccentrically located impeller. Such non-typical arrangements have not yet been described in the literature. In order to apply them to industrial processes, it is necessary to know corresponding hydrodynamic characteristics. The following dimensionless criteria are mostly determined experimentally:

1. Power number defined as:

$$Po = P/\varrho\, n^3\, d^5 \qquad (1)$$

2. Flow-rate number, describing impeller pumping capacity $\dot{V}$ in dimensionless form:

$$Kp = \dot{V}/n\, d^3 \qquad (2)$$

If the impeller is eccentrically located in the vessel, it is worth observing also the so-called induced flow $\dot{V}_i$ to be able to determine a portion of the liquid flow not passing through the body of the impeller. Attention is usually focused on the total flow-rate in the vessel $\dot{V}_c$, i.e. on the volumetric flow-rate of the agitated liquid in the plane of the lower edges of impeller blades[5,6]. Then the relation

$$\dot{V}_c = \dot{V} + \dot{V}_i \qquad (3)$$

is valid. The corresponding dimensionless numbers are defined as follows:

$$Kp_c = \dot{V}_c/n\, d^3 \qquad (4)$$

for the total flow of the agitated liquid, and

$$Kp_i = \dot{V}_i/n\, d^3 \qquad (5)$$

for the induced flow (see Fig. 2).

In case the values of the quantities Po, Kp, Kp_c and Kp_i are known from experiments made, it is easy to compare the different geometric arrangements from the view-point of energy consumption (Po) or circulating ability of the impeller (Kp, Kp_c). At the prevailing axial liquid-flow in the vessel, it is also possible to define another dimensionless number as a proportion of the impeller power input necessary for pumping to the total impeller power input:

$$E_{ax} = P_{ax}/P \tag{6}$$

For the determination of the quantity P_{ax} it is necessary to assume that at the flow of the agitated liquid through the body of the impeller no mechanical energy is dissipated, so that at a constant height of the liquid in the vessel at rest for pressure drop Δp over the body of the rotating impeller, the following equation is valid:

$$\Delta p = \rho \, (\overline{v}^2/2) = \frac{1}{2} \rho \left(\frac{4 \dot{V}}{\pi d^2}\right)^2 \tag{7}$$

The value of the mean velocity $\overline{v}$ is calculated from the measured pumping capacity of the impeller $\dot{V}$ passing through the body of the rotating impeller. Then for the impeller power input necessary for the pumping, the simple relation is valid:

$$P_{ax} = \dot{V} \Delta p \tag{8}$$

Combination of Eqs. (7) and (8(gives the following relation:

$$P_{ax} = \frac{8}{\pi^2} \, Kp^3 \, \rho \, n^3 \, d^5 \tag{9}$$

Let us introduce the quantity:

$$E = Kp^3/Po \tag{10}$$

and similarly, the quantity for total volumetric flow-rate in the vessel:

$$E_c = Kp_c^{\,3}/Po \tag{11}$$

If the maximum liquid circulation at the minimum energy consumption is to be reached, the values of the impeller energetic efficiency E or E_c must be at their maximum irrespective of the absolute values of the Po and Kp (Kp_c). The maximum values correspond to the most suitable geometric arrangement with respect to the impeller pumping efficiency. However, with regard to the asumption introduced, the criteria E and E_c can be used only for comparison of individual geometric arrangements of the agitated systems, e.g. at different eccentricities of the impeller in the vessel.

2. EXPERIMENTAL

The following variants of geometrical arrangements of the agitated systems were observed experimentally:

1) Agitated system with four baffles and eccentrically located impeller in the plane of symmetry between two adjacent baffles (see Fig. 3).

2) Agitated system with one baffle and an eccentrically located impeller between the baffle and the vessel axis (see Fig. 4).

3) Agitated system with two baffles and an eccentrically located impeller in the plane of symmetry between the baffles (see Fig. 5).

4) Agitated system with three baffles and eccentrically located
impeller in the plane of symmetry between two adjacent baffles
(see Fig. 6).

5) Agitated system without baffles with impeller eccentrically
e = D/3.

The above systems had a six blade impeller with inclined plane blades[1]
producing prevailing axial flow in the baffled vessel. The ranges of
geometric simplexes for geometric arrangements used at the experiments are summarized in Tab. I. All experiments were carried out at
a turbulent flow of the agitated liquid with the mixing Reynolds
number $Re_m > 1.10^4$. The diameter of the vessel D = 445 mm. In all
the observed cases no depression liquid surface appeared, so that
the inference of the mixing Froude number could be neglected.

For the determination of the corresponding dimensionless number
/see Eqs. (10) and (11)/, the following quantities were observed
experimentally: impeller power input (P), impeller pumping capacity
($\dot{V}$), and total volumetric flow-rate in the vessel /$\dot{V}c$/. The impeller
power input was calculated from the torque measured on the rotatably
located electromotor, the rotation frequency of the impeller being
measured simultaneously. The method of the impeller power input
measurement is one of the usual experimental procedures and has been
sufficiently described in the literature [2,3,4].

The pumping capacity of the impeller and total volumetric flowrate of the agitated liquid were measured by means of the mean
circulation times of the flow-follower (the indicating particle)[8].
Under the assumption that the flow-follower follows the stream lines
in the agitated liquid, the quantity $\dot{V}$ was measured from the mean
time of primary circulations between two passages through the body
of the rotating impeller, and the quantity $\dot{V}c$ from the mean time
of total circulations between the two passages through plane of
the lower edges of the impeller blades in the closen direction
(downwards or upwards). The number of circulations for the calculation of the mean value was always 1 000, only at a great impeller
eccentricity, owing to a very long time of the primary circulation,
the quantity $\dot{V}$ was calculated as a mean value from 300 passages
through the body of the rotating impeller. Always the individual
passages were observed visually in a transparent vessel. Water at
room temperature was used as agitated liquid.

3. RESULTS

The results of the experiments made are drawn tabularly for
the individual geometrical arrangements in Figs. 3, 4, 5 and 6.
The quantities Kp, Kpc and Po are plotted as functions of the impeller eccentricity e. For the arrangement of the agitated system without baffles, only one impeller position was observed, e = D/3; the
values measured are listed in Tab. II. Calculated values of dimensionless numbers E and Ec for geometrical arrangements examined are
listed in Tab. III.

4. DISCUSSION

The equipment and techniques used in the experiments were
verified by at set of previous experiments, whose results were
compared with the data obtained from the literature. In all cases
the results were congruent with the data and thus it is possible to
conclude that the measured values in the present contribution are
quite real. During the observations carried out visually, no

tangential vortex appeared on the liquid surface at the ^{all} arrangements.
Therefore the influence of the Froude number for mixing was elimi-
nated. At a greater impeller eccentricity occurs a significant
change of the liquid macroflow in the vessel in contrast to the flow
-pattern at the "standard arrangement". The original shape of the
stream-lines changes into one loop circulation as shown in Fig. 8.
The change of the stream-lines appears at the eccentricity e = D/4.
The resultant shape being less energy consuming, which was verified
experimentally. A decrease of the values Kp and Kp$_C$ in comparison
with the "standard arrangement" is caused by a more significant
tangential component of the liquid flow in the mixing vessel
especially at a smaller number of baffles. As the most effective
arrangement from the view-point of the energetic efficiency E or Ec
appears the system with three baffles; see Tab. III. From the experimental results is obvious that at the greater impeller eccentricity less of the total volumetric flow-rate of the agitated liquid
passes through the body of impeller. This phenomenon is caused by
the influence of the vessel wall upon the impeller pumping ability,
which may sometimes increase the impeller power input. Therefore there always
exists on the optimum of the impeller eccentricity when transformed
into the pumping intensity of the macroflow in the mixing vessel.

References

1. Czechoslovak Standard "Mixing" ON 691 000-37. Chepos, Brno 1959.

2. Uhl, V.W., Gray, J.B.: Mixing - Theory and Practice. Academic
Press, New York and London, 1966.

3. Nagata, S.: Mixing - Principles and Applications. Kodasha Ltd.,
Tokyo, 1976.

4. Strek, F.: Mixing and Mixing Equipments (in Czech). SNTL,
Prague, 1977.

5. Medek, J.: Ph.D. Thesis. Technical University Brno. Brno, 1976.

6. Medek, J., Fořt, I.: Chemický průmysl 30, 12, 1980.

7. Medek, J., Fořt, I.: Chemický průmysl 29, 3, 1979.

8. Porcelli, J.V., Marr, G.R.: Ind. Chem. Eng. Chem., Fundam 1,
172, 1962.

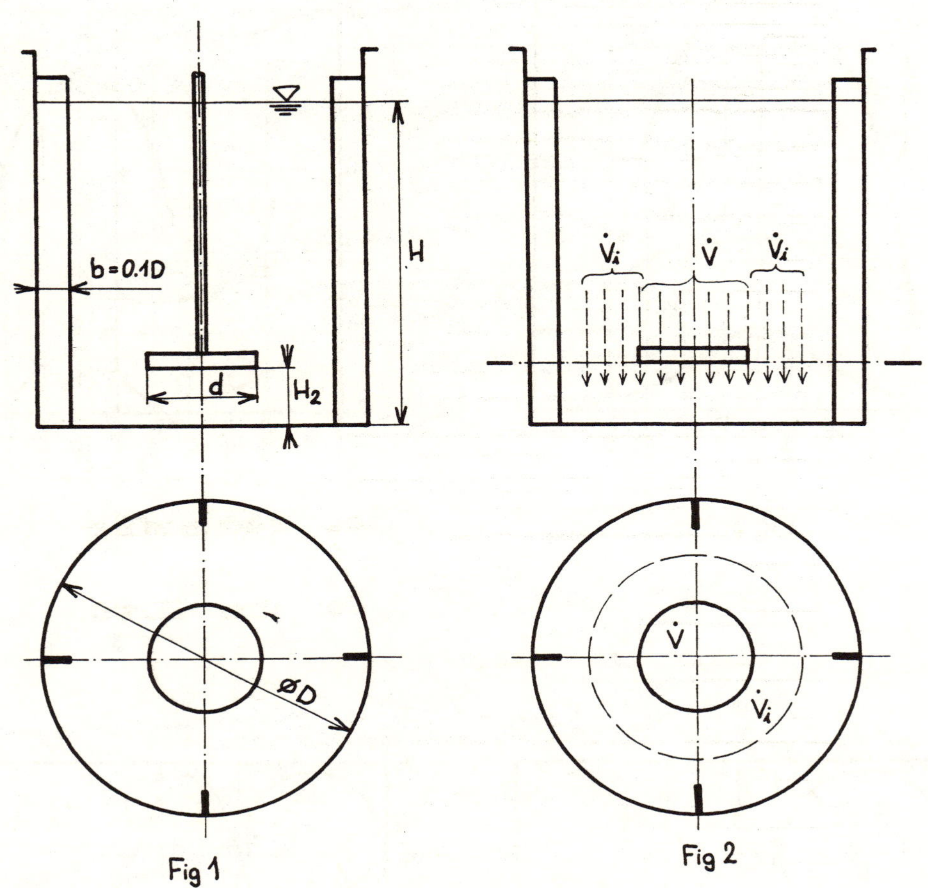

TABLE I. GEOMETRICAL ARRANGEMENTS OF AGITATED SYSTEMS
(" STADARD ARRANGEMENT ")

H/D	D/d	H_2/d	NUMBER OF BAFFLES	b/D
0.8 – 1.2	2.5 – 5.0	0.5 – 1.5	4	0.10 – 0.12
1.0	4.45	1.0	4	0.10

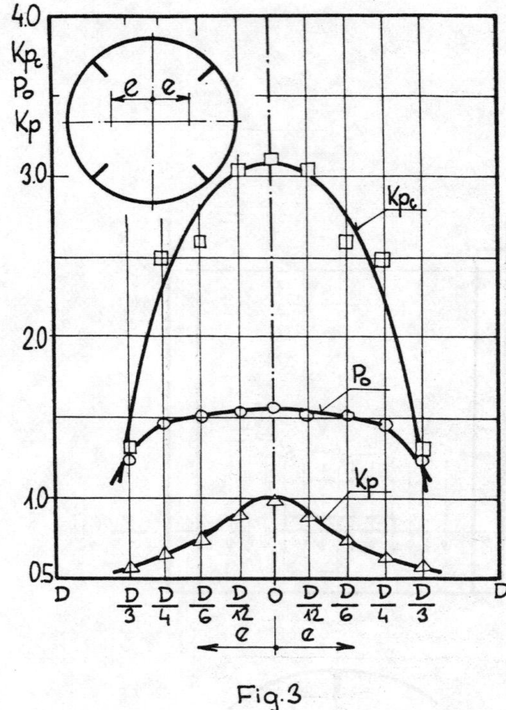

Fig. 3

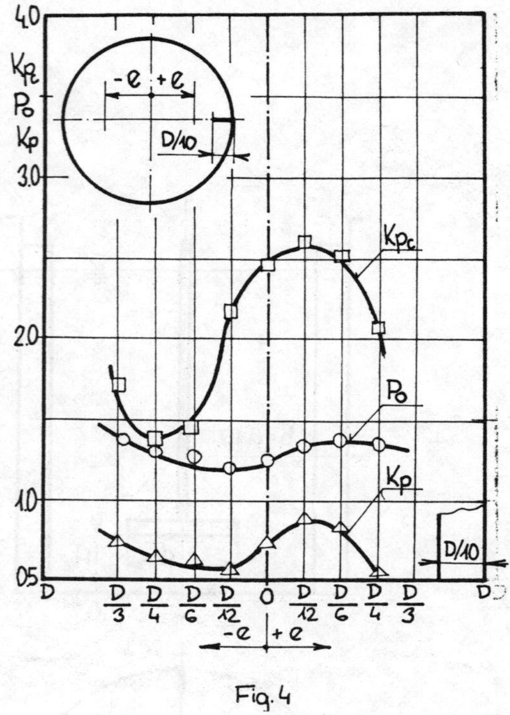

Fig. 4

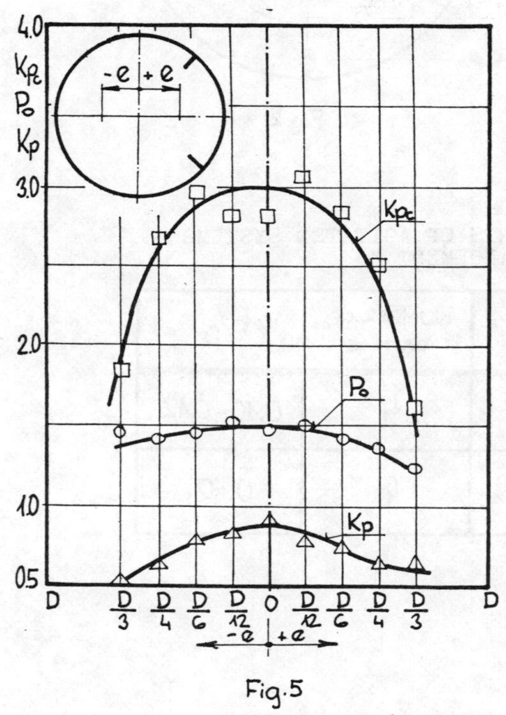

Fig. 5

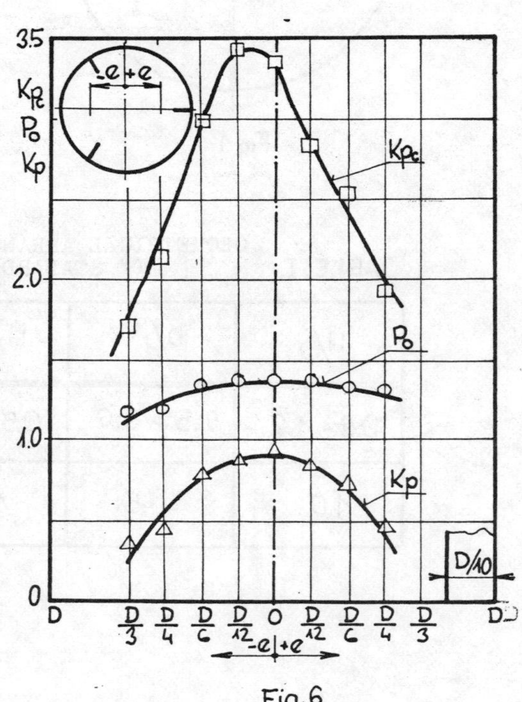

Fig. 6

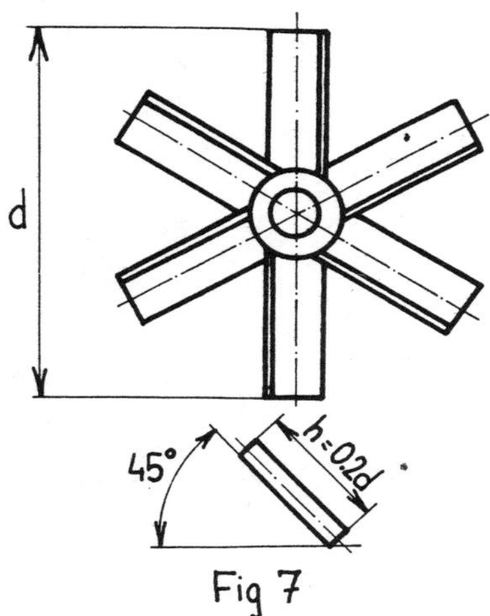

Fig 7

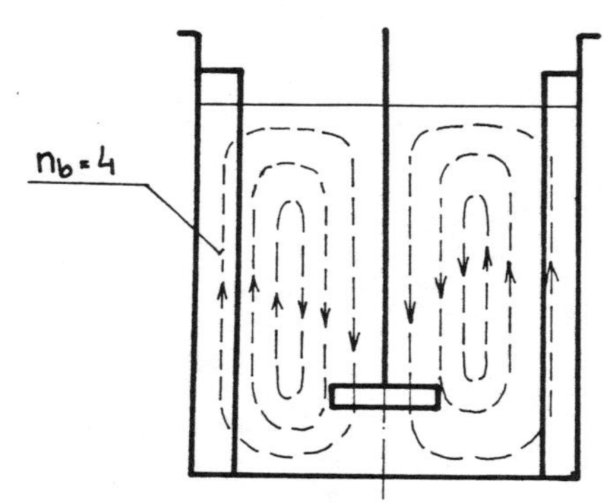

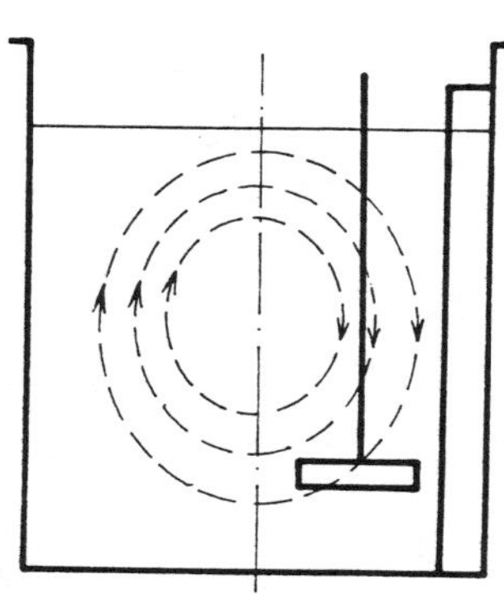

Fig 8

TABLE II. RESULTS OF EXPERIMENTS WITHOUT BAFFLES

e	kp_c	kp	P_o	
$\dfrac{D}{3}$	1.72	0.58	1.54	e

TABLE III. RESULTS OF IMPELLER ENERGETIC EFFICIENCY

n_b		e								
		$-D/3$	$-D/4$	$-D/6$	$-D/12$	0	$+D/12$	$+D/6$	$+D/4$	$+D/3$
0	E_c	3.30	—	—	—	—	—	—	—	3.30
	E	0.13	—	—	—	—	—	—	—	0.13
1	E_c	3.69	2.06	2.35	8.24	12.29	13.02	5.85	2.85	—
	E	0.33	0.19	8.18	0.13	0.31	0.52	0.40	0.10	—
2	E_c	4.02	13.14	18.07	14.41	14.74	19.06	15.86	11.22	3.19
	E	0.15	0.22	0.24	0.27	0.50	0.35	0.31	0.17	0.05
3	E_c	4.31	8.28	18.63	29.37	28.92	16.45	12.54	5.08	—
	E	0.03	0.07	0.32	0.44	0.60	0.40	0.23	0.05	—
4	E_c	1.76	10.51	11.33	18.55	19.41	18.55	11.33	10.51	1.76
	E	0.16	0.19	0.23	0.45	0.66	0.45	0.23	0.19	0.16

SCALE-UP OF AGITATED VESSELS FOR DIFFERENT MIXING PROCESSES

A. Mersmann, H.D. Laufhütte
Technical University Munich
Federal Republik of Germany

Summary

In agitated vessels some mixing processes such as macro-mixing and the suspension of particles are strongly dependant upon the mean fluid velocities whereas others such as micro-mixing and the dispersion of fluid are principally influenced by the fluctuating velocities, i.e. the specific power input. Subsequently design of stirred vessels require scale-up rules based upon knowledge of the fields of both the mean and the fluctuating velocity and of the specific power input. The scale-up rule for mixing and dispersing processes are well established but considerable confusion exists in dealing with suspensions. Explanations are given for the existence of contradictory scale-up rules in association with the specific power input required for the suspension of solid particles in liquids.

Held at Wurzburg, 10-12 June, 1985.

Organised by DVCV· Deutsche Vereinigung für Chemie- und Verfahrenstechnik
(German Association of Chemical and Process Engineering).

Organisation: GVC·VDI-Gesellschaft Verfahrenstechnik und Chemieingenieurwesen. 

Nomenclature

C_1, C_2, C_3	constants
D_{AB}	diffusivity
D	pipe diameter
D_1	overall agitator diameter
d_p	particle diameter
d_{32}	Sauter-mean diameter
g	acceleration due to gravity
H	total liquid height
k_o	lowest wave number
l_K	Kolmogoroff scale
N_V	discharge coefficient
n	agitator speed
P	mixing power
r	radius
T	inside diameter of stirred tank
u	fluid velocity
u'	fluctuating velocity
V	volume of the vessel
$\dot{V}$	volumetric discharge flow
w_{min}	minimum settling velocity
w_s	settling velocity of a single particle
w_{ss}	settling velocity in a swarm

<u>Greek symbols</u>

ε	specific power
η	dynamic viscosity
θ	mixing time
λ_F	micro-scale of turbulence
Λ_F	macro-scale of turbulence
ν	kinematic viscosity
ρ	density
$\Delta\rho$	density difference
σ	interfacial tension
τ	time interval
τ	shear stress
φ	tangential coordinate
φ_V	volumetric hold-up
Index M	for Model

<u>Dimensionless numbers</u>

$$Fr = \frac{w_{min}^2 \, \rho}{d_p \, \Delta\rho \, g} \qquad \text{Froude number of pipe flow}$$

$$Fr^* = \frac{D_1^2 \, \pi^2 \, n^2 \, \rho}{d_p \, \Delta\rho \, g} \qquad \text{Froude number of stirred vessel}$$

$$Fo = \frac{\theta_{micro} \, D_{AB}}{l_K^2} \qquad \text{Fourier number of micro-mixing}$$

$$Re = \frac{n \, D_1^2}{\nu} \qquad \text{Stirrer Reynolds number}$$

1. Mixing tasks in industry

This paper deals with the main mixing tasks such as homogenization, suspension and dispersion of particles in Newtonian liquids of low viscosity. Usually liquid (L), solid-liquid (S/L), gas-liquid (G/L), liquid-liquid (L/L) and gas/liquid-solid (G/L/S) systems are to be agitated, see Fig. 1. Two different types of stirrers are depicted in this figure: For L and S/L systems the vessel is equipped with a marine type propeller with a low drag. In G/L and L/L systems as well as in G/L/S systems a fluid (gas or liquid) has to be dispersed into bubbles or drops. This can be done advantageously by a one- or multiblade flat turbine with a high drag.

Homogenization of chemically non-reacting liquids and suspensions of solid or liquid particles are influenced to a high degree by the mean velocity, $\bar{u}_z$, which is dominant throughout the entire vessel. The settling velocity, w_{ss}, of such particles increases with the density differences, $\Delta\rho = \rho_d - \rho$, of the two phases and must be compensated for by the mean velocity $\bar{u}_z$ of the continuous phase (density ρ). Micro-mixing of chemically reacting liquids as well as dispersion (break-up) of fluids depend upon the fluctuating velocities, u', or upon the local energy dissipation rates, ε.

Knowledge of the mean velocities, $\bar{u}$, and the fluctuating velocities, u', is essential for the projection of mixing operations. The velocity field of a homogeneous single-phase system can be employed as a design basis for two-phase systems.

1.1. Single-phase systems

Sometimes two non-reacting liquids, different in concentration or temperature, have to be mixed. In association herewith it is common practice to distinguish between macro-mixing (macro scale) and micro-mixing (mixing on the molecular scale) [1], [2]. The chemical engineer is interested in the mixing time, θ, required for a certain degree of mixing. As it is, things become more difficult in the presence of chemical reactions. According to Bourne [3] inhomogeneity on a molecular scale develops depending on the ratio of half times, encountered in mixing with chemical reaction and in micro-mixing without chemical reactions. Micro-mixing is governed by molecular diffusion within small fluid elements. In this context expressions such as "scale of segregation" and "intensity of segregation" are helpful [4], [5]. Nevertheless, the situation becomes quite complicated when a consecutive reaction develops in which the product can react with one of the reactands [6].

1.2. Two- and three-phase systems

First systems with chemically inert solid or liquid particles exchanging mass with the continuous phase shall be considered. In suspensions and emulsions a certain homogeneity of the dispersed phase is desired without having material disposed at the bottom of the reaction vessel. Of prime interest is the functional relationship between stirrer speed and the average size of the fluid particles in gas-liquid dispersions and emulsions. In general similarity experiments are performed in a scaled-down model vessel with corresponding stirrer speeds, and the question arises: what is an appropiate scale-up rule for the design of the full scale agitated vessel? The answer can only be given by understanding the scale-up rules governing turbulent flow.

2. Turbulent flow in an agitated vessel

Here the fluctuating velocities are most important since the specific energy dissipation rate, ε, is closly related to these velocities.

In turbulent flow regimes the fluid velocity, $u(t)$, can be written as a superposition

$$u(t) \;=\; \bar{u} + u'(t) \tag{1}$$

$\bar{u}$ is the time averaged velocity in the radial (r) and axial (z) as well as in the tangential (φ) direction. The fluctuating velocity, $u'(t)$, depends on time and is usually measured in terms of Root Mean Square (RMS) values (i.e. $\bar{u}'$) of the radial, axial and tangential velocity components.

2.1. Pumping capacity and mean velocity

The pumping capacity of the stirrer is according to [7], [8]

$$\dot{V} \;=\; N_V\, n\, D_1^3 \qquad\qquad N_V \approx 0.8 \;\; \text{for marine type} \tag{2}$$
$$\text{impeller}$$
$$N_V \approx 0.75 \;\; \text{for flat blade}$$
$$\text{impeller}$$

and the local mean velocity is expressed as

$$\bar{u}_{loc} \;=\; k\, n\, \pi\, D_1 \qquad\qquad \text{with } \; 0 < k < 1 \tag{3}$$
$$\text{and } \; k = f(r^*,\, z^*,\, \varphi,\, Re)$$

Sometimes the characteristic velocity [23]

$$\bar{u}_{car} \;\sim\; n\, D_1\, \frac{D_1}{\sqrt[3]{V}} \;\sim\; n\, D_1 \left(\frac{D_1}{T}\right) \qquad \text{for } \frac{H}{T} = \text{const} \tag{4}$$

is useful for the computation of the slip velocity on occasion of mass transfer in suspensions [27]. Recalling the power consumption, P, of the stirrer

$$P \;=\; Ne\, \rho\, n^3\, D_1^5 \tag{5}$$

and the specific mean power input, $\bar{\varepsilon}$, for $\frac{H}{T} = \text{const}$

$$\bar{\varepsilon} \;\sim\; n^3\, D_1^2 \left(\frac{D_1}{T}\right)^3 \tag{6}$$

and a constant local or averaged mean velocity, $\bar{u}$, in the vessel the expression for the mean specific power input becomes

$$\bar{\varepsilon} \;\sim\; \frac{1}{D_1} \qquad \text{or} \qquad \bar{\varepsilon} \sim \frac{1}{T} \;\; \text{for } \frac{D_1}{T} = \text{const.} \tag{7}$$

2.2. Fluctuating velocities

The definition of the mean fluctuating velocity is

$$\text{RMS} \;=\; \sqrt{\overline{u'^2}} \;=\; \sqrt{\frac{1}{\tau} \int\limits_{t}^{t+\tau} u'^2 dt} \;\;. \tag{8}$$

Measurements show that the turbulence is nearly isotropic, ($\overline{u'_r} \approx \overline{u'_z} \approx \overline{u'_\varphi} \approx \overline{u'}$) throughout the entire vessel with the except for the immediate vicinity of the stirrer [9]. Under isotropic conditions the shear stress becomes

$$\tau \;=\; -\,\rho\, \bar{u}'^2 \;\;. \tag{9}$$

Laufhütte et.al. [9] have demonstrated the validity of the expression

$$\bar{u}' \sim \pi \, n \, D_1 \tag{10}$$

leading to

$$\tau \sim \rho \, n^2 \, D_1^2 \; . \tag{11}$$

Considering micro-mixing and break-up processes of the fluids the scale of an eddy, l_K, according to Kolmogoroff is

$$l_K = \sqrt[4]{\frac{\nu^3}{\varepsilon}} \; , \tag{12}$$

and the micro-scale, λ_F, of turbulence due to Taylor [10] is given by

$$\lambda_F \sim \sqrt{\frac{\Lambda_F \cdot \nu}{\bar{u}'}} \tag{13}$$

in which the macro-scale, Λ_F, of turbulence according to Brodkey [1] is

$$\Lambda_F \approx 0{,}75 \, \frac{1}{k_o} \tag{14}$$

Herein the denominator, $k_o \approx 4/D_1$, is the lowest wave number in agitated vessels.
The preceding scaling diameters describing the behaviour of the fluid elements are useful properties for the modeling of turbulent flow.

2.3. The field of specific power input
In turbulent flow the local specific power input can be calculated with [10]

$$\varepsilon = 15 \, \nu \, \frac{\bar{u}'^2}{\lambda_F^2} \tag{15}$$

A combination of these equations (13), (14) and (15) yields

$$\varepsilon = C_1 \, \frac{\bar{u}'^3}{D_1} \; . \tag{16}$$

For a constant mean power input $\bar{\varepsilon}$ the relation

$$u' \sim D_1^{1/3}$$

is obtained.
Measurements by Laufhütte [11] and the theoretical predictions given by Brodkey [1] are in agreement for the range $6.0 < C_1 < 8.6$ of the constant in equation (16). The ratio, $\varepsilon/\bar{\varepsilon}$, depends upon the dimensionless coordinates (r^*, z^*, φ) and reaches its maximum value of approximately 15 in the discharge flow caused by a flat blade turbine $(1 < 2r/D_1 < 1.5)$.

3. Relation between mixing operation and turbulent flow
According to the previous statements some processes (macro-mixing in absence of chemical reactions and suspensions) require a certain mean velocity, $\bar{u}$, in the vessel, whereas for other processes (micro-mixing with chemical reactions or break-up of fluid particles) a certain fluctuating velocity, $\bar{u}'$, or energy dissipation rate, ε, is decisive for the result. Combination of equations (2), (5) and (11) yields the ratio, $\dot{V}/\tau$, between the discharge flow rate, $\dot{V}$, and the shear stress τ

$$\frac{\dot{V}}{\tau} \sim \left(\frac{N_V}{\varepsilon}\right)^{1/3} \frac{(D_1)^{5/3}}{\rho} \left(\frac{D_1}{T}\right) \quad . \tag{17}$$

The power, $P \sim \dot{V}\tau$, dissipated by the stirrer can either be converted into a high discharge flow ($\dot{V}/\tau$ high) or strong shear rates ($\dot{V}/\tau$ low). On
the other hand the pumping capacity is high in large vessels fitted with
a small drag marine type propeller.

3.1. Processes dependent on the mean velocity
   In agitated vessels high average velocities promote macro-mixing
in inert systems and also very efficient suspension operations.

3.1.1. Macro-mixing
   Many macro-mixing experiments have proven that the total number of
revolutions or the dimensionless mixing time $n \cdot \theta$ depends upon the stirrer
related Reynolds number and the geometry of the vessel. The shortest
macro-mixing time, θ_{macro}, for optimal parameters in the turbulent flow
regime can be calculated for a desired quality of mixing according to
[12] with

$$\theta_{macro} = 7.3 \sqrt[3]{\frac{T^2}{\bar{\varepsilon}}} \quad . \tag{18}$$

The equation is valid for height, H, to diameter, T, ratio of the vessel
equal to unity. Mixing time increases with the tank diameter at constant
specific power input but it increases only slightly with $\bar{\varepsilon}$ for a given
vessel. Operating at the optimal Reynolds number, Re_{opt}, the question
arises whether the constant, 7.3, is universal for all vessels. To this
end a simple theoretical convective model for macro-mixing was tested
using experimentally determined fluid velocities [13]. Good agreement
exists with equation (18) which is a correlation of extensive experimental results by various authors.

3.1.2. Suspensions
   Next the scale-up procedure for stirred vessels in which solid
particles are suspended will be discussed. Unfortunately, this is a very
confusing task illustrated by the "harp" published in 1975 [12] and included in the book "Fluid Mixing Technology" by J.Y. Oldshue [14], (see
Fig. 2). According to this "harp" the scale-up recommendation ranges from
$\bar{\varepsilon} \sim 1/T$ [15, 18] to $\bar{\varepsilon} \sim T^{1/2}$ [16]. Application of these expressions to
suspensions in large agitated vessels results in stupendous differen-
ces in power inputs (see Fig. 3). Considering a tank with a diameter
T = 15 m, and volume V = 2700 m^3 a power input of 10 000 kW is calculated with the equation given in [17], whereas application of another
equation [18] yields as low a power dissipation as 10 kW. These low powers now appear to be adequate according to an information received from
an industrial company which performed large scale experiments.
   What can be the reason for this confusing situation? Clarification
is a prerequisite for a reliable scale-up rule for the dimensioning of
large vessels used in industry. At first the question has to be answered,
whether it is reasonable to compare these different scale-up rules since
some authors employ the "one second criterion" considering only particles
resting at the bottom of the vessel, whereas others utilize the layer
height criterion regarding the height of the suspension in the vessel.
   According to recent publications [19], [9] the mean specific power
input, $\bar{\varepsilon}$, is dissipated by two superimposed processes: generation of the
discharge flow rate in the vessel and the consumption of power to coun-
teract the sinking of the particles:

$$\bar{\varepsilon} \;=\; \bar{\varepsilon}_{flow} + \bar{\varepsilon}_{sink} \;=\; \xi\,\frac{\bar{u}^3}{T} + \left(\frac{w_{ss,turb}}{w_{ss}}\right) w_{ss}\cdot g\cdot\frac{\Delta\rho}{\rho}\,\varphi_v \tag{19}$$

Dealing with large tanks the contribution $\bar{\varepsilon}_{sink}$ prevails. However, in small laboratory vessels, with a high ratio of wall area to vessel volume a minimum velocity $\bar{u}_r$ at the bottom is needed for particles motion with the result that $\bar{\varepsilon}_{flow}$ dominates.

During the sixties and seventies many authors used the "one second criterion" considering only the motion of the particles at the bottom. An analoguous two-phase flow problem is the hydraulic transport of a suspension in pipes. For safe transportation the liquid has to exceed a certain minimum velocity, w_{min}, called the settling velocity. According to Brauer [20] the dimensionless minimum settling velocity depends on the ratio, d_p/D, (diameter d_p of particles to pipe diameter D), see Fig. 4.

In case of suspended particles in a stirred vessel the highest velocity of the agitator is $D_1 \pi n$ and the largest dimension is the tank diameter T. Experimental results obtained for pipe flow [20] and for suspended particles in stirred vessels are displayed in Figure 4 which has two ordinates (Froude numbers) and two abscissae: $Fr = w_{min}^2\,\rho/(d_p\,\Delta\rho\,g)$ versus d_p/D for hydraulic transport and $Fr^* = D_1^2\,\pi^2\,n_{min}^2\,\rho/(d_p\,\Delta\rho\,g)$ versus d_p/T for stirred vessels. This figure depicts results for suspended particles in vessels fitted with a pitched six blade turbine, Ne = 1.8 [21] and with an eight blade flat turbine, Ne = 6 [22]. In both cases the dimensionless minimum velocity or speed respectively decrease with increasing values of the diameter ratio.

In this context the crucial fact is now that by plotting the Froude numbers versus the diameter ratio on log-log paper the experimental data fit on a straight line only over a minor range of the plot as indicated in Figure 4. The deviation from the power law was confirmed by data published in the literature and by own preliminary investigations.

From the results of all these experiments it was concluded that the Froude number remains approximately constant at small values of the diameter ratio (large vessels) but that this number decreases with increasing d_p/T-values i.e., decreasing size of the vessel.

The common scale-up rule for stirred suspension vessels with $D_1/T = const$ (with the slope m to be taken from Fig. 4) is given by:

$$\bar{\varepsilon}_{flow} \;\sim\; \left(\frac{\Delta\rho^3\,g^3\,d_p}{\rho^3}\right)^{1/2} \left(\frac{d_p}{T}\right)^{\frac{3m+2}{2}} \quad \text{with } m = f(d_p/T). \tag{20}$$

With the expression for large vessels $(m \rightarrow 0)$

$$\frac{D_1^2\,\pi^2\,n^2\,\rho}{\Delta\rho\cdot g\cdot d_p} \;=\; const$$

reflecting the flow contribution the specific power input becomes

$$\bar{\varepsilon}_{flow} \;\sim\; \left(\frac{\Delta\rho^3\cdot g^3\cdot d_p}{\rho^3}\right)^{1/2} \left(\frac{d_p}{T}\right) \tag{21}$$

In large tanks the power dissipation in the process of suspending particles in a fluid is no longer limited to the flow contribution only but now it is necessary to include the contribution caused by the sinking particles, $\bar{\varepsilon}_{sink}$, in equation (19).

$$\bar{\varepsilon}_{sink} \approx \left(\frac{w_{ss,turb}}{w_{ss}}\right) w_{ss} \cdot g \cdot \frac{\Delta\rho}{\rho} \varphi_V \qquad (22)$$

with $0.5 < \dfrac{w_{ss,turb}}{w_{ss}} < 1$ due to [23].

Equation (22) yields reasonable results for the example presented in Figure 3 (Tank-diameter T = 15 m).

From figure 4 it becomes obvious that for small vessels the decrease of the dimensionless stirrer speed is the most salient feature for the scale-up rule. With a slope, m = - 1/3, from Fig. 4 or

$$\frac{D_1^2 \pi^2 n^2 \rho}{\Delta\rho \cdot g \cdot d_p} \sim \left(\frac{d_p}{T}\right)^{-1/3} \qquad (23)$$

equation (20) becomes

$$\bar{\varepsilon}_{flow} \sim \left(\frac{\Delta\rho^3 \cdot g^3 \cdot d_p}{\rho^3}\right)^{1/2} \left(\frac{d_p}{T}\right)^{1/2} \qquad (24)$$

Moreover with a slope, m = - 1/2, the scale-up rule changes to $\bar{\varepsilon}_{flow} \sim T^{-1/4}$, and for m = - 2/3 the term $\bar{\varepsilon}_{flow}$ is independent of T.

Another crucial point of the new findings is that the scale-up rule now includes the ratio of particle to vessel diameter, d_p/T. In order to suspend particles of identical diameter, d_p, both in a model and in a full scale vessel the scale-up rule shows a dependency on the diameter, T, of the vessel. Very likely this dependency was not taken into account in the relevant publications of the sixties and seventies. Hence it is no longer advisable to extrapolate the earlier experimental results for the design of large vessels and to use them as a scale-up criterion.

This statement is illustrated by Fig. 5 in which the contributions $\bar{\varepsilon}_{flow}$ and $\bar{\varepsilon}_{sink}$ as well as the sum $\bar{\varepsilon} = \bar{\varepsilon}_{flow} + \bar{\varepsilon}_{sink}$ are plotted versus the tank diameter.

The upper curves are valid for a particle diameter d_p = 1000 μm. The lower lines describe the relation for d_p = 100 μm. In case of small tanks and high d_p/T-ratios the mean specific power input according to [21, 22] is

$$\bar{\varepsilon} \approx \bar{\varepsilon}_{flow} \approx 2.2 \left(\frac{\Delta\rho^3 \ g^3 \ d_p}{\rho^3}\right)^{1/2} \qquad \left(\text{for } \frac{H}{T} = 1\right) \qquad (25)$$

In very large vessels with low d_p/T-ratios equation (19) simplifies to

$$\bar{\varepsilon} \approx \varepsilon_{sink} \approx w_{ss} \cdot g \frac{\Delta\rho}{\rho} \varphi_V \qquad (22a)$$

In the relation $\bar{\varepsilon} \sim T^b$ the exponent b ranges from 0 to -1, compare Fig. 2.

Suspensions exhibit another characteristic feature: the distribution of the solid particles in liquids of small viscosity is not uniform. This degression of uniformity possibly leads to high particle concentrations in the vicinity of the stirrer. For that reason neither the liquid density, ρ nor the mean density of the suspension $\bar{\rho} = (1-\varphi_V)\rho + \varphi_V\rho_d$, is the effective density "felt" by the stirrer. This effective density depends on the local and quite often high particle concentration close to the impeller. The density distribution varies with height and volume of the vessel for given particle diameter and stirrer speed. Both factors the decrease of the Froude number in Figure 4 and the inhomoge-

neity of the particle distribution in agitated vessels lend credence to
the applicability of the diameter ratio, d_p/T, in scaling-up procedures.

3.2. Stirring operations depending in the fluid fluctuating velocity or the local specific energy dissipation

Briefly the process of micro-mixing and dispersion of a fluid shall
be discussed.

3.2.1. Micro-mixing with chemical reaction

In eddies micro-mixing takes place by convection and diffusion.
Bourne and coworkers [5] developed models to describe the progress of
mixing and to predict the time required for micro-mixing which is much
smaller than the time required for macro-mixing, θ_{macro}, [24]. It is
noteworthy that time for micro-mixing can be of the same order of magnitude as the reaction time. Pure diffusion is described by Fourier numbers Fo formed with the diameter of the small eddy according to Kolmogoroff

$$Fo = \frac{\theta_{micro} \cdot D_{AB}}{l_K^2} \tag{26}$$

A combination of the equations (26), (12) and (16) yields

$$Fo \sim \theta_{micro} D_{AB} \sqrt{\frac{\varepsilon}{\nu^3}} \sim \theta_{micro} D_{AB} \sqrt{\frac{\bar{u}'^3}{D_1 \nu^3}} \tag{27}$$

These equations indicate the dependency of micro-mixing on the fluctuating velocity, u', or the local specific power input, ε. In scale-up procedures the value of ε should remain constant at the point of immission
of the reactands.

3.2.2. Break-up of fluids

In association with the dispersion of fluids into bubbles or
droplets it is essential to know the ε- and u'-fields because the mean
Sauter diameter, d_{32}, depends on the parameters containing these properties:

$$d_{32} = C_2 \frac{\sigma^{0.6}}{\rho^{0.6} \varepsilon^{0.4}} = C_3 \frac{D_1^{0.4} \sigma^{0.6}}{\rho^{0.6} (u')^{1.2}} \tag{28}$$

However, up to now discrepancies exist in regard to the values of the
constants C_2 and C_3 and to the range of validity of equation (28). Some
authors claim that the constants C_2 and C_3 depend on time and that steady state is achieved only after several hours of operation [25], [26].
Equation (28) is only valid in a medium range of fluctuating velocities and of particle sizes, see Fig. 6. Larger drops tend to settle,
while smaller drops display a strong tendency towards coalescence.
The resulting drop-size distribution is governed by the nature of the
break-up and coalescence processes and it is a function of the mean spe-
cific power input, $\bar{\varepsilon}$, and of the ε-field as well. Figure 6 shows the general drop size distribution of drops according to Grossmann [26] on a
normalized logarithmic distribution plot. The two straight lines can be
explained by break-up and coalescence processes.

Literature

[1] Brodkey: Turbulence in Mixing Operations, Academic Press, Inc.
     New York, San Francisco, London (1975) p. 80-110
[2] Hinze, J.O.: Turbulence, McGraw Hill, New York (1975)
[3] Bourne, J.R.: Chem. Engng. Sci. 38(1983), p. 5-8
[4] Danckwerts, P.V.: Chem. Engng. Sci. 1958, p. 5
[5] Baldyga, S.; J.R. Bourne: Chem. Engng. Commun. 28(1984), pp.231-
     241
[6] Angst, W.H.: Doctoral Thesis, Technical university Zürich (1982)
[7] Kipke, K.-D.: Chemie-Technik 5(1976), p. 169-174
[8] Revill, B.K.: Fourth European Conference on mixing, paper B1,
     Noordwijkerhout, 1982
[9] Laufhütte, H.D.; A. Mersmann: Chem. Ing. Techn. 56(1984), pp.862-
     863
[10] Taylor, G.I.: Proc. Royal Soc. A 151 (1935) pp 421-478
[11] Laufhütte, H.D., A. Mersmann: Fifth European Conference on Mixing,
     Würzburg, 1985
[12] Mersmann, A.; M. Käppel; W.-D. Einenkel: Chem. Ing. Techn. 47(1975),
     pp. 953-964
[13] Voit, H.; H.D. Laufhütte; A. Mersmann: Internal report, Lehrstuhl B
     für Verfahrenstechnik of the Technical university Munich, 1984
[14] Oldshue, J.Y.: Fluid Mixing Technology, McGraw Hill, New York, 1983
[15] Müller, W.; E.K. Todtenhaupt: Aufbereitungstechnik (1972), pp.38-42
[16] Kneule, F.; P.M. Weinspach: Verfahrenstechnik 1(1967), pp. 531-540
[17] Zlokarnik, M.; H. Judat: Chem. - Ing.-Techn. 41(1969), pp. 1270-1273
[18] Niesmak, G.: Doctoral Thesis, Technical university Braunschweig
     1982
[19] Kneule, F.: Chem. Ing. Techn. 55(1983), pp. 275-281
[20] Brauer, H.: Grundlagen der Einphasen- und Mehrphasenströmung;
     Sauerländer Aarau, 1971
[21] Ditl, P.; F. Rieger: Proceedings of the 8th Chisa Congress, paper
     No. V3-61, Praha, 1984
[22] Baldi, G.; R. Conti; F. Alaria: Chem. Engng. Sci. 33(1978),
     pp. 21-25
[23] Schwartzberg, H.G.; R.E. Treybal: I. & E. Fundamentals 7(1968),
     pp. 1/12
[24] Mersmann, A.: Turbulence and Mixing; paper presented at the GVC-
     working party on mixing, Bad Driburg, 1983
[25] Langner, F.; H.V. Moritz; K.H. Reichert: Chem. Ing. Techn.
     51(1979), pp. 746-747
[26] Grossmann, H.: Doctoral Thesis, Technical university Munich, 1982.
[27] Herndl, G.: Fluiddynamics and Mass Trnasfer in Stirred Suspensions.
     Chem. Eng. Commun. 13(1981), p. 23-27.

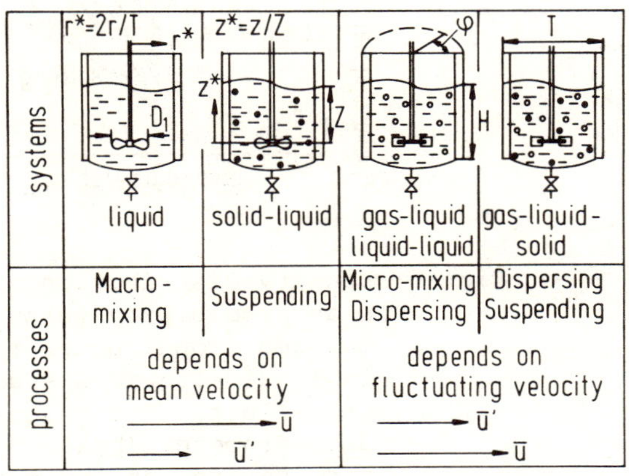

Fig. 1: Systems and processes
        in stirred vessels

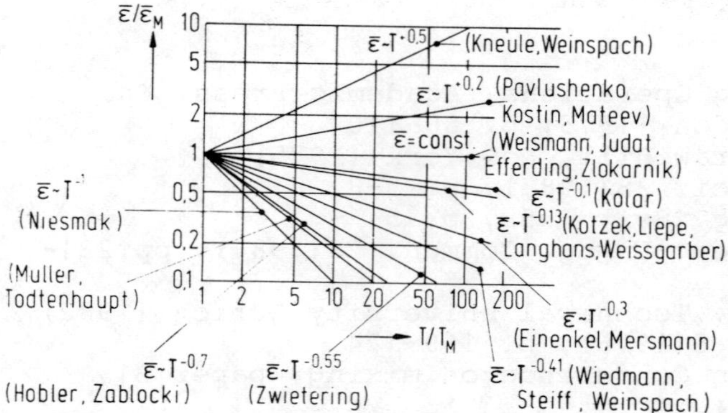

Fig. 2: Scale-up rules for suspending of solid particles in stirred vessels

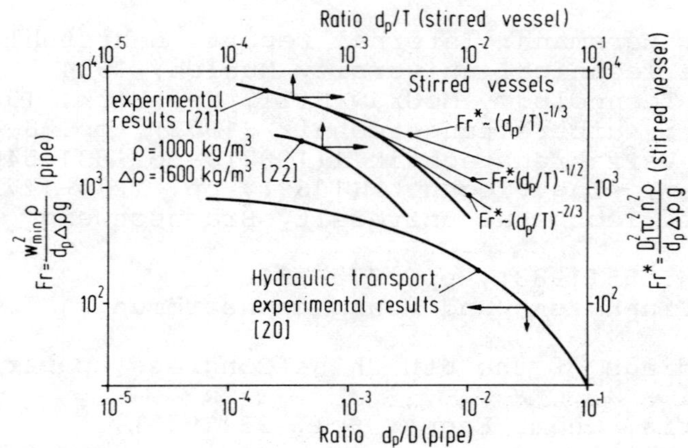

Fig. 4: Froude number versus d_p/D for the hydraulic transport in pipes (lower curve) and Froude number Fr* versus d_p/T for stirred vessels (upper curves)

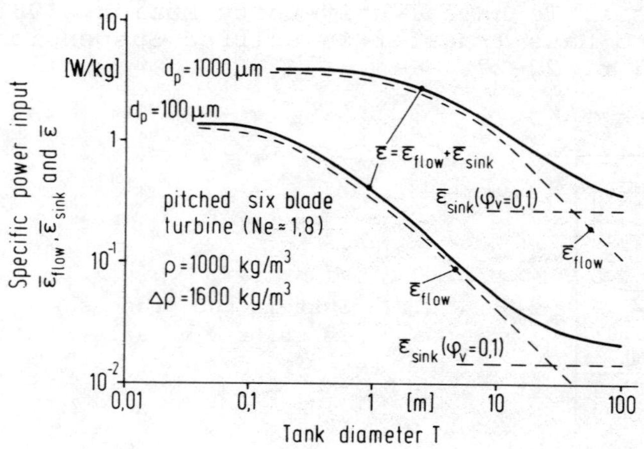

Fig. 5: Mean specific power inputs $\bar{\varepsilon}_{flow}$, $\bar{\varepsilon}_{sink}$ and $\bar{\varepsilon} = \bar{\varepsilon}_{flow} + \bar{\varepsilon}_{sink}$ versus tank diameter T. Upper curve for $d_p = 1000$ µm and lower curve for $d_p = 100$ µm ($\varphi_V = 0.1$)

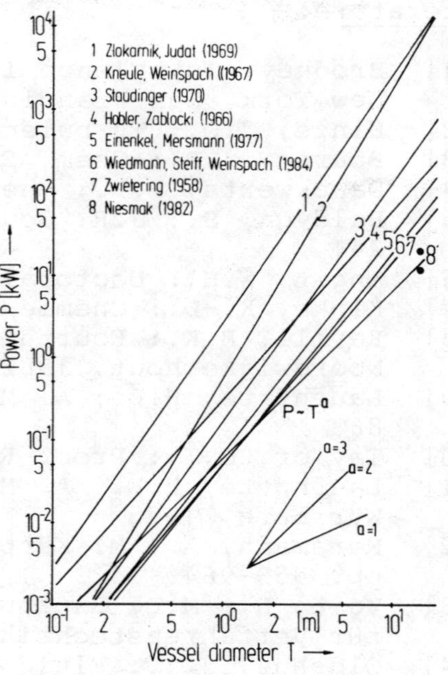

Fig. 3: Power input versus tank diameter for suspending of solid particles

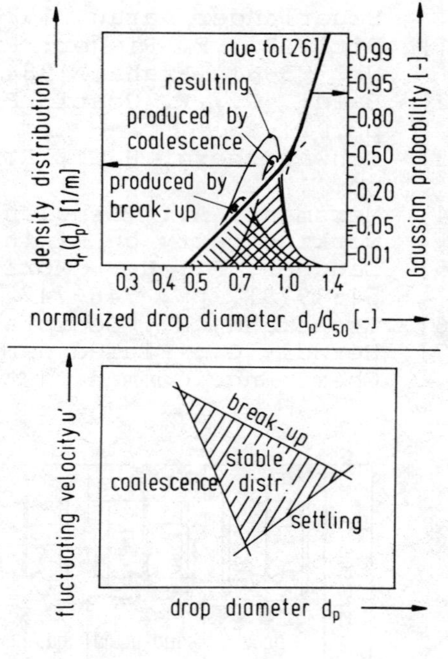

Fig. 6: Density distribution and Gaussian probability versus normalized dropdiameter for drops in liquid-liquid systems (above). Fluctuating velocity versus drop diameter (below) depicting the region for stable drop size distribution

"FLUID LOADING AND POWER MEASUREMENTS ON AN ECCENTRICALLY MOUNTED PITCHED BLADE IMPELLER

R KING AND M J MUSKETT

SUMMARY

A single pitched blade impeller was tested in a vessel of 2.67m diameter at one eccentricity. Impeller height from the vessel base was varied, and reference tests carried out with the impeller mounted centrally. The power numbers obtained with eccentric mounting were similar to the fully-baffled tank, whilst the shaft bending moments were considerably higher for eccentric mounting due to asymmetries in the vessel flow patterns.

Held at Wurzburg, 10-12 June, 1985.

Organised by DVCV· Deutsche Vereinigung für Chemie- und Verfahrenstechnik
(German Association of Chemical and Process Engineering).

Organisation: GVC·VDI-Gesellschaft Verfahrenstechnik und Chemieingenieurwesen. 

NOMENCLATURE

Symbol	Explanation	Units
C	Impeller clearance from bottom of tank	m
C_H	Force coefficient	–
D	Impeller diameter	m
e	Eccentricity	m
F_B	Forces producing bending	N
F_H	Hydraulic Force	N
F_T	Forces producing torque	N
f_n	Natural frequency of shaft and impeller in liquid	s^{-1}
L	Shaft length	m
m	Number of impeller blades	–
N	Shaft rotational speed	rev/s
N_R	Ratio of blade passing frequency to natural frequency	–
p	Parameter defining line of action of F_T	–
P	Power drawn by impeller	W
T	Tank diamter	m
σ_i	Standard deviation of signal from ith set of strain gauges	Nm
σ_{N-1}	Standard deviation of N results	–
μ	Effective shaft bending moment	Nm
ν	Kinematic viscosity of liquid	m^2/s
λ_{MAX}	Maximum value of force coefficient	–
λ_1	Value of force coefficient at high reduced speed	–
ρ	Liquid density	kg/m^3
τ	Torque on impeller shaft	Nm

<u>DIMENSIONLESS GROUPS</u>

Fr   Froude number $= \dfrac{N^2 D}{g}$

Po   Power number $= \dfrac{P}{\rho N^3 D^5}$

Re   Reynolds number $= \dfrac{N D^2}{\nu}$

λ Force coefficient $= \dfrac{\mu D}{\tau L}$

1.   INTRODUCTION

There are situations which demand the use of stirred vessels without baffles, such
as in the food and paint industries where absolute cleanliness is required.   To
maximise the power consumption, the impeller may be mounted eccentrically on a
vertical shaft.   This arrangement may also be used in the contacting of floating
solids which do not agglommerate, since a large vortex is created which draws
material from the liquid surface.   This phenomenon has been studied for different
baffle arrangements by Edwards and Ellis (Ref. 1) and Joosten, Schilder and Broere
(Ref. 2).

There is little data in the literature on even the power consumption of
eccentrically mounted agitators other than the work of Novak, Ditl and Rieger
(Ref. 3), who also considered surface aeration due to the formation of a vortex.
Novak et al did not study the effect of impeller bottom clearance when mounted
eccentrically, and the present study aims to fill that gap.   Kamei, Nagata and
Yoshida (Ref. 4) produced some data for a flat paddle, indicating that the impeller
power consumption is maximised for eccentricities greater than a quarter of a tank
diameter.   Also, no large-scale information is available for power consumption in
unbaffled vessels.   In addition, shaft bending loads were measured to produce data
relevant to the design of shafts for eccentric mounting.

2.   EXPERIMENTAL METHOD

2.1   Test Facility

All the tests were carried out in BHRA's 2.67m diameter vessel, shown schematically
in figure 1.   One impeller type was used, a four-bladed 45° pitched blade turbine
consisting of plates bolted to a central hub (Figure 2):-

TABLE 1   DETAILS OF IMPELLER

	(m)	
Impeller diameter	0.914	0.343T
Blade width	0.209	0.078T
Projected blade width	0.147	0.055T
Blade thickness	0.010	

The impeller was driven by a 37.5 kW electric motor, controlled by a thyristor drive
unit to give variable speeds up to 100 RPM.

Strain gauges were fixed to the top of the shaft to measure torque and bending loads
in two perpendicular planes.   The signal was pre-amplified and transmitted by
telemetry as an FM radio signal to aerials fixed to the vessel framework.   The
telemetry equipment was manufactured by Astech Electronics Ltd, UK, who also
manufactured the de-modulation units shown in the schematic representation of the
instrument chain given in figure 3.   The voltage-modulated signal then passed
through an amplifier to utilise the full range on the computer analog-to-digital
converter, and was finally filtered to prevent signal aliasing (Ref. 5) before being
sampled by a DEC PDP 11/23 laboratory computer.

Sampling was carried out over a 20 second period, giving 2048 samples on each of the
3 channels.   For each impeller speed this was repeated 9 times.

For each test, the flow patterns in the tank were allowed to establish themselves,
so that only steady state data were recorded.   In the absence of baffles, this may
take some time as the tank contents are accelerated from rest.   All the tests were
carried out in water, with a Reynolds number range of 1.6×10^5 to 1.0×10^6.

2.2   Configurations tested

For reasons of convenience, only one eccentricity was used, with $e = \dfrac{T}{4}$.

There were also some tests carried out with the impeller centrally located for reference purposes. Table 2 summarises the configurations. Impeller clearance is defined as the distance from the impeller centre-line to the bottom of the dished end of the tank.

TABLE 2:   CONFIGURATIONS TESTED

Configuration	Eccentricity e	Impeller Clearance, C	Baffles
1	T/4	T/3	No
2	T/4	T/2	No
3	T/4	0.176T	No
4	0	T/3	No
5	0	T/3	4 x T/12

3.   THEORY AND BACKGROUND

3.1   Power Number

It is very well established that in baffled, turbulent vessels, power number is independent of Reynolds number (Ref. 6, 7 and 8). In unbaffled vessels dimensional analysis implies that, since a free surface is present:-

$$Po = f\,(Re,\ Fr)$$

In contrast to Rushton (Ref. 9), Nagata (Ref. 10) showed that the effect of Fr is negligible, except for very large values of D/T. If Reynolds number is sufficiently high, power number becomes independent of Reynolds number, as shown by Novak et al (Ref. 3) regardless of the eccentricity of the mounting.

3.2   Shaft Bending Loads

Pollard (Ref. 11) and Kipke (Ref. 12) have described the nature of forces on an impeller shaft due to the action of the fluid on the impeller. These forces are unsteady and give rise to fluctuating bending moments on the shaft. The method of non-dimensionalising these bending moments used by BHRA is to consider the relationship between forces producing bending and forces producing torque.

The force producing bending is given by;

$$F_B = \frac{\mu}{L}$$

The force producing torque is given by;

$$F_T = \frac{\tau}{pD}$$

p is a parameter accounting for the fact that the line of action of the force F_T does not pass through the impeller diameter, but coincides with some point on the impeller blade which is characteristic of the impeller geometry. For geometrically similar impellers, p should not be a function of scale, and the following dimensionless group is defined:-

$$\lambda = \frac{\mu D}{\tau L}$$

Extensive work at BHRA and on site has shown that this quantity does not depend on scale for a particular geometry.

Since, in normal operation, the stresses induced in the shaft by the action of the fluid do not exceed the yield stress of the shaft material, the bending moments are considered in the context of shaft fatigue, leading to the use of only the fluctuating component of shaft bending in the definition of λ. There is also a steady component due to shaft imbalance or run out which is removed from the experimental data. The root-mean-square is then calculated for each bending plane and μ is calculated as follows;

$$\mu^2 = \sigma_1{}^2 + \sigma_2{}^2$$

λ is plotted against a quantity termed the <u>reduced speed</u>, defined as;

$$N_R = \frac{mN}{f_n} = \frac{\text{Blade-passing frequency}}{\text{Natural frequency}}$$

This is a method of normalising the frequency response of the shaft. Fourier analysis of the bending moment signal shows that the most significant excitation frequency is the blade-passing frequency, mN, and when this equals shaft natural frequency, a local maximum in the fluctuating bending moment is found, giving a sharp peak in λ. For reduced speeds in excess of unity for fully turbulent flow but below the critical speed, λ is independent of speed implying that in this region, for turbulent flow;

$$\mu \propto N^2$$

The form of this relationship is also predicted by Kipke (Ref. 12), who defined a force coefficient;

$$C_H = F_H / \rho\, N^2 D^4 .$$

C_H was found to be constant for speeds below the critical speed. Pollard (Ref. 11) has discussed the similarities between the two approaches and pointed out some geometrical differences in test apparatus which produced some disagreement in the data obtained.

4.   <u>RESULTS</u>

4.1 <u>Off-centre Tests</u>

The power number data for configurations 1 to 3 are given in figures 4-6 respectively. Table 3 gives the averaged values of power number, together with the standard deviation of the sample and the number of separate determinations for each configuration.

TABLE 3:   POWER NUMBERS: OFF CENTRE CONFIGURATIONS

Configuration	Impeller Clearance	Po_{MEAN}	σ_{N-1}	No. of Samples
1	T/3	1.47	0.1	19
2	T/2	1.75	0.13	15
3	0.176T	1.85	0.15	12
4	T/3	0.54	0.11	15
5	T/3	1.58	0.07	16

Figures 7-9 give the non-dimensionalised shaft loading results for the eccentric configurations.

Identifying the peak value of λ as λ_{MAX}, and the value of λ at high reduced speed as λ_1, Table 4 gives a summary of the data.

TABLE 4:   NON-DIMENSIONAL SHAFT LOADING: OFF-CENTRE CONFIGURATIONS

Configuration	Impeller Clearance	λ_{MAX}	λ_1
1	T/3	2.55	0.93
2	T/2	1.65	0.94
3	0.176T	2.03	0.93

4.2   Centred Impeller

Fig. 10 shows the power number data for an unbaffled vessel. At Re = 5 x 10^5 there is no dependance of Po on Re, and in this region the average power number is 0.49, which compares with a value of 1.58 obtained in this study using 4 flat baffles of width T/12. Table 3 includes these data. The bending data are given in figures 11 and 12.

5.   DISCUSSION

5.1   Power Numbers

Comparing the fully baffled result with the off-centre mounting, C = T/3, shows that the unbaffled, eccentric configuration gives 93% of the power drawn in the fully baffled vessel. This is in line with Novak et al. (Ref. 3), who found this ratio to be 0.85 for a 3 bladed pitched blade turbine and 0.96 for a 6 bladed pitched blade turbine.

The off-centre results do not show a consistent trend, in that power number is not a monotonic function of impeller clearance as is found in baffled vessels with pitched blade turbines (Ref. 6). The results at low speeds show considerable scatter in power number, and analysis of the raw data shows that at low speed there are higher standard deviations than at high speed. This indicates:-

    (i)   The higher standard deviations may be due to instrument noise, since signals are relatively small.

    (ii) There are instabilities in the flow patterns which give rise to large fluctuations in torque.

The latter point is borne out by visual observation, which shows that except for the lowest impeller position, the large vortex is very unsteady at low speeds. The lowest impeller position did not give such large scatter in the low speed power numbers.

For the impeller in a central position, the fully-baffled power number is rather high in comparison with the smaller scale work of Bates et al (Ref. 6) and with data obtained in the Fluid Mixing Processes (FMP) programme at BHRA in a vessel of 0.61m diameter. However, the fully-baffled result agrees well with the off-centre power number at a corresponding impeller clearance.

The unbaffled data with the impeller mounted centrally show no influence of either Froude number or Reynolds number, as would be expected in view of the very high Reynolds numbers encountered at this scale.

5.2   Bending Moments

The off-centre data show that the bending moments experienced by the shaft exceed those for the centred tests by a factor of about 4.5. This is not surprising in view of the highly asymmetric nature of the flow, which tends to put a lateral force on the shaft. Very sharp peaks in λ are found at reduced speeds of around unity, which is often the case in fully baffled tanks. Clearly there is a fluctuating component in the flow which causes the shaft to vibrate at the blade-passing frequency. This tendency is far less clear in the highest impeller position, but this may be due to the very narrow bandwidth of the peak, which could lead to it being missed. The centred impeller results show a far less marked peak for flat baffles, in spite of the attention paid to the peak region during testing. This behaviour is not found in other impellers which differ only in the ratio of blade width to impeller diameter, and leads one to the conclusion that geometric factors such as this play a large part in the response of impellers at reduced speeds of around unity. In the flat region of the λ curve, however, the results agree very well with other data taken at BHRA, and it is in this region that vessels should be operated, to avoid small changes in shaft speed producing large changes in shaft bending moment.

In general, the centred, unbaffled results show that there is less bending and less torque than for a centred, baffled tank operating at the same speed. This is consistent with the high degree of swirling observed in unbaffled tanks, which implies a relatively small amount of interaction between fluid and impeller. Also, there is no tendency for λ to show coherent peaks; the nature of the fluid-impeller interaction is more random than for baffled vessels.

5.3   Aeration

When using the eccentric configurations, aeration through the vortex must be considered since this could impair process performance. Unfortunately, it was not possible to study aeration in this series of tests since the tank contents could not be viewed during operation. Also, a rigorous definition of aeration is necessary before correlations may be produced, since aeration can vary from a few bubbles in the surface region to the whole vessel containing a gas-liquid dispersion. The study of surface aeration is further complicated by the fact that the process will probably follow different scaling laws to the aeration, and data taken on one particular scale are not likely to be of great value, unless the scaling law for aeration is known. In this study, no sudden drops in power number were observed, which would occur with the onset of very severe aeration, but the work of Warmoeskerken, Speur and Smith (Ref. 13) and Chapman, Nienow, Cooke and Middleton (Ref. 14) indicate that the gassed power curves for pitched blade turbines in fully-baffled tanks are relatively flat at the low gassing rates corresponding to mild surface aeration.

The phenomenon of surface aeration is likely to be dependent on liquid phase properties such as viscosity or ionic strength, so it would seem better to consider aeration for the particular process fluid to be used, than generalising from data obtained in relatively pure water, which is unrepresentative of many process fluids.

6.   CONCLUSIONS

For the impeller mounted at T/4 from the vessel axis, the following conclusions were drawn:

1.   No substantial reduction in power number is observed relative to the fully baffled, centred impeller case. Variation of power number with impeller-base clearance did not follow a consistent trend, thought to be due to unstable flow patterns created by the formation of a large vortex, particularly at low speeds.

2.   Fluctuating bending moments caused by fluid loading increased by a factor of about 4.5 relative to the centred agitator tests. This feature should be included in initial stress evaluations at the impeller-shaft design stage.

3.   The formation of a large, persistent air entraining vortex makes this arrangement potentially suitable for drawing down solids through the liquid surface, although the accompanying aeration may not be acceptable from process considerations. Furthermore, the performance of the vessel in suspending solids should be considered if material denser than water is involved at any stage in the process. There are some indications that the eccentric mounting reduces the suspension performance in comparison to an axially mounted impeller in an unbaffled tank.

<u>REFERENCES</u>

1. Edwards M.F. and Ellis D.I.
    "The drawdown of floating solids into mechanically agitated vessels"
    Fluid Mixing II, I.Chem.E. Symposium Series No. 89
    Bradford, UK, April 1984, p.1.

2. Joosten G.E.H., Schilder J.G.M. and Broere A.M.
    "The suspension of floating solids in stirred vessels"
    Trans. I. Chem. E., Vol. 55, p.220, 1977.

3. Novak V., Ditl P. and Rieger F.
    "Mixing in unbaffled vessels. The influence of an eccentric impeller position
    on power consumption and surface aeration"
    4th European Conf. on Mixing, BHRA, Cranfield, UK, p.57, 1982.

4. Kamei S, Nagata S. and Yoshida, N.
    "Studies on power requirements for paddle agitators in cylindrical vessels",
    Japan Sci. Rev. Mech. and Elec. Vol. 1, p. 43, 1950.

5. Pollard G.J.
    "Digital Spectral Analysis - a brief introductory guide"
    BHRA report TN 1545, May 1979.

6. Bates R.L., Fondy P.L. and Corpstein R.R.
    "An examination of some geometric parameters of impeller power"
    I & E.C.,Proc. Des. & Dev., Vol. 2, No. 4, 1963.

7. Nienow A.W. and Miles D.
    "Impeller power numbers in closed vessels"
    I & E.C., Proc. Des. & Dev.,Vol. 10, No.1, 1971.

8. O'Kane K.
    "The effect of geometric parameters on the power consumption of turbine
    impellers operating in non-viscous liquids"
    1st European Conference on Mixing, BHRA. Cranfield, UK, 1974.

9. Rushton J.H., Costich, E.W. and Everett H.J.
    "Power characteristics of mixing impellers"
    Chem. Eng. Prog., Vol. 46, Nos.8 & 9, 1950.

10. Nagata S.
    "Mixing: Principles and Applications"
    Halsted Press, 1975.

11. Pollard G.J.
    "Hydraulic bending loads on mixer shafts"
    4th European Conf. on Mixing, BHRA, Cranfield, UK, 1982, Paper K3.

12.   Kipke K.D.
      "Hydraulic forces on different impeller types"
      3rd European Conference on Mixing, BHRA,Cranfield, UK, 1979, Paper El.

13.   Warmoeskerken M.M.C.G., Speur J. and Smith John M.
      "Gas-liquid dispersion with pitched blade turbines"
      Chem. Eng. Comm., Vol. 25, p.11, 1984.

14.   Chapman C.M., Nienow A.W. Cooke M. and Middleton
      "Particle-gas-liquid mixing in stirred vessels - Part II"
      Chem. Eng. Res. Des., Vol 61, p.82, 1983.

ACKNOWLEDGEMENTS

The authors are indebted to Mr N Pendall of BHRA for his ceaseless enthusiasm in conducting the experimental work reported here.

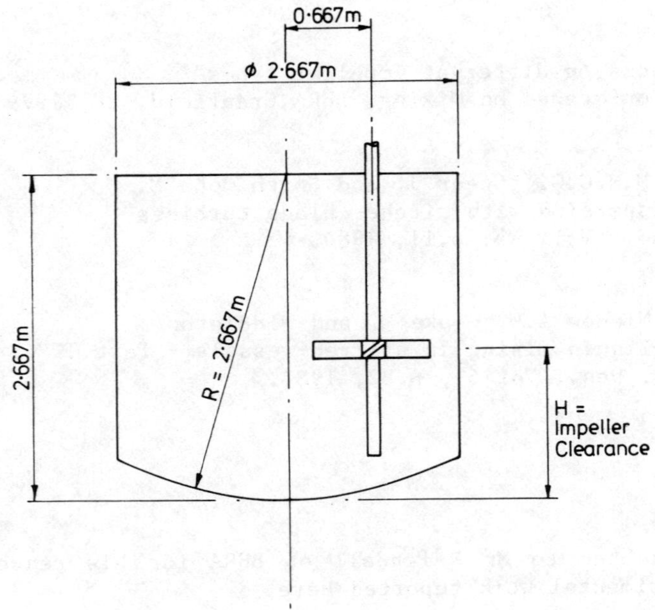

Standard Fill Level Shown In Drawing

FIG. 1 TEST VESSEL IN SIDE ELEVATION SHOWING
IMPELLER MOUNTED ECCENTRICALLY.

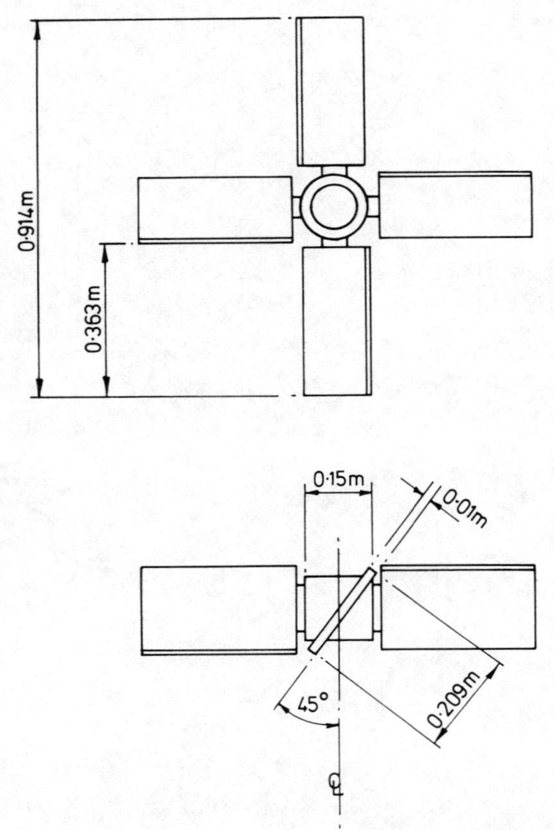

FIG. 2   IMPELLER USED IN TESTING

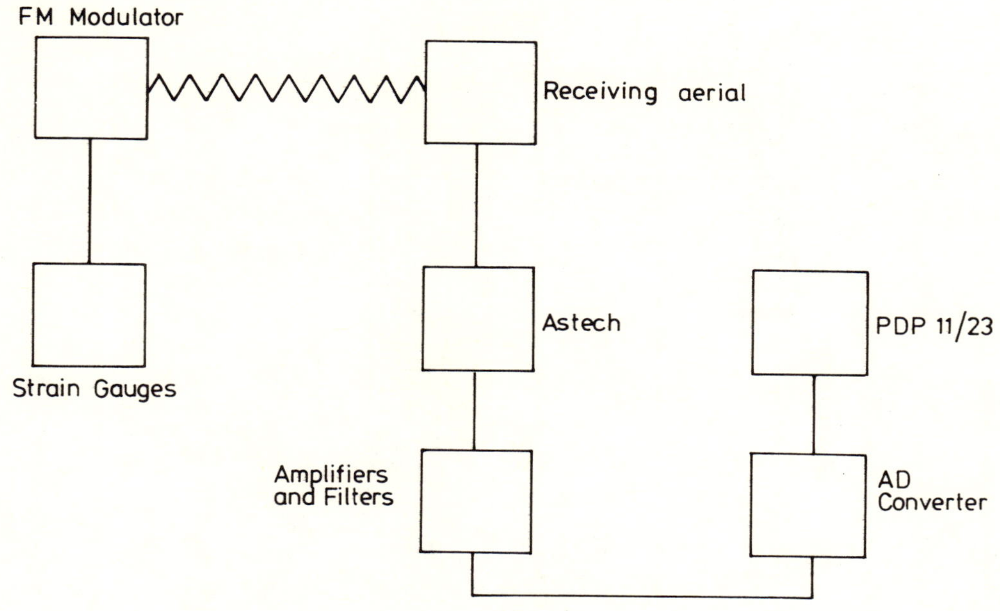

FIG. 3   DIAGRAMATIC REPRESENTATION OF THE INSTRUMENT CHAIN

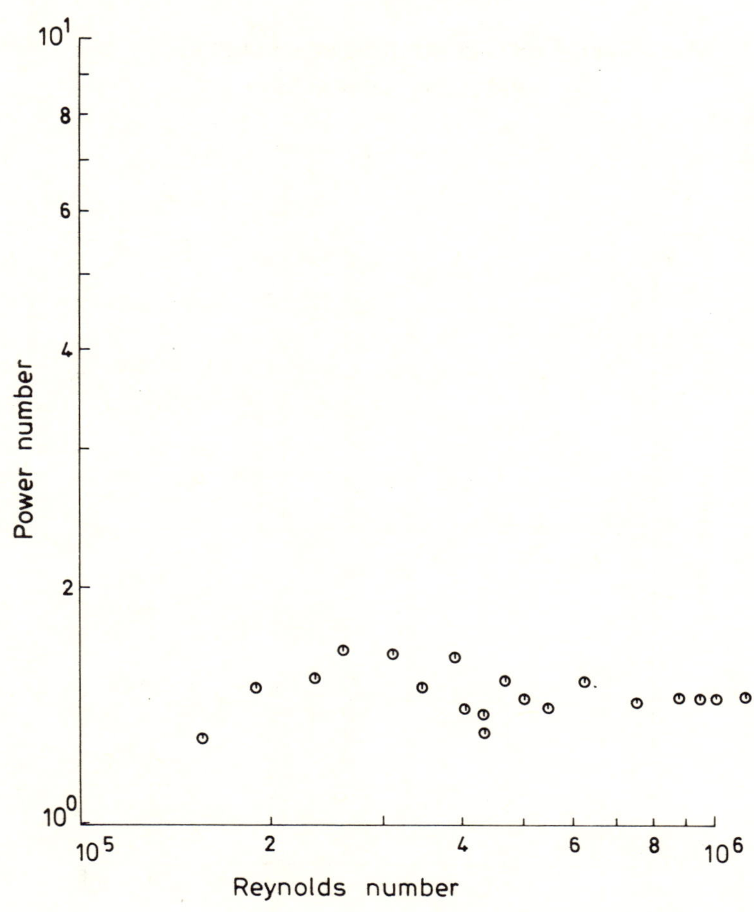

FIG.4 OFF-CENTRE POWER NUMBER
IMPELLER CLEARANCE = T/3

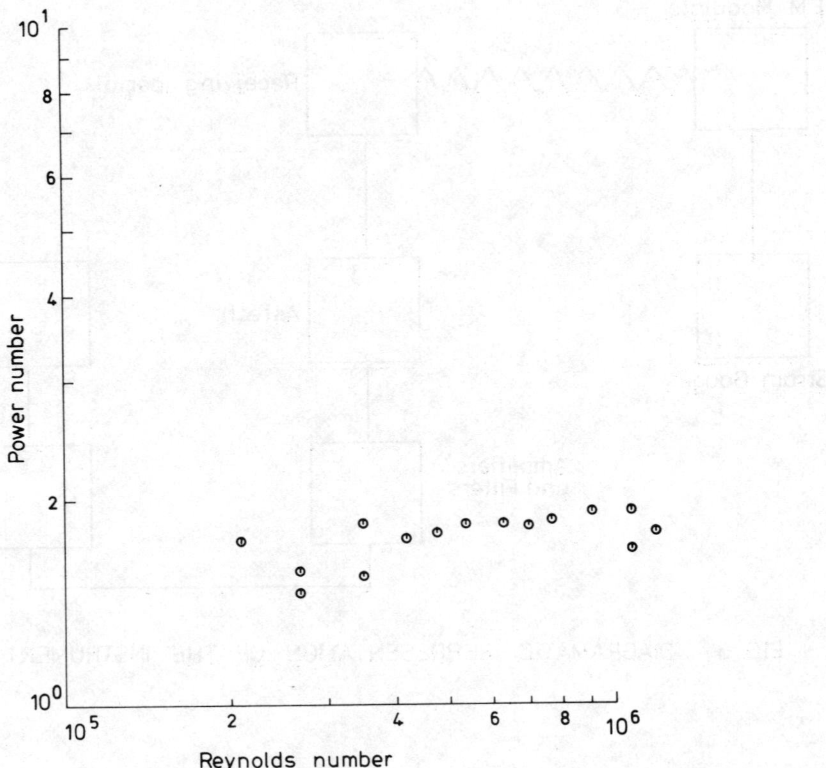

FIG.5 OFF-CENTRE POWER NUMBER.
IMPELLER CLEARANCE = T/2

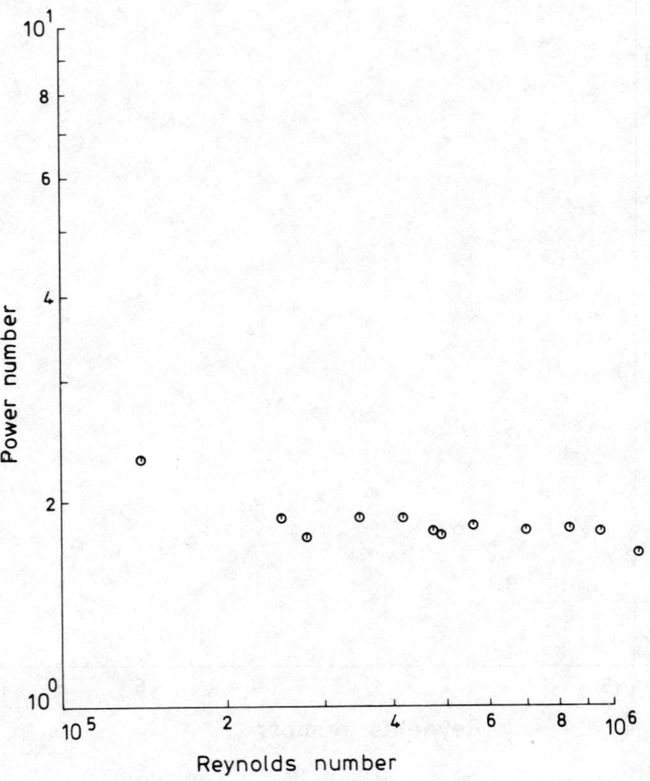

FIG.6 OFF-CENTRE POWER NUMBER
IMPELLER CLEARANCE = 0·176T

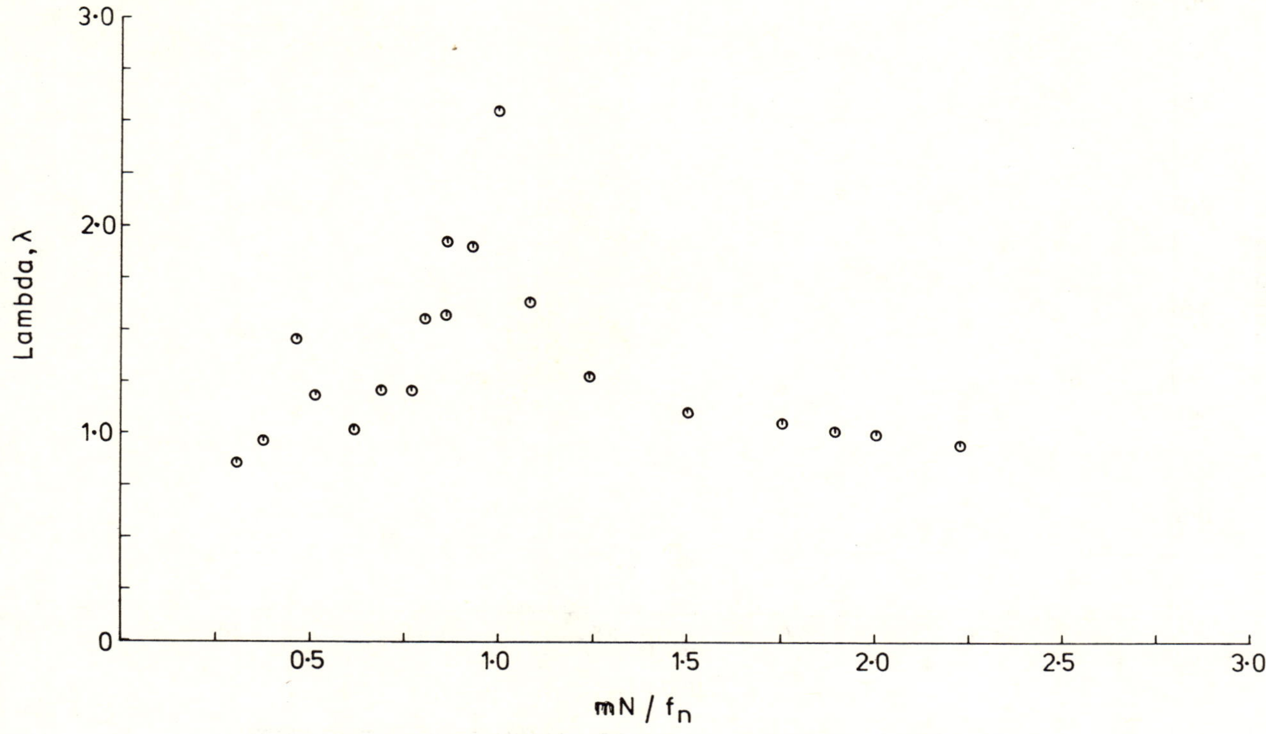

FIG.7   BENDING LOADS , ECCENTRIC MOUNTING , C = T/3

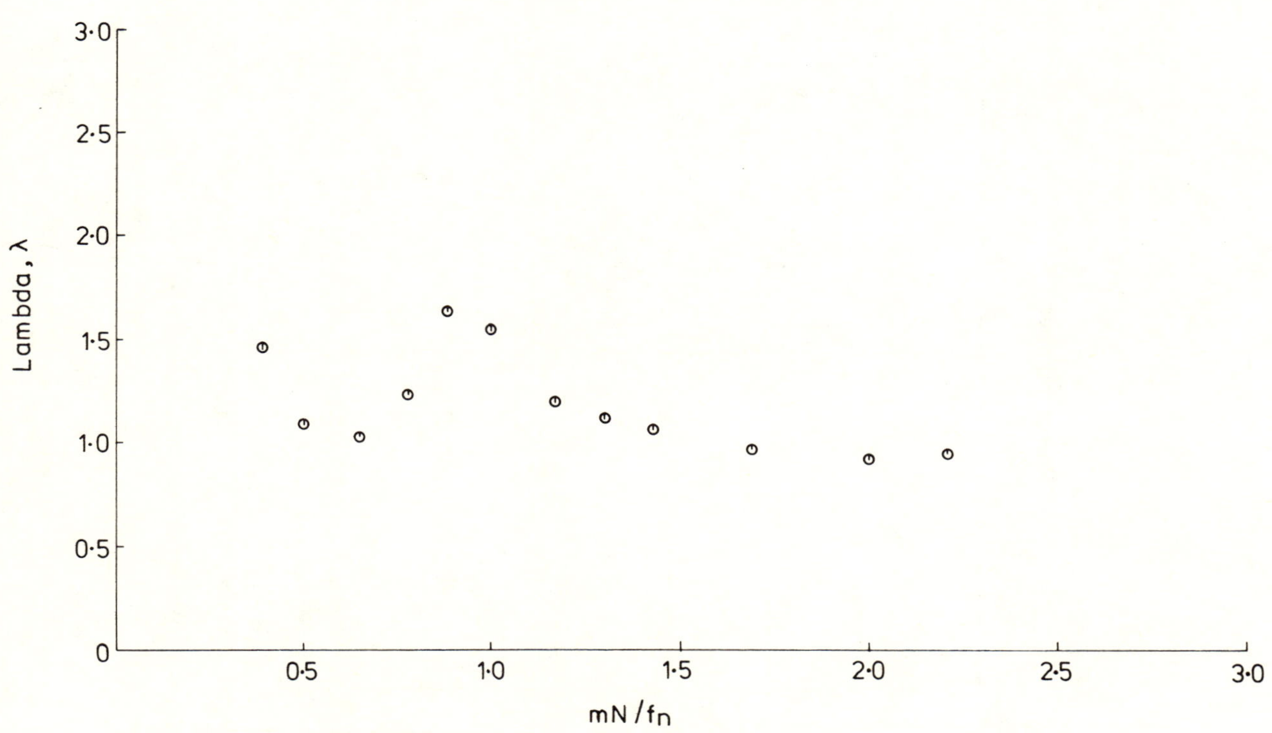

FIG.8   BENDING LOADS ECCENTRIC MOUNTING , C = T/2

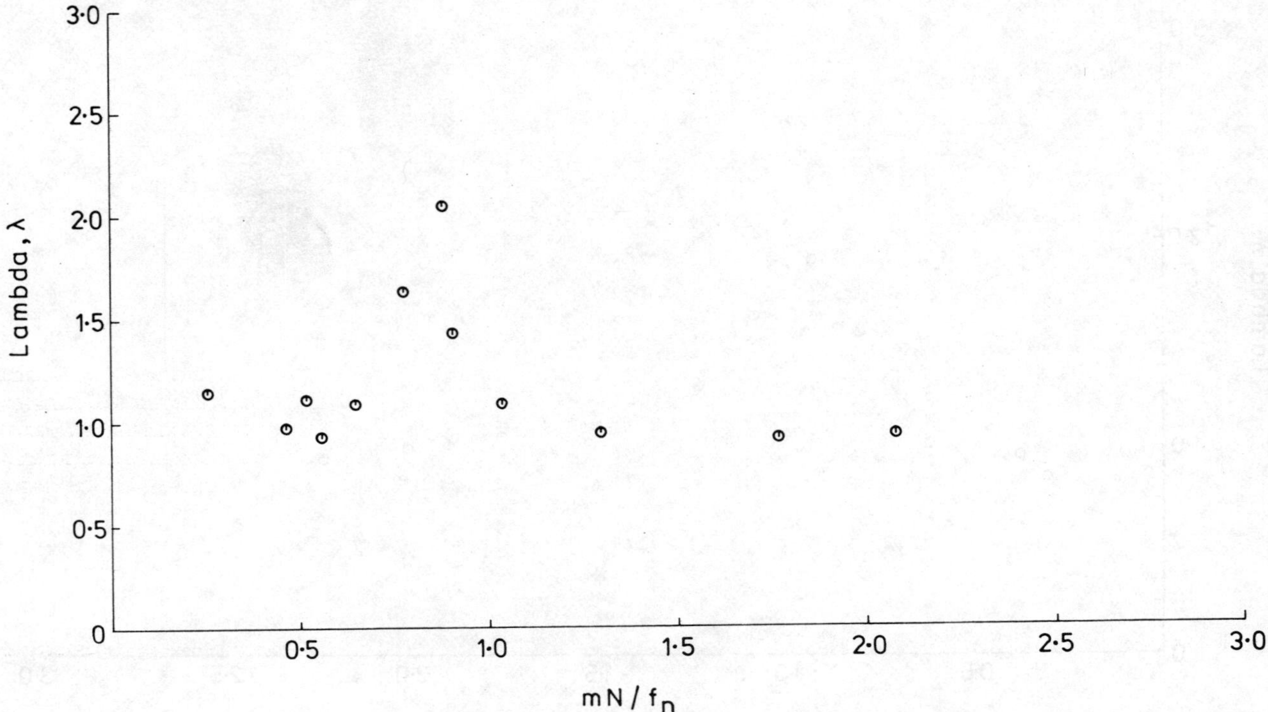

FIG.9   BENDING LOADS , ECCENTRIC MOUNTING , C=0·176 T

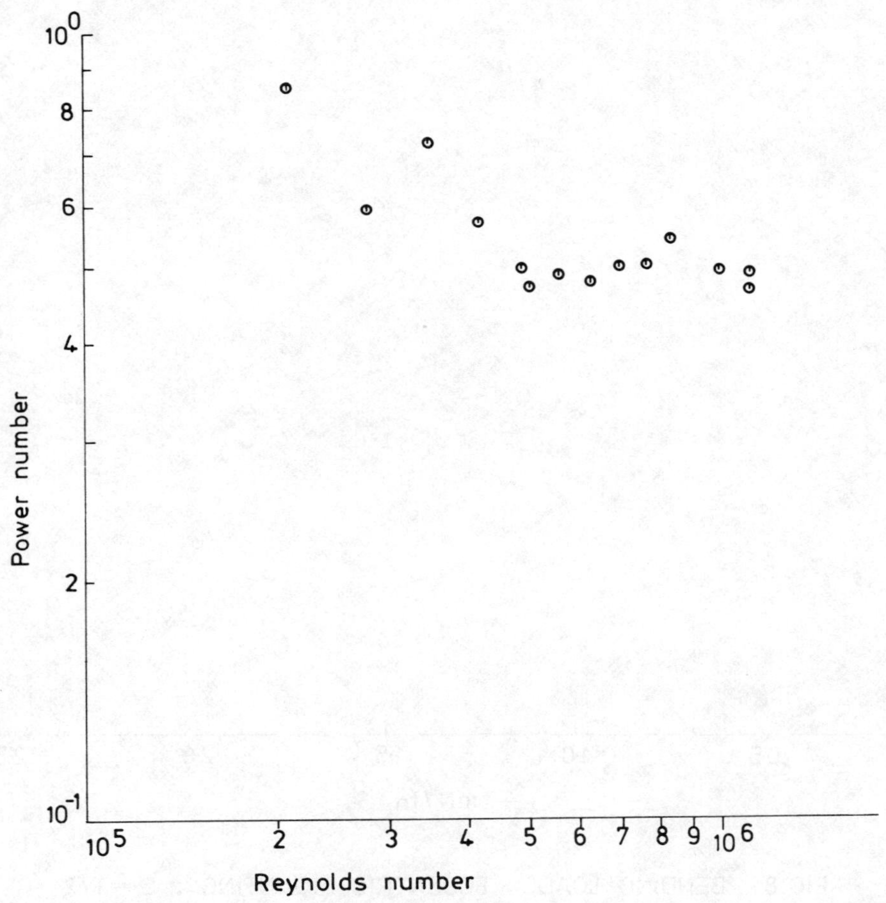

FIG.10   AXIALLY MOUNTED , POWER NUMBER , NO BAFFLES

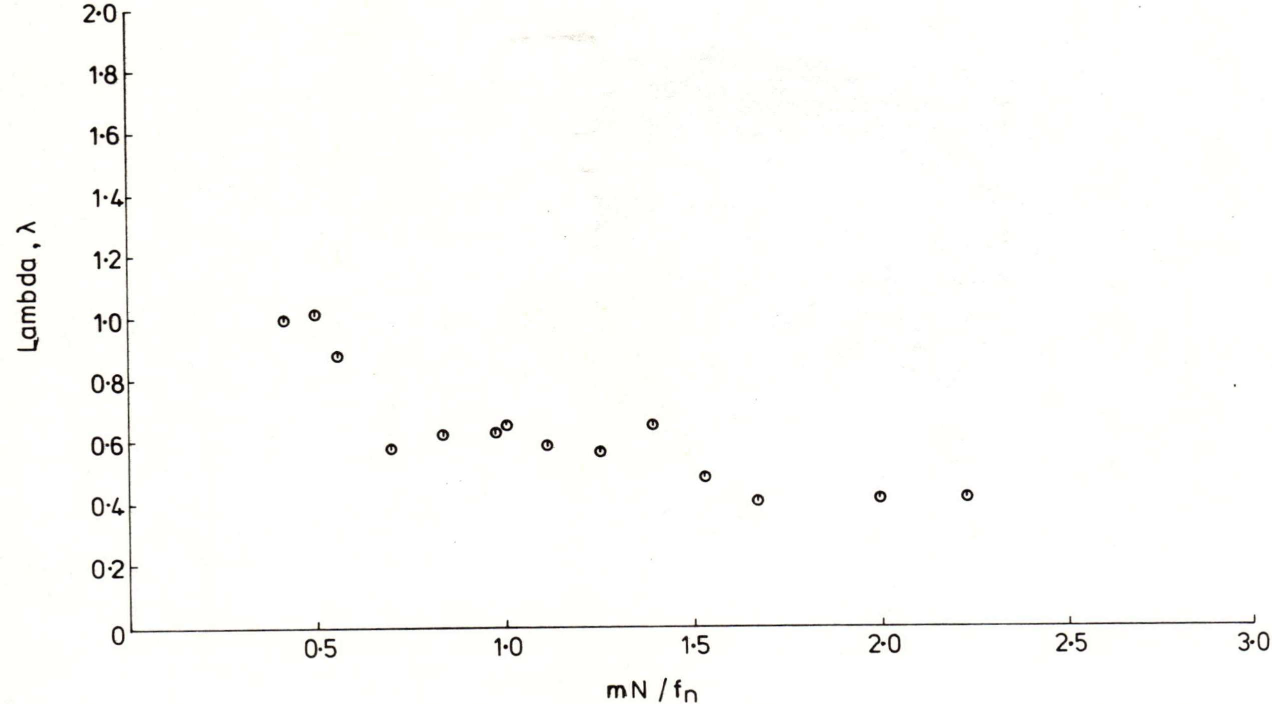

FIG.11   BENDING LOADS , AXIAL MOUNTING , NO BAFFLES

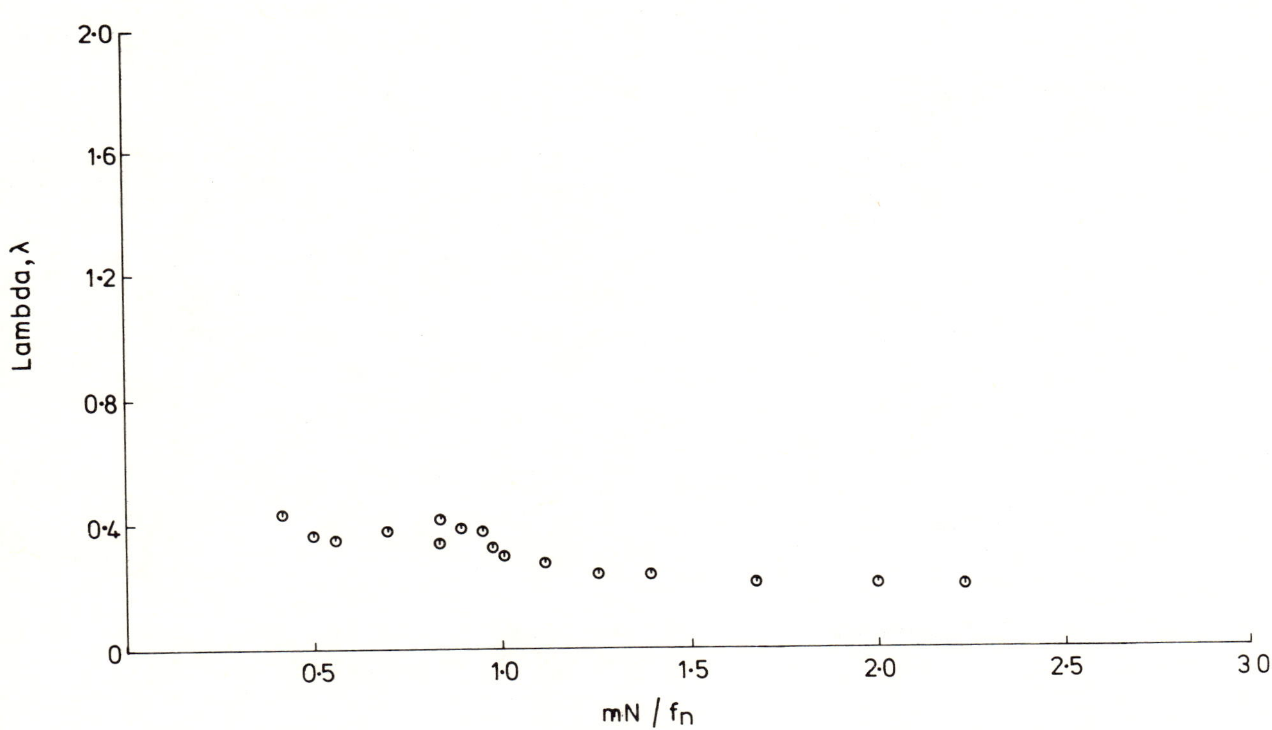

FIG.12   BENDING LOADS , AXIAL MOUNTING , 4 FLAT BAFFLES

Mixing

SCALE-UP OF MASS TRANSFER IN HIGHLY VISCOUS LIQUIDS

H.-J. Henzler, J. Kauling

Central Research / Applied Physics
BAYER AG, Wuppertal,
West Germany

Summary

Sorption characteristics for the systems air/water and air/viscous liquids with Newtonian and non-Newtonian flow behaviour are developed from measurements published in literature (reactor diameter T = 0.06-0.4 m) and from our own measurements (T = 0.1-2.6 m). The presentation of our results is based on concepts founded on the theory of similarity, which is introduced in this article and is also valid for non-Newtonian liquids. To enable the Π-relationships to be applied usefully to non-Newtonian liquids, a concept of apparent viscosity, identical for both mixing vessels and bubble columns, is introduced. This makes the sorption characteristics of the two types of reactor directly comparable. The use of this concept has the added advantage that it eliminates the so far observed difference, which can hardly be explained in physical terms, between mixing vessels and bubble columns with regard to the viscosity functions of the $k_L a$ values.

The case of bubble columns differs from that of mixing vessels because, as may be seen from the scale-up rules, an additional influence of the column diameter T is observed at T ≤ 0.6 m.

Held at Wurzburg, 10-12 June, 1985.

Organised by DVCV· Deutsche Vereinigung für Chemie- und Verfahrenstechnik
(German Association of Chemical and Process Engineering).

Organisation: GVC·VDI-Gesellschaft Verfahrenstechnik und Chemieingenieurwesen. 

NOMENCLATURE

A = cross sectional area of reactor

d = aperture diameter

D = impeller diameter

$\mathbb{D}$ = diffusion coefficient of air in various liquids

g = acceleration due to gravity

H = liquid height

K = consistency coefficient (for definition see Equation (9))

$k_L a$ = mass transfer coefficient

m = flow exponent (for definition see Equation (9))

n = rotational speed of the impeller

P = power consumption of the impeller

P_B = total power consumption in gas-liquid contacting

q = gas throughput

S_i = material parameters for description of coalescence behaviour

S_i^+ = dimensionless numbers for description of coalescence behaviour

Sc = $\nu/\mathbb{D}$, Schmidt number

T = reactor diameter

V = reactor volume

v = superficial gas velocity

γ = shear gradient

η = dynamic viscosity

ν = kinematic viscosity

$\bar{\gamma}, \bar{\eta}, \bar{\nu}$ = representative quantities

γ_1, η_1 = material constants

ρ = density of the liquid

σ = surface tension

σ^+ = $\sigma/\rho \, (\nu^4 \, g)^{1/3}$, material number

ABBREVIATIONS

SP = sintered plate

PP = perforated plate

RS = ring sparger

SH = single hole

1. INTRODUCTION

Gas-liquid contacting with the aim of introducing oxygen into viscous liquids and liquids with non-Newtonian flow behaviour is a problem that arises in industries where fermentation is employed. To enable suitable dimensions to be chosen for fermentation vessels, and aerobic fermentation processes to be carried out in the most efficient manner, it must be possible to predict the gas-liquid mass transfer. Reliable precalculation is only possible if suitable design data for a number of empirical quantities, such as the $k_L a$ value, are available for use in the scaling-up process. In this article, calculation data are developed for the mass transfer coefficient $k_L a$ on the basis of measurements obtained for model systems (air/water and air/viscous Newtonian and non-Newtonian liquids).

2. THEORETICAL BASIS

For a given apparatus geometry the mass transfer coefficient $k_L a$ is influenced by the operating conditions gas throughput q and impeller power P and by the material parameters viscosity η, density ρ, coalescence parameters S_i of the liquid and diffusion coefficient D of the gas in the liquid. The material parameters S_i, which should describe the coalescence behaviour, are not yet known. For this reason the effect of the surface tension σ is also unknown because σ can only be changed through an alteration of the system, which, on the other hand, may lead simultaneously to an alteration of the coalescence behaviour. In spite of this we have included σ in the list of relevant factors because in certain cases a material parameter with the dimensions $\sigma \cdot /N/m/$ is needed to enable the measurements to be presented in a dimensionless form.

If the behaviour of the liquid is non-Newtonian, the results can only be correlated if the material function determining the process is known. In general the material function

$$\frac{\eta}{\eta_1} = f_r \left[\frac{\gamma}{\gamma_1} , \quad \Pi_r \right] \tag{1}$$

should be valid.

The dimensional material parameters η_1 and γ_1 present in Equation (1), and also additional dimensionless numbers Π_r needed to describe the rheological function of state, have to be substituted for η in the list of limiting quantities, so that the list of relevant factors is now:

$$k_L a = f_1 \ (\rho, \ \eta_1, \ \gamma_1, \ D, \ \sigma, \ S_i, \ v, \ \frac{P}{V}, \ g, \ \Pi_r) \tag{2}$$

Because the target function $k_L a$ itself is a related quantity we have chosen the superficial velocity $v = q/A$ and the specific impeller power P/V as operational limiting quantities here.

According to the Π-theorem 7 dimensionless numbers follow from the 10 dimensional quantities in the above list:

$$k_L a \ (\frac{\nu_1}{g^2})^{\frac{1}{3}} = f_2 \left[\frac{v}{(g\nu_1)^{1/3}} , \ \frac{P}{V\rho(\nu_1 g^4)^{1/3}} , \ \frac{1}{\gamma_1}(\frac{g^2}{\nu_1})^{\frac{1}{3}}, \ \frac{\nu_1}{D}, \ \frac{\sigma}{\rho(\nu_1^4 g)^{1/3}}, \ S_i^{+}, \ \Pi_r \right] \tag{3}$$

The viscosity behaviour of the liquid is described here by the $ $ -quantities that contain γ_1 and ν_1. The taking into account of elastic properties necessitates additional numbers from the set Π_r. To enable these elasticity numbers to be formulated, the definition of suitable additional dimensional material parameters (Ref. 1-3), the measurement of which is often very problematical, is necessary.

We will confine our attention to the properties of viscous liquids, since most of the available measurements have been obtained for CMC solutions, whose elastic properties can evidently be neglected.

The concept of an apparent viscosity has been used in the literature to present the measurements in a simple way. This concept was introduced by Metzner and Otto to describe the power input of impellers in ungassed liquids in the range of laminar flow

(Ref. 4) and has subsequently been applied to the heat transfer in bubble columns
(Ref. 5) and to mass transfer in mixing vessels (Ref. 6) and in bubble columns (Refs.
7, 8). As with Newtonian liquids, the set of dimensionless numbers is reduced by one
number ($g^{2/3}/\gamma_1 \cdot \nu_1^{1/3}$ becomes irrelevant), so that

$$k_L a \left(\frac{\nu}{g^2}\right)^{\frac{1}{3}} = f_2 \left[\frac{\nu}{(g\nu)^{1/3}} \, , \, \frac{P}{V\rho(\nu^4 g)^{1/3}} \, , \, \frac{\nu}{D} \, , \, \frac{\sigma}{\rho(\nu^4 g)^{1/3}} \, , \, S_i^+ \right] \tag{4}$$

In the case of non-Newtonian liquids ν is replaced by $\bar{\nu}$, which is obtained from the
flow curve at a representative shear gradient $\bar{\gamma}$. For calculation of $\bar{\gamma}$ the equation

$$\bar{\gamma} = A \cdot n \tag{5}$$

was used for mixing vessels and

$$\bar{\gamma} = B \cdot v \tag{6}$$

for bubble columns, A and B being empirical constants. As may be seen from the following, these definitions largely contradict the physical conditions

1. The useful concept $\bar{\gamma} \sim (P/\rho V\bar{\nu})^{1/2} \sim n$, contained in Equation (5), does not hold for
   the gassing of a stirred liquid because the law of the range of laminar flow $P/V \sim$
   ηn^2, on which Equation (5) is based, appears not to apply to the total power consumption in gas-liquid contacting (Equation (8)).

2. Equation (6) does not describe the physical circumstances in their entirety, since
   B is a dimensional constant. To achieve dimensional consistency, $1/B$ would have to
   be replaced by a characteristic length, e.g. by the bubble diameter. But every con-
   ceivable characteristic length is itself dependent on the superficial velocity,
   which means that the relation $\bar{\gamma} \sim v$ cannot be expected.

To enable the measurements to be correlated more appropriately, we would like to introduce a common concept for mixing vessels and bubble columns. The state of motion in
the reactor and hence the representative shear gradient will depend on the specific
power input $P_B/\rho V$ $[m^2/s^3]$. For the sake of a dimensionally consistent formulation we
have used, as a further limiting quantity, the kinematic viscosity, which likewise
influences the state of motion. With these limiting factors the following dimensionally consistent formulation of the representative shear gradient is arrived at:

$$\bar{\gamma} = \left(\frac{P_G}{V\rho} \cdot \frac{1}{\nu} \right)^{1/2} \tag{7}$$

where P_B is the total power consumption of the gas/liquid system:

$$P_G = \rho g H q + P = g v + \frac{P}{V\rho} \tag{8}$$

If the flow curve can be described by the Ostwald-de Waele equation:

$$\eta = K \gamma^{m-1} \tag{9}$$

then

$$\bar{\gamma} = \left(\frac{P_G}{V} \, \frac{1}{K}\right)^{\frac{1}{m+1}} \tag{10}$$

follows from Equations (7) and (8) for the representative shear gradient and

$$\bar{\nu} = \left(\frac{K}{\rho}\right)^{\frac{2}{m+1}} \left(\frac{P_G}{V\rho}\right)^{\frac{m-1}{m+1}} \tag{11}$$

follows for the apparent viscosity.

Correspondingly modified Π-numbers are thus obtained from the numbers of the Π-equa-
tion (4). For purely formal reasons the flow exponent m should, apart from this, be

incorporated in the Π-equation as an additional number.

3. RESULTS OBTAINED FOR MIXING VESSELS

If, to begin with, certain material systems are considered at a constant process temperature, the material numbers $Sc = \nu/D$, $\sigma^+ = \sigma/\rho\,(\nu^4 g)^{1/3}$ and S^+ are omitted because D, σ and the coalescence behaviour do not change. The available experimental results can then be expressed by power equations such as :

$$k_L a \left(\frac{\nu}{g^2}\right)^{1/3} = C_1 \left(\frac{P}{V\rho\,(\nu g^4)^{1/3}}\right)^a \left(\frac{\nu}{(g\nu)^{1/3}}\right)^b \qquad (12)$$

The constants C, a and b are dependent on the system. With all the investigated systems, however, $b \approx 1-a$. This leads finally to the following simple formulation, by which the measurements can be described virtually as accurately as they can according to the concept of Equation (12):

$$\frac{k_L a}{\nu} \left(\frac{\nu^2}{g}\right)^{1/3} = C_2 \left(\frac{P}{V\rho}\,\frac{1}{\nu g}\right)^a \qquad (13)$$

Figure 1 gives a corresponding correlation of various measurements for water on the

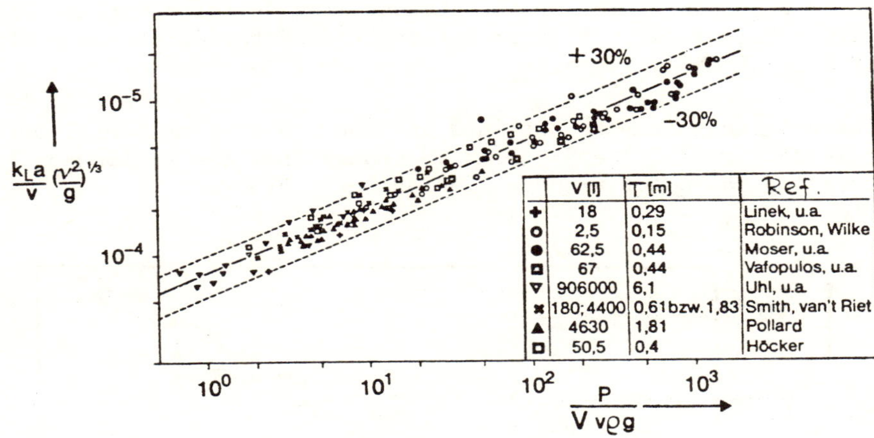

	V [l]	T [m]	Ref.
+	18	0,29	Linek, u.a.
o	2,5	0,15	Robinson, Wilke
●	62,5	0,44	Moser, u.a.
▣	67	0,44	Vafopulos, u.a.
▼	906000	6,1	Uhl, u.a.
✕	180; 4400	0,61 bzw. 1,83	Smith, van't Riet
▲	4630	1,81	Pollard
◰	50,5	0,4	Höcker

Figure 1 Sorption characteristics for water, disc turbine, D/T = 0.14-0.5; H/T = 0.5-1, (after Ref. 10)

scale of V = 2.5 l to 906 m^3 (cf. Figure 6 in Ref. 9). As the measurement points for different volumes are very close together, the scaling-up rule $k_L a \sim (P/V\rho)^a\,\nu^b$ may be considered valid.

The $k_L a$ values for viscous liquids can be correlated in the same way (see Figure 2, Ref. 10). Although the viscosity has changed by more than 1 to 2 orders of magnitude and the diffusion coefficient (in the case of millet jelly and glucose solutions) by more than 1 order of magnitude, a summary presentation of the measurements is possible without Sc and σ^+ being taken into consideration. This is so because the influence of the viscosity is evidently accounted for correctly by the sorption number. It is not necessary to use σ^+ and Sc also, since, where oxygen is absorbed by a particular liquid, σ^+ and Sc are changed almost exclusively by the viscosity. (With the material systems considered here the surface tension changes only very slightly and the change of the diffusion coefficient is tied firmly to the viscosity where a particular material system is concerned: $D \sim \nu^{-a}$ (see Ref. 8)). It is not necessary to include the Sc number unless one requires sorption characteristics for calculating the mass transfer of different gases that have different diffusion coefficients at a given viscosity of a given liquid. The use of the Sc number then necessarily entails that of σ^+ because the influence of viscosity additionally arising from Sc has to be compensated.

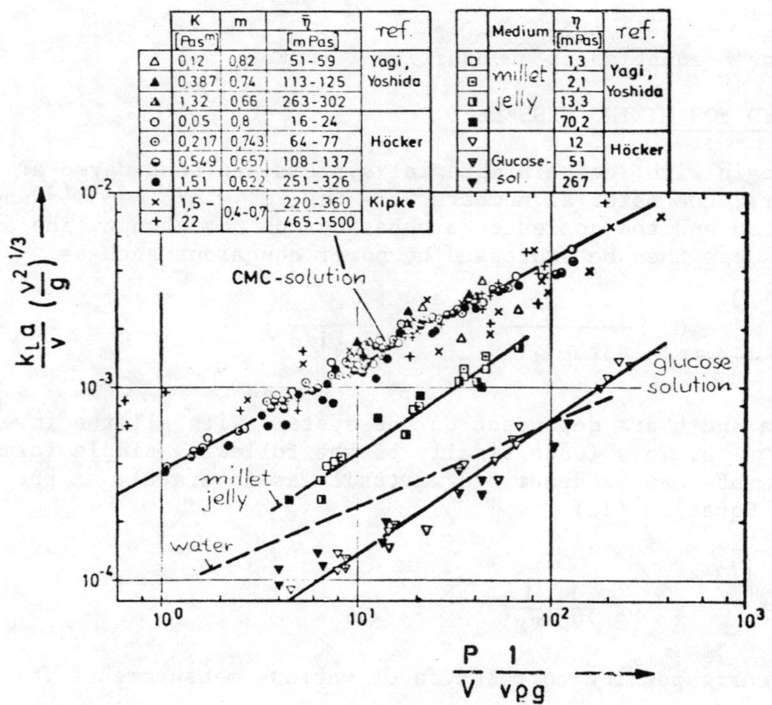

Figure 2   Sorption characteristics of air/viscous liquids, disc turbine D/T = 0.24–
0.4, H/T = 1, V = 12–50 1 (from Ref.10) (presentation by means of $\bar\gamma$ = 11.5 n
in the case of CMC solutions)

Figure 2 contains only measurements obtained in laboratory experiments. Our own measurements on the scale of V = 47.5 1 and V = 13 m^3 showed that the scaling-up laws given by Equation (13) are also valid for viscous liquids – see Figure 3.

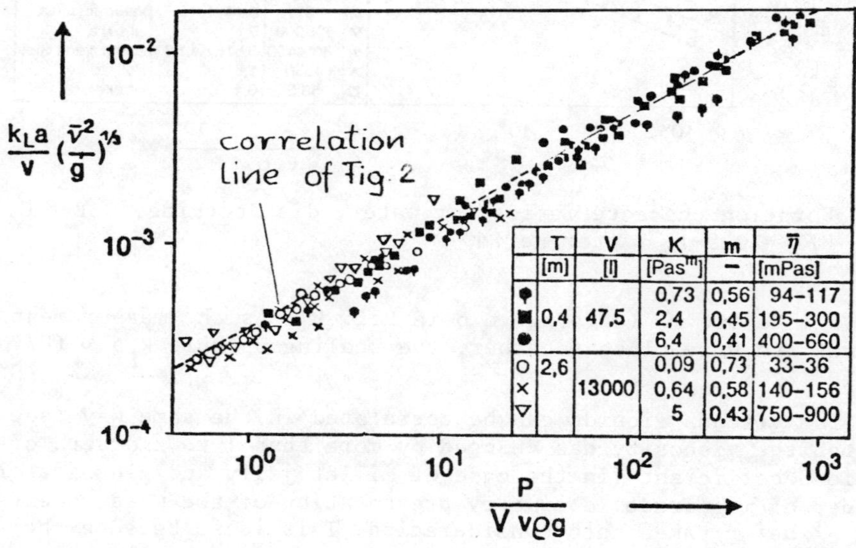

Figure 3   Sorption characteristics for air/CMC solutions, Inter–MIG impeller, D/T =
0.65, H/T = 1 (presentation by means of $\bar\gamma$ = 17.5 n)

For the non-Newtonian CMC solutions the old concept of the apparent viscosity according to Equation (5) was used. For comparison, the measurements for V = 13 m^3 have been presented in Figure 4 according to the concept introduced above on the basis of Equation (7). Not only are the measurement points shifted downwards, but their progression for $P/V v \rho g$ <1 follows the straight line more closely.

308

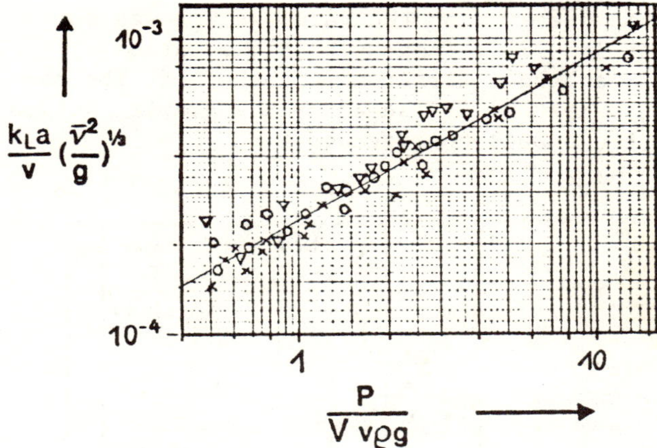

Figure 4

Sorption characteristics for air/ CMC solutions (presentation according to the concept of Equation (7)). For explanations see Figure 3.

4. RESULTS FOR THE BUBBLE COLUMNS

The available measurements show that in the case of technically suitable gas distributors there is a direct proportionality between the $k_L a$ value and v at low superficial velocities v. Not until certain velocities are exceeded does the ratio $k_L a/v$ decrease, though this is not perceptibly connected with the viscosity. If we again consider a particular material system, the following dimensionless formulation can be used to present the results:

$$\frac{k_L a}{v}\left(\frac{v^2}{g}\right)^{1/3} = f_3\left(v\left(\frac{\rho}{\sigma g}\right)^{1/4}\right) \tag{14}$$

Using the same sorption number as has been used for the mixing vessel permits a simple comparison of the two apparatuses. The independent number $v(\rho/\sigma g)^{1/4}$ is obtained by combining the numbers $v/(gv)^{1/3}$ and $\sigma/\rho(gv^4)^{1/3}$ (see Equation (3) or (4)).

Figures 5 to 8 show the dependence $k_L a/v = f(v)$ which has been described above. According to the material system and gas distributor, the differences may be merely quantitative.

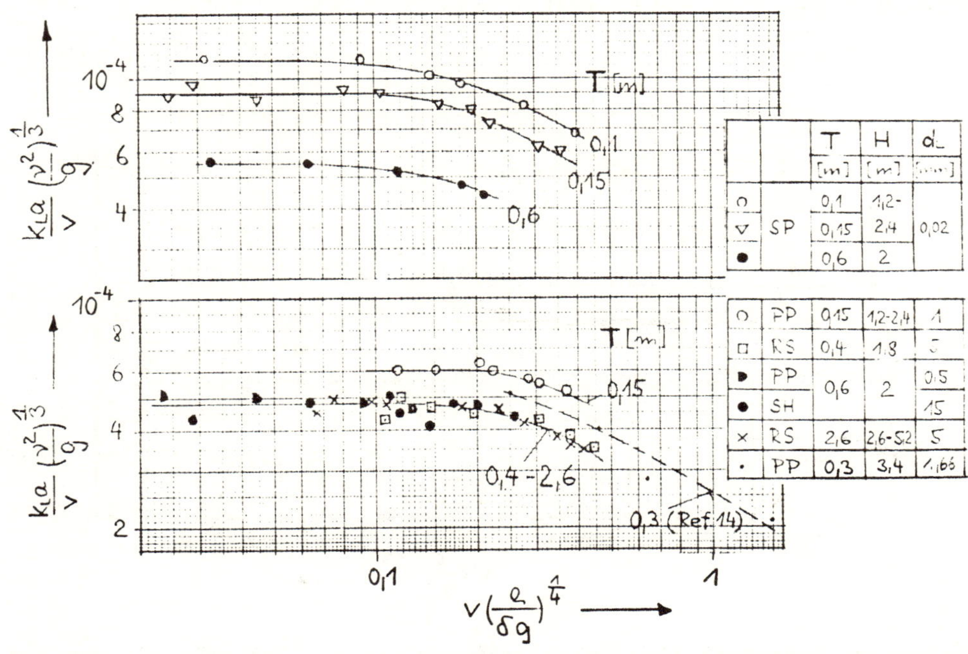

		T [m]	H [m]	d_L [mm]
○		0,1	1,2-	
▽	SP	0,15	2,4	0,02
●		0,6	2	
○	PP	0,15	1,2-2,4	1
□	RS	0,4	1,8	5
◗	PP	0,6	2	0,5
●	SH			15
×	RS	2,6	2,6-5,2	5
·	PP	0,3	3,4	1,65

Figure 5   Sorption characteristics for air/water

In the case of water an influence of the column diameter on the value of k_La was observed at T < 0.6 m (fine-bubble aeration) and T < 0.4 m (coarse-bubble aeration). This influence is particularly marked where fine-bubble aeration is concerned. The reduction of the k_La value at increasing column diameters is caused by the growing motion of the liquid, which leads to an increase of the bubble coalescence.

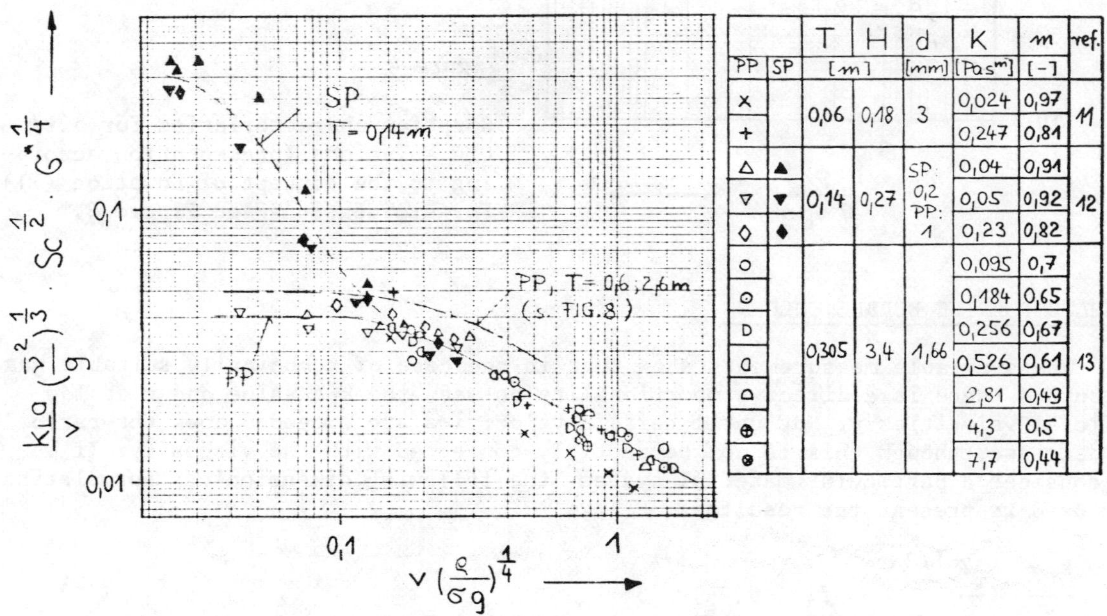

Figure 6 Sorption characteristics for CMC solutions (presentation by means of $\gamma = 1500\ v\ [m/s]$)

As shown by the comparison of the results published in the literature (see Figures 6 and 7) with our own results (see Figure 8), the diameter evidently has an opposite influence where the non-Newtonian CMC solutions and fairly high superficial velocities $v \gtrsim 0.03$ m/s ($v(\rho/\sigma g)^{1/4} \gtrsim 0.1$) are concerned. This is probably a consequence of the slug flow which has been observed at fairly high velocities v where T $\leq$ 0.3 m (Ref. 11 to 13).

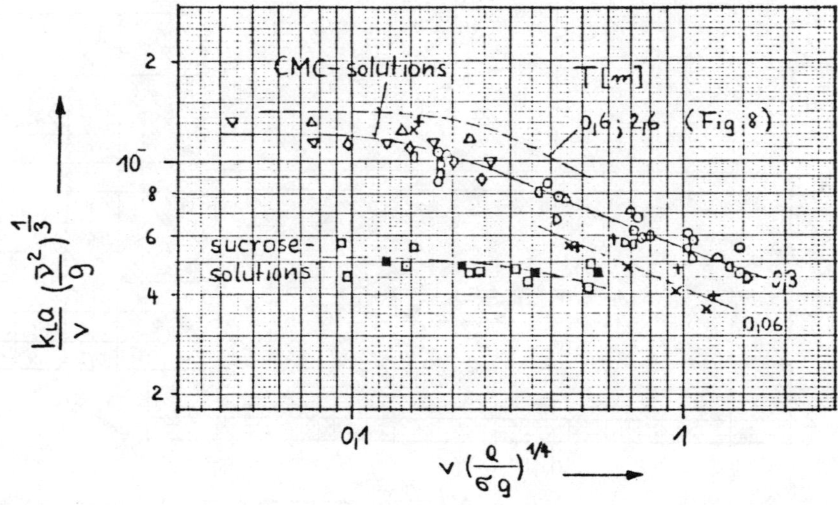

Figure 7 Sorption characteristics for air/CMC solutions. For explanations see Fig. 6 (presentation according to the concept of Equation (10))

The results obtained for viscous sucrose and CMC solutions can likewise be correlated
with just two dimensionless numbers (see Figures 7 and 8). Where the non-Newtonian CMC
solutions are concerned, though, that is only possible if the concept introduced in
the form of Equation (7) or (10) and (11) is used. If the apparent viscosity is taken
according to the formula presented in Equation (6), as applied until now, the additional numbers Sc and σ^+ have to be used, too, to correlate the measurements (see
Figure 6). The correlation expressed in Figure 6 leads, finally, to a more pronounced
viscosity dependence $k_La \sim \nu^{5/6}$ than that seen in the case of mixing vessels, for
which $k_La \sim \nu^{2/3}$ was found. As, however, the same flow zones exist in the mixing
vessel and bubble column (because the same liquid is used and the values of $P/V\rho$ are
comparable), this difference cannot be explained in physical terms and must be attributed to the unsuitability of the definition of the apparent viscosity given by
Equation (6).

	K $[Pa\,s^m]$	m $[-]$		K $[Pa\,s^m]$	m $[-]$
△	0,047	0,88	○	0,055	0,8
△	0,1	0,81	◑	0,2	0,72
▲	0,33	0,67	◐	0,52	0,64
▲	1,6	0,53	○	0,56	0,56

Figure 8

Sorption characteristics for air/CMC
solutions,
△ D = 0.6 m,
o D = 2.6 m,
(presentation according to the concept of Equation (7))

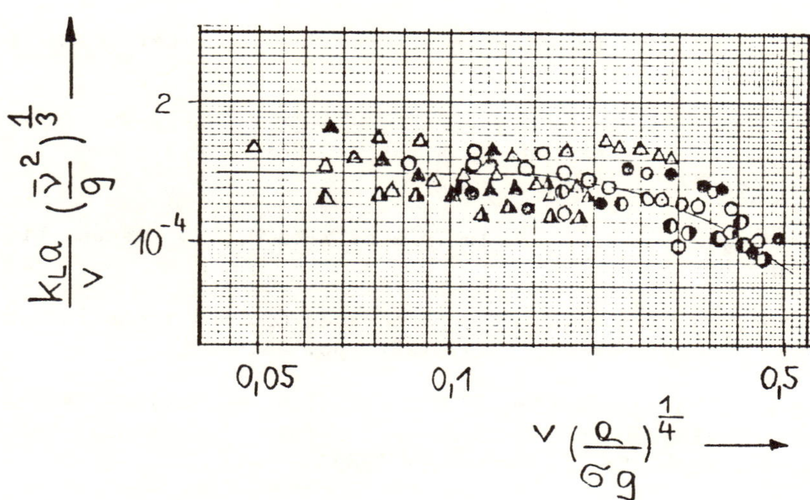

A direct comparison of the mixing vessel and bubble column is made possible by
Figures 4 and 8, for which the same definition of viscosity was used to present the
measurements. As was expected, the sorption numbers of the mixing vessel for
$P/Vv\rho g \ll 1$ approach those for the bubble column.

5. REFERENCES

1 Yagi, H. and Yoshida, F.: "Gas absorption by Newtonian and non-Newtonian fluids in sparged agitated vessels". Ind. Eng. Chem. Process Des. Dev. 14, 4, 1975, pp. 488-493

2 Ranade, V.R. and Ulbrecht, J.J.: "Influence of polymer additives on the gas-liquid mass transfer in stirred tanks". AIChE Journal 24, 5, September 1978, pp. 796-803

3 Nakanoh, Yoshida: "Gas Absorption by Newtonian and Non-Newtonian Liquids in a Bubble Column", Ind. Eng. Chem. Process. Des. Dev. 19, 1, 1980, pp. 190-195

4 Metzner, A. and Otto, R.: AIChE Journal 3, 1957, pp. 3-9

5 Nishikawa, M. et al.: "Heat Transfer in Aerated Tower Filled with Non-Newtonian Liquids". Ind. Eng. Chem. Process. Des. Dev. 16, 1, 1977, pp. 133-137

6 Höcker, H.: "Mass transport in aerated mixing vessels filled with non-Newtonian liquids". (Zum Stoffübergang begaster Rührer in nicht-Newtonischen Flüssigkeiten). Paper presented at the GVC specialist Committee meeting at Bad Kreuznach, 1978.

7 Henzler, H.-J.: "Aeration of viscous liquids". (Begasen höher viskoser Flüssigkeiten). Preprints of conference entitled "Verfahrenstechnische Fortschritte beim Mischen, Dispergieren und Wärmeübertragen in Flüssigkeiten" held by the GVC in Düsseldorf on 4/5 December 1978

8 Henzler, H.-J.: "Aeration of viscous liquids". (Begasen höherviskoser Flüssigkeiten). Chem.-Ing.-Tech. 52, 8, August 1980, pp. 643-652

9 Judat, H.: "Gas/Liquid Mass Transfer in Stirred Vessels", Ger. Chem. Eng. 5, 1982, pp. 357-363

10 Henzler, H.-J.: "Design data for stirred vessels as fermenters". (Verfahrenstechnische Auslegungsunterlagen für Rührbehälter als Fermenter). Chem. Ing. Tech. 54, 5, May 1982, pp. 461-476

11 Schumpe, A. and Patwari, A.N.: "Bubble Columns with Highly Viscous Liquids". Chem. Eng. Commun.

12 Deckwer, W.D.: "Oxygen Mass Transfer into Aerated CMC Solutions in a Bubble Column", Biotechnology and Bioengineering, XXIV, 1982, pp. 461-481

13 Schumpe, A.; et al.: "Oxygenization of CMC solutions in a wide-diameter bubble column". (Sauerstoffeintrag in CMC-Lösungen in einer Blasensäule großen Durchmessers". Chem.-Ing.Tech. 55, 9, September 1983

14 Godbole, S.P., et al.: "Hydrodynamics and Mass Transfer in a Bubble Column with an Organic Liquid". The Canadian Journal of Chem. Eng. 62, June 1984, pp. 440-445

AGITATION OF HIGHLY VISCOUS FLUIDS BY GATE-AGITATORS

J. BERTRAND and J.P. COUDERC
Institut du Génie Chimique
U.A. CNRS 192
Chemin de la Loge
31078 TOULOUSE Cédex (France)

Summary

For a simple geometry and a laminar flow, the partial differential motion equations have been integrated to obtain the velocity and stress distributions in the case of newtonian and non-newtonian liquids. Numerical analysis is a rather inexpensive way of predicting the hydrodynamic behaviour of such systems, avoiding the major part of experimental measurements at the laboratory or pilot plant scale. The fluids are highly viscous newtonian and pseudoplastic liquids of the CARREAU type. They are agitated by gate-agitators in flat bottomed cylindrical vessels without baffles : the height of the blades is equal tothat of the liquid in the vessel.

Examples of the results detailing the stream function, the vorticity function, the radial and tangential velocity profiles, shear and normal stresses, the dissipation function, are presented. At least, the power number is determined.

Results of velocity are compared with experimental ones which are obtained by hot film anemometry. The power consumption is also measured by an appropriate device. Both kinds of results are in good agreement.

The numerical procedure developed here can be used for designing new impellers in all cases where no axial motion occurs.

Held at Wurzburg, 10-12 June, 1985.

Organised by DVCV· Deutsche Vereinigung für Chemie- und Verfahrenstechnik
(German Association of Chemical and Process Engineering).

Organisation: GVC·VDI-Gesellschaft Verfahrenstechnik und Chemieingenieurwesen.  GVC

<u>NOMENCLATURE</u>

r, θ, z = radial, angular and axial position respectively in cylindrical coordinates

D = agitator diameter (m)

$f(\eta)$ = function of the derivatives of the apparent viscosity

n = flow behaviour index (–)

N = agitator rotational speed (S^{-1})

Ne = power number ($Ne = P/\rho N^3 D^5$)

P = power input (W)

Re = Reynolds number (defined by equations 15 and 16)

T = tank diameter (m)

V = tank volume (m^3)

γ = shear rate (S^{-1})

ϕ_V = viscous dissipation (S^{-2})

ψ = stream function ($m^2 S^{-1}$)

$\psi_P, \psi_G, \psi_{G'}$ = particular values of the stream function ($m^2 S^{-1}$)

η = apparent viscosity (kg $m^{-1} S^{-1}$)

η_o = viscosity at rest for a Carreau fluid ($kg m^{-1} S^{-1}$)

λ = characteristic time (S)

ρ = fluid density ($kg m^{-3}$)

ω = vorticity (S^{-1})

τ = stress (Pa)

1. INTRODUCTION

Mixing or agitation of highly viscous newtonian or non-newtonian fluids takes place in various fields of the chemical or food industries. The corresponding scientific problems to be solved by chemical engineers are difficult and only a few results can be found in the literature.

This paper presents a numerical method to solve the basic continuity and motion equations in the case of a gate-agitator and of laminar flow.

The height of the impeller is equal to the depth of the liquid in the tank. The tank is cylindrical, flat bottomed and the depth of the liquid is equal to the tank diameter (for all other dimensions, see figure 1).

2. GENERALITIES

2.1 Numerical method

The numerical method has already been presented in detail elsewhere (Ref. 1,2). It can briefly be recalled that the equations to be solved are the continuity, motion and rheology equations, this last one in the case of a pseudoplastic fluid, the rheology of which is represented by the Carreau model.

Experimental observations have suggested that the flow is only horizontal : all planes are identical and the axial velocity is equal to zero. So, the problem to be solved is only two-dimensional. The equations have been written in terms of the stream and vorticity functions and apparent viscosity :

$$\omega = -\nabla^2 \psi \tag{1}$$

$$\frac{1}{r}\, Re\, \frac{\partial(\psi,\omega)}{\partial(r,\theta)} = \eta\,\nabla\omega + f(\eta) \tag{2}$$

For a newtonian fluid, η is known in every point. For a pseudoplastic fluid represented by the Carreau model, η is written as :

$$\eta = \left\{ 1 + \lambda^2 \left[\tfrac{1}{2}(\gamma:\gamma) \right] \right\}^{\frac{n-1}{2}} \tag{3}$$

The shear rate γ is given by

$$\tfrac{1}{2}(\gamma:\gamma) = 4\left(\frac{1}{r}\frac{\partial^2\psi}{\partial r\,\partial\theta} - \frac{1}{r^2}\frac{\partial\psi}{\partial\theta}\right)^2 + \left(2\frac{\partial^2\psi}{\partial r^2} - \omega\right)^2 \tag{4}$$

These equations are solved with the following boundary conditions :

. Shaft :

$$\psi = 0 \tag{5}$$

$$\omega = -\frac{\partial^2\psi}{\partial r^2} \tag{6}$$

$$\eta = (1 + \lambda^2\omega^2)^{\frac{n-1}{2}} \tag{7}$$

. tank wall :

$$\Psi = \Psi_P \tag{8}$$

$$\omega = 1 - \frac{\partial^2 \Psi}{\partial r^2} \tag{9}$$

$$\eta = \left| 1 + \lambda^2 (\omega + 2)^2 \right|^{\frac{n-1}{2}} \tag{10}$$

. impeller blades :

$$\Psi = \Psi_G \quad \text{for} \quad \frac{D_{YO}}{T} < r < \frac{D_{PO}}{T} \tag{11}$$

$$\Psi = \Psi_{G'} \quad \text{for} \quad \frac{D_Y}{T} < r < \frac{D}{T} \tag{12}$$

$$\omega = - \frac{1}{r^2} \frac{\partial^2 \Psi}{\partial \theta^2} \tag{13}$$

$$\eta = \left| 1 + \lambda^2 \omega^2 \right|^{\frac{n-1}{2}} \tag{14}$$

The constants Ψ_P, Ψ_G and $\Psi_{G'}$ are determined, by writting the conservation of angular momentum (Ref. 1,2).

Finally the product power number-Reynolds number is obtained :

- for a newtonian fluid :

$$NeRe = C_1 \iiint_V \left| \tfrac{1}{2} (\gamma : \gamma) \right| \, dV, \quad \text{with } Re = \frac{\rho ND^2}{\eta} \tag{15}$$

- for a Carreau fluid :

$$NeRe = C_1 \iiint_V \left\{ 1 + \lambda^2 \left| \tfrac{1}{2} (\gamma : \gamma) \right| \right\}^{\frac{n+1}{2}} dV, \quad \text{with } Re = \frac{\rho ND^2}{\eta_0} \tag{16}$$

The constant C_1 is characteristic of the system geometry.

2.2 Experimental conditions :

The experimental tank is 0.2 m^3 in volume

The selected fluids have been a viscous oil (newtonian) and a carbopol solution (pseudoplastic liquid).

Measurements of velocity components were carried out by hot film anemometry : the probe was rotating with the impeller. In every point in the tank, the three velocity components were obtained : it was shown that the axial velocity component was negligible. The radial and tangential velocity components were in good agreement with the ones obtained by the numerical procedure.

Then the power consumption P was determined by torque measurement on the shaft.

3. <u>RESULTS</u>

The numerical procedure provides very numerous results. Only few results will be presented and discussed here.

3.1   <u>Stream function</u>

Figure 2 shows streamlines for Re = 0.1 and a newtonian fluid. Three regions can be distinguished :

- the first one, within the line ψ = 0, is a region of eddies where the streamlines are closed on themselves. These eddies are rotating with the impeller shaft. Mixing within this region is not efficient,

- the second one is a region of circular streamlines, showing a tangential motion,

- the third one is a region of local eddies clung close to the blades and limited by the lines ψ = - 0.0251 for the first blade and ψ = -0.0538 for the second one.

3.2   <u>Vorticity</u>

Examples of variation of vorticity function ω versus radius r are shown on figure 3, for different Reynolds numbers and for a newtonian fluid. Characteristical values are observed at the impeller tips. They are maximum values at the internal tips and minimum values at the external tips. It can be noted that these tips are the highest shear points in the tank.

It is hard to correlate the influence of Reynolds number on the variations of the vorticity function. Generally, the flow remaining laminar, no variations of local parameters are observed.

3.3   <u>Velocity field</u>

The velocity field is presented on figures 4 and 5. Figure 4 shows the variations of the two velocity components v_θ and v_r versus the radius r for an angle θ equal to 6° ahead the blades. The tangential component has two local maximum values corresponding to the width of the blades. It can be noted that the maximum value (v_θ = 0.65) has decreased since the angle θ equal to 0°, where the angular velocity is equal to the blade velocity (v_θ = 0.83).

The variations of the radial component can be summarized by :

- the first blade is pushing the liquid till the shaft (v_r is negative),

- the second one is pushing the liquid till the wall (the maximum value of v_r around 0.3). A local pumping action is so characterized.

The variations of v_r are able to describe a secondary flow in all the tank. For instance, as the maximum value of v_r is increasing, the absolute minimum value of v_θ (near the wall) is increasing too.

Figure 5 presents the same variations as previously, but for an angle equal to 45° ahead the blades. The variations are no more influenced by the location of the blades.

Only a maximum value is observed for v_θ and v_r is closed to zero.

3.4   Stresses

Figure 6 gives an example of shear stress $\tau_{r\theta}$ and normal stress τ_{rr} for Re = 0.1. This figure shows the variations of the stresses versus the radius r for an angle θ equal to 6° ahead the blades. The maximum values of $\tau_{r\theta}$ and τ_{rr} are respectively equal to 6.5 and 4.7 ; they are obtained in the vicinity of the radius corresponding to the second impeller tip. It has yet been noted that this region is the highest shear one.

For an angle θ equal to 45° ahead the blades, the stresses are varying between −0.5 and 1.5 (see figure 7). This example means that the bulk is not a stressed region.

3.5   Dissipation function

The dissipation function is an index of the overheating in the tank due to shear rates. It is obvious that the tips are very sensible regions : figure 8 shows that the overheating is at these points more than 500 times the overheating in the other parts of the tank. This fact has to be known when products sensible to thermic effects are agitated.

3.6   Power consumption

For low values of Reynolds numbers, the product NeRe is a constant depending only on the geometry and flow behaviour index. Figure 9 presents the results obtained in that work for a newtonian fluid. A rather good agreement is observed between experimental and numerical values.

These results are conveniently correlated by the following equation :

$$NeRe = 169 \tag{17}$$

For a pseudoplastic fluid, the results are correlated by,

$$NeRe = 169(1 + 3.05\,\lambda^2)^{\frac{n-1}{2}} \tag{18}$$

Both equations (17) and (18) are very useful for the chemical engineer. They are only valid for the impeller described on figure 1. For other geometries, the results obtained in this work show that for a newtonian fluid, the product NeRe is given by :

$$NeRe = \frac{65}{\frac{D}{T} - (\frac{D}{T})^3} \tag{19}$$

4. CONCLUSIONS

By a convenient numerical procedure, very detailed results are obtained on the flow generated by a gate-agitator agitating newtonian or pseudoplastic liquids.

Local results as velocities field, stresses field, viscous dissipation have been obtained ; global results as power consumption have been obtained too.

Unfortunately these results cannot be compared with literature data and there are no other references. Indeed this impeller was not studied although it is industrially used, particularly in food industry for the production of melt cheese.

This numerical procedure developed here is able to give very useful informations in order to design new gate-agitators (Ref. 3).

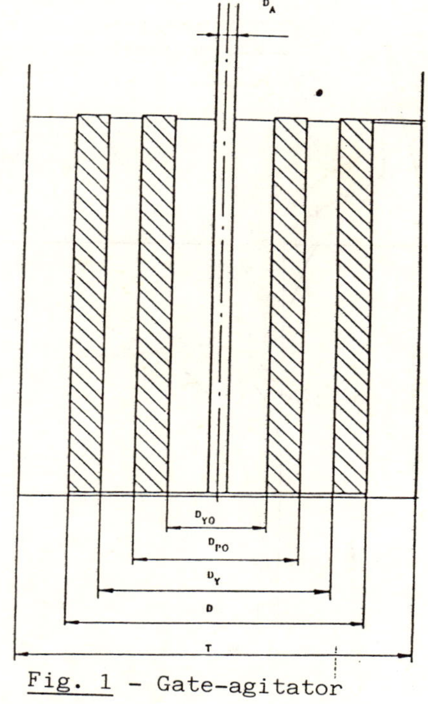

$D_A/T = 0.0417$

$D_{YO}/T = 0.361$

$D_{PO}/T = 0.509$

$D_Y/T = 0.681$

$D/T = 0.828$

Fig. 1 - Gate-agitator

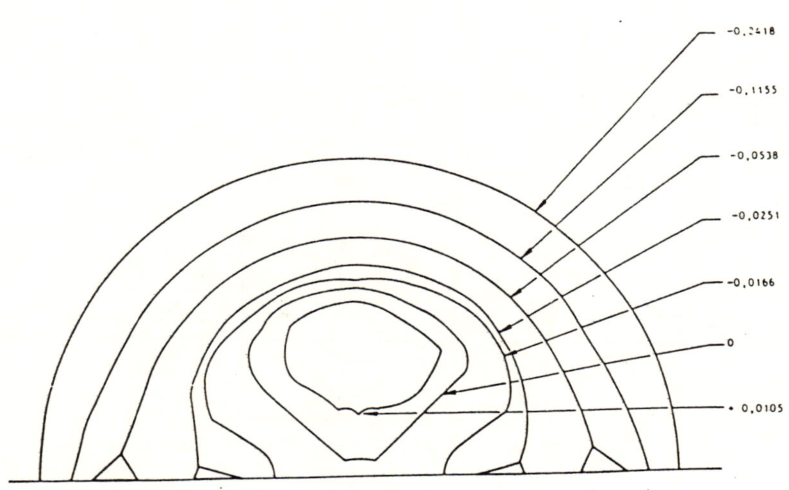

Fig. 2 - Streamlines for Re = 0.1 and n = 1

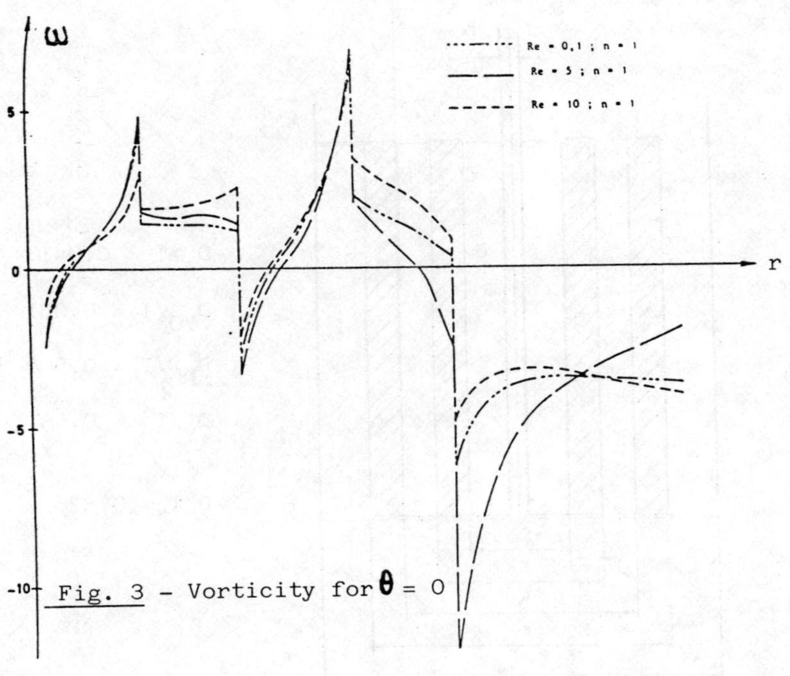

Fig. 3 - Vorticity for $\theta = 0$

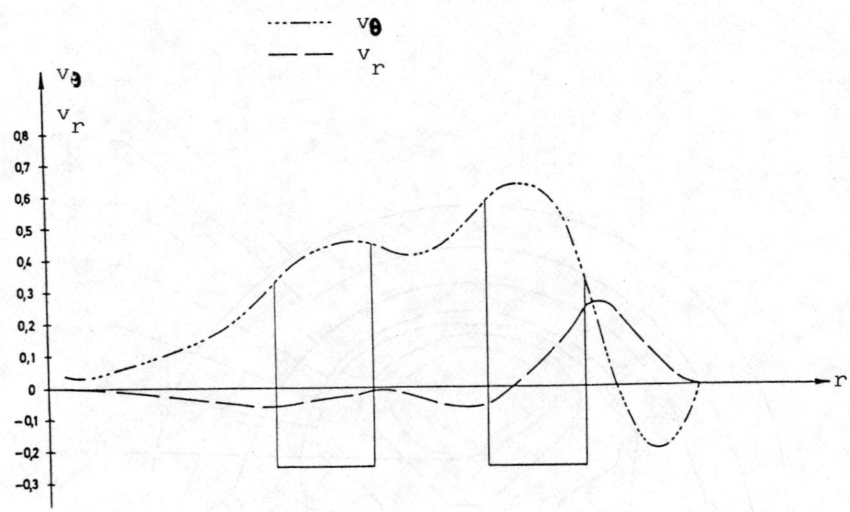

Fig. 4 - Radial and tangential velocities for Re = 0.1, n = 1 and $\theta = 6°$

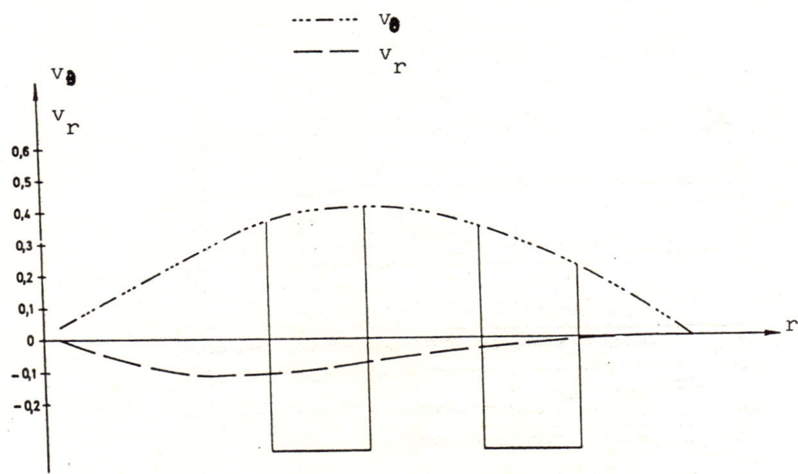

Fig. 5 – Radial and tangential velocities for Re = 0.1, n = 1 and θ = 45°

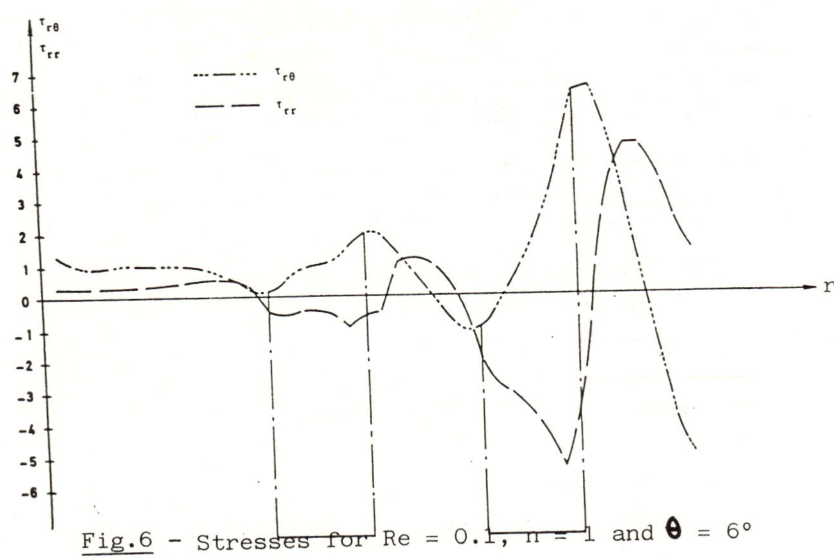

Fig.6 – Stresses for Re = 0.1, n = 1 and θ = 6°

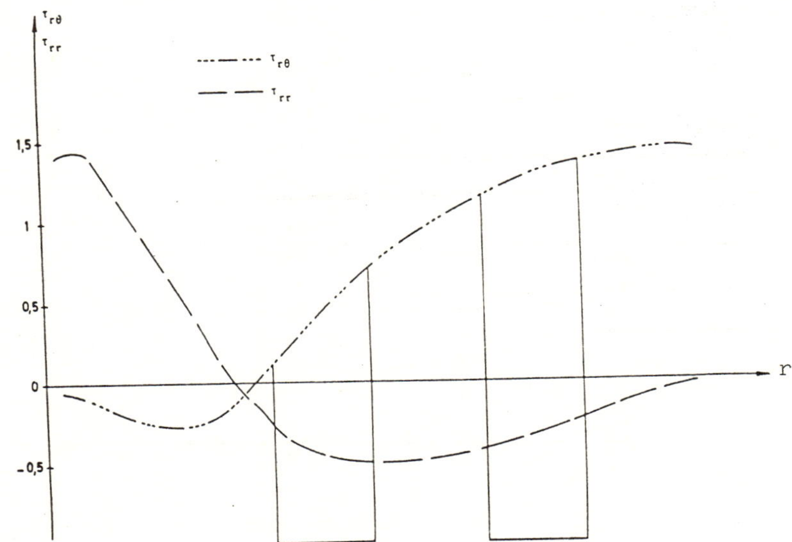

Fig. 7 – Stresses for Re = 0.1, n = 1 and θ = 45°

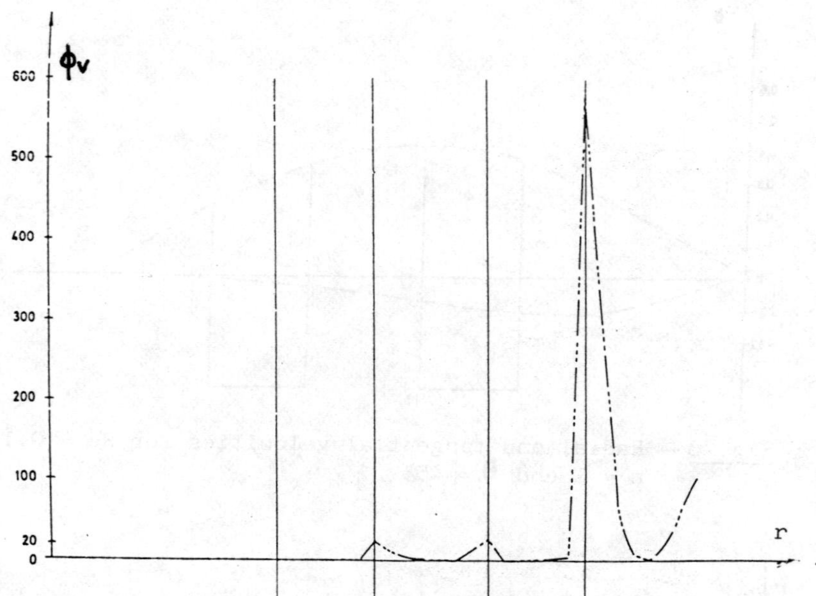

Fig. 8 - Viscous dissipation for Re = 0.1, n = 1 and θ = 0°

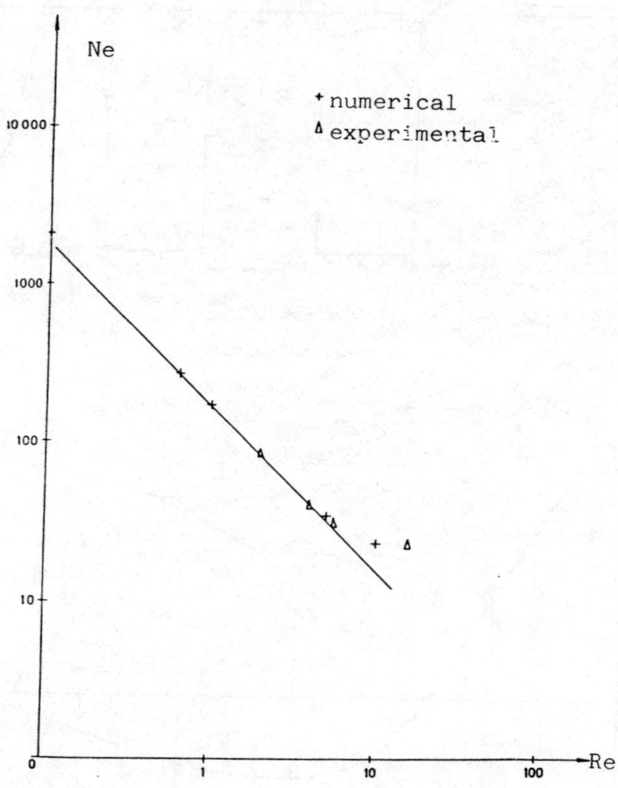

Fig. 9 - Power consumption

MIXING OF POWER LAW FLUIDS IN LOOP REACTORS

K.H.Tebel and P.Zehner

BASF Aktiengesellschaft
6700 Ludwigshafen
F R Germany

Summary

Numerous processes in chemical and bioengineering involve
fluids with non-Newtonian properties. Mainly there are polymer melts
and polymer solutions, and with increasing importance, fermentation
products which exhibit a complex non-Newtonian behaviour. The processing of such products in mixing apparatus or reactors, as for
instance stirred tanks or loop reactors, is often dominated by viscous flow regimes. Therefore it is necessary to determine a shear
rate that is representative of the hydrodynamics in mixing processes.
Generally this average shear rate is different from that one recommended by Metzner and Otto in connection with power correlations.
Once the representative shear rate is known, the flow curve leads to
a representative viscosity. With that the Newtonian mixing characteristic can applicated to both Newtonian and non-Newtonian fluids and
scale up rules can be developed.

A detailed description of the hydrodynamics of pseudoplastic
non-Newtonian fluids in loop reactors that leads to a new concept of
a representative viscosity for mixing processes will be presented.
Assuming the power law equation according to Ostwald-de Waele a representative shear rate can be predetermined theoretically. A comparison with known data from literature and with own measurements of
the liquid circulation flow rate in a loop reactor confirms the physical model.

Held at Wurzburg, 10-12 June, 1985.
Organised by DVCV· Deutsche Vereinigung für Chemie- und Verfahrenstechnik
(German Association of Chemical and Process Engineering).

Organisation: GVC·VDI-Gesellschaft Verfahrenstechnik und Chemieingenieurwesen.  GVC

MEANINGS OF THE SYMBOLS EMPLOYED

d	diameter
F	surface area
k	consistency factor
l	length
m	fluidity index
N	flow rate ratio
Re	Reynolds number
t	mixing time
v	velocity
Σ	resistance coefficient
$\dot{\gamma}$	shear rate
μ	diameter ratio
ϵ	proportionality factor
λ	coefficient of friction of the tube
η	viscosity
ρ	density

INDEX SYMBOLS

a	annular space
e	draft tube
h	hydraulic
t	driving jet

1. INTRODUCTION

The processing of polymer melts or polymer solutions, and to an increasing extent the processing of products derived from bioengineering processes, almost always creates flow conditions which are influenced by a stress-dependent viscosity. In order to be able to assess the consequences arising from this in a scale-up procedure, a comprehensive description of the hydrodynamics of non-Newtonian fluids in such apparatus is absolutely essential.

An analysis of the results known from the literature which relate to the mixing of non-Newtonian fluids shows that no consideration is given to the flow condition which is of decisive importance for mixing or for circulation. The critical resistance coefficient representing the flow conditions in the loop is usually correlated on a formal basis with the Reynolds number of the driving jet and not with the Reynolds number applicable to the flow in the loop. Although it is indeed the case that this procedure provides useful results as regards the turbulent and the laminar flow range of a Newtonian fluid, the procedure of extension to non-Newtonian fluids which is in general undertaken with the aid of the concept of the representative viscosity according to the Metzner and Otto resistance characteristic, as reproduced by way of example in Fig. 1, cannot result in a consistent description of the conditions. Indeed, detailed consideration of the field of flow shows that the stress, which is critical for mixing purposes, is to a great extent determined by the rate of circulation and the diameter of the reactor or the diameter of the circulation tube. On the basis of this finding, in order to describe the hydrodynamics of non-Newtonian fluids under conditions of ordered flow in a loop, the derivation and application of a new concept of a viscosity which is representative for this purpose are presented.

2. THEORY

The hydrodynamic properties of a jet loop reactor are characterized in that the pulse delivered by means of the driving jet causes an increase in the hydrostatic pressure in the draft tube [2], which under turbulent conditions substantially compensates for the deflection losses and under conditions of laminar flow provides similar compensation for the actual friction losses affecting the flow in the tube and the annular space. In addition to assuming that the velocity profile in the nozzle outlet cross-section is piston-shaped, it is in general also possible to assume that the ratio of the area of the draft tube to that of the driving jet nozzle will be large. The association of these assumptions with the pulse energy balance leads to equation (1). The resistance coefficient which is critical for the purposes of the flow conditions in the loop is not broken down any further and is, in the form of an appropriate mean value, associated with the mean velocity in the draft tube. Consideration of the definitive equation (2) for the flow rate ratio leads to the necessity of determining the resistance characteristic of the flow taking place in the loop within the laminar flow range in accordance with equation (3), due consideration being given to a suitable representative viscosity. If it is assumed that the flow properties are those of a non-Newtonian fluid and can be described with reference to the Ostwald-de Waele power law and the pertinent representative shear rate according to equation (4), then the final result is the descriptive equation (5) which is generally valid for Newtonian and non-Newtonian fluids. In this connection, the representative shear rate is determined on the basis of the illustrative principle that the characteristic of a Newtonian fluid and that of a non-Newtonian fluid are identical. Comparison of this theoretical analysis with the flow rates determined experimentally by Marquardt [1] using aqueous solutions of CMC and glucose leads to the graphical representation shown in Fig. 3. Furthermore, equation (5) leads to the result that the concept which is customary in the literature cannot lead to bunching of the family of curves for different fluidity indices in dependence upon a Reynolds number determined with reference to the characteristic data of the driving jet.

In the case of non-Newtonian fluids, it is observed that as the fluidity index decreases the rises in the laminar characteristic increase. Since, in addition to this, the coefficient also changes on the basis of the corresponding root function, a descriptive concept based on the Reynolds number applicable to the flow conditions in the loop is in general recommended.

If the flow losses following application of the pressure causing flow in the loop are reduced to pure friction losses, and deflection and inlet losses are ignored, then it is possible to formulate explicitly the representative shear stresses in the draft tube and in the annular space. As has been stated on several occasions in the literature [3], the equating of the Newtonian and non-Newtonian resistance characteristic in accordance with equations (6) to (8) with due consideration of the known pressure drop relation (9) applicable to the laminar flow, within the confines of a tube, of a non-Newtonian fluid with a constant fluidity index gives rise to the representative shear rate evident from equation (10). On this basis, any additional dependence upon the fluidity index can be disregarded without the introduction of large errors. On the basis of equations (11) to (13), and with the use of an explicit approximate equation (14) for the pertinent pressure drop [4], analogous application to annular flow gives rise to a comparable expression (15) for the representative shear rate in the annular space.

The deviation from the value applicable to the draft tube is restricted, within a wide range of the fluidity index, to a factor dependent on the ratio of diameters. As can be seen from Fig. 5, the two shear stress values are of the same order of magnitude in the case of diameter ratios which are applicable for industrial purposes. The same is approximately true for the ratio of the two pertinent Reynolds numbers which are applicable for the determination of the circulation characteristic. Moreover, comparison of the shear rate derived from the approximate equation (14) with the solution presented by Hoffmann [5] for the integral equation gives a high level of agreement.

In summary, it can be stated that these considerations give rise to the simplified principle that the viscosity which is critical for the purposes of the Reynolds number applicable to flow in the loop should be determined from the flow curve on the basis of the known representative shear rate within the draft tube.

3. EXPERIMENT

In order to assess the errors associated with these assumptions in relation to the reproduction of experimental results, the volume flow rates of fluid circulating in a jet loop reactor were measured under different operating conditions and with the use of materials with different properties. Fig. 5 shows the principal geometric dimensions, and Fig. 6 shows the flow curves which were produced. In contrast to the CMC solutions which were used for measurement purposes by Marquardt [1] and which show a constant fluidity index within the pertinent range of shear rate values, the HEC solution which was investigated in the course of our own experiment is characterized by a maintained zero viscosity value. The zero viscosity range for the Luviskol solution extends over several decades. Since the properties of slight elasticity which are also present are of subsidiary importance under conditions of laminar flow in a

loop, the result may immediately be classified as following the Newtonian characteristic determined on the basis of the glycerin solutions. Additional measurements carried out with pure water show the limiting case for turbulence with a constant flow rate ratio. Fig. 7 shows the closed form of the circulation flow characteristic determined with the aid of the newly developed concept of a representative viscosity for flow within a loop. The observed deviations amount to ±10% are in accord with the accuracy of measurement which can be achieved; the latter is essentially determined by the optical determination of the flow rate profile within the draft tube.

4. CONCLUSION

On the basis of the new concept, and with the application of the representative viscosity applicable to the conditions of flow within a loop, the mixing of Newtonian and non-Newtonian fluids can be described on a common basis. As is indicated in [6], this approach can be extended without difficulty to the flow rate ratios found in propeller loop reactors and to the dimensionless mixing time characteristics determined with the use of conventional mixing units. An appropriate assessment with reference to suitable experimental investigations will be undertaken in a subsequent paper.

Furthermore, it is now for example possible to state explicitly the effect of various hydrodynamic boundary conditions, such as constant driving jet velocity, constant driving jet Reynolds number or constant, volume-specific energy input, in the course of a scale-up procedure involving complete geometric similarity, on the mixing time in dependence upon the non-Newtonian viscosity. On the assumption that the fluidity index is constant over the entire range under consideration, the result shown in Fig. 8 is obtained for jet loop reactors. The transfer (scale-up) ratios applicable to the condition of turbulent flow are also given, for comparison purposes. In principle, of course, any change in the condition of flow in any individual case occurring as a result of the scale-up procedure is to be determined independently.

REFERENCES

[1] Marquardt, R.: Verfahrenstechnik 13 (1979) 6, 527

[2] Zehner, P.: Chem.-Ing.-Techn. 52 (1980) 11, 910

[3] Schümmer, P.: Chem.-Ing.-Tech. 41 (1969) 21, 1021

[4] Mishra, P.; Mishra, I: AIChE J. 22 (1976) 3, 617

[5] Hoffmann, L.; Schümmer, P.; Schwerdt, H.: Rheol. Acta 14 (1975) 7, 626

[6] Tebel, K.H.; Zehner, P.: Chem.-Ing.-Techn. 57 (1985) 1, 49

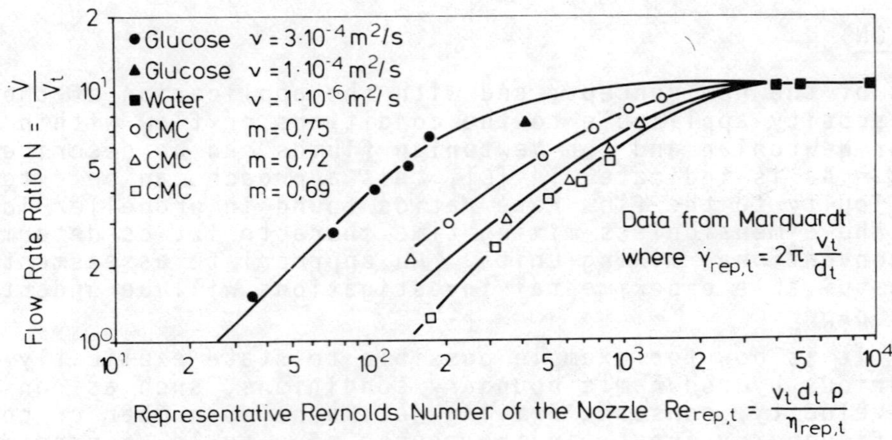

Fig. 1:   Conventional form of representation of the flow rate ratio as a function of the Reynolds number of the nozzle

$$\rho\, v_t^2\, F_t \cong \Delta p\, F_e \qquad \text{where} \qquad \Delta p = \zeta\, \frac{\rho}{2}\, v_e^2 \qquad (1)$$

$$N = \sqrt{\frac{2}{\zeta}\, \frac{F_e}{F_t}} \qquad \text{where} \qquad N = \frac{v_e\, F_e}{v_t\, F_t} \qquad (2)$$

$$\zeta = \frac{C}{Re_{rep}} \qquad \text{where} \qquad Re_{rep} = \frac{v_e\, d_e\, \rho}{\eta_{rep}} \qquad (3)$$

$$\eta_{rep} = k\, \dot{\gamma}_{rep}^{\,m-1} \qquad \text{where} \qquad \dot{\gamma}_{rep} = \varepsilon\, \frac{v_e}{d_e} \qquad (4)$$

$$N = \frac{2}{C}\, \frac{F_e}{F_t}\, Re_{rep}^{1/2} = \left(\frac{d_e}{d_t}\right)^{3m} \sqrt{\frac{2}{C}\, \frac{F_t}{F_e}}\; Re_{rep,t}^{1/m} \qquad (5)$$

Fig. 2:   Equations describing the hydrodynamic properties of the system

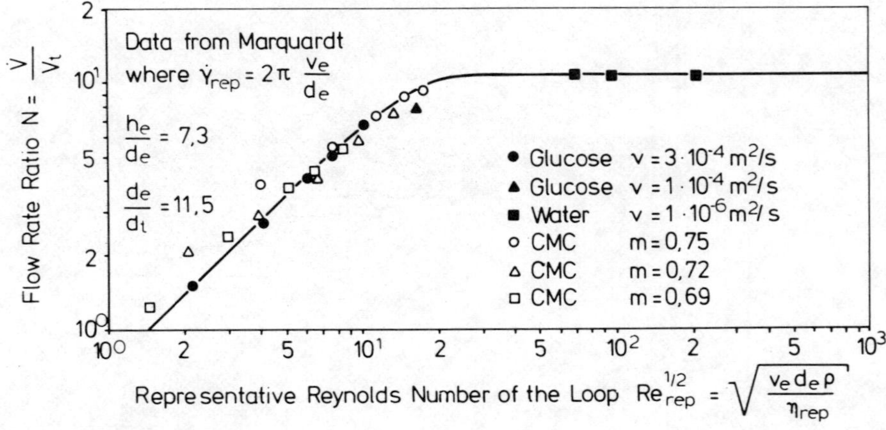

Fig. 3:   New form of representation of the flow rate ratio as a function of the Reynolds number of the loop

DRAFT TUBE

$$\lambda_e^{N} = \lambda_e^{NN} \qquad (6)$$

$$\lambda^{N,NN} = \frac{\Delta p_e}{\frac{\rho}{2} v_e^2} \frac{d_e}{l_e} \qquad (7)$$

$$\lambda_e^{NN} = \frac{64}{Re_{rep,e}} \qquad (8)$$

$$\Delta p_e^{NN} = 4k \frac{l_e}{d_e} \left(\frac{v_e}{d_e}\right)^m f_e(m) \qquad (9)$$

ANNULAR SPACE

$$\lambda_a^{N} = \lambda_a^{NN} \qquad (11)$$

$$\lambda_a^{N,NN} = \frac{\Delta p_a}{\frac{\rho}{2} v_a^2} \frac{d_{h,a}}{l_a} \qquad (12)$$

$$\lambda_a^{NN} = g(\varkappa) \frac{64}{Re_{rep,a}} \qquad \varkappa = \frac{d_e}{d_a} \qquad (13)$$

$$\Delta p_a^{NN} = 4k \frac{l_a}{d_a} \left(\frac{v_a}{d_{h,a}}\right)^m f_a(\varkappa, m) \qquad (14)$$

$$\dot{\gamma}_{rep,e} = 2\pi \frac{v_e}{d_e} \qquad (10) \qquad 0{,}1 \leq m \leq 1$$

$$\dot{\gamma}_{rep,a} = 2\pi \frac{v_e}{d_e} \cdot 0{,}005 \cdot e^{9{,}2\varkappa} \qquad (15) \qquad 0{,}1 \leq m \leq 1 \quad 0{,}4 \leq \varkappa \leq 0{,}7$$

Fig. 4: Concept of the representative viscosity of the fluid flowing in the loop

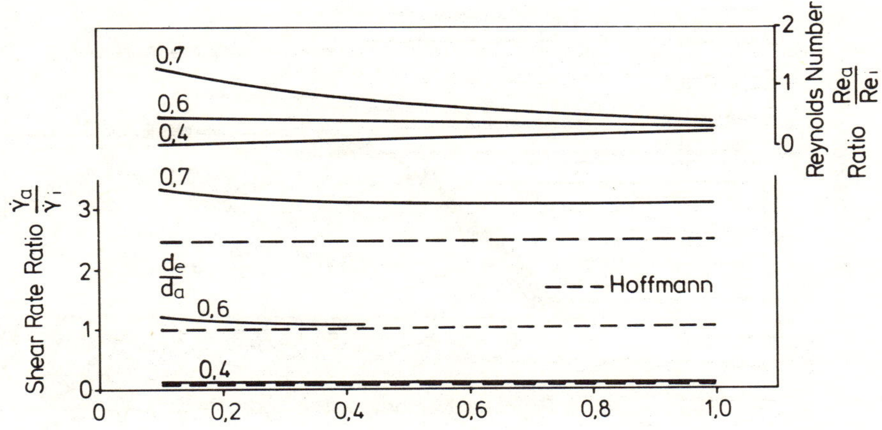

Fig. 5: Shear stress and Reynolds number within the draft tube and within the annular space in the case of non-Newtonian fluids

JET LOOP REACTOR DESIGN

l_a = 1380 mm	d_a = 140 mm	l_a/d_a = 9,9
l_e = 1300 mm	d_e = 90 mm	l_d/d_e = 14,4
a_o = 49 mm		a_o/d_a = 0,35
a_u = 31 mm		a_u/d_a = 0,22
a_t = 18 mm	d_t = 10 mm	d_t/d_e = 0,11
s = 5 mm		

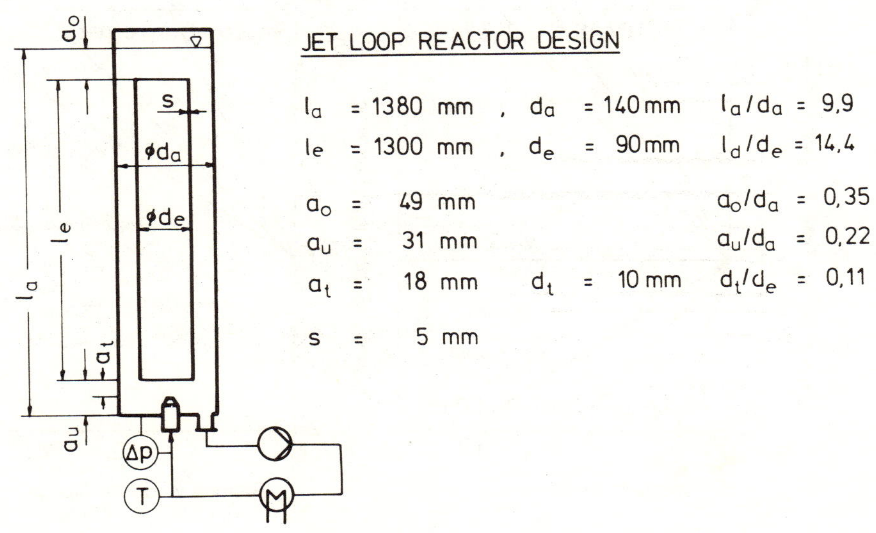

Fig. 6: Principal dimensions of the jet loop reactor

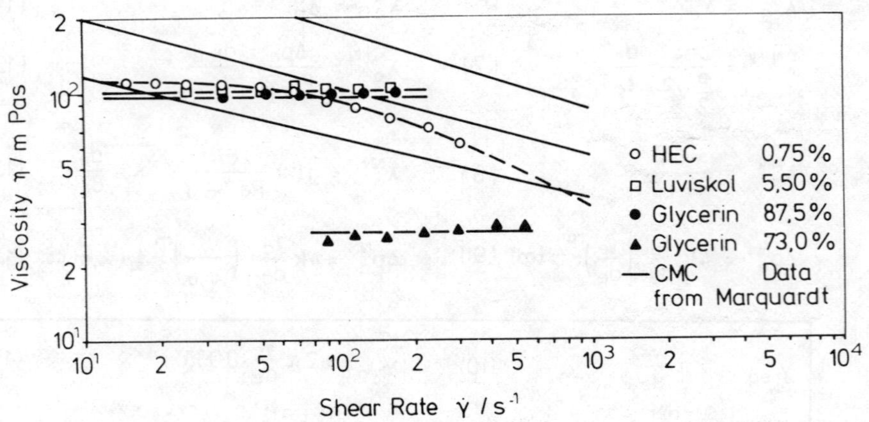

Fig. 7:   Flow curves for aqueous solutions

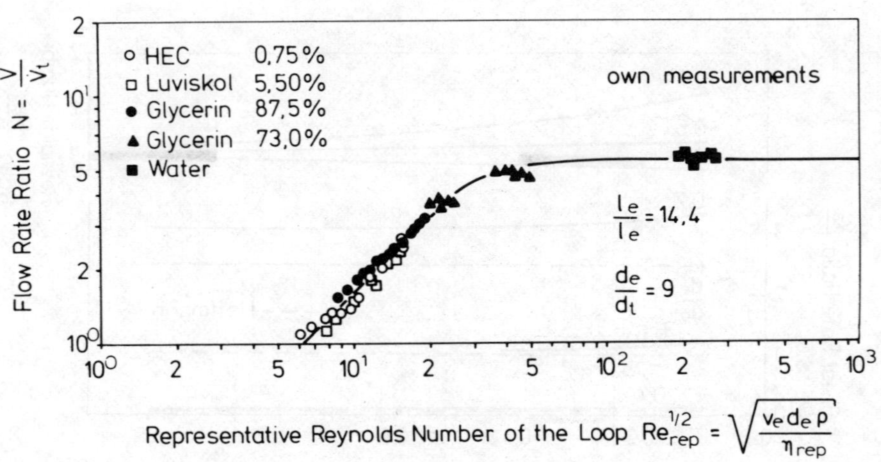

Fig. 8:   Results of the measurement of the flow rate ratio

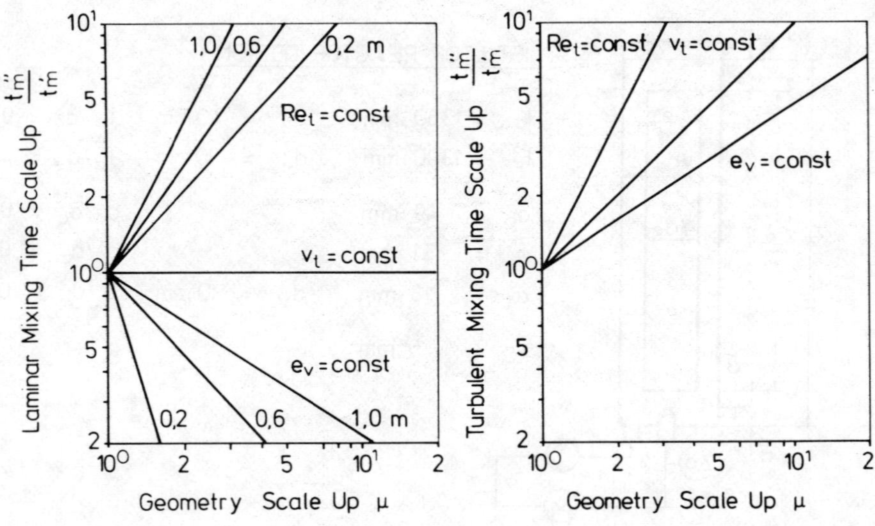

Fig. 9:   Effect of the scale-up procedure on the mixing time

DISSIPATION OF POWER IN STIRRED VESSELS

H.D. Laufhütte, A.B. Mersmann
Lehrstuhl B für Verfahrenstechnik
Arcisstr. 21, D-8000 München 2

Summary

The local energy dissipation rates of a stirred fluid found in the literature are based on insufficient measuring techniques. In this paper the laser-Doppler-velocimetry is used to measure the velocity field. It is shown that only the turbulent fluctuating velocity is to take into account to determine the local energy dissipation in the turbulent flow range. The fields of the fluctuating velocities and of the energy dissipation rate in a stirred vessel equipped with a Rushton turbine are discussed.

Held at Wurzburg, 10-12 June, 1985.

Organised by DVCV· Deutsche Vereinigung für Chemie- und Verfahrenstechnik
(German Association of Chemical and Process Engineering).

Organisation: GVC·VDI-Gesellschaft Verfahrenstechnik und Chemieingenieurwesen.  GVC

Nomenclature:

C_1, C_2, C_3 constants

T inside diameter of stirred tank

D_1 overall agitator diameter

d_{32} Sauter-mean-diameter

$E(f)$
$E(k)$ one-dimensional-energy spectrum

f frequency

g acceleration due to gravity

k wave number

L typical length

n agitator speed

P mixing power

r radius

u fluid velocity

$u' \triangleq$ RMS   fluctuating velocity (Root-Mean-Square-value)

V volume

z axial distance

Dimensionless numbers:

$Re = \dfrac{n \cdot D_1^2}{\nu}$ Reynolds number

$Ne = \dfrac{P}{\rho_c \, n^3 \, D_1^5}$ Newton number

Greek symbols:

$\gamma(f)$ standard energy spectrum

ε energy dissipation rate per unit mass

λ_f, λ_f' micro scale of turbulence

Λ_f macro scale of turbulence

μ, μ^* friction factors

ν kinematic viscosity

ρ density

σ surface tension

φ tangential coordinate

Subscripts:

c continuous phase

l laminar

per   periodic

t turbulent

tot   total

z z-coordinate

"——"   averaged value of space or time

1. Introduction

Processes in stirred vessels are influenced strongly by the distribution of the local energy dissipation rates. According to the turbulence theory of Kolmogorov the Sauter-mean-diameter of drops can be calculated from

$$d_{32} \sim \frac{\sigma^{0.6}}{\rho_c^{0.6} \, \varepsilon^{0.4}} \tag{1}$$

in the case of a dispersion or emulsification process [1]. Other research activities stress the micromixing process. Different micromixing models are based on a known ε-field, see for instance Bourne [2].

2. State of research

In the literature different statements are made concerning the distribution of local energy dissipation rates. Fig. 1 gives an overlook of the somewhat contradictory statements about the local energy dissipation in the outflow region of a Rushton turbine. In all cases the flow was fully turbulent. In the vicinity of the stirrer there are differences of the ratios $\varepsilon/\bar{\varepsilon}$ in the range of the factor 70. In the following chapter the measuring techniques and the theories employed by the various authors will be discussed briefly.

Cutter [3] determined the velocity of the flow by a photographic method. He calculated the kinetic energy contents of the flow of radial sectors from the mean velocities and the fluctuating velocities according to

$$-\frac{dE}{dr} = 2 \, \pi r \int_{-\infty}^{+\infty} \varepsilon \, dz$$

$$= \frac{2 \, \pi \, dr}{dr} \int_{0}^{\infty} ([u_r'^2 + u_z'^2 + u_\varphi'^2 + \bar{u}_r^2 + \bar{u}_z^2 + \bar{u}_\varphi^2]2 \cdot u_r$$

$$+ 2 \, \bar{u}_\varphi \cdot \overline{u_z' \, u_\varphi'}) \, dz \tag{2}$$

Because of the wide scattering of the data of the fluid velocities the statements of Cutter were criticized by Günkel and Weber [4].

The equation of Liepe [5] for the distribution of the local energy dissipation rate is also based on measurements of the fluid velocities. This author employed a hot film anemometer. He calculated the local energy dissipation rate by

$$\varepsilon = 30 \, \nu \, \frac{\bar{u}'^2}{\lambda_f^2} \tag{3}$$

with the micro scale λ_f of turbulence according to

$$\lambda_f = \frac{\bar{u}}{2 \, \pi} \cdot \frac{1}{(\int_0^\infty f^2 \, \gamma(f) \, df)^{0.5}} \tag{4}$$

Komasawa [6] used a high-speed cine-camera to measure the fluctuating velocities and calculated an one-dimensional energy spectrum of these fluctuating velocities. Equ. (3) was used for the calculation of the ε distribution with the micro scale λ_f of turbulence according to the following equation different from equ. (4)

$$\lambda'_f = \frac{\bar{u}}{\sqrt{2} \, \pi} \frac{1}{(\int_0^\infty f^2 \, \gamma(f) \, df)^{0.5}} \tag{5}$$

The statements of Okamoto et al. [7] are based on measurements carried out with hot film anemometers.

In the equation

$$\varepsilon_{measured} = 15 \, \nu \int_0^\infty k^2 \, E(k) \, dk \tag{6}$$

the wave number is calculated from the turbulence frequency according to

$$k = \frac{2 \, \pi \, f}{\bar{u}}$$

Due to a recommendation of Lumley [8] values drawn from equ. (6) will be
corrected by the relationship

$$\varepsilon = \frac{\varepsilon_{measured}}{(1 + 5\ u'^2/\overline{u}^2)} \tag{7}$$

It is assumed that this correction takes into account the anisotropy of
turbulence.

Last not least Barthole et al. [9] measured the flow velocities with a
hot film anemometer and determined the energy distribution using the obtained data. The local energy dissipation rate ε was calculated from

$$\varepsilon = 30\,\nu \int_0^\infty k_1^2\ E(k)\ dk \tag{8}$$

Besides the different theoretical equations for the calculation of the
ε-field also the measuring techniques often not exact enough for this
purpose are responsible for different and partly contradictory statements
concerning the distribution of the local energy dissipation rates.

3. Experimental set-up

The following statements are based on flow measurements carried out with
a laser-Doppler-velocimeter. Information about the set-up is given in
literature [10]. The measurements were carried out in a standard agitated
vessel equipped with a Rushton turbine and four baffles. The diameter of
the vessel is 0.19 m. Water, aqueous solutions of glucose and mixtures
of glycerol and water were used as liquids. By means of a thermostat the
temperature of the vessel was hold constant and by this way the kinematic viscosity was constant, too.

4. Laminar and turbulent flow

In the case of an one-phase flow in geometrically similar vessels the
power consumption P of a stirrer depends on

$$P = f(D_1,\ n,\ \nu,\ \rho) \quad or \quad P = Ne\ \rho\ n^3\ D_1^5 \tag{9}$$

and is independant of the gravity acceleration g when the vortex does
not reach the agitator. Fig. 2 shows the relationship between the Newton
number and the stirrer Reynolds number for the vessel investigated by us.
For the laminar flow region (Re < 10) the Newton number is inversely proportional to the Reynolds number. The Newton number remains constant in
the turbulent range (Re > 10 000). In this chapter we will try a physical
explanation of the whole Ne = f(Re)-curve. For this purpose we separate
the Newton number in a laminar and a turbulent contribution (Index 1 and
t rsp.):

$$Ne = Ne_1 + Ne_t \tag{10}$$

According to the calculation of the power consumption in a pipe flow the
following expression is suggested for the laminar contribution:

$$P_1 = \mu^*\ T \cdot T \cdot \rho \cdot \overline{u}_z^3 \tag{11}$$

In this equation $\overline{u}_z$ is the mean axial velocity in the whole vessel and
can be determined from measured velocity profiles. μ^* is a factor like
the friction factor of a pipe flow. Since in the laminar range the Newton number of a stirred vessel as well as the friction factor of a pipe
flow are inversely proportional to the Reynolds number we obtain

$$Ne_1 = \mu_1 \cdot \left(\frac{T}{D_1}\right)^2 \cdot \left(\frac{\overline{u}_z}{\pi n D}\right)^3 \cdot \frac{1}{Re} \tag{12}$$

The turbulent contribution can be described by theoretical results found
for the turbulent flow by Taylor [11]

$$\varepsilon = \frac{P}{\rho V} = 15\ \nu\ \frac{u'^2}{\lambda_f^2} \tag{13}$$

According to this author a relationship exists between the micro scale λ_f and macro scale Λ_f of turbulence

$$\lambda_f \sim \sqrt{\Lambda_f \, \nu/u'} \tag{14}$$

In a stirred vessel operating in the turbulent range the macro scale Λ_f is proportional to the impeller producing turbulence

$$\Lambda_f = C_1 \cdot D_1 \tag{15}$$

Assuming that the filling height H of the vessel is equal to the tank diameter T we obtain

$$Ne_t = \mu_t \left(\frac{T}{D_1}\right)^3 \left(\frac{\overline{u'}}{\pi n D_1}\right)^3 \tag{16}$$

In this equation $\overline{u'}$ is the mean value of the fluctuating velocity for the whole vessel and can be calculated from the measured velocity profiles. The mean velocity as well as the fluctuating velocity is written in dimensionless form in equs. (11) and (16) since both velocities are proportional to the circumferential velocity of the stirrer. Now it is possible to describe the Ne = f(Re)-curve with the help of equs. (10), (12) and (16). The friction factors $\mu^* = \mu_1 \cdot Re = 3 \cdot 10^6$ and $\mu_t = 223$ can be found by adjustment of these parameters to the experimental results. At low Reynolds number the Newton number is proportional to 1/Re. The flattening of the curve for Re > 10 can be explained by an increase of $\overline{u}_z$. In this range the flow is spreading over the whole liquid within the vessel. For Re > 10 000 flow takes place all over the tank and as a consequence $\overline{u}_z$ is proportional to the circumferential velocity of the impeller. Then the contribution Ne_1 can be neglected. The contribution Ne_t effects power consumption in the range Re > 800 when the turbulent fluctuatings are spreading and remains constant for Re > 10 000. According to this model developped here it can be seen that the mean velocity and influences of fluid friction contribute only the energy dissipation to a neglecting degree in the turbulent range. As a consequence it is sufficient to take only into account the fluctuating velocities according to the equs. (13), (14) and (15).

5. Determination of the fluctuating velocities in a stirred vessel with a Rushton turbine

As mentioned already various measuring techniques deliver different fluctuating velocities. However the calculation of the distribution of energy dissipation essentially depends on exact measurements of the fluctuating velocities since we have the relationship $\varepsilon \sim (u')^3$.
Fig. 3 shows a comparison of fluctuating velocities (as Root-Mean-Square-values) determined with different measuring equipment. These results are all valid for the discharge flow of a Rushton turbine. The data of Liepe [5] who used a hot film anemometer are smaller than the results of other authors.
Our own data obtained with laser-Doppler-velocimetry (LDV) agree very well with the LDV-values of Reed et al. [12]. Fig. 3 contains axial component velocities u_z' as well as radial component velocities u'_r. Data measured by Cutter [3] using a photographic method are lying above our points and are the largest found in literature. Taking into account an insufficient resolution of photographic methods and the disturbance of the flow using a probe it seems that laser-Doppler-velocimetry is the most objective method. As a consequence further calculations of the energy dissipation rates should be based on fluctuating velocities obtained by this method.
Fig. 4 shows the distribution of the axial fluctuating velocities in the whole vessel. We have shown (see [10]) that the field of turbulence with the exception of the vicinity of the stirrer is nearly isotropic. In the neighbourhood of the impeller blade there are high RMS-values u'_z. According to van't Riet et al. [13, 14] and Günkel, Weber [4] these values are induced by periodic fluctuations caused by the passing of the im-

peller blades which move relatively to a fixed measuring volume. In this
region the fluctuating velocity u' must be separated in a periodic contribution u'_{per} and a turbulent part u'_t which does not correlate with
u'_{per}:

$$u'_{tot} = u'_{per} + u'_t \tag{17}$$

Strictly speaking the statements of the turbulence theory can only
applied on the statistical value u'_t. This was recognized by various
authors. So Günkel, Weber [4] and van der Molen, van Maanen [16] suggest
to cut the peaks of the energy spectrum E(k) for the purpose to obtain
the statistical turbulent velocity u'_t without periodic contributions.
For illustration fig. 5 shows the energy spectrum E(k) versus wave number
for different distances of the measuring volume from the blade according
to Möckel [15]. However van't Riet et al. [14] point out that the periodic contribution does not only occur in the peaks at the one- and two-
fold stirrer frequency but can be detected up to the tenfold frequency.
As a result of this the whole energy spectrum is shifted continuously to
higher frequencies.
In the case of probes (hot film anemometer etc.) it is possible to let
rotate the probe with the stirrer. In this way only statistical fluctuations are measured but these data are only valid for rotating coordinates and cannot be simply compared with values obtained at a fixed point.
Statements of Bertrand et al.[17] are only qualitative since there are
contradictions between the measured u'-values in [17] to those in later
publications [18] of the authors.
There is another possibility to evaluate the periodic contribution: It is
assumed that u'_{per} is proportional to the circumferential stirrer velocity since the periodic contribution is induced by the passage of the
stirrer:

$$u'_{per} \sim C_2 \cdot n \cdot D_1 \tag{18}$$

This may be valid in the whole Reynolds number range, see fig. 2.
The fluctuating velocities occuring in the laminar range in the vicinity
of the stirrer are identical with the periodic contribution. Equ. (18)
then allows to extrapolate this velocity u'_t into the turbulent range.
Corresponding calculations were carried out, and the results are plotted
in fig. 6. This figure shows measured values u' (here dimensionless as
$u'/(\pi nd)$ or $RMS/(\pi nd)$) depending on the axial distance z from the stirrer
as well as the periodic contribution u'_{per} (as $u'_{per}/(\pi nd)$ or $RMS/(\pi nd)$).
All data are valid for the immidiate vicinity of the stirrer ($2r/T = 0.39$
or $2r/D_1 = 1.18$).
According to this evaluation the periodic contribution near the stirrer
amounts to a maximum of 30 % of the whole value and decreases rapidly
with the radius. In the region $2r/T > 0.6$ a noticable periodic contribution was not detectable. Our evaluation is supported by the fact that
according to fig. 5 for $2r/T > 0.6$ there are no peaks indicating a periodic contribution.

6. Distribution of local energy dissipation rates in stirred vessel with Rushton turbine

Equs. (13), (14) and (15) deliver the relationship

$$\varepsilon = C_3 \cdot \frac{u'^3}{D_1} \tag{19}$$

The constant C_3 must be calculated for the purpose to determine the
ε-field.
Remembering the statements in chapter 4 roughly the total power input is
dissipated in turbulent flow of the fluid. Assuming this the constant C_3
can be determined from an integral power balance. For this purpose the
total volume of the vessel is divided into part volumina of equal energy
dissipation. On the other hand the total power input of the whole system
is determined.

Now we will distinguish between two cases:
a) For the immediate vicinity of the stirrer the total measured fluctuating velocity u' is inserted in the power balance. By this way we obtain

$$\varepsilon_1 = 6.0 \cdot \frac{u'^3_{tot}}{D_1} \qquad (20a)$$

b) Only the turbulent fluctuating velocity u'_t is taken into account as was explained in chapter 5. Then we obtain

$$\varepsilon_2 = 6.5 \cdot \frac{u_t'^3}{D_1} \qquad (20b)$$

In fig. 7 lines of equal ratios $\varepsilon_2/\overline{\varepsilon}$ - so-called isoenergetic lines - are plotted.

The differences of the constants C_3 is smaller than 10 % but the two models deliver different results especially in the discharge flow near the stirrer blades as can be seen from fig. 8. In this figure the specific power input ratio $\varepsilon/\overline{\varepsilon}$ is plotted versus the dimensionless radius. Corresponding to equation (20b) the local energy dissipation rate increases to a value of $\varepsilon/\overline{\varepsilon} \approx 15$ at the dimensionless radius $2r/T = 0.5$. Then the dimensionless energy dissipation rate decreases rapidly. Considering that the fluid is not rotating with the circumferential velocity of the stirrer it's to assume that near the stirrer blades the periodical part of the fluctuating velocities becomes important. In equation (20a) this periodical part is treated as "pseudo turbulence". Accordingly, the graph of equation (20a) in fig. 8 is to interpret as the extremity of the local energy dissipation rate.
The question arises if the constant C_3 is universal. I.e. is there an influence of the geometrical configuration on the constant C_3 in equations (20a, 20b)? In this context a theoretical estimation of C_3 by Brodkey [19] is helpful. The following equation for ε based on some simplifications was obtained

$$\varepsilon = (3.68 \div 4.4) \, \frac{u'^3}{L} \qquad (21a)$$

In the case of a stirred vessel Brodkey obtains independently of the type of the stirrer with $L = D_1/2$

$$\varepsilon = (7.36 \div 8.8) \, \frac{u'^3}{D_1} \qquad (21b)$$

With regard to the simplification of this model the agreement of the constants both experimentally and theoretically determined is satisfactory. Accordingly, the constant C_3 in equation (20a,b) seems to be independent of the type and the geometry of the stirrer.
Consequently, we succeeded in determining the local energy dissipation rate in stirred vessels by an useful equation.
The distribution of the local energy dissipation rate in a stirred vessel equipped with a Rushton turbine was discussed in detail. Based on these results a solution of more complex mixing tasks should be possible.

7. References

[1] Grassmann, P.: Physik. Grundlagen der Verfahrenstechnik. 3. Aufl., Verlag Sauerländer, Aarau (1983) p. 692

[2] Baldyga, J.; Bourne, J.R.: Chem. Eng. Commun. 28 (1984) p. 231-280

[3] Cutter, L.: AIChE Journal 12 (1966) p. 35-45

[4] Günkel, F.; Weber, M.: AIChE Journal 21 (1975) 5, p. 931-948

[5] Liepe, F.; Möckel, H.O.; Winkler, H.: Chem. Techn. 23 (1971) 4/5 p. 231-237

[6] Komasawa, I.; Kuboi, R.; Otake, R.: Chem. Eng. Sci. 29 (1974) p. 641-650

[7] Okamoto, Y.; Nishikawa, M.; Hashimoto, K.: Int. Chem. Eng. 21 (1981) p. 88-94

[8] Lumley, J.L.: Phys. Fluids 8 (1965) p. 1059

[9] Barthole, J.; Maisonneuve, J.; Gence, J.; David, R.: Chem. Eng. Fund. 1 (1983) p. 17-26

[10] Laufhütte, H.D.; Mersmann, A.: Chem. Ing. Tech. 11 (1984) p. 862-863

[11] Taylor, G.I.: Proceedings of the Royal Society A 151 (1935) p. 421-428

[12] Reed, X.; Princz, M.; Hartland, S.: Sec. Europ. Conf. on Mixing (1977) Paper B1, p. 1-26

[13] Van't Riet, K.; Smith, J.: Chem. Eng. Sci. 30 (1975) p. 1093-1105

[14] Van't Riet, K.; Bruijn, W.; Smith, J.: Chem. Eng. Sci. 31 (1976) p. 407-412

[15] Möckel, H.O.: Doct. Thesis TH Köthen (1978)

[16] Van der Molen, K.; van Maanen, H.: Chem. Eng. Sci. 33 (1978) p. 1161-1168

[17] Bertrand, J.; Couderc, J.P.; Angelino, H.: Chem. Eng. J. 19 (1980) p. 113-123

[18] Costes, J.; Couderc, J.P.: Fourth Europ. Conf. on Mixing, Paper B2 (1982) p. 25-34

[19] Brodkey, R.S.: Turbulence in Mixing Operations. Academic Press, Inc. New York, San Francisco, London (1975) p. 80-100

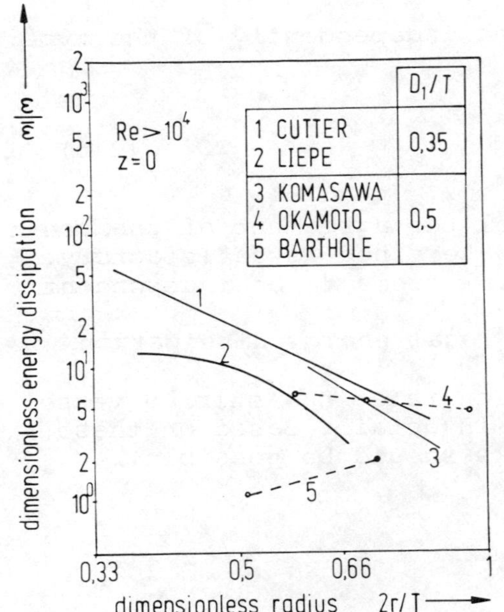

Fig. 1: Energy dissipation in the outflow of a Rushton turbine

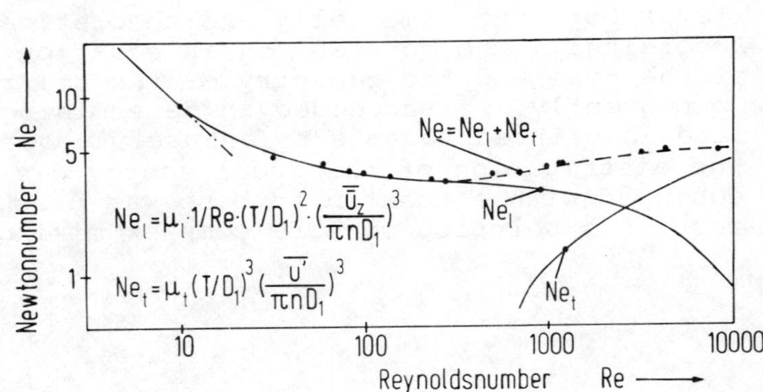

Fig. 2: Measured and calculated Newton number for a Rushton turbine ($D_1/T = 0.33$)

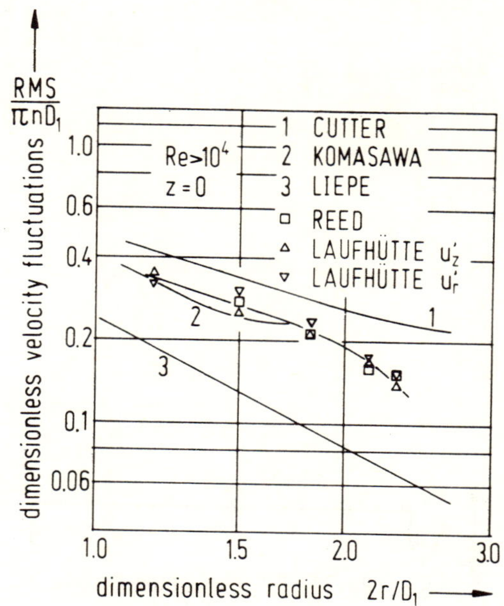

Fig. 3: Fluctuating velocity RMS
in the outflow of a Rush-
ton turbine

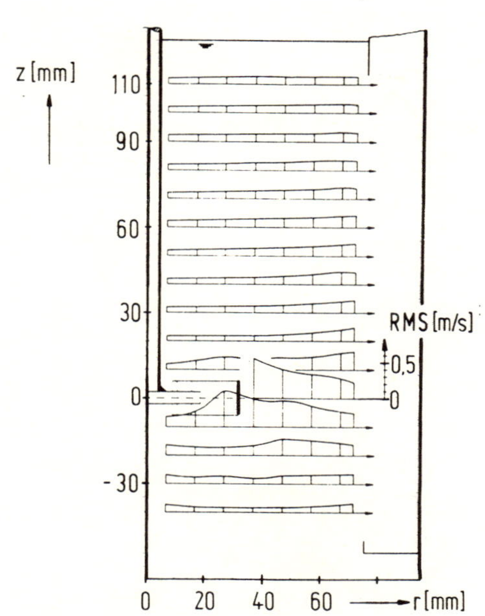

Fig. 4: Axial RMS velocity distribution in the $\varphi = 0°$ plane
$(Re = 3 \cdot 10^4)$

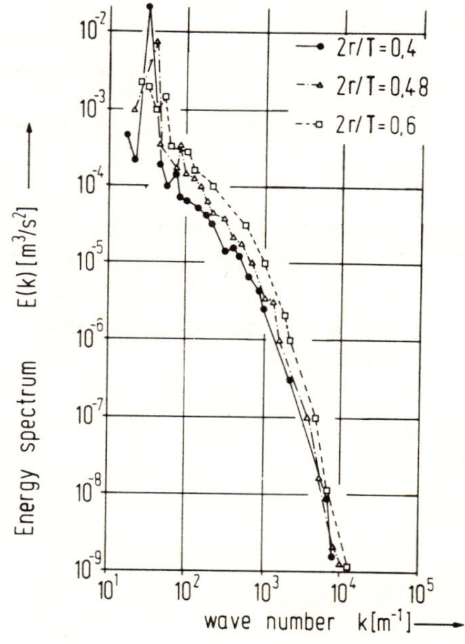

Fig. 5: One dimensional energy
spectra in the impeller
stream (from [15])

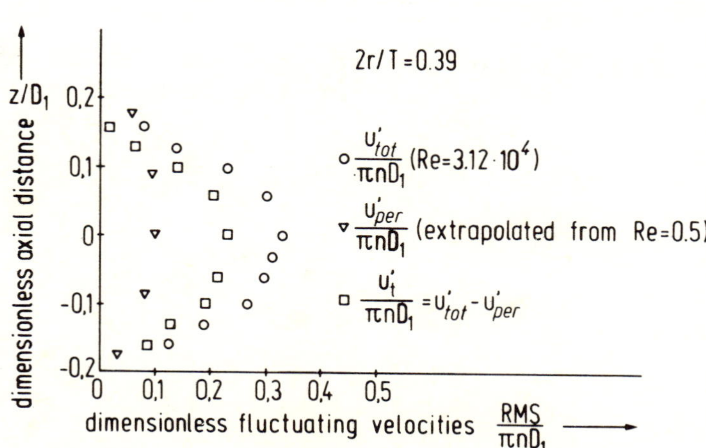

Fig. 6: Calculation of u'_t
near the impeller
blade $(2r/T = 0.39)$

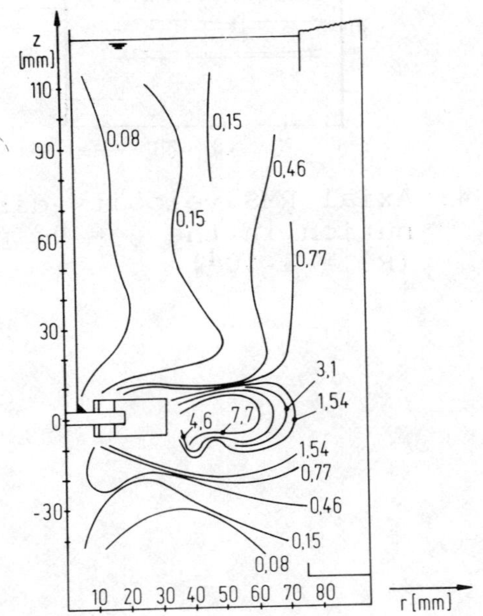

Fig. 7: Map of iso $\varepsilon_2/\bar{\varepsilon}$-curves in one half of the vessel

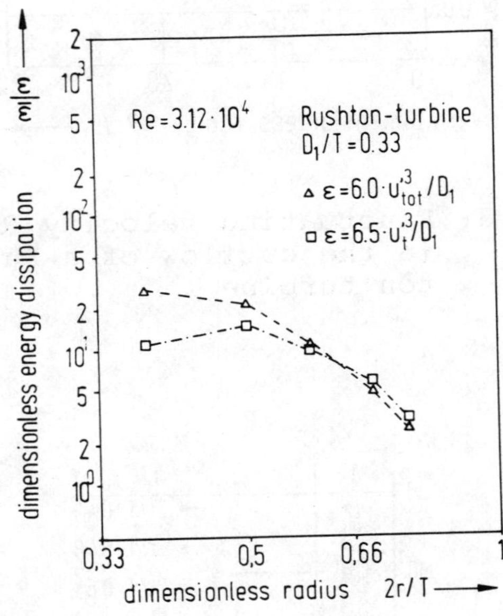

Fig. 8: Energy dissipation in the outflow of a Rushton turbine (consideration of the total and turbulent RMS-values)

STUDY BY LASER DOPPLER ANEMOMETRY OF THE FLOW INDUCED BY A PROPELLER IN A STIRRED TANK - INFLUENCE OF BAFFLES

P. PLION, J. COSTES, J.P. COUDERC
Laboratoire de Génie Chimique
U.A. CNRS 192
Chemin de la Loge
31078 TOULOUSE Cédex (FRANCE)

Summary

The hydrodynamical characteristics of the flow generated by a three blades propeller in a flat bottomed cylindrical vessel have been determined by Laser Doppler Anemometry.

The vessel had a diameter of 0.445 m and was filled with water up to 0.445 m of the bottom (volume of liquid equal to 70 l). The marine propeller had a diameter of 0.2 m and was located at 0.150 m above the bottom of the tank.

The power consumption and the power number have been determined. Mean velocities and velocity fluctuations have been measured, first without baffled then with four baffles of width equal to 0.044 m and displayed at 90° one of the other. The pumping capacity and the non dimensional pumping coefficient have been calculated.

A comparison of the results obtained will give information on the influence of the baffles on flow patterns in such a tank.

Held at Wurzburg, 10-12 June, 1985.
Organised by DVCV· Deutsche Vereinigung für Chemie- und Verfahrenstechnik
(German Association of Chemical and Process Engineering).
Organisation: GVC·VDI-Gesellschaft Verfahrenstechnik und Chemieingenieurwesen. 

<u>NOMENCLATURE</u>

r, Θ , z = radial, angular an axial position respectively in cylindrical coordinates

b = baffle width (m)

D = agitator diameter (m)

f = frequency (s^{-1})

f_D = Doppler frequency (s^{-1})

h = agitator level in the tank (m)

H = water level in the tank (m)

n = agitator rotational speed (s^{-1})

Ne = power number

N_Q = pumping coefficient

P = power input (W)

Q_p = pumping capacity ($m^3.s^{-1}$)

T = tank diameter (m)

Tu = turbulence intensity

v = velocity ($m.s^{-1}$)

v' = velocity fluctuation ($m.s^{-1}$)

η = viscosity ($kg.m^{-1}.s^{-1}$)

ρ = density ($kg.m^{-3}$)

The symbol "$\tilde{\ }$" indicates a time smoothed value.

1. INTRODUCTION

Mechanical agitation is very often used in chemical engineering for operations such as blending, heat transfer, mass transfer, chemical reaction, ...

In the case of moderately viscous liquids, measurements inside stirred vessels are very difficult, in particular because flow is three dimensional and highly turbulent. A lot of experimental technics have been suggested, but mainly two of them seem really promising : hot film anemometry and laser doppler anemometry ; these methods provide a very detailed information on the flow structure inside stirred vessels. In fact, recent analysis demonstrate that laser anemometry, which provides directly one component of the velocity with its sign, is more precise and must be recommended when its use is possible. Data are presented in the case of a three blades propeller in a flat bottomed cylindrical vessel filled with 70 liters of water at room temperature. Mean velocities and velocity fluctuations have been measured by laser doppler anemometry in the volume of the vessel, first without baffles and then with four baffles.

2. EXPERIMENTALS

2.1 Agitated vessel (Ref. 1)

The experiments have been carried out in a vessel filled with 0.070 m^3 of water (Fig. 1).

The diameter of this glass cylindrical tank was equal to : T = 0.445 m, and the level of water at rest in the tank H was equal to the tank dimater : H = T. The tank had a flat bottom and it was possible to organize four vertical baffles displayed at 90° one of the other ; the baffle width being b = 0.0445 m.

The propeller used here had three blades, with an overall diameter D = 0.200 m and a pitch of 0.130 m. The propeller was located at a distance h = 0.150 m from the bottom of the vessel. It was mounted at the end of a tubular shaft, with a diameter of 0.025 m, axially located in the vessel and was rotated by an electrical motor with a variable speed and a system of pulleys and belts.

The cylindrical tank was put inside a cubic tank, also filled with water, to decrease the effects of light refraction.

2.2 Laser Doppler Anemometry (Ref. 2, 3, 4)

A laser gives a beam of monochromatic light which is divided in two and focalized, at the measuring point, by a convenient optical equipment. The two beams interfere to produce interference fringes. If now, a very little solid particle, carried by the liquid flow, enters the measuring volume, it will move through these interference fringes, scattering light with a modulated intensity corresponding to its travel through the successives fringes. This light is then collected by a photomultiplier and analyzed in frequency, so giving an information on the velocity of the particle, that is to say of the fluid, perpendicular to the planes of interference.

The optical arrangement is shown on Figure 2. We have used a 0.015 W Helium-Neon Laser (Spectra Physics, model 124 B) with a red beam of wavelength λ_o = 632.8 nm. The photomultiplier (PM) converts the light signal into an electrical signal, which is then analysed by the frequency shifter (DISA 55 N 10 and 55 N 20).

The distance between fringes Δf, can be related to the angle θ between the two beams and the wavelengh of the laser light λ_o by

$$\Delta f = \frac{\lambda_o}{2 \sin \frac{\theta}{2}}$$

A particle moving across the fringe pattern with a velocity component v_z (for example) perpendicular to the fringe planes will scatter light at a frequency

$$f_D = \frac{v_z}{\Delta f} = \frac{2\, v_z \sin \Theta/2}{d_o}$$

3.   RESULTS AND DISCUSSION

We present here the hydrodynamical characteristics of the flow in the tank for an agitator rotational speed : $n = 3.33\ s^{-1}$ with and without four baffles.

3.1   Power consumption

The power input corresponding to a given agitator is of importance for the design of the necessary mechanical assembly and electrical motor. In each case, power consumption has been measured by determining the torque on the shaft. For that, the motor was mounted vertically between two ball bearings so that it could turn freely on itself. A counterweight system allowed then the measure of the resisting couple necessary to maintain the motor motionless. Starting from these data, the power input P, and the power number, Ne

$$Ne = \frac{P}{\rho\, n^3 D^5}$$

have been determined.

It has been found, in agreement with previous works, that when the Reynolds number,

$$Re = \frac{ND^2 \rho}{\eta}$$

is greater than 10^4, the power number Ne remains constant and independant of Re.

The values of the power number obtained are given in the following table. This power number can be compared with values given by other authors (Ref. 5) and the agreement is good.

n (s^{-1})	Ne		Re	
	without baffle	with baffles		
3.3	0.29	0.39	128 000	this work
	0.27	0.34	100 000–200 000	Literature (5)

3.2   Characteristics of the flow in the tank without baffles

In an unbaffled vessel, the rotation of the agitator induces the formation of a vortex, the depth of which increases with the speed of rotation.

The shape of the vortex has been determined by a photographic method ; these observations being in agreement with NAGATA's theories (Ref. 5) that distinguish two zones : in the central zone, called the forced vortex, the liquid rotates with the same angular velocity as that of the agitator, while the flow in the outer part is similar to a free vortex.

a/ Time-smoothed velocity

Figure 3 presents the spatial distribution of the resultant of the radial and axial components :

$$\bar{v}_r + \bar{v}_z$$

It can be seen that the discharge flow is neither radial nor axial, but inclined in direction of the bottom of the vessel, with an angle varying between 45° and 90° with the horizontal plane. The discharge flow strikes the vessel wall and bottom, turns upwards generating two axial flows : the first one, induces an upward movement at the vessel wall ; the second flow induces a loop under the porpeller and returns to it near the central axis of the tank.

Figure 4 presents the angular velocity measured in an horizontal plane at 0.03 m from the bottom of the tank. The two zones which correspond to the forced vortex (central zone) and the free vortex can clearly be seen, in agreement with the theory of NAGATA. In this unbaffled tank, the angular velocity component is high relative to radial and axial components.

For such a propeller, the pumping capacity can be calculated using equation :

$$Q_p = \frac{1}{2} \int_S (\vec{v}.\vec{n})\ dS$$

and then the non dimensional pumping coefficient :

$$N_Q = \frac{Q_P}{nD^3}$$

The results are presented in the following table :

n (s^{-1})	Q_p (m^3/s)	N_Q
3.33	$5.1\ 10^{-3}$	0.19

b/ Velocity fluctuations and turbulence

The radial $(\overline{v_r'^2})^{1/2}$, axial $(\overline{v_z'^2})^{1/2}$ and angular $(\overline{v_\theta'^2})^{1/2}$ root mean square velocity fluctuations distribution in the agitated vessel are presented on figures 4, 5, and 6.

As is evident from figures 5 and 6, the radial and axial r.m.s. velocities are very large in the vicinity of the propeller, then rapidly decreasing to reach a quasi uniform distribution far from the agitator.

Near the propeller, the r.m.s. value is higher than the mean velocity. Using all the data obtained, the radial, angular and axial turbulence intensities have been calculated :

$$Tu_r = \frac{(\overline{v_r'^2})^{1/2}}{\bar{v}_r} \quad , \quad Tu_\theta = \frac{(\overline{v_\theta'^2})^{1/2}}{\bar{v}_\theta} \quad , \quad Tu_z = \frac{(\overline{v_z'^2})^{1/2}}{\bar{v}_z}$$

and then the global turbulence intensity, when possible,

$$Tu = \frac{(\overline{v_r'^2} + \overline{v_\theta'^2} + \overline{v_z'^2})^{\frac{1}{2}}}{(\overline{v_r}^2 + \overline{v_\theta}^2 + \overline{v_z}^2)^{\frac{1}{2}}}$$

All the results are presentes in table 1.

It can be seen that the radial and axial turbulence intensities are higher than 100 % at many points. But angular turbulence intensity is smaller than 100 % and so is the global turbulence intensity.

3.3 Characteristics of the flow in the tank with four baffles

a/ Time-smoothed velocity

Figures 7 and 8 present the spatial distribution of the resultant of the radial and axial components for two vertical planes : one at $\theta = 0°$, in the plane of one baffle, and the other at $\theta = 45°$ in the middle plane between two baffles.

In this case, the discharge flow is parallel to the agitator shaft : the circulation flow is high and as with a jet, entrainment of the surrounding liquid occurs, below the propeller. The upward flow is near the vessel wall.

In the middle plane between the baffles the velocities are smaller than in the plane of a baffle.

The values of the angular velocity are non negligible with changes in the sign showing some reversal flow (figure 9). The pumping capacity and the pumping coefficient for each plane have been calculated ; the results are presented in the following table :

n	Q_P (m^3/s)		N_Q	
(s^{-1})	$\theta = 0°$	$\theta = 45°$	$\theta = 0°$	$\theta = 45°$
3.33	$11.40\ 10^{-3}$	$12.60\ 10^{-3}$	0.43	0.47

The pumping capacity is higher in the mid plane between baffles showing that the baffles exert a non negligible influence on the discharge flow of the propeller.

b/ Velocity fluctuations and turbulence

The radial, axial and angular root mean square velocity fluctuations distribution are presented on figures 9, 10 and 11. It can be seen that velocity fluctuations are more homogeneous in the volume of the tank than for the unbaffled vessel. The maximum value of fluctuations occurs near the propeller. The turbulence intensities, axial, radial, angular and global have been calculated, and the results are presented in the following table 2.

In the zones where flow is mainly radial (at the bottom of the tank) or axial (at the vessel wall) the corresponding turbulence intensity is smaller than 100 %. But in the other parts, the turbulence intensity is always greater than 100 %.

4. CONCLUSIONS

This paper simply presents experimental data obtained by laser anemometry, without any attempt to model the results. It shows clearly the interest of this

anemometric method which provides a very detailed information on the flow structure
in the vessel.

The particular point of interest is the comparison between the velocity distributions in a baffled or an unbaffled vessels. It appeared clearly that the mainly angular movements in an unbaffled tank turn to a more axial discharge flow near the propeller in a baffled tank. It has also been observed that the turbulence distribution is more homogeneous in a baffled vessel.

Our conclusions is that laser doppler anemometry is able to provide to everybody interested in agitation, the basic hydrodynamic information necessary to study in details any physical or chemical process such as mixing or homogeneous chemical reaction : the use of this technic should then develop in laboratories specialised in agitation.

5.   <u>REFERENCES</u>

1.   <u>PLION P.</u> "Study by Laser Doppler Anemometry of the flow induced by a propeller in a cylindrical tank", Thesis of Docteur-Ingénieur, INPT (1981).

2.   <u>PLION P., LAULAN A., and COUDERC J.P.</u>"Study by Laser Anemometry of the discharge flow of a propeller in an unbaffled tank", Information Chimie, n° 210, pp. 161 (1981).

3.   <u>REED X.B., PRINCZ M. and HARTLAND S.</u> "Laser Doppler Measurements of turbulence in a standard stirred tank", Second European Conference on Mixing, Cambridge (England), March 1977.

4.   <u>Mc COMB W.D. and RABIE L.H.</u> "Laser Doppler Measurements of turbulent structure", A.I.Ch.E. Journal, <u>28</u>, n° 4, 1982, pp. 558.

5.   <u>NAGATA S.</u>, "Mixing - Principles and application", John Wiley, New York, 1985.

Table 1   –   Turbulence intensity in unbaffled vessel

Each cell lists, stacked top-to-bottom, Tu_r / Tu_z / Tu_θ (%). In the bottom row ($z = 0.03$) the bold figure is Tu.

For a given point (r,z):

$$\begin{array}{|ccc|} \hline Tu_r & & \\ Tu_z & Tu & (\%) \\ Tu_\theta & & \\ \hline \end{array}$$

z (m)	$r=0.01$	$r=0.05$	$r=0.10$	$r=0.15$	$r=0.20$
0.25		114 175 –	200 100 –	200 67 –	200 67 –
0.20		∞ 240 –	400 475 –	185 175 –	240 50 –
0.15				433 288 –	243 65 –
0.14		240 175 –	157 2920 –		
0.09	– 380 –	200 180 –	400 363 –	113 360 –	240 263 –
0.06	– 300 –	117 700 –	240 267 –	280 367 –	333 189 –
0.03	– 225 35	60 100 **15** 11	250 200 **22** 12	425 188 **36** 17	400 200 **40** 23

bottom

Table 2 : Turbulence intensity in baffled vessel

Each cell lists, stacked top-to-bottom, Tu_r / Tu_z / Tu_θ (%). In the bottom row ($z = 0.03$) the bold figure is Tu.

$$\begin{array}{|ccc|} \hline Tu_r & & \\ Tu_z & Tu & \% \\ Tu_\theta & & \\ \hline \end{array}$$

$\theta = 45°$

z (m)	$r=0.05$	$r=0.10$	$r=0.15$	$r=0.20$
0.40	167 367 –	367 325 –	575 200 –	550 ∞ –
0.30	600 150 –	242 150 –	97 167 –	163 117 –
0.25	320 97 –	256 142 –	211 1700 –	350 57 –
0.20	333 95 –	385 525 –	∞ 1100 –	667 70 –
0.15	– – –	1000 282 –	1050 192 –	1500 69 –
0.14	460 48 –	533 188 –	– – –	– – –
0.03	77 71 **87** 100	80 112 **114** 135	55 185 **110** 2800	68 82 **73** 71

$\theta = 0°$

z (m)	$r=0.05$	$r=0.10$	$r=0.15$	$r=0.175$
0.40	240 433 –	200 675 –	260 467 –	700 700 –
0.30	517 145 –	200 400 –	1700 138 –	1800 131 –
0.25	1800 114 –	650 194 –	717 180 –	525 90 –
0.20	475 68 –	157 171 –	∞ 350 –	550 98 –
0.15	– – –	1550 311 –	– – –	– – –
0.14	∞ 54 –	1700 150 –	1200 1250 –	460 96 –
0.03	40 106 **72** 135	50 158 **41** 31	88 98 **97** 48	115 83 **78** 75

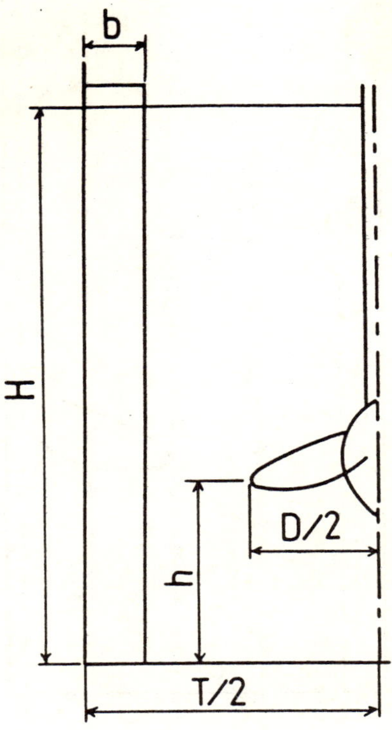

Fig. 1 – Characteristics of
agitated vessel and propeller

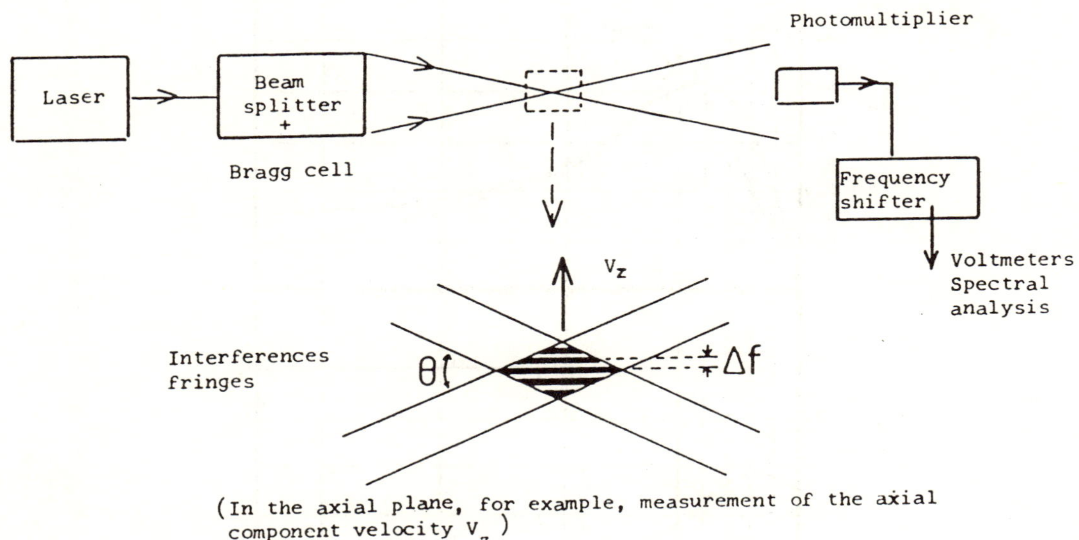

Fig. 2 – Laser Doppler Anemometry

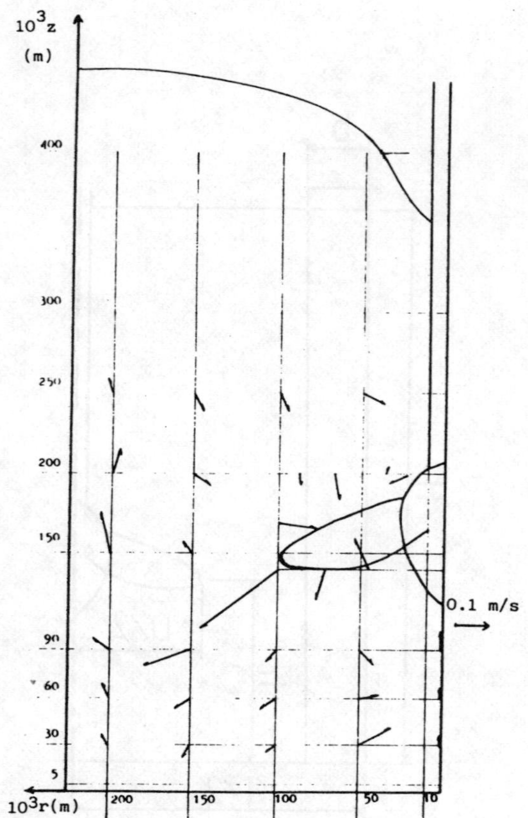

<u>Fig. 3</u> - Mean resultant velocity :
$\overline{v}_r + \overline{v}_z$ (unbaffled tank)

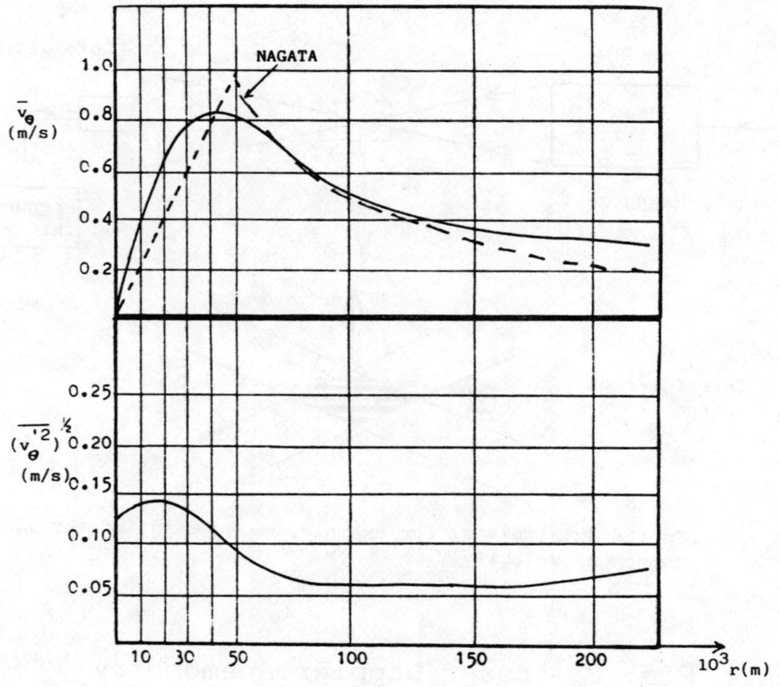

<u>Fig. 4</u> - Mean angular and r.m.s. fluctuation
velocity at z = 0.03 m (unbaffled tank)

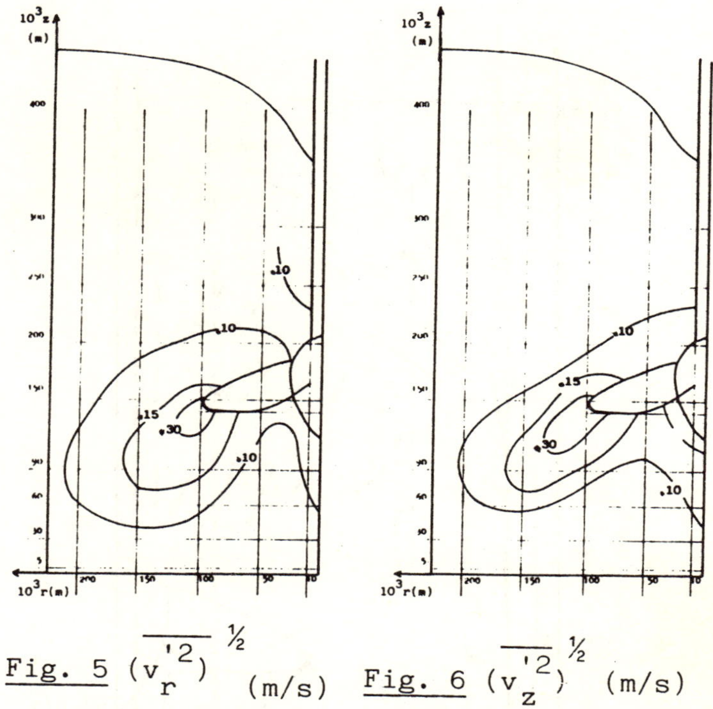

Fig. 5 $(\overline{v_r'^2})^{1/2}$ (m/s)   Fig. 6 $(\overline{v_z'^2})^{1/2}$ (m/s)

Radial and axial r.m.s. fluctuation velocity (unbaffled tank)

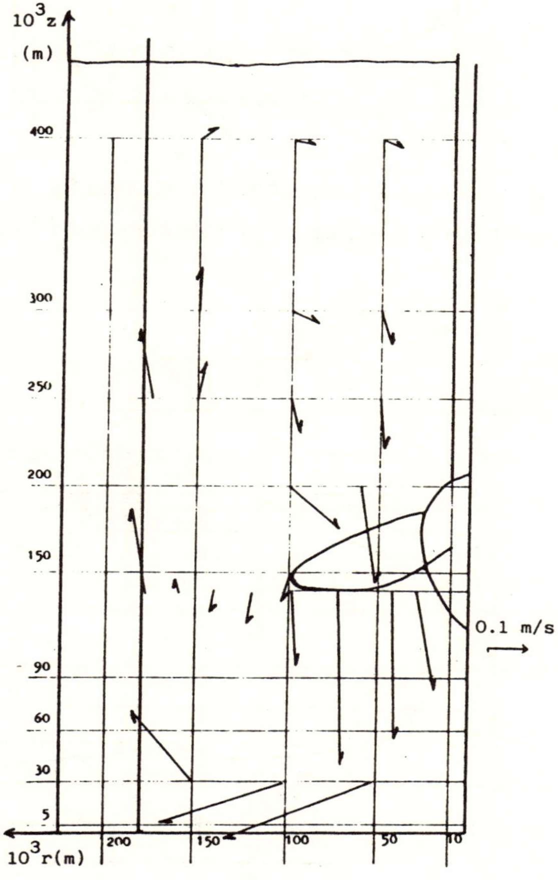

Fig. 7 - Mean resultant velocity : $\overline{v}_r + \overline{v}_z$ in the plane of baffles ($\Theta = 0°$)

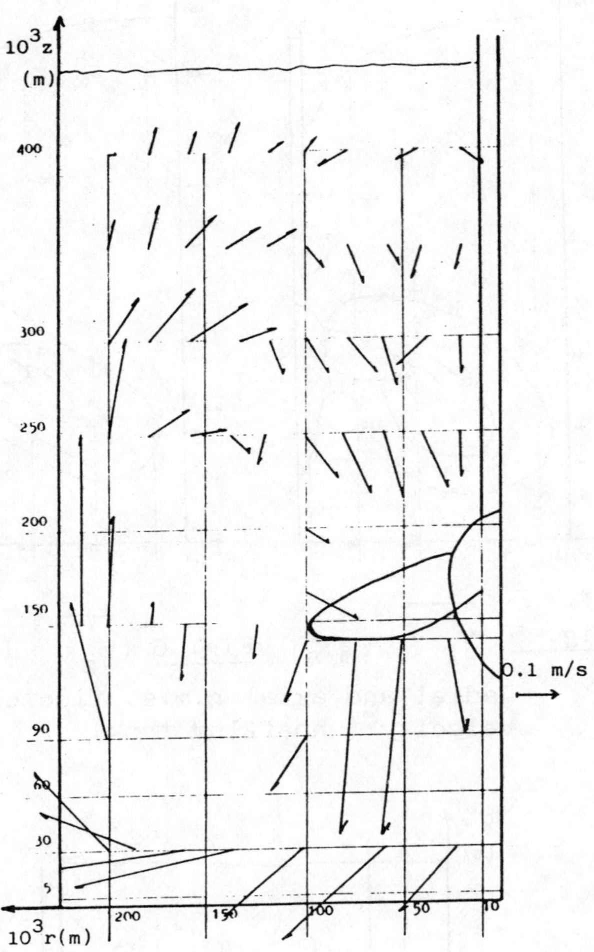

Fig. 8 – Mean resultant velocity : $\overline{v}_r + \overline{v}_z$ in the middle plane between baffles ($\theta = 45°$)

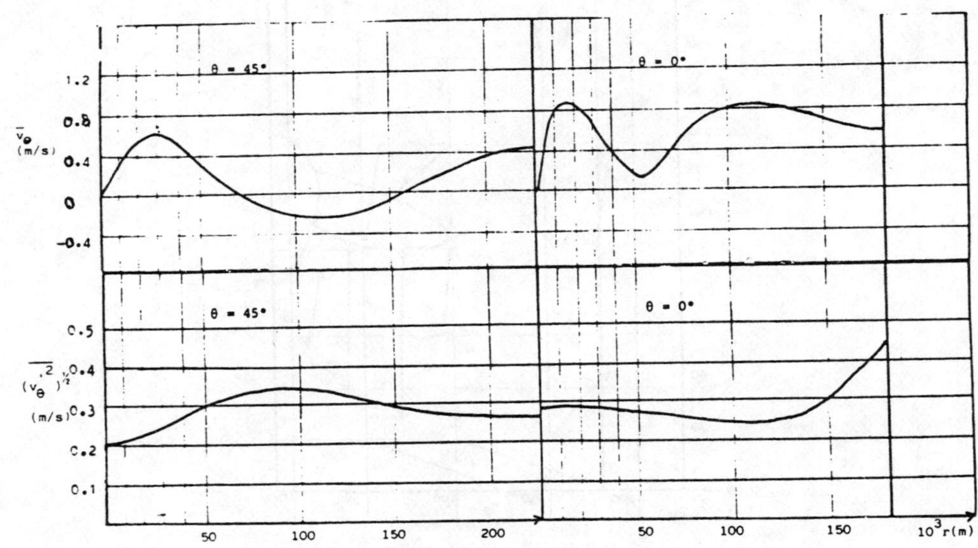

Fig. 9 – Mean angular and r.m.s. fluctuation velocity at z = 0.03 m (baffled tank)

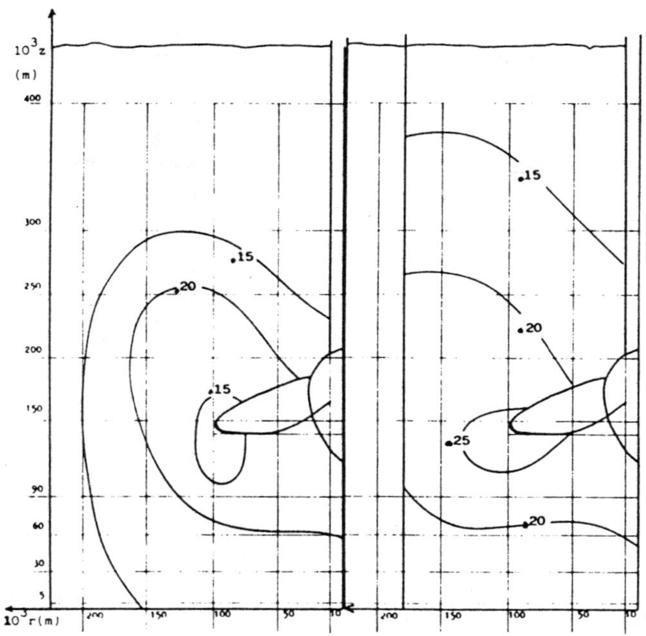

$$\Theta = 45° \qquad \Theta = 0°$$

$$(\overline{v_r'^2})^{1/2} \ (m/s)$$

<u>Fig. 10</u> - Radial r.m.s. fluctuation velocity (baffled tank)

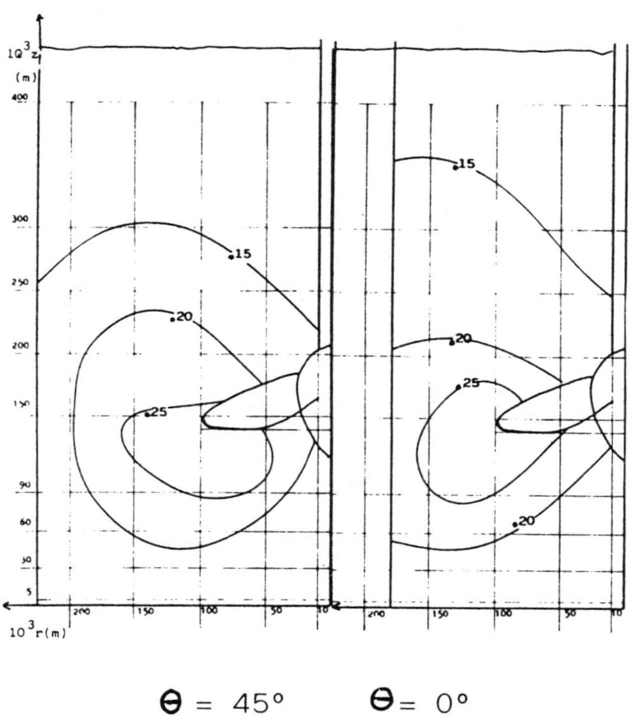

$$\Theta = 45° \qquad \Theta = 0°$$

$$(\overline{v_z'^2})^{1/2} \ (m/s)$$

<u>Fig. 11</u> - Axial r.m.s. fluctuation velocity (baffled tank)

DISTRIBUTION OF TURBULENCE ENERGY DISSIPATION RATES IN MIXERS

by

G. K. Patterson and H. Wu
University of Arizona

An approximate method of measuring the turbulence energy dissipation rate
(ε) in mixers by use of laser-Doppler measurements of the velocity
autocorrelation and turbulence energy was successful in yielding useful
values. The dissipation rate is one of the most important parameters in
modelling mixing rates in turbulent mixing. With the integral length scale of
segregation the dissipation rate forms a parameter, $(\varepsilon/L_s^2))^{1/3}$, which is
essentially proportional to mixing rate.

Values of ε measured in a stirred tank with a disk turbine yield total
power dissipations comparable to those expected. Also the distributions vary
spatially very similarly to measurements made in other ways. The method
reported here is much easier than other methods available.

Held at Wurzburg, 10-12 June, 1985.

Organised by DVCV· Deutsche Vereinigung für Chemie- und Verfahrenstechnik
(German Association of Chemical and Process Engineering).

Organisation: GVC·VDI-Gesellschaft Verfahrenstechnik und Chemieingenieurwesen. 

NOMENCLATURE

c	fluctuating concentration, $C - \bar{C}$
D	impeller diameter
$\mathscr{D}$	molecular diffusivity
k	turbulence energy ($= q$)
L_s	macroscale of segregation
L_x	integral velocity scale
N	impeller rotation rate
N_{Sc}	Schmidt number ($\nu/\mathscr{D}$)
q	turbulence energy ($= k$)
r	radial distance from impeller shaft
R	impeller radius
U, V, W	velocity in x, y, z directions
u, v, w	velocity fluctuations
u_r, u_θ, u_z	fluctuating velocities in radial, tangential, and axial (shaft) directions
U_r, U_θ, U_z	velocities
z	axial distance from impeller disk
ε	turbulence energy dissipation rate
μ	viscosity

OBJECTIVE

The objective of the experimental work reported here is to develop and prove the usefulness of an easy method for obtaining approximate values of the turbulence energy dissipation rate in mixers. Mixers typically have very high intensities of turbulence and in some locations poorly defined flow directionality. Such conditions make measurements very difficult unless the transducer response is linear. The laser-Doppler anemometer meets that criterion, so quite reliable values of average and fluctuating velocities may now be measured. The turbulence energy dissipation rate (ε), however, is not a directly measurable quantity.

Even though ε is a scalar and is manifested by local generation of thermal energy, a thermal energy generation rate probe is not available. Temperature is, of course, the variable measured to indicate thermal energy, but the temperature changes are much too small to sense with any known method. Combinations of vector quantities must consequently be used to determine ε. All the correlation terms which arise in the turbulence energy balance dissipation term may be measured and summed, as was done by Laufer (1) for a pipe flow. That approach is much too laborious for determinations of large numbers of values. A simpler method is sought which can be used to obtain spatial distributions of ε in mixers at many different conditions.

MOTIVATION

Corrsin (2) has shown that the local turbulence energy dissipation rate and the integral scale of the segregation are the strongest variables determining mixing rate in turbulent mixers. He showed that

$$-\frac{\overline{dc^2}}{dt} = \frac{2\overline{c^2}}{4.1(L_s^2/\varepsilon)^{1/3} + (\mu/\varepsilon)^{1/2}\ln N_{Sc}} \tag{1}$$

for fluids of high Schmidt number. The Schmidt number term is usually much smaller than the other term so

$$-\frac{\overline{dc^2}}{dt} \approx \frac{\overline{c^2}}{2\,(L_s^2/\varepsilon)^{1/3}} \tag{2}$$

These relationships can be used as the segregation decay (mixing rate) term in balance equations for $\overline{c^2}$.

In order to use (1) or (2), measured or computed values of L_s and ε must be available. They can be generated for most turbulent flow systems by use of a multi-equation turbulence model such as the k-ε (3) model. In order to assure that proper values are used, however, data on model systems must be obtained by measurement. The availability of many such sets of data will be very valuable in the development of better mixing models.

PREVIOUS WORK

Previous measurements of ε have been reported by Cutter (4), Rao and Brodkey (5), Antonia, Satyaprakash, and Hussain (6), Sato et al. (7), and Sato and Yamamoto (8). Cutter's measurements were based on velocities inferred from streak lengths on photographs of suspended particles. He then computed ε as follows using a simplified version of Laufer's approach:

$$\varepsilon \approx \frac{1}{2r}\frac{\partial}{\partial r}\,[(2q + U^2 + V^2 + W^2 + 2\overline{W}\overline{vw})r] \tag{3}$$

where $q = \frac{1}{2}\,(\overline{u^2} + \overline{v^2} + \overline{w^2})$.

Cutter found that most of the power was dissipated in the impeller stream for
a disk turbine stirred vessel.

Rao and Brodkey computed ε as follows:

$$\varepsilon = 15 \, \nu \, \overline{u^2}/\lambda^2$$

$$\text{where } \lambda = \overline{u^2}/\overline{(du/dr)^2} \tag{4}$$

λ is the microscale of turbulence. They made their velocity measurements in
the impeller stream with a hot-film anemometer aligned with the direction of
the resultant velocity vector.

Antonia, Satyaprakash, and Hussain used a method suggested by Batchelor (9),
Tennekes (10), and Townsend (11). Batchelor proposed that
$\varepsilon \approx A \, (\overline{u^2})^{3/2}/L_x$, where L_x is the integral scale. Tennekes and Townsend
suggested a way to account for non-isotropy by the use of the following:

$$\varepsilon = Aq^{3/2}L$$

$$\text{where } q = \frac{1}{2} \, (\overline{u^2} + \overline{v^2} + \overline{w^2}) \tag{5}$$
$$L = L_x + L_y + L_z$$

Each integral scale is obtained from equations of the form:

$$L_x = \frac{U}{\overline{u^2}} \int_o^\infty \overline{u(t)u(t+\tau)} \, d\tau \tag{6}$$

The basis for the method of equations (5) and (6) is as follows:
1. Energy cascade from large to small eddies exists.

2. Local equilibrium exists between turbulence production (at the large,
   <u>measureable</u> scales) and dissipation (at the small, <u>unmeasureable</u>
   scales).

3. Local isotropy of small scales exists even through large scales are
   very anisotropic.

Antonia, Satyaprakash, and Hussain made measurements of $\overline{u^2}$ and L_x in
round and plane jets, then computed ε using Batchelor's equation. They found
that A should be 1.0 for a round jet and 0.5 for a plane jet.

EXPERIMENTAL WORK

The apparatus used in this work consisted of a stirred-tank mixer and a
laser-Doppler anemometer. A schematic drawing of the stirred tank is shown in
Figure 1. The tank was 27 cm inside diameter and 27 cm high. The top was in
contact with the fluid, which was water. The 4 baffles were 2.7 cm wide. The
6-bladed disk turbine used was of standard design, of diameter approximately
equal to one-third the tank diameter. The shaft ran the full length of the
tank and the impeller was centered between the top and bottom.

The laser-Doppler anemometer used was a standard one-direction model by
DISA with coaxial optics operating in the dual-beam differential mode. The

anemometer was used to measure velocities in all three directions in all portions of the tank. The highest density of measurement points was near the impeller tip where the radial jet was formed. Corrections were made for the periodic non-turbulent velocity pulsations produced by the turbine blades. The turbulence intensity near the blade tips was reduced by about 10%. The correction diminished rapidly away from the impeller.

The length scales were measured by integrating autocorrelations, then multiplying by average velocity, as in Equation (6). Only small effects of measurements in different directions were noticed, since turbulence in most locations was not strongly anisotropic. In the case of a Pfaudler impeller (results to be reported later), the turbulence was strongly anisotropic, so Equation (5) would be necessary.

RESULTS

Figure 2 shows smoothed curves for the radial turbulence intensity values, u'_r/U_b, for locations at v/R_b between 1.08 and 2.15. There was very little difference between u'_r, u'_θ and u'_z in the impeller stream. As indicated, turbulence levels outside the impeller stream were very low. Impeller rotation rates were 100, 100, 300 rpm.

Figure 3 shows a typical set of smoothed plots for radial and tangential average velocities in the impeller stream. The radial velocities divided by tip speed varied little with impeller rotation rates, but normalized tangential velocities were higher at higher rotation rates.

Figures 4-8 show the distributions of measured turbulence energy dissipation rates. Small effects of impeller rotation rate were observed for the normalized values, $\varepsilon/N^3 \cdot D^2$, because of the variations of tangential average velocity. The peak values in the impeller stream were an order of magnitude greater than the regions just outside the stream. That supports the idea that most mixing occurs in the impeller stream.

The measurements of $\varepsilon' = q^{3/2}/L$ were spatially integrated over the entire tank to obtain the total energy dissipation rate. When the impeller power was divided by $\rho \int \varepsilon' dV$, the resulting value of A, the constant in Equation (5), was found to be 0.4. That is remarkably close to Antonia, Satyaprakash, and Hussain's value for a plane jet.

REFERENCES

1. Laufer, J., NACA Tech. Report 1174, 1953.

2. Corrsin, S., "The Isotropic Tubulent Mixer: Part II. Arbitrary Schmidt Number", A.I.Ch.E. J., 10, 870 (1964).

3. Launder, B., and D.B. Spalding, Mathematical Models of Turbulence, Academic Press, London, 1972.

4. Cutter, L.A., "Flow and Turbulence in a Stirred Tank", A.I.Ch.E. J., 12, 35 (1966).

5. Rao, M.A., and R.S. Brodkey,"Continuous Flow Stirred Tank Turbulence Parameters in the Impeller Stream", Chem. Eng. Sci., 27, 137 (1972).

6. Antonia, R.A., B.R. Satyaprakash, and A.K.M.F. Hussain, "Measurements of Dissipation Rate and Some Other Characteristics of Turbulent Plane and Circular Jets", Phys. Fluids, 23, 695 (1980).

7. Sato, Y., M. Kamiwano, and K. Yamamoto, "Turbulent Flow in a Stirred Vessel-Effects of Impeller Types", Kagaku Kogaku, 34, 104 (1970).

8. Sato, Y., and K. Yamamoto, "Two-Equation Model of W.C. Reynolds for
    Isotropic Turbulence", A.I.Ch.E. J., <u>30</u>, 831 (1984).

9. Batchelor, G.K., <u>The Theory of Homogeneous Turbulence</u>, Cambridge Univ.
    Press, 1953.

10. Tennekes, H., "Turbulence: Diffusion, Statistics, Special Dynamics",
    <u>Handbook of Turbulence</u>, Vol. 1, 127, Ed. by Frost and Moulden, Plenum
    Press, 1977.

11. Townsend, A.A., <u>The Structure of Shear Turbulent Flow</u>, 2nd Ed., Cambridge
    Univ. Press, 1976.

ACKNOWLEDGEMENTS

The authors are indebted to Maartin von Doorn, who did most of the
initial measurements, and to Imperial Chemicals Industries, who provided part
of the financial support.

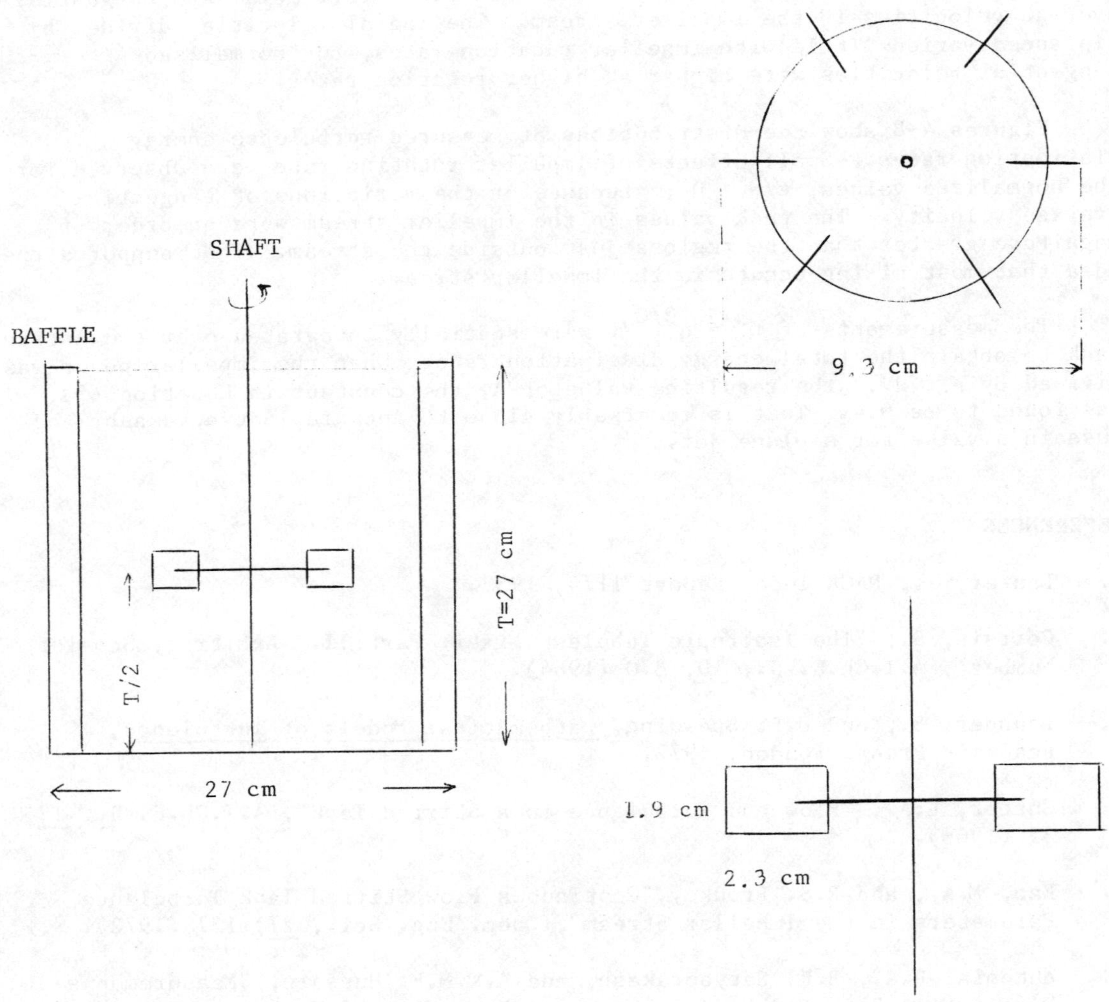

Figure 1. Tank and the 6-flat-blade Disk Turbine

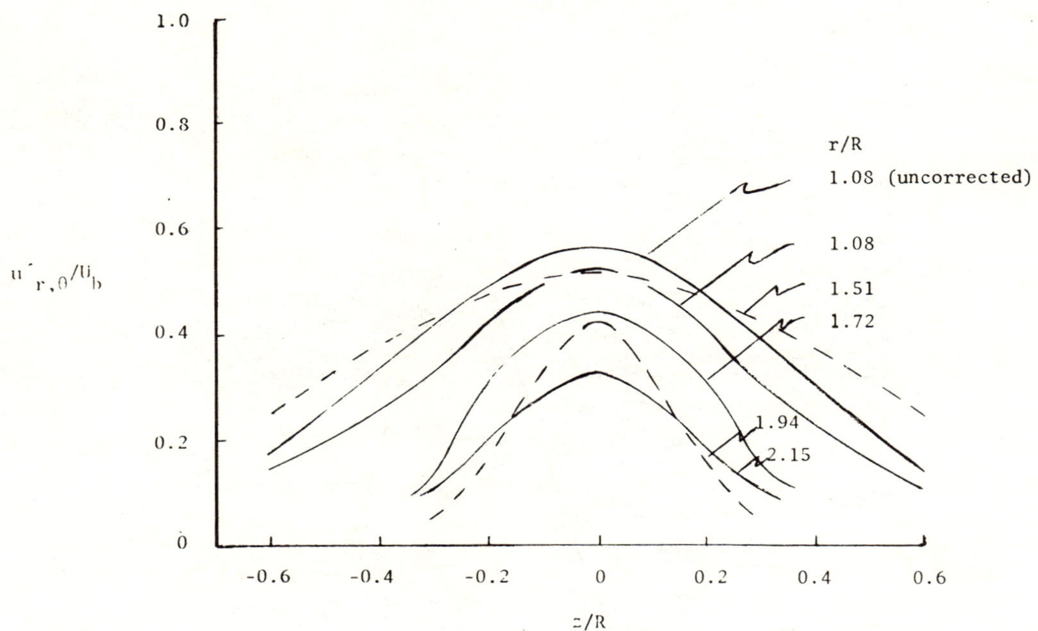

Figure 2.   Turbulence intensities in the impeller stream
normalized by impeller tip speed;
$100 < N < 300$ rpm.

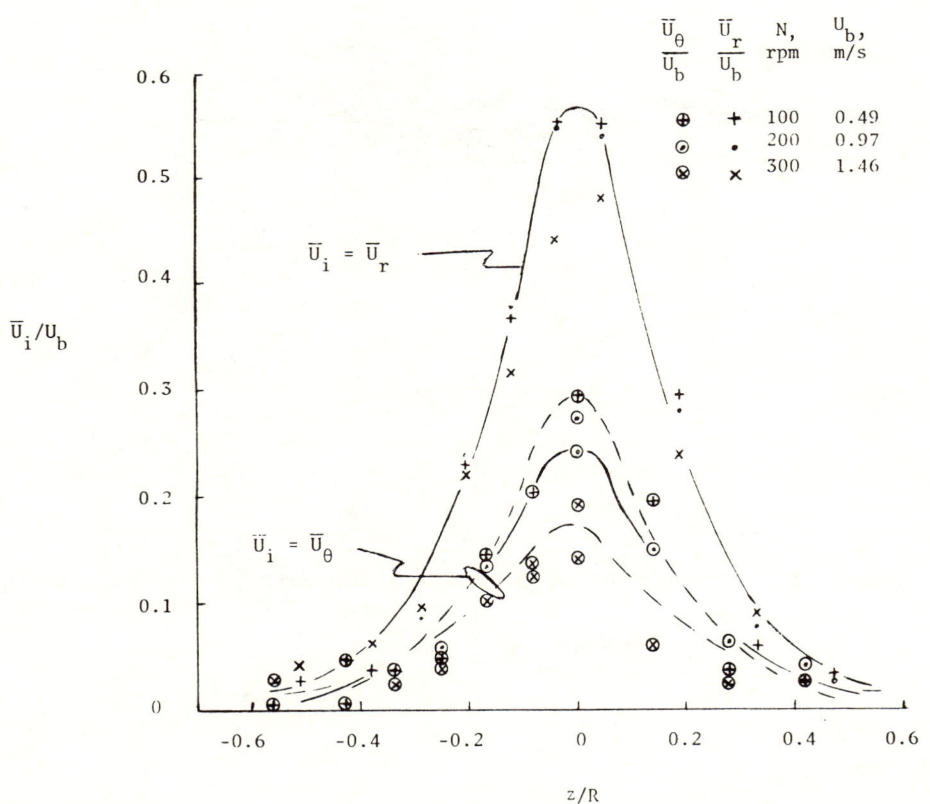

Figure 3.   Average velocities in the impeller stream
normalized by impeller blade tip speed.

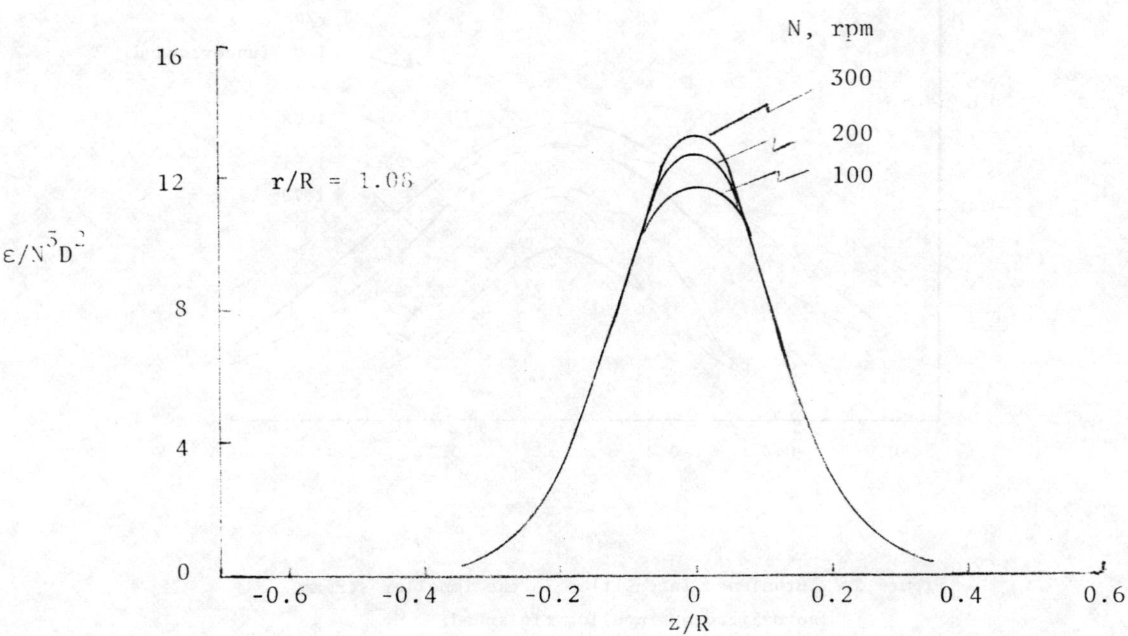

Figure 4. Normalized turbulence energy dissipation rate in the impeller stream.

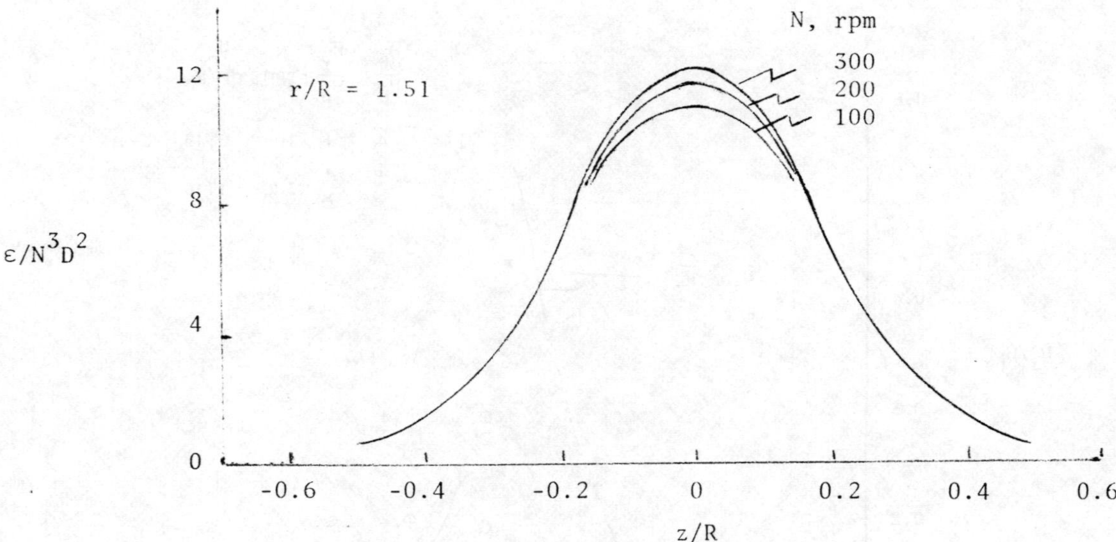

Figure 5. Normalized turbulence energy dissipation rate in the impeller stream.

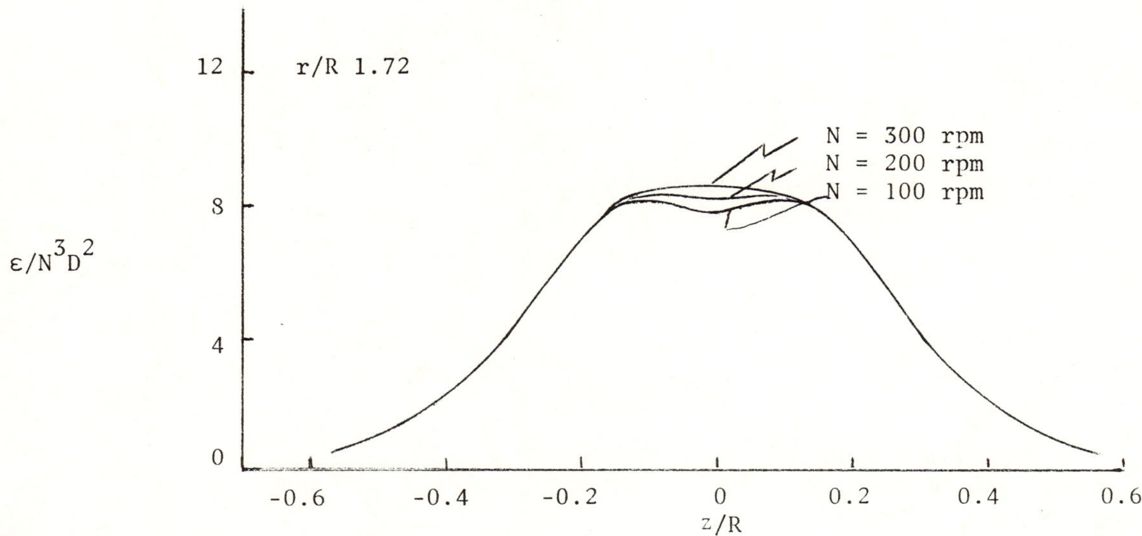

Figure 6. Normalized turbulence energy dissipation rate in the impeller stream.

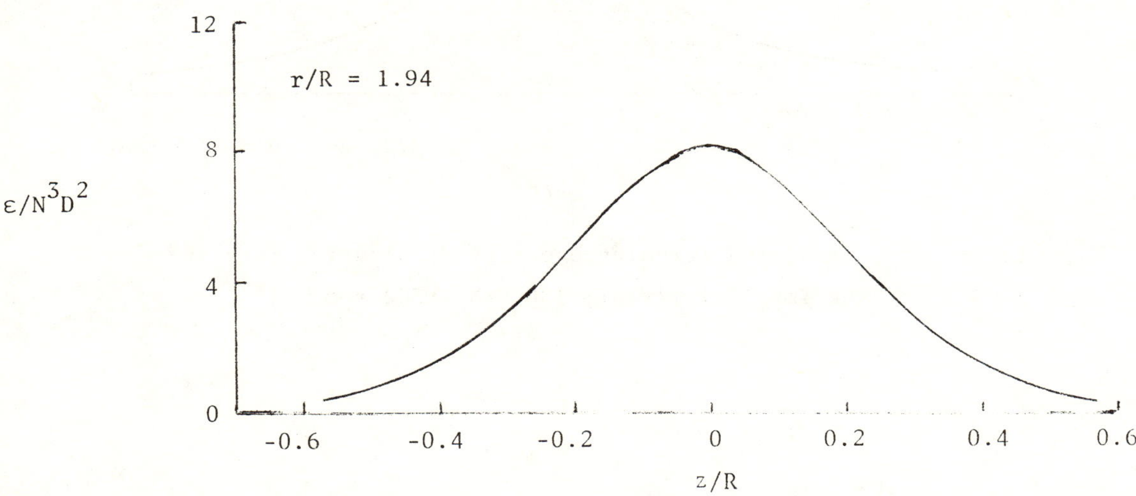

Figure 7. Normalized turbulence energy dissipation rate in the impeller stream:
100 < N < 300 rpm.

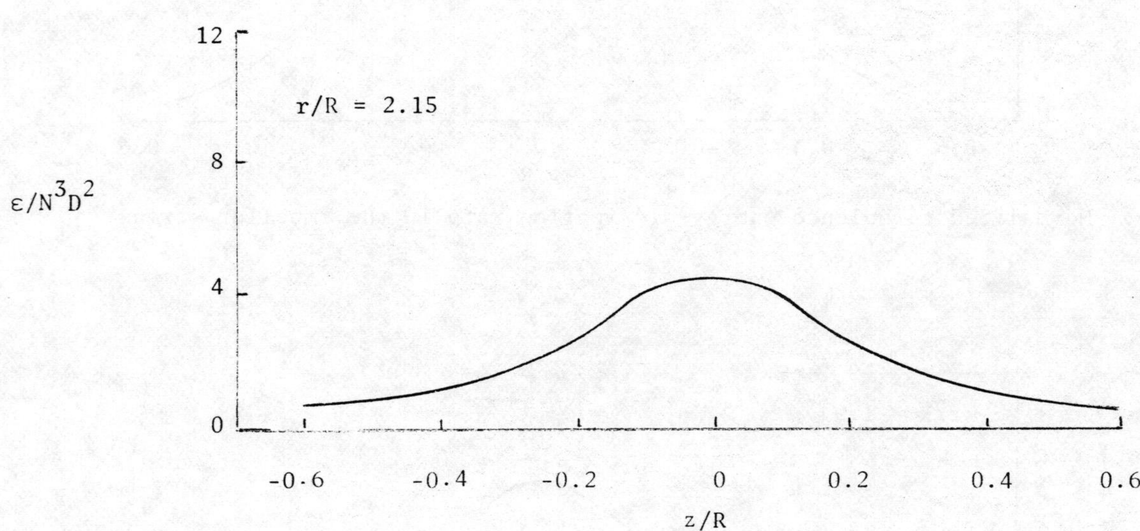

Figure 8. Normalized turbulence energy dissipation rate in
the impeller stream; 100 < N < 300 rpm.

MIXING TIMES FOR PASSIVE TRACERS IN STIRRED TANKS

C.D. Rielly & R.E. Britter
Wolfson Group for Fluid Flow and Mixing in Industrial Processes
Cambridge University Engineering Department
Trumpington Street, Cambridge CB2 1PZ.

SUMMARY

The $N\theta$ = constant scale up rule for the blending of two miscible, initially separate, liquid phases is not successful in predicting mixing times in large scale stirred tanks. This work is an experimental investigation, using the conductivity method, of the mixing time for a salt solution tracer in an agitated vessel and aims to show the importance of other length and time scales in mixing experiments. The effect of Reynolds number has been examined by changing both the impeller speed and the liquid viscosity; the effect of the probe length scale and the initial length scale of the tracer source have been measured. It has been been shown that the mixing time measured in these types of experiments is only a time to homogeneously distribute the tracer about the tank and the importance of properly and uniquely defining the criteria for mixedness in these type of experiments has been discussed.

Held at Wurzburg, 10-12 June, 1985.

Organised by DVCV· Deutsche Vereinigung für Chemie- und Verfahrenstechnik
(German Association of Chemical and Process Engineering).

Organisation: GVC·VDI·Gesellschaft Verfahrenstechnik und Chemieingenieurwesen. 

©BHRA, The Fluid Engineering Centre, Cranfield, Bedford MK43 0AJ, England.

1. INTRODUCTION

Although the mixing or blending of two miscible fluids is one of the most common unit operations in the chemical and process industries and although many studies of the mixing of tracers have been carried out there are still a number of problems associated with the prediction of mixing times in stirred tanks. It is commonly found that mixing times do not simply scale on impeller speed and diameter and that mixing times predicted on the basis of pilot plant or laboratory scale trials do not agree with full scale results. This work investigates experimentally the prediction of mixing times to achieve a certain degree of homogeneity of two miscible, initially separate, liquid phases and looks at the influence of two length scales not generally regarded as important in laboratory scale experiments.

Kipke [1] estimated that some 50% of mixing operations involve the blending of two miscible liquids, however, a successful non-dimensional scale up rule does not exist. Kipke reported that at Reynolds numbers in the turbulent regime ($Re = ND^2/\nu > 10^4$) the empirical equation

$$N\theta = \text{constant}$$

(1)

predicts equal mixing times, θ, in both large and small tanks of similar geometry at the same impeller speed. Kipke presented experimental measurements of mixing times (using the conductivity method) and showed that eqn.(1) does not hold for scale up from pilot plant to full scale. In these experiments the volume of the passive tracer was geometrically scaled with the volume of the tank but the size of the conductivity probe used was not scaled.

Middleton [2] measured average circulation times (which may be taken as proportional to mixing times) in 0.61, 0.91 and 1.81 m diameter tanks using a "radio-pill" follower. The average circulation time, t_c, was found to depend on the tank volume, V_T, and was correlated by:

$$Nt_c = 0.5V_T^{0.30}\left(\frac{T}{D}\right)^3$$

(2)

Mersmann et al [3] also noted that mixing time increased with increasing tank diameter, T, viz:

$$\theta \sim T^{2/3}$$

In this work mixing times were measured experimentally using the conductivity method: a pulse of electrolyte, passive tracer was released into the stirred tank and the change in the resistance of a small volume of the fluid was measured using a conductivity probe (the resistance of the fluid is proportional to the concentration of the electrolyte: Gibson & Schwarz [4]).

Dimensional analysis of the experimental variables shows that for geometrically similar vessels -

$$\theta = f(N,D,\nu,L_s,L_p,K)$$

(3)

where L_s and L_p are the characteristic length scales of the tracer source and the probe and K is the molecular diffusivity of the electrolyte in the working fluid. Gravitational acceleration has not been included in eqn.(3) : g only influences behaviour at the free surface. In fully baffled stirred tanks the free surface is relatively flat and it may be assumed that the mixing processes are not governed by phenomena at the surface. However, in unbaffled tanks a combined Rankine vortex forms (Nagata [5]) and the shape of the free surface depends on gravitational acceleration; in this case g may exert some influence on mixing times.

In these experiments the smallest length scale of concentration fluctuations that can be detected depends on the resolution of the conductivity probe. Clearly, concentration fluctuations exist on a length scale less than the resolution of the probe and are dissipated by molecular diffusion at the Batchelor scale which is given by:

$$\eta_B = (\nu \kappa^2/\varepsilon)^{1/4}$$

$$(4)$$

where ε is the energy dissipation per unit mass. Consequently, the mixing time measured in these experiments is the time for concentration fluctuations to be on a length scale smaller than the length scale of the probe, L_p, and is a <u>macro-mixing time</u>. It is assumed that molecular diffusivity has little effect on concentration gradients in time scales of order θ. Forming dimensionless groups from eqn.(3) :

$$N\theta = f(Re, L_s/D, L_p/D)$$

$$(5)$$

where

$$Re = ND^2/\nu$$

$$(6)$$

The experiments described in the following section investigated the effect of these three dimensionless groups on $N\theta$. Water and glycerol/water solutions were used as the working fluids and the effects of N, ν, L_s and L_p on mixing time were determined.

2. EXPERIMENTAL

The mixing vessel used in these experiments was 29 cm in diameter and had the standard geometry shown in figure 1. The impeller was a standard six bladed Rushton turbine with a diameter T/3 at a clearance of T/3 from the bottom of the tank. The turbine was driven by a TASC unit variable speed drive and the shaft rotation rate was measured using a calibrated tachogenerator. A torque transducer in line with the motor and shaft enabled the power input to be determined (since the final bearing was after the transducer a "no-load" power input had to be subtracted from any reading to allow for frictional losses in the bearing). Small fluctuations in the torque were observed at constant impeller speed (possibly due to the feedback control loop of the TASC unit). An average torque value was calculated by integraing over a period of approximately 60 seconds using the statistical programs of a microprocessor voltmeter.

The mixing time was measured by examining the concentration fluctuations of a pulse of passive tracer as a function of time from its relesase at t = 0. In these experiments θ was defined as the time from release of the tracer until concentration fluctuations at the probe were within $\pm 10\%$ of the final concentration difference. A known volume of strong salt solution (300 g/litre) was released at the surface of the mixing tank, midway between the baffles, as shown in figure 1. A marker was superimposed on the conductivity signal (by shorting the probe across a large resistance) as the source was released to indicate t = 0. The source release device is illustrated in figure 1. The bottom plate was sealed against the cylinder using an 'O-ring' and the clamp locked. The device was charged with a known volume of salt solution and the free surface of the tracer aligned with the free suface of the mixing tank. When the clamp was released the compression spring rapidly removed the cylinder from the tank and released the volume of tracer. Little dispersion of the tracer resulted from the release device itself. Two sizes of release device (20 mm and 45 mm inside diameter) allowed source volumes in the range 5 - 120 cm^3 to be investigated.

A conductivity probe located on the opposite side of the tank, as shown in figure 1, detected changes in salinity. The tips of the smaller probes were coated with platinum black to increase active surface area and sensitivity (the technique is described by Gibson & Schwarz [4]). The second electrode, which completed the probe circuit, was located on the opposite side of the tank to the probe: a strip of stainless steel was attached to one of the baffles and a connection made to the earth of the probe circuitery. The probes were fed with an a.c. voltage of 0.2 v rms at 10 kHz. Gibson & Schwarz [4] examined such a system (where one electrode is much larger and distant from the other) and showed that the conductivity of the fluid between the electrodes is determined primarily by the restitivity of the solution around the tip of the probe. They estimated that the output signal from conductivity fluctuations with a wavelength of ten probe length scales would be attenuated by about 50% and so we may assume that the smallest spatial scales that may be detected by our probes are of order $10L_p$.

A further limitation on the resolution of the probe is the frequency response. Some measurements of the temporal response of the probes used in this work was made by dropping them through a clean salt/water interface. The response of the probe to this step change in concentration was recorded on a digital storage oscilloscope and the time, τ, for the output to rise by $(1-e^{-1})$ was measured. The velocity of the probe as it passed through the interface was measured using a Scmitt trigger and timer/counter. Clearly the response time of the probes depends on this velocity and their length L_p. An approximate calculation of the mean velocity close to the probe in the tank showed that parcels of fluid are convected passed the tip at about 0.5 m/s. At this velocity it was estimated that the 0.3 mm probe was capable of measuring concentration fluctuations on a scale greater than about 3 mm.

The output signal from the probe was amplified and rectified and, in view of the measured frequency response, was low-pass filtered at 1 kHz to remove high frequency noise (some noise remained on the signal, possibly due to inadequate RF screening). The filtered analogue output was stored on magnetic tape and was later digitised and stored on magnetic discs for analysis on the CED/Alpha mini computer.

The standard experimental conditions were : a probe of length 0.3 mm, a tracer volume of 20 cm^3 (which was 0.10% of the tank volume) and an impeller speed of 6.7 s^{-1}. The working fluid was tap water in the standard case; glycerol/water solutions were used to vary the viscosity in other experiments. Four series of experiments were carried out in the same tank geometry:

1. to change Re by varying N in the range 2.5 - 6.7 s^{-1} at constant ν, L_s and L_p

2. to change Re by varying ν in the range 1×10^{-6} - 6.3×10^{-6} m^2/s at

constant N, L_s and L_p

3. to vary L_p in the range 0.3 - 20 mm at constant N, ν, L_s

4. to vary L_s in the range 5 - 120 cm^3 at constant N, ν, L_p

The 90% mixing time is a statistical quantity and so each experiment was repeated ten times to obtain an average mixing time, $\bar{\theta}$. Each data point has been plotted with an error bar of one standard deviation: in all cases this was between 5 - 10% of the mean value.

3. DISCUSSION OF RESULTS

It is useful to initially consider an idealised system in which there is no molecular diffusion and there is no limit to the resolution of the probe. The effect of molecular diffusion is to smear out concentration gradients and to reduce the magnitude of concentration fluctuations with time ie. fluctuations are dissipated by molecular diffusivity. At t = 0 the initial length scale of concentration fluctuations is the length scale of the tracer source; at later times turbulent mixing produces finer and finer scales of concentration fluctuations (see figure 2). For the idealised case the scale of concentration fluctuations decreases indefinitely with time and there is a change in the spectrum towards higher frequencies.

In the real case, molecular diffusion dissipates concentration fluctuations so that there is a cut-off point in the spectrum (at the Batchelor scale). The magnitude of the fluctuations would therefore decay as a function of time and diffusion rate. The resolution of the probe plays an important role in determining the mixing time: the probe does not respond to the full spectrum of concentration fluctuations and the output signal from high frequency fluctuations is attenuated. Therefore the fluctuations in the output signal from the probe decay with time as the scales of concentration fluctuations become smaller than the scales that may be detected by the probe. Although the output becomes constant and appears to indicate that the tracer is well mixed, concentration gradients still exist in the tank but on scales too small to be seen by the probe.

In these experiments the smallest probe size used was 0.3 mm and it was estimated from its frequency response and size that it could measure concentration fluctuations above a scale of 3 mm. The Batchelor microscale at which concentration fluctuations are dissipated by diffusion is estimated as 1 μm and clearly a large part of the spectrum cannot be detected by the probe. Therefore the mixing time measured in this and many other works is only a <u>gross homogenisation time</u> for the tracer to be distributed on a scale less than the resolution of the probe. In such cases two criteria are necessary to properly define mixedness:

1. the percentage, of the final concentration difference, to which output fluctuations must decay

2. the scale of resolution of the probe

3.1 Effect of Reynolds Number

Reynolds number was changed in the range 1×10^4 to 6.3×10^4 by varying both N and ν . The experimental results are presented in figure 3 for these cases. The results show a slight decrease in dimensionless mixing times at lower Re in the range 1×10^4 to 3×10^4 but are constant at higher Re. Some preliminary measurements obtained with a 4 mm probe also indicated a decrease in Nθ at lower Re. In each of these series of experiments the power number remained constant throughout.

Several previous workers have noted some dependency of Nθ on N (Norwood & Metzner [6]; Biggs [7]; and van de Vusse [8]) but have mistakenly attributed this to a Froude number effect ($Fr = ND^2/g$). It has already been stated that θ should not depend on free surface phenomena and hence there should be no effect of Fr. In our experiments at constant N the Froude number is also constant yet Nθ is not a constant. More recently Zlokarnik [9] and Gramlich & Lamade [10] found that Nθ increased slowly with increasing Re for Re > 10^4. Such a Reynolds number effect would be consistent with observations that mixing times are longer in large scale equipment (Re$\propto$D^2) at the same impeller speed.

Some caution must be exercised in interpreting the results of varying Re. In the experiments which varied ν using glycerol solutions the two criteria for mixedness are not the same in all cases. The frequency response of the probe depends on two factors: the speed of the electronics (which is usually very fast) and the rate at which liquid can drain from the probe tip and be replenished. As the viscosity increases it takes longer for the probe to drain and replenish and hence the frequency response decreases. Consequently the probe may not be able to detect as small concentration fluctuations as before and there is an apparent decrease in mixing times. At this stage in the work it is not clear whether it is this effect that causes a decrease in mixing times at lower Reynolds numbers.

3.2 Effect of the Initial Length Scale of the Source

The initial length scale of the source was varied by changing the volume of the salt tracer released. Figure 4 shows the effect of changing V_s in the range 5 – 120 cm^3, corresponding to $1.7 < L_s < 4.9$ cm. There appears to be no significant effect of changing L_s in these experiments; it was not practicable to use sources smaller than 5 cm^3 (0.025% of the tank volume).

The source size might be expected to have an effect when L_s becomes smaller than the integral length scale of velocity fluctuations in the flow. In this case initially the main source of concentration fluctuations is the distortion or stretching of the tracer fluid by large scale velocity fluctuations. Results obtained for elevated plumes in a turbulent boundary layer indicate that meandering is the main source of concentration fluctuations for source sizes less than the integral length scale (Fackrell & Robins [11]).

Rao & Brodkey [12] made measurements of mean and turbulent velocities in a stirred tank of the same diameter as that used in this work. They calculated that the integral length scale in their tank was less than 1 cm. If we assume that the integral length scale in our tank is of the order 1 cm then the initial length scale of the source was <u>always</u> greater and no effect should have been observed.

3.3 Effect of the Length Scale of the Probe

The physical size of the probe was changed by varying the length of exposed platinum wire (see figure 1): lengths of 0.3,4,5,10 and 20 mm were used. It has already been discussed that the decay of output signal fluctuations depends on the resolution and response of the probe. Figure 5 shows the effect of L_p on the mixing time: there is some slight decrease in θ with larger probe sizes. The 20 mm probe averages over a much larger volume than the 0.3 mm probe and only responds to large scale concentration fluctuations. Although the large probes indicate a well mixed condition the smaller probes can still detect concentration fluctuations at the same point and consequently shorter mixing times are observed with larger probes.

Figure 6 shows a comparison of the output from the 0.3 and 20 mm probes under identical experimental conditions. The difference in structure between the output signals verifies that the large probe only responds to large scale concentration fluctuations and measures a shorter mixing time.These results show the importance of the second criterion for mixedness viz: the scale to which the probe can measure concentration fluctuations.

4. Conclusions

The Reynolds number of the flow was observed to affect the dimensionless mixing time and would account for longer mixing times being reported in larger scale equipment. However, at this stage of the work it is not possible to state whether this is truly an effect of Re or an effect of a change in the resolution of the probe. The tracer volume was not found to have any influence on mixing times. It may be expected that so long as the initial length scale of the source is greater than the velocity integral length scale there should be no effect on θ . Shorter mixing times were observed with larger probes which demonstrates that probes should be carefully scaled to measure similar scales in different sizes of tank. It should be noted that the resolution of the probe is determined by both its size and its frequency response.

This work has shown that the criterion for mixedness that concentration fluctuations should decay by some percentage of the final concentration difference is not sufficient to uniquely define a mixing time. The scale of resolution of the probe determines the amount by which the output signal is attenuated and thus how quickly concentration fluctuations appear to decay. Concentration gradients still exist for a time after the probe shows a well mixed condition until they are dissipated by molecular diffusion. Scale up of mixing experiments is not simply a matter of geometrically scaling the probe and source and care must be taken to ensure that the criteria for mixedness are identical at different scales of operation.

5. REFERENCES

1. Kipke, K.: "problems in Mixing Technology", Ger. Chem. Eng., 6, 1983, pp.119-128

2. Middleton, J.C.: "Measurement of Circulation within Large Mixing Vessels", Third Eur. Conf. Mixing, BHRA Fluid Eng., York 1979, paper A2, pp.15-36

3. Mersmann, A, Einenkel, W.-D. & Kappel, M.: "Design and Scale Up of Mixing Equipment", Int. Chem. Eng., 16, 1976, pp.590-603

4. Gibson, C.H. & Schwarz, W.H.: "Detection of Concentration Fluctuations in a Turbulent Flow Field", J. Fluid Mech., 16, 1963, pp.357-364

5. Nagata, S.: "Mixing. Principles and Applications", Halstead Press, John Wiley, 1975, Ch.1

6. Norwood, K.W. & Metzner, A.B.: "Flowpatterns and Mixing Rates in Agitated Vessels", A.I.Ch.E.J., 6, 1962, pp.432-437

7. Biggs, R.D.: "Mixing Rates in Stirred Tanks", A.I.Ch.E.J., 9, 1963, pp.636-640

8. van de Vusse, J.G.: "Mixing by Agitation of Miscible Liquids", Chem. Eng. Sci., 4, 1955, pp.178-200

9. Zlokarnik, M.: "Suitability of Stirrers for the Homogenisation of Liquid Mixtures", Chem. Ing. Tech., 39, 1967, pp.539-548

10. Gramlich, H. & Lamade, S.: "Comparison of Enamelled Agitator Systems for Homogenisation of Liquid Mixtures", Chem. Ing. Tech., 45, 1973, pp.116-122

11. Fackrell, J.E. & Robins, A.G.: "Concentration Fluctuations and Fluxes in Plumes from Point Sources in a Turbulent Boundary Layer" J. Fluid Mech., 117, 1982, pp.1-26

12. Rao, M.A. & Brodkey, R.S.: "Continuous Flow Stirred Tank Turbulence Parameters in the Impeller Stream", Chem. Eng. Sci., 27, 1972, pp.137-156

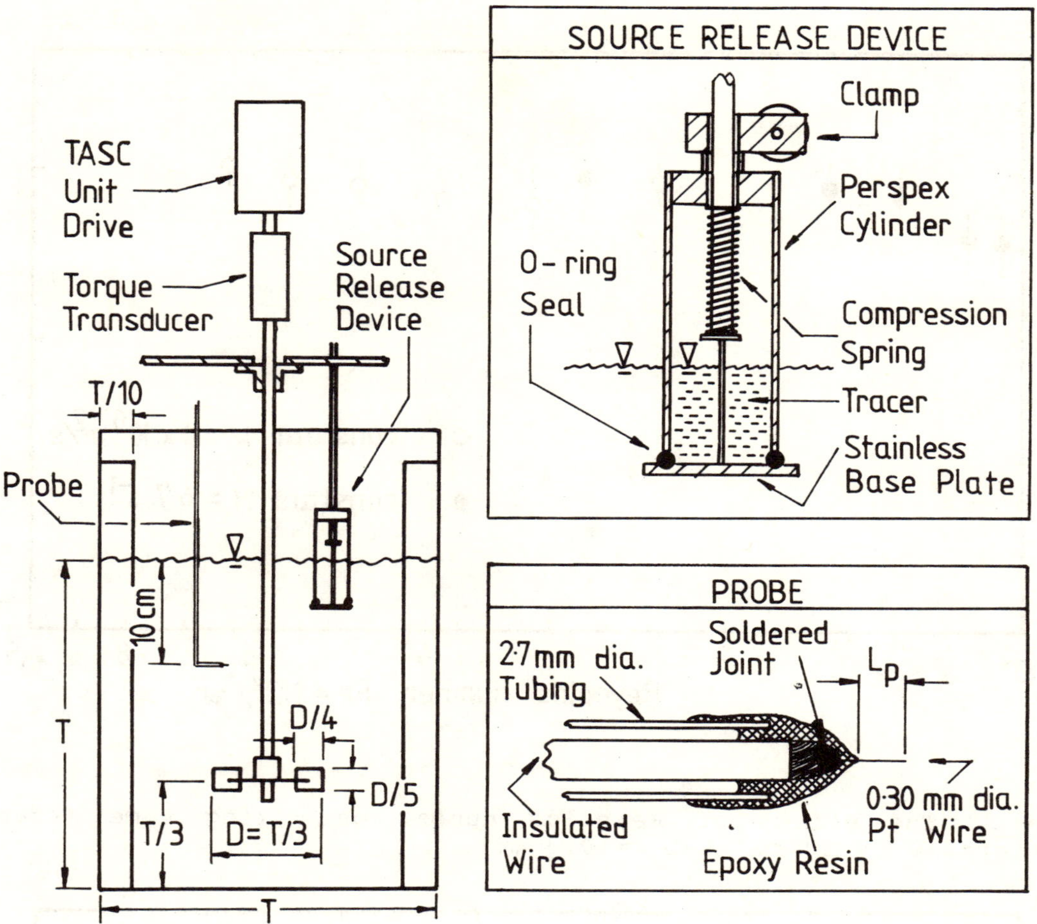

Figure 1. Experimental mixing tank, source release and conductivity probe. The source release device operates by clamping the perspex cylinder against the base and sealing with the O-ring. When the clamp is released the spring removes the cylinder from the tank.

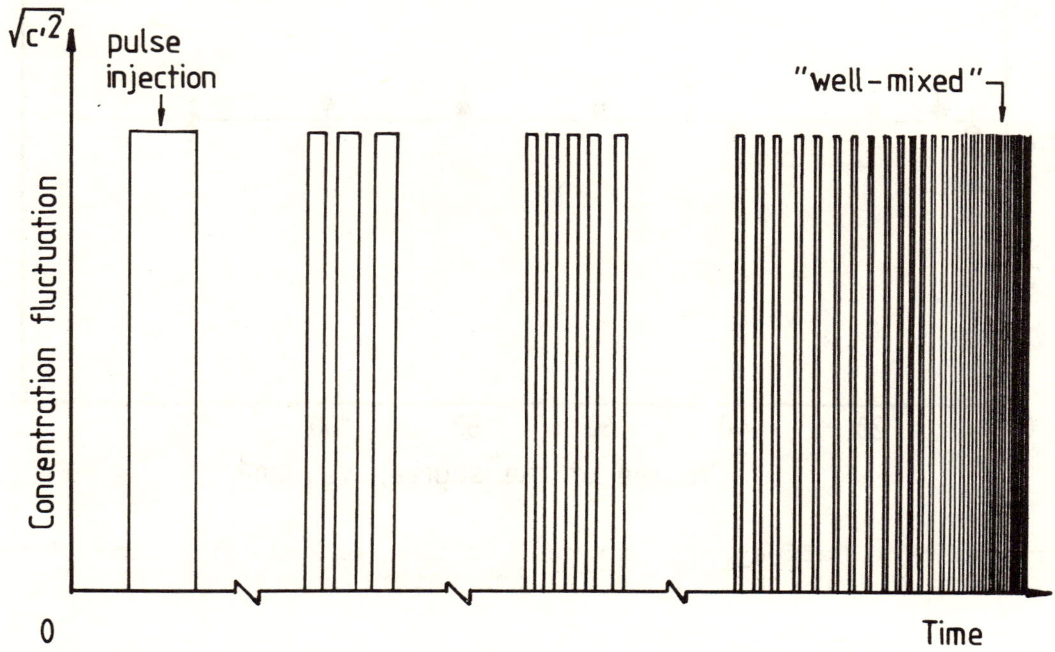

Figure 2. The production of fine scale concentration fluctuations in an ideal system.

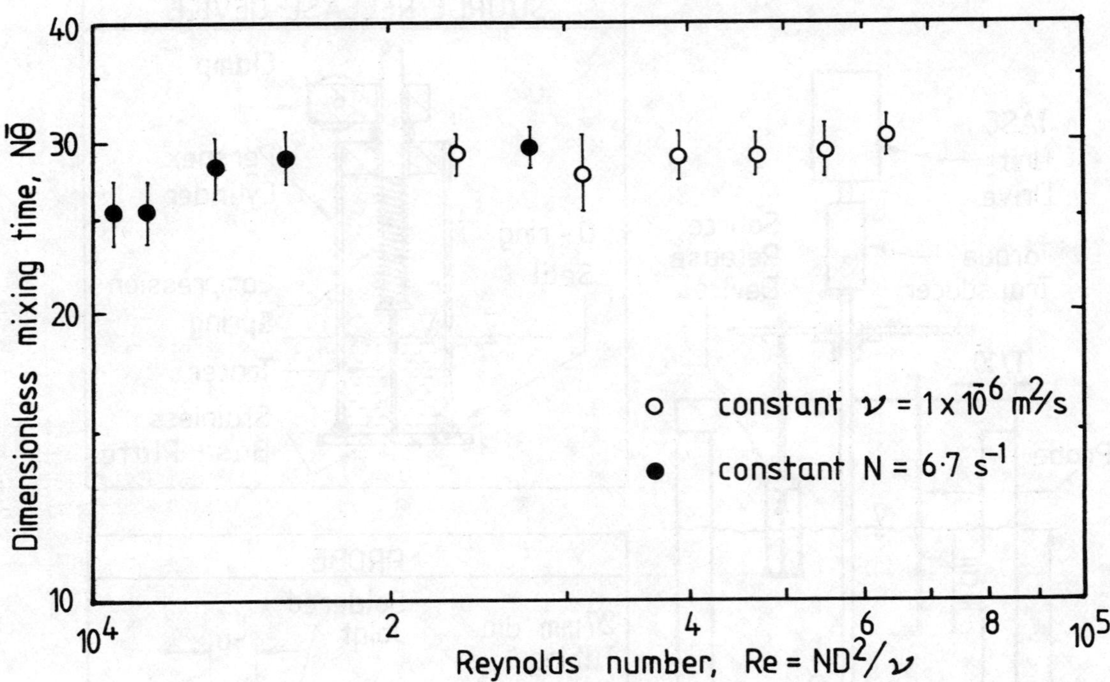

Figure 3. The effect of Reynolds number on mixing time (standard conditions: V_s = 20 cm³ & L_p = 0.3 mm).

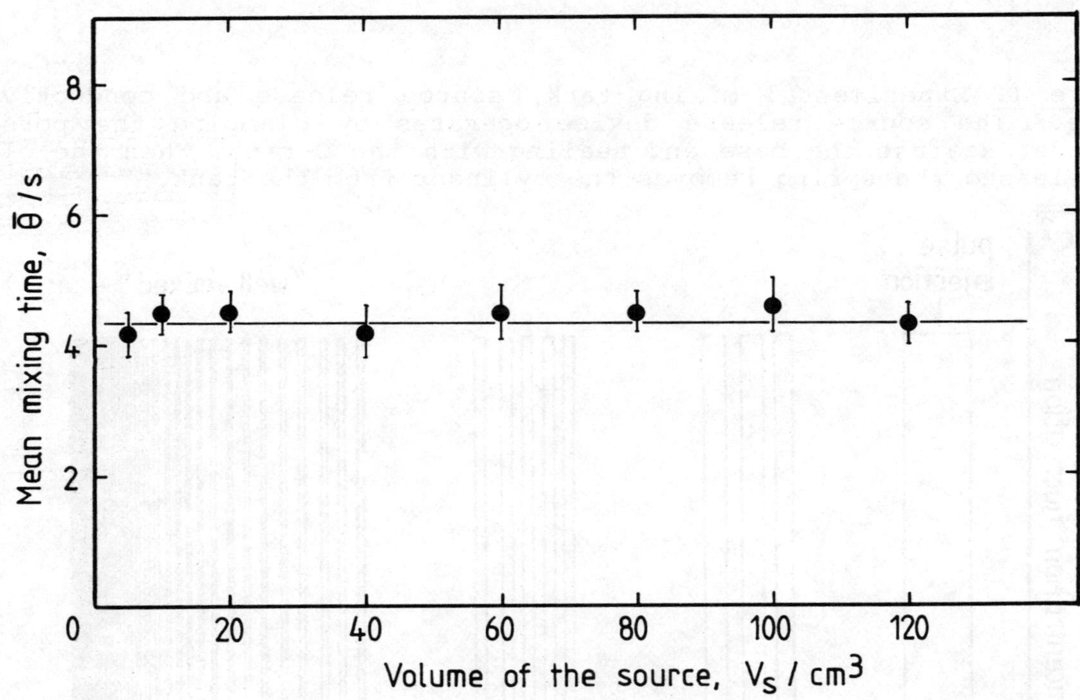

Figure 4. The effect of source volume on mixing time (standard conditions: N = 6.7 s⁻¹, L_p = 0.3 mm, water).

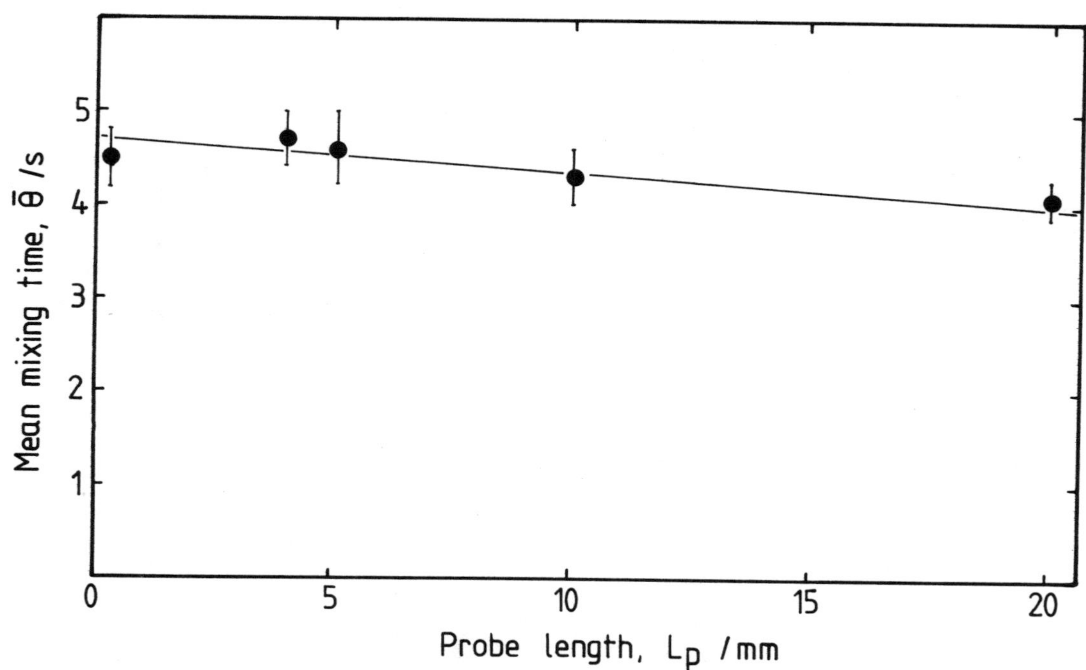

Figure 5. The effect of probe length on mixing time (standard conditions: N = 6.7 s^{-1}, V_s = 20 cm^3, water).

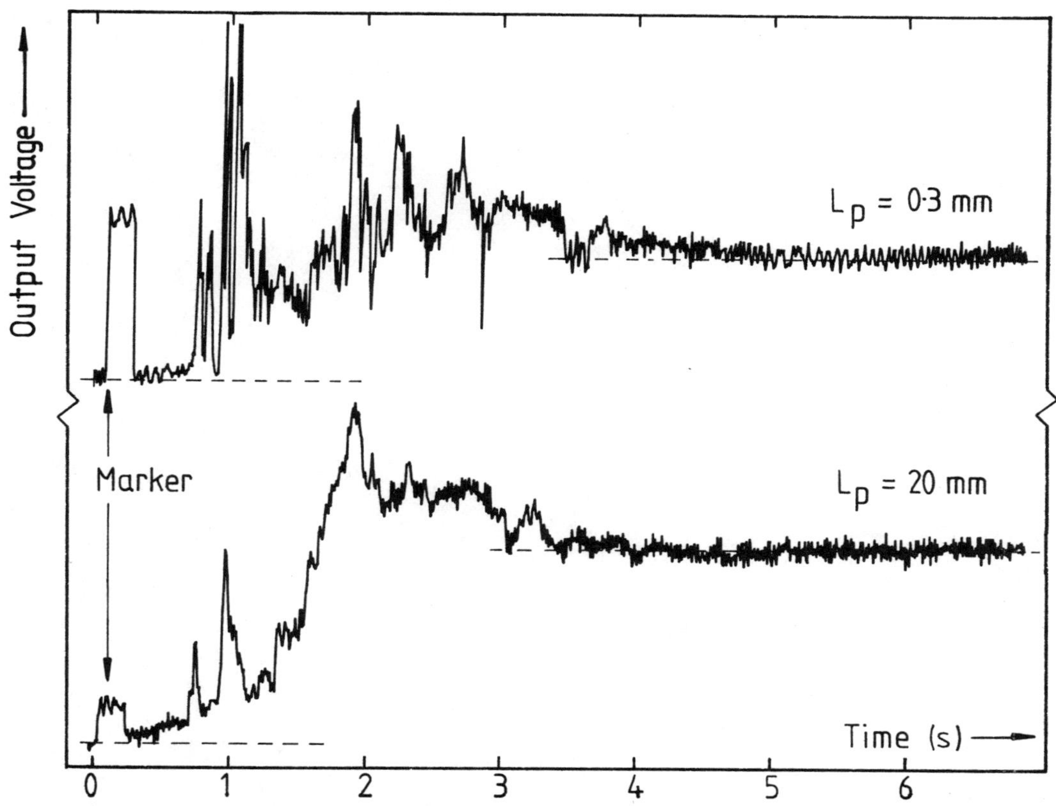

Figure 6. Comparison of the output from the 0.3 and 20 mm probes for the standard conditions: N = 6.7 s^{-1}, V_s = 20 cm^3, water.

NUMERICAL SIMULATION OF LIQUID HOMOGENIZATION
IN A STIRRED TANK WITH THE DISC TURBINE AGITATOR

Z. Jaworski*, I. Fořt** and F. Stręk*

* Technical University of Szczecin
71-065 Szczecin, Al. Piastów 42
Poland

**Institute of Chemical Technology
166-28 Prague 6, Suchbatarova 5
Czechoslovakia

Summary

Some details of the numerical solution of the homogenization model
for a stirred tank have been presented. The results of simulation of
the homogenization course have been evaluated in terms of three indices.
Their values have been compared with the published experimental data
for three agitator sizes. The computed values for two indices agreed
closely with the corresponding experimental data. The applicability of
the proposed simulation method has been demonstrated.

Held at Wurzburg, 10-12 June, 1985.

Organised by DVCV· Deutsche Vereinigung für Chemie- und Verfahrenstechnik
(German Association of Chemical and Process Engineering).

Organisation: GVC·VDI-Gesellschaft Verfahrenstechnik und Chemieingenieurwesen. 

NOMENCLATURE

A = constants in Eq./6/

c = tracer concentration

C = dimensionless tracer concentration = $(c - c_{min})/(c_{max} - c_{min})$

$\overline{C}$ = mean dimensionless tracer concentration

C_μ = constant in Eq./7/

D = agitator diameter

D_e = effective diffusion coefficient

D_t = turbulent diffusion coefficient

E = dimensionless variable = $4D_e R/nT^2$

G = dimensionless variable = $D_e R/nH^2$

h = agitator distance from bottom of vessel

H = liquid level height

I = homogeneity indices defined in Eqs./8,9,10/

$\overline{K}$ = kinetic energy of turbulence

n = agitator speed

r = radial coordinate

R = dimensionless radial coordinate = $2r/T$

t = time

T = tank diameter

U = dimensionless variable = $2v_r/nT$

v_r = radial component of mean velocity

v_z = axial component of mean velocity

w = width of agitator blades

W = dimensionless variable = $v_z R/nH$

z = axial coordinate

Z = dimensionless axial coordinate = z/H

ε = dissipation rate of turbulence energy

Θ = dimensionless time = nt

1. INTRODUCTION

Mathematical models for the description of homogenization process in
a stirred tank are based either on the dynamic and stochastic characteristics of the apparatus or on the differential mass balance of a tracer.
The recent models (Refs. 1,2) of the latter group referred to batch turbulent mixing in a baffled tank with the disc turbine agitator. They
were formulated for three independent variables due to the assumption of
fluid velocities and of tracer concentrations. A similar model was also
used by Placek and Tavlarides (Ref. 3) to predict circulation time in a
stirred tank. Liepe (Ref. 4) presented a special numerical procedure to
avoid errors caused by false diffusion.

2. MATHEMATICAL MODEL

The general concept of modifications in the primary model developed
in (Ref. 2) and solved numerically in (Ref. 5) was presented elsewhere
(Ref. 6). The basic model equation can be written in a dimensionless
version as

$$\frac{\partial(RC)}{\partial\Theta} + \frac{\partial(UC)}{\partial R} + \frac{\partial(WC)}{\partial Z} = \frac{\partial}{\partial R}\left(R\frac{\partial C}{\partial R}\right) + \frac{\partial}{\partial Z}\left(G\frac{\partial C}{\partial Z}\right) \qquad /1/$$

The boundary conditions determine:
- the initial regions of high /C=1/ and low /C=0/ concentration of the
  tracer

$$C = 1 \qquad \text{for} \quad \Theta = 0, \ R_{min} < R < R_{max}, \ Z_{min} < Z < Z_{max} \qquad /2/$$

$$C = 0 \qquad \text{for} \quad \Theta = 0, \ 0 < R < R_{min}, \ R_{max} < R < 1, \ 0 < Z < Z_{min}, \ Z_{max} < Z < 1 \qquad /3/$$

- axial symmetry of concentration field and isolation from mass transfer
  with its environment /batchwise operation/

$$\frac{\partial C}{\partial R} = 0 \qquad \text{for} \quad \Theta > 0, \ R=0, \ R=1, \ 0 < Z < 1 \qquad /4/$$

$$\frac{\partial C}{\partial Z} = 0 \qquad \text{for} \quad \Theta > 0, \ 0 < R < 1, \ Z=0, \ Z=1 \qquad /5/$$

3. NUMERICAL PROCEDURE

The numerical approach was based on the method by Patankar and Spalding (Refs. 6,7). The calculation domain in the dimensionless coordinates
R,Z,Θ was divided into N "control volumes" by the grid adopted. Owing to
some assumptions on the C profile inside every control volume the discretized form of Eq./1/ became a linear algebraic equation. This equation
can be written for the i,j-th control volume as

$$C_{i,j} = A_1 C_{i,j+1} + A_2 C_{i,j-1} + A_3 C_{i+1,j} + A_4 C_{i-1,j} + A_u C_u \qquad /6/$$

Eq./6/ links the C value for the i,j-th grid node with the four neighbouring ones and with its value C_u for the former time step. The A coefficients were derived from the local values of U,W,E,G and from the grid size. To calculate the A values the distribution of mean velocity components and of the turbulent diffusion coefficient D_t must be known.

The approximate velocity distributions for a baffled tank with the disc turbine agitator proposed in literature (Refs. 8,9) led to small deviations from the continuity requirements. Therefore another approximate distribution of mean velocity components had to be developed, which obeyed strictly the difference mass balance of the mixture (Ref. 10).

The D_t values were calculated in two ways. For the regions of the agitator discharge stream and of the wall jet the experimental data cited by Liepe (Ref. 11) were adopted. For the remaining area of the tank the data on local energy dissipation rate ε (Ref. 12) and on fluctuating velocities (Refs. 13,14) were applied. The D_t values were computed by means of Eq./7/.

$$D_t = C_\mu \bar{k}^2/\varepsilon \qquad /7/$$

The numerical solution of Eq./1/ is reduced to the solution of a set of N algebraic, fully implicit Eqs./6/ formulated for N control volumes. Patankar (Ref. 6) recommended to solve them iteratively by means of a line-by-line method which is a combination of the tri-diagonal-matrix-algorithm and the method of Gauss-Seidel.

A computer programme based on the described procedure was prepared. A matrix of C values for the grid nodes was derived iteratively for the consecutive steps of dimensionless time. The iterations were considered as converged when the individual C values did not differ more than by 0,02% from the former ones and when the actual mean concentration did not differ more than by 0,002% from the initial one.

By means of the obtained distributions of the tracer concentration within the stirred tank the values of different homogeneity indices were chosen for which their experimental dependency on Θ was known. These indices were defined in terms of the applied dimensionless concentration C by the following equations:
- the index I_{PL} by Prochazka and Landau (Ref. 16)

$$I_{PL} = \frac{1}{N} \sum_{i=1}^{N} \left| \frac{C_i - \bar{C}}{C_{\alpha_i} - \bar{C}} \right| \qquad /8/$$

- the index I_{HL} by Henzler and Liepe (Ref. 17)

$$I_{HL} = \frac{C_{max} - C_{min}}{\overline{C}} \qquad\qquad /9/$$

- the index I_{KL} by Khang and Levenspiel (Ref. 18)

$$I_{KL} = \frac{C_{max} - \overline{C}}{\overline{C}} \qquad\qquad /10/$$

4. RESULTS AND DISCUSSION

A baffled, cylindrical tank with the disc turbine agitator was chosen to test the proposed model of homogenization. The main geometrical proportions of the mixer were as follows: $T=0,3m$, $H/T=h/D=1$, $w/D=1/5$. Several computer simulations were made for the turbulent flow of the agitated liquid.

The obtained profile of the indices defined in Eqs./8,9,10/ for $D/T=1/3$ was compared with the course of their experimental values. This is presented graphically in Figs. 1,2,3. The most interesting range for a practical purpose is the one where the indices fall below 0,1. The predicted course appeared to converge to the experimental one either poorly /Fig. 1/, fairly /Fig. 2/ or well /Fig. 3/ in dependence of the data source and of the index applied.

A comparison of the time necessary to reach the commonly used value of 0,05 of the applied indices was also made for different D/T ratios. The model allowed to perform simulations for $D/T=0,25 \div 0,50$. The $I=f(\Theta,D/T)$ correlations (Refs. 16,17,18) were also valid for this range. The computation results are presented in Table 1. The predicted values of time Θ_p to reach $I_{PL}=0,05$ differ again from the experimental values Θ_e by Prochazka and Landau (Ref. 16). This is probably caused by the influence of the applied probe size on the experimental results which was proved by others (Ref. 15). The values of Θ_e and Θ_p for the I_{HL} and I_{KL} indices are rather consistent for every D/T ratio applied /see Table 1/. This can be recognized as a proof of the applicability of the model described.

The possible errors which can arise in the modelling can be attributed to:
- model simplifications,
- discretization procedure,
- approximate distributions of mean velocity components and turbulence characteristics,
- numerical approach accompanied by false diffusion, truncation errors and incomplete convergence.

<u>5. CONCLUSIONS</u>

- Further investigation is required to explain some differences between
the predicted and experimental values of the homogeneity index I_{PL}.
- The presented model and the method of its solution can be recognized
as a useful tool to predict a homogenization course in a stirred tank
with the disc turbine agitator.

<u>6. REFERENCES</u>

1. <u>Liepe F.</u>: "On the synthesis of models for flow processes in process
engineering". /Zum Aufbau von Modellen für Strömungsprozesse der Ver-
fahrenstechnik/. Maschinenbautechnik <u>32</u>,7,1983,pp.313-316./In German/.

2. <u>Jaworski Z.</u>: "Modelling of liquid homogenization in stirred tank.
I. Mathematical model and computational algorithm". /Modelowanie
homogenizacji cieczy w mieszalniku. I.Model matematyczny i algorytm
obliczeń/. Prace Naukowe Politechniki Szczecińskiej No.270,14,1984,
pp.23-36. /In Polish/.

3. <u>Placek J.</u>, <u>Tavlarides L.L.</u>: "A turbulent flow model applied to fluid
flow in a stirred tank". 1982 Annual AIChE Meeting,/Los Angeles,USA,
Nov.14-19,1982/, AIChE, 1982, Paper 101C, 13pp.

4. <u>Liepe F.</u>, <u>Sperling R.</u>, <u>Kawa P.</u>: "Modelling of homogenization processes". 8th International Congress CHISA`84,/Prague,Czechoslovakia,
Sept.3-7,1984/ Czechoslovak Academy of Sciences,1984,Paper V.3.63,6pp.

5. <u>Jaworski Z.</u>: "Simulation of mixing in an agitated vessel". 8th International Congress CHISA`84,/Prague,Czechoslovakia, Sept. 3-7,1984/
Czechoslovak Academy of Sciences, 1984, Paper V.3.43, 7 pp.

6. <u>Patankar S.V.</u>: "Numerical heat transfer and fluid flow". Washington,
Hemisphere Publishing Co., 1980, pp. 25-109.

7. <u>Patankar S.V.</u>, <u>Spalding D.B.</u>: "A calculation procedure for heat,
mass and momentum transfer in three-dimensional parabolic flows".
International Journal of Heat and Mass Transfer <u>15</u>,10,1972, pp.
1787-1806.

8. <u>Platzer B.</u>, <u>Noll G.</u>: "An analytical approximate solution for the flow
in baffled tanks with radially pumping agitators". /Eine analytische
Näherungslösung für die Strömung in bewehrten Behältern mit radial-
fördernden Rührern/. Chemische Technik <u>33</u>,8,1981, pp. 430-432 /In
German/.

9. <u>Fořt I.</u>, <u>Obeid A.</u>, <u>Březina V.</u>: "Flow of liquid in a cylindrical
   vessel with a turbine impeller and radial baffles". Collection of
   Czechoslovak Chemical Communications <u>47</u>,1,1982, pp. 226-239.

10. <u>Jaworski Z.</u>: Unpublished results.

11. <u>Liepe F.</u>: "Some applications of the theory of turbulence to problems
    of mixing". Advances in Mechanics /edited in Warsaw, Polish Academy
    of Sciences/ <u>4</u>,4,1981, pp. 3-45. /In Russian/.

12. <u>Barthole J.P.</u>, <u>Maisonneuve J.</u>, <u>Gence J.N.</u>, <u>David R.</u>, <u>Mathieu J.</u>,
    Villermaux J.: "Measurement of mass transfer rates, velocity and
    concentration fluctuations in an industrial stirred tank".
    Chemical Engineering Fundamentals <u>1</u>,1,1982, pp. 17-26.

13. <u>Fořt I.</u>, <u>Rogalewicz V.</u>, <u>Richter M.</u>: "Simulation of two-phase flow
    in a mechanically agitated gas-liquid charge". Collection of
    Czechoslovak Chemical Communications /In press/.

14. <u>Schwartzberg H.G.</u>, <u>Treybal R.E.</u>: "Fluid and particle motion in tur-
    bulent stirred tanks". Industrial and Engineering Chemistry, Fun-
    damentals <u>7</u>,1,1968, pp. 1-12.

15. <u>Thyn J.</u>, <u>Novak V.</u>, <u>Pock P.</u>: "Effect of the measured volume size on
    the homogenization time". The Chemical Engineering Journal <u>12</u>,1976,
    pp. 211-217.

16. <u>Prochazka J.</u>, <u>Landau J.</u>: "Homogenation of miscible liquids in the
    turbulent region". Collection of Czechoslovak Chemical Communications
    <u>26</u>,12,1961, pp. 2961-2973.

17. <u>Henzler H.J.</u>, <u>Liepe F.</u>: "Investigations on substance joining in
    fluid phase". /Untersuchungen zum Stoffvereinigen in flüssiger Phase.
    Teil 2. / Chemische Technik <u>26</u>,8,1974, pp. 516-517, and Depotbericht
    No. 85/74. /In German/.

18. <u>Khang S.J.</u>, <u>Levenspiel O.</u>: "New scale-up and design method for
    stirrer agitated batch mixing vessels". Chemical Engineering Science
    <u>31</u>,7,1976, pp. 569-577.

Table 1.   Values of the dimensionless time Θ necessary to obtain $I=0,05$ in experiments Θ_e and in simulations Θ_p.

D/T	$I_{PL}=0,05$		$I_{HL}=0,05$		$I_{KL}=0,05$	
	Θ_e (Ref.16)	Θ_p	Θ_e (Ref.17)	Θ_p	Θ_e (Ref.18)	Θ_p
0,250	51,1	24,4	44,3	46,0	44,7	44,0
0,333	24,4	13,8	24,9	25,4	23,1	24,5
0,500	8,6	6,3	11,1	11,5	9,1	10,1

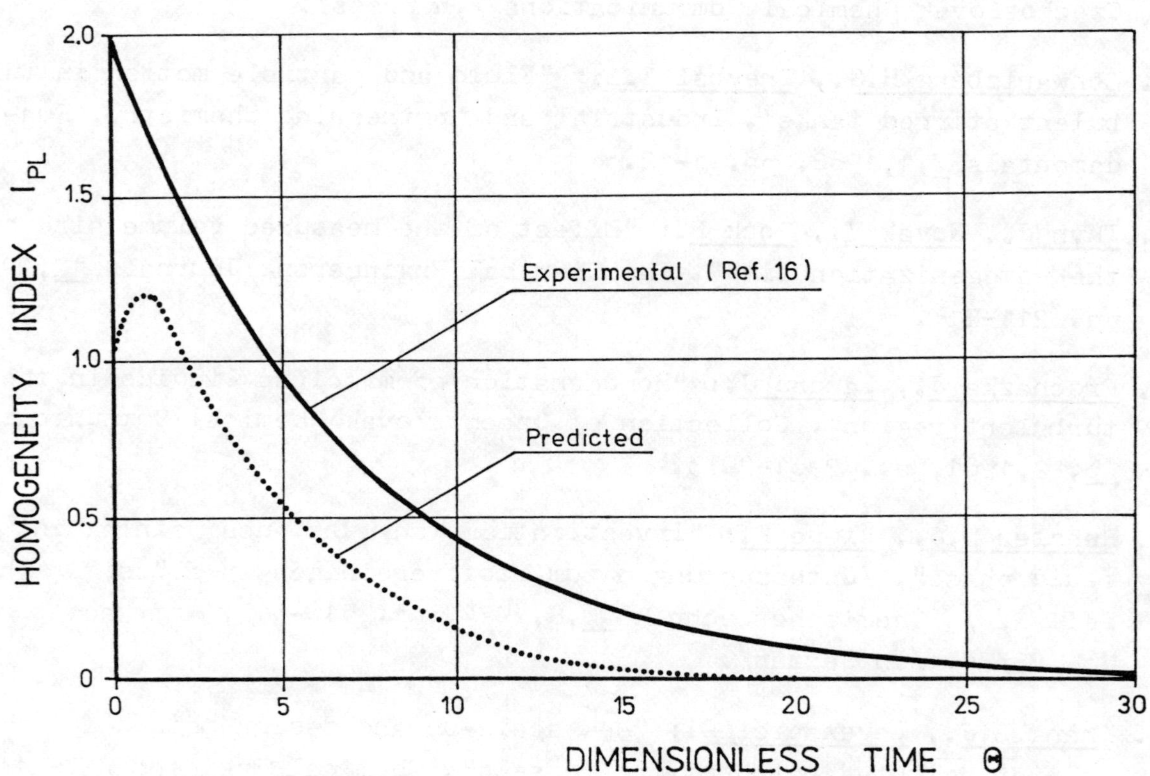

Fig. 1.   Experimental and calculated profiles of the I_{PL} index.

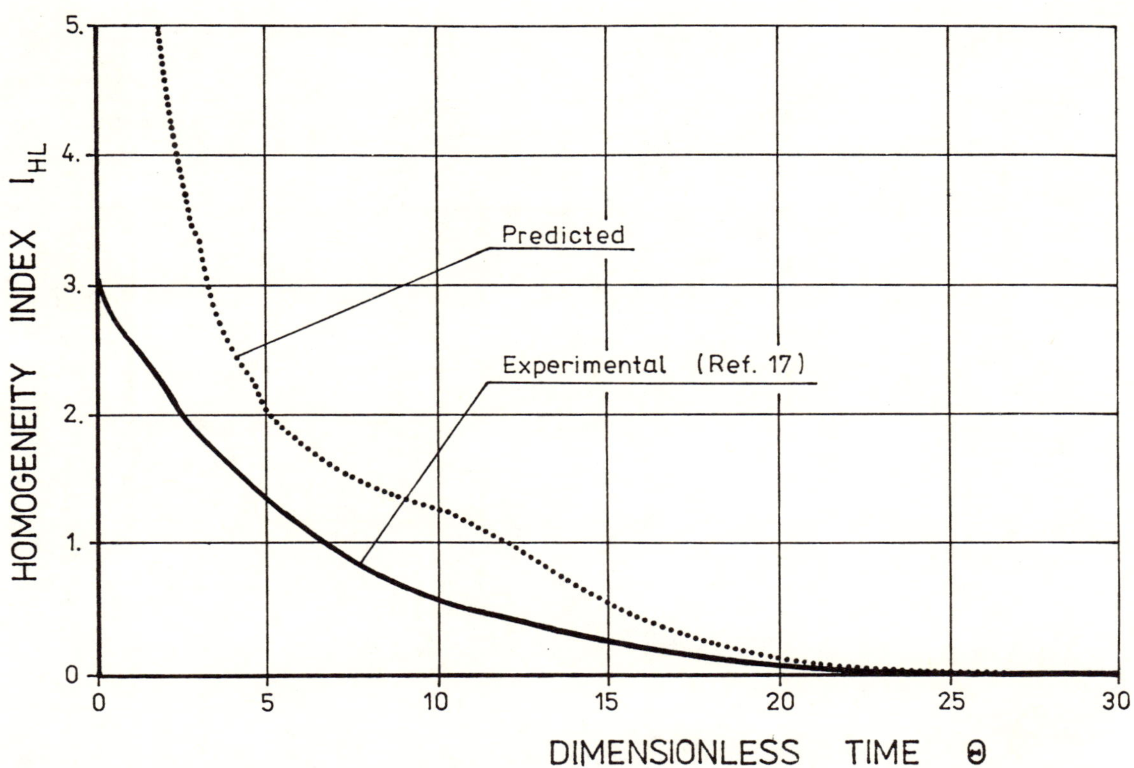

Fig. 2.   Experimental and calculated profiles of the I_{HL} index.

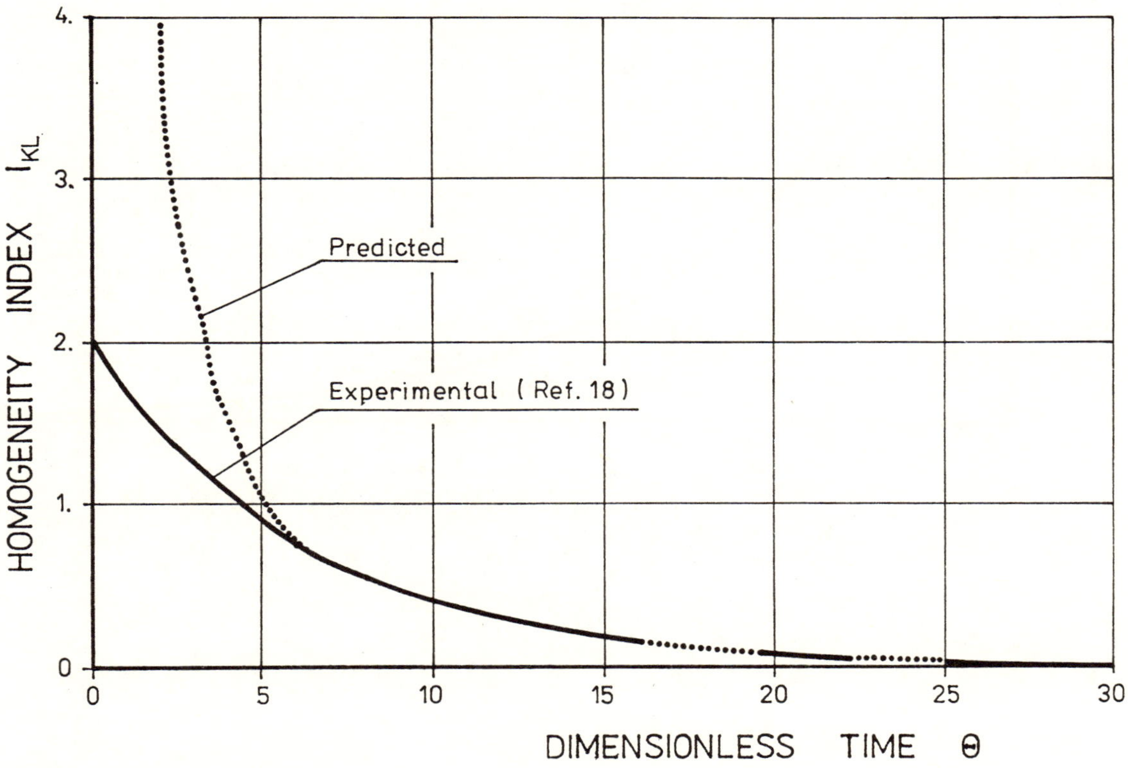

Fig. 3.   Experimental and calculated profiles of the I_{KL} index.

Mixing in Twin Screw Extruders

David B. Todd
Baker Perkins Inc.
Saginaw, MI   48601

Held at Wurzburg, 10-12 June, 1985.

Organised by DVCV· Deutsche Vereinigung für Chemie- und Verfahrenstechnik
(German Association of Chemical and Process Engineering).

Organisation: GVC·VDI-Gesellschaft Verfahrenstechnik und Chemieingenieurwesen. 

<u>Introduction</u>:

The mixing process used in the plastics industry to incorporate additives and fillers into very viscous polymers is known as <u>compounding</u>. The requirements for effective compounding are:

a. complete fluxing of the base resin.
b. uniform dissolution of all soluble components in the base polymer matrix.
c. uniform distribution of all solid ingredients, including dispersion of agglomerates where present.
d. such other modifications of grafting, alloying, vis-breaking, etc., as may be desired.

Economics dictate that the above tasks be achieved at the minimum possible temperature, otherwise energy is needlessly put into the process and must be paid for twice, since the stored thermal energy must also be removed from the polymer.

Melting is best accomplished by conversion of work energy, as the thermal properties of most polymers preclude effective heat transfer through surfaces. Heat transfer at the surface can be important, however, in controlling the wall film properties such as friction coefficient, viscosity, and wetability, all of which can affect the amount of work energy imparted by the compounder.

Supplying the energy required for fluxing mechanically, such as in an extruder, frequently can achieve the dispersion and distribution requirements by the time that melting has occurred.

<u>Twin Screw Compounders</u>:

This work concerns intermeshing, co-rotating, self-wiping twin screw compounders. This type equipment offers some unique characteristics that can be particularly advantageous in compounding. Comparisons of twin co-rotating equipment with single screw and with counter-rotating twin screw have been made previously (1,2).

With the self-wiping specification the shape of
the flight channel cannot be arbitrarily varied, and
is fixed primarily by the centerline to diameter ratio
(Cl/D). Booy (3) has described the geometric relations required for self-wiping for various screw
combinations.

Screw profiles and cross-sections are shown for
single, double, triple, and quadruple start screws in
Figure 1. With an odd number of starts, the screws
are in phase. With two starts, the profiles are rotated
90° with respect to each other, and with four starts, 45°.

The volume of the figure-8 barrel available for
compounding is shown in Figure 2 as a function of the
Cl/D ratio for conventional single, double and triple
start screw profiles. As Cl/D ratio is decreased, the
amount of the cross-section filled by metal also decreases, but so also does the torque capability of the
rotors. The selection of an appropriate Cl/D ratio
involves a compromise between the ability to supply the
necessary work energy through the shafts while retaining
enough depth to the flights to have sufficient forwarding
capacity without the need for crammer feeders.

The relative surface areas of self-wiping twin
screws are shown in Figure 3. The surface of the screws
is divided into two categories: the land, or screw-
to-barrel rubbing surface (L), and the flank (F) which
never contacts the barrel wall. If the rotors are cored
for heat transfer, the responses of the L and F surfaces
will be different.

Kneading Paddles:

Paddles with the same cross-section as the screws
shown in Figure 1 are available, and can be considered
as screws with 90° helix. Usually they are used in
arrays of short lengths, staggered along a shaft. An
individual pair of paddles has no axial forwarding
capability, and each three rotations of the shafts
causes the barrel contents to undergo four stages of
compression and expansion.

This is shown in Figure 4, where the shaded material between paddle 1 and the barrel goes through expansion over the top saddle point, and then gets compressed again between paddle 2 and the barrel.

Unless the barrel is operating starved, the material being compressed will want to squish over the tips, or axially forward or aft along the barrel. This kneading action is unique to co-rotating self-wiping twin shaft systems.

Also unique is the continual transfer of material from one barrel half to the other, and the smearing change of direction as the material passes over the two saddle points.

The extent of axial movement will depend upon the offset position of the next upstream and downstream paddles. With multi-keyed slip-on paddles on a keyed shaft, a great variety of forwarding or retarding arrays are feasible.

If the barrel is running partially starved, this compression-expansion effect disappears. Placing a variable resistance at the end of series of paddles, such as a barrel valve (5), allows control over the number of paddle pairs that are working in a completely filled condition, and thus permits control over the mixing energy imparted.

Twin Screw Processing:

Because of the kneading action of paddle sections, energy for completely fluxing polymer can be imparted in a very short axial length, and a variety of compounding functions can be staged.

Figure 5 shows a twin screw compounder in schematic form. Free-flowing ingredients can be metered into top ports. The forwarding capacity of deep flighted twin screws is usually so great that crammer feeders are not required. Barrel valves can be used to assure complete fluxing of a polymer before addition of fillers. A side entering auxiliary feeder can be used instead of the top entering port. The rotor configuration is arranged with

feed screws following kneading sections to present a
partially starved condition at the point where subsequent
addition is to take place.

Liquid ingredients can be pumped into the barrel at
various ports along the barrel. Staging of liquid addition is frequently very desirable. For example, in a
polymer letdown process, adding all of the solvent initially
will tend to cause the polymer to "liver", to slip around
with very slow absorption of the solvent into the polymer.
Staging allows the mass to absorb solvent progressively.

Vent ports can be used to evict air which accompanies
addition of powdered ingredients, or to remove by-product
volatile components.

Designing a Rotor Configuration:

The condition of each feed stream needs to be
identified, with particular attention paid to the soften-
ing points of the polymers and the viscosity after melting. Are there any critical temperatures which must not
be exceeded? Does one of the solid additives need to be
deagglomerated?

Multi-stage mixing is used for handling stages of
varying viscosities or for alloying polymers having
vastly different melt temperatures or viscosities.

Paddles are better mixing components than continuous
screw sections, as there is a dispersing, smearing, action
in addition to the compression-expansion kneading. Screws
provide more scraping, and perhaps a rolling action, on the
material being forwarded.

The extent of the work energy to be imparted is also
controlled by the number of mixing paddles and the degree
of conveying imparted by the amount of offset. Paddles
at 90° alternating orientation provide no net forwarding,
but still will be largely self-emptying when the feed is
turned off. Forward conveying is achieved by each downstream paddle lagging its predecessor by some lesser offset, such as 60°, 45°, or 30°. If the downstream paddle
leads, then a reverse or holdback action occurs.

Because of the ability to effect holdback by barrel valve and kneading paddle configuration, throughput rate is not necessarily proportional to rotor speed. Thus, more work energy can be imparted at higher rotor speeds without incurring a proportionate reduction in residence time.

Typical single stage and two-stage rotor configurations are shown in <u>Figure 6</u>. The two-stage configuration is particularly advantageous for addition of a fragile filler like fiberglass, or of a lower melting point polymer after complete fluxing of the higher melting polymer in the first zone.

With a two-stage rotor configuration such as shown in Figure 6, the extent of process control was demonstrated with an 80mm twin screw compounder at various barrel valve settings. An engineering resin was being compounded at 150 kg/h at a rotor speed of 300 RPM. The final product temperatures could be varied between 213 to 252°C as a function of the downstream barrel valve closure.

<u>Compounding Performance:</u>

Mixtures of polymers such as ABS, SAN, PS, were compounded into a base of semi-rigid PVC to demonstrate the mixing action of kneading paddles. The 100mm twin screw compounder was dead-stopped and rapidly opened, and the contents allowed to solidify. Carcasses were removed from around the kneading blocks, and eventually sliced to illustrate the extent of mixing.

The sliced sections indicate a rapid progression from a barely cohesive mixture to a fully compounded condition in less length than 1.5 barrel diameters. (Colored slides of the sections will be presented). Close examination of the cross-sections illustrates the rapid elongation smearing, and dissolution of the higher melting polymer pellets into the PVC matrix.

<u>Conclusions:</u>

The unique compression/expansion and dispersion features of kneading paddles in co-rotating, self-

wiping, twin screw extruders permit rapid compounding
of polymer systems under the adiabatic conditions which
lead to optimum utilization of energy.

Staging of processing operations along sections of
the barrel is readily accommodated by various arrays of
kneading paddles and screw components. For any given
array, additional process control can be exercised by
use of a barrel valve which can be automatically adjusted
to hold a given energy input or product temperature output.

References:

1. Kroll, W. Kunststoffe 67 576 (1977).

2. Hermann, H; U Burkhardt, S. Jakopin, SPE 33rd
    ANTEC, Preprints 481 (1977).

3. Booy, M.L., Polym, Eng. Sci. 18, 973 (1978).

4. Loomans, B.A., U.S. Patent 3,900,187 (1975).

5. Todd, D.B., SPE ANTEC, Preprints 25, 220 (1979).

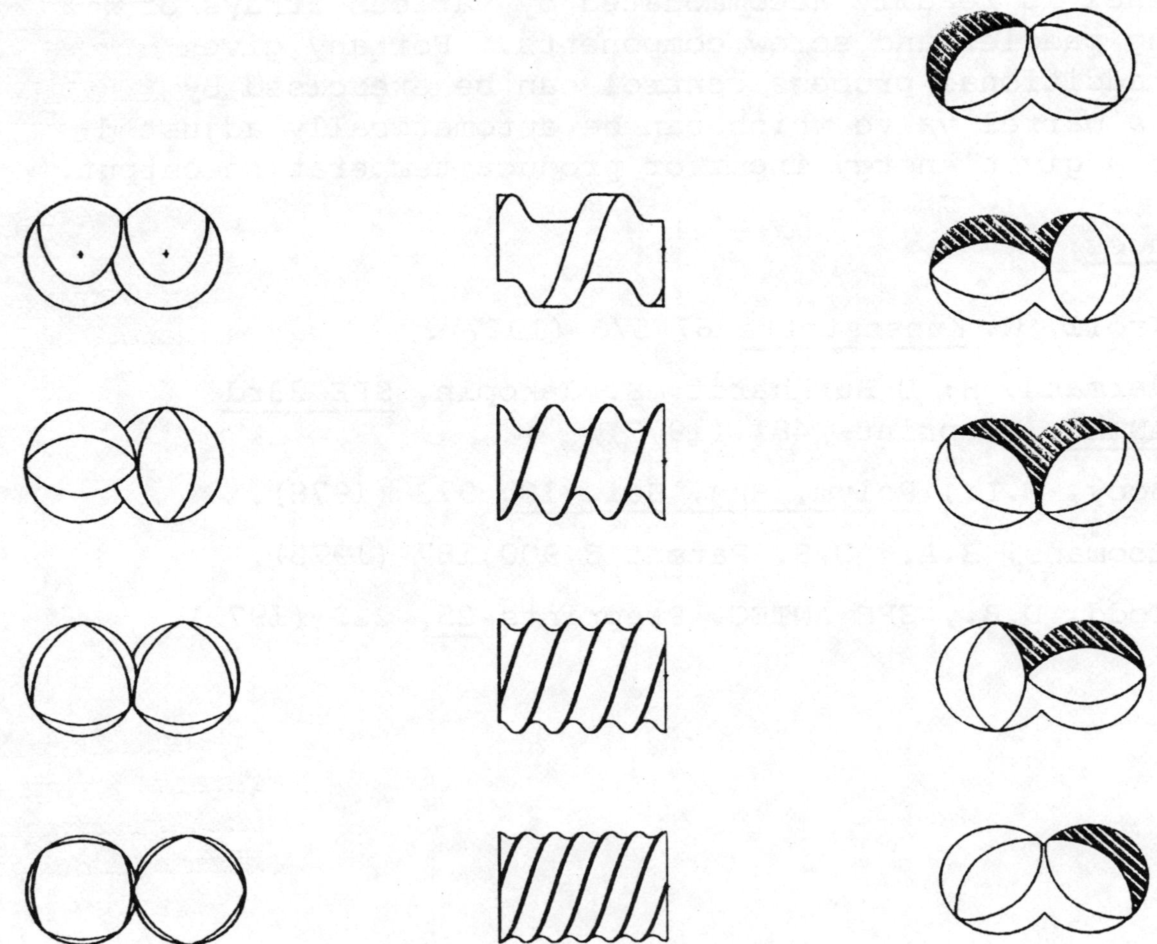

FIG. 1 SCREW PROFILES   FIG.4 COMPRESSION EXPAN-
SION EFFECT

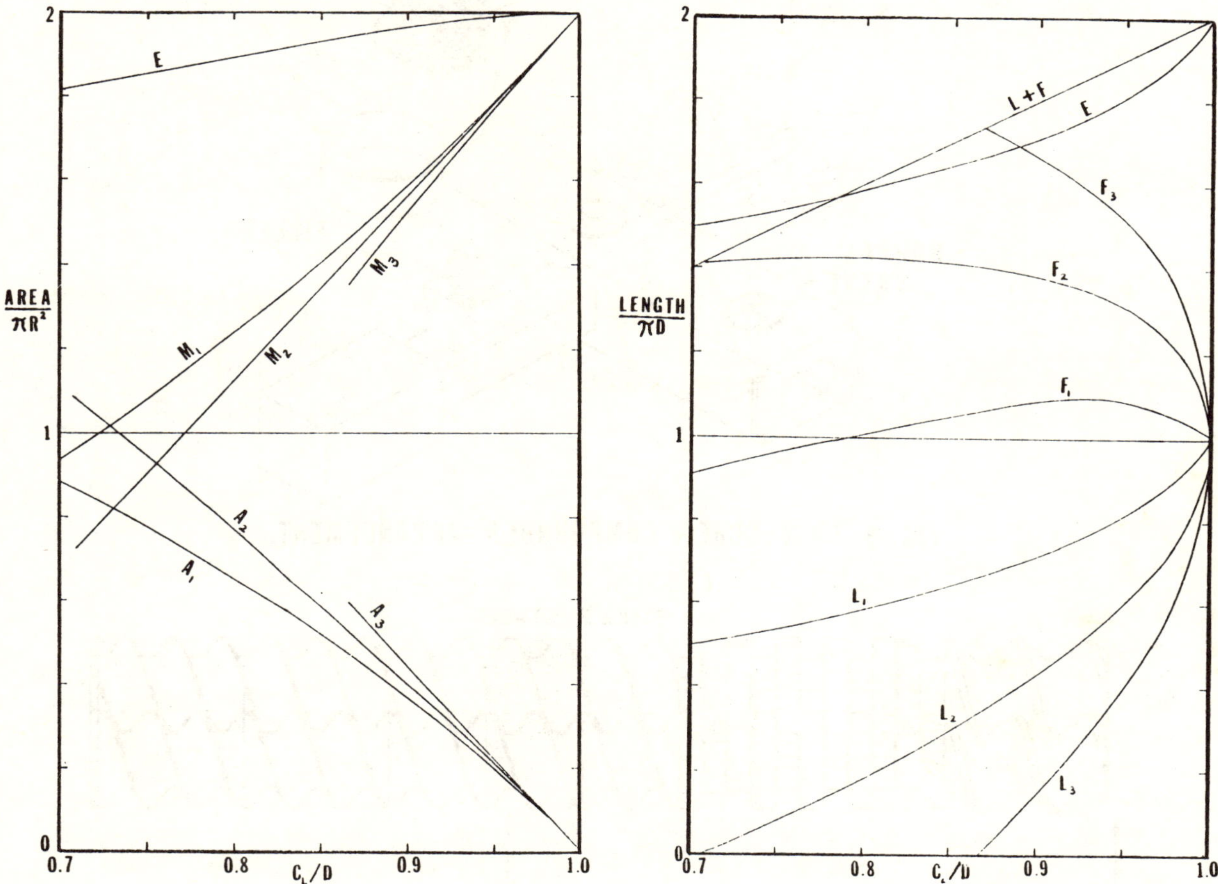

FIGURE 2. RELATIVE VOLUME PER UNIT LENGTH, SINGLE, DOUBLE, AND TRIPLE START SCREWS. E EMPTY BARREL. M METAL. A AVAILABLE FOR PROCESS.

FIGURE 3. RELATIVE SURFACE PER UNIT LENGTH, SINGLE, DOUBLE, AND TRIPLE START SCREWS. E EMPTY BARREL. F FLANK OF SCREW. L LAND OF SCREW CONTACTING BARREL.

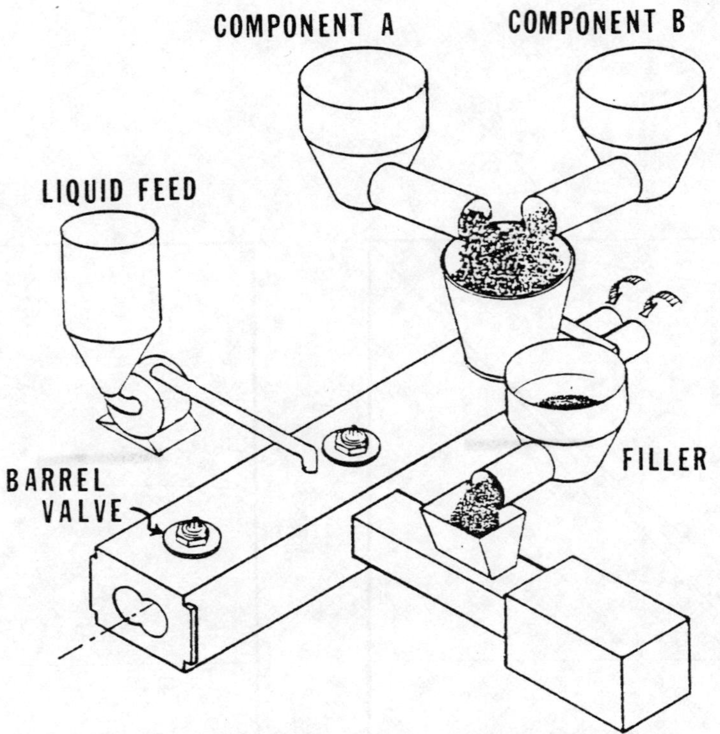

FIG. 5 TWIN-SCREW COMPOUNDER ARRANGEMENT

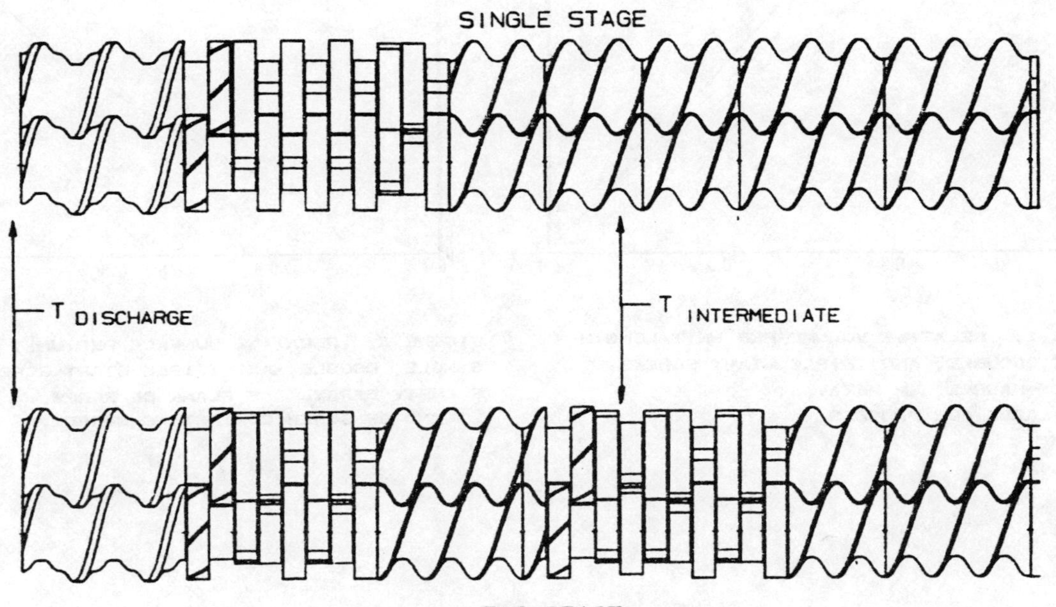

FIG. 6
TYPICAL ROTOR CONFIGURATIONS

COMPARATIVE EFFICIENCIES FOR TURBULENT PIPELINE MIXING
FOR DRAG REDUCING AND OTHER AQUEOUS SOLUTIONS

P.V. Bartels, R. Van Beelen and John M. Smith,
Laboratory for Physical Technology, Delft University of Technology,
Prins Bernhardlaan 6, Delft, The Netherlands.

<u>Summary</u>

Turbulent pipeline mixing in aqueous solutions has been studied using techniques developed by Hiby, based on an acid–base reaction, though here with an iodine – thiosulphate reaction. The two techniques are compared. The iodine conversion, which is controlled by turbulent diffusion, has been measured along the tube. Aqueous ionic solutions have been used and the effects of drag reducing quantities of polyacrylamide studied.

Held at Wurzburg, 10-12 June, 1985.

Organised by DVCV· Deutsche Vereinigung für Chemie- und Verfahrenstechnik
(German Association of Chemical and Process Engineering).

Organisation: GVC·VDI-Gesellschaft Verfahrenstechnik und Chemieingenieurwesen. 

Symbols

C_o	Initial concentration of iodine	mol/1
C_e	Final concentration of iodine	mol/1
C_n	Normalised concentration of iodine	mol/1
L_m	Mixing length	m
f	Friction factor	-
R	Conversion rate	1/s
Re	Reynolds number	-
t	Fluid residence time	s
V	Superficial velocity	m/s
X	Ratio acid/base or thiosulphate/iodine	-
ε	Energy dissipation needed to homogenise a given mass	J/kg

<u>Introduction</u>

Much is known about the mechanism of drag reduction caused by polymers like polyethyleneoxide or polyacrylamide (PAAm). In several pipe flow studies, for instance those of Mizushina and Usui (1), Thielen (2) and of McComb and Rabie (3), it has been shown that in pipe flow the boundary layer is thickend and the maximum turbulence intensity is located further from the wall with increasing concentrations of polymer. Whilst the absolute maximum of intensity is reduced, the intensity in the centre is not affected, however there is a relative shift towards lower turbulence frequencies. As a result, it can be expected that the modification in turbulence structure resulting from the presence of polymer in the solution will not only decrease the friction at the wall but also the rate of homogenisation in the fluid bulk.

In this study the effect of the concentration of polymer on turbulent mixing has been investigated using fast chemical reactions. Because of the low concentrations used, the elastic effects produced by the polymer molecules will be local, whilst the fluid as a whole is virtually Newtonian.

<u>Experimental</u>

The apparatus

The apparatus for the reaction experiments consists is a precision bore glass tube, 9 m long with an inner diameter of 16.4 mm (figure 1). Aqueous solutions are prepared in two tanks of 250 1. These solutions were pumped through the pipe in equal volumetric flows, controlled by maintaining zero pressure difference between the identical entrances. Initially the two flows are separated by a partition, 0.42 m long, in the tube.

It was possible to traverse a light absorption detector along the tube. Red light from an LED travelled across a diameter and fell on a photodetector (figure 2). The beam was about 3 mm in diameter. The extent of mixing could be followed from measurements of the decoloration of the blue indicators used, starch for an iodine – thiosulphate reaction and bromothymol blue for acid neutralisation. The signal detected by the photodiode was processed by a HP-1000 computer. Samples were taken with a frequency of 33 Hz, and every measurement was the average of 100 samples.

The pressure drops have been measured over .5 and 5 m length of the 16.4 mm pipe with a Depex DP15 pressure transducer. The average velocity was determined by an inductive flow meter. These meters are also coupled to the computer.

Laser Doppler Anemometry

Turbulence profiles across the tube were measured with a laser doppler anemometer. A reference beam method has been used with a rotating grating preshift and tracking system. Measurements were made at 5 m downstream of the tube entrance. A short thin wall section of plastic tube surrounded with a rectangular chamber filled with the experimental fluid served to minimize the optical distortion. It was possible to measure in the axial, radial and tangential directions. The apparatus is similar to that used by Mizushina and Usui (1).

The results of the turbulence measurements will be presented in detail in a separate publication.

Reactions

The turbulent dispersion has been measured by fast chemical reactions. The acid – base reaction, suggested by Danckwerts (4) and developed by Hiby (5), has been compared with the iodine – thiosulphate reaction in solutions without PAAm. Because of the changes in behaviour of PAAm at different acidities, another reaction must be used. The iodine –

thiosulphate reaction in combination with a starch derivate is suitable. In one tank iodine is added (0.5 mmol/l) to the solution with the starch (0.2 g/l), giving a strong blue colour. The other tank contains sodium thiosulphate in a concentration up to 1.0 mmol/l. The ratio thiosulphate/iodine (X) is always larger than 1. This is necessary to obtain complete decoloration. This redox reaction is very fast (6).

It can be presumed that the conversion is diffusion controlled. As a result of the high energy input, the turbulent component of dispersion dominates.

Fluids used

The solutions were prepared from Dow Chemical polyacrylamide (PAAm) (Separan 30). A 2.5 % bulk solution was diluted to appropiate concentrations below 500 ppm. Measurements were made at least 48 hours after dilution to allow the solutions to stabilize, and to reduce the danger of changes due to degradation during the measurements. The viscosity was measured with Ubbelohde viscometers. The viscosities of all the solutions were below 3 mPas.

<u>Results and discussions</u>

Conversion along the pipe

The use of bromothymol – blue as an indicator for the HCl – NaOH reaction is well established as a means of following mixing and reaction in ionic solutions without polymer (7). The course of the decoloration has the same general pattern (figure 3 and 4). In figure 4, which is a semi – logarithmic plot of the tracer concentration against the elapsed time available for the mixing process, the local tracer concentration (iodine) C_n has been normalised by the initial concentration C_o and the concentration at the end of the tube C_e according to:

$$C_n = \frac{C_t - C_e}{C_o - C_e} \qquad (1)$$

In figure 3 the same has been done for the blue color which is proportional to the concentrations of volume elements with an excess of base. For the results used for figure 3 the signal to noise level is about 30. For the iodine – thiosulphate reaction the ratio is about 1000 at the data used in figure 4. Included in the noise is the light absorption by the glass wall. This figure shows the practical value of the starch indicator in following mixing processes. The colour – concentration relationship is not linear at very low concentrations. This means that it is difficult to use the final stages decoloration as a measure of the rate of mixing.

A distinction can be made between two regions. Following the mixing of the two potential cores a reduction in the blue color can be seen in the first few centimeters. The second region can be characterized as the mass transfer region for pipe flow.

In figure 4 $\ln(C_n)$ decreases linearly as the elapsed residence time t of a fluid element in the pipe increases. The assumption of first order kinetics is supported reasonably by this figure. For this reason the turbulent mass transfer can be described by a conversion rate R in analogy with the segregation reduction used by Hiby (5):

$$R = \frac{C_n}{t} = \frac{C_n L_t}{V} \qquad (2)$$

With V the superficial velocity and L_t the distance downstream of the partition reached by the fluid at time t.

In this form the conversion rate will be useful for studying the downstream dispersion effects of static mixers, like the Kenics mixer.

Influence polymer additives

The mixing length L_m, defined as the pipe length necessary to obtain a certain reduction in the scale of segregation (figure 6), is associated with the conversion rate R, according to equation 2, exept that the entry region has its own conversion rate. This mixing length has been used to study the effect of drag reducing agents on pipeline mixing.

The presence of polyacrylamide has a significant effect on the mixing (figure 5). Higher polymer concentrations have a much greater mixing length. Conversion rates are reduced because of a decrease in radial transport and an increased Kolmogoroff scale. The last effect is associated with the drag reduction of polymer solutions (figure 6). The mixing length increases almost exponentially with polymer concentration at the same zero shear viscosity base Reynolds number (figure 8).

Efficiency

As a measure of the efficiency of a mixing process, the energy dissipation needed to obtain a given rate of homogenisation, can be used. The pressure drop over the mixing length per fluid mass, gives the specific mixing energy ε directly (figure 7).

At the low polymer concentrations, the energy needed for mixing can be less than in solutions without polyacrylamide (figure 9). At higher concentrations than the concentration of PAAm at which the Virk asymptote (8) is reached (figure 6) for the given Reynolds number, specific mixing energy consumption increases rapidly. This can be explained in terms of the differing mechanism of drag reduction and turbulent mixing. Drag reduction is primarily caused by repression of the turbulence in the buffer region, and it is possible that this effect is more important than the bulk mixing. From the velocity measurements it can be shown that the distribution of velocity fluctuations between radial and axial directions is dependent on the concentration of PAAm. Generally radial dispersion is reduced more rapidly than that in the axial direction as concentration is increased. This can result in a greater fall in the pressure drop than the proportional increase in mixing length, thus reducing the total energy demand. At high concentrations of PAAm the axial velocity fluctuations are considerably reduced.

Conclusions

The iodine – sulphate reaction is convenient for examining mixing in turbulent pipeline flow. Mixing rates are strongly repressed in drag reducing polyacrylamide solutions, though the coupling with pressure drop phenomea is not straightforward.

References

1   Mizushina, T. and Usui, H.: Phys. of Fluids, $\underline{20}$, (1977), 100
2   Thielen, W.: Thesis, TH Aachen 1977
3   McComb, W.R. and Rabie, L.H.: A.I.Ch.E. J., $\underline{28}$, (1982), 547
4   Danckwerts, P.V.: Appl. Sci. Res. Sect., A 3, (1952), 279
5   Hiby, J.W., Verfahrenstechnik: $\underline{4}$, (1970), 538
6   Dodd, G. and Griffith, R.O.: Trans. Faraday Soc., $\underline{45}$, (1949), 546
7   Hiby, J.W.: Chem. Ing. Tech., $\underline{51}$, (1979), 704
8   Virk, P.S.: A.I.Ch.E. J., $\underline{21}$, (1975), 625

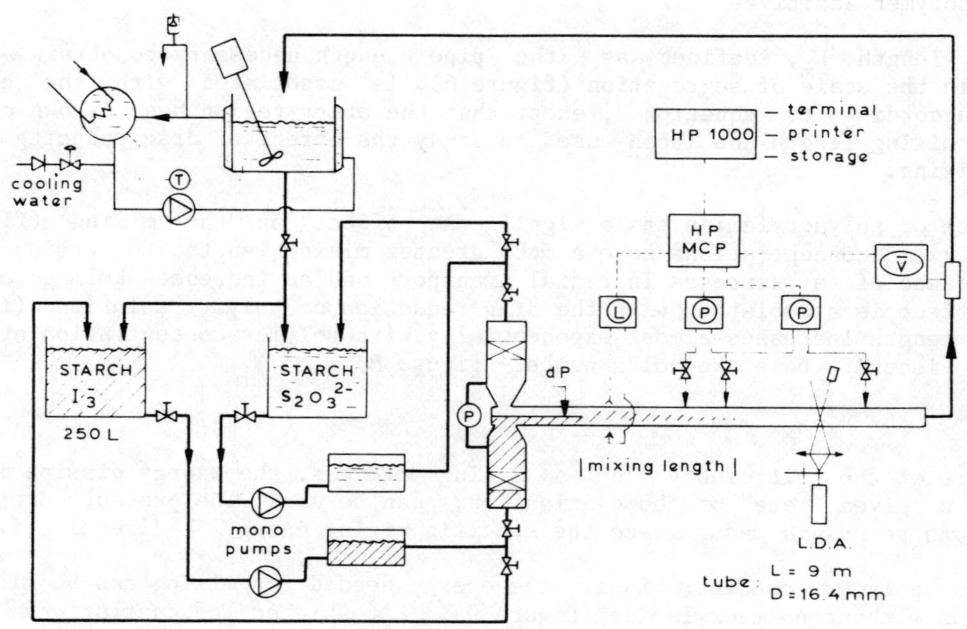

Figure 1   The general arrangement for the mixing experiments

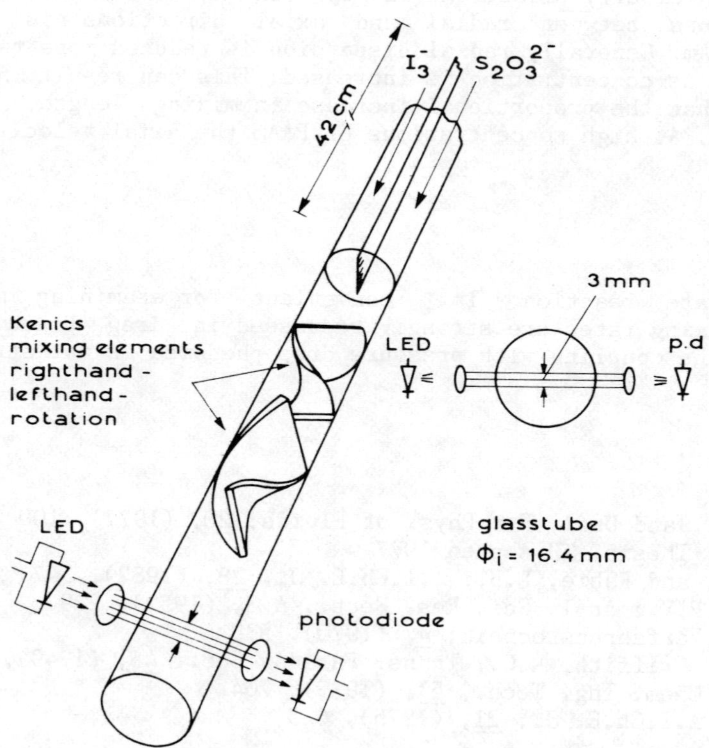

Figure 2   The initial part of the tube with the light absorption meter

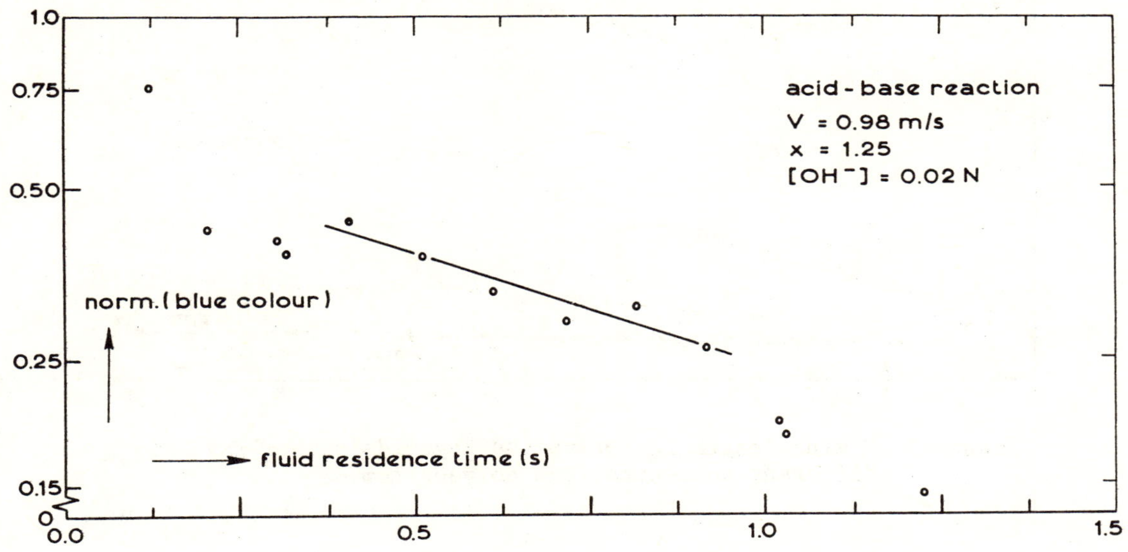

Figure 3 Bromothymol blue colour conversion from the acid–base
neutralisation reaction along the tube

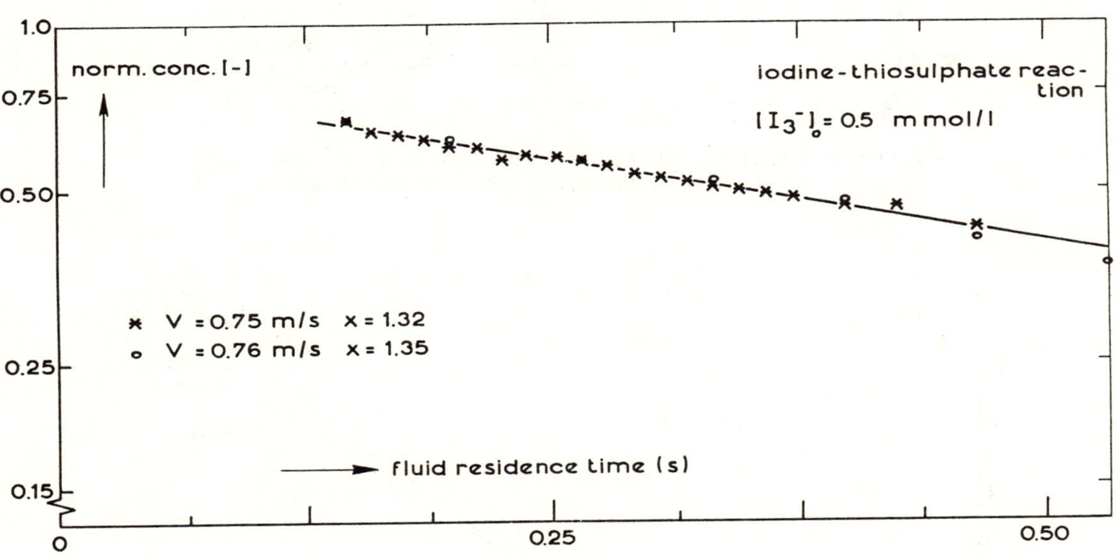

Figure 4 The iodine conversion along the tube

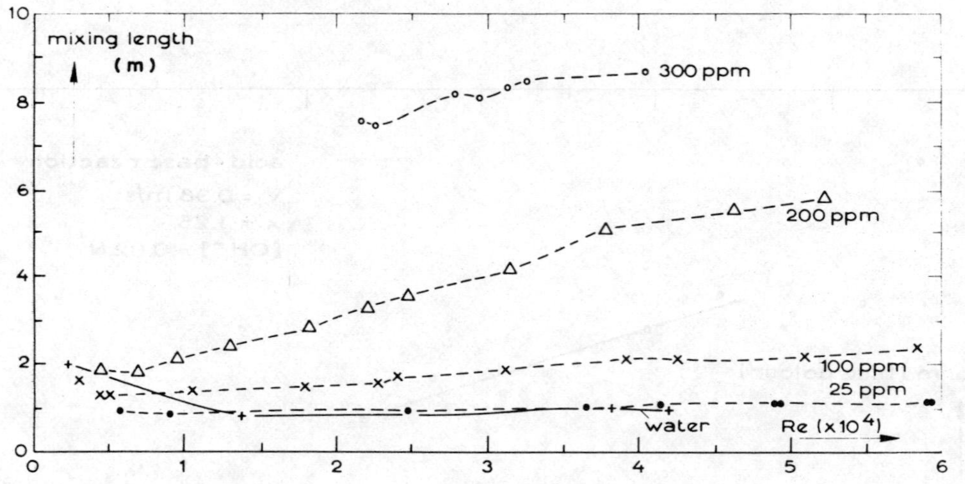

Figure 5   Mixing length L_m versus the Reynolds number for different concentrations polyacrylamide

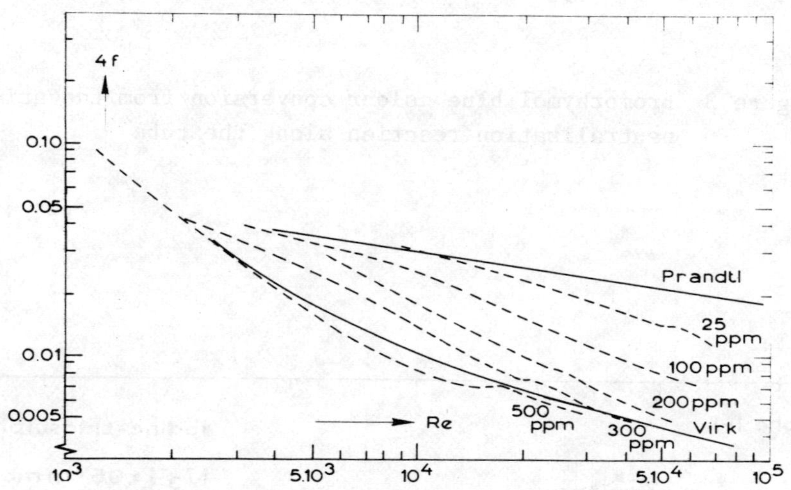

Figure 6 Friction factor versus Reynolds number for different concentrations polyacrylamide

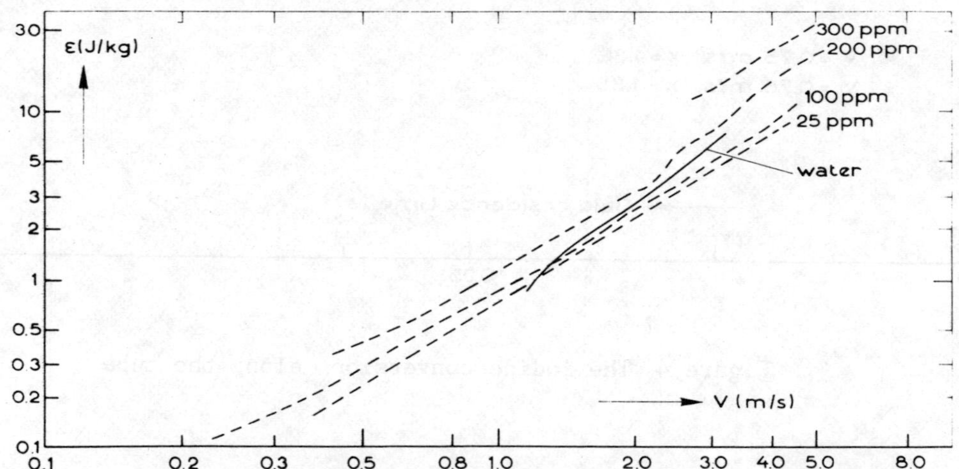

Figure 7   Specific energy dissipation ε versus superficial velocity for different concentrations polyacrylamide

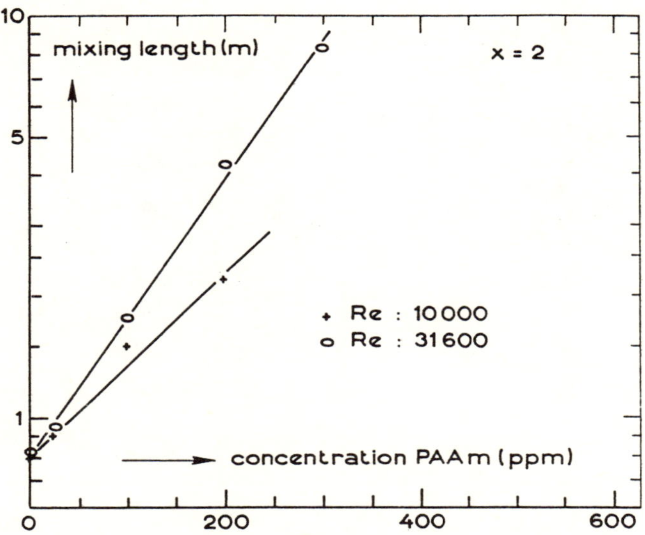

Figure 8 Mixing length versus concentration polyacrylamide for two Reynolds numbers

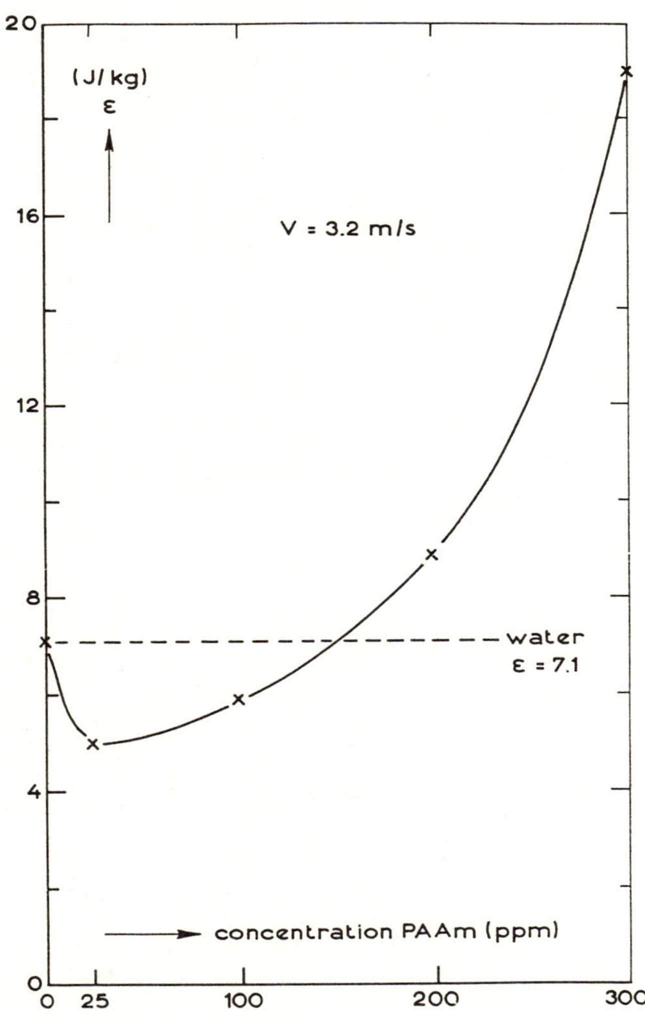

Figure 9 Specific energy dissipation ε for different concentrations polyacrylamide

405

RESIDENCE TIME DISTRIBUTION OF OSTWALD-DE WAELE FLUID IN KENICS STATIC MIXER

Dr. Piotr Pustelnik
Dr. Jerzy Petera

Institute of Chemical Engineering
Lodz Technical University
ul.Żwirki 36, 90-924 Lodz, Poland

Summary

The paper presents a model of residence time distribution in tubes filled with static KSM mixers (Kenics Corp., Denvers, USA).

The assumptions of the proposed model were formulated on the basis of literature and our own optical investigations of flow hydrodynamics in the system. Two model cases of flow were considered, i.e. dispersion-free flow and the flow with superimposed longitudinal dispersion. As a result the differential function and cumulative distribution function of RTD of Ostwald-de Waele fluid in a laminar flow regime were obtained. The final forms of these functions were obtained using a distribution convolution theorem. In the proposed model the influence of Reynolds number, the number of elements of KSM applied and rheological properties of the fluid were taken into account.

An experimental verification of the proposed model was carried out. The model RTD curves and our own experimental data, compared here, were in good agreement.

Held at Wurzburg, 10-12 June, 1985.

Organised by DVCV· Deutsche Vereinigung für Chemie- und Verfahrenstechnik
(German Association of Chemical and Process Engineering).

Organisation: GVC·VDI-Gesellschaft Verfahrenstechnik und Chemieingenieurwesen. 

©BHRA, The Fluid Engineering Centre, Cranfield, Bedford MK43 0AJ, England.

NOMENCLATURE

$F(r), F(\ )$ = cumulative distribution function

k = parameter of the flow model

L_E = length of KSM element

L^* = equivalent length of model tube

N = number of transformation planes

P_E = parameter of the flow model

Q = volumetric flow rate

q_i^*, q_N^* = total propagation along the i-th (N-th) transformation plane

q^* = maximum total propagation

R = radius of model tube

U = dimensionless flow velocity

b = parameter of the flow model (cf. eq.(15))

m, m_N, m_o = exponent in the model of velocity profile

n_{ow} = exponent in Ostwald-de Waele model

r_N^* = radius of velocity distributions' intersection; undisturbed from those existing along the N-th transformation plane

$t(r)$ = residence time of fluid in the system

$u(r), u_i(r), u_N(r)$ = local linear fluid velocity in the system

u_{max}, u_{omax} = fluid velocity in the model tube axis

v = mean linear velocity of fluid in the system

Θ = dimensionless residence time

Re = Reynolds number after Metzner and Reed

1. INTRODUCTION

Static mixers are an example of unconventional way of carrying out unit operations of chemical engineering. They have been used in modern industry for many years. They have been applied in many mixing processes and in heat and mass transfer processes. Bor (Ref. 1) observed applicability of static mixers in tubular flow reactors. In this case they are a useful tool for performing reactor processes. Danckwerts (Ref. 2) proved that the yield of first order reaction is explicitly related to residence time distribution of reaction mass in the system. Hence, it is evident that the determination of this distribution shape or introducing of a proper model is of basic importance in the problems of chemical reaction engineering.

2. ASSUMPTIONS OF THE MODEL

The basic assumptions of the model of RTD in Kenics static mixers are as follows:
1) the fluid flow is steady, isothermal and laminar;
2) the fluid is incompressible;
3) the fluid is a generalized Newtonian shear-thinned fluid and can be described by the rheological model of Ostwald-de Waele;
4) the RTD is strongly affected by the number of KSM elements, flow parameters and longitudinal dispersion in the system.

3. MODEL OF FLOW

Complex flow phenomena that occur in Kenics static mixers (Ref. 3) were replaced by the following simplified model of flow.

The flow takes place in a hypothetical cylindrical empty pipe of the same diameter and length as a real reactor. In the pipe considered (Fig. 1) there are only hypothetical transformation planes of velocity distribution perpendicular to pipe axis. They are at the distance L_E equal to the length of single KSM elements. On these planes there is an instantaneous change of flow velocity distribution. Between the planes the velocity distribution does not change.

It was also assumed that the fluid flowing to the first transformation plane had a fully developed, laminar and non-Newtonian velocity profile. On this plane a partial flattening of the velocity profile occurs but next the profile can be still described by a parabolic dependence.

Naturally, on subsequent transformation planes the fluid must flow from the central part of the pipe to the wall region (Fig. 2). This phenomenon is called fluid propagation. As shown in Fig. 2 a total propagation on the i-th transformation plane Q^*_i is this part of volumetric flow rate in the central part of the pipe which is contained between velocity distribution before the first plane and the distribution on the i-th transformation plane. Total propagation Q^* reaches the maximum value when for the number of planes $N = \infty$ the transformed velocity profile is flat (Fig. 3). We can write

$$Q^*_\infty = \lim_{N \to \infty} Q^*_N \qquad (1)$$

All the discussed velocity profiles can be described by the following general dependence

$$u(r) = u_{max} \left[1 - \left(\frac{r}{R}\right)^m \right] \qquad (2)$$

Consider first the flow in which longitudinal dispersion in the system
is neglected.

In the given model residence time of the fluid in the pipe
containing N transformation planes may be described as follows

$$t(r) = \sum_{i=1}^{N} \frac{L_E}{u_i(r)} \tag{3}$$

In the above model a simple form of a cumulative distribution function
(Ref. 4) was used

$$F(r) = \frac{2\pi}{Q} \int_{0}^{r} u_{max} [1-(\frac{r'}{R})^{m}] r'dr' \tag{4}$$

In eq. (4) the argument r should be replaced by the inverse function of
residence time t(r) determined by eq. (3). To avoid problems resulting
from the presence of the family of velocity profiles in eq. (3) a substitute definition of residence time was proposed

$$t(r) = \frac{L^*}{u_N(r)} \tag{5}$$

where L^* is related to the N number of transformation planes distant
from each other by L_E

$$L^* = K \cdot N \cdot L_E \tag{6}$$

After introducing a dimensionless residence time   and after all transformations the form of function F for a hypothetical pipe containing N
transformation planes of velocity profile is obtained

$$F_N(\theta) = \frac{m_N+2}{m_N} [1- \frac{m_N}{K\theta(m_N+2)}]^{\frac{2}{m_N}} - \frac{2}{m_N} [1- \frac{m_N}{K\theta(m_N+2)}]^{\frac{m_N+2}{m_N}} \tag{7}$$

The above equation for N = 0 and $m_N = 2$ has the form

$$F(\theta) = 1 - \frac{1}{4\theta^2} \tag{8}$$

while for $N = \infty$ has the forms

$$F(\theta) = \begin{cases} 0 & \theta < 1 \\ & \text{for} \\ 1 & \theta > 1 \end{cases} \tag{9}$$

According to the model assumptions the shape of F curves is
significantly affected by the number of Kenics elements equal to transformation planes, flow parameters and rheological properties of fluid.

The aim of further considerations is to derive a functional
dependence between the parameter m_N in eq. (7) and the above mentioned
values. In order to do this a model which will enable a dependence
between the values of Q_N^* and the number N of transformation planes (number of elements) N = 1, 2,... to be obtained is proposed.

Note that after each transformation plane N total propagation
of part of the volume Q_N^* out of the whole volume Q^* occurs. Thus, there
can be a probability of an event P_N which consists in the fact that on
the N-th transformation plane the $Q^* - Q_N^*$ fluid will not propagate

$$P_N(Q_\infty^* - Q_N^*) = \frac{Q_\infty^* - Q_N^*}{Q_\infty^*} \tag{10}$$

Therefore a conditional probability may also be determined

$$P_{N-1,N} = P(Q_\infty^* - Q_N^* \mid Q_\infty^* - Q_{N-1}^*) = \frac{Q_\infty^* - Q_N^*}{Q_\infty^* - Q_{N-1}^*} \tag{11}$$

Assume that

$$1^\circ \quad P(Q_\infty^* - Q_N^* \mid Q_\infty^* - Q_{k_1}^* \cap Q_\infty^* - Q_{k_2}^* \cap \ldots \cap Q_\infty^* - Q_{k_n}^*) = P(Q_\infty^* - Q_N^* \mid Q_\infty^* - Q_{k_n}^*),$$

$$2^\circ \qquad 0 \leqslant k_1 < k_2 < \ldots < k < N \qquad Q_0^* = 0$$

and

$$P_{N-1,N} = a = \mathrm{const}, \quad 0 < a < 1, \quad N = 1,2,\ldots$$

$$P_{0,1} = \frac{Q_\infty^* - Q_1^*}{Q_\infty^*} = P(Q_\infty^* - Q_1^*)$$

Condition 1° means that the event of an arbitrary number N_0 contains the whole stochastic information that describes the behaviour of the system for the numbers of events $N > N_0$. So defined sequence of dependent experiments is a homogeneous Markov chain.

Taking the above assumptions we have

$$P((Q_\infty^* - Q_0^*) \cap (Q_\infty^* - Q_1^*) \cap \ldots \cap (Q_\infty^* - Q_N^*)) = 1 \cdot P_{0,1} \cdot P_{1,2} \cdots P_{N-1,N} =$$

$$= a^N = 1 \cdot \frac{Q_\infty^* - Q_1^*}{Q_\infty^* - Q_0^*} \cdot \frac{Q_\infty^* - Q_2^*}{Q_\infty^* - Q_1^*} \cdot \ldots \cdot \frac{Q_\infty^* - Q_N^*}{Q_\infty^* - Q_{N-1}^*} = \frac{Q_\infty^* - Q_N^*}{Q_\infty^*} \tag{12}$$

One can find such a constant b that

$$a^N = e^{-bN} \tag{13}$$

where: $b = -\ln a > 0$.
Hence we have

$$\frac{Q_\infty^* - Q_N^*}{Q_\infty^*} = e^{-bN} \tag{14}$$

or

$$Q_N^* = Q_\infty^* (1 - e^{-bN}) \tag{15}$$

According to the definition, the total propagation Q_N^* can be calculated from the equation

$$Q_N^* = 2\pi \left\{ \int_0^{r_N^*} u_{o\max} \left[1 - \left(\frac{r}{R}\right)^{m_o}\right] r\, dr - \int_0^{r_N^*} u_{N\max} \left[1 - \left(\frac{r}{R}\right)^{m_N}\right] r\, dr \right\} \tag{16}$$

where m_o is a value of the exponent of laminar non-Newtonian velocity profile related to the exponent n in the Ostwald-de Waele rheological model by the dependence

$$m_o = \frac{n_{ow} + 1}{n_{ow}} \tag{17}$$

and the local velocity on the axis of hypothetical pipe is

$$u_{o\max} = \frac{3 n_{ow} + 1}{n_{ow} + 1} \cdot v \tag{18}$$

The value r_N^* corresponds to intersection points of undisturbed profile and the profile occurring after the N-th transformation plane. Hence

$$u_{o\max}\left[1 - \left(\frac{r_N^*}{R}\right)^{m_o}\right] = u_{N\max}\left[1 - \left(\frac{r_N^*}{R}\right)^{m_N}\right] \tag{19}$$

The system of equations (15)-(19) determines from the theoretical point of view the dependence of the exponent m_N and the number of transformation planes N:

$$m_N = g(N) \tag{20}$$

In practice this system of equations can be solved numerically.

Constant b in eq. (15) determines the rate of changes in total propagation Q_N^* and also the rate of changes of velocity profile. These

changes are a result of two opposite phenomena: profile flattening due
to the operation of KSM elements and profile rebuilding due to formation
of a new boundary layer. For low values of the Reynolds number the
velocity profile will rebuild to a large extent and thus the rate of its
changes is small. For high values of the Reynolds number the profile rebuilding is small which means a high rate of changes. Therefore it
follows that constant b in eq. (15) is linked functionally with the
Reynolds number.

In the second part of the considerations concerning the model
the authors discuss the case of fluid flow with longitudinal dispersion.
The phenomenon of dispersion is caused by an additional particle motion
in the flow which can be induced by diffusion. We assume that the
resultant velocity profile is a sum of the dispersion-free flow profile
(main profile) described above and the independent profile resulting
from longitudinal dispersion. Therefore, basing on the theorem of distribution of the sum of two independent random variables the function of
resultant velocity density distribution as a convolution of density functions of distributions of two component velocity profiles

$$\hat{f}_{CON}(U) = (f_1 * f_2)(U) = \int_0^\infty \hat{f}_1(U - U') \, d\hat{F}_2(U') \tag{21}$$

where: U – dimensionless velocity
$\hat{f}_{CON}(U)$ – density function of resultant velocity distribution
$\hat{f}_1(U)$ – density function of the main profile distribution
$\hat{F}_2(U)$ – distribution function of velocity due to longitudinal
                dispersion.

In the description of dispersion the well-known density distribution function of residence time (Ref. 5) for the case of axial
dispersed plug flow model was used

$$f_2(\theta) = \frac{1}{2} \left(\frac{P_E}{\Pi \theta} \right)^{\frac{1}{2}} \exp \left[- \frac{P_E (1 - \theta)^2}{4\theta} \right] \tag{22}$$

To take advantage of eq. (21) the variables in the equations
defining density functions f_j, $j = 1,2$ should be replaced

$$F_j(\theta) = \int_0^\theta f_j(\theta') \, d\theta' \qquad , \; j = 1,2 \tag{23}$$

using the dependence of dimensionless velocity and dimensionless residence time

$$U = \frac{1}{\theta} \tag{24}$$

Substituting eq. (23) to (24) and using normalizing condition

$$\int_0^\infty \hat{f}_j(U) \, dU = 1 \tag{25}$$

one obtains

$$\hat{F}_j(U) = 1 - \int_0^U f_j\left(\frac{1}{U'}\right) \left(\frac{1}{U'}\right)^2 dU' \quad , \; j = 1,2 \tag{26}$$

The respective density functions of velocity distributions are
as follows

$$\hat{f}_j(U) = \frac{d\hat{F}_j(U)}{dU} = - f_j\left(\frac{1}{U}\right) \left(\frac{1}{U}\right)^2 \quad , \; j = 1,2 \tag{27}$$

The obtained functions (27) for $j = 1,2$ can be substituted to eq. (21)
thus giving the above mentioned convolution of density functions of
velocity distribution. After another replacement of variables using eq.
(24) the convolution of density functions of residence times is obtained

$$f_{CON}(\theta) = \int_0^\infty f_2(\theta')\, f_1\left(\frac{1}{\dfrac{1}{\theta}-\dfrac{1}{\theta'}}\right) \frac{1}{\left(\dfrac{1}{\theta}-\dfrac{1}{\theta'}\right)^2}\, d\theta' \tag{28}$$

The cumulative function F is respectively

$$F(\theta) = \int_0^\theta \int_0^\infty f_2(\theta')\, f_1\left(\frac{1}{\dfrac{1}{\theta'}-\dfrac{1}{\theta''}}\right) \frac{1}{\left(\dfrac{1}{\theta'}-\dfrac{1}{\theta''}\right)^2}\, d\theta'\, d\theta'' \tag{29}$$

Upon substitution to eq. (29) of detailed formulae for the functions f_1 and f_2, respective differential forms of functions (7) and (22) can be obtained after a few transformations

$$F_N(\theta) \int_0^\theta \int_0^\infty \left(\frac{P_E}{\Pi\,\theta'}\right) \exp\left[-\frac{P_E(1-\theta')^2}{4\theta'}\right] \frac{\theta'-K\theta''}{K\theta\theta''(m_N+2)-m_N(\theta'-K\theta'')} \cdot$$
$$\cdot \left[1-\frac{m_N(\theta'-K\theta'')}{K\theta\theta''(m_N+2)}\right]^{\frac{2}{m_N}} d\theta'\, d\theta'' \tag{30}$$

This is the cumulative distribution function of residence times of the fluid at the outlet from the hypothetical pipe containing N transformation planes with superimposed longitudinal dispersion.

4. EXPERIMENTAL VERIFICATION

The functional interrelations between model parameters and the parameters of the flow system can be determined in two ways, i.e. employing theoretical considerations or using correlations. The authors chose the latter way having rich experimental material and using the multi-parameter optimization method. Particular parameters of the model were functionally related to the parameters of the system in the following way:

a) the parameter K could be assumed constant in the whole range of system parameters, hence

$$k = \text{const} \tag{31}$$

b) the parameter b was correlated depending on the Reynolds number using the equation

$$b = b_0 + c_0 \log Re \tag{32}$$

c) the parameter P_E was determined depending on the number of Kenics elements, in the form

$$P_E = 1 - N^{n_1} \tag{33}$$

where: b_0 — constant
c_0 — constant
$n_1 = f(Re)$.

Figures 4, 5 and 6 present a comparison of model curves and experimental data. The range of system parameters was as follows: the number of Kenics elements ranged from 2 to 30, the Reynolds number based on empty pipe was 0.72 to 442.5 and the exponent in the rheological model of Ostwald-de Waele fluids ranged from 0.63 to 1.

5. CONCLUSIONS

The above proposed mathematical model of RTD in a tubular chemical reactor filled with Kenics mixers operating in laminar flow regime can be a basis for the determination of proper methods for designing of these reactors.

The application of a convolution of distributions as an element linking non-dispersive and dispersive flows enables us to avoid
troublesome solution of dispersion differential equations.

Despite the application of a simplified optimization method
the agreement of model curves of RTD with the experimental data is good.

It should be noted that the shape of RTD curves was most
strongly influenced by the number of Kenics mixers applied.

6. REFERENCES

1. Bor, T.: "The static mixer as a chemical reactor". British Chemical
    Engineering, July 1971, pp. 56-57.

2. Danckwerts, P.V.: "Continuous flow systems: distribution of resi-
    dence times". Chem. Eng. Sci., 2, 1953, pp. 1-18.

3. Tung, T.T.: "Low Reynolds number entrance flows: a study of a
    motionless mixer". Dissertation, University of Massachusetts,
    Amherst, April, 1976.

4. Wein, O., Ulbrecht, J.: "Residence time distribution in laminar
    flow systems". Coll. Czech. Chem. Commun., 37, 1972, pp. 412-428.

5. Levenspiel, O.: "Chemical reaction engineering". London, John Wiley
    & Sons, Inc., 1972, pp. 276-277.

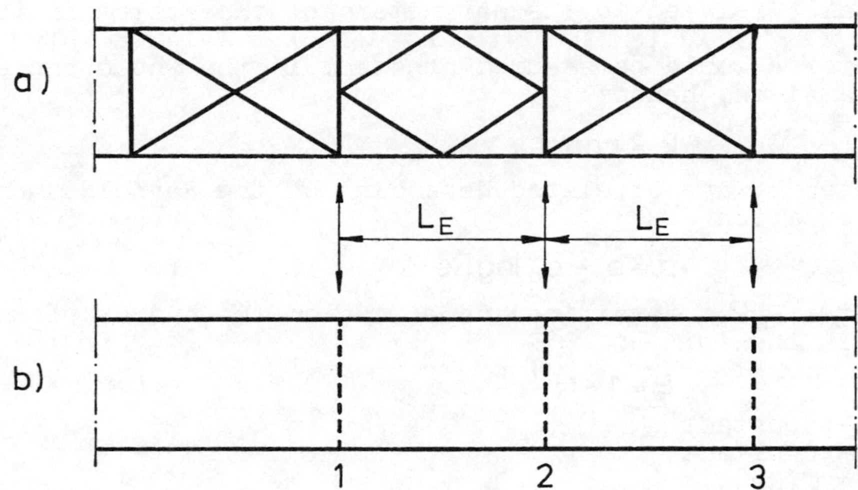

Fig.1. Model geometry of flow through KSM mixers
       a) real tube filled with KSM elements of length L_E
       b) hypothetical pipe with transformation planes (1,2,3)

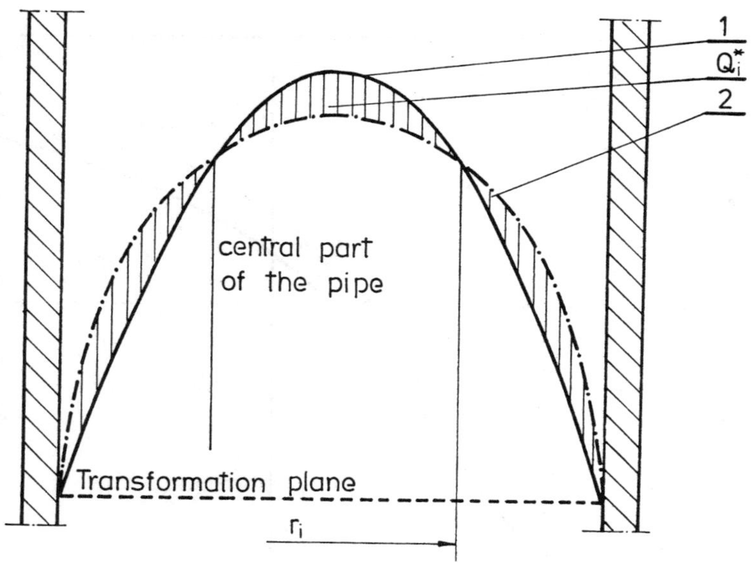

Fig. 2 Total propagation Q_i^*

1 - fluid velocity profile before the first transformation plane;

2 - fluid velocity profile after the i-th transfromation plane;

r_i - radial coordinate of velocity profiles intersection before the first and after the i-th transformation planes.

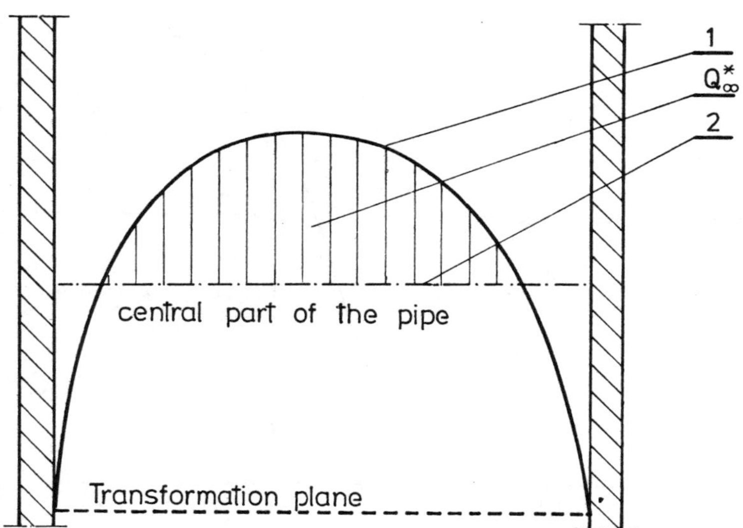

Fig. 3 Maximum total propagation Q_∞^*

1 - fluid velocity profile before the first transformation plane.

2 - flat velocity profile.

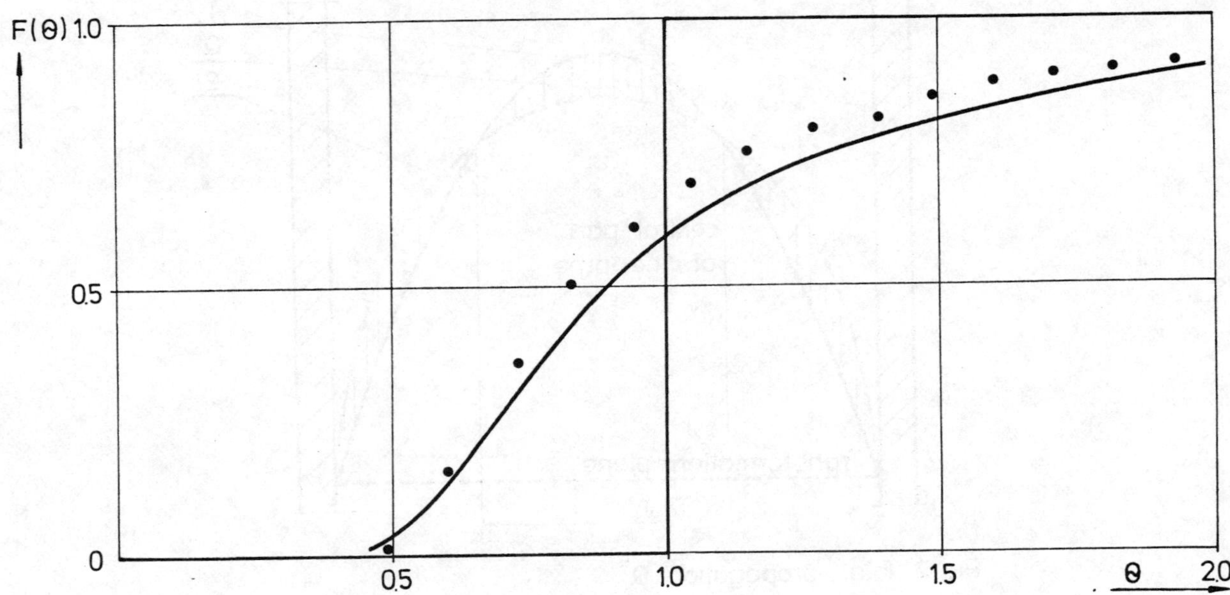

Fig. 4 Comparison of model distribution function F (eg. 30) and experimental
data for glycerol.
(D=21mm ; Re =0.72 ; N=4 ; P= 0.305 ; m_N = 2.25) – experimental points.

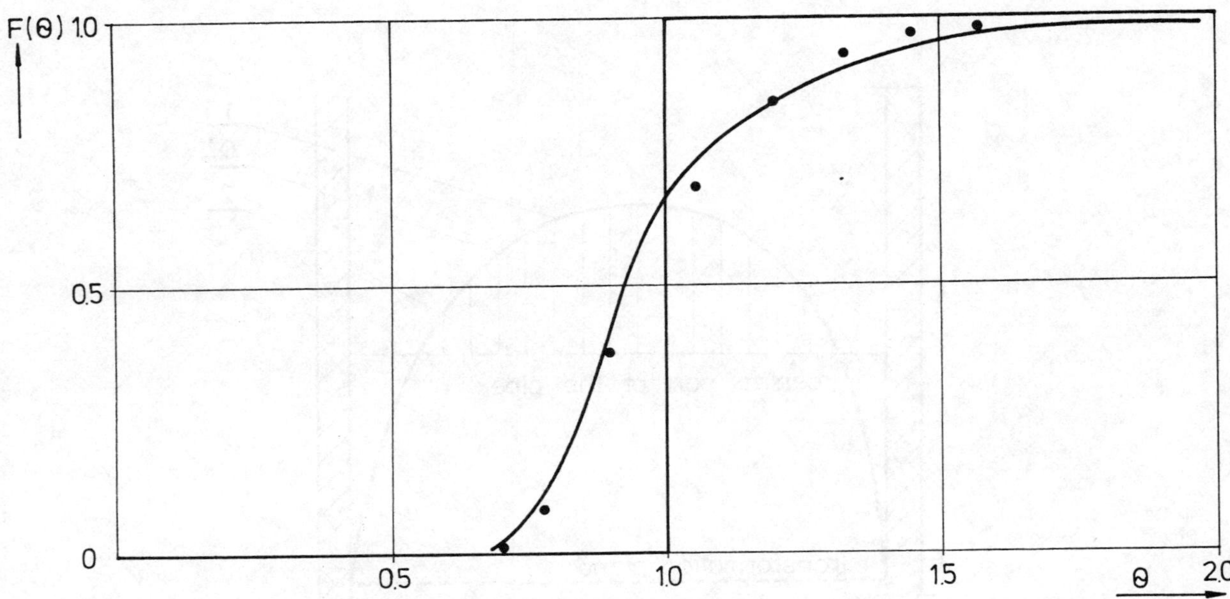

Fig. 5 Comparison of model distribution function F (eg. 30) and experimental
data for glycerol.
(D=21mm ; Re =0.72 ; N=30 ; P=0.425 m_N =14.2) - experimental points.

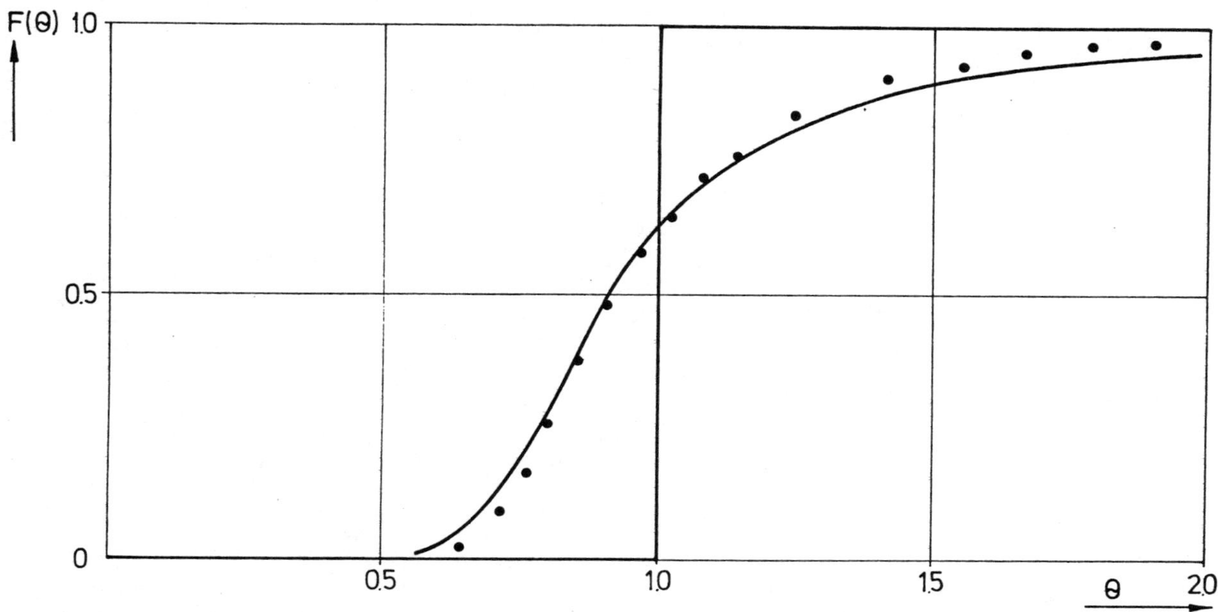

Fig.6 Comparison of model distribution function F (eg.30) and experimental data for glycerol.

(D = 21mm ; Re = 0.72 ; N = 16 ; P = 0.395 m_N = 6.56) − experimental points.

"Gas-Liquid Mass Transfer Data in an Three Phase Stirred Vessel"

A. Brehm* and H. Oguz, D-Oldenburg; B. Kisakürek, TR-Ankara

* Author to whom correspondence should be adressed at Department of Chemistry,
  P.O.Box 25 03, D-2900 Oldenburg, W.-Germany

Held at Wurzburg, 10-12 June, 1985.

Organised by DVCV· Deutsche Vereinigung für Chemie- und Verfahrenstechnik
(German Association of Chemical and Process Engineering).

Organisation: GVC·VDI-Gesellschaft Verfahrenstechnik und Chemieingenieurwesen.  **GVC**

Symbols

A	surface area
a, b, c, d	exponents for correlation
c_E	actual concentration of solute gas
c^*	equilibrium concentration of solute gas
D	agitator diameter
$k_L a$	mass transfer coefficient
$k_L a_o$	mass transfer coefficient of solid-free solution
n	agitator speed
P	power input
Q_G	gas flow rate
T	inside diameter of vessel
T_E	time constant of pO_2-probe
t	time
V_{SL}	volume of slurry-phase
v_G	superficial gas velocity
X_G	volume fraction of gas
X_S	volume fraction of solid
η	dynamic viscosity
η_o	dynamic viscosity of solid-free solution
ρ_{SL}	density

1. Introduction

Reactions in three-phase systems (gas-liquid-solid) are usually carried out in
gas-sparged stirred vessels. They permit to carry out processes, in which gaseous components react with a suspended solid or on the surface of a suspended
catalyst. For the design of such reactors and for analysing conversion data, it
is necessary to know the gas-liquid mass transfer characteristics.
In recent years, several investigations on gas-liquid mass transfer have been
published. It has been observed, that the addition of solids has an influence
on gas-liquid mass transfer characteristics, even if these particles are not
reactive and do not show increased adsorption of dissolved gases /1-4/. This
particle effect is remarkably dependent on the type of solids. For instance, it
has been found that an addition of kieselgur, alumina und glass beads in small
volume fractions influences the volumetric mass transfer in different ways. Thus
the addition of five percent of alumina results in an increase of the $k_L a$-values
by the factor of three in a stirred autoclave while a sharp decline is observed
when Kieselgur is added /4/. On the other hand, in a standard stirred vessel the
$k_L a$ data increase in the presence of alumina (5%) by a factor of only 1.5 /6/.
These results were obtained with an isochoric method which does not permit introduction of gas through a sparger. Therefore, another method was applied to
determine $k_L a$-data under sparging conditions.

2. Experimental set up

The vessel used in this study was of standard configuration. The proportions
were as given by Chapman, Gibilaro and Nienow /7/. The vessel has a diameter of
$T = 0,145$ m. Its total volume was 4 l. The four flat-blades impeller ($D = 0,5$ T)
was mounted with a off-bottom clearance of 0.3 T. The gas was sparged by a ring
with 12 holes located at vessel bottom.
All measurments were done at 25°C under variation of gas flow rate (50 to 1000
l/h), stirrer speed (500 to 800 rpm) and particle concentration (0 to 10 Vol.%).
The following particles were used: alumina, iron oxide, Kieselgur, glass beads
(50 μm and 300 μm) and see sand (~ 300 μm and pulverize). The $k_l a$ values were
determined from the transient oxygen concentration in the slurry phase. The
oxygen partial pressure was measured by a fast response oxygen probe (WTW EO 90).
The experimental procedure was as follows: The liquid phase was made O_2 free
by stripping with N_2. If $po_2 \approx 0$ air flow and stirring was started and the
oxygen partial pressure was monitored. Due to the fast response of the probe $k_L a$
was determined from

$$\ln(C^* - C_E) = k_L a' t + \text{const.} \qquad /8, 9/ \qquad (1)$$

Linek /10/ has studied the influence of nitrogen desorption on oxygen adsorb-
tion. This author reported that simultaneous N_2 desorption results in lower
$k_L a$-values for O_2, if the residence time of bubbles is nearly the same as
the time of the whole measurments.

When using the dynamic method in slurries it is difficult to detect the effect
of N_2 desorption. Therefore, measurements were done with water and aqueous solutions of polyethylenglycol in order to find out the influence of N_2 desorption
on the k_La values for O_2 absorption. Fig. 1 shows the relative differences in
k_La measured in the presence of desorbing N_2 and with gas free liquid (degassed
by evacuation).

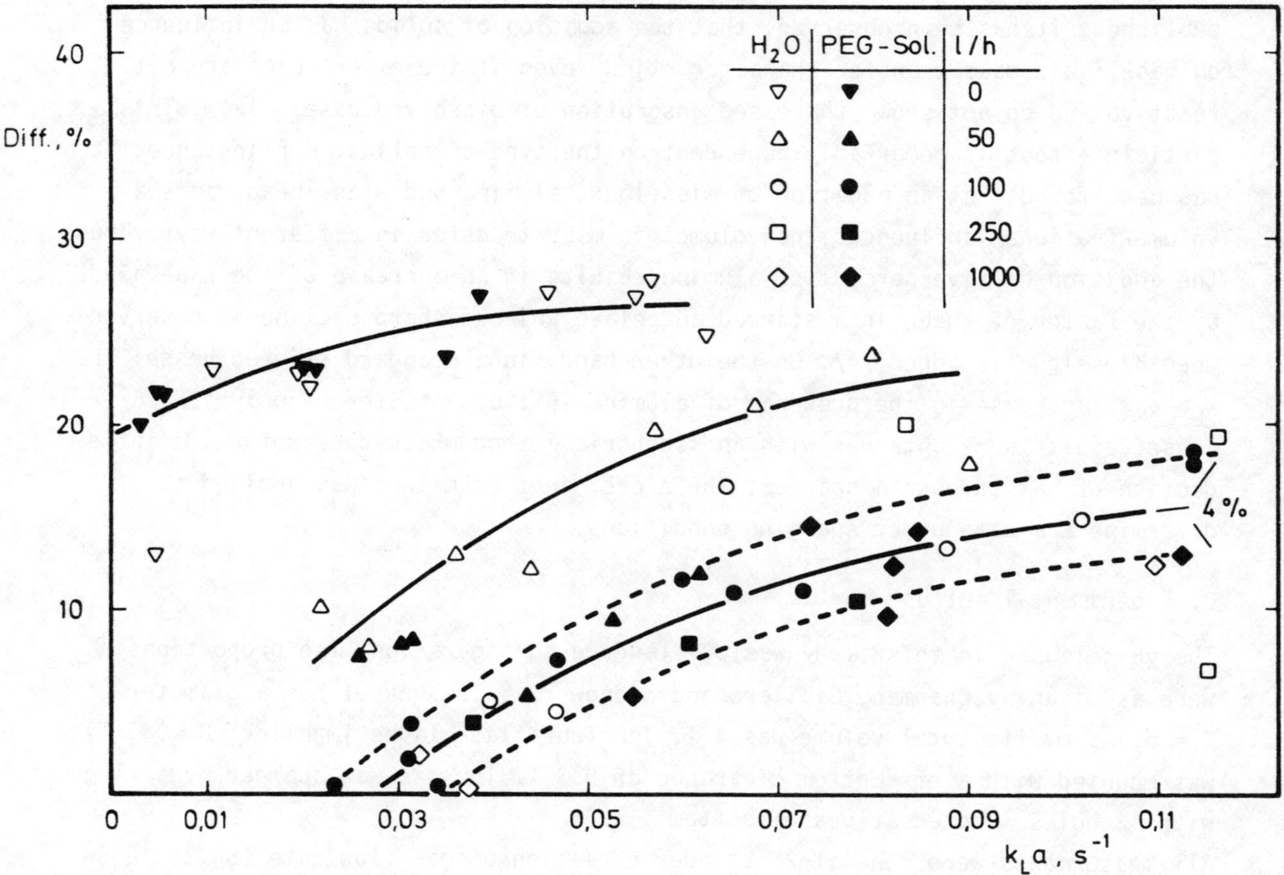

Fig. 1: Differences in k_La measured in the presence of desorbing N_2 and with
gas free liquid (degassed by evacuation).

All the k_La data for the various slurries were measured under the conditions of
N_2 desorption and were corrected according to the findings shown in Fig. 1. The
corrected data agree well with those k_La-values obtained from an isochoric
method starting with gas free liquid.

The k_La' values evaluated from eq. 1 are refered to the liquid volume. By multiplication with (1 - X_S) k_La values are obtained which are refered to the slurry
volume. These values are reported in the following.

3. Results

Small alumina particles increase the $k_L a$-values by a factor of 1,5 if the concentration is about 5%. For higher solid concentrations the $k_L a$ values decrease. Introduction of gas through a ring sparger results in larger effects if absolute values are compared (Fig. 2). On the other hand the relative mass transfer coefficient $(k_L a / k_L a_o)$ is hardly influenced by gas-sparging.

An increase in $k_L a$ as shown in figure 2 for alumina particles was not observed for all kinds of solid particles. For instance glass beads did not show such an effect.

Disregarding the effects observed at low particle concentration it was thought possible to describe the data of different particles by a mutual correlation of the general type

$$k_L a = a \left(\frac{P}{V_{SL}}\right)^b \cdot \left(\frac{\eta}{\eta_o}\right)^c \cdot v_G^d \tag{2}$$

which considers only volumetric power input, relative viscosity and superficial gas velocity.

Indeed this could be achieved by taking into account only those measurements where the apparent viscosity (η/η_o) due the presence of solids was larger than 1,3.

The volumetric power input was calculated from the relation reported by Hassan and Robinson /11/ which for our setup is given by

$$\frac{P}{V_{SL}} = \frac{n^{2,94} \cdot D^{5,48} \cdot \rho_{SL}}{Q_G^{0,38} \cdot (1 - \varepsilon_G) \cdot V_{SL}} \tag{3}$$

The regression analysis considering 185 experimental points leads to final correlation

$$k_L a = 0,5612 \cdot \left(\frac{P}{V_{SL}}\right)^{0,65} \cdot \left(\frac{\eta}{\eta_o}\right)^{0,47} \cdot v_G^{0,40} \tag{4}$$

which describes the data by an average error of only 11,5%.

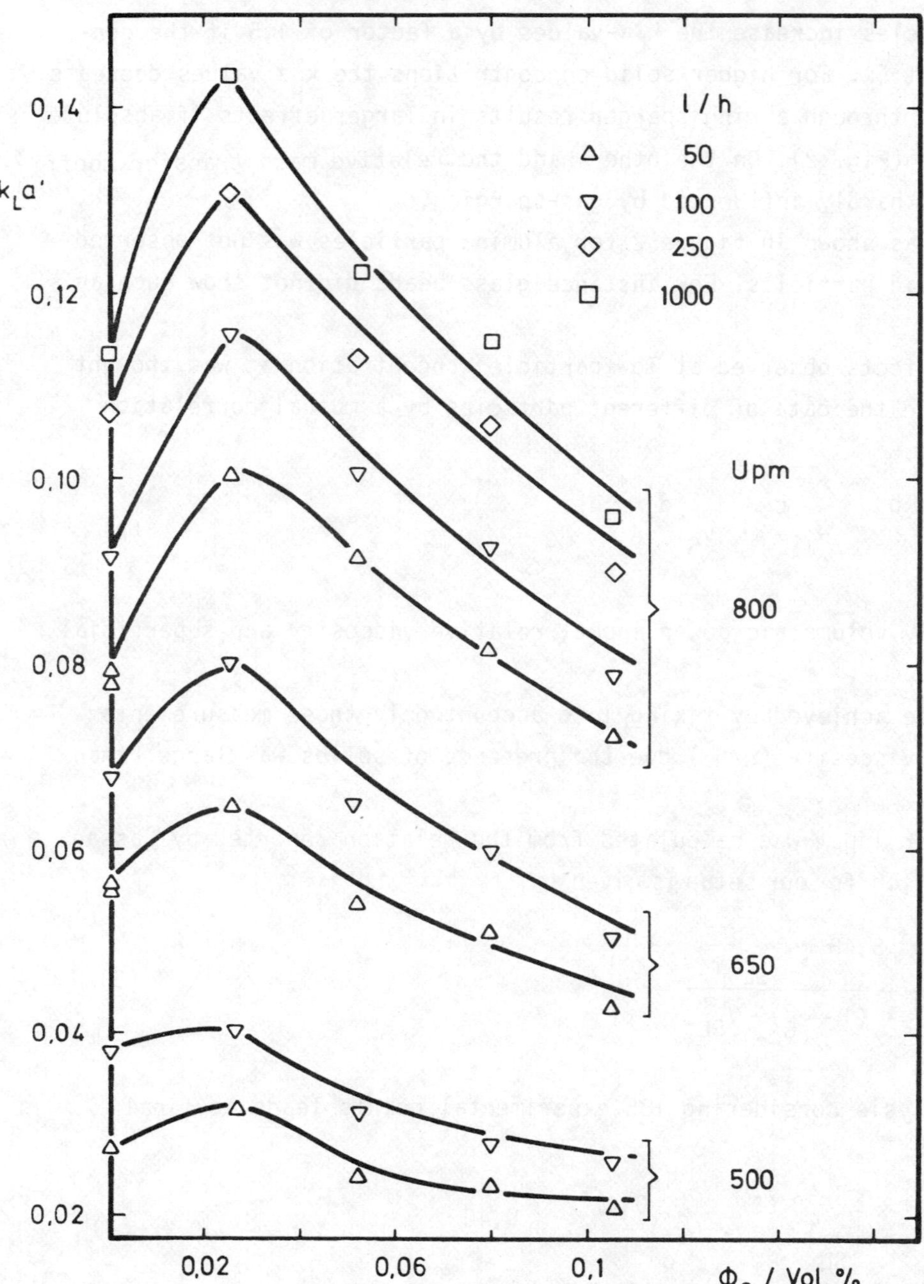

Fig. 2
effect of an
addition of
alumina par-
ticles on k_La'

4. Discussion and conclusions

The effect of suspended particles on k_La was investigated in a stirred vessel under variation of the kind of particles and their concentration, the power input and the gas flow rate by the dynamic methode. Taking into account the influence of nitrogen desorption, the k_La-values of this study agree with those measured independently by a different method.

It was possible to correlate the mass transfer coefficient measured at relative viscosity > 1,3 independent from the kind of solid particles by

$$k_L a = 0,5612 \; (\frac{P}{V_{SL}})^{0,65} \cdot (\frac{\eta}{\eta_0})^{-0,47} \cdot v_G^{0,40}$$

where the power input is calculated from

$$\frac{P}{V_{SL}} = \frac{n^{2,94} \cdot D^{5,48} \cdot \rho_{SL}}{Q_G^{0,38} \cdot (1 - X_G) \cdot V_{SL}}$$

References

/ 1/ G.E.H. Joosten, J.G.M. Schilder and J.J. Janssen, Chem.Eng.Sci., 32 (1977) 563.

/ 2/ V.D. Mehta and M.M. Sharma, Chem.Eng.Sci. 26 (1971) 461.

/ 3/ M. Schmitz, A. Steiff and P.-M. Weinspach, GVC-Fachausschuß "Mischvorgänge", Bad Driburg (1983)

/ 4/ R.S. Albal, Y.T. Shah, A. Schumpe and N.L. Carr, Chem.Eng.J., 27 (1983) 61.

/ 5/ S. Ledakowicz, R. Kokuun, A. Brehm and W.-D. Deckwer, Chem.-Ing.-Tech. 55 (1983), 648.

/ 6/ A. Brehm, S. Ledakowicz and R. Kokuun, 8th. Intern. Congress CHISA '84 Praha.

/ 7/ C.M. Chapman, L.G. Gibilaro and A.W. Nienow, Chem.Eng.Sci. 37 (1982), 891.

/ 8/ M. Nakanoh and F. Yoshida, Ind.Eng.Chem.Process Des.Dev. 19 (1980), 190.

/ 9/ A. Schumpe, Chem.-Ing.-Techn. (submitted)

/10/ V. Linek, P. Beues and F. Hovorka, Biotechnol. Bioeng. 23 (1981) 301.

/11/ I.T.M. Hassan and C.W. Robinson, AIChE J. 23 (1977) 48.

THE SENSITIVITY OF THE GAS LIQUID CSTR DESIGN
TO COEFFICIENTS EVALUATION

Alberto Brucato, Valerio Brucato
and
Lucio Rizzuti

Istituto di Ingegneria Chimica
Università di Palermo
Viale delle Scienze, 90128 Palermo, Italy

Summary

The influence of the liquid side mass transfer coefficient and of the interfacial area on the volume of a mechanically stirred gas-liquid tank reactor has been investigated by using a trial and error design procedure. The correlation used in this procedure for the estimation of the interfacial area has been multiplied by a factor, whose range was 0.4-2.5 and the reactor volume was calculated. This procedure has been repeated for various input design quantities, as diffusivities, gas solubility, kinetic constant, concentrations and flow rates. A power-law relationship has been obtained between the reactor volume and the multiplying factor for each calculation run. The exponents obtained in this way have been related to dimensionless numbers and reported in diagrams. The same calculations performed by varying the value of the liquid side mass transfer coefficient led to diagram similar to the previous ones. All those diagrams allow the definition of the uncertainty range on the calculation of the reactor volume due to uncertainties on the estimation of those two mass transfer parameters. Moreover they characterize fields of influence of both interfacial area and liquid side mass transfer coefficient in terms of dimensionless numbers.

Held at Wurzburg, 10-12 June, 1985.
Organised by DVCV· Deutsche Vereinigung für Chemie- und Verfahrenstechnik
(German Association of Chemical and Process Engineering).

Organisation: GVC·VDI-Gesellschaft Verfahrenstechnik und Chemieingenieurwesen. 
©BHRA, The Fluid Engineering Centre, Cranfield, Bedford MK43 0AJ, England.

NOTATION

A   = reagent present in the liquid phase

B   = gaseous reagent

a_i = interfacial area per unit volume of liquid, m^{-1}

C_A = concentration of A into the liquid bulk, $Kmol\ m^{-3}$

D_B^o = diffusivity of the gaseous reagent into the liquid, m^2s^{-1}

Ha = Hatta modulus

E   = enhancement of mass transfer by reaction

E_i = enhancement factor for instantaneous reaction

k_c = kinetic constant, $Kmol^{-1}m^3s^{-1}$

P   = agitation power, W

v_{sg} = gas superficial velocity, $m\ s^{-1}$

V_L = volume of liquid into the reactor, m^3

V_1 = volume of liquid into the reactor for $\Omega_a,\ \Omega_\beta = 1$, m^3

Z   = dimensionless parameter defined by eqn(4)

Greek

β_L = liquid side mass transfer coefficient, $m\ s^{-1}$

ε_a = exponent in eqn(2) for a_i

ε_β = exponent in eqn(2) for β_L

Ω_a = multiplying coefficient for a_i calculation

Ω_β = multiplying coefficient for β_L calculation

1. <u>INTRODUCTION</u>

Within the area of polyphasic reactors, the mechanically stirred reactors are widely used in the chemical industry for various reactions as fermentations, oxidations, hydrogenations etc. The design of these apparatuses is very complex as many mechanical, physical and chemical processes are involved in this particular type of gas liquid contacting device.

In dealing with the design, one can be surprised of the great number of empiric correlations for the mass transfer parameters that are found in the pertinent literature. These correlations are often very different one from the other even when they involve the same quantities. The reason of this lies in the complexity of the phenomena taking place into the reaction vessel. Excellent reviews on this matter can be found in recent literature (Refs. 1,2,3,4).

Two design procedures have recently been proposed for these two-phase reactors and for second order reactions (Refs. 5,6). Despite the calculation precision that can be achieved, the uncertainty affecting the determination of the values of the above referred parameters obviously reflects on the design results.

Aim of this paper is the study of the sensitivity of the design results to errors in the determination of the mass transfer coefficient, β_L , and of the interfacial area per unit volume of liquid phase, a_i . This allows the easy estimation of the uncertainty affecting the design results and the identification of the fields where the influence of each parameter is critical or negligible.

2. <u>APPROACH TO THE PROBLEM</u>

From the designer's point of view, the reactor dimensions are the most important design target. For this reason the quantity on which the attention has been focused in this study is the liquid volume in the reactor, that is strictly related to the reactor dimensions. Among all the parameters which affect the reactor volume and that are graved by the largest uncertainties, the interfacial area per unit volume and the mass transfer coefficient have been chosen to perform this investigation.

In order to investigate the influence of the β_L and a_i one of the previously referred design procedure has been utilized (Ref. 5). This trial and error procedure requires the specification of a complete set of geometrical, physical, chemical and process data. It eventually leads to the definition of reactor dimensions, required stirring power and stirrer speed, which allow the design to further proceed.

The correlation utilized in this procedure for the determination of the interfacial area was (Ref.4):

$$\beta_L \, a_i \;=\; 0.002 \, (P/V_L)^{0.7} \, v_{sg}^{0.2} \qquad\qquad (1)$$

from which a_i is easily obtained once the value of β_L is fixed. With regard to β_L , as its value normally ranges between $2 \cdot 10^{-4}$ and $4 \cdot 10^{-4}$ ms^{-1}, a value of $3 \cdot 10^{-4}$ has been chosen, and used in the previous equation.

The design procedure was necessarily performed by trial and error because the equations that link the reactor volume to a_i and β_L are quite complex. This implies that an analytical study of the problem is not straightforward. For this reason a numerical approach has been used: the value of a_i obtained from eqn (1) and the value of β_L have been multiplied respectively by the coefficients Ω_a and Ω_L and the design program has been changed accordingly. The range in which both coefficients Ω were varied was $0.4 - 2.5$. The design program was run and at the end of each run a value of the reactor volume was obtained.

These values have been correlated with the Ω values by a regression analysis performed on the following power-law relationship:

$$V_L = V_1 \, \Omega^\varepsilon \qquad (2)$$

that was found to be very effective in correlating the results. This procedure has been repeated for several values of the physical, chemical and process input data, leading each time to the determination of the exponents ε_a and ε_β.

These exponents have then been plotted against various combinations of the design parameters. Various fields of influence have been individuated, and in most of them a dimensionless parameter, which only the exponent is related to, has been picked out.

3. RESULTS AND DISCUSSION

3.1 Influence of the interfacial area.

It is obvious that, due to the used calculation approach, the results can be represented by continuous curves. The results obtained by varying the interfacial area coefficient Ω_a are represented by the curves reported in Fig. 1. In this figure the abscissa is the Hatta modulus,

$$Ha = \frac{\sqrt{k_c \, C_A \, D_B^o}}{\beta_L} \qquad (3)$$

and the ordinate is the exponent ε_a.

It can be seen that the influence of a_i can be divided into three fields, whose boundaries can be put in terms of the Hatta modulus and of the parameter Z , that will be defined later.

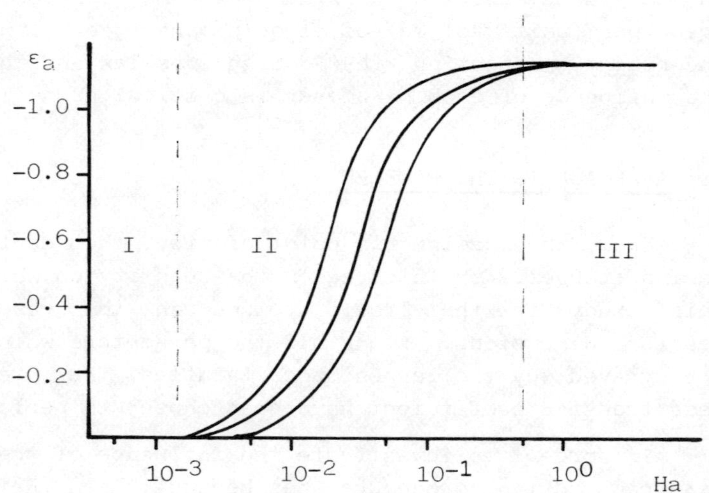

Fig 1. Hatta modulus vs. ε_a

I) Ha < 0.5 and Z < 0.05
   In this field the exponent is always close to zero. In other words, the reactor volume required is not influenced by the interfacial area.

II) Ha < 0.5 and Z > 0.05
   In this field the exponent ranges approximately from zero to -1.1538 . It can also be seen that it depends not only on the Hatta modulus but also on some other parameters, in fact various curves have been obtained for the various input design data. For the sake of clarity only a limited number of curves has been reported.

III) Ha > 0.5
   In this field the exponent is always close to the value of 1.1538 for all the data sets. This means that the reactor volume has a dependence sligthly more than linear on the reciprocal of the interfacial area.

The results obtained in field I are not surprising, as this field corresponds to the kinetic regime in which the concentration of the gaseous reactant is close to that of saturation and the total reaction rate is simply proportional to the reactor volume, and not dependent on the mass transfer parameters.

The field II corresponds to the diffusional regime in which the bulk concentration of the gaseous reactant starts to decrease because of mass transfer limitations. A parameter can be introduced by which only a curve can represent all the different design specifications for the three regions. This parameter has been found to be

$$Z = \frac{k_c\, c_A}{\beta_L\, a_i} \qquad (4)$$

the ratio between the product of the kinetic constant and the bulk concentration of the liquid reagent and the liquid mass transfer product.

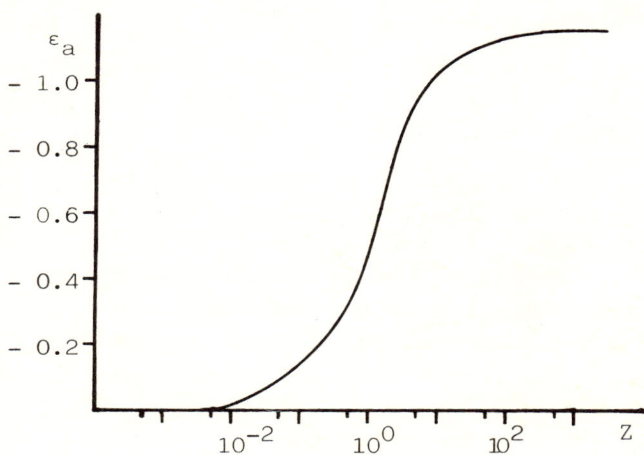

Fig 2. The parameter Z vs. ε_a

In Fig.2 Z has been plotted against the exponent for all the design specifications and only one curve can be observed.

Calling γ the exponent of the gas superficial velocity in the correlation for the determination of the interfacial area, it can be easily found that the limiting value of ε_a in region III can be written as

$$\varepsilon_a = \frac{1}{1 - \gamma\, 2/3} \qquad (5)$$

As in our case the value of γ is 0.2 , substitution in the previous equation gives ε_a = 1.1538. This value corresponds exactly to that found in Figs 1 and 2.

3.2 Influence of the mass transfer coefficient.

The results obtained by varying β_L are shown in Fig 3 where the abscissa is the parameter Z and the ordinate is the exponent ε_β .

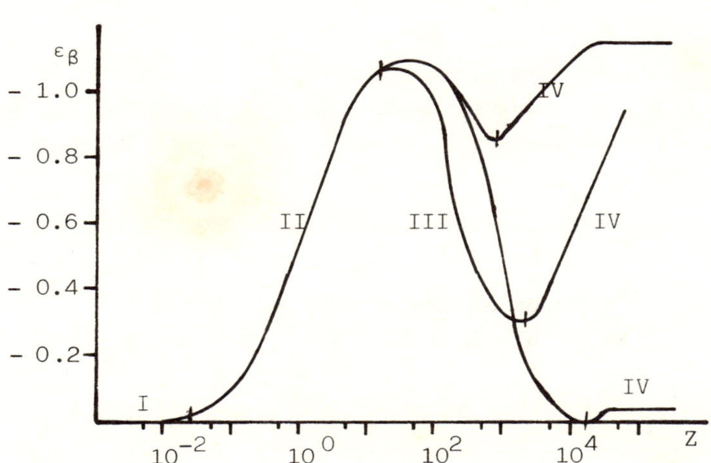

Fig 3. The parameter Z vs. ε_β

Also in this case each curve corresponds to different design specifications and can be divided into four regions as shown in Fig 3.

- Regions I and II). These regions coincide exactly with the corresponding ones in Fig 1, so that the same considerations can be applied here. This is not surprising because in these regimes of mass transfer the two parameters β_L and a_i always appear in the product $\beta_L\, a_i$, hence their influence on the reactor volume must be the same.

- Region III). In this region both the reaction occurring in the film and that occurring in the liquid bulk are important.

- Region IV). This region corresponds to a regime in which the reaction takes place only in the liquid film. The various curves correspond to different sets of starting design data.

The existence of several curves in regions III and IV shows that the previously defined parameter Z does not correlate the various design situations. It was not possible so far to correlate all these situations in region III, while it was possible to introduce a parameter that represents all the design situations by a single curve in region IV. This region is defined by

$$Ha > 4 \sqrt{\frac{E_i - E}{E_i - 1}} \qquad (6)$$

and the parameter is the ratio E/E_i where E is the enhancement factor and E_i is that of an instantaneous reaction.

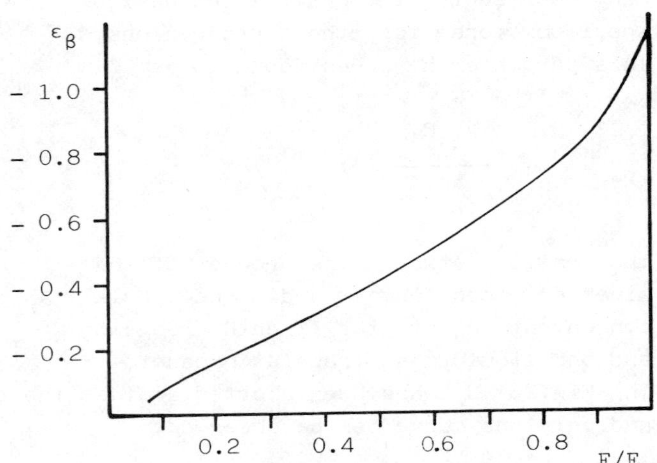

Fig 4. E/E_i vs. ϵ_β

In Fig 4 ϵ_β has been plotted against E/E_i for the same design conditions of Fig 3, and only one curve is observed.

To the limiting value of -1.1538 found for ϵ_β when E/E_i approaches unity, the same considerations made on the limiting value of ϵ_a , can be applied.

4.   REFERENCES

1.   Midoux, N. and Charpentier, J-C.: "Mechanically stirred gas-liquid tank reactors. Part 1 : Hydrodynamics". .(Les réacteurs gaz-liquide à cuve agitée mécaniquement. Partie 1 : Hydrodynamique). Entropie, n.88, 1979, pp.5-38. (in French)

2.   Midoux, N. and Charpentier, J-C.: "Mechanically stirred gas-liquid tank reactors. Part 2 : Interfacial areas". .(Les réacteurs gaz-liquide à cuve agitée mécaniquement. Partie 2 : Aires interfaciales). Entropie, n.101, 1981, pp.3-31. (in French)

3.   Joshi, J.B., Pandit, A.B. and Sharma, M.M.: "Mechanically agitated gas-liquid reactors". Chem. Engng Sci., 37, 1982, pp.813-843.

4.   Van't Riet, K.: "Review of measuring methods and results in nonviscous gas-liquid mass transfer in stirred vessels". Ind. Eng. Chem. Proc. Des. Dev., 18, 1979, pp.357-364.

5.   Brucato, A. and Rizzuti, L.: "Design procedure for gas-liquid mechanically stirred reactors". .(Procédure de projet pour les réacteurs gaz-liquide à cuve agitée mécaniquement). Entropie, n.115, 1984, pp.52-57. (in French)

6.   Barona, N.: "Design CSTR reactors this way". Hydrocarbon Processing, July 1979, pp.179-194.

INTERPRETATION OF MICROMIXING EXPERIMENTS INVOLVING
CONSECUTIVE COMPETITIVE REACTIONS IN A STIRRED TANK
BY A SIMPLE INTERACTION MODEL

R. DAVID, J.P. BARTHOLE* and J. VILLERMAUX

Laboratoire des Sciences du Génie Chimique
CNRS - ENSIC
1, rue Grandville, F 54042 Nancy, France

* Present address :   Rhône - Poulenc,
Centre de Recherche Technique
9, quai Jules-Guesde, F 94400 Vitry

SUMMARY :

The precipitation of barium sulphate from a basic EDTA-complex upon addition of a small volume of acid in a semi-batch stirred tank of 0.14 m^3 has been used to study the local state of micromixing. A model for the interpretation of previously published data obtained with this consecutive-competitive reaction system is presented in this paper. The model assumes that the injected acid is surrounded by a growing volume of fluid which circulates within the reactor. The tank is divided into two zones where the intensities of mixing are different. The reacting cloud passes successively through these zones. The cloud itself is divided into small eddies which interact and react according to the IEM (Interaction by Exchange with the Mean) model. The model accounts well for the influence of experimental parameters : location of injection points, agitation speed, injected volume and initial concentrations of reactants.

Held at Wurzburg, 10-12 June, 1985.

Organised by DVCV· Deutsche Vereinigung für Chemie- und Verfahrenstechnik
(German Association of Chemical and Process Engineering).

Organisation: GVC·VDI-Gesellschaft Verfahrenstechnik und Chemieingenieurwesen. 

NOMENCLATURE :

c_j concentration of reactant j, $mol.m^{-3}$

c'_{jo} initial concentration before mixing of reactant j, $mol.m^{-3}$

k , D_T turbulent diffusivities, $m^2.s^{-1}$

k_2 rate constant of reaction step 2, $m^5.mol^{-5/3}.s^{-1}$

L linear dimension of reaction cloud, m

n initial concentration ratio of barium and hydroxide ions

N agitation speed, s^{-1}

r_1 rate of reaction 1, $mol.m^{-3}.s^{-1}$

t_i age of volume i, s

t_c, t_{c_1}, t_{c_2} internal circulation times (overall, in zone 1, in zone 2), s

t_m micromixing time, s

V volume of reaction cloud $(V = \sum_i V_i)$, m^3

V_i incremental volume of rank i in the cloud, m^3

V_o initial volume of mixture before addition of acid, m^3

V_1 volume of injected acid, m^3

V_T tank volume $(V_T = V_o + V_1)$, m^3

X_S segregation index

Z_i integration variable in volume i, $mol.m^{-3}$

ε dissipated power per unit mass in the tank, $m^2.s^{-3}$

ν_{j1} stoichiometric coefficient of reactant j in reaction 1

superscripts $^-$ or $^=$ denote averaging over the cloud or over the tank, respectively

It is well known $|1-3|$ that consecutive-competitive reactions are very sensitive to the micromixing state of the fluids in which they occur. The precipitation of barium sulfate (S) from a basic EDTA barium-complex (A) by addition of hydrochloric acid (B) has been used to study local states of micromixing in a $V_O = 140$ dm^3 tank stirred by a Rushton turbine $|4-5|$.

1. REACTION SYSTEM

The reactions occur via three steps :

$$A + B \rightarrow R + 2W \quad \text{(instantaneous)} \qquad (1)$$

$$nU + R + 2nB \rightarrow nS\downarrow + nY\,H_2 \quad \text{(fast)} \qquad (2)$$

$$2A + YH_2 \rightarrow 2R + Y^{2-} + 2W \quad \text{(instantaneous)} \qquad (3)$$

where : $A = (Ba^{++},Y^{2-})_n OH^-$; $R = (Ba^{++},Y^{2-})_n$; $U = SO_4^{--}$; $YH_2 = $ EDTA-acid

$Y^2 = $ EDT-acetate ; $W = H_2O$; $B = H_3O^+$; $n = (Ba^{++})_o/(OH^-)_o$

The kinetics of the second stage were carefully determined in a batch reactor free from micromixing effects $|6|$ as :

$$r_2 = k_2 C_R C_U (C_S + nC_R + nC_A)^{2/3} \qquad (4)$$

with $k_2 = 1.5 \times 10^{-2}$ m^5.mol$^{-5/3}$.s^{-1} at 15°C.

The reaction was initiated by letting $|4|$ a volume $V_1 = 0.1$ dm^3 of HCl (concentration $C_{Bo} = 1.2$ N) burst into the mixture of the other reactants at different locations in the tank (semi-batch regime). Species A being in excess, if the acid B is instantaneously neutralized after addition, no S is produced. Conversely, if B eddies remain unmixed even for a short time (in the same order of magnitude as stage 2) S will be formed by reaction of R on B. The redissolution of S is very slow, so that its final amount, after completion of neutralization via stages 1 and 3, can be measured by turbidimetry $|4-5|$. The larger the amount of S, the higher the segregation. A segregation index X_S may thus be defined according to BOURNE et al. $|2|$ as :

$$X_S = (2n + 1)C_S V_T/(nV_1 C'_{Bo}) \qquad (5)$$

The aim of the present paper is to present a model accounting for the data published in ref. $|4|$. The basic assumptions of this model are the following.

2. MACROSCOPIC FLOW

The macroscopic flow pattern in the tank is described in detail in ref $|8|$. The injected volume V_1 is initially associated to a volume V_2 of surrounding fluid to form a reacting cloud which starts growing by turbulent diffusion. Simultaneously, the cloud moves along recirculation streams. As reported by NAGATA $|7|$, the linear dimension L of the cloud increases with cloud's age according to :

$$\frac{dL^2}{dt_i} = k \quad , \quad \text{or} \quad \frac{dV}{dt_i} = \frac{3}{2} k\, V^{1/3} = D_T V^{1/3} \qquad (6)$$

if $V = L^3$ is the cloud's volume (assumed cubic here).

The cloud's growth can be simulated by adding to the cloud incremental volumes $V_i = D_T V^{1/3}\Delta t$ from the bulk of the tank at Δt time intervals. Starting from the injection point, the cloud is convected along a trajectory determined by the average velocity pattern. It goes through two zones, one close to the stirrer (high mixing intensity), the other far from it (low mixing intensity) and follows an internal recirculation loop during circulation time t_c (fig. 1). Coefficient D_T is supposed by NAGATA to be proportional to the fluctuating velocity u'. From results published in ref. $|8|$, D_T is thus assumed to be 3-times larger in the stirrer zone than in the rest of the tank. The injection point is defined by the time required by the cloud to reach the next zone (fig. 1), namely t_{c1} (injection in quiet zone 1), or t_{c2} (injection in turbulent zone 2). The time for passing through zone 2 was set to 0.1 t_c $|8|$.

3. MICROMIXING PROCESS

Micromixing within the cloud is represented by the IEM model which assumes mass exchange between eddies and chemical reaction. Volumes V_i consist of small eddies which interact by exchange with a fictitious average concentration over the cloud $|9|$. For reactant j :

$$\frac{dc_{ij}}{dt_i} = \sum_{l=1}^{3} \nu_{jl}\, r_{il} + \frac{\overline{C}_j - C_{ij}}{t_m} \tag{7}$$

$$c_{ij}(0) = C'_{jo} \tag{8}$$

The micromixing time t_m is different in zones 1 and 2. One may assume with CORRSIN $|10|$ and VILLERMAUX $|9|$ that $t_m \sim \varepsilon^{-1/3}$. It was found in ref $|8|$ that $\varepsilon_2/\varepsilon_1 \simeq 10$ from which the ratio t_{m1}/t_{m2} was set equal to 2.2. $\overline{C}_j$ is calculated as :

$$\overline{C}_j = (\sum_i C_{ij}\, V_i)/V \tag{9}$$

The overall concentration in the tank is then :

$$j \neq B \quad \overline{\overline{C}}_j = \overline{C}_j\, V + (V_T - V)C'_{jo}$$
$$j = B \quad \overline{\overline{C}}_B = \overline{C}_B\, V/V_T \tag{10}$$

Integration of system (7) – (10) is made by using variables $z_{1i} = C_{Bi} + 2C_{YH_2i} - C_{Ai}$, $z_{2i} = C_{Ri} + C_{Ai}$, $z_{3i} = C_{YH_2i} + C_{Y2-i}$, in order to eliminate the rates of instantaneous steps (1) and (3).

4. RESULTS

Free parameters of the model are t_{m1} and D_T. From photographs, V_2 was estimated to about 1 dm^3 ; t_c , t_{c1} and t_{c2} were determined from the hydrodynamic flow pattern $|8|$. The initial concentration of reactants were $C'_{Ao} = 10$ mols.m^{-3} ; n = 0.2 ; $C'_{Uo} = 5$ mols.m^{-3} if not otherwise specified. Results are summarized in fig. 2-4. Fig 2 shows the spatial distribution of X_S vs. t_{c1} , both calculated and experimental, for different values of D_T , and a constant agitation speed N = 1.35 s^{-1} . The best fit is obtained with $t_{m1}/t_c \sim 0.1$ and $D_T t_c \simeq 0.175$ m^2 , i.e. $t_{m1} = 0.175$ s and $D_T = 0.1$ m^2.s^{-1} (in zone 1). Fig. 3 represents experimental and calculated variations of X_S vs. N. As $t_c \sim N^{-1}$, $t_m \sim \varepsilon^{-1/3} \sim N^{-1}$ $|9|$ and $u' \sim N$ $|8|$, t_m/t_c and $D_T t_c$ may be assumed to be independent on N. The fair agreement observed in fig. 3 confirms these assumptions. Fig. 4 shows the influence of the initial volume V_1 of acid, keeping $V_1 C'_{Bo} = 0.12$ mols and $V_2/V_1 = 10$. Experiments with different concentrations of BaCl$_2$ and Na$_2$SO$_4$, keeping the other concentrations constant, were also performed at two points in the tank (point 1 : $t_{c1} = 0.95\, t_c$, point 2 : $t_{c1} = 0.15\, t_c$). In each case, the agreement between model predictions and experiments is excellent. Besides, the time for total consumption of B ($\overline{C}_B = 0$) is also well predicted by the model (this time is about 0.5 t_c).

5. CONCLUSIONS

This simple model excludes spatial variation of concentration in the eddies and the set-up of a network of mixing cells within the tank. But is takes into account the macroscopic growth of a reacting cloud and the difference in turbulence intensity near and far from the stirrer. These are especially important when the time for completion of the reaction makes it possible for the cloud to pass through both zones. Finally all the parameters could be related to hydrodynamic data.

The calculation time corresponding to one experiment is about 2 min. with a middle-class computer.

Work is in progress for applying the model to other reactions and other mixing conditions.

6. <u>REFERENCES</u>

1.   <u>Levenspiel, O.</u>: "Chemical Reaction Engineering", New York, John Wiley and Sons,
     2nd ed. 1972, pp. 343-345

2.   <u>Bourne J.R., Kozicki F. and Rys P.</u>: "Mixing and fast chemical reaction. I - Test
     reactors to determine segregation". Chem. Eng. Sci., <u>36</u>, oct. 1981, pp. 1643-1648

3.   <u>David R., Lintz H.G. and Villermaux J.</u>: "Investigation of the mixing state of
     continuous reactors with chemical reactions". (Untersuchung der Durchmischung in
     kontinuiertich betriebenen Reaktoren mit Hilfe von chemischen Reaktionen).
     Chemie Ingenieur Technik, <u>56</u>, Feb. 1984, pp.104-110

4.   <u>Barthole J.P., David R. and Villermaux J.</u>: "A new chemical method for the study
     of local micromixing conditions in industrial stirred tanks". ACS Symp. Ser., <u>196</u>,
         1982, pp. 545-554

5.   <u>Barthole J.P., David R. and Villermaux J.</u>: "A new chemical method for investiga-
     tion of the micromixing state in stirred tanks". (Eine neue chemische Methode zur
     Untersuchung des Mikrovermischungsgrades in Rührbehältern). Chemie Ingenieur
     Technik, <u>56</u>, Jan. 1984, p. 63

6.   <u>Barthole J.P., David. R., Molleyre J.F., Bourret P. and Villermaux J.</u>: "Macrosco-
     pic kinetics of barium sulfate precipitation in the presence of EDTA". (Cinétique
     macroscopique de la précipitation du sulfate de barium en présence d'EDTA). J. de
     Chimie Physique, <u>79</u>, oct. 1982, pp. 719-724

7.   <u>Nagata S.</u>: "Mixing", New York, John Wiley and Sons, 1975, p. 181

8.   <u>Barthole J.P., Maisonneuve J., Gence J.N., David R., Mathieu J. and Villermaux J.</u>:
     "Measurement of mass transfer rates velocity and concentration fluctuations in an
     industrial stirred tank". Chem. Eng. Fundamentals, <u>1</u>, Jan. 1982, pp. 17-26

9.   <u>Villermaux J.</u>: "Mixing in Chemical Reactors", ACS Symp. Ser., <u>226</u>, 1983, pp. 138-
     144

10. <u>Corrsin S.</u>: "The isotropic turbulent mixer. Part II. Arbitrary Schmidt Number".
     AIChE J., <u>10</u>, nov. 1964, pp. 870-877

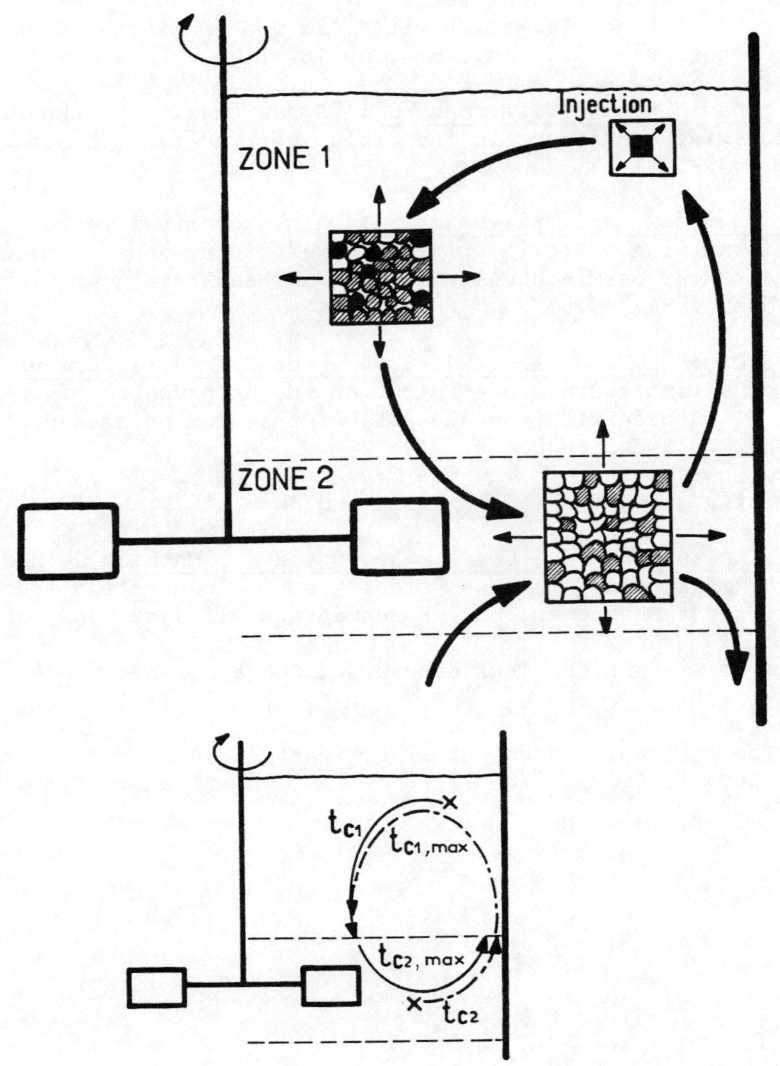

1. Description of model :   ○ A eddies ,   ● ⦸ B eddies (more or less concentrated)

Definition of t_{c_1}, t_{c_2} : ——— addition zone 1   ------- addition zone 2

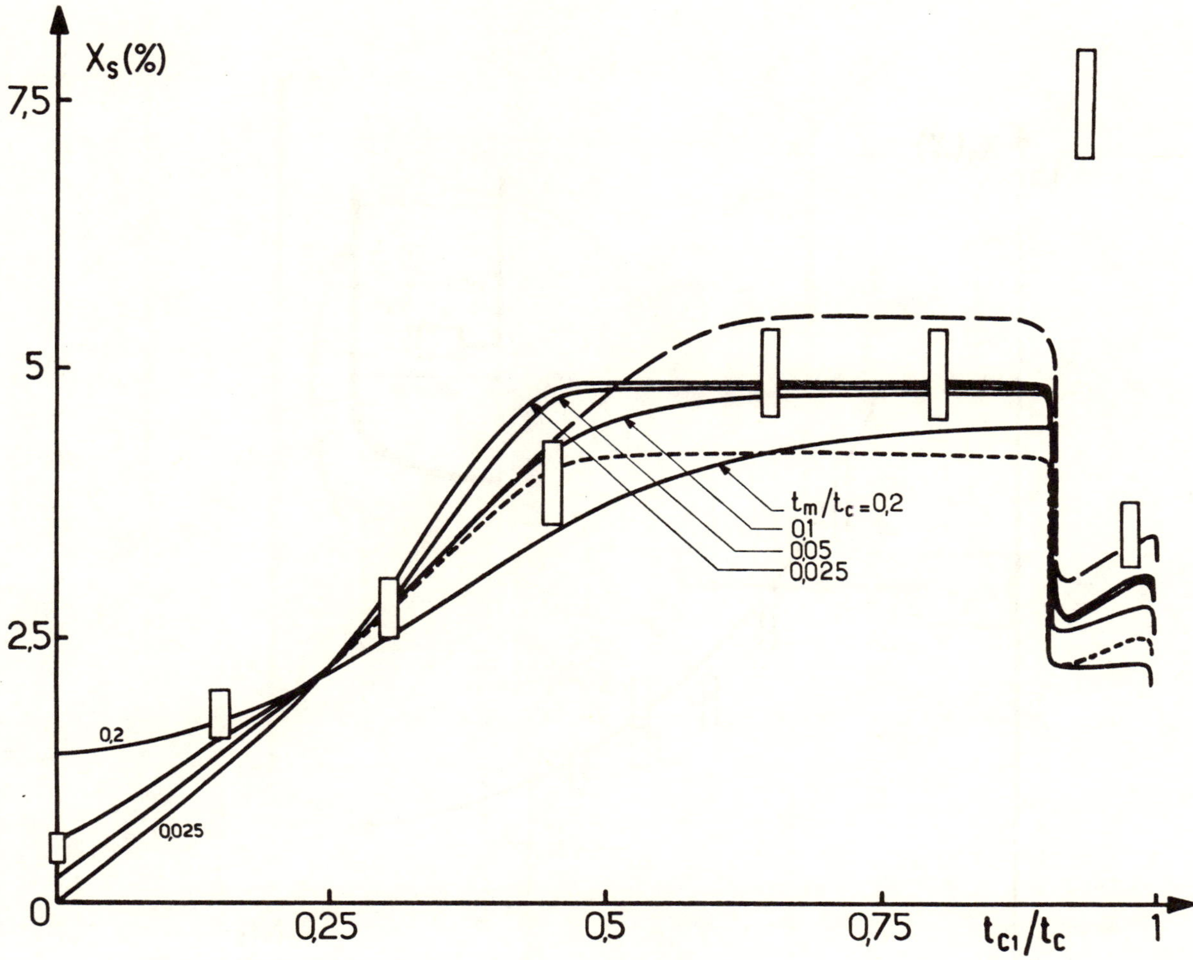

2. Segregation index X_S vs. t_{c_1}/t_c at $N = 1.35$ s^{-1} and $t_c = 1.75$ s ($t_{c_1}/t_c > 0.9$ indicates injection in zone 2) ; ▯ experiments ; ——— iso.t_m at $D_T t_c = 0.175$ m^2 ; ----- iso $D_T t_c = 0.2$ m^2 and $t_{m_1}/t_c = 0.1$; — — — iso $D_T t_c = 0.15$ m^2 and $t_{m_1}/t_c = 0.1$

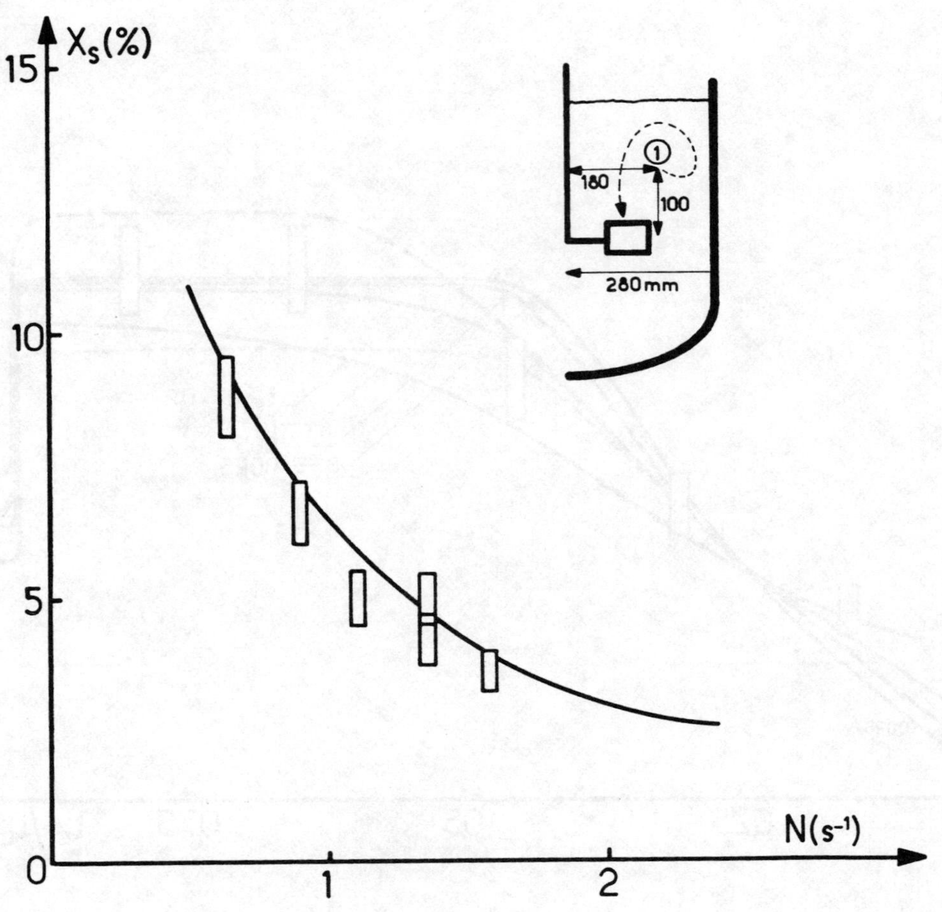

3. Segregation index X_S vs. Notation speed N at point n° 1
experiments ———— model ($t_m/t_c = 0.1$; $D_T t_c = 0.175$ m^2)

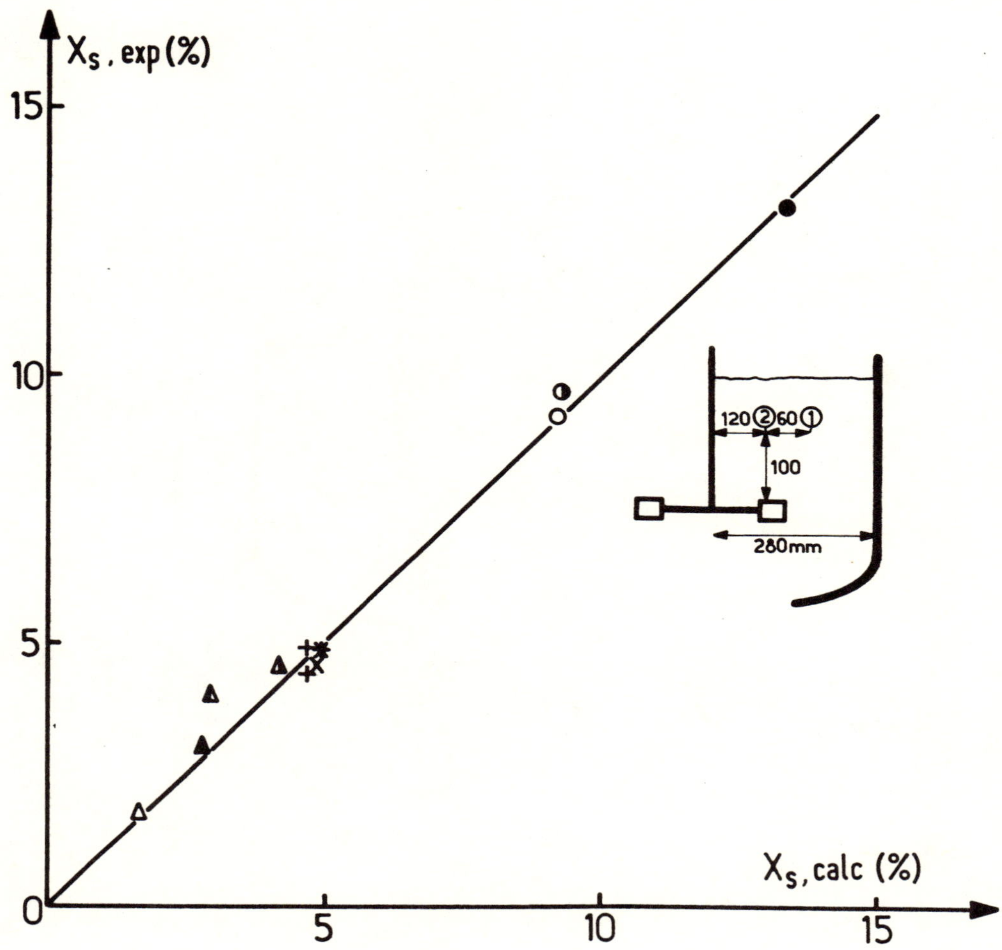

4. Influence of initial volume of acide V_1 and initial concentration on segregation index X_S.

Symbol	$V_1 (cm^3)$	Point No.	C'_{Uo} (moles/m^3)	n
+	100	1	5	0.2
×	50	1	5	0.2
✳	10	1	5	0.2
○	100	1	10	0.2
●	100	1	15	0.2
◑	100	1	5	0.4
△	100	2	5	0.2
▲	100	2	10	0.2
◭	100	2	15	0.2
◮	100	2	5	0.4

PERFORMANCE OF A CATALYTIC BASKET REACTOR FOR
KINETIC STUDIES OF A GAS-LIQUID-SOLID SYSTEM.

A. Gianetto, G. Baldi, S. Sicardi, R. Conti,
V. Specchia

Dipartimento di Scienza dei Materiali e Ingegneria Chimica
Politecnico di Torino, 10100 Torino, Italy

Summary

Although frequently used for kinetic studies, the catalytic basket reactor has
been very seldom examined in the case of a three-phase reacting system. Recent results
have indicated a partial utilization of the catalyst which has been not satisfactorily explained.

In this work, a catalytic basket reactor, equipped with a static basket and a 6-
blade disk turbine, was examined; as a test reaction, the catalytic oxidation of ethanol with gaseous oxygen was employed. Three sets of experiments were carried out, with
a different catalyst loading.

The results have shown a partial utilization of the catalyst that increased as the
stirrer speed increased and the catalyst loading decreased. This was explained as due
to both solid-liquid mass transfer and O_2 depletion in the catalytic layer.

Held at Wurzburg, 10-12 June, 1985.

Organised by DVCV· Deutsche Vereinigung für Chemie- und Verfahrenstechnik
(German Association of Chemical and Process Engineering).

Organisation: GVC·VDI-Gesellschaft Verfahrenstechnik und Chemieingenieurwesen. 

C_t total liquid concentration, $kmol/m^3$

C_1, C_2 oxygen and ethanol concentration in the liquid, $kmol/m^3$

K composite kinetic constant, $kmol/(s.m^3)$

k_0 2-nd order kinetic constant, $m^3/(s.kmol)$

k_2 constant in the kinetic rate equation, $m^3/kmol$

N revolution speed, rps

n_G gas phase flow rate, kmol/s

Q liquid rate flowing through the catalyst layer, m^3/s

Q_F liquid feed rate, m^3/s

R_{exp} experimental conversion rate, $kmol/(s.m^3_{cat})$

R_{kin} theoretical conversion rate,   "   "

R_I intrinsic reaction rate per unit cat. volume, $kmol/(s.m^3_{cat})$

T temperature, K

V_c catalyst layer volume, m^3

x_1^0 oxygen molar fraction in the outlet liquid, dim.less

x_2^1, x_2^0 ethanol molar fraction in inlet and outlet liquid respectively, dim.less

y_2^0 ethanol molar fraction in the outlet gas, dim.less

η catalyst effectiveness factor, dim.less

1. INTRODUCTION

The kinetic study of a gas-liquid-solid reaction is usually rather complicated, because of the several elemental phenomena involved (interphase and intraphase mass and heat transfer) and the influence of hydrodynamics. Sometimes, the three-phase system can be reduced to a two-phase system, by using a liquid phase previously saturated with the gaseous reactant (Ref. 1, 2, 3). This makes easier the analysis of the results, because there is a wide literature on laboratory catalytic reactors with one fluid phase (Ref. 4, 5, 6, 7). But generally a three-phase reactor must be used, to test large conversions of the liquid reactant; the analysis of the kinetic results is then more difficult, since for reactors of this type very few results are available in the literature.

Among the laboratory three-phase reactors, the so called "catalytic basket reactor" (CBR) is particularly interesting for kinetic studies with catalyst pellets. This reactor can be equipped either with a rotating basket containing the catalyst, or with a static basket. In the first case, the basket itself works like a stirrer; in the second case there is a stirrer promoting the gas and liquid motion.

Although the CBR, at first glance, seems a very simple reactor (the fluid phases can be assumed as perfectly mixed), in some papers recently published (Ref. 8, 9, 10, 11), a partial utilization of the catalyst was pointed out; this was explained as due to a blanketing of the pellet by the gas bubbles entrapped in the catalytic bed (Ref. 9, 11).

The aim of this work is to examine the performance of a CBR with a static basket for kinetic studies of three-phase systems. The reaction examined was the calatytic oxidation of ethanol with gaseous O_2; a 0.5% Pd on alumina catalyst was employed. The intrinsic reaction rate per unit catalyst volume R_I was previously determined (Ref.1); The following equation was found to hold:

$$R_I = - k_o C_1 C_2 \ / \ (1 + k_2 C_2) \qquad\qquad (1)$$

where:

$$k_o = 9.564.10^8 \ \exp(-6030/T) \qquad [m^6/s.m^3_{cat}.kmol]$$

$$k_2 = 2.126.10^5 \exp(-2982/T) \qquad [m^3/kmol]$$

2. EXPERIMENTAL APPARATUS

Fig. 1 provides the schematic layout of the experimental apparatus. The glass jacketed reactor of about $6.10^{-3} \ m^3$, had a spherical bottom and was equipped with 4 baffles, a 6-blade disk turbine and a stationary basket, whose major dimensions are shown in fig. 2.

The reactor was continuous and isothermal. The liquid feed was an aqueous solution of ethanol and K_2CO_3; this was added because in a basic medium there are no side reactions (Ref. 1, 12). Pure oxygen was fed from bottles in a large excess through a sparger; the gas leaving the reactor was passed through a condenser, cooled with brine at -5 °C to condense most of the stripped ethanol.

The concentrations of ethanol in the liquid entering and leaving the reactor, and in the gas leaving the condenser were determined by chromatographic analysis. The concentration of oxygen in the outlet liquid was measured by a polarographic electrode. The observed values were practically equal to those in equilibrium with the O_2 partial pressure in the reactor.

Three sets of experiments were carried out:

- Set A : the basket was completely filled with catalyst pellets (.37 kg of catalyst).

- Set B : the outer part of the basket contained .152 kg of catalyst; the inner part was filled with glass cylinders of the same size of the catalyst.

- Set C : as set B as far as the catalyst is concerned; the inner part of the
  basket was instead completely empty.

Table 1 shows the range of the operating conditions.

3. RESULTS

The experimental conversion rates of ethyl alcohol were calculated from the following mass balance equation:

$$R_{exp} \cdot V_c = Q_F C_t (x_2^o - x_2^i) + n_G y_2^o \qquad (2)$$

The values of R_{exp} were compared with the theorethical ones R_{kin} calculated by assuming that the concentrations of ethyl alcohol and oxygen at the external surface of the catalyst pellets are those measured in the liquid leaving the reactor:

$$R_{kin} = - \eta \, k_o C_t^2 x_1^o x_2^o / (1 + k_o C_t x_2^o) \qquad (3)$$

The catalyst effectiveness factor η was evaluated by taking into consideration that the active component is distributed in a thin superficial layer. O_2 was considered the controlling reactant, due to its low solubility.

Typical results are shown in fig. 3. The ratio R_{exp}/R_{kin} is always less than 1, indicating a certain degree of partial utilization of the catalyst. It increases as the stirrer speed increases, and furthermore is markedly affected by the arrangement of the apparatus: set C was more effective than set B, and this more effective than set A. This can be also seen from fig. 4, as far as setA and B are concerned.

It was observed that the gas bubbles did not seem to enter the packed layer, so that through the catalytic bed there was very likely a liquid flow only.

The influence of the hydrodynamic conditions is generally explained as due to the relevance of mass transfer phenomena. Here, solid-liquid mass transfer must be considered, because the liquid surrounding the basket is always at equilibrium with the gas phase. The solid-liquid mass transfer in the catalytic bed is affected by the liquid velocity, and hence by the stirrer speed.

The higher conversion rates in the runs with set C could really be due to a higher liquid flow rate through the catalyst, and hence to a higher mass transfer coefficient between solid and liquid, because of the smaller thickness of the bed. The data of set C are also an indirect confirmation that the blanketing of the pellets due to the gas bubbles does not occur in our system. In fact at high N values the probability for the bubbles to enter the catalytic layer is high; but for these runs, on the contrary to the findings of other Authors (Ref. 10, 11), the values of R_{exp}/R_{kin} is close to 1.

Another parameter that may affect considerably the reactor performance is the depletion of oxygen in the bed due to the reaction. Let Q be the liquid flow rate pumped through the bed by the stirrer; the microscopic mass balance of O_2 gives:

$$C_t Q \, dx_1 = R \, dV_c \qquad (4)$$

Bearing in mind the low solubility of oxygen, the concentration of ethanol may be considered constant along the bed, so that the reaction rate can be written as follow:

$$R = -K \, x_1 \qquad (5)$$

K is a composite "kinetic constant", which takes account of both the solid-liquid mass transfer and the intrinsic reaction rate.

Integration of eq. 4 results in an exponential decay of x_1 along the basket as a function of V_c. Therefore, the mean concentration of O_2 in the bed, and hence the average reaction rate, at the same hydrodynamic conditions in the bed, would decrease as the thickness of the catalytic layer is increased:

$$x_{1,o} = x_{1,e} \cdot \exp(-KV_c/Q) \tag{6}$$

The effect of oxygen depletion is shown by the fact that the conversion rate in set A is generally lower than that   in set B at the same conditions. It may be stressed that the two sets are equivalent from an hydrodynamic point of view, because the thickness of the solid layer in the two cases is the same. They only differ in the catalyst loading.

4. CONCLUSIONS

During the runs, it was observed that no bubbles enter the catalytic bed. On the contrary of the findings of other Authors (Ref. 10, 11), the gas phase does not seem to affect the wetting of the catalyst, at least in the range of the stirrer speed examined.

The performance of the CBR can on the other hand be affected by the mass transfer between the liquid and the solid particles. Furthermore an important effect may also have the depletion of the key reactant in the liquid flow through the bed, so that the catalyst load may play an important influence.

Aknowledgements: This work was carried out with the financial support of C.N.R., Chimica Fine e Secondaria, grant n° 82.00605.95.

5. REFERENCES

1. Tukac, V., Mazzarino, I., Baldi, G., Gianetto, A., Sicardi, S., and Specchia, V.: "Conversion rates in a laboratory trickle-bed reactor during the oxidation of ethyl alcohol". Submitted to Chem. Eng. Sci.

2. Baldi, G., Goto, S., Chow, C.K. and Smith, J.M.: "Catalytic oxidation of formic acid in water. Intraparticle diffusion in liquid-filled pores". Ind. & Eng. Chem. Proc. Des. Devel., 13, 1974, pp. 447-452.

3. Levec, J. and Smith, J.M.: "Oxidation of acetic acid solutions in a trickle-bed reactor". AIChE J., 22, 1976, pp. 159-168.

4. Weekman, V.W. Jr.: "Laboratory reactors and their limitations". AIChE J., 20, 1974, pp. 833-840.

5. Dorayswamy, L.K., and Taijbl, D.G.: " Laboratory catalytic reactors". Cat. Rev. - Sci. Eng., 10, 1974, pp. 177-219.

6. Shah, Y.T.: "Gas-liquid-solid reactor design". McGraw-Hill Inc.,1979, New York.

7. Carberry, J.J.: "Chemical and catalytic reaction engineering". McGraw-Hill Inc., 1976, New York.

8. Myers, E.C.,and Robinson, K.K.: "Multiphase kinetic studies with a spinning basket reactor". Proc. 5-th Int. Symp. Chem. Reaction Eng., Weekman V.W. Jr. and Luss D. Editors, ACS Symposium Series n° 65, 1978, pp. 447-458.

9. Ohta, H., Goto, S., and Tashima, H.: "Liquid phase oxidation of phenol in a rotating catalytic basket reactor". Ind. & Eng. Chem. Fundam., 19, 1980, pp.180-185.

10. Njiribeako,A.I., Silveston, P.L., and Hudgins, R.R.: "A laboratory spinning catalyst basket reactor for multiphase contacting". Can. J. Chem. Eng., 56, 1978, pp. 643-645.

11.   Pavko, A., Misic, D.M., and Levec, J.: "Kinetics in three-phase reactors". Chem. Eng. J., 21, 1981, pp. 149-154.

12;   Hsu, S.H., and Ruether, J.A.: "Kinetics of the liquid phase oxidation of ethanol by oxygen over a Palladium-alumina catalyst". Ind.& Eng. Chem. Process Des. Devel., 17, 1978, pp 524-527.

<u>Table 1</u>: Operating conditions.

Liquid flow rate	$1.315.10^{-6}$ m³/s
Oxygen flow rate	$2.23.10^{-5}$ Nm³/s
Catalyst loading	0.152-0.37 kg
Stirrer speed	3.33-13.33 rps
Temperature	45-80 °C
Ethanol concentration	0.013-0.041 kmol/m³
Oxygen partial pressure	$5.10^{4} - 8.8.10^{4}$ Pa
K_2CO_3 concentration	$1.45.10^{-2}$ kmol/m³

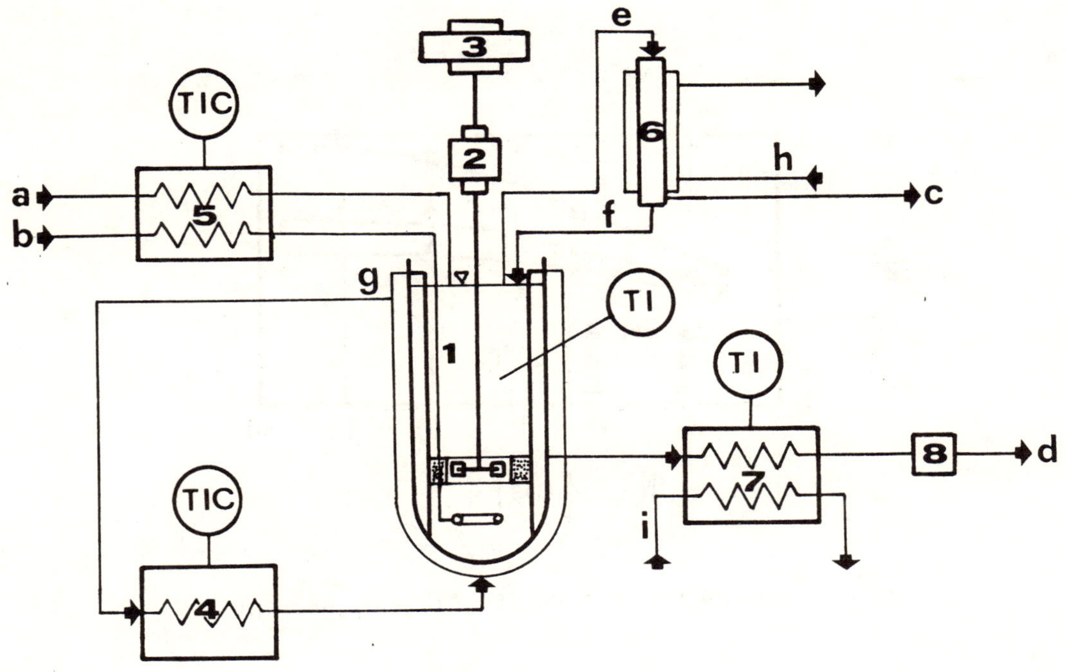

Fig. 1: Schematic diagram of the experimental set-up; 1: reactor; 2: torquemeter; 3: d.c. electric motor; 4, 5: thermostatic baths; 6: condenser; 7: cooler; 8: cell for O_2 analysis; a: liquid feed ; b: gas feed; c: gas out; d: liquid out

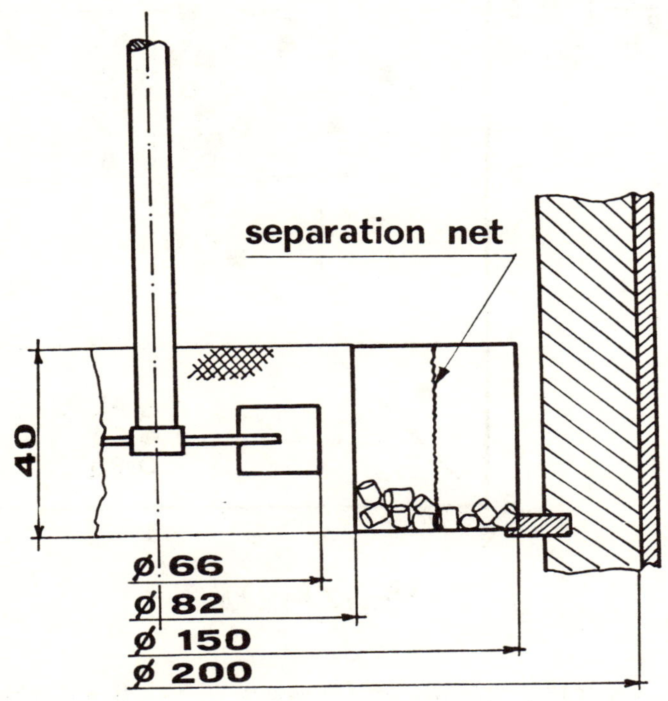

Fig. 2: Major dimensions of the basket reactor.

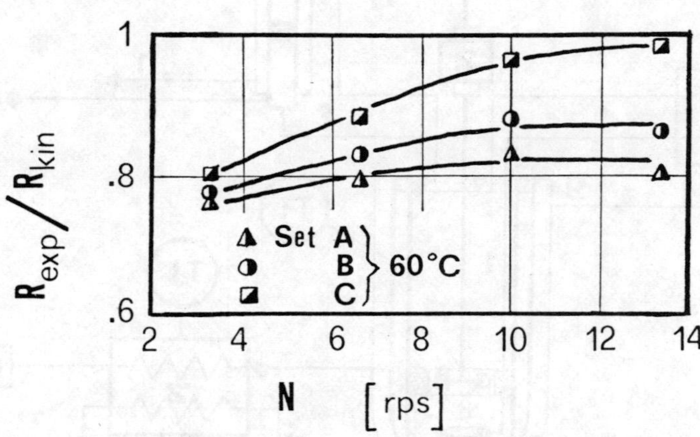

Fig. 3: Influence of N on the ratio R_{exp}/R_{kin}.

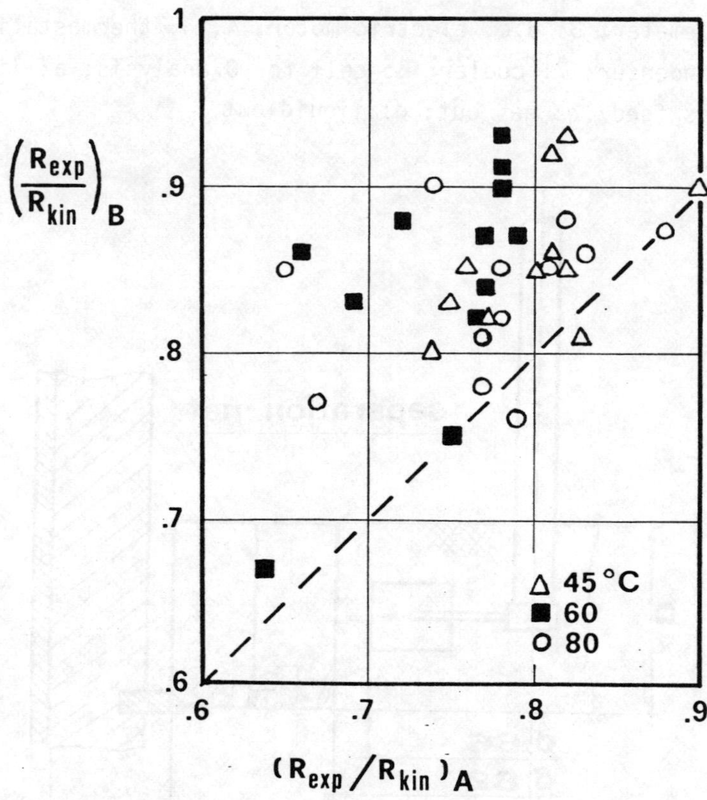

Fig. 4: Comparison of the performance of set B and that of set A.

EFFECT OF HIGH SOLIDS CONCENTRATIONS ON MASS TRANSFER AND GAS HOLD-UP IN THREE-PHASE MIXING

M. Greaves and V.Y. Loh

School of Chemical Engineering
University of Bath, Bath BA2 7AY, U.K.

Summary

Gas-liquid mass transfer and gas hold-up have been investigated at high concentration levels of solids in a 0.2 m stirred reactor. Glass ballotini and ion exchange resin particles at concentrations up to $X = 50\%$ w/w and $X = 30\%$ w/w, respectively, were used. For oxygen absorbing into water, k_La values were accurately determined by means of a continuous flow steady-state technique.

At $N > N_{JSg}$, k_La is found to decrease rapidly beyond a certain level of solids concentration. On a volumetric basis, this corresponds to approximately $X' = 15\%$ v/v, for both particle types. In both coalescing and 'non-coalescing' air-water systems, the mass transfer coefficient tends to a limiting value ($k_La \sim 0.005$ s^{-1}) as the highest solids concentrations are approached. Under these conditions, k_La values are not noticeably affected by changes in gassing rate or impeller speed.

For glass ballotini particles, ε_G generally decreases as X increases, the overall reduction being approximately 30% at $X = 50\%$ w/w. By comparison, very little change is observed when suspending ion exchange resin particles. It is concluded that, there are two main factors responsible for the lowering of k_La when solids are present at high concentrations. These effects are explained in terms of enhanced bubble coalescence arising from turbulence dampening and rheological-viscous behaviour.

Held at Wurzburg, 10-12 June, 1985.

Organised by DVCV· Deutsche Vereinigung für Chemie- und Verfahrenstechnik
(German Association of Chemical and Process Engineering).

Organisation: GVC·VDI-Gesellschaft Verfahrenstechnik und Chemieingenieurwesen. 

© BHRA, The Fluid Engineering Centre, Cranfield, Bedford MK43 0AJ, England.

NOMENCLATURE

a	bubble interfacial area ($m^2 m^{-3}$)
d_p	particle diameter (μm)
C	impeller clearance (m)
D	impeller diameter (m)
k_L	mass transfer coefficient (ms^{-1})
$k_L a$	specific mass transfer coefficient (s^{-1})
N	impeller speed (s^{-1})
N_{CD}	impeller speed for complete dispersion of gas throughout vessel (s^{-1})
N_{JSg}	impeller speed for just suspending solid particles (s^{-1})
P/V	power per unit volume (wm^{-3})
Q_G	gas sparging rate ($m^3 s^{-1}$)
Q_L	liquid flow rate ($m^3 s^{-1}$)
T	vessel diameter (m)
v_s	superficial gas velocity (ms^{-1})
X	solids concentration (w/w%)
X'	solids concentration (v/v%)
ε_G	gas hold-up
ρ_s	particle density ($kg\ m^{-3}$)

1   INTRODUCTION

Three-phase mixing is important in a wide range of industrial processes, which involve hydrogenation reactions, oxidation reactions, polymerisation, biological fermentation, and mineral froth flotation. Additional complexity arises in three-phase mixing because of the simultaneous requirement for suspension of solid particles and dispersion of gas into the liquid. Normally, it is considered sufficient for the solid particles to be just-suspended ($N \geqslant N_{JSg}$), so that all of the particle surface is exposed and available for mass transfer, or reaction. At the same time, the gas must be sufficiently dispersed throughout the vessel ($N \geqslant N_{CD}$) in order that gas-liquid mass transfer can proceed efficiently. The gassed power and energy dissipation required have been extensively investigated by Weidmann et al[1] and Chapman et al[2,3]. Nienow[4] has also recently analysed the effectiveness of various agitator geometries for three-phase mixing. It is concluded that the disc turbine is desirable from the view-point of stability, but may not be the most energy efficient.

The emphasis in this paper is on gas-liquid mass transfer at high solids concentrations (X up to 50% w/w). At such high particle loadings, there will be intense particle-bubble interaction causing changes in the rates of bubble coalescence and bubble size. Rheological-viscous effects also become important, producing turbulence dampening. The resulting effect on k_L, a or $k_L a$ is complex, since bubble surface area or gas hold-up may increase, or decrease, depending on the extent of particle-bubble inter-action and type of suspension rheology.

2   MEASUREMENT OF $k_L a$

In practice, it is generally more convenient to use a physical, rather than chemical absorption method, for the measurement of the specific mass transfer coefficient, $k_L a$. Chapman et al[5] have described a novel dynamic response technique which eliminates the need to make any assumptions about the state of gas-phase mixing. However, an oxygen probe with a very fast response is required. The steady-state method was selected for the three-phase experiments for two main reasons. Firstly, there is no restriction concerning the establishment of gas hold-up, and, secondly, it is not necessary to have the probe located at any particular position in the vessel. The time to attain steady-state in the three-phase systems at high solids concentrations, is much longer compared with gas-liquid mixing[6]. Therefore, under dynamic conditions, the time required to displace all of the nitrogen, say, (from t = 0) will also be longer. This influences the initial part of the response curve, which critically affects the accuracy with which $k_L a$ can be determined. On the second point, if the probe is placed directly in the vessel, there will inevitably be interference from colliding solid particles (and bubbles). Even at quite low solid concentrations, this can cause the probe measurement to fluctuate by $\pm$ 20%.

For the steady-state method, $k_L a$ is determined from the following equation:

$$k_L a = \frac{Q_L \, (p - p_i)}{V \, (p^* - p)} \tag{1}$$

Full details of the derivation are given by Loh[7].

3   EXPERIMENTAL

3.1 Procedure

The measurements of the $k_L a$ were carried out in a 0.2 m stainless steel vessel, using the arrangement shown in Fig. 1. Three sizes of disc turbine impeller were used (D/T = 0.375, 0.5 and 0.66), set at a clearance of C/T = $\frac{1}{4}$. A liquid height of 1.25 T was used throughout. Solid particles were retained in the vessel by fitting a fine mesh on the exit pipe. The second vessel, of 0.9 m diameter, was used to deoxygenate the water supplied to the stirred vessel.

Initially, the water in the large holding tank was sparged with air to saturate it with oxygen. The liquid was then circulated to the mixing vessel, where it was further aerated at a selected impeller speed and gassing rate. At steady-state, a 100 cm^3 sample was taken and the dissolved oxygen concentration measured using a Beckman 0260 DO probe (p*%). The sample was agitated with a magnetic stirrer to ensure that there was a constant supply of fresh liquid at the probe membrane. The gas flow to the holding tank was then switched to nitrogen and deoxygenated water (typically 8% O_2) pumped to the stirred vessel where it was aerated at the same conditions as previously. A number of samples were taken at the inlet and exit of the mixing vessel to determine when steady-state had been attained. These readings are, respectively, p_i and p in Eqn.(1). All of the experiments were carried out at atmospheric pressure using either filtered tap water or a solution of 0.05 M K_2SO_4, maintained at 20 ^{0}C. The properties of the solid particles are given in Table 1.

3.2 Liquid flow rate and gas phase depletion

The effect of liquid flow rate on the hydrodynamic regime in the stirred vessel is shown in Fig. 2. This was determined for the gas-liquid case, but very similar behaviour was exhibited by the particle-liquid-gas systems. For the larger impeller sizes, there is a significant effect on k_La at a flow rate greater than $50 \times 10^{-6} m^3 s^{-1}$, approximately. It is desirable to have as large a flow rate as possible in order to minimise the value of p (*i.e.* approach to saturation). The liquid flow rate was therefore fixed at $Q_L = 50 \times 10^{-6} m^3 s^{-1}$.

The stead-state method of measuring k_La assumes that there is negligible variation of the oxygen concentration in the gas phase, *i.e.* constant p* throughout the vessel. Under the worst conditions, *i.e.* low Q_G and high N, the maximum error incurred in k_La (without correction for gas depletion) was only 6%. Therefore, the value of p* can be taken as essentially constant.

4 EXPERIMENTAL RESULTS

4.1 Effect of solids concentration on k_La

The trends exhibited in Figs. 3 to 6 are very similar in each case. k_La is not greatly affected by the presence of solids up to X = 15% for ion exchange resin and up to 30% for glass ballotini. Thereafter, a rapid reduction in k_La occurs, eventually tending towards a limiting value of approximately 0.005 s^{-1}. The effect of gassing rate also diminishes at high solids concentrations.

k_La first begins to fall when the volumetric concentration of solids reaches a value of about 15% v/v, for both ion exchange resin and glass ballotini particles. This appears to be mainly a volumetric effect. The results generally confirm the trends reported by other workers[8,9]. However, the first lowering of k_La occurs at lower values of X than those reported by Joosten *et al*[9]. In the size range studied (glass ballotini), there is no significant effect of particle size. The reducing effect on k_La as X increases is logical, since the spherical shape of the particles would not completely restrict the diffusional path in the liquid film surrounding the bubble, nor totally block the area available for mass transfer. This may also explain why gas flow rate has little effect on k_La at very high solids concentrations. Both Joosten *et al* and Chapman *et al* noticed a slight increase in k_La at very low solids concentrations, which is the opposite of Lee's *et al*[10] findings for ionic systems. Neither of these trends are exhibited by the present results.

A comparison of k_La data with the correlation of Van't Riet[11] for coalescing water systems is shown in Fig. 7:

$$k_La = 0.026 (P/V)^{0.4} (v_s)^{0.5} \tag{2}$$

At low gassing rates, there is good agreement with Eqn.(2). The agreement at $Q_G = 1.79$ vvm is less good, but still falls well within the accuracy limits of Van't Riet's correlation. For X < 16% v/v, the variation of k_La with P/V is very similar to that for gas-liquid systems, but at X = 25% v/v there is a reduced proportionality.

4.2 Gas hold-up

The effect of impeller speed on ε_G is shown in Figs. 8 and 9. For N < N_{JSg}, ε_G is very small, mainly due to poor dispersion of the gas. All of the mass transfer data presented in this paper refers to conditions with N > N_{JSg}.

Figs. 10 and 11 show that for ion exchange particles, there is hardly any effect of solids concentration on ε_G. However, in the case of glass ballotini (Figs. 12 to 15), there is a continuous decrease in ε_G as X increases. This cannot be explained solely in terms of a physical interference effect, *i.e.* "crowding-out" of the gas by solid particles, since no such effect is observed for the ion exchange particles, even though the highest values of X (on a volumetric basis) are virtually the same in both cases. The suggestion by Lee *et al*[10] that the solids act to dampen turbulence, thereby reducing bubble break-up, seems to be applicable. Density also has an influence, since ε_G is only affected by the heavier glass particles. It has been observed for this case that the glass ballotini concentrated in the lower section of the vessel, even for $N \geqslant N_{JSg}$. This reduces the size of the high shear zone, so that the turbulent forces further away from the impeller zone are markedly reduced. The combined effect of reduced gas dispersion and increased bubble coalescence decreases the residence time for the bubbles, which in turn reduces the gas hold-up.

In very concentrated suspensions, the effective viscosity is increased and this effect is probably the main reason for the large reduction in ε_G beyond about X = 25% w/w (glass ballotini). Rheological-viscous effects of this type will tend to promote bubble coalescence, with consequent reduction in $k_L a$ (Figs. 3 to 6).

4.3 Electrolyte solution

Results for 'non-coalescing' electrolyte solution (0.05 M K_2SO_4) exhibit similar trends to those obtained for tap water. However, as shown in Figs. 16 to 19, both $k_L a$ and ε_G are higher at low solids concentrations, but they eventually approach those of tap water when X is greater than 40% w/w. This converging trend is further evidence of increasing turbulence dampening and rheological-viscous behaviour at high solids concentrations. The coalescence inhibition produced by the electrolyte solution is continually diminished and eventually suppressed altogether.

5 CONCLUSIONS

The steady-state absorption technique provides an accurate means of determining gas-liquid mass transfer for three-phase mixing in stirred vessels, when the $k_L a$ values are comparatively low.

At low solids concentrations, $k_L a$ is not significantly different from values obtained for gas-liquid systems. $k_L a$ starts to fall rapidly, however, when X exceeds 15% v/v. At solids concentrations higher than this, further reduction of $k_L a$ occurs due to increasing effective viscosity of the suspension and turbulence dampening effects. A limiting value of $k_L a \sim 0.005$ s^{-1} is approached when X reaches very high concentrations of 30% w/w (ion exchange particles and 40% w/w (glass ballotini). Under these conditions, $k_L a$ is not significantly affected by either gassing rate or impeller speed.

Gas hold-up in three-phase suspensions is dependent on particle density. However, the lowering effect on ε_G is probably due to poor gas dispersion resulting from increased solids concentrations in the lower regions of the vessel. For ionic solution, $k_L a$ and ε_G values decrease rapidly as X increases, eventually approaching those of tap water at X = 40% w/w.

REFERENCES

1   Wiedmann, J.-A., Stieff, A. and Weinspach, P.A., "Fluid Dynamics of Stirred Three-phase Reactors", *Ger. Chem. Eng.*, 4, 1981, 125-136

2   Chapman, C.M., Nienow, A.W., Cooke, M. and Middleton, J.C., "Particle-gas-liquid Mixing in Stirred Vessels. Part III. Three-phase Mixing", *Chem. Eng. Res. & Dev.*, 61, (3), 1983, 167

3   Chapman, C.M., Nienow, A.W., Cooke, M. and Middleton, J.C., "Particle-gas-liquid Mixing in Stirred Vessels. Part IV. Mass Transfer and Final Conclusions", *Chem. Eng. Res. & Dev.*, 61, (3), 1983, 182

4   Nienow, A.W., "Particle-gas-liquid Mixing in Stirred Vessels", Paper presented at A.I.Ch.E. Annual Meeting, San Franciso, November 1984, (Session 1296), pp 1-15

5   Chapman, C.M., Gibilaro, L.G. and Nienow, A.W., *Chem. Eng. Sci.*, 37 (6), 1982, 891

6   Greaves, M. and Loh, V.Y., *I.Chem.E. Symposium Series, No. 89*, 1984, pp 69-96

7   Loh, V.Y., Ph.D. Thesis, University of Bath, 1983

8   Chapman, C.M., Ph.D. Thesis, University of London, 1981

9   Joosten, G.E.H., Schilder, J.G.M. and Janssen, J.J., *Chem. Eng. Sci.*, 32, 1977, 563

10   Lee, J.C., Ali, S.S. and Tasakoru, P., Fourth European Conference on Mixing, Leeuweuhorst, Netherlands, 1982, 379 (BHRA)

11   Van't Riet, K., I.E.C. (PDD), 18, 1979, 357

Address:

Dr V.Y. Loh is now at the Food Research Institute, Colney Lane, Norwich NR4 7UA

TABLE 1

Particle Properties

Particle Type	Size d_p, μm	Density ρ_s, kg m^{-3}
'Amberlite' ion exchange resin	780 ± 70	1260
Lead glass ballotini	655 ± 55 1300 ± 100	2950 2950

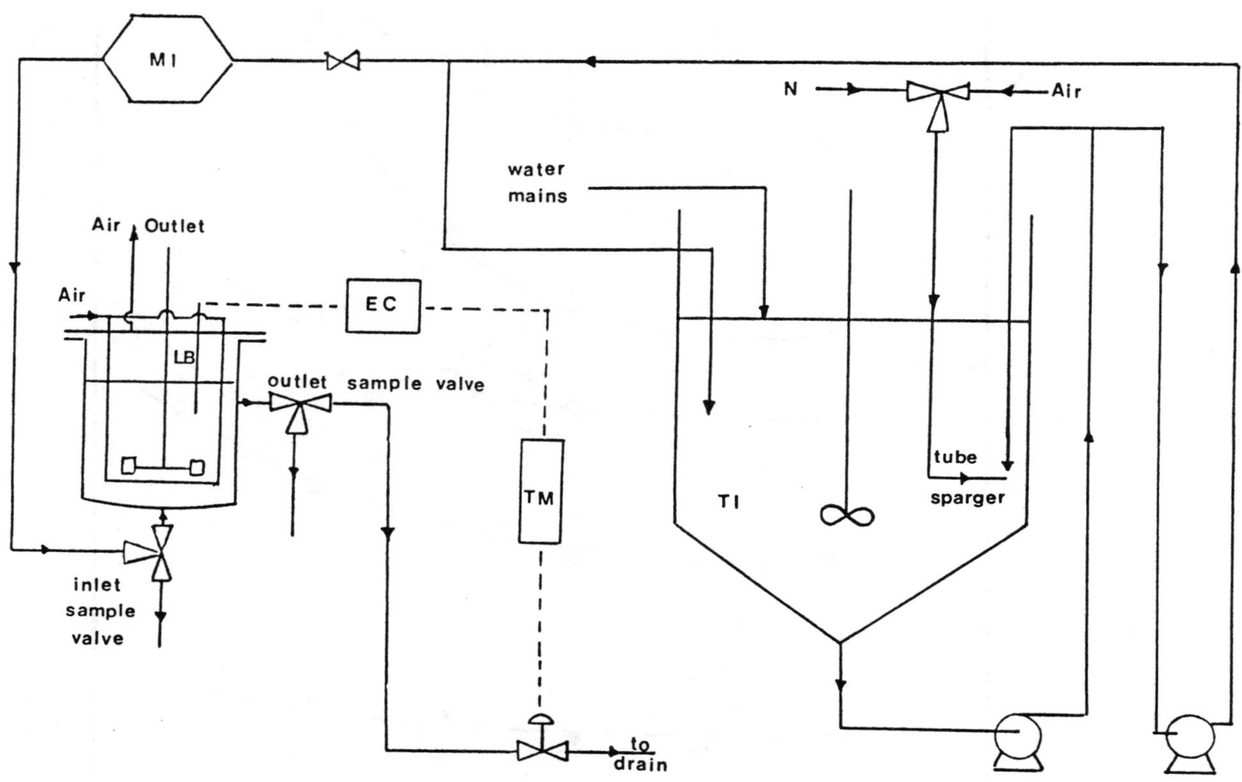

Figure 1   Flow Circuit for $k_L a$ Measurement

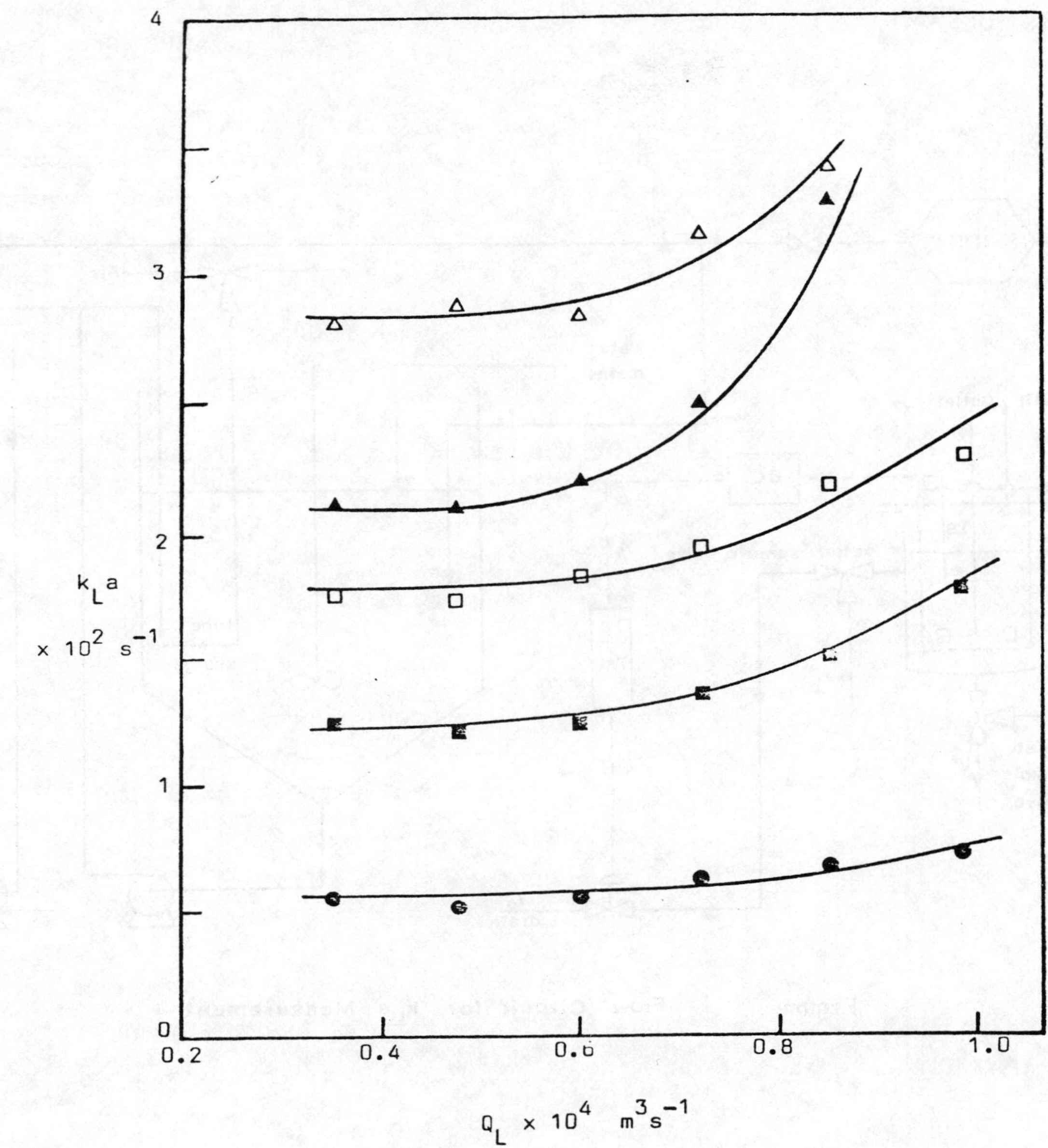

Figure 2   Effect of liquid flow rate on $k_L a$

	D (m)	N (s^{-1})	Q_G (vvm)
●	0.0762	5	1.79
■	0.1016	5	1.79
▲	0.135	5	1.79
□	0.1016	6	1.07
△	0.135	6	1.07

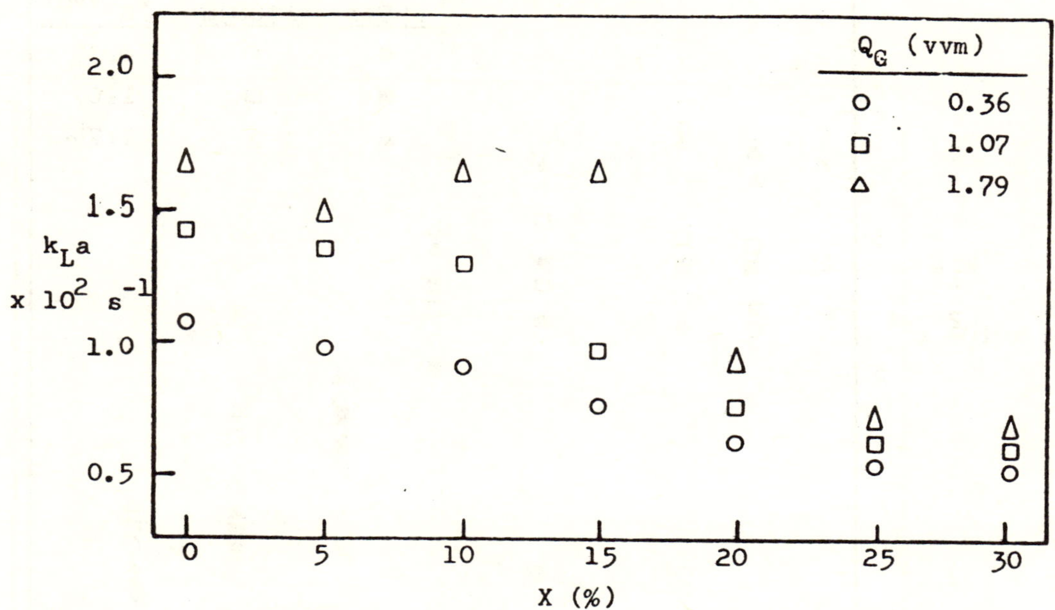

Figure 3   **Effect of Solids on $k_L a$ (D=0.135 m, N=4 s^{-1})**
            (Ion Resin Particles, d_p=780 μm)

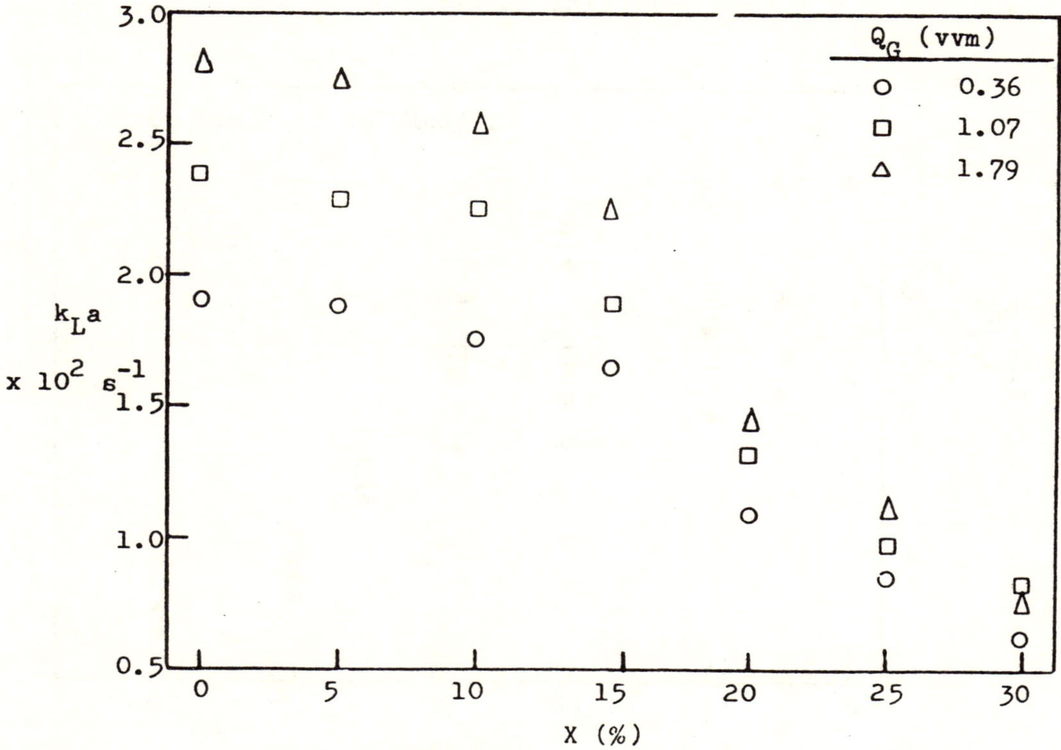

Figure 4   **Effect of Solids on $k_L a$ (D=0.1016m, N=7 s^{-1})**
                (Ion Resin Particles, d_p=780 μm)

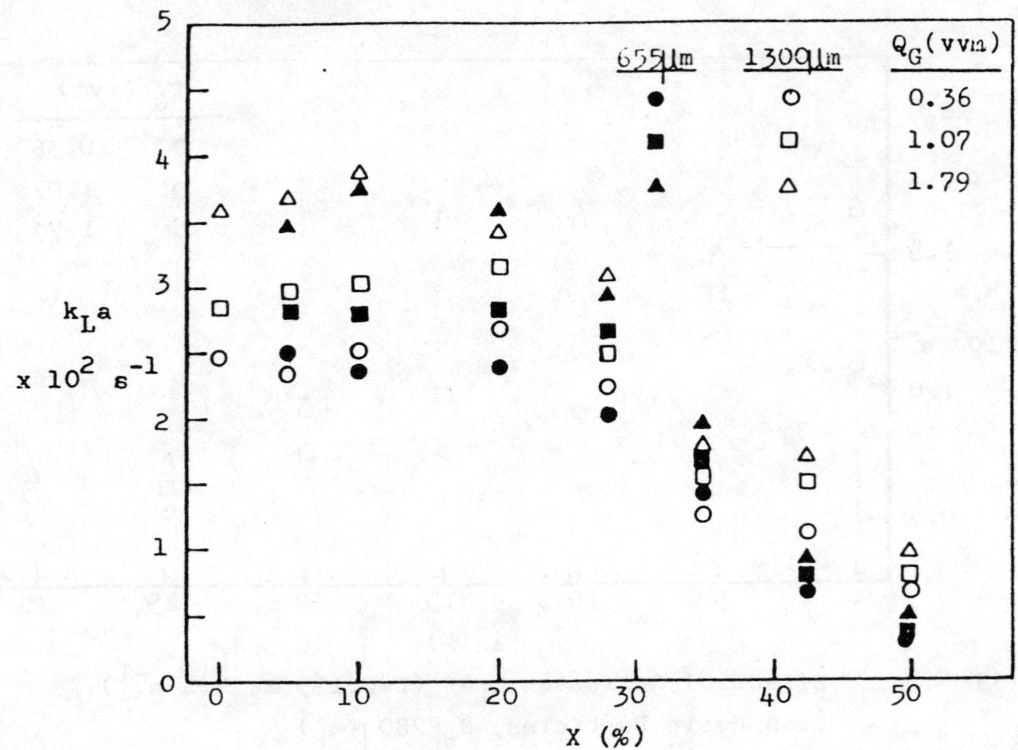

Figure 5   Effect of Solids on k_La (D=0.135 m, N=6 s^{-1})
(Glass Ballotini Particles)

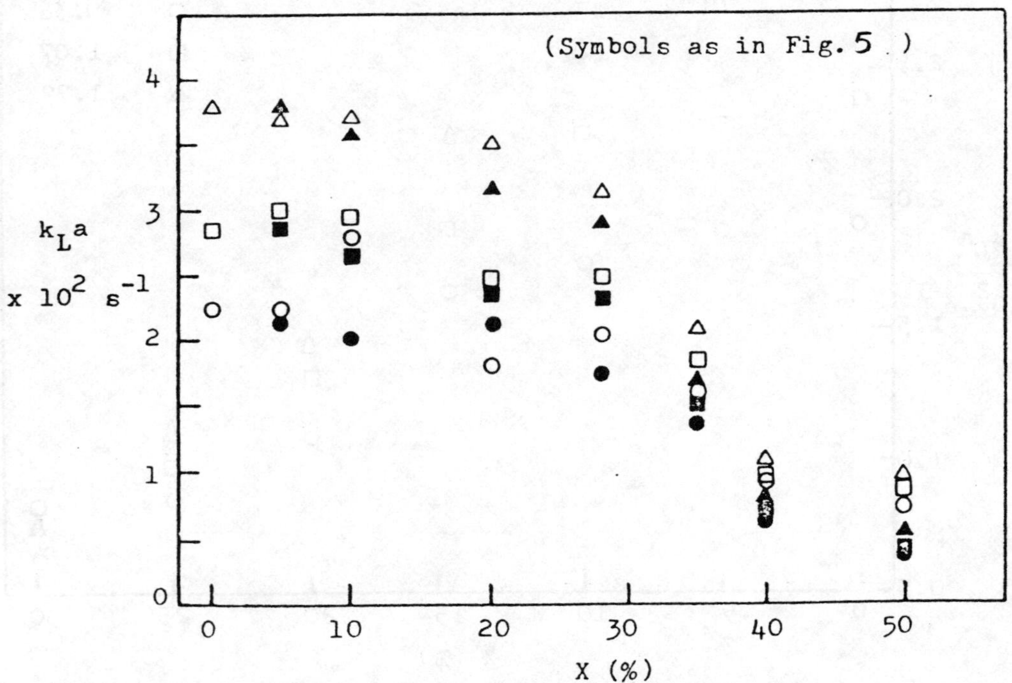

Figure 6   Effect of Solids on k_La (D=0.1016 m, N=9 s^{-1})
(Glass Ballotini Particles)

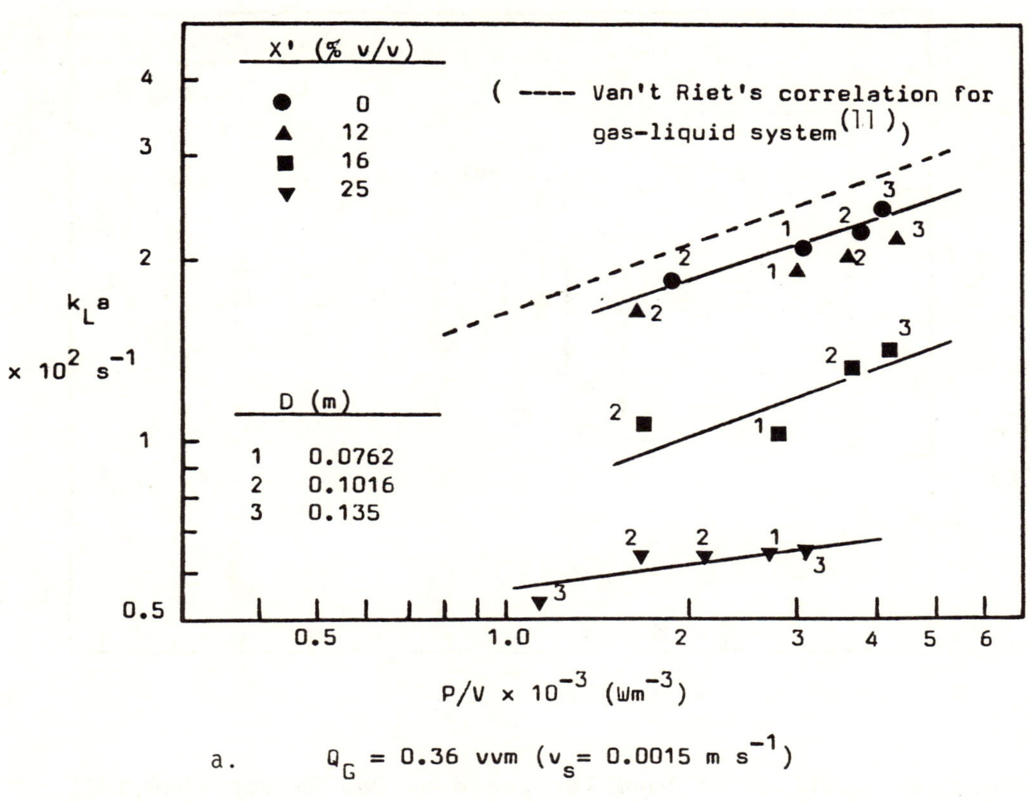

a. Q_G = 0.36 vvm (v_s = 0.0015 m s^{-1})

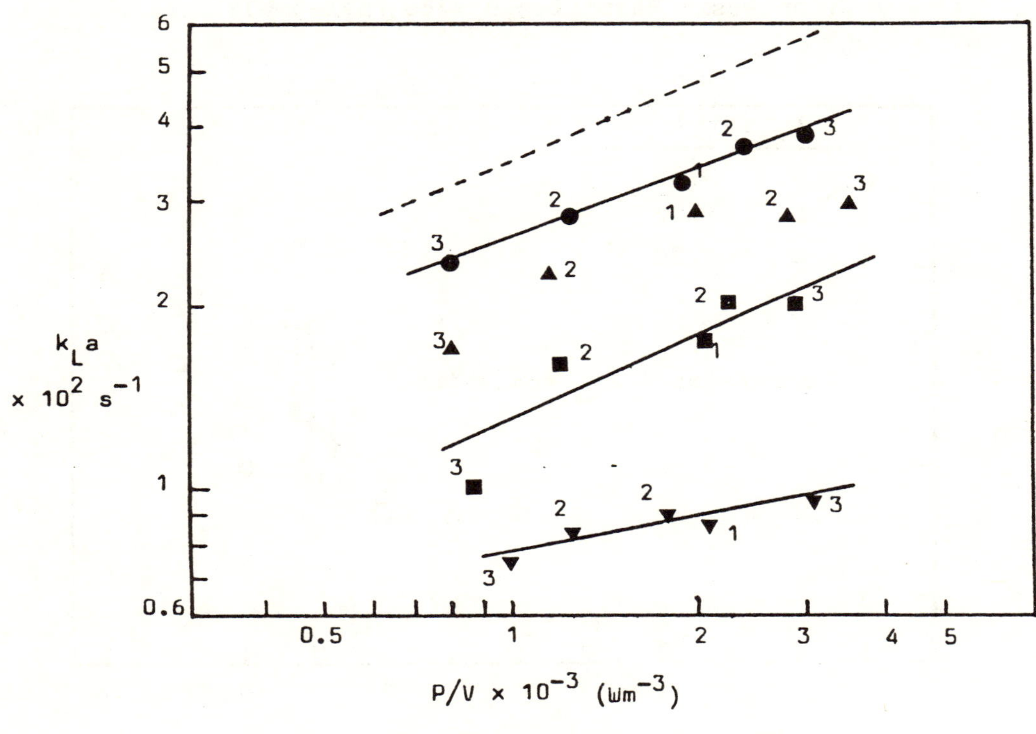

b. Q_G = 1.79 vvm (v_s = 0.0077 ms^{-1})

Figure 7 $k_L a$ versus P/V for various Solids Concentrations.

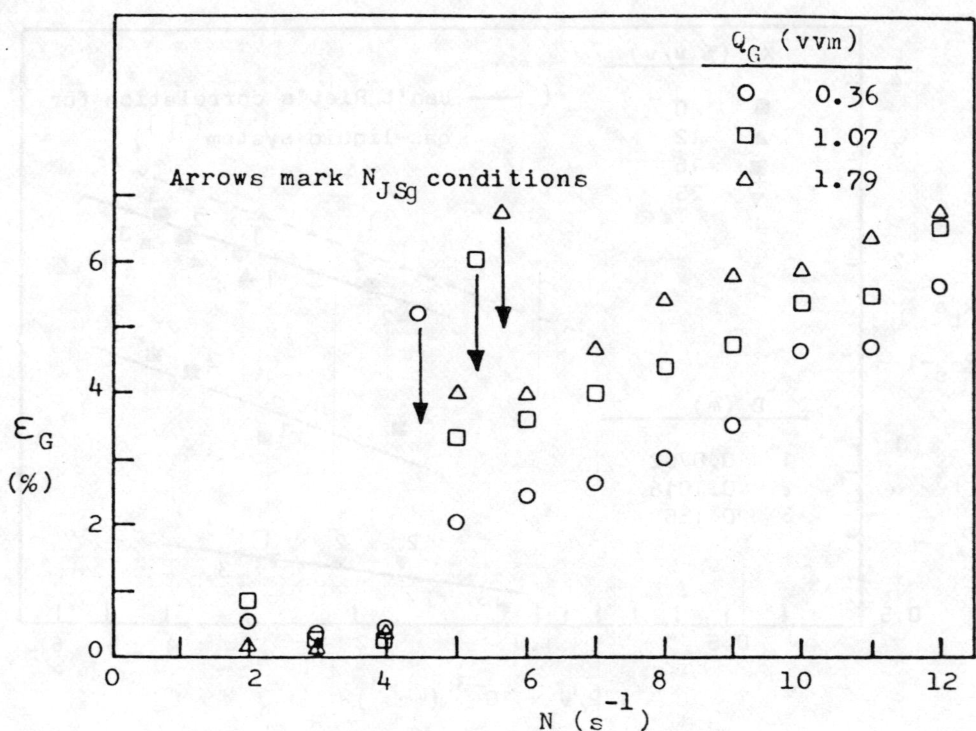

Figure 8   Effect of Impeller Speed on Gas Holdup (D=0.1016 m)
(Ion Resin Particles, d_p=780 μm, X=30%)

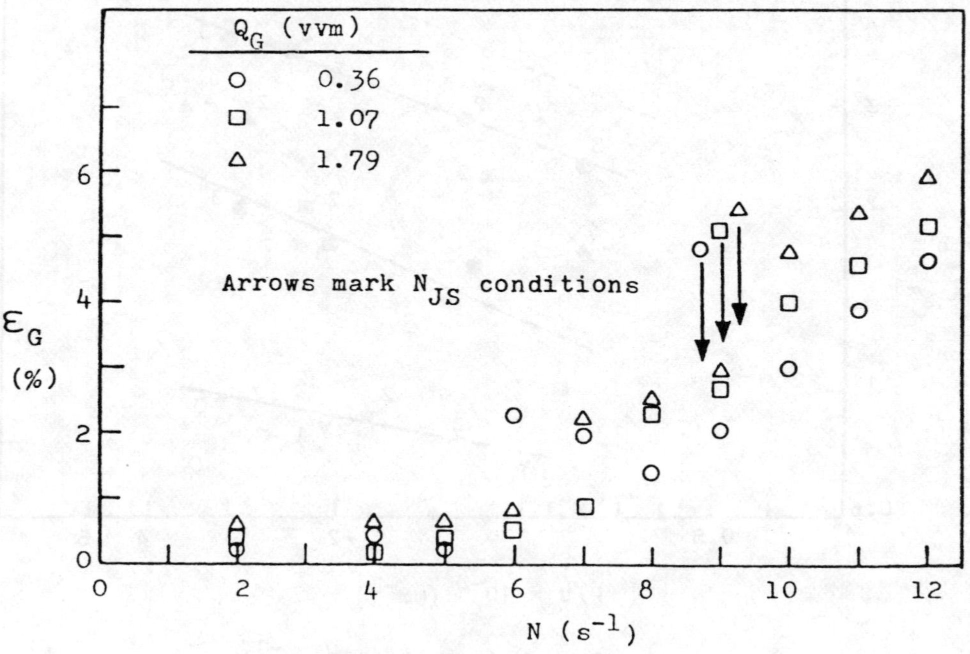

Figure 9   Effect of Impeller Speed on Gas Holdup (D=0.1016 m)
(Glass Ballotini Particles, d_p=655 μm, X=40%)

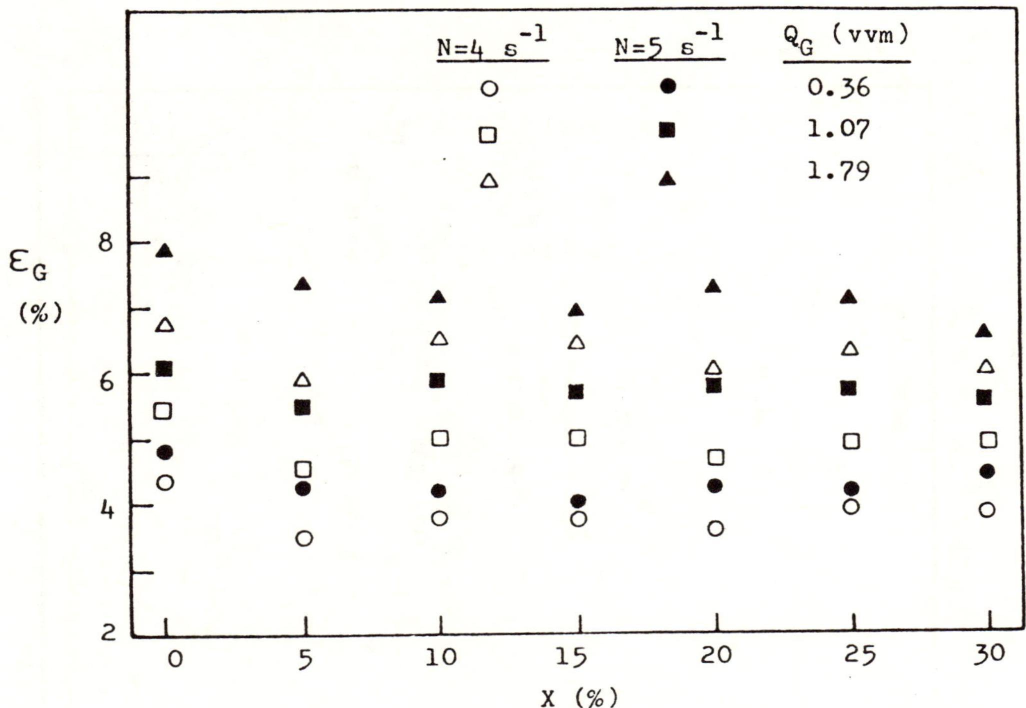

Figure 10   Effect of Solids on Gas Hold p (D=0.135 m)
(Ion Resin Particles, d_p=780 μm)

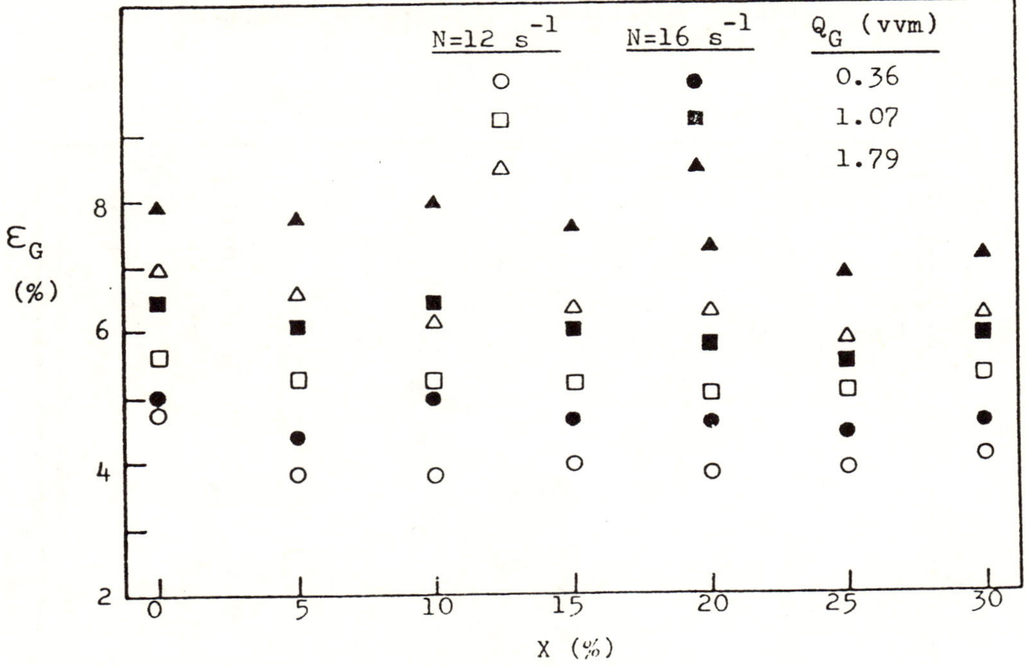

Figure 11   Effect of Solids on Gas Holdup (D=0.0762 m)
(Ion Resin Particles, d_p=780 μm)

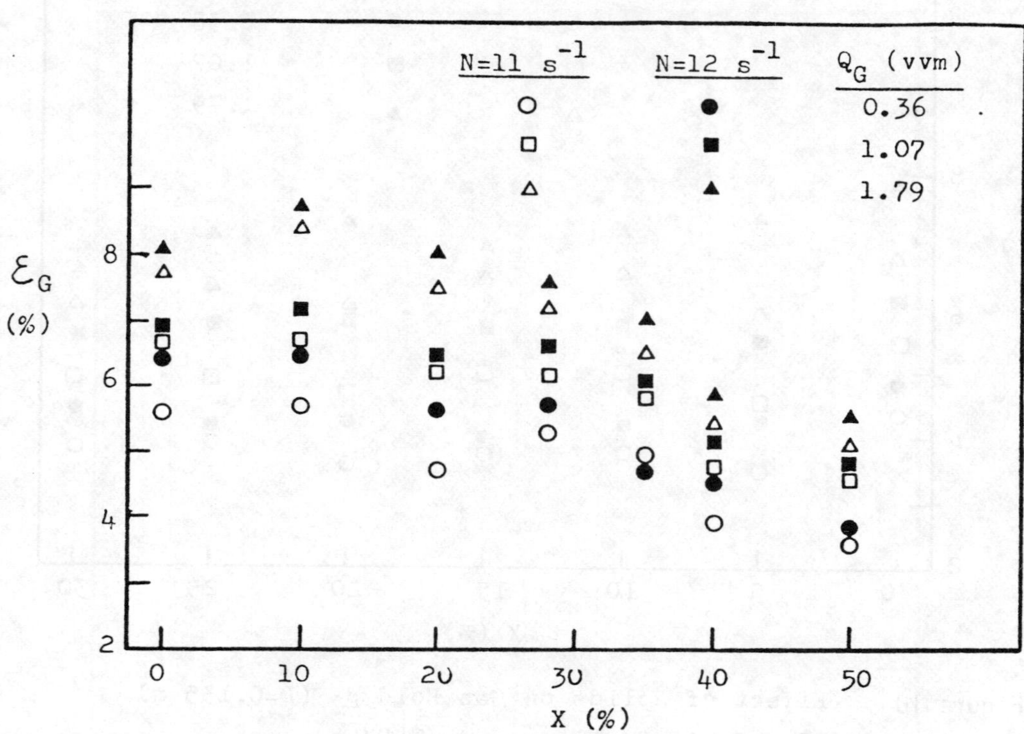

Figure 12   Effect of Solids on Gas Holdup   (D=0.1016 m)
(Glass Ballotini Particles, d_p= 655 μm)

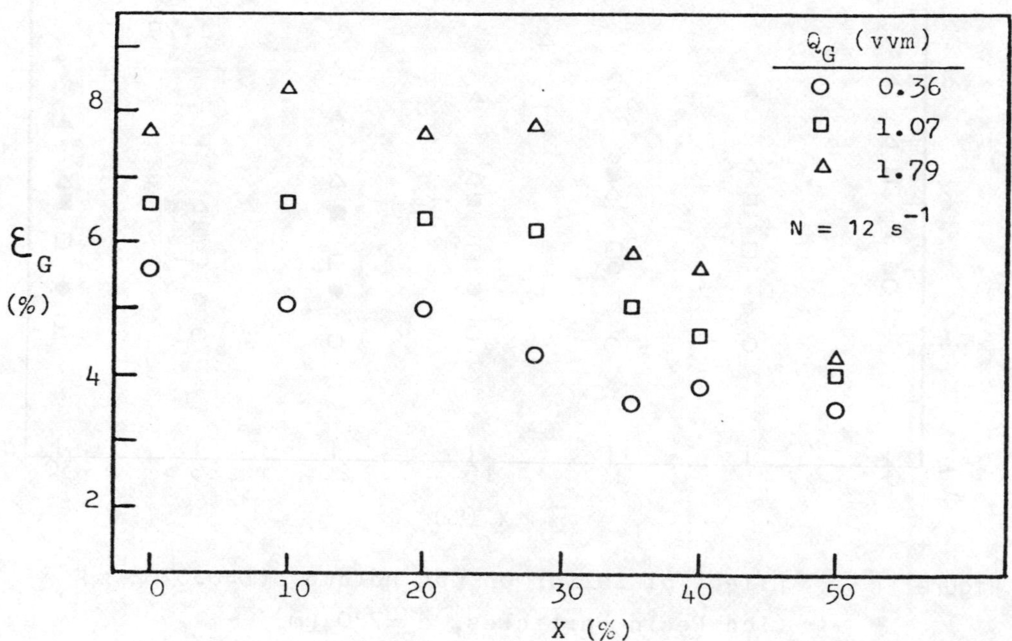

Figure 13   Effect of Solids on Gas Holdup   (D=0.1016 m)
(Glass Ballotini Particles, d_p=1300 μm)

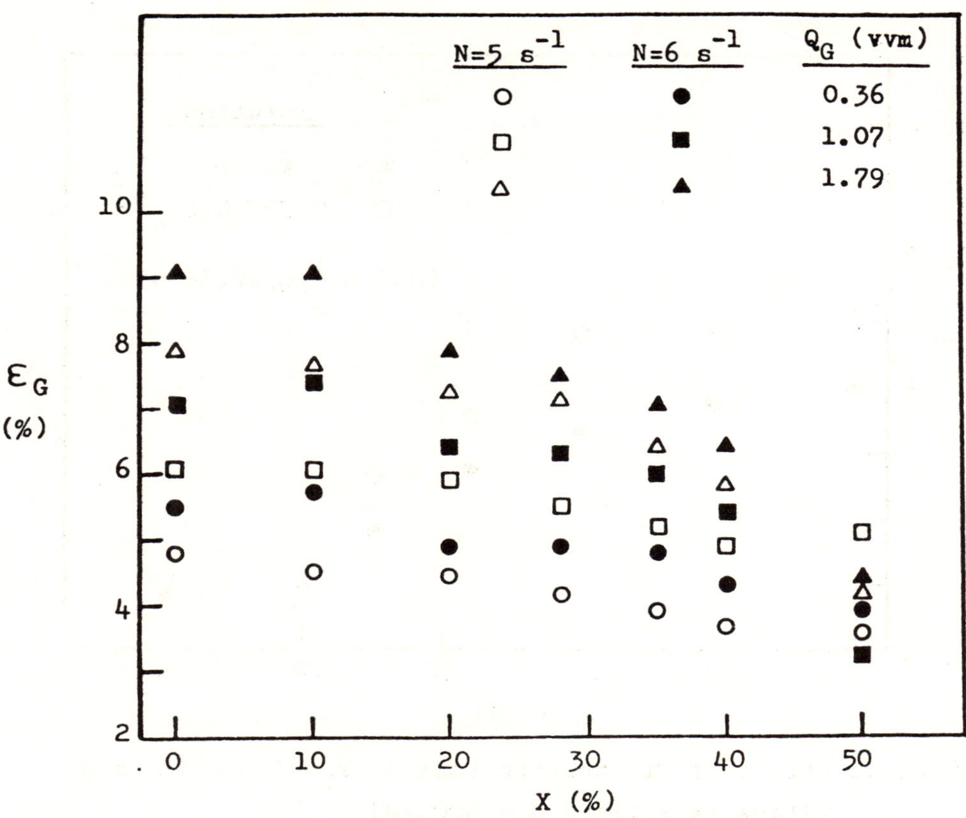

Figure 14   Effect of Solids on Gas Holdup   (D=0.135 m)
(Glass Ballotini Particles, d_p=655 μm)

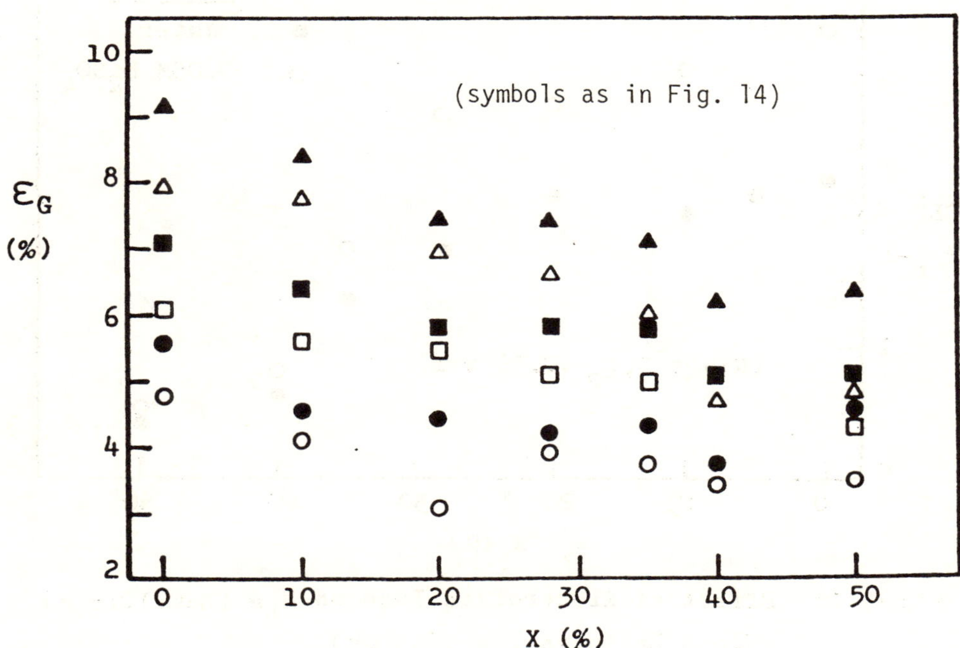

Figure 15   Effect of Solids on Gas Holdup   (D=0.135 m)
(Glass Ballotini Particles, d_p=1300 μm)

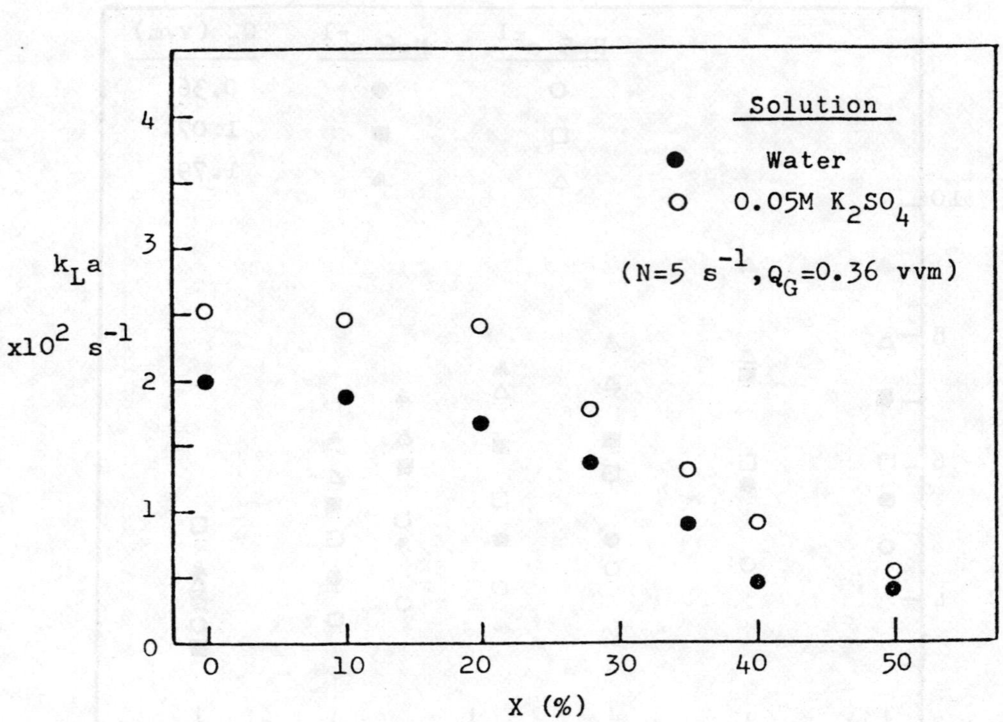

Figure 16 **Effect of Electrolyte Ions on** k$_L$a (D=0.135 m)
(Glass Ballotini, d$_p$=1300 μm)

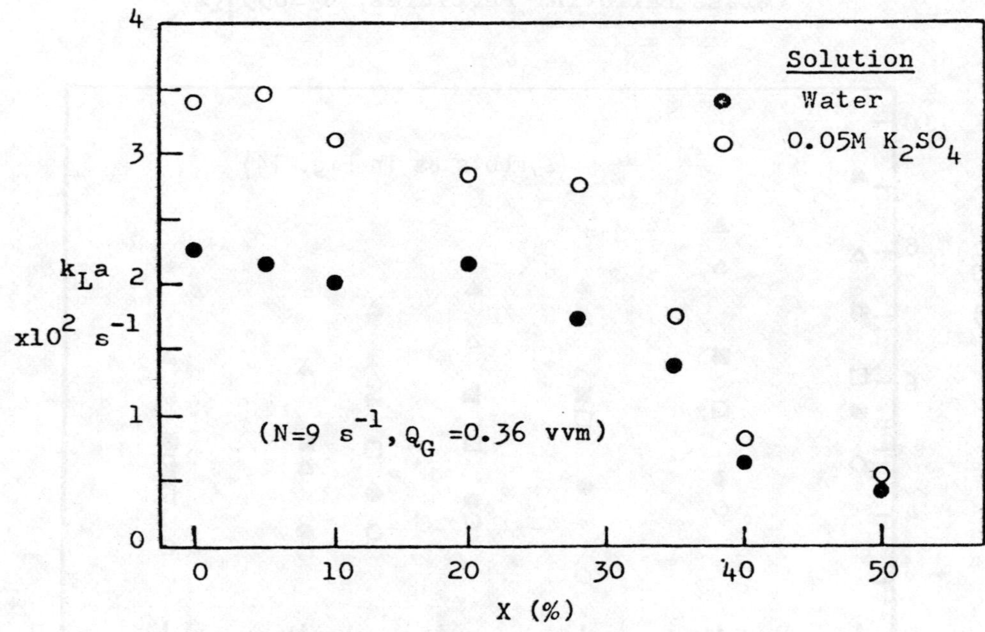

Figure 17   Effect of Electrolyte Ions on k$_L$a (D=0.1016 m)
(Glass Ballotini, d$_p$=655 μm)

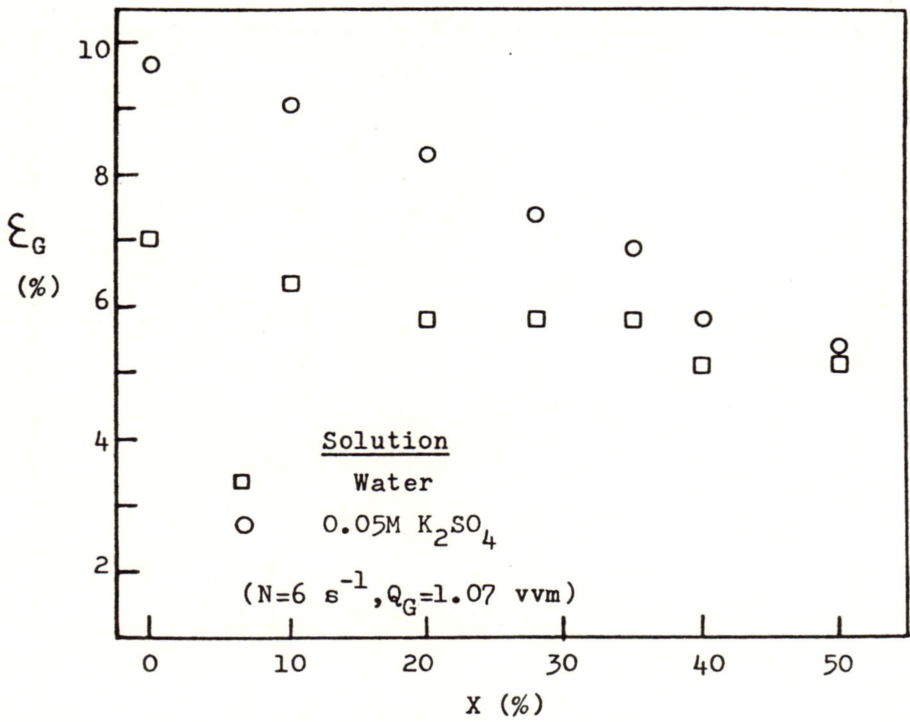

Figure 18   **Effect of Electrolyte Ions on Gas Holdup**
(D=0.135m, Glass Ballotini, d_p=1300μm)

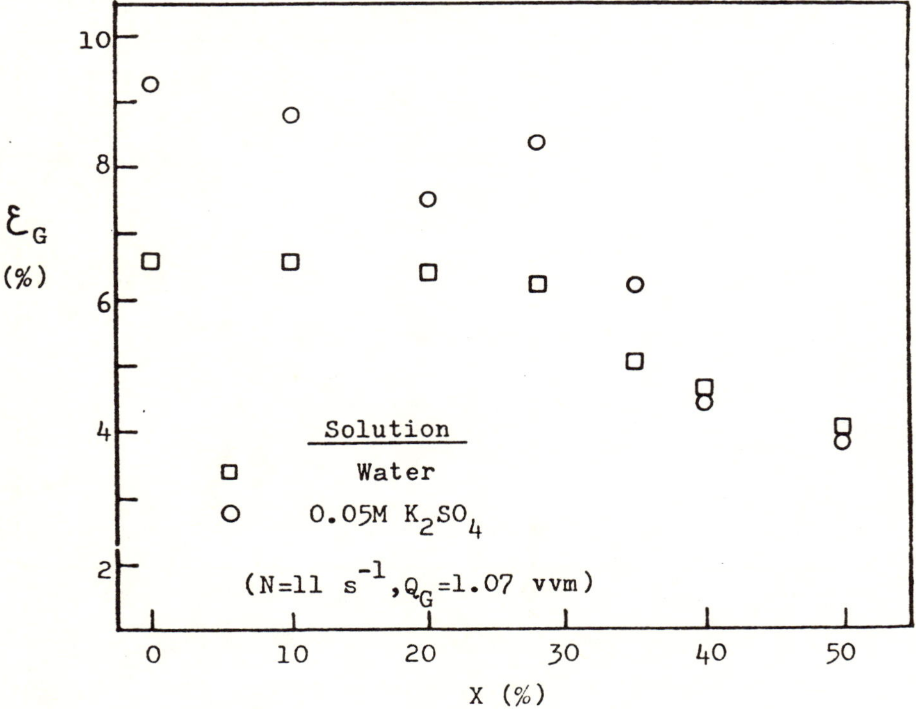

Figure 19   Effect of Electrolyte Ions on Gas Holdup
(D=0.1016m, Glass Ballotini, d_p=1300μm)

HOMOGENIZATION OF LIQUID IN STIRRED TANKS.
MODIFIED MODEL AND NUMERICAL SOLUTION

Z. Jaworski

Technical University of Szczecin
71-065 Szczecin, Al. Piastów 42
Poland

Summary

The modified model of homogenization in a stirred tank has been described. Some remarks on the introduced modifications in the previous model and numerical approach have been made. The fundamental steps of the computer programme based on the model have been briefly outlined. An example of the course of the local tracer concentration during homogenization has been presented.

Held at Wurzburg, 10-12 June, 1985.

Organised by DVCV· Deutsche Vereinigung für Chemie- und Verfahrenstechnik
(German Association of Chemical and Process Engineering).

Organisation: GVC·VDI-Gesellschaft Verfahrenstechnik und Chemieingenieurwesen. **GVC**

©BHRA, The Fluid Engineering Centre, Cranfield, Bedford MK43 0AJ, England.

c = tracer concentration

C = dimensionless tracer concentration defined in Eq./3/

D = agitator diameter

D_e = effective diffusion coefficient

D_t = turbulent diffusion coefficient

E,G = dimensionless variables defined in Eqs./9,10/

n = agitator speed

r = radial coordinate

R = dimensionless radial coordinate

t = time

T = tank diameter

U,W = dimensionless variables defined in Eqs./7,8/

v_r = radial component of mean velocity

v_z = axial component of mean velocity

z = axial coordinate

Z = dimensionless axial coordinate

θ = dimensionless time defined in Eq./4/

1. INTRODUCTION

A number of papers on mixing processes was devoted to the theoretical
and experimental investigations of homogenization of miscible liquids.
The recent models (Refs. 1,2) describing homogenization in stirred tanks
were based on the tracer differential mass balance formulated for three
independent variables. The author's primary model (Ref. 2) was adopted
to the numerical simulation of homogenization process (Ref. 3). This
model was solved using a procedure of numerical solution based on the
Patankar's method (Ref. 4). Mean velocity profiles and turbulence characteristics of the stirred liquid were adopted from literature. However,
the velocity values did not obey strictly the difference mass balance of
the mixture. This resulted in some discrepancies between computation
results and experimental data.

2. MODIFIED MODEL AND NUMERICAL SOLUTION

The basic equation of the models are obtained on the assumption of
constant liquid density and of axial symmetry of the tracer concentrations c as well as of velocity components. In the cylindrical coordinates r,z this equation takes the form of Eq./1/.

$$\frac{\partial c}{\partial t} + v_r \frac{\partial c}{\partial r} + v_z \frac{\partial c}{\partial z} = \frac{1}{r} \frac{\partial}{\partial r}\left(D_e \frac{\partial c}{\partial r}\right) + \frac{\partial}{\partial z}\left(D_e \frac{\partial c}{\partial z}\right) \qquad /1/$$

For turbulent flows the values of c, mean velocity components v_r
and v_z, effective diffusion coefficient D_e are to be the expected ones.
The other relation which must be fulfilled is the continuity equation.
With the above assumptions it can be expressed as follows:

$$\frac{\partial (rv_r)}{\partial r} + \frac{\partial (rv_z)}{\partial z} = 0 \qquad /2/$$

Introduction of the dimensionless variables defined in Eqs./3-10/

$$C = \frac{c - c_{min}}{c_{max} - c_{min}} \quad /3/ \qquad \theta = n t \quad /4/ \qquad R = \frac{2r}{T} \quad /5/$$

$$Z = \frac{z}{H} \quad /6/ \qquad U = \frac{2v_r R}{nT} \quad /7/ \qquad W = \frac{v_z R}{nH} \quad /8/$$

$$E = \frac{4D_e R}{nT^2} \quad /9/ \qquad G = \frac{D_e R}{nH^2} \quad /10/$$

allows one to derive the modified generalized form of differential
Eqs./1,2/.

$$\frac{\partial(RC)}{\partial\Theta} + \frac{\partial(UC)}{\partial R} + \frac{\partial(WC)}{\partial Z} = \frac{\partial}{\partial R}\left(E\,\frac{\partial C}{\partial R}\right) + \frac{\partial}{\partial Z}\left(G\,\frac{\partial C}{\partial Z}\right) \tag{11}$$

$$\frac{\partial U}{\partial R} + \frac{\partial W}{\partial Z} = 0 \tag{12}$$

The model is completed by boundary conditions which define the initial regions of high $/c_{max}/$ and low $/c_{min}/$ tracer concentrations as well as isolation of the system from its environment. The governing Eq./11/ should be solved numerically as its analytical solution is not known. To develop the effective numerical procedure the discretization method of Patankar and Spalding (Refs. 4,5) was adopted. Due to the modified form of Eq./11/ fewer simplifications in the procedure were required as compared to the previous approach (Ref. 3). Another alteration was the application of an approximate velocity distribution (Ref. 6) that fulfilled the discretized form of Eq./12/.

The effective diffusion coefficient D_e is usually calculated as a sum of the turbulent D_t and molecular coefficients. The applied distribution of D_t in a stirred tank with disc turbine agitator was compiled from some papers (Refs. 7,8,9). Because of false diffusion introduced by the numerical approach a correction of D_e values could be provided. The false diffusion coefficient was estimated from an approximate expression cited in (Ref. 4).

As a result of the discretization of Eq./11/ the set of linear and fully implicit N algebraic equations for C value in N grid points was obtained. These equations were solved iteratively by means of a line-by-line method (Ref. 4) for the consecutive steps of dimensionless time Θ. The iterations were considered as converged when the last iteration did not change the individual C`s more than by 0,02% and when the current mean concentration did not differ from the initial one more than by 0,002%. A computer programme based on the described method was developed. It contained the calculation of different indices of mixture homogeneity obtained for the given time Θ. A typical profile of the predicted local concentration C during the homogenization period is shown in Fig.1. Further results of numerical simulations obtained by means of the programme are described elsewhere (Ref. 10).

3. REFERENCES

1. Liepe F.: "On the synthesis of models for flow processes in process engineering". /Zum Aufbau von Modellen für Strömungsprozesse der Verfahrenstechnik/. Maschinenbautechnik 32,7,1983,pp.313-316. /In German/.

2. Jaworski Z.: "Modelling of liquid homogenization in stirred tank.
   I. Mathematical model and computational algorithm". /Modelowanie
   homogenizacji cieczy w mieszalniku. I.Model matematyczny i algorytm
   obliczeń/. Prace Naukowe Politechniki Szczecińskiej No.270,14,1984,
   pp.23-36. /In Polish/.

3. Jaworski Z.: "Simulation of mixing in an agitated vessel". 8th
   International Congress CHISA`84 /Prague,Czechoslovakia,Sept.3-7,
   1984/ Czechoslovak Academy of Sciences,1984,Paper V.3.43, 7 pp.

4. Patankar S.V.: "Numerical heat transfer and fluid flow". Washington,
   Hemisphere Publishing Co.,1980, pp. 25-109.

5. Patankar S.V., Spalding D.B.: "A calculation procedure for heat,
   mass and momentum transfer in three-dimensional parabolic flows".
   International Journal of Heat and Mass Transfer $\underline{15}$,10,1972, pp.
   1787-1806.

6. Jaworski Z.: Unpublished results.

7. Liepe F.: "Some applications of the theory of turbulence to problems of mixing". Advances in Mechanics /edited in Warsaw, Polish
   Academy of Sciences/ $\underline{4}$,4,1981, pp. 3-45. /In Russian/.

8. Barthole J.P., Maisonneuve J., Gence J.N., David R., Mathieu J.,
   Villermaux J.: "Measurement of mass transfer rates, velocity and
   concentration fluctuations in an industrial stirred tank".
   Chemical Engineering Fundamentals $\underline{1}$,1,1982, pp. 17-26.

9. Fořt I., Rogalewicz V., Richter M.: "Simulation of two-phase flow
   in a mechanically agitated gas-liquid charge". Collection of
   Czechoslovak Chemical Communications /In press/.

10. Jaworski Z., Fořt I., Stręk F.: "Numerical simulation of liquid
    homogenization in a stirred tank with the disc turbine agitator".
    In. Proc. 5th European Conference on Mixing /Würzburg,West Germany,
    June 10-12,1985/ DVCV, /In press/.

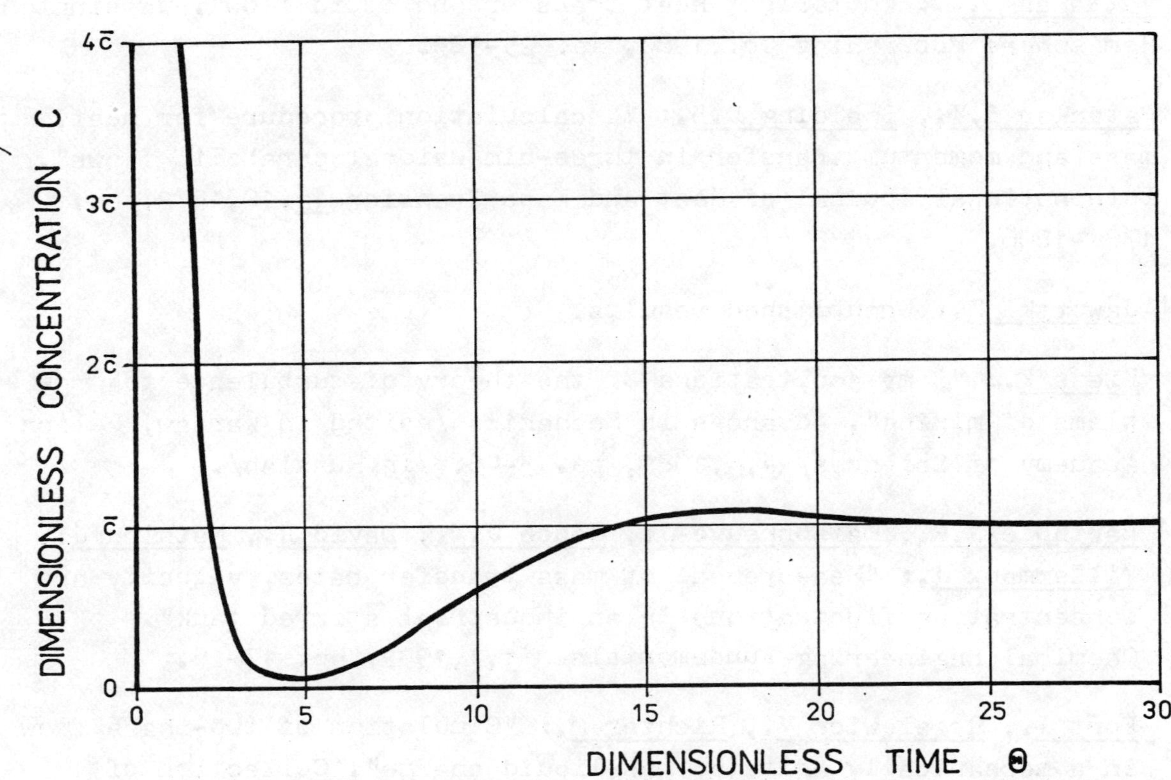

Fig. 1.   The profile of the tracer concentration
in the agitator disc plane.

TO DETERMINE THE PUMPING CAPACITY OF A DISK TURBINE IMPELLER AT
DIFFERENT REYNOLDS NUMBERS

D. B. A. Keller BSc. Hons

National Engineering Laboratory, East Kilbride, Glasgow, UK

Summary

Measurements of velocity profiles around the impeller of an agitated vessel at
Reynolds numbers in the range of 4×10^4 to 7×10^4, were taken using Laser Doppler
Anemometry in Backscatter Mode. It was shown that as the impeller speed increased, the
peak mean radial velocity fell due to aeration on the impeller surface. When applying
the models for pumping capacity, these values obtained were found to be higher than the
current recommended design values.

Held at Wurzburg, 10-12 June, 1985.

Organised by DVCV· Deutsche Vereinigung für Chemie- und Verfahrenstechnik
(German Association of Chemical and Process Engineering).

Organisation: GVC·VDI-Gesellschaft Verfahrenstechnik und Chemieingenieurwesen. 

©BHRA, The Fluid Engineering Centre, Cranfield, Bedford MK43 0AJ, England.

<u>NOMENCLATURE</u>

B = Baffle width   m

D_1 = Impeller diameter   m

h_1 = Height of impeller   m

H = Height of liquid   m

L = Length of blade   m

N = Impeller speed   r/min

Re = Reynolds number $\rho ND^2/\eta$

T = Tank diameter   m

U_r = Mean radial velocity   m s^{-1}

U_t = Impeller tip velocity   m s^{-1}

W_1 = Blade width   m

η = Viscosity   kg/m s

κ = Pumping capacity

ρ = Density of water   kg/m^3

1. INTRODUCTION

Over the past few years, our understanding of the radial flow around the impeller and in agitated vessels has increased. Cutter (Ref. 1) and a number of authors (Ref. 2) have shown that the radial velocity profiles in the impeller stream has a jet-like behaviour. By knowing this profile, it is possible to determine the impeller's pumping characteristic (Ref. 3) (ie the pumping capacity).

Most of the data which have been collected on the impeller stream, have been obtained using hot wire or hot film anemometry and pitot probes. The main problem with these techniques, is that the probe is introduced into the flow; this will affect the flow behaviour. Also pitot probes cannot measure the fluctuating components of velocity and therefore cannot adequately measure the turbulence intensity of the flows. Laser Doppler Anemometry (LDA) does not suffer from these advantages.

Reed (Ref. 4) and Oldshue (Ref. 5) are at present the only authors who have published velocity data in agitated vessels using the LDA. Both of the researchers' measurements were in forward scatter mode. This provides very high data rates however exact optical layouts both for the sending and receiving optics are required, making the application on large scales impractical. At NEL, it was decided to develop a backscatter technique so that it would be possible to investigate flow patterns in large vessels up to say 2 m in diameter.

This paper presents data giving the mean velocity around the shaft and impeller and also gives a determination of pumping capacity of the impeller at different impeller speeds.

2. EXPERIMENTAL

The stirred vessel was standard 4 baffled glass vessel (Fig. 1) with a six-bladed disc turbine impeller. To prevent optical distortion of the probe volume, the vessel was surrounded by a glass jacket so that the beams entered the medium through a flat surface. The mixing fluid was water.

There were facilities in the rig to divert the probe volume in three dimensions (Fig. 2). All the measurements were taken at a radius 59 mm at the centre of the vessel. The probe volume length was 3 mm and was moved along the vertical axis.

3. LASER DOPPLER ANEMOMETER AND ASSOCIATED EQUIPMENT

The anemometer used in the tests is a DISA LDA 10 model. This is a three beam, two component anemometer with the facility for forward or backward scatter measurements. By means of a Bragg Cell the fringes of the probe volume can be shifted. The laser is a 5 W argon ion laser. This amount of power is required since the scattered light in backward scatter is considerably less than that obtained in forward scatter. The intensity of the scattered light was measured using a photomultiplier connected to a counter linked to an HP 9836 computer. The number of readings taken was 4000 approximately. The time to take a set of readings was dependent on the data rate. The system was only used to measure in one-dimensional components and was calibrated accordingly. This has been reported by Keller (Ref. 6).

4. MEASUREMENT OF VELOCITY

A series of measurements were taken over a range of impeller speeds, Reynolds numbers covered were in the range between 4×10^4 to 7×10^4, it was impossible to measure velocities at Reynolds numbers above 7×10^4 due to surface aeration. Below 4×10^4, the data acquisition rate was too low to be used with the counter employed in the system. Ideally velocity measurements should be taken with a frequency tracker as reported by Reed (Ref. 3). Seeding of the fluids was achieved using neutrally buoyant latex particles (diameter approximately 10 μm). Natural contamination of water, and the addition of drops of milk did not produce adequate data rates. For the backscatter method to be effective, it is vital that the seeding has a higher refractive index than the fluid.

5.   RESULTS

Radial velocities were taken at Reynolds numbers in the range 4×10^4 to 7×10^4. The measurements were taken along the axis of the shaft at a radius of 59 mm. The results are presented in Figs 3 and 4. In Fig. 3, the results are plotted, as a function of the ratios of the height of the (h) to liquid height (H) against mean radial velocity (U_r) to tip impeller speed (U_t). In Fig. 4, they are a function of h/H against the ratio of standard deviation (fluctuating component of the flow) to tip speed (U_t).

6.   VELOCITY PROFILE

From the data presented in the two plots, it can be seen that at h/H = 0.33, there is as might be expected, a peak in the profile of the radial velocity. However at the higher impeller speeds, it can be noted that the value of the peak velocity unexpectedly drops. During the test it was noticed that there was surface air entrainment and it is considered that viscous decoupling effects produced this unexpected fall in peak velocity. In the impeller stream, the plot of h/H against the standard deviation follows a similar pattern to the velocity profile.

7.   PUMPING CAPACITY

The pumping capacity coefficients for the impeller are shown in Table 1. The coefficient is defined as

$$K = Q/ND^3$$

where

$$Q = \pi D \int_{-W/2}^{W/2} U_r dz.$$

The flowrate Q was calculate by determining the area of the profile graphically.

The pumping capacity values are higher than the recommended design value (which is 0.75 ± 0.15). A possible reason for this difference is that the profile recorded was not close enough to the impeller. This will be investigated in future studies. A large number of points will also be obtained along the impeller blade height. Reville (Ref. 3) has commented on the effect of measuring the profile at distances along the radius.

8.   CONCLUSIONS

An LDA technique using backscatter has been developed to measure radial mean velocity components in a stirred vessel. It was found that as the Reynolds number increased (ie impeller speed increases), the peak radial mean velocity fell; this was due to the surface air entrainment and viscous decoupling effects noted during the tests. The fluctuating component profile followed a similar pattern to the velocity profile. The pumping capacity coefficients for the four velocity profiles were higher than the values recommended for design.

9.   REFERENCES

1.   Cutter, L. A.: "Flow and turbulence in a stirred tank". Amer. Inst. of Chem.
      Engng, $\underline{V12}$, 1966, pp. 35-45.

2.   Cooper, R. G. and Wolf, D.: "Velocity profiles and pumping capacities for turbine
      type impellers". Can. Journ. of Chem. Engng, $\underline{V46}$, 1962, pp. 94-100.

3.   Reville, B.: "Pumping capacity of disc turbine agitators - A literature review".
      Fourth European Conference on Mixing, 27-29 April 1982, Paper B1, pp. 11-24.

4.   Reed, C. B., Princz, M. and Hartland, S.: "Laser Doppler measurements of
      turbulence in a standard stirred tank". Second European Conference on Mixing,
      30 March-1 April 1977, pp. B1-1 to B1-26.

5.   Oldshue, J. Y.: "Fluid mixing technology chemical engineering series". McGraw
      Hill, 1983, pp. 166-175.

6.   Keller, D. B. A.: "Single measurements in a stirred vessel using Laser Doppler
      Anemometry". NEL Internal Report in process of being published.

T A B L E   1

PUMPING CAPACITY COEFFICIENTS

Reynolds No $Re \times 10^4$	Pumping capacity K
4.2	1.200
5.0	1.242
5.8	1.394
6.7	1.013

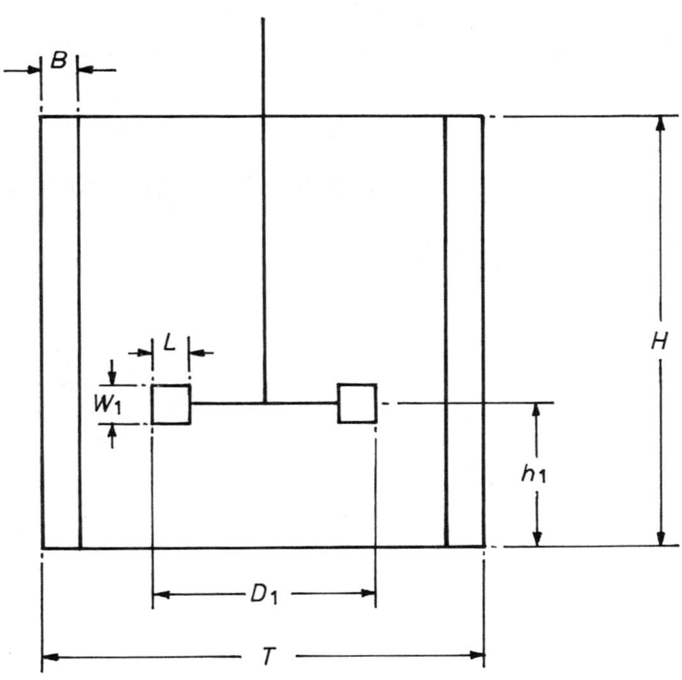

Fig 1   Vessel Configuration

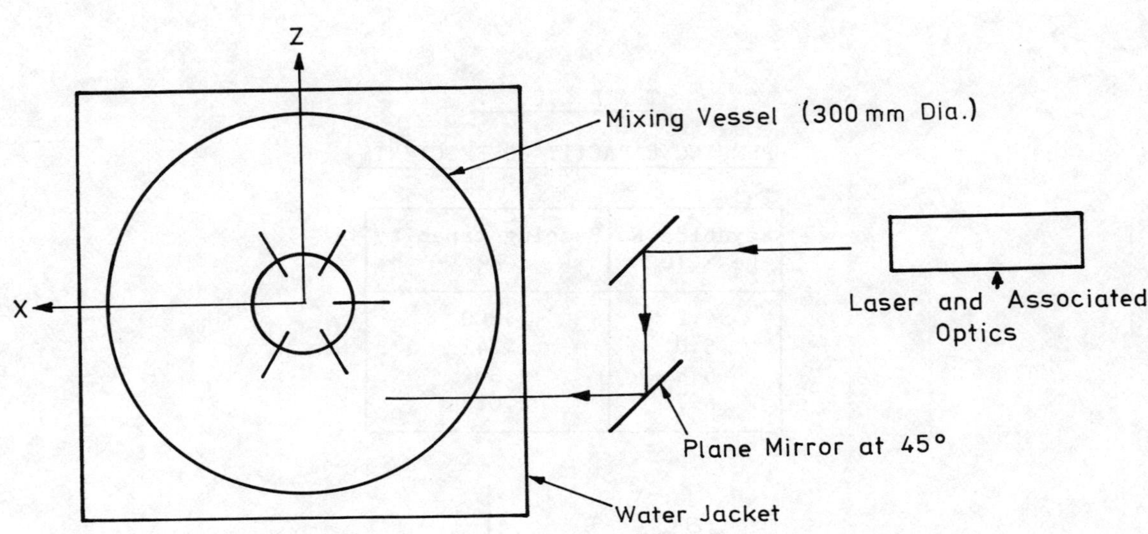

Fig 2   General Arrangement

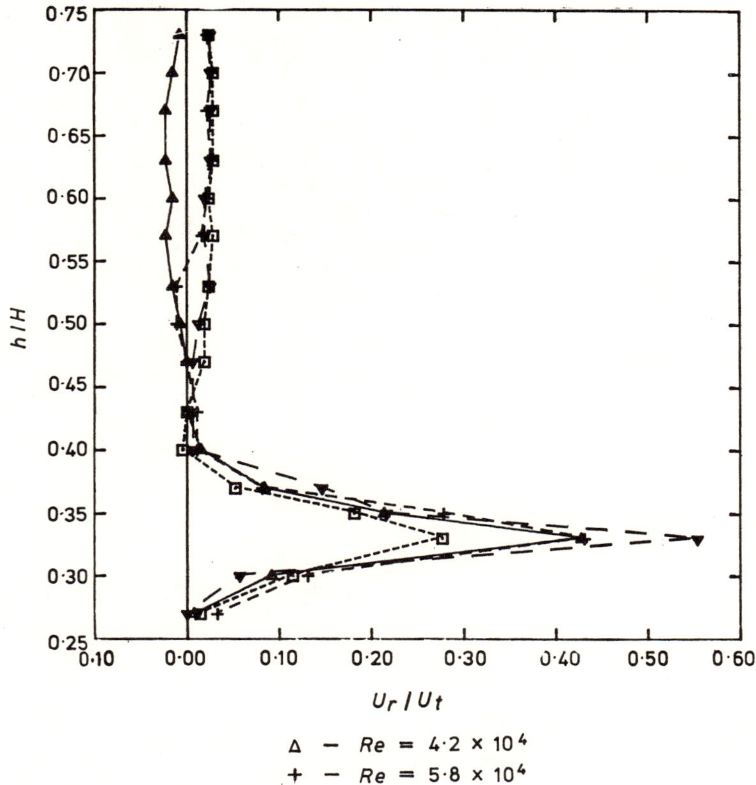

Fig 3 U_r / U_t against h/H

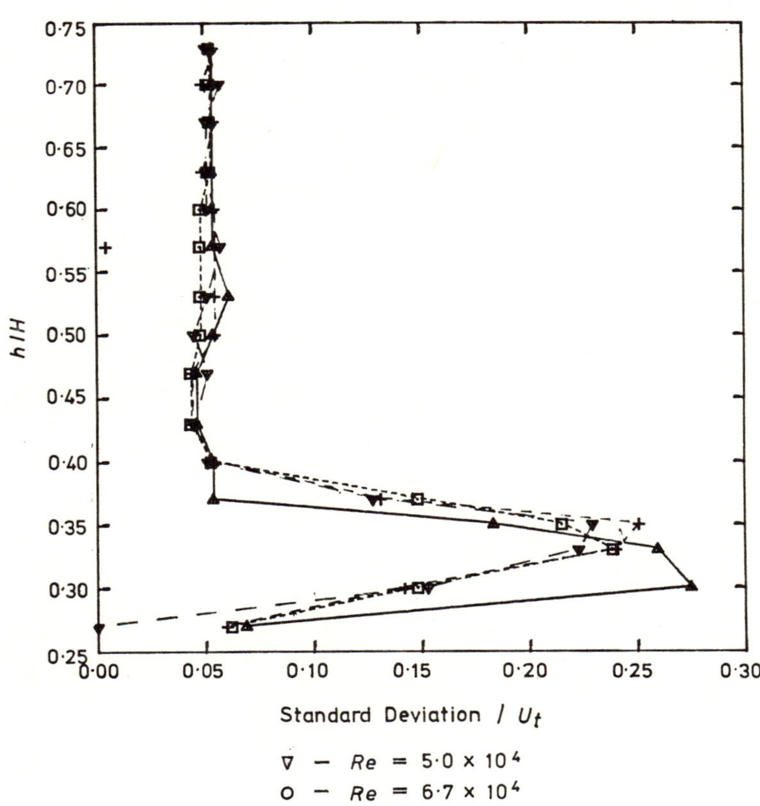

Fig 4 Standard Deviation $/ U_t$ against h/H

THE SWIRLING RADIAL JET MODEL AND ITS

APPLICATION TO A RADIAL IMPELLER STREAM

V.Kolář, P.Filip and A.G.Curev

Institute of Hydrodynamics
166 12   Prague, Czechoslovakia

Summary

The aim of this contribution is to present a survey of the hitherto theoretical and experimental results of the investigation of the swirling radial jet velocity field in connection with its typical realization represented by a high speed radial impeller. The swirling radial jet problem is considered from two different viewpoints :
1/ as a purely hydrodynamic problem supposing ideal flow conditions,
2/ as a stream generated by a radial impeller in an agitated baffled vessel.

The development of the theoretical solution of the mean-time radial jet velocity field is outlined in the frame of a solely hydrodynamic approach. The solutions of the radial jet without swirl, with a weak swirl and with an arbitrary one are successively discussed. These descriptions deal with both a laminar regime and - significant from a practical point of view - a turbulent one characterized by simple turbulence models. Further, this part of the contribution presents mean-time velocity field description of complex swirling radial jets, i.e. multiple and antisymmetrical swirling ones.Finally, the experimental results concerning the above stated problems are briefly mentioned.

The other part of the contribution introduces the hitherto theoretical approach to the high speed radial impeller and its linkage to the results stated in the preceding part. This connection has provided an improved model characterized by the parameters of clear physical meaning (further expressed by mixing conditions). A good agreement was found between this theoretical approach and the experimental data of other investigators.

Held at Wurzburg, 10-12 June, 1985.

Organised by DVCV· Deutsche Vereinigung für Chemie- und Verfahrenstechnik
(German Association of Chemical and Process Engineering).

Organisation: GVC·VDI-Gesellschaft Verfahrenstechnik und Chemieingenieurwesen. **GVC**

<u>NOMENCLATURE</u>

c_e = dimensionless swirl parameter
c_R = dimensionless radius of a source
$C_{d/D}$ = dimensionless momentum flux rate parameter
d = impeller diameter
D = vessel diameter
e = swirl parameter
k = spreading rate parameter
M_∞, N = integral invariants
n = rotational speed of an impeller
R = radius of a circumference-point-source
s = natural number
u = time-averaged radial velocity component
$\overline{u^2}$ = time-averaged square power of an instantaneous radial velocity component
$\overline{uw}$ = time-averaged product of instantaneous radial and peripheral velocity components
v = time-averaged axial velocity component
w = time-averaged peripheral velocity component
$\overline{w^2}$ = time-averaged square power of an instantaneous peripheral velocity component
x = radial coordinate
y = axial coordinate
$Y_1, \ldots, Y_s$ = axial locations of individual sources

β = angle (declination of an outflow from the radial direction)
δ = jet width
η = similarity variable (scaled x and y coordinate)
η_0 = axial displacement parameter
$\eta_1, \ldots, \eta_s$ = similarity variables
ρ = liquid density
τ_{xy} = xy-component of shear stress tensor
$\tau_{\phi y}$ = ϕy-component of shear stress tensor
ϕ = peripheral coordinate

The swirling radial jet is treated in two following steps. First, the problem is considered from a purely hydrodynamic viewpoint supposing ideal flow conditions, i.e. the jet issuing into a large stagnant surroundings and expanding under zero pressure gradient, since there is no external force involved (consequently, the jet momentum flux is preserved). The attention is also paid to complex swirling radial jets (multiple and antisymmetrical swirling ones). Then, a radially discharging impeller stream in an agitated baffled vessel is modelled by the swirling radial jet.

The radial jet with swirl is governed by the equations of time-averaged motion (written in cylindrical coordinates x,ϕ,y – see Fig.2a)

$$u\frac{\delta u}{\delta x} + v\frac{\delta u}{\delta y} - \frac{w^2}{x} = \frac{1}{\rho}\cdot\frac{\delta\tau_{xy}}{\delta y} \quad , \tag{1}$$

$$u\frac{\delta w}{\delta x} + v\frac{\delta w}{\delta y} + \frac{uw}{x} = \frac{1}{\rho}\cdot\frac{\delta\tau_{\phi y}}{\delta y} \quad , \tag{2}$$

by the continuity equation

$$\frac{\delta}{\delta x}(xu) + \frac{\delta}{\delta y}(xv) = 0 \quad , \tag{3}$$

by the conditions of symmetry (with respect to the plane y=0), and by the boundary ones expressing stagnant surroundings.

The papers dealing with the above stated problem may be divided approximately into three groups. The first one solves the non-swirling case ($w\equiv0$, $\tau_{\phi y}\equiv0$). Riley (Ref.1), Schwarz (Ref.2) present the solution for a laminar regime; Schwarz (Ref.2) and the present authors (Ref.3) for a turbulent regime that is experimentally studied by Heskestad (outflow from a peripheral slot formed by the edges of two disks (Ref.4)), Tanaka (outflow from a cylindrical nozzle (Ref.5)), Witze and Dwyer (radial jet flow formed by either two opposing round jets or using a peripheral nozzle (Ref.6)). The so-called "weak" swirl problem (neglecting the term of centrifugal acceleration $-w^2/x$ in the first equation of motion) characterizing the second group of authors is solved in Chanaud (Ref.7) for a laminar regime, and in Vulis and Kashkarov (Ref.8) for both regimes of flow. The swirling radial jet problem governed by (1)-(3) is treated by the third group, for a laminar regime in Riley (Ref.9), O'Nan and Schwarz (Ref.10); for a turbulent case in Chanaud (Ref.11), Fallenius (Ref.12), Smirnov (Ref.13), and in the papers (Refs.14,15), experimentally in Chanaud (flow beyond a rotating disk (Ref.11)), Muhe (outflow from an orifice between disks one of which rotates (Ref.16)).

The considered three cases of a radial jet (non-swirling, with "weak" swirl, swirling) differ mainly by their space flow geometry as shown in Fig.1 (top view).
The space flow geometry of the radial jet
- without swirl is obviously represented by the system of radial planes (strictly said half-planes);
- with "weak" swirl is formed by the system of cylindrical manifolds parallel to the axis of jet symmetry y, the projection of a cylindrical manifold to the plane y=Const is a hyperbolic spiral;
- with swirl is characterized by the system of the tangential half-planes to the cylinder of the radius e with the axis coinciding with the axis of jet symmetry.

The Görtler-type solution (using Prandtl´s Second Hypothesis) of a radial turbulent jet with swirl is derived in the paper (Ref.14) in the form

$$u(x,y) = \left(\frac{3M_\infty}{4k\rho}\right)^{1/2}\cdot\frac{1}{x}\cdot\left(\frac{x^2-e^2}{x^2-R^2}\right)^{1/2}\cdot(1-\tanh^2\eta(x,y)) \quad , \tag{4}$$

$$w(x,y) = \frac{e}{\sqrt{x^2-e^2}} \cdot u(x,y) \qquad , \tag{5}$$

$$\eta(x,y) = \frac{1}{2k} \cdot (x^2-e^2)^{1/2} \cdot (x^2-R^2)^{-1} \cdot y \qquad \left(\equiv \frac{y}{\delta(x)}\right) \tag{6}$$

where

$$M_\infty = \lim_{x \to +\infty} \int_0^\infty \rho x u^2 dy \qquad , \qquad N = \int_0^\infty \rho x^2 uw \, dy \qquad . \tag{7}$$

The integral conditions (7) express the asymptotic behaviour of the radial flux of radial momentum and the conservation of the radial flux of angular momentum, respectively. The so-called swirl parameter e is given by

$$e = N/M_\infty \qquad . \tag{8}$$

This solution obviously holds for an arbitrary swirl parameter e and in a limiting case $e \to 0$ reduces to the solution of flow without swirl.

In the following we present the Reichardt-type similarity solutions of multiple and antisymmetrical turbulent radial jets with swirl (using 3-D formulation of Reichardt's Hypothesis).

The multiple (antisymmetrical) swirling radial jet is formed by a system of s (two) single ones swirling in the same (opposite) sense whose axes of rotation coincide and which are characterized by the same flow parameters, see Fig.2b,c.

The advantage of Reichardt's procedure (introducing the time-averaged magnitudes of the type $\overline{u^2}, \overline{w^2}, \overline{uw}$, etc.) over those in the above mentioned papers (dealing with Reynolds equations of motion) is the possibility to apply the superposition principle to the single jet solutions. Hence, this method enables to obtain an analytical description of the mean velocity field of complex turbulent jets (the detailed analysis will be presented in the prepared paper (Ref.17)).

If we denote Y_i, i=1,...,s, the axial (i.e. related to the y-coordinate) locations of sources of individual single swirling radial jets, the mean velocity field description of the multiple swirling radial jet is given in the form

$$\overline{u^2}(x,y) = \frac{M_\infty}{\sqrt{\pi}k\rho} (x^2-e^2)^{1/2} \cdot x^{-2} \cdot \left[x-R+e \cdot \ln\left(\frac{x+(x^2-e^2)^{\frac{1}{2}}}{R+(R^2-e^2)^{\frac{1}{2}}}\right)\right]^{-1} \cdot F(\eta_1,\ldots,\eta_s) \tag{9}$$

$$\overline{w^2}(x,y) = e^2 \cdot (x^2-e^2)^{-1} \cdot \overline{u^2}(x,y) \qquad , \tag{10}$$

$$\overline{uw}(x,y) = e \cdot (x^2-e^2)^{-1/2} \cdot \overline{u^2}(x,y) \tag{11}$$

where

$$F(\eta_1,\ldots,\eta_s) = \sum_{i=1}^{s} \exp(-\eta_i^2) \qquad , \tag{12}$$

$$\eta_i(x,y) = \left[2k(x-R) + 2ke \cdot \ln\left(\frac{x+(x^2-e^2)^{\frac{1}{2}}}{R+(R^2-e^2)^{\frac{1}{2}}}\right)\right]^{-1} \cdot (y-Y_i) \qquad , \tag{13}$$

$$M_{\infty} = \lim_{x \to +\infty} \int_0^{\infty} \rho x \overline{u^2} \, dy \quad , \quad N = \int_0^{\infty} \rho x^2 \overline{uw} \, dy \quad , \quad e = N/M_{\infty} \quad . \qquad (14)$$

In the case of the antisymmetrical swirling radial jet we obtain for $\overline{u^2}, \overline{w^2}$ the same relations as for the multiple swirling radial jet, i.e. (9),(10) with s=2. The relation for the magnitude $\overline{uw}$ is given by

$$\overline{uw}(x,y) = \frac{N}{\sqrt{\pi} k \rho} \cdot x^{-2} \cdot \left[x-R+e \cdot \ell n \left(\frac{x+(x^2-e^2)^{\frac{1}{2}}}{R+(R^2-e^2)^{\frac{1}{2}}} \right) \right]^{-1} \cdot \left[\exp(-\eta_1^2) - \exp(-\eta_2^2) \right]$$
$$(15)$$

Turbulent jet model describing the high speed turbine impeller stream in agitated baffled vessels was first used by Nielsen (Ref.18) who introduced the so-called tangential jet model. This model based on an a priori assumption of a space flow geometry was taken as a basis for further investigation by DeSouza and Pike (Ref.19). The tangential jet model is discussed in the paper (Ref.20) where the above mentioned swirling radial jet model (4)-(8) is proposed for the description of a radially discharging impeller stream. This application means the replacement of a radial impeller by an equivalent circumference-point-source of momentum flux with definite parameters (location, radius, direction and momentum flux of outflow), see Fig.3. Hence, the parameters appearing in the velocity profile relations (4), (5) have clear physical meaning and are expressed (using outflow approximations, e.g. Uhl and Gray (Ref.21)) by mixing conditions (impeller and vessel diameters, rotational speed)

$$u(x,y) = \left(\frac{C_{d/D} n^2 d^5}{ke} \right)^{1/2} \cdot \frac{1}{x} \cdot \left(\frac{x^2-e^2}{x^2-R^2} \right)^{1/2} \cdot \left[1-\tanh^2 \left(\frac{(x^2-e^2)^{\frac{1}{2}}}{2k(x^2-R^2)} \cdot y - \eta_0 \right) \right] \quad , \qquad (16)$$

$$w(x,y) = \frac{e}{\sqrt{x^2-e^2}} \cdot u(x,y) \qquad (17)$$

where

$$e = c_e \cdot d/2 \quad , \quad R = c_R \cdot d/2 \quad . \qquad (18)$$

For the given geometry of a radial impeller the parameters $C_{d/D}$, c_e, c_R, k, η_0 are constants or functions only of the impeller-to-vessel diameter ratio. Different types of radial impellers may be characterized by the introduced parameters with respect to the generated flow field.

A good agreement was found between the theoretical and experimental data both for the case of standard six-flat-blade and six-curved-blade turbine impellers as shown in the paper (Ref.20).

REFERENCES

1. Riley,N.:"Asymptotic expansions in radial jets".J.Math.Phys.,41,
   1962,pp.132-146.

2. Schwarz,W.H.:"The radial free jet".Chem.Eng.Sci.,18,1963,pp.
   779-786.

3. Kolář,V.,Filip,P. and Curev,A.G.:"Comments on the radial turbulent
   jet".Acta Techn.ČSAV,27,1982,pp.563-567.

4. Heskestad,G.:"Hot-Wire Measurements in a Radial Turbulent Jet".
   Trans.ASME,Ser.E,J.Appl.Mech.,33,1966,pp.417-424.

5. Tanaka,T. and Tanaka,E.:"Experimental Study of a Radial Turbulent
   Jet (1st Report, Effect of Nozzle Shape on a Free Jet)".Bull.JSME,
   19,133,July 1976,pp.792-799.

6. Witze,P.O. and Dwyer,H.A.:"The turbulent radial jet".J.Fluid
   Mech.,75,1976,pp.401-417.

7. Chanaud,R.C.:"The radial free jet with weak swirl".Chem.Eng.Sci.,
   19,1964,p.933.

8. Vulis,L.A. and Kashkarov,V.P.:"The Theory of Jets of Viscous
   Fluid".Moscow,Nauka,1965,Chaps.4,11.(In Russian).

9. Riley,N.:"Radial jets with swirl,Part I. Incompressible flow".
   Quart.J.Mech.Appl.Math.,15,1962,pp.435-458.

10. O´Nan,M. and Schwarz,W.H.:"The swirling radial free jet".Appl.sci.
    Res.,A15,1965,pp.289-312.

11. Chanaud,R.C.:"Measurements of Mean Flow Velocity Beyond a Rotating
    Disk".Trans.ASME,J.Basic Engng.,93,1971,pp.199-204.

12. Fallenius,K.:"On the Swirling Radial Jet".Acta Polytechn.Scand.,
    Appl.Phys.Ser.No.Ph 120,1970,pp.3-17.

13. Smirnov,E.M.:"The similarity solution of the radial free jet with
    an arbitrary swirl".Zh.Prikl.Mekh.Tekhn.Fiz.,5,1977,pp.71-75.
    (In Russian).

14. Kolář,V.,Filip,P. and Curev,A.G.:"The swirling radial jet".Appl.
    Sci.Res.,39,1982,pp.329-335.

15. Filip,P.,Kolář,V. and Curev,A.G.:"Space flow geometry of the
    radial free, wall and liquid jets with swirl".Appl.Sci.Res.,42,
    1985,in press.

16. Muhe,H.:"The Swirling Radial Free Jet".In:Proc.Symp.IUTAM
    (Marseille,France:Aug.31-Sept.3,1982) Berlin/Heidelberg/New York,
    Springer-Verlag,1983,pp.229-239.

17. Filip,P.,Kolář,V. and Curev,A.G.:"Complex swirling radial jets".
    To appear.

18. Nielsen,H.:"Flow and turbulence from a flat blade turbine mixing
    impeller".Ph.D.Thesis,Illinois Inst.Technol.,Chicago,1958,107 pp.

19. DeSouza,A. and Pike,R.W.:"Fluid Dynamics and Flow Patterns in
    Stirred Tanks with a Turbine Impeller".Can.J.Chem.Eng.,50,1972,
    pp.15-23.

20. Kolář,V.,Filip,P. and Curev,A.G.:"Hydrodynamics of a radially
    discharging impeller stream in agitated vessels".Chem.Eng.Commun.,
    27,1984,pp.313-326.

21. Uhl,V.W. and Gray,J.B.:"Mixing - Theory and Practice".New York-
    London,Academic Press,1966,pp.182-183.

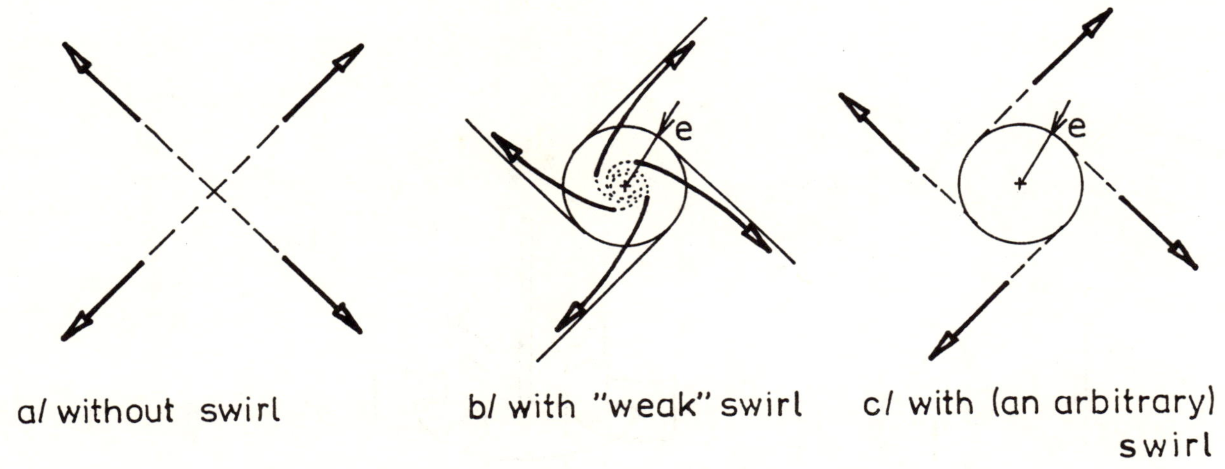

a/ without swirl   b/ with "weak" swirl   c/ with (an arbitrary) swirl

Fig.1   Space flow geometry of a radial jet.

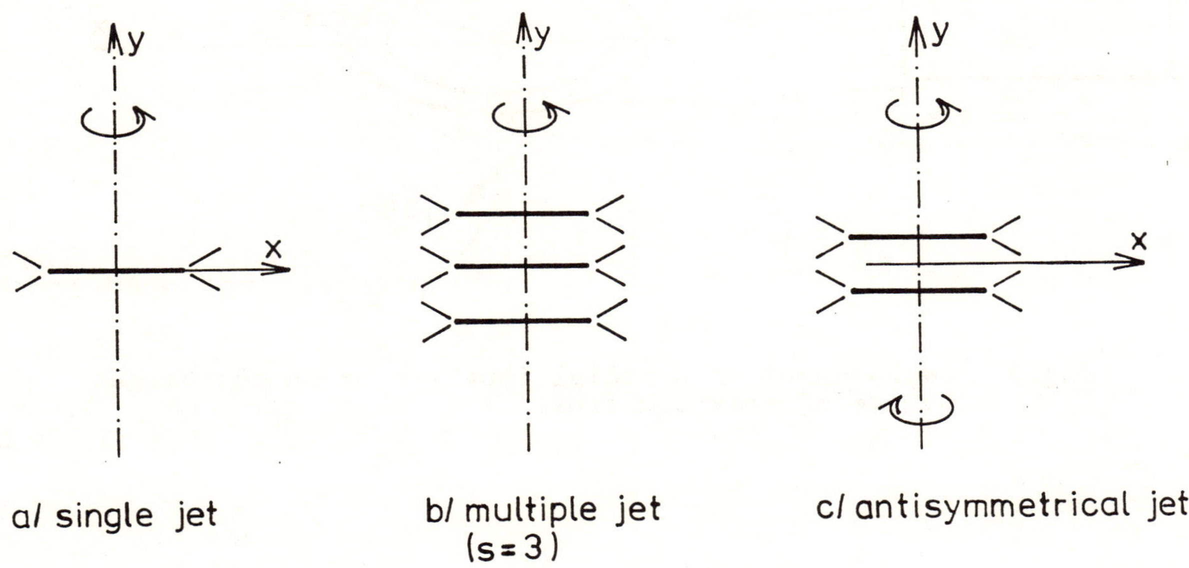

a/ single jet   b/ multiple jet (s=3)   c/ antisymmetrical jet

Fig.2   Sketch of complex radial swirling jets.

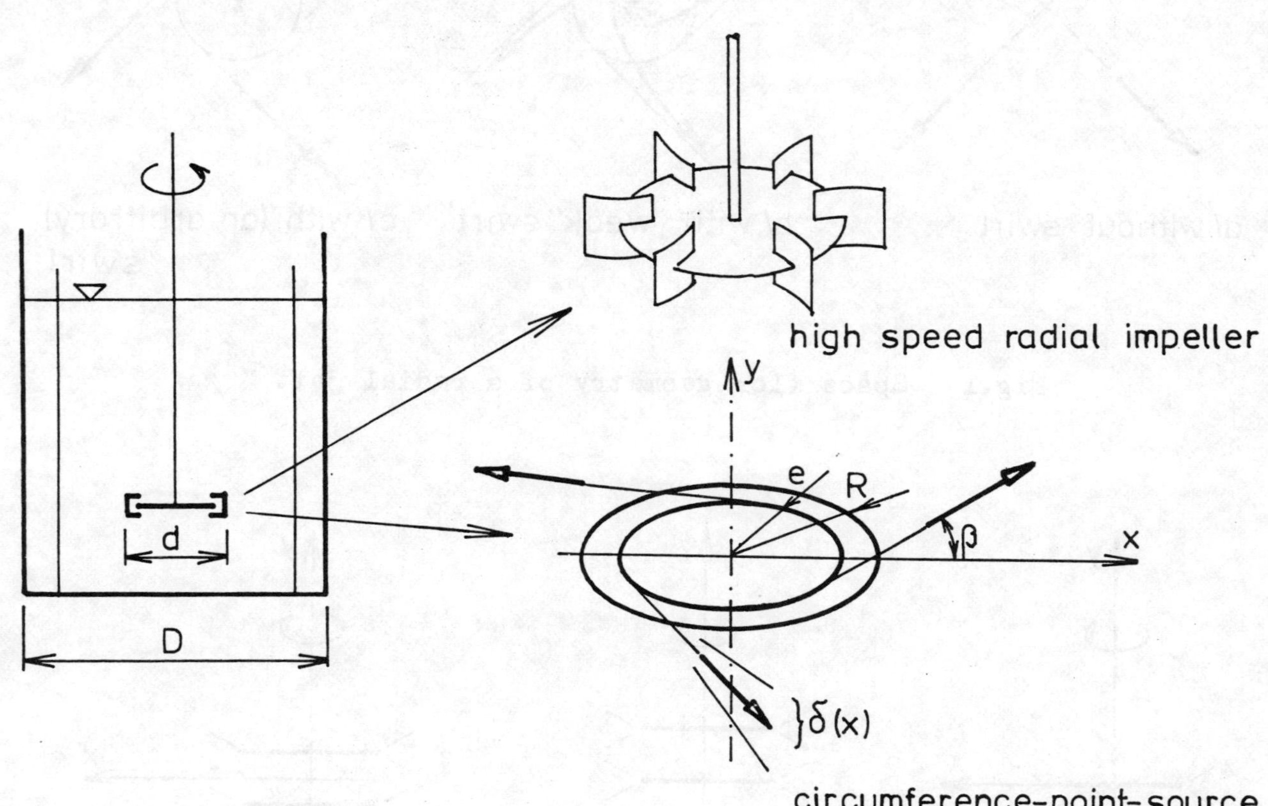

Fig.3   Replacement of a radial impeller by an equivalent
        source of momentum flux.

PUMPING CAPACITY OF SCREW AGITATORS
WITH A DRAUGHT TUBE

F. Rieger and V. Novák

Czech Technical University
166 07 Prague 6
Czechoslovakia

Summary

The aim our research was to estimate the effect of geometrical parameters on pumping capacity of screw agitators with a draught tube.

The main geometrical parameters of screw agitators were: the screw pitch, screw length, screw root diameter, the clearance between the screw and the draught tube and the ratio of the screw to the vessel diameter.

The mathematical model for calculation of pumping capacity of screw agitator operated in the draught tube was proposed. This model, which is valid for the creeping flow regime only was tested experimentally.

The good agreement between calculated and experimental values was found. On the basis of results presented in this paper the effect of main geometrical parameters on pumping capacity of screw agitators with a draught tube was estimated.

Held at Wurzburg, 10-12 June, 1985.

Organised by DVCV· Deutsche Vereinigung für Chemie- und Verfahrenstechnik
(German Association of Chemical and Process Engineering).

Organisation: GVC·VDI-Gesellschaft Verfahrenstechnik und Chemieingenieurwesen. **GVC**

©BHRA, The Fluid Engineering Centre, Cranfield, Bedford MK43 0AJ, England.

NOMENCLATURE

a = dimensionless constant in Eq. (1)

b = dimensionless constant in Eq. (1)

D = screw agitator diameter

D_1 = screw root diameter

D_t = draught tube diameter

e = screw flight thickness

F_d = flight correction factor for drag-flow

F_{dc} = curvature correction factor for drag flow

F_p = flight correction factor for pressure flow

F_{pc} = curvature correction factor for pressure flow

k = dimensionless constant in Eq. (4)

k_d = leakage correction factor for drag flow

k_p = leakage correction factor for pressure flow

L = length of screw

L_t = length of draught tube

n = agitator speed

p = pressure

s = screw pitch

T = tank diameter

$\dot{V}$ = volumetric flow rate

φ_t = helix angle on diameter D_t

η = dynamic viscosity

1. INTRODUCTION

Screw agitators with a draught tube are often employed for mixing of highly **viscous** liquids (Ref. 1). The pumping capacity is one of the most important parameters which influences the homogenization effect or heat transfer rate. The aim of our research was to estimate the effect of geometrical parameters on pumping capacity of screw agitators with a draught tube.

The screw agitator with a draught tube is shown in Fig. 1. The main geometrical parameters of screw agitator with a draught tube are: the agitator diameter D, the screw pitch s, the screw root diameter D_1, the screw lenght L, the diameter of draught tube D_t, the draught tube length L_t and the vessel diameter T.

The volume of the mixing vessel can be divided into two main regions. The region 1 includes the screw rotating in a draught tube which operates as a pump which transports the liquid into the region 2 - see Fig. 1. The region 2 can be subdivided into regions with flow reversals 2a an 2c and annular space between the draught tube and the vessel wall 2b.

2. MATHEMATICAL MODEL OF SCREW AGITATOR WITH A DRAUGHT TUBE

The mathematical model of a screw agitator with a draught tube can be obtained by simultaneous solution of equation of pumping characteristic of the screw and equation of hydraulic characteristic of the system.

The equation of pumping characteristic of the screw is linear in the creeping flow regime and may be written in the form:

$$\frac{\dot{V}}{nD^3} = a - b\,\frac{\Delta p}{\eta\,n} \tag{1}$$

Constants a and b in Eq. (1) depend on geometry of the screw and can be calculated from relations:

$$a = \frac{\pi}{4}\left(\frac{s}{D} - \frac{e}{D}\right)\left(\frac{D_t}{D} - \frac{D_1}{D}\right)\frac{D_t}{D}\cos^2\varphi_t\,F_d\,F_{dc}\,(1 - k_d) \tag{2}$$

$$b = \frac{1}{96}\left(\frac{s}{D} - \frac{e}{D}\right)\left(\frac{D_t}{D} - \frac{D_1}{d}\right)^3\frac{D}{L}\,\sin\varphi_t\cos\varphi_t\,F_p\,F_{pc}\,(1 + k_p) \tag{3}$$

where the correction factors F_d, F_p, F_{dc}, F_{pc}, k_d and k_p can be calculated from relations published in (Refs. 2 and 3).

The equation of hydraulic characteristic of the system
(region 2) was reduced to the equation for region 2b because there is
a lack of reliable data for pressure drop calculations in the regions
of flow reversals 2a and 2c. The equation of hydraulic characteristic
of annular region 2b can be written in the form see e.g. (Ref. 4)

$$\Delta p = \frac{128 \frac{L}{D}}{\pi \left(\frac{T}{D}\right)^4} \left\{ \left[1 - \left(\frac{D_t}{T}\right)^4 - \frac{\left[1 - \left(\frac{D}{T}t\right)^2 \right]^2}{\ln\left(\frac{T}{D_t}\right)} \right] \right\}^{-1} \frac{\eta \dot{V}}{D^3} = k \frac{\eta \dot{V}}{D^3} \qquad (4)$$

The relationship for dimensionless pumping capacity of screw agitator
rotating in a draught tube can be obtained by elimination Δp from
Egs. (1) and (4)

$$\frac{\dot{V}}{nD^3} = \frac{a}{1 + bk} \qquad (5)$$

For estimates of the effect of agitator diameter up on circulation
capacity it is more convenient to use vessel diameter T as characteristic length in the equation for dimensionless pumping capacity:

$$\frac{\dot{V}}{nT^3} = \frac{\dot{V}}{nD^3} \left(\frac{D}{T}\right)^3 = \frac{a}{1 + bk} \left(\frac{D}{T}\right)^3 \qquad (6)$$

3. EXPERIMENTAL

In formulating of the mathematical model (5) the influence
of regions 2a and 2c has been neglected and therefore its experimental
verification was necessary. Experiments were carried out for 22 geometries of screw agitator with a draught tube in the vessel of diameter T = 150 mm. Aqueous solutions of glycerol and corn syrup were
used in the measurements. The pumping capacity was calculated from
liquid velocity in the annulus which was measured using a flow
follower having approximately the same density as the mixed liquid.
The typical shape of the plot of dimensionless pumping
capacity $\dot{V}/nD^3$ vs. Reynolds number is shown in Fig. 2. From this
Fig. it follows, that in the creeping flow regime the dimensionless
pumping capacity does not depend on the Reynolds number. The values
of $\dot{V}/nD^3$ obtained experimentally in the creeping flow regime are
listed (with their 95 % confidence limits) in Tab. 1. Tab. 1 also
contains the main geometrical parameters of agitators and the values
of $\dot{V}/nD^3$ calculated on the basis of mathematical model (3). From
Tab. 1 it is obvious that   most   experimental values do not
differ significantly from calculated values. There are only 2 excep-

tions, geometry No 22 characterized by high screw pitch causing that
the screw flight does not form one full thread and geometry No 18
characterized by extremely high values of screw root diameter and the
clearance between screw and draught tube. For both these geometries
are experimental values $\dot{V}/nD^3$ lower than calculated ones.

4. THE EFFECT OF GEOMETRY ON PUMPING CAPACITY

From the preceding paragraph it may be concluded that
proposed mathematical model is suitable for estimation of the effect
of geometrical parameters on pumping capacity of screw agitators with
a draught tube. As an example of its application the dependence of
dimensionless circulation capacity on the ratio of agitator to vessel
diameter D/T is shown in Fig. 3. From this figure it is obvious
that the circulation capacity of screw agitator increases   with
relative agitator size up to a maximum value, whereafter decreases
rapidly. This can be explained in the following way. Hydraulic resistance of annular region is negligible at small values of D/T ratio,
dimensionless pumping capacity $\dot{V}/nD^3$ is constant and circulation
capacity given by $\dot{V}/nT^3$ increases with third power of D/T according
to Eq. (6). With D/T ratio further increasing, the hydraulic resistance of annular region starts to increase significantly, the value of
$\dot{V}/nD^3$ starts to decrease and the rate of the increase of $\dot{V}/nT^3$ is
getting lower up to critical value. With increasing D/T above critical value hydraulic resistance of annular region increases rapidly
and circulation capacity decreases very rapidly. As it can be expected
the circulation capacity decreases with increasing clearance between
agitator and draught tube also the position of maximum value of $\dot{V}/nT^3$
shifts to smaller values of D/T ratio.

5. CONCLUSIONS

This paper deals with the effect of geometry on the pumping
capacity of screw agitator with a draught tube. The mathematical model suggested earlier (Ref. 5) was confirmed experimentally and a
good agreement was found in contrast with conclusions presented by
Chavan and Ulbrecht (Ref. 6). The proposed mathematical model can be
used for the estimation of the effect of geometry on pumping capacity
of screw agitator with a draught tube in the creeping flow regime.

REFERENCES

1. <u>Novák,V. and Rieger,F.</u>: "Homogenization with helical screw agitators". The Transactions of the Institution of Chemical Engineers, <u>47</u>, 10, Dec 1969, pp. T 335 - 340.

2. <u>Rieger,F.</u>: "Process characteristics of screws and screw agitators". ("Procesní charakteristiky šroubových rotorů a míchadel") DrSc Thesis ČVUT Prague 1984 (in Czech).

3. <u>Boy,L.M.</u>:"On isothermal flow of viscous liquids through screw pumps". Thesis. Technische Hoogeschool Delft 1964.

4. <u>Bird,R.B., Stewart,W.E. and Lightfoot,E.N.</u>: "Transport phenomena", New York, John Wiley & Sons, Inc, 1960, pp. 51 - 54.

5. <u>Rieger,F.</u>: "Polymerization reactor with screw agitator". ("Polymerizátor se šnekovým míchadlem") MSc Thesis. ČVUT Prague 1964 (in Czech).

6. <u>Chavan,V.V., Ulbrecht,J.</u>: Internal Circulation in Vessels Agitated by Screw Impellers. The Chemical Engineering Journal, <u>6</u>, 1973, pp. 213 - 223.

TABLE 1 - Experimental and calculated values of dimensionless pumping capacity $\dot{V}/nD^3$

geometry No	1	2	3	4	5	6	7	8	9	10	11
T/D	2.0	2.0	2.0	2.0	2.0	2.0	1.59	1.59	1.59	1.59	1.59
D_1/D	0.2	0.2	0.2	0.2	0.4	0.6	0.2	0.2	0.2	0.2	0.4
s/D	1.0	1.0	1.0	1.0	1.0	1.0	1.0	1.0	1.0	1.0	1.0
D_t/D	1.03	1.06	1.10	1.15	1.10	1.10	1.02	1.06	1.10	1.14	1.02
$\dot{V}/nD^3$ experiment	0.417 ±0.078	0.422 ±0.051	0.392 ±0.028	0.378 ±0.047	0.356 ±0.018	0.253 ±0.013	0.350 ±0.024	0.330 ±0.023	0.272 ±0.008	0.203 ±0.024	0.334 ±0.046
$\dot{V}/nD^3$ Eq.(5)	0.382	0.379	0.373	0.364	0.327	0.244	0.330	0.311	0.276	0.241	0.310

geometry No	12	13	14	15	16	17	18	19	20	21	22
T/D	1.59	1.59	1.59	1.59	1.59	1.59	1.59	2.0	2.0	2.0	2.0
D_1/D	0.4	0.4	0.4	0.6	0.6	0.6	0.6	0.2	0.4	0.6	0.2
s/D	1.0	1.0	1.0	1.0	1.0	1.0	1.0	1.5	1.5	1.5	2.0
D_t/D	1.06	1.10	1.14	1.02	1.05	1.10	1.14	1.10	1.10	1.10	1.10
$\dot{V}/nD^3$ experiment	0.309 ±0.038	0.279 ±0.009	0.228 ±0.022	0.266 ±0.030	0.259 ±0.025	0.214 ±0.008	0.144 ±0.016	0.481 ±0.015	0.426 ±0.039	0.323 ±0.041	0.416 ±0.055
$\dot{V}/nD^3$ Eq.(5)	0.297	0.267	0.234	0.242	0.237	0.223	0.206	0.466	0.423	0.324	0.485

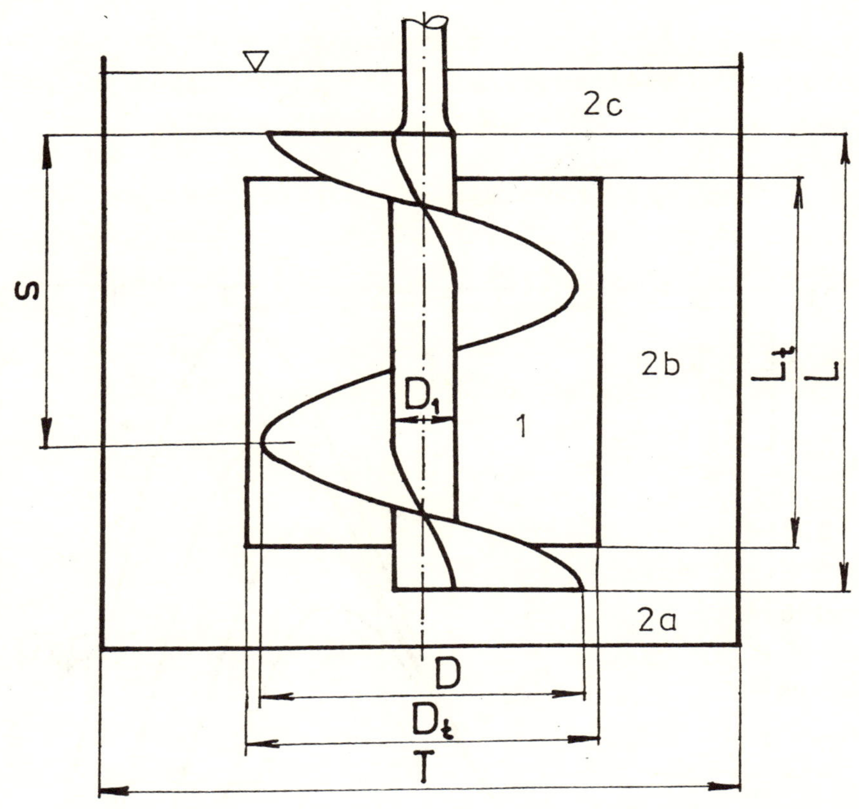

FIG. 1 - Screw agitator with a draught tube

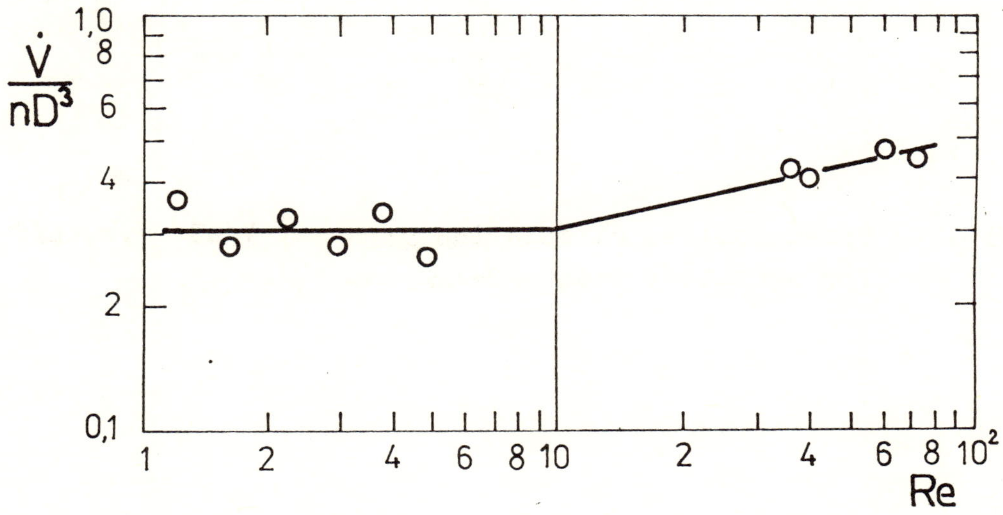

FIG.2 - Pumping characteristic of screw agitator No 12

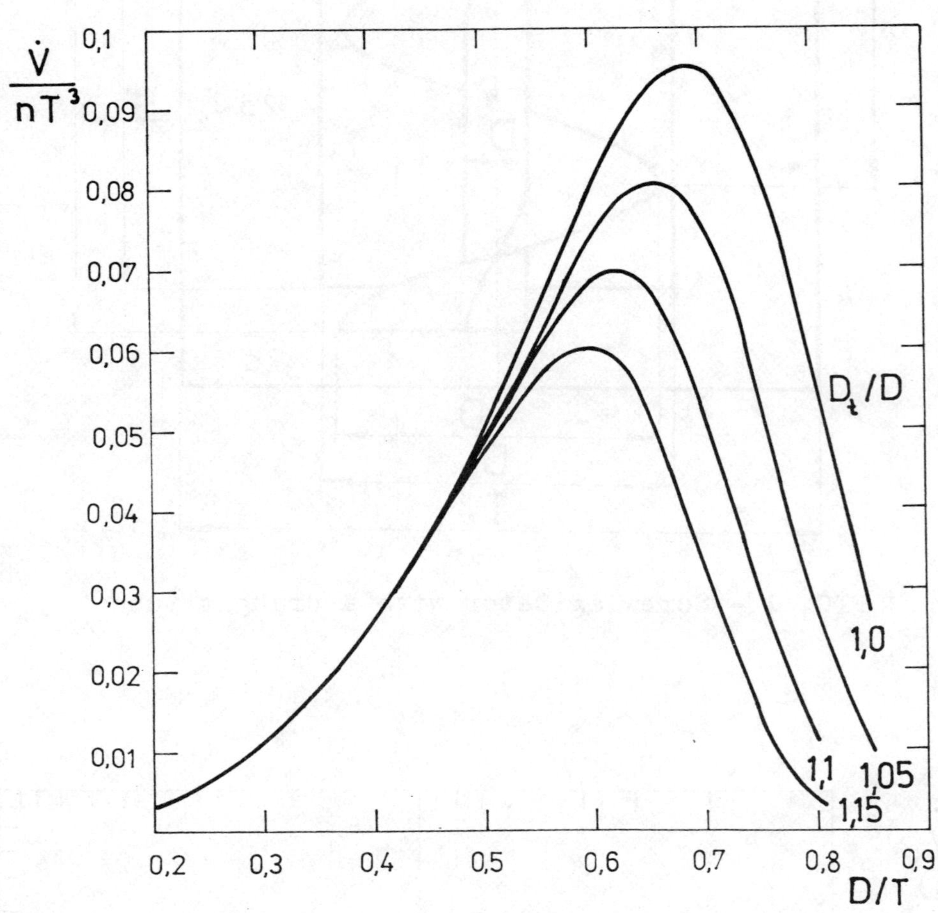

FIG.3 – Dimensionless circulation capacity $\dot{V}/nT^3$ v.s. D/T for agitators with s/D=1.0 and D_1/D=0.2

Dispersion Efficiency of a Pitch-Blade Stirrer in a Three-Phase (G-L-L) System

Authors:

A.J.F. Simons, J. van Hinsberg, L.L. van Dierendonck

DSM/Central Laboratory

Geleen, Netherlands

Held at Wurzburg, 10-12 June, 1985.
Organised by DVCV· Deutsche Vereinigung für Chemie- und Verfahrenstechnik
(German Association of Chemical and Process Engineering).

Organisation: GVC·VDI-Gesellschaft Verfahrenstechnik und Chemieingenieurwesen. **GVC**

NOMENCLATURE

D	Impeller diameter	m
F	Surface area of the horizontal cylinder at overflow height	m^2
Fr	Froude number $= \dfrac{n^2 D^2}{g.(H-h)}$	-
G-L-L	Gas-Liquid-Liquid	-
H	Liquid depth	m
h	Impeller clearance from base	m
L-L	Liquid-Liquid	-
n	Impeller rotation speed	s^{-1}
P	Total energy input	W
P_o	Power number	-
P_v	Specific power input	$W.m^{-3}$
Q	Volumetric feed rate	$m^3.s^{-1}$
Re	Reynolds' number $(= \dfrac{\rho.n.D^2}{\eta})$	-
v_g	Gas superficial velocity	$m.s^{-1}$
α	Dispersed phase hold-up	-
α_F	Hold-up of phase to be dispersed in total feed $(\alpha_F = Q_{aq}/(Q_{aq} + Q_{org}))$	-
η	Liquid viscosity	$N.s.m^{-2}$
ρ	Liquid density	$kg.m^{-3}$

INDICES

aq	aqueous
cal	calculated
d	dispersed
g	gas
md	measured
org	organic

1. INTRODUCTION

Information in literature about dispersion of liquid-liquid systems focuses on modes of realization (use of mixer-settler, RDC, pulsation column, etc.). A procedure that finds little application makes use of a horizontal cylindrical reactor in which a constant flow is maintained.

In most cases the horizontal cylindrical reactor is divided into individually stirred compartments. For the dispersion of immiscible liquids use is often made of a Rushton turbine stirrer. For the multi-phase system here dealt with a <u>pitch-blade</u> stirrer is employed (four blades at 45 °).

In another little-used process operation, in which an aqueous phase is dispersed in an organic phase with which it is immiscible, reaction heat is removed through evaporation of the organic solvent. This leads to a complicated problem if the resulting dispersion has to satisfy a strict requirement of uniformity.

The phase ratio is primarily controlled through the mass flows of the feed to the reactor. On account of insufficiency of the stirring energy and/or of stirring effects in combination with other influences, such as the evaporation of the solvent referred to above, the phase ratio in the reactor or in individual reactor compartments may come to differ considerably from the feed ratio chosen. In the case here considered, we have to do with an extraction involving a chemical reaction whose selectivity depends in part on the hold-up of the aqueous phase.

A reaction component A is contained in the <u>continuous organic phase</u> (the solvent) and is extracted in the reactor by the <u>disperse aqueous phase</u>. In the aqueous phase component A is converted into several products. The formation of the main product and of undesirable components (i.e. the selectivity) depends on the hold-up of the aqueous phase. The reaction products are re-extracted into the organic phase. The overall reaction may be rendered as:

$$A_{org} \xrightarrow{\quad OH^- \quad} B_{org} + C_{aq} \qquad (1).$$

For the reasons set forth above, a good translation of laboratory to plant conditions is required. In this translation a central role is played by the stirring energy needed in industrial-scale application, in view of the requirement of dispersion uniformity. The large quantity of vapour formed in the reaction mixture makes adequate translation a difficult problem.

The extractive reaction forms part of an industrial process which in a previous stage was studied in a pilot plant in order to find ways towards process improvement.

2. PILOT-PLANT STUDY AND SCALE-UP

In the pilot-plant the extractive reaction was studied in a CSTR. The reaction was carried out at a pressure at which the liquid did not boil. The heat of reaction was carried off through the reactor wall.

In this study an examination was undertaken of the influence of the aqueous-phase hold-up on:
- selectivity;
- rate of the reaction.

At the same time a determination was made of the minimum impeller speed of the <u>Rushton turbine</u> needed to obtain complete dispersion.

Afterwards the reaction had to be carried out on plant scale in an existing horizontal cylindrical reactor (pressure vessel). Fig. 1 is a diagram of this reactor.

In consultation with a stirrer manufacturer, an investigation was made into the minumum dispersion energy required in a batch-operated scale model of the first reactor compartment, with a view to the scale-up. For a good L-L dispersion (in the absence of a vapour phase) a stirring energy of about 0.3 kW/m^3 was found to be needed if use was made of a four-blade, downward-pumping stirrer. Fig. 2 gives a diagrammatical representation of the scale model.

The Rushton turbine was found not to produce a good dispersion with the same energy input. Therefore a choice was made in favour of the 'pitch-blade' stirrer, and scaling up was done on the basis of a constant specific energy input.

3. STATEMENT OF THE PROBLEM

After the industrial reactor had been started up, it was found that the
amount of vapour generated caused a gas hold-up of 15 - 20 %. This was in reasonable
agreement with predictions based on literature correlations (Van Dierendonck, 1968).
The energy input measured was 0.33 kW/m^3 liquid for the first compartment
and 0.41 kW/m^3 for the second.
The aqueous-phase hold-up, α, in the first compartment proved to be much
too high ($\alpha/\alpha_F > 5$), pointing to non-uniformity of the dispersion, resulting in low
selectivity.
DSM's Central Laboratory was commissioned to give further advice to the
plant people about the power input requirement and/or about installation of a stirrer
adapted to realize the objective originally envisaged for the reactor vessel.

4. INVESTIGATION BY MEANS OF A MOCK-UP

In a mock-up of the existing reactor (Fig. 2) the influence of the <u>pitch-blade</u>
stirrer on the dispersion behaviour was subjected to closer investigation – for batch
as well as continuous operation – through variation of stirrer speed and stirrer
diameter. The following aspects were considered:
- the gas velocity in the vessel, defined as $v_g = Q_g/F$;
- the gas flow rate Q_g in relation to the pump capacity of the stirrer (nD^3), with
  upward- as well as with downward-pumping stirrer.
The liquids used were the components of the feed to the existing reactor.
Vapour generation was simulated by feeding N_2 into the reactor under the stirrer.
Use was made of two model reactors, with volumes of 6 litres and 60 litres.
Measurements were made explicit of:
- the minimum stirrer speed required for effecting homogeneous dispersion;
- the energy input;
- the speed at which gas is drawn in through the surface without gas supply to the
  reactor.

5. RESULTS

5.1   The minimum stirrer speed for liquid dispersion.

5.1.1   Continuous flow reactor without gas loading.

The minimum stirrer speed at which the aqueous phase was uniformly
dispersed thoughout the vessel has been determined visually. This speed was dependent
on the pumping direction of the pitch blade stirrer (see also Section 5.1.3).
The measured hold-up α was signnificantly different from values calculated
with correlations given in literature (e.g. Weinstein, Treybal, 1973; Thornton,
Bouyatiotis, 1967); see Table 1.

Table 1

$\dfrac{n}{s^{-1}}$	$\left(\dfrac{\alpha}{\alpha_F}\right)_{md}$	$\left(\dfrac{\alpha}{\alpha_F}\right)_{cal}$
5	0.3	1.7
7	0.5	1.6

Surprisingly the aqueous phase hold-up ws found to be systematically lower
than the hold-up related to the feed flows ($\alpha < \alpha_F$).
The difference in behaviour noted with respect to the situation in standard
stirred vessels and with standard stirrers is attributed to the following causes:
- the strongly deviant geometry of the stirred vessel;
- the strong circulation lengthwise in the vessel, with droplets being preferentially
  passed across the overflow weir;
- the high rate of gas draw-in.

The results of the experiments carried out by the stirrer supplier – in which the criterion was that no non-dispersed water phase should be visual on the vessel bottom – cannot, on the strength of our new observations, be used for scale-up.

5.1.2   Continuous-flow reactor with gas loading.

When gas is being sparged into the vessel, the dispersing effect of the stirrer is negatively affected to the extent that homogeneous L-L dispersion does not result until also the gas passed through is seen to be well dispersed in the liquid.

In our experiment, the linear gas velocity v_g was about 0.03 m/sec.

5.1.3   Batch reactor without gas loading.

The criterion applied was that, judged visually, the water phase distribution should be as nearly uniform as possible. The stirrer speeds observed were:
- n = 305 rpm, for downward pumping,
- n = 375 rpm, for upward pumping.

The specific energy input values were, respectively:
P_v = 0.35 kW/m^3 and 0.65 kW/m^3.

5.1.4   Batch reactor with gas loading.

With downward-pumping action and $v_g \approx$ 0.03 m/s, the stirrer speed has to be raised to n > 500 rpm to obtain a visually homogeneous dispersion; the specific energy input then is: P_v > 3 kW/m^3.

With upward-pumping action it was found that a value of n > 600 was needed for homogeneous dispersion, at a specific energy input of P_v > 4   kW/m^3.

5.2   The minimum stirrer speed for gas aspiration.

As a result of vortex formation, gas is drawn into the liquid from the vapour-filled space. The manner in which this happens depends on the direction of the pumping action.

We did not succeed in finding an unequivocal criterion, analogous to that for a turbine or propeller stirrer in a vertical cylindrical vessel (Fr = 0.2).

The stirrer speeds for gas aspiration with the stirrer employed at a level of h = ½ H was:
- n = 140 – 160 rpm, for downward-pumping action,
- n = 286 – 375 rpm, for upward pumping action,
corresponding to Fr = 0.1 and Fr = 0.3, respectively.

5.3   Energy input.

5.3.1   Characteristic value of the energy input without gas aspiration.

The characteristic value of the energy input P_o ($= \dfrac{P}{\rho \cdot n^3 D^5}$) was determined under conditions where the horizontal cylindrical reactor was completely filled, to prevent gas aspiration.

For upward as well as downward pumping, we found:
P_o = 1.8 $\pm$ 0.2, at Re > 10^5.

5.3.2   The stirring energy with gas aspiration.

The energy input is influenced by gas draw-in owing to vortex formation.

The measuring results, which are subject to large fluctuations are shown in Fig. 3 for downward and upward pumping action.

5.3.3   The stirring energy with gas loading.

When the plant situation is simulated by passing N_2 through the liquid it is seen that the power input changes with increasing gas load at constant stirrer speed (Fig. 4). There is found to be remarkable difference between the required energy inputs with downward and with upward pumping of the stirrer. The overall picture does not agree with the observations given in literature. (Warmoeskerken, 1982).

6. CONCLUSIONS

- The energy input is strongly affected when gas is drawn in or sparged into the
liquid.
With gas aspiration we see a decrease of the energy input, which effect is much
stronger with a downward-pumping than with an upward-pumping stirrer.
When gas is passed through, the energy input rises with downward pumping action and
decreases with upward pumping action of the stirrer.
- With downward pumping action of the stirrer, gas aspiration is observed at relatively low stirrer speeds.
- In a continuous-flow compartment (with spherical front) the hold-up is not at any
stirring energy equal to the feed value.
- Literature data do not provide sufficient information for predicting the dispersion
behaviour in pilot-plant or commercial scale installations.
- For the geometry we are concerned with here, no unequivocal scale-up rule can be
given.

7. RECOMMENDATIONS FOR THE COMMERCIAL PLANT

The following advice was given to the plant people:
- Raise stirring energy substantially. Limitations imposed by the mechanical strength
of the vessel did not permit of raising the energy beyond $P_v \approx 0.75$ kW/m^3. This
value was attained by increasing the stirrer diameter by 33 %.
- Place an additional baffle in the spherical front of the cylinder in order to
obtain a stabler flow pattern and so increase the energy input.
The mechanical construction made it impossible for this advice to be
followed.

8. RESULTS IN THE COMMERCIAL PLANT

The results in plant operation were the following:
- stabler energy input;
- reduction of aqueous-phase hold-up to about 8 % of the vessel volume (as against
the required 4 %).

LITERATURE

1. Van Dierendonck L.L., Fortuin J.M.H., Venderbos D., Proc. 4th Symposium Chem.
React. Eng., Brussel 1968.

2. Warmoeskerken M.M.C.G., Smith J.M., Proc. Symp. Physical aspects of chemical
reactors; The Institution of Chemical Engrs., European Branch, Utrecht 1982.

3. Weinstein B., Treybal R.E., A.J.Ch.E.J., 19, 851 (1973).

4. Thornton J.D., Bouyatiotis B.A., Inst. Chem. Engrs., Symp. Ser. no. 26, 43
(1967).

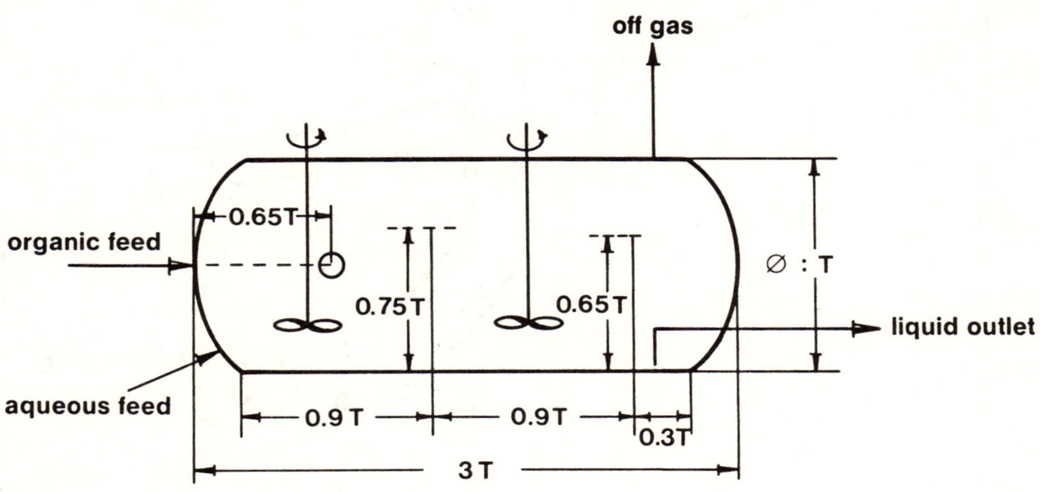

Fig. 1 Plant-reactor

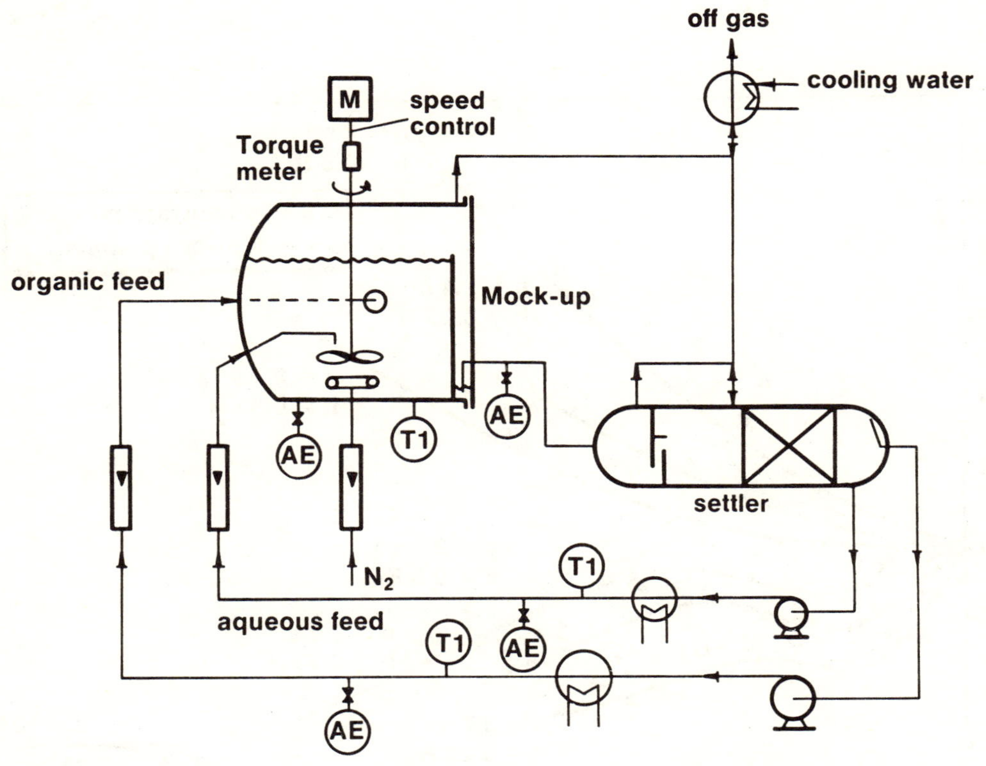

Fig. 2   The experimental set up

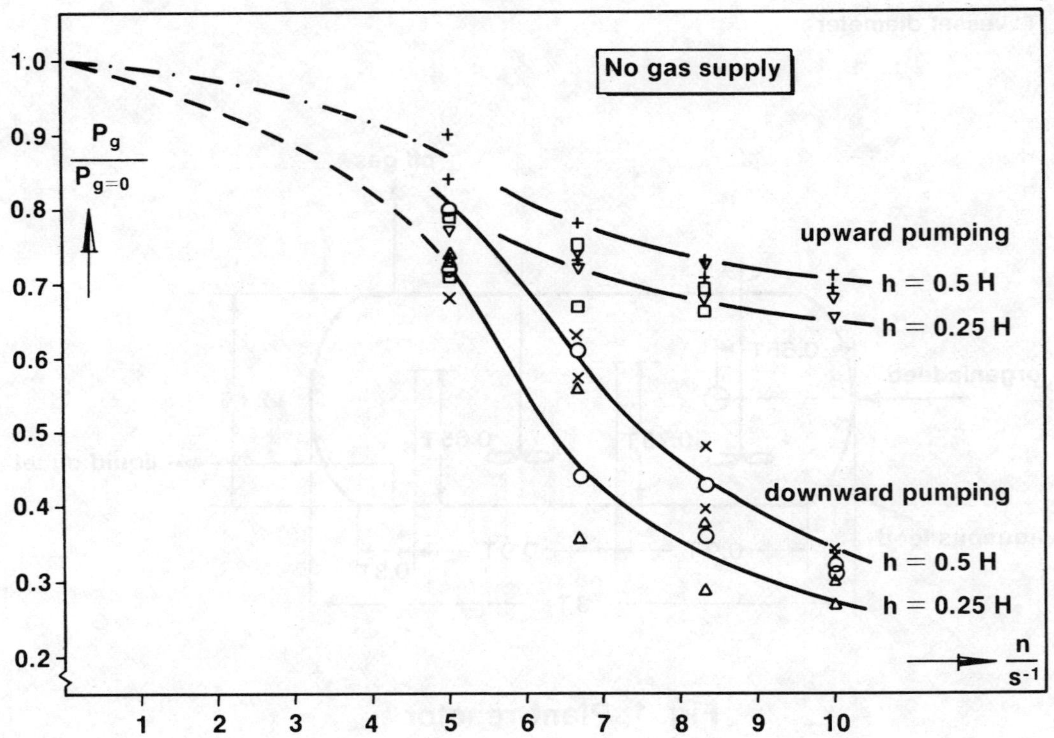

Fig. 3 Gassed power ratio against stirring rate

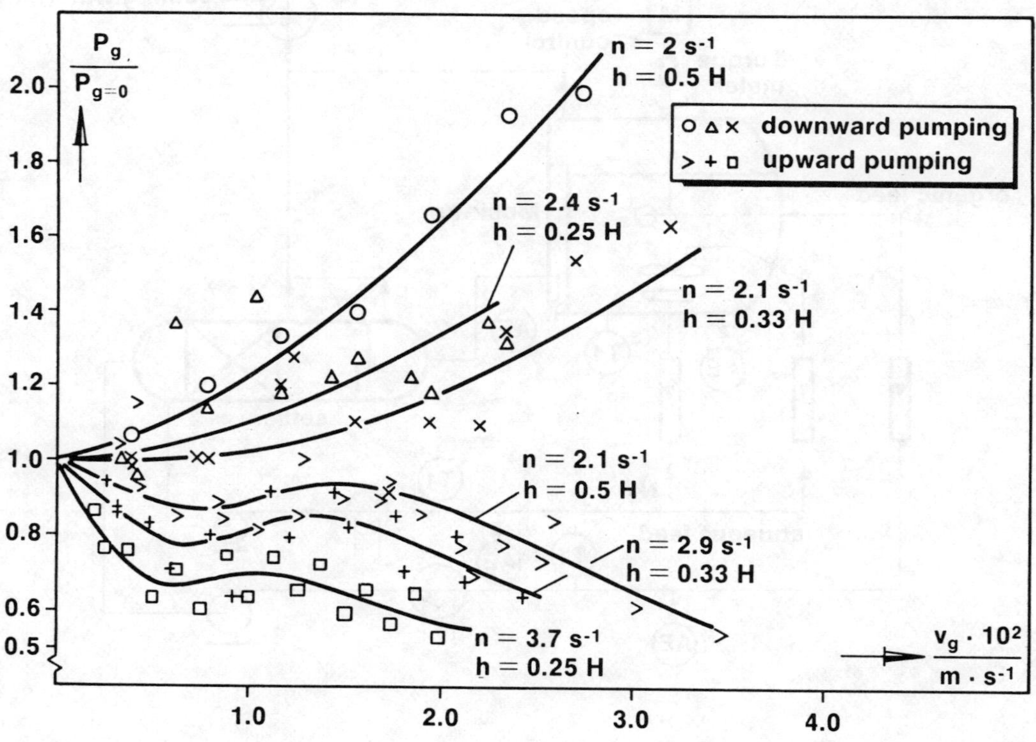

Fig. 4 Gassed power ratio against gas rate

MIXING AND FLOW IN A SPINNING DISC BLENDER

A. J. P. Spragg*, S. J. Maskell[+] and M. A. Patrick[+]

*National Engineering Laboratory, East Kilbride, Glasgow, UK
[+]Centre for Industrial and Geophysical Fluid Dynamics, University of Exeter, Exeter, UK

Summary

The problem of predicting flow fields in stirred vessels continues to attract much research attention because of the highly complex hydrodynamic processes involved and their implications for vessel design. The work reported here addresses a simpler problem, that of predicting the steady laminar Newtonian flow due to a rotating disc, mounted coaxially on a shaft, in a cylindrical unbaffled vessel. This is intended as a first step towards a more complicated and realistic model. The predictions (which are in stream function/vorticity formulation) are compared with photographic results from a parallel experimental investigation.

The computationally predicted steady laminar flow fields are then used to investigate the mixing capability of the system by introducing bands of coloured fluid into the steady flow and solving the unsteady convection/diffusion equation for a passive scalar until 95 per cent uniformity of concentration is reached.

Held at Wurzburg, 10-12 June, 1985.

Organised by DVCV· Deutsche Vereinigung für Chemie- und Verfahrenstechnik
(German Association of Chemical and Process Engineering).

Organisation: GVC·VDI-Gesellschaft Verfahrenstechnik und Chemieingenieurwesen. 

© BHRA, The Fluid Engineering Centre, Cranfield, Bedford MK43 0AJ, England.

NOMENCLATURE

c = Concentration

D = Impeller diameter

D_{ab} = Diffusion coefficient

d = Shaft diameter

H = Vessel height

h = Impeller distance from vessel bottom

N = Total number of grid points

N_p = Pumping number = $2\pi(\psi_{max} - \psi_{min})/(\rho n D^3)$

n = Rotational speed of impeller, revolutions per second

$\underline{q}$ = Velocity vector

Re = Reynolds number = $\rho n D^2/\eta$

r = Radial coordinate

Δr = Radial mesh size

Sc = Schmidt number = $\eta/(\rho D_{ab})$

T = Vessel diameter

t^* = Mixing time, defined in text

t_1 = $1/n$

u = Radial velocity component

v = Tangential velocity component

w = Axial velocity component

z = Axial coordinate

Δz = Axial mesh size

ζ = Vorticity component perpendicular to (r, z) plane

η = Viscosity

θ = Tangential coordinate

ρ = Density

ϕ = Mixing index, defined in text

ψ = Stream function

1.   INTRODUCTION

A commonly used geometry for mixing of a fluid is that of a bladed Rushton turbine mounted coaxially on a shaft in a cylindrical baffled vessel. Here, a simplified system is considered in an attempt to perform comparisons between experimental observations and theoretical calculations both for the fluid flow field and for the mixing properties. The flow is generated by a spinning disc blender in an unbaffled cylindrical tank (Fig. 1).

2.   CALCULATION OF THE FLOW FIELD

The flow field is taken as steady and axisymmetric, with the fluid being incompressible and Newtonian. The equation for mass continuity is then

$$\frac{1}{r}\frac{\partial}{\partial r}(\rho r u) + \frac{\partial}{\partial z}(\rho w) = 0 \tag{1}$$

which can be used to define a stream function ψ via

$$u = -\frac{1}{\rho r}\frac{\partial \psi}{\partial z}, \quad w = +\frac{1}{\rho r}\frac{\partial \psi}{\partial r}. \tag{2}$$

The component of the vorticity perpendicular to the (r, z) plane can be expressed in terms of ψ via

$$\zeta = \frac{\partial u}{\partial z} - \frac{\partial w}{\partial r} = \frac{\partial}{\partial z}\left(-\frac{1}{\rho r}\frac{\partial \psi}{\partial z}\right) - \frac{\partial}{\partial r}\left(\frac{1}{\rho r}\frac{\partial \psi}{\partial r}\right) \tag{3}$$

which may be regarded as a Poisson type equation for ψ, given ζ.

A transport equation for ζ can be obtained by taking the curl of the equation for momentum conservation, and is

$$r^2\left\{\frac{\partial}{\partial z}\left(\frac{\zeta}{r}\frac{\partial \psi}{\partial r}\right) - \frac{\partial}{\partial r}\left(\frac{\zeta}{r}\frac{\partial \psi}{\partial z}\right)\right\} - \frac{\partial}{\partial z}\left\{r^3\frac{\partial}{\partial z}\left(\frac{\eta \zeta}{r}\right)\right\} - \frac{\partial}{\partial r}\left\{r^3\frac{\partial}{\partial r}\left(\frac{\eta \zeta}{r}\right)\right\} - \rho r\frac{\partial(v^2)}{\partial z} = 0. \tag{4}$$

An equation is needed for the swirl velocity, v, and the $\hat{\theta}$ component of the momentum conservation equation can be used to obtain

$$\left\{\frac{\partial}{\partial z}\left(rv\frac{\partial \psi}{\partial r}\right) - \frac{\partial}{\partial r}\left(rv\frac{\partial \psi}{\partial z}\right)\right\} - \frac{\partial}{\partial z}\left\{\eta r^3\frac{\partial}{\partial z}\left(\frac{v}{r}\right)\right\} - \frac{\partial}{\partial r}\left\{\eta r^3\frac{\partial}{\partial r}\left(\frac{v}{r}\right)\right\} = 0. \tag{5}$$

The problem for the flow field thus consists of solving equations (3)-(5) subject to suitable boundary conditions. The method of Gosman et al (Ref. 1) is used to replace these differential equations by five-point difference formulae, with all three variables, v, ζ and ψ, stored at locations of the type denoted by dots in Fig. 1.

The boundary conditions chosen are

a   rv: rv = 0 at tank bottom and outer tank wall

$\qquad\qquad\qquad$ rv = $2\pi n r^2$ on shaft and disc

$$\frac{\partial v}{\partial z} = 0 \text{ at free surface (ie neglible shear)} \tag{6}$$

$\qquad\qquad\qquad \dfrac{\partial v}{\partial r} = 0$ on axis of symmetry b ψ: $\qquad\qquad\qquad\qquad \psi = 0$ on the boundary of the fluid $\hfill$ (7)

c ζ: $$\left.\frac{\zeta}{r}\right|_W = -\frac{1}{2}\left.\frac{\zeta}{r}\right|_{NW} - \frac{3\psi_{NW}}{\rho r_W^2(\Delta z)^2}$$ $\hfill$ (8a)

at walls of constant z

$$\left.\frac{\zeta}{r}\right|_W = -\frac{r_{NW}}{r_W}\left\{\frac{4r_W - 3\Delta r}{5\Delta r - 8r_W}\right\}\left.\frac{\zeta}{r}\right|_{NW} + \frac{24\psi_{NW}}{\rho(\Delta r)^2 r_W(5\Delta r - 8r_W)}$$ $\hfill$ (8b)

at the outer wall

$$\left.\frac{\zeta}{r}\right|_W = -\frac{r_{NW}}{r_W}\left\{\frac{4r_W + 3\Delta r}{5\Delta r + 8r_W}\right\}\left.\frac{\zeta}{r}\right|_{NW} - \frac{24\psi_{NW}}{\rho(\Delta r)^2 r_W(5\Delta r + 8r_W)}$$ $\hfill$ (8c)

at shaft and edge of disc

where the subscripts W, NW indicate wall and near-wall grid points respectively.

These conditions on ζ are obtained by assuming that near the boundaries tangential derivatives are negligible by comparison with normal gradients.

3.   CALCULATION OF CONCENTRATION FIELD

Once the flow field has been solved to obtain a steady velocity field $\underline{q}$, the transient evolution of the concentration c of a tracer species in the flow field can be determined from the equation

$$\frac{\partial c}{\partial t} + \nabla\cdot(c\underline{q} - D_{ab}\nabla c) = 0$$ $\hfill$ (9)

subject to the boundary condition

$$\frac{\partial c}{\partial n} = 0 \text{ on the boundary,}$$ $\hfill$ (10)

ie no mass transfer takes place through the walls. Thus a constraint on the calculation is that

$$\int_{\substack{\text{volume}\\\text{of field}}} c\,dv = M_o, \text{ constant, for all time.}$$ $\hfill$ (11)

The method of integration used is an explicit time-stepping scheme, with the concentration c stored at locations of the type denoted by crosses in Fig. 1.

4.   FLOW FIELD PREDICTIONS

The reference length is taken as D and the reference speed as nD, giving a Reynolds number of the form $Re = \rho nD^2/\eta$. The flow pattern, Figs 4, 3a and 2b, consists of a primary tangential flow with a superimposed toroidal secondary flow, and will be seen to agree well with observation, Fig. 2a. Computational runs were performed to show how the predicted flow field varies with Re in a fixed geometry, and with geometry for a fixed Re.

The results are compared via a pumping number, N_p, defined as the total mass recirculating in the (r, z) plane per unit time, non-dimensionalised by $\rho n D^3$. Fig. 5 presents the results obtained at eight values of Re in the range (3, 100) for three different values of D/T at constant h/H = 0.333. To investigate how the secondary flow varies with h/H, a series of runs were performed at fixed D/T = 0.333 for varying h/H, Fig. 6. The results show consistently that the highest secondary flows are induced by the value h/H $\simeq$ 0.5.

5. CONCENTRATION FIELD PREDICTIONS

The mixing of a 'band' of tracer fluid was investigated computationally, illustrating the mixing effects of the secondary flow. That is, with the steady velocity field present, a band of marked fluid is initialised at time t = 0, and its subsequent dispersion computed. Typical results can be seen in Fig. 7.

To quantify the mixing of the system, a mixing index $\phi(t)$ is defined by

$$\phi(t) = \frac{1}{N} \sum_i \sum_j (c_{ij}^n - c_\infty)^2 \tag{12}$$

where c_∞ is the uniform concentration when the contents are perfectly mixed. Fig. 8 shows non-dimensionalised $\phi(t)$, $\phi^* = \phi(t)/\phi(0)$ plotted against non-dimensional time, $t^* = t/t_1$, where t_1 is the time for one revolution of the disc, for various Sc at constant Re.

6. CONCLUSIONS

As an initial step, the flow induced by a rotating disc has been investigated numerically and experimentally. The computations predict that the highest recirculation rates are induced by a disc placed halfway up the tank. The mixing effects of the secondary flow have also been investigated numerically. It was found that the model could not distinguish between Schmidt numbers greater than 300. The mixing calculations await experimental investigation.

The next stage of investigation will be to compare the Newtonian results with non-Newtonian flow and mixing in the same geometry.

7. REFERENCES

1. Gosman, A. D. et al: "Heat and mass transfer in recirculating flows". Academic Press, 1969.

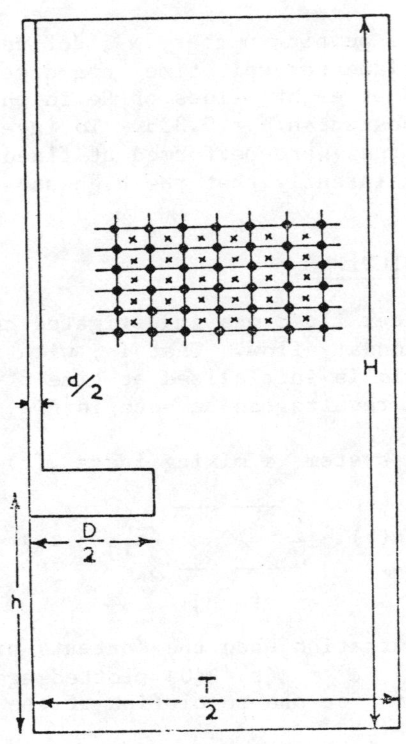

Fig. 1   Geometry and grid
      o   Flow variables
      x   Concentration

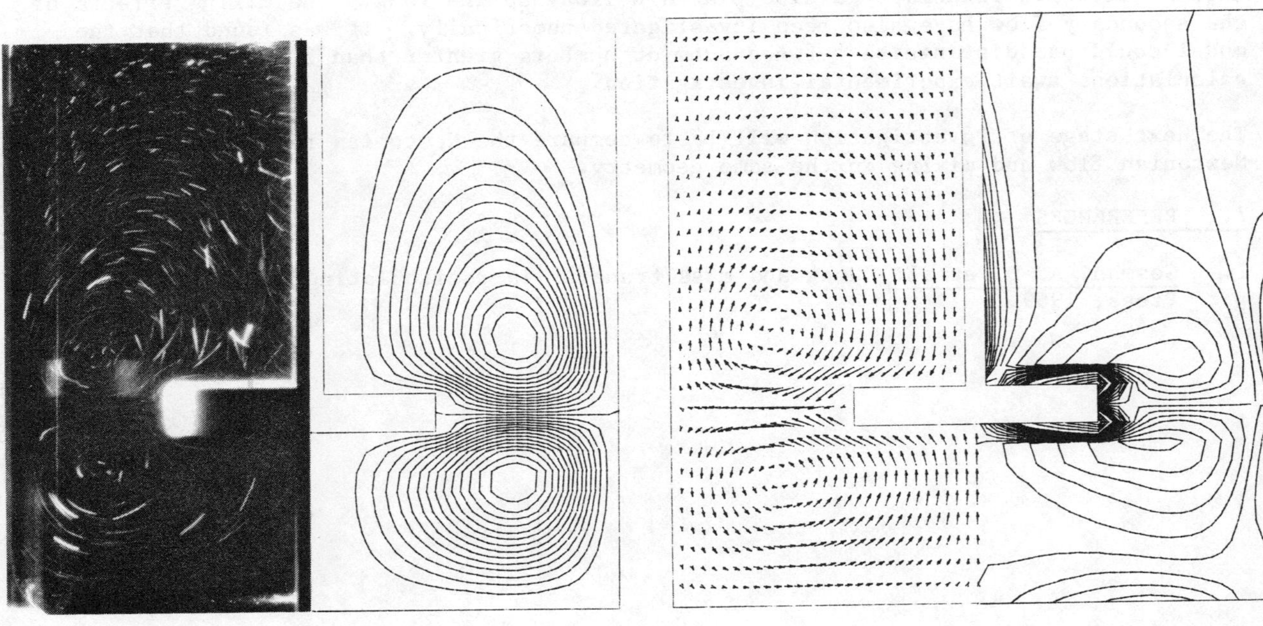

Fig. 2a   Fig. 2b   Fig. 3a   Fig. 3b

Photograph of flow   Contours of   Velocity vector   Vorticity
pattern   stream function   distribution in   contours
Re = 66,   T/H = 1   vertical plane flow
D/T = 0.4,   h/H = 0.333

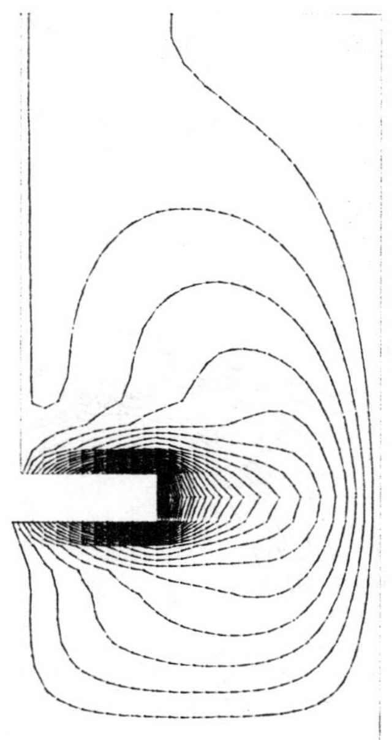

Fig. 4   Swirl-velocity contours

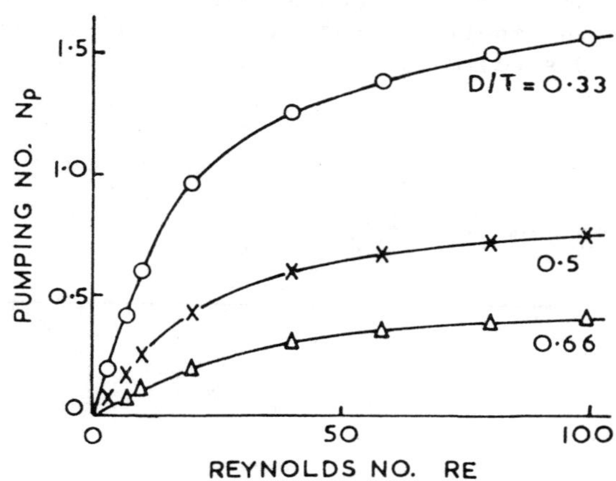

Fig. 5   Pumping No vs Reynolds No

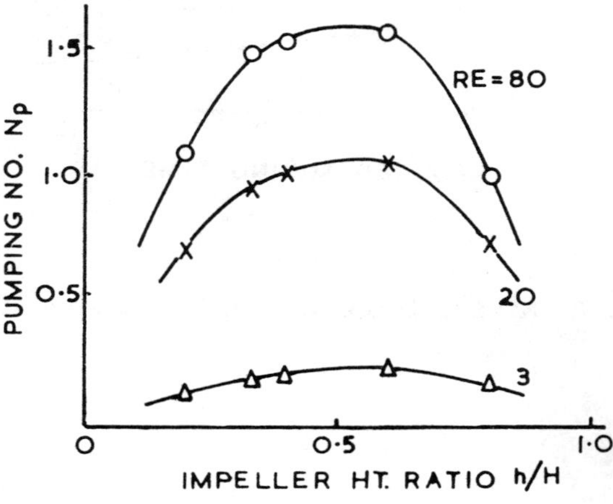

Fig. 6   Pumping No vs impeller height ratio

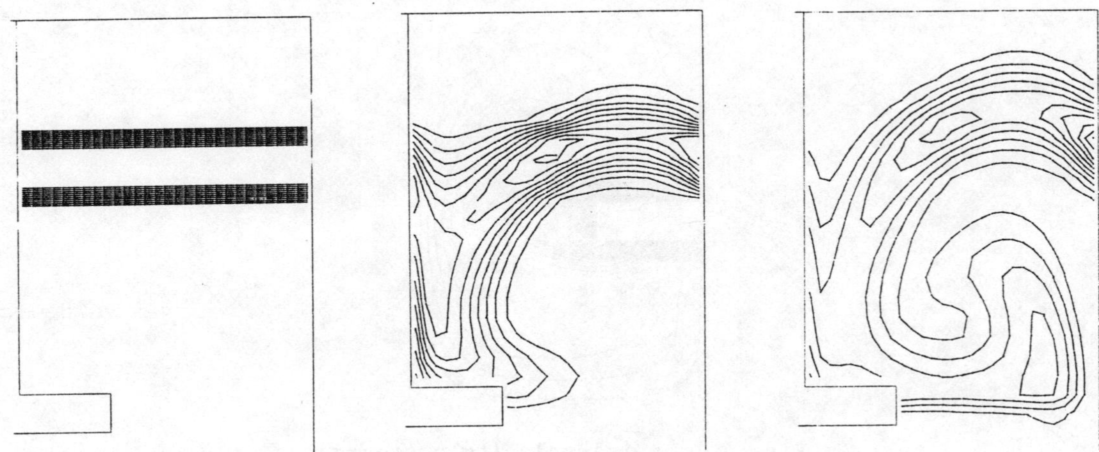

Fig. 7   Time sequence showing concentration distribution in region of
tank above impeller: (a) initially (b) after 5 impeller
revolutions (c) after 20 revolutions

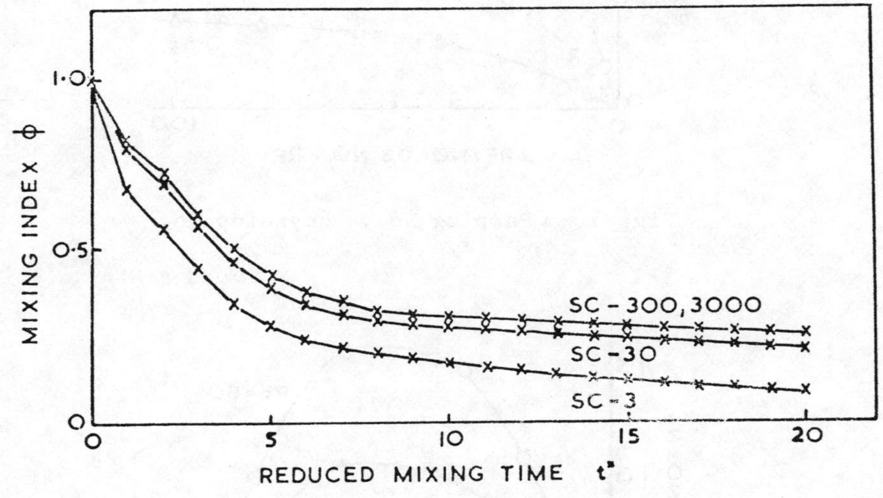

Fig. 8   Mixing index vs dimensionless mixing time

514

OPTIMIZATION OF GEOMETRICAL PARAMETERS OF THE TURBINE WITH THE
PITCHED BLADES IN AGITATED VESSEL FOR HEAT TRANSFER

F. Strek and J. Karcz

Technical University of Szczecin
71-065 Szczecin, Al. Piastów 42
Poland

Summary

In this paper the optimum geometrical parameters of the pitched blade agitator for heat transfer in the jacketed stirred tank have been
evaluated experimentally. The coefficient K has been applied as the optimization criterion and calculated on the basis of the results of the
investigations of both heat transfer and mixing power. The following
optimal geometrical parameters of the pitched blade agitator have been
proposed: $D/T = 0.5$; $w/D = 0.13$; $z = 8$ and $\alpha = 45°$. This geometry differs from the usually applied, so called standard agitator geometry:
$D/T = 0.33$; $w/D = 0.2$; $z = 6$ and $\alpha = 45°$.

Held at Wurzburg, 10-12 June, 1985.

Organised by DVCV· Deutsche Vereinigung für Chemie- und Verfahrenstechnik
(German Association of Chemical and Process Engineering).

Organisation: GVC·VDI-Gesellschaft Verfahrenstechnik und Chemieingenieurwesen.  **GVC**

NOMENCLATURE

A, B, E = exponents in equation /4/

A = surface of heat transfer

b = baffle width

C_1, C_2 = constants in equations /3/ and /4/

c_p = specific heat at constant pressure

D = overall agitator diameter

H = liquid level height

h = agitator distance from bottom of vessel

J = baffles number

K = optimization criterion

n = agitator speed

P = mixing power

$\dot{Q}$ = flow rate of heat

T = inside diameter of stirred tank

t = temperature

V = liquid volume

z = number of blades on agitator

α = blade pitch angle to the horizontal

α = heat transfer coefficient

η = dynamic viscosity of liquid

λ = thermal conductivity

ρ = liquid density

$$Ne = \frac{P}{n^3 D^5 \rho} \text{ – Newton number;} \qquad Nu = \frac{\alpha T}{\lambda} \text{ – Nusselt number;}$$

$$Pr = \frac{c_p \eta}{\lambda} \text{ – Prandtl number;} \qquad Re = \frac{nD^2 \rho}{\eta} \text{ – Reynolds number}$$

The purpose of the investigations presented in this paper was an experimental determination of the optimum geometrical parameters of the pitched blade agitator for heat transfer in the jacketed stirred tank. The coefficient K, defined in the paper (Ref. 1) on the basis of the connection between thermal effect $\dot{Q}/V$ and an agitation energy P/V

$$\frac{\dot{Q}}{V} = f\left(\frac{P}{V}\right) = K\varphi\left(\frac{P}{V}\right) , \tag{/1/}$$

was applied as the optimization criterion. In order to receive the detailed form of the equation /1/ the following relationships were taken into account:

$$\dot{Q} = \alpha A \Delta t \quad /2/; \qquad P = C_1 n^3 D^5 \rho \quad /3/; \qquad \alpha = C_2(\lambda/T)Re^A Pr^B(\eta/\eta_w)^E \quad /4/$$

Transforming the equations /2 - 4/ we obtained equation /5/

$$\frac{\dot{Q}}{V} = K \frac{4(\pi/4)^{A/3} Pr^B(\eta/\eta_w)^E \lambda \rho^{2A/3} \Delta t}{\eta^A T^{(6 - 4A)/3}} \left(\frac{P}{V}\right)^{A/3} \tag{/5/}$$

where the coefficient K was defined by equation /6/

$$K = \left(H/T\right)^{A/3}\left[C_2/\left(C_1 \frac{T}{D}\right)^{A/3}\right] \tag{/6/}$$

and was calculated on the basis of the results of the investigations of both heat transfer and mixing power. This coefficient depends only upon the geometrical parameters of the stirred tank.

The influence of the agitator diameter D, width of the agitator blade w, number of blades z and angle of inclination of the blade α /Fig. 1/ upon both heat transfer and mixing power has been examined in the stirred tank of inner diameter T = 0.45m. A total 35 different agitator geometries has been tested. The measurements were performed under the turbulent regime of the flow of a homogeneous Newtonian fluid, applying the planned experiment.

The mixing power was measured using a strain gauge technique. The heat transfer was investigated under conditions of steady state. The scheme of the experimental apparatus is shown in Fig. 2. The fundamental parts of this equipment were: tank /1/ with baffles /2/ and agitator /3/, driving system /4/, system for measurements mixing power /5/ and liquid cooler /6/.

The results of the experiments have been approximated analytically.

On the basis of this analytical function the optimum agitator geometry
for heat transfer has been determined. The following optimal geometrical
parameters of the pitched blade agitator have been proposed: D/T = 0.5;
w/D = 0.13; z = 8; α = 45°. This geometry differs from the usually applied, so called standard agitator geometry: D/T = 0.33; w/D = 0.2; z =6
and α = 45°. In the stirred tank equipped with the optimal agitator the
required flow rate of heat can be obtained at minimum mixing power.

Reference
1. Stręk F. and Karcz J.:"Optimization of geometrical parameters of baffles in agitated vessel for heat transfer". 8th International Congress
of Chemical Engineering, Chemical Equipment Design and Automation,
CHISA'84 /Prague, Czechoslowakia: Sept. 3-7, 1984/, Prague, ČSAV and
ČSVTS, 1984, Paper G8.6, 10 pp.

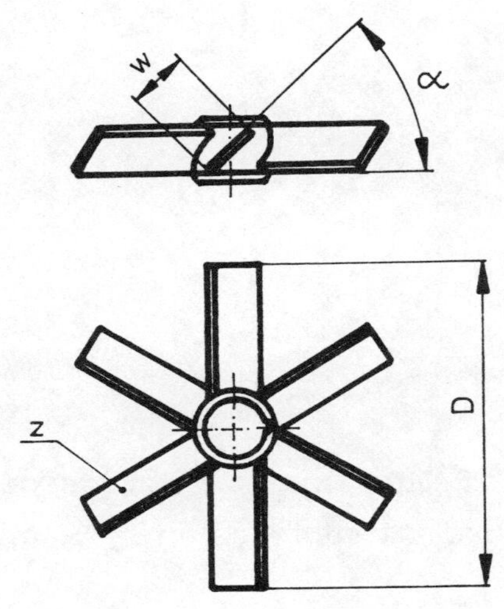

Fig. 1. The pitched blade
agitator

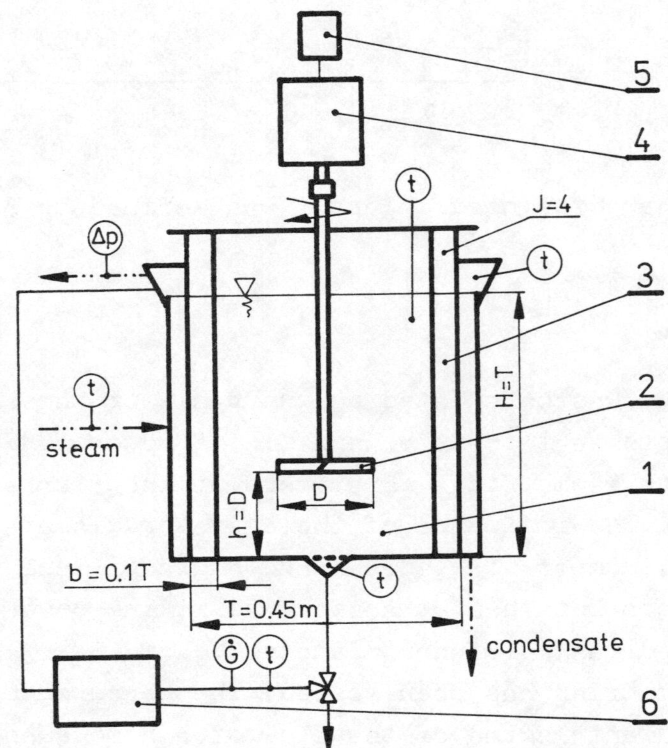

Fig. 2. Scheme of experimental equipment
1-stirred tank, 2-agitator with shaft,
3-baffle, 4-driving system, 5-system for
measurements of mixing power, 6-liquid
cooler

EMULSIFICATION OF TWO IMMISCIBLE LIQUIDS

UNDER THE ACTION OF A GAS JET

E. ARATO, M. GAGGERO, M. GIAMA$^{(*)}$, G. LODI, P. COSTA, B. CANEPA

Istituto di Scienze e Tecnologie dell'Ingegneria Chimica
Facoltà di Ingegneria dell'Università di Genova
Via dell'Opera Pia 15 - 16145 Genova - ITALY

(*) Università di Mogadiscio

Summary

In contacting two immiscible liquids by means of an impinging gaseous jet, the dynamic agitation due to the jet can be effective for the formation of a liquid-liquid emulsion only if a strong interaction between the jet and the lower liquid is achieved.

In this paper some data are presented on such three-phase systems in terms of the depth of the cavity induced by the jet, and especially in terms of the bulge height which occurs at the liquid-liquid interface at different working conditions (lance height, jet momentum, thickness of the upper liquid layer, viscosity ratio of the two liquids, etc...).

The critical conditions for the onset of emulsification are studied: they depend firstly on the cavity depth in comparison with the thickness of the upper liquid, but are also affected by the bulge height: the deformation of the liquid-liquid interface decreases the actual thickness of the upper layer, so that the interaction between the gas and the lower liquid becomes easier and the emulsification occurs under softer conditions. For this reason only sufficient conditions for emulsification can be obtained from two-phase data (gas-one liquid systems).

On the other hand a semi-empirical correlation of bulge height data leads to more realistic estimations of the working conditions at which emulsification occurs.

Held at Wurzburg, 10-12 June, 1985.

Organised by DVCV· Deutsche Vereinigung für Chemie- und Verfahrenstechnik (German Association of Chemical and Process Engineering).

Organisation: GVC·VDI-Gesellschaft Verfahrenstechnik und Chemieingenieurwesen. 

©BHRA, The Fluid Engineering Centre, Cranfield, Bedford MK43 0AJ, England.

NOMENCLATURE

F = jet momentum

H_b = bulge height

H_c = cavity depth relative to undisturbed liquid surface

H_{cc} = cavity depth at the onset of spraying

H_o = nozzle-liquid distance (lance height)

K = a constant, see equation 9

Mo $= (g \, \mu_1^4) \, / \, (\rho_1 \, \sigma^3)$, Morton number

S_e = actual thickness of the upper liquid layer

S_o = thickness of the undisturbed upper liquid layer

d = nozzle diameter

g = acceleration due to gravity

k_u = momentum transfer coefficient

u = axial gas velocity

z = distance downstream from the nozzle exit

μ = dynamic viscosity

ν = kinematic viscosity

ρ = density

σ = surface tension

Indexes

g = gas

l = liquid

1 = upper liquid

2 = lower liquid

1. INTRODUCTION

In the reactions involving two immiscible liquids and a gas the conctact among the phases can be obtained by means of a gaseous jet impinging on the two stratified liquids with the formation of a two or three-phase emulsion.

This method is particularly useful in metallurgical processes (e. g. the LD process) where any mechanical agitation is out of question.

However, the dynamic agitation due to the jet is effective only if a strong interaction between the jet and the lower liquid is achieved. In other words, the depth of the cavity induced by the jet must be greater than the actual thickness of the upper liquid layer. Moreover, in emulsionating three-phase system, a significant role is played by the spraying.

So the behaviour of a three-phase system can be analyzed just as a two-phase system in terms of the depth of the cavity induced by the jet and in terms of the conditions for the onset of spraying.

In addition, for three-phase systems, we have to investigate the interactions at the liquid-liquid interface, which in particular, can strongly decrease the actual thickness of the upper liquid layer.

2. TWO-PHASE SYSTEMS

As an introduction to the discussion on three-phase system, our work on two-phase system is here briefly summarized. The related experimental results have been reported elsewhere (Ref. 1) in some more details.

Much work (Ref. 2-6) has been done on the interactions between a gaseous jet and a single liquid phase, especially for what concerns the cavity depth and shape, and the onset of spraying.

The results of these authors have been confirmed and, for some topics, completed and corrected by our investigations.

By using a blow-down system that provides a gas jet impinging on a liquid surface we have studied, in particular, the effects of the following variables on the cavity depth and on the onset of spraying:
- the nature of the two phases (water, ethylene glycol, siliconic oils; air, carbon dioxide, helium);
- the geometrical parameters in the ranges: $d = .7$, 1, $1.5 \ 10^{-3}$ m (nozzle diameter); $H_o = 1-20 \ 10^{-2}$ m (lance height);
- the jet momentum (that is the flow rate for a given nozzle) in the range: $F = 3.5 \ 10^{-3} - 3.3$ N.

The experimental data have been obtained by means of photografic methods.

The cavity depth has been correlated by the know equation (Ref. 3, 4):

$$(H_c/H_o) (1+H_c/H_o)^2 = (2 k_u^2 F)/(\pi \rho_1 g H_o^3) \tag{1}$$

In our work the validity of this correlation has been verified also for systems where the density of the jet and the density of the surrounding atmosphere are different ($\rho_g(0) = \rho_g(z)$). This means that the central velocity of the jet can be espressed as:

$$u(z) = k_u u(0) d/z (\rho_g(0)/\rho_g(z))^{\frac{1}{2}} \tag{2}$$

So, in the range of our measurements, the pressure distribution at the impingiment seems to be unaffected by the density gradients inside the gas. This experimental finding disagrees with the theoretical results presented by Ito (Ref. 7, 8) and based on an erroneous balance equation, which suggest a remarkable increase of the cavity depth as a consequence of the temperature gradients existing in a steelmaking converter.

The cavity depth at the onset of spraying can be correlated by:

$$H_{cc} = (\sigma/\rho_1 \, g)^{\frac{1}{2}} \, f(Mo) \tag{3}$$

where Mo is the Morton number. The observed dependence on Mo is slightly different from the one found by other Authors (Ref. 4). Two ranges can be individuated: at low Mo (10^{-10}–10^{-6}) we have f(Mo)=const, while at high Mo (10^{-6}–10^{6}) there exists an important effect of the liquid viscosity: the momentum which is necessary for the onset of spraying is quickly increasing with the viscosity.

3. THREE-PHASE SYSTEMS

The above correlations can also be useful in order to describe the behaviour of three-phase systems in the range of conditions where liquid-liquid emulsification does not occur. On the other hand the interaction between the two liquids which induces the emulsification depends not only on the cavity depth, but also on the deformation of the liquid-liquid interface (see Fig. 1). Such fenomena were firstly reported by Turkdogan (see Ref. 9).

In other words the effective thickness of the upper layer which must be bored by the impinging jet is :

$$S_e = S_o - H_b \leq S_o \tag{4}$$

where the bulge height H_b is strongly affected by the operating conditions.

3.1 Experimentals

Three-phase tests have been carried out by using an air jet, water as lower liquid and a siliconic oil as upper liquid.

Three different oils have been used, with equal densities (ρ_1 =960 kg m^{-3}) and surface tensions (σ=2.25 10 N m^{-1}), but quite different viscosities (ν_1=.5, 1, 3.5 St).

For each air-oil-water system tests have been performed by varying the jet momentum at different lance heights and at different thickness of the upper layer. The bulge at the liquid-liquid interface was measured by means of a photografic method: for each condition a high velocity sequence of pictures was made and an averaged value of the bulge height was derived.

It is also to observe that the cavity formed on the liquid surface by the impinging jet is unstable and has two main motions: a rotation around the jet axis and a radial oscillation: Molloy (Ref. 10) pointed out that these motions come from the instability and the oscillatory nature of the impinging jet. The instability of the jet and the connected instability of the cavity induce fluctuations on the liquid-liquid interface, rotations and deformations of the bulge as well. So a certain scattering of the experimental points is unavoidable: for example the measurements can be altereted by very small deviations from the radial simmetry.

The ranges of experimental conditions on three-phase systems were:

- tank radius: 1.5 10^{-1} m;
- nozzle diameter: 1 10^{-3} m;
- lower liquid thickness (water): .15 m
- upper liquid thickness (oil): 2 10^{-2} –6 10^{-2} m
- lance height: 4 10^{-2} –2 10^{-1} m
- gas flow rate (air): 7.6 10^{-5} –9.7 10^{-4} kg s^{-1}
- jet momentum: 7 10^{-3} –5 10^{-1} N

3.2 Data correlation and discussion

The bulge height data have been correlated by means of the following empirical equation based on dimensional analysis:

$$H_b/(H_o+H_c) = \alpha \; ((H_o+H_c)/S_o)^\beta \; (F/((\rho_{12}-\rho_{11}) \; g \; (H_o+H_c)^3))^\gamma \; (\nu_{11}/\nu_{12})^\delta \qquad (5)$$

The best fit values of the parameters

$$\alpha = .023; \qquad \beta = 1.02 \cong 1; \qquad \gamma = .5; \qquad \delta = .29 \cong .3; \qquad (6)$$

present small standard deviations and a confidence intervals of about 10%. Fig. 2 shows the experimental points for three different water-oil systems together with the correlating straight line 5 in a bilogaritmic plot.

Our theoretical understanding of the dependence 5 is not yet completely satisfactory. As the bulge height is mainly connected to a ricirculating flow and to the pressure drop induced in the upper liquid by the impinging jet, in order to evaluate the influence of the various parameters, it should be necessary to describe the flow patterns of this phase. Such a goal could be attained through the rigorous solution of the Navier-Stokes equations in turbulent regime (Ref.11). However, in our opinion, the agreement between computed and measured values should always result only qualitative because of the uncertainties on the boundary conditions and on the cavity shape.

Therefore, in a previous work (Ref. 1, 12) a simplified description was used from which a direct proportionality between the bulge height and the jet momentum was deduced. With reference to our experimental conditions $(\rho_{11}/\rho_g=\text{const.})$ this semi-theoretical correlation is:

$$H_b/S_o = \text{const.} \; f \; F/((\rho_{12}-\rho_{11}) \; g \; S_o^3) \qquad (7)$$

where f is a shape factor and the dependence on viscosities is implicit.

Now, by taking the rounded values $\beta =1$, $\delta = .3$ equation 5 can be rearranged in a form which is in agreement with equation 6 only under the assumption that the shape factor f is proportional to $(H_o+H_c)/H_b$.

3.3 Critical emulsification condition

However, by substituting equation 1 in equation 5 and by putting equal to zero the actual thickness of the upper liquid layer $(S_e=0; \; H_b+H_c=S_o)$ we obtain a single relation between the reduced lance height (H_o/H_c) and the reduced thickness of the upper layer (S_o/H_c) which describes the critical conditions (onset) of emulsification:

$$(S_o/H_c)^2 \; (S_o/H_c-1)^2 = K \; (H_o/H_c+1)^3 \qquad (8)$$

where

$$K = (.023/k_u) \; (\pi \rho_{11}/(2 \; (\rho_{12}-\rho_{11})))^{\frac{1}{2}} \; (\nu_{11}/\nu_{12})^{.3} \qquad (9)$$

is a constant for each liquid-liquid system.

The comparison (see Fig. 3) between the sufficient condition for emulsification

$$(S_o/H_c) < 1 \qquad (10)$$

and the necessary condition

$$(S_o/H_c) < (S_o/H_c)_{critical} \qquad (11)$$

where the right hand side term is given by equation 8, shows the effects of the liquid-liquid interaction (bulge height) in promoting the emulsification.

In Fig. 3 some experimental points with reference to $\nu_{11}=3.5$ St are also reported: the figure shows that the measuraments have been carried out up to the onset of emulsification.

A summary of our results on the onset of emulsification and some previsional

extrapolations are reported in Fig. 4 together with a plot of the reduced force
against the reduced lance height: for each reduced force the reduced lance height
and then the reduced layer can be quickly estimated on the curve corresponding to
the appropriate value of K. The effects of the nature of the liquid-liquid system
on the emulsification is so put in evidence, while, on the contrary, the sufficient
condition 10 is not affected by the characteristics of the lower liquid.

4. REFERENCES

1. De Marchi, G., Arato, E., Costa, P. and Canepa, B.: "The critical conditions
    of emulsification of two immiscible liquids under the action of a gas jet".
    In: Proc. 4th Italian-Yugoslavian-Austrian Chemical Engineering Conference (Grado,
    I: Sep. 24-26, 1984) Trieste, Alessi, P. and Kikic, I. Editors, 1984, pp. 214-221.

2. Chatterjee, A.: "On some aspects of supersonic jets of interest in LD steelmaking".
    Part 1. Iron and Steel, Dec. 1972, pp. 627-634.

3. Chatterjee, A.: "On some aspects of supersonic jets of interest in LD steelmaking".
    Part 2. Iron and Steel, Feb. 1973, pp. 38-40.

4. Chatterjee, A. and Bradshaw, A. V.: "Break-up of a liquid surface by an impinging
    gas jet". Journal of the Iron and Steel Institute, 210, Mar. 1972, pp. 179-187.

5. Banks, R. B. and Chandrasekhara, D. V.: "Experimental investigation of the
    penetration of a high-velocity gas jet through a liquid surface". Journal of Fluid
    Mechanics, 15, 1, 1963, pp.13-34.

6. Cheslak, F. R., Nicholls, J. A. and Sichel, M.: "Cavities formed on liquid
    surfaces by impinging gaseous jets". Journal of Fluid Mechanics, 36, 1, 1969,
    pp. 55-63.

7. Ito, S. and Muchi, I.: "Transport phenomena of supersonic jet in oxygen top
    blowing converter". British Iron and Steel Industry Translation Service,
    Dec. 1973, BISI 11562 (Translation from: Tetsu to Hagane, 55, Nov. 1969,
    pp. 1152-1163).

8. Ito, S. and Muchi, I.: "Effects of characteristics of supersonic jet in the
    operations of oxygen top blowing converter". British Iron and Steel Industry
    Translation Service, Mar. 1971, BISI 8664 (Translation from: Tetsu to Hagane, 55,
    Nov. 1969, pp.1164-1175).

9. Szekely, J.: "Fluid flow phenomena in metals processing". New York, Academic
    Press, 1979, p. 344.

10. Molloy, N. A.: "Impinging jet flow in a two-phase system: the basic flow pattern".
    Journal of the Iron and Steel Institute, 208, Oct. 1970, pp.943-950.

11. Szekely, J. and Asai, S.: "Turbulent fluid flow phenomena in metal processing
    operations: mathematical description of the fluid flow field in a bath caused by
    an impinging gas jet". Metallurgical Transactions, 5, Feb. 1974, pp. 463-467.

12. Costa, P., Canepa, B. and De Marchi, G.: "Flow analysis of gas-liquid reactors
    stirred by a central gas injection". Poster presented at: Engineering Foundation
    Conference on Chemical Reaction Engineering (Santa Barbara, U.S.A.: Oct. 23-28,
    1983).

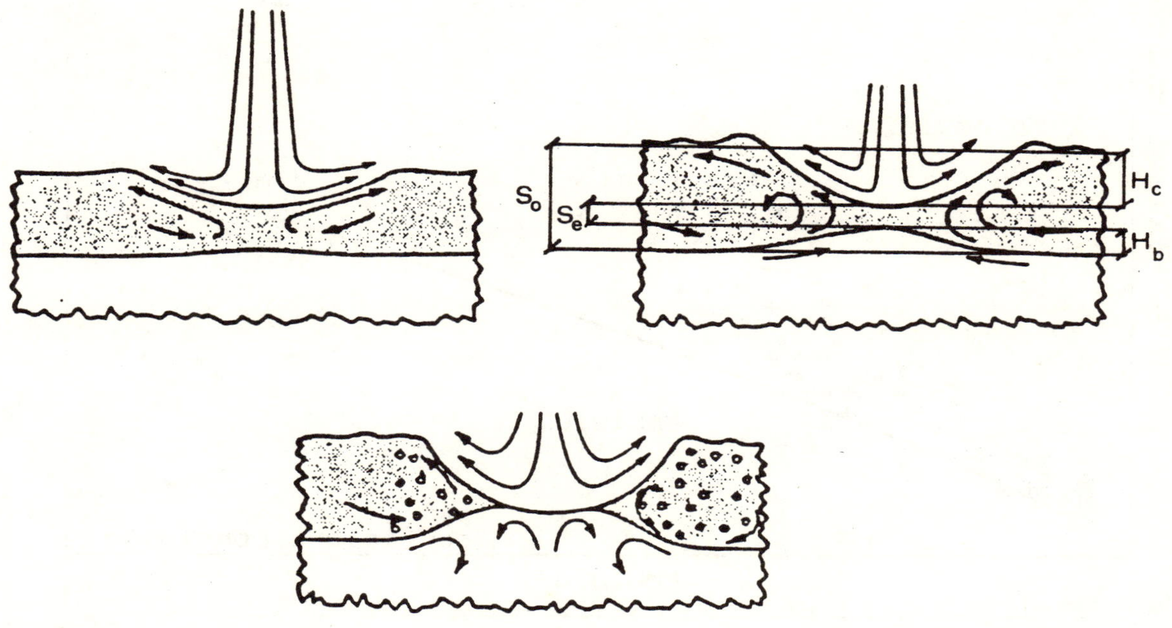

Fig. 1 —Effect of jet momentum increasing on bulge formation.

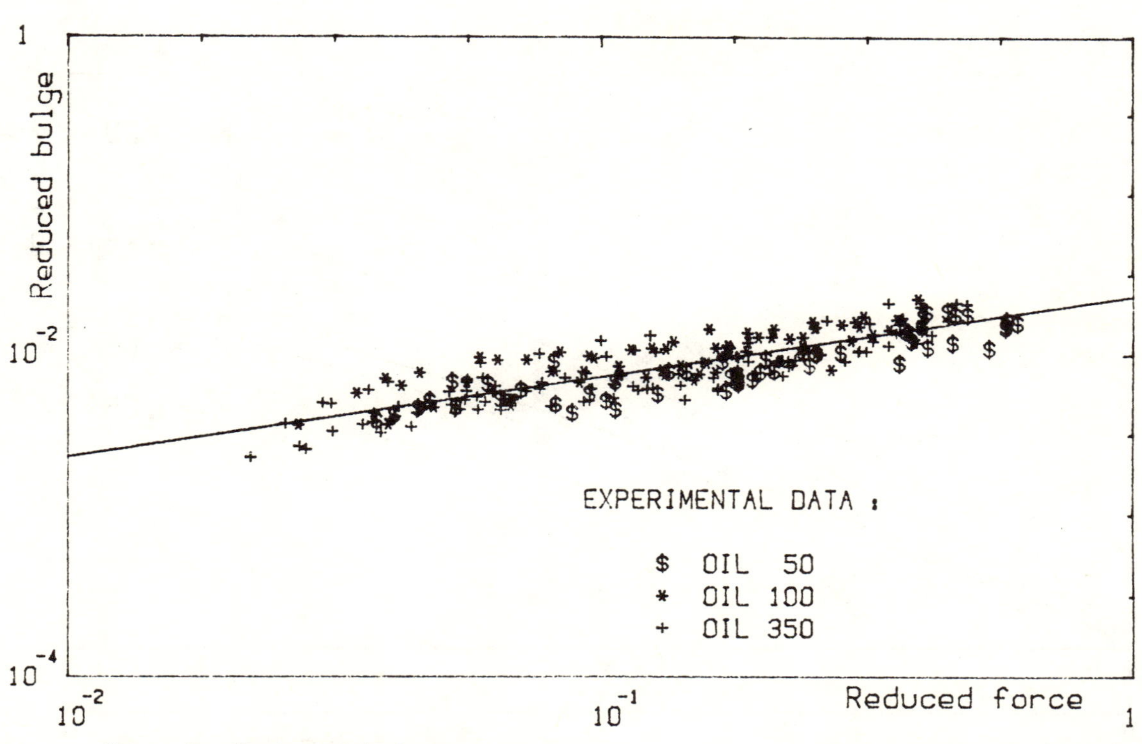

Fig. 2 —Reduced bulge $(H_b S_o/((H_o+H_c)^2(\nu_{11}/\nu_{12})^{.3})$ versus reduced force $(F/((\rho_{12}-\rho_{11}) g (H_o+H_c)^3))$ for oils with different viscosity (ν_{11} = .5, 1., 3.5 St).

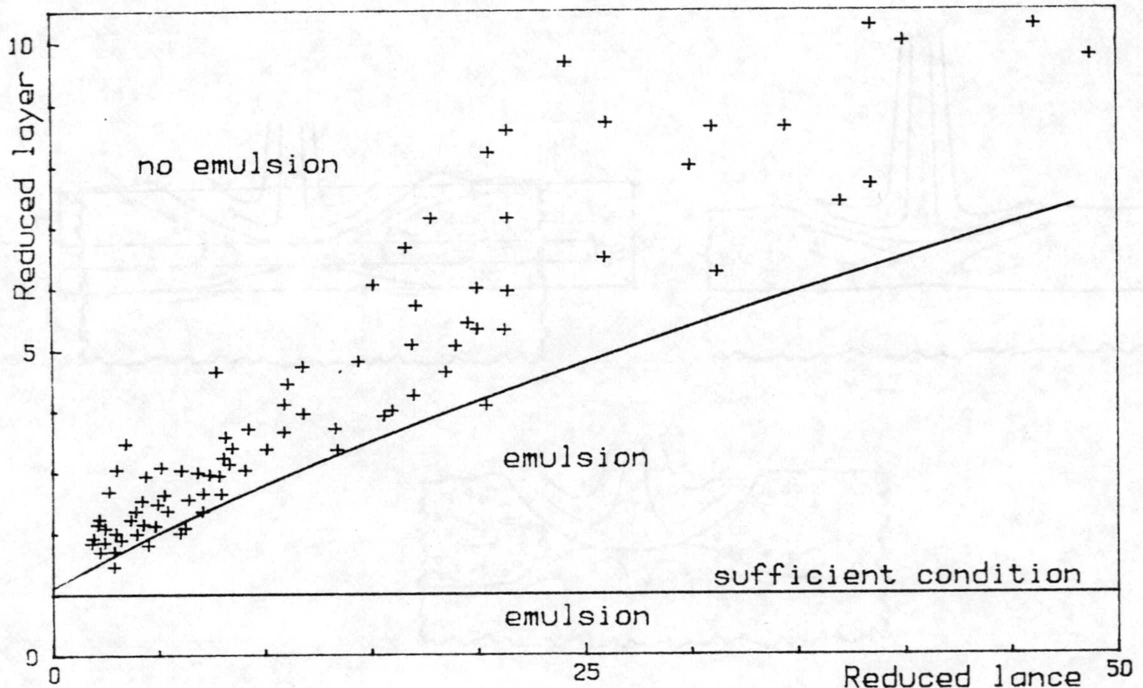

Fig. 3 —Reduced layer (S_o/H_c) at the onset of emulsification versus reduced lance (H_o/H_c) . Experimental points (+) refer to oil 350 are also reported.

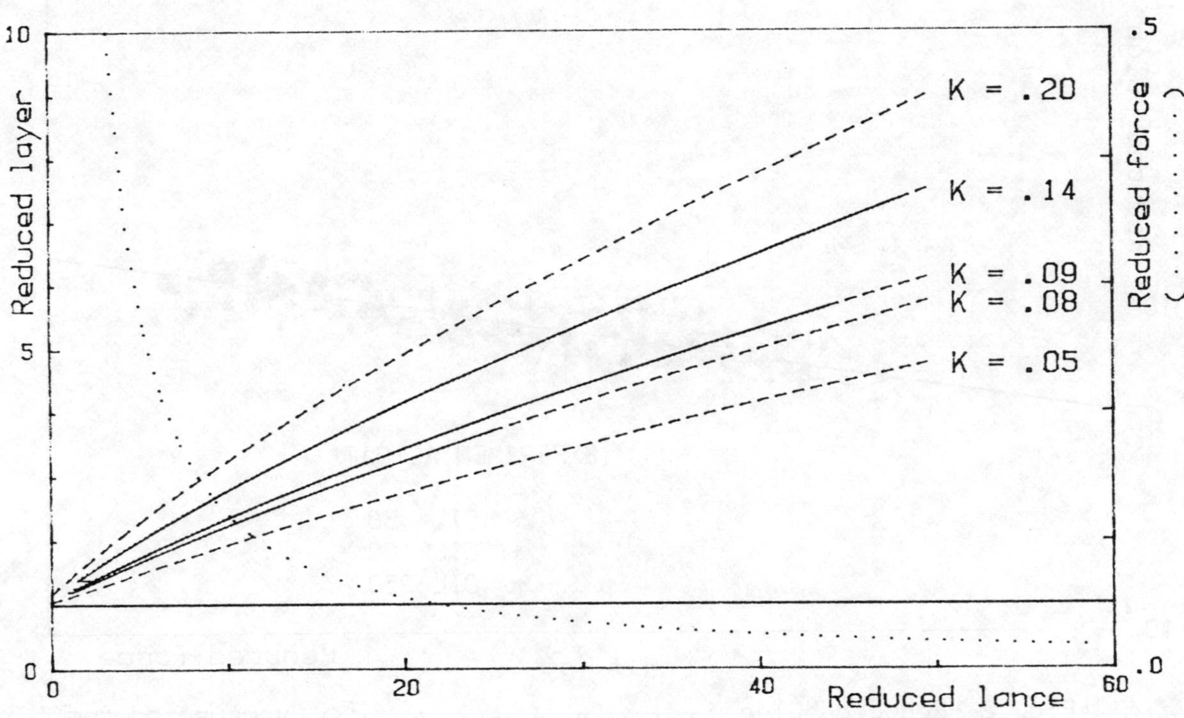

Fig. 4 —Reduced force $((F\ 2\ k_u^2)/(\rho_1 H_o^3 \pi g))$ versus reduced lance (H_o/H_c) (dotted curve) and reduced layer (S_o/H_c) at the onset of emulsification versus reduced lance for different systems. Continous curves are experimental, dashed are extrapolations.

INVESTIGATIONS OF MATERIAL FLOW DYNAMICS
BY MEANS OF RADIOTRACER METHODS

Andrzej G. Chmielewski

Institute of Nuclear Chemistry and Technology
Department of Nuclear Methods for Process
Engineering
Dorodna 16, 03-195 Warszawa
Poland

Summary

The radiotracer technique is sometimes an exclusive tool for
the investigations of material flow dynamics in the industrial in-
-stallations and environment. The sophisticated equipment and
suitable techniques of labelling enable skilled researchers to solve
almost every problem in field.
To illustrate the methodology of radiotracer measurements,
spme recently carried out investigations have been discused in the
paper.
The flow dynamics of solid, liquid and gaseous phases was
examined in the instalation for coal liquefaction. Introduction of
kinetic data has given the possibility to relate degree of conversion with stream recycle ratio. Similar experiments were performed on
digester for the continuous cooking of pinewood chips, where the
solid and liquid flow were investigated.
Mixing parameters for metallic and slug phases in the copper production flash process were determined. As the results of radiotracer
investigations of the glass melting tank furnace, the main streams,
surface current, bulk load current and backmixing draining current
have been found.
The possibilites of application of radiotracer for investigations of wastewater plants and effluents transport have been
discussed later on in this paper.
Finally, some hydrological applications connected with the
study of the bedload sediment in rivers and marine breaker zone
have been presented.

Held at Wurzburg, 10-12 June, 1985.

Organised by DVCV· Deutsche Vereinigung für Chemie- und Verfahrenstechnik
(German Association of Chemical and Process Engineering).

Organisation: GVC·VDI-Gesellschaft Verfahrenstechnik und Chemieingenieurwesen. 

©BHRA, The Fluid Engineering Centre, Cranfield, Bedford MK43 0AJ, England.

1. INTRODUCTION

The radiotracer methods are sometimes an exclusive tool for
research. The multichanel, mobile microcomputer monitoring system
assures simultaneous data collecting from several detectors.
Very fast /resolution 1 ms/ and slow processes can be observed.
The tracers for all three phases: gaseous, solid and liquid have been
worked out. The particular results of the investigations will not be
discussed in this paper. The author would like to show a variety of
problems in which the radiotracer methods can be use.

2. INDUSTRIAL INVESTIGATIONS

2. 1. Loop reactor for coal liquefaction.

The high-pressure catalytic hydrogenization process of coal
extract have been investigated. Gas phase /hydrogen/ was labelled
with Kr-85, liquid phase was labelled with Br-82 /in chemical form
of bromobenzene, bromotoluene or bromonaphtalene/ and solid phase
/coal or ash/ was traced with La-140 /ash impregnated with lanthanum
nitrate/. In similar way catalyst was labelled during the measurements of the bed expansion. The points in which scintillation detectors were placed during liquid phase flow experiments, are marked in
Fig. 1. From the point of view of liquid dynamics, the reactor can be
divided in three regions described by transfer functions:

$$G_1/s/ = \frac{2 \exp\left[Pe_1/2\ (1 - B_1)\right]}{1 - B_1}, \quad Pe_1 = \frac{u_1\ H}{E_1}, \quad B_1 = /1 + \frac{4sT_1}{Pe_1}/^{1/2}$$

Dispersion coefficient $E_1 = 0,289u_L + 0,575 \cdot 10^{-2}$

$$G_2/s/ = \frac{\exp\left[Pe_2/2\ (1 - B_2)\right]}{B_2}, \quad Pe_2 = \frac{u_2/L - H/}{E_2}, \quad B_2 = /1 + \frac{4sT_2}{Pe_2}/^{1/2}$$

$$E_2 = 0,137u_L + 0,426 \cdot 10^{-2}$$

$$G_3/s/ = \frac{\exp/-sT_3/}{/sT_4 + 1/} \qquad u_L - \text{liquid superficial velocity}$$

u_1, u_2 – real liquid velocities.
By introducing the kinetic data it was possible to relate the degree
of conversion with recycle number – Fig. 2. /Ref. 1, 2/

2. 2. Digester for the continuous cooking of pinewood chips.

The wooden chips were impregnated with lantanum compounds and
irradiated in nuclear reactor/tracer La-140/. Na-24 in the form of
Na_2CO_3 was used as a tracer of liquid. Points in which detectors
were placed are marked in Fig. 3. Some response curves for chips and
liquid are presented in Fig. 4. The dynamics data were related with
temperature and conversion factor along the reactor /Ref. 3, 4/

2. 3. Tank furnace for glass melting.

Dynamics of the single-machine tank furnace with cullet
recycle for sheet-glass production has been determined. Sc-46 in
form of scandium oxide was used as a tracer. The tracer was introduced into the tank at the bulk inlet and also at different depths of
the product take up part of the tank. The main streams of molten
glass have been found, including surface current, bulk load current
and backmixing draining current. The response curve is shown in
Fig. 5. On its base the contribution of the stagnation zone have
been determined. /Ref. 5/.

2. 4. Flash furnace for copper production.

Residence times of the particles in the furnace shaft and in the whole system of the flash furnace have been determined. Tracers were obtained by means of direct neutron activation of copper concentrate and La_2O_3 of appropriate granularity.

Mixing parameters for metallic and slag phases were also determined. As a tracer for metallic phase Au-198 was used and La-140 in the form of La_2O_3 for slag phase. Homogenization parameters were determined by means of sampling and radiometric testing of copper and slag samples collected during subsequent tappings./Ref. 6/

3. ENVIRONMENT PROTECTION

3. 1. Wastewater treatment plants.

Mixing conditions have a big impact on efficiency of particular installations of the wastewater treatment plants. Mainly inlet-outlet radiotracer measurements are performed. However in some cases detectors are placed inside the tank to observe liquid or solid streams. Both methods were applied to investigate an equalizer. Detectors placed under the liquid level at different depths were moving with air mixer. Br-82 in the form of KBr solution was used as a tracer. Beside the flow dynamics model which can be described by the transfer function:

$$G/s/ = \frac{1}{/1 + sT_1//1 + sT_2/}$$

the main streams in the tank were determined. On the base of transfer function the frequency response can be calculated. Gain, as it was stated elsewhere /Ref. 7/, determines the equalization efficiency. /Ref. 8/

3. 2. Effluent dispersion.

Br-82 in the form of KBr solution is used as a tracer for investigations of transport of effluents in rivers, lakes and sea. The impulse or continuous tracing can be applied. However, from the plume tracing the prediction of the dispersion of continuously discharged sewage can be obtained /Fig. 6/. From the continuous injection the profiles of longitudinal and transversial /Fig. 7/ concentration changes are measured. Further the appropriate dispersion coefficients can be calculated. /Ref. 9, 10/

4. HYDROLOGICAL APPLICATION

4. 1. Bedload sediment transport study.

The beginning parameters of the bedload sediment movement have been studied in rivers and marine breaker zone. For stone fractions the natural stones were bored and tantalum wire inserts fixed in /Ta-182 was a tracer/. Artificial sand and gravel fractions were made of a sodium-free glass to which iridium oxide was added /tracer Ir-192 obtained by direct neutron activation of the glass sample/. The observation of the counting curve by a detector placed in water over investigated fraction /Fig. 8/ allows to relate the beginning of fraction transport with hydrological conditions. The quantitative mass transport was calculated from the data obtained for traced sand plume by boat-mounted monitoring system as shown in Fig. 9. /Ref. 11,12/

5. REFERENCES

1. Iller E., Dobrowolski A. and Przybytniak W.:
"Investigations of hydrodynamics of reactor for coal liquefaction"

Badanie hydrodynamiki reaktora uwodorniania węgla, Internal Reports
of Inst. Nuclear Res., 104/XVI/82, 1982 /In Polish/.

2. Iller E., Dobrowolski A. and Klimkiewicz T.:

"Investigations of dynamis of reagents flow in the installations for
coal liquefaction" /Badanie dynamiki przepływu reagentów w instala-
cjach upłynniania węgla/. Internal Reports of Inst. Nuclear Chem.
and Technology, 116/XVI/83, 1983 /In Polish/.

3. Iller E. and Klimkiewicz T.:

"Radiotracer investigations of kinetics and dynamics of the pinewood
chips cooking process" /Radioznacznikowe badania kinetyki i dynamiki
procesu roztwarzania zrębków w warniku/. Jbid., 136/XVI/83, 1983,
/In Polish/.

4. Iller E., Klimkiewicz T., Młodzianowski W., Kalinowski J. and
Czapiewska L.:

"Radiotracer investigations of kinetics and dynamics of the pinewood
chips cooking process   Part II" /Radioznacznikowe badania kinetyki
i dynamiki procesu roztwarzania zrębków w warniku Część II/,
Ibid., 13/VI/84, 1984, /In Polish/.

5. Harasimowicz M., Palige J. and Masiak A.:

"Determination of some parameters of glass melting tank furnace".
/określenie niektórych parametrów pracy wanny szklarskiej/, Ibid.,
80/VI/84, 1984, /In Polish/.

6. Palige J. and Chamer R.:

"Investigations of mixing parameters of slag and copper in the flash
furnace" /Badanie parametrów mieszania żużla i miedzi w piecu za-
wiesinowym/ Ibid., 47/VI/84, /In Polish/.

7. Chmielewski A. G.:

"Method for efficiency estimation of wastewater equalizers" /Metoda
oceny działania zbiorników uśredniających stężenie zanieczyszczeń w
ściekach". Gosp. Wodna. 40/5/ 110-113, 1980, /In Polish/.

8. Chmielewski A. G., Zwoliński K. and Dobrowolski A.:

"Radiotracer investigations of the equalization basin" /Radioznaczni-
kowe badania zbiornika uśredniającego/, Ibid., 2/VI/85, 1985,
/In Polish/.

9. Szpilowski St.:

"Investigations of initial sewage dispersion of effluents in natural
water reservoirs" /Badanie wstępnej fazy rozcieńczania się zanieczyszczeń w naturalnych zbiornikach wodnych/, Ibid, 97/XVI/82, 1982,
/In Polish/.

10. Owczarczyk A., Wierzchnicki R., Szpilowski St. and Basiński T.:

"Application of radiotracers for bedload sediment transport investigations in the marine breaker zone" /Zastosowanie radioznaczników do
badania ruchu osadów w morskiej strefie brzegowej/, Ibid.,
105/VI/84, 1984, /In Polish/.

11. Owczarczyk A. et al:

"Radiotracer investigations of bedload sediment transport in the
mountain river", /Radioizotopowe badania pomiaru transportu rumowiska
wleczonego w rzece rejonu podgórskiego/ Ibid., 62/VI/84,1984/In Polish

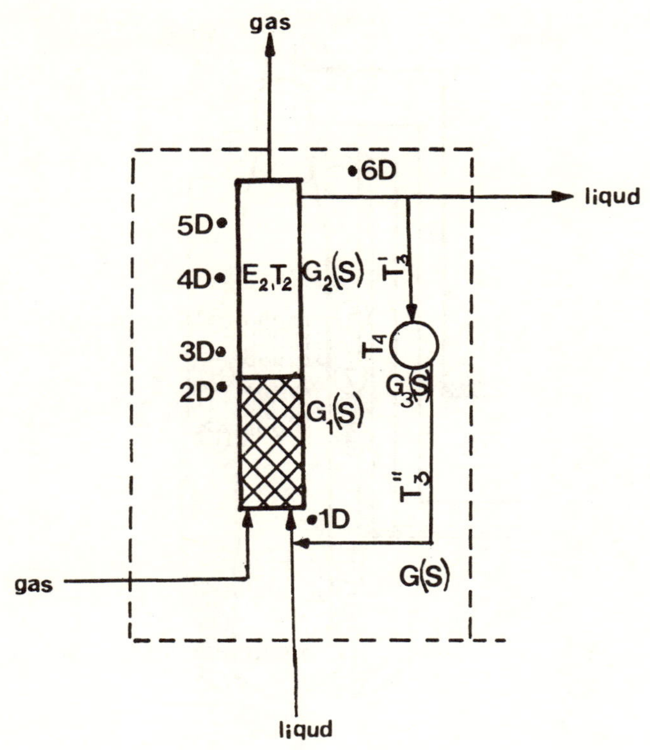

Fig. 1. Scheme of loop reactor for coal liquefaction.
Detectors marked as 1D, 2D, etc.

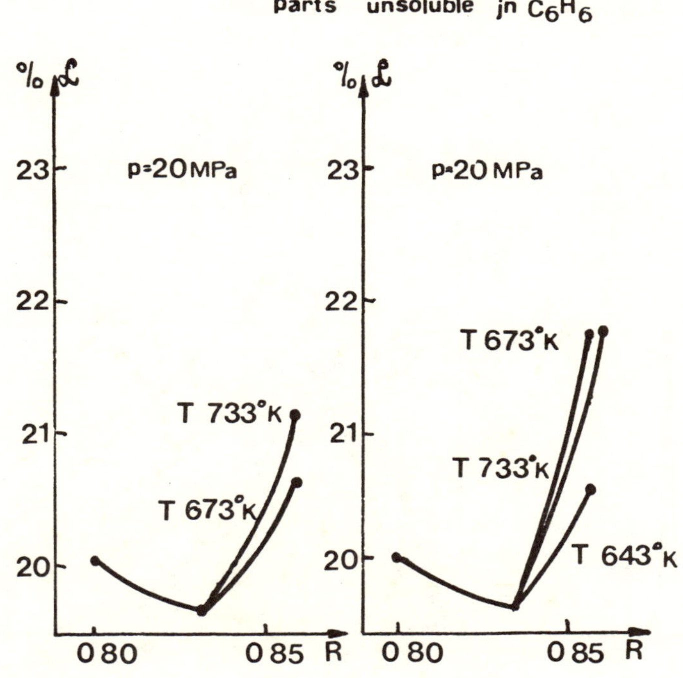

Fig. 2. Relationship between conversion coefficient and
recycle ratio.

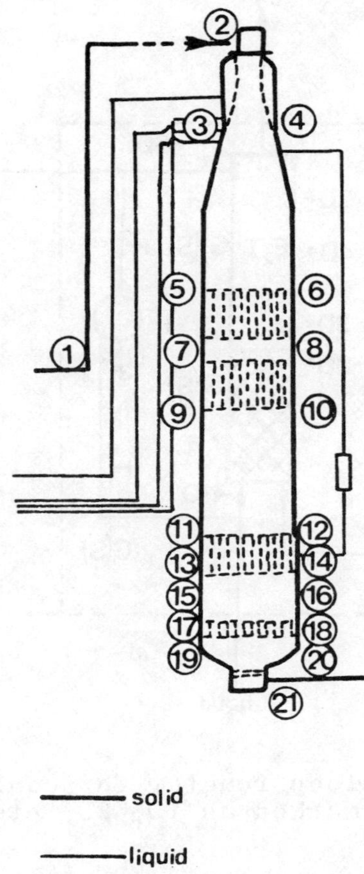

Fig. 3. Scheme of digester. Detectors marked as 1, 2, etc.

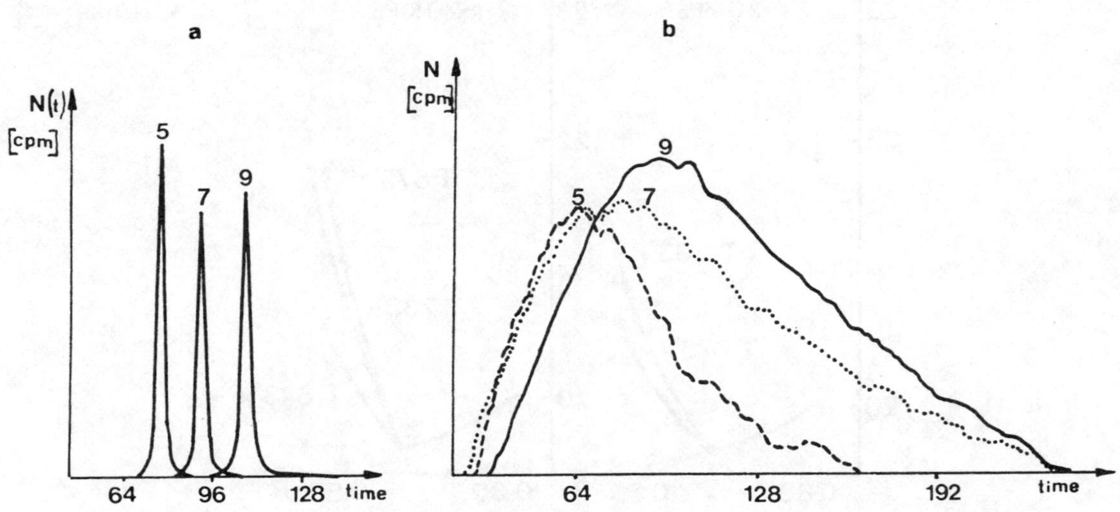

Fig. 4. Response curves. a — for labelled chips, b — for liquid.

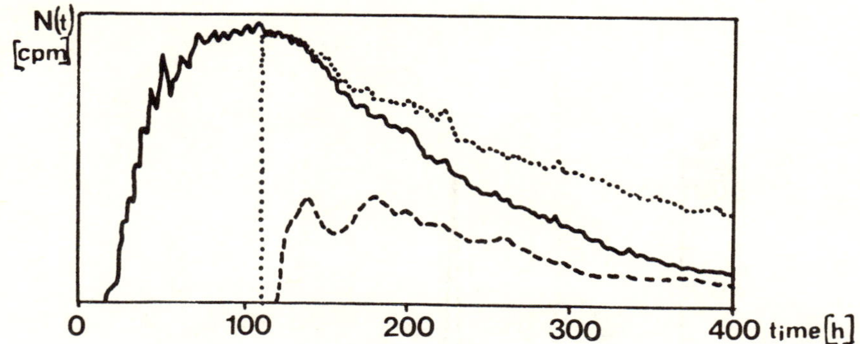

Fig. 5. Response curves for glass melting furnace.
a – total, b – for recycled active stream.

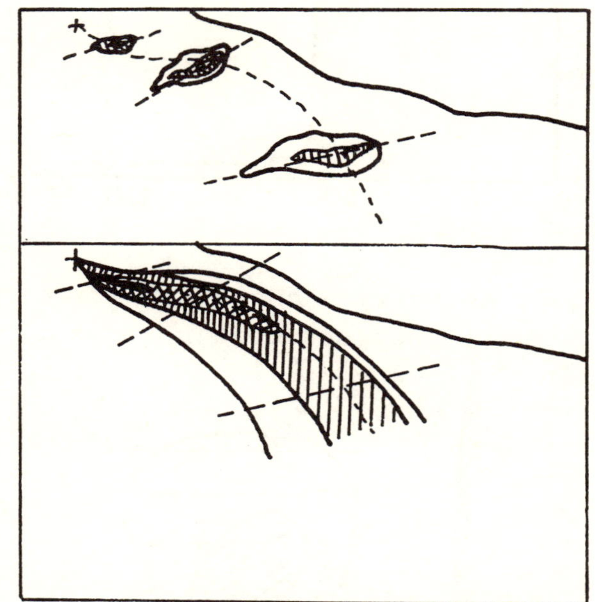

Fig. 6. Effluents transport scheme. a – plume tracing,
b – predicted continuous sewage discharge.

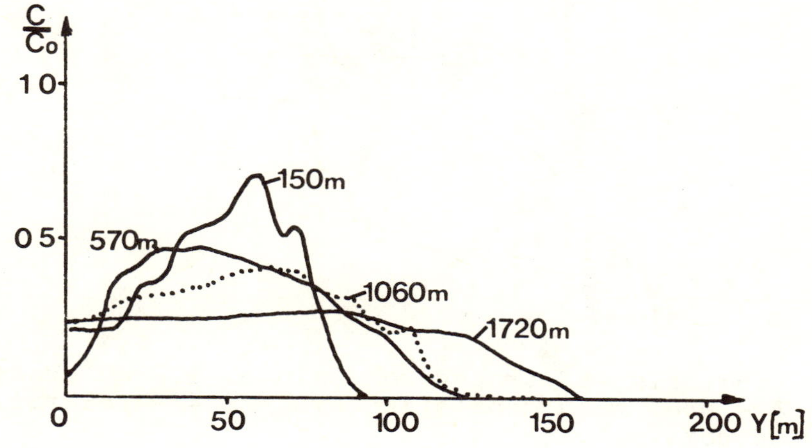

Fig. 7. Transversial tracer concentration changes in river.

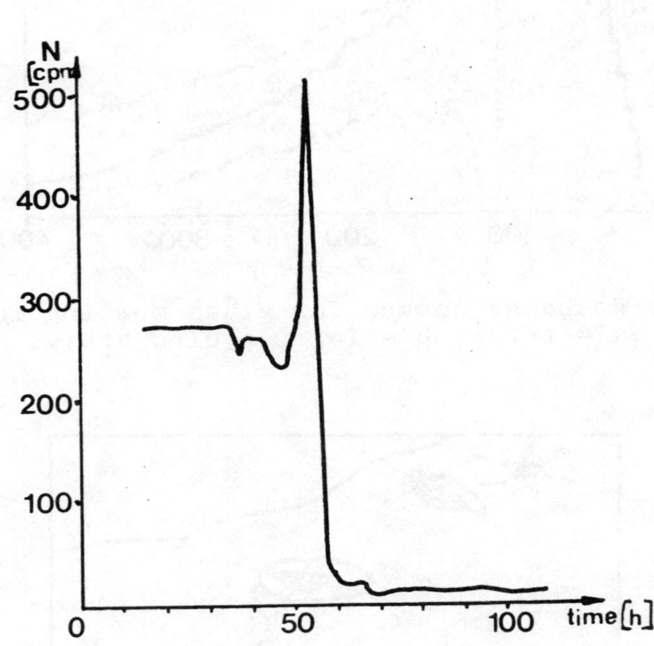

Fig. 8. Activity changes for start of
bedload sediment movement.

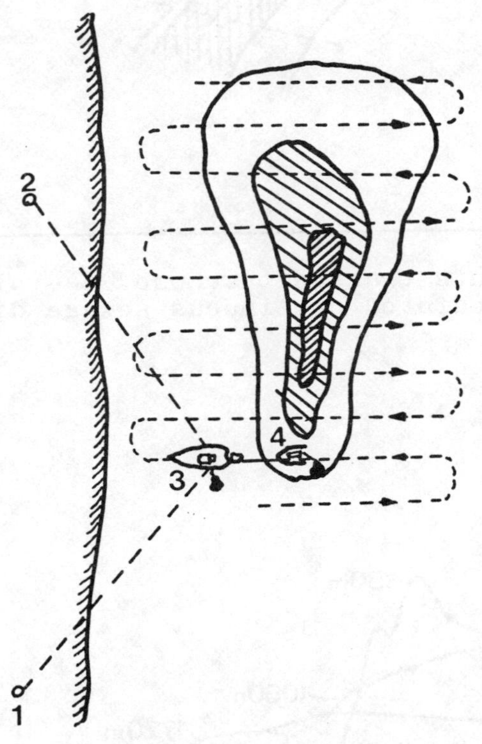

Fig. 9. Monitoring of traced sand plume. 1,2 - theodolits,
3 - boat with monitoring system, 4 - scintillation
detectors.

A STUDY OF THE PERFORMANCE OF A STATIC MIXER IN PULSATILE FLOW

J.R.E. Christy, N. Macleod

Chemical Engineering Department
University of Edinburgh
The King's Buildings
Edinburgh EH9 3JL
Scotland, U.K.

Summary

It is required to provide a means by which two reagents, supplied at equal rates by metering pumps, can be uniformly mixed on-line with a pulsatile stream of milk having fifty times their combined flow rates, at the entry to a rig for evaluating the thrombogenic tendency of prosthetic heart valves. Results are reported for the uniformity of composition obtained under various conditions of operation in trial experiments, with model fluids, using a particular type of commercial static mixer proposed for this application.

The uniformity of mixing depends on three factors:

1. The configuration of the injection system.

2. The pulse amplitude, shape and frequency of the main flow, both absolutely and relative to the pulsatile characteristics of the injected stream.

3. The performance of the static mixer, as measured in steady flow experiments.

In the experiments reported here, injections are made through a nozzle at the axis of the main fluid flow. The reagent injection pumps and the main flow pulsatile pump (whose flow characteristics can be made to resemble those of the human heart) are independent, i.e., not phase or frequency linked. We have observed the performance of a Chemineer Kenics KMS static mixer as a function of the frequency and pulse shape of the main flow in these circumstances, and have compared these pulsatile results with those obtained in steady flow. The method used has been to inject electrolyte into water, making measurements of the resulting conductivity immediately downstream of the static mixer.

A dimensional analysis enabled an assessment to be made of the importance of the various operating parameters. The best mixing was achieved in pulsatile flow with a dimensional pulse number greater than two and the frequencies and phase of the main fluid flow and injection flow being matched, the Reynolds Number and main fluid pulse shape having limited effects.

Held at Wurzburg, 10-12 June, 1985.

Organised by DVCV· Deutsche Vereinigung für Chemie- und Verfahrenstechnik
(German Association of Chemical and Process Engineering).

Organisation: GVC·VDI-Gesellschaft Verfahrenstechnik und Chemieingenieurwesen. 

©BHRA, The Fluid Engineering Centre, Cranfield, Bedford MK43 0AJ, England.

NOMENCLATURE

Cmax, Cmin, $\overline{C}$	=	Maximum, minimum and average salt concentration	kmol/m^3
D	=	Static mixer internal diameter	m
f	=	injection pulse frequency	Hz
F	=	main pulse frequency	Hz
fr	=	required resolution frequency	Hz
Nc	=	dilution number = $\dot{v}/\dot{V}$	–
Nf	=	frequency number = f/F	–
Nld	=	static mixer length/diameter ratio = $4Vsm/\pi D^3$	–
Np	=	pulse number = $Vsmf/\dot{V}$	–
Nr	=	resolution number = $Vcc\ fr/\dot{V}$	–
Ns	=	static mixer shape factor	–
Re	=	Reynolds number = $4\dot{V}\rho/\eta\pi D$	–
$\dot{v}$	=	volumetric flowrate of injection fluid	m^3/s
$\dot{V}$	=	volumetric flowrate of main fluid	m^3/s
Vcc	=	conductivity cell volume	m^3
Vsm	=	static mixer volume	m^3
w	=	injection pulse shape factor	–
W	=	main fluid pulse shape factor = systolic/diastolic duration ratio	–
Δ	=	degree of segregation = $(Cmax - Cmin)/2\overline{C}$	–
η	=	dynamic viscosity of water	Ns/m^2
ρ	=	density of water	kg/m^3
ϕ	=	phase angle between injection and main flow pulses	–
ψ	=	general function	–

1. INTRODUCTION

In studies of the causes of blood clotting (thrombosis) in artificial heart valves it has been shown, by Lewis (Ref. 1), that a freshly-prepared, homogeneous mixture of milk, rennet and calcium chloride, forms a clot at the same sites on currently used heart valves as those affected by thrombosis in clinical practice. Such a coagulable milk mixture can therefore be used in investigations of the causes of valve-related thrombosis and in evaluating the relative clotting propensities of artificial heart valves of different design.

To facilitate such investigations, an experimental apparatus is required capable of supplying to any heart valve under test a pulsating flow of milk containing 1% by volume each of rennet and saturated aqueous calcium chloride solutions. As the mixture is unstable, the two reagents must be introduced into the pulsating pumped flow of milk as soon as possible before entry to the valve under test. However, for valid tests of the influence of local flow characteristics at the valve on clotting tendency, the composition of the test mixture must be uniform both spatially (across the valve diameter) and temporally (at every phase of the pulsatile pumping cycle).

The experiments described here have been performed to investigate how well a commercial static mixer is able to satisfy these requirements for homogeneity in the kind of pulsating flow required. Electrolyte was injected into steady or pulsatile flows of pure water, the resulting conductivity being measured immediately downstream of the static mixer.

2. DIMENSIONAL ANALYSIS

The measured degree of segregation, Δ [defined as $(Cmax - Cmin)/2\bar{C}$], was assumed to depend on the following operating variables:-

a) The main fluid flowrate ($\dot{V}$), frequency (F), pulse shape (W = systolic : diastolic ratio), viscosity (η) and density (ρ).
b) The injection fluid flowrate ($\dot{v}$), frequency (f), pulse shape (w) and phase angle (ϕ).
c) The static mixer volume (Vsm), diameter (D) and element type and arrangement denoted by a combined shape factor (Ns).
d) The conductivity cell volume (Vcc).

Thus, $\Delta = \psi$ (F, $\dot{V}$, W, η, ρ, f, $\dot{v}$, w, ϕ, Vsm, D, Ns, Vcc).

The dimensionless segregation thus depends on thirteen variables in three dimensions including four dimensionless quantities, implying the need for six dimensionless combinations of the dimensional quantities. The groups chosen were:-

1) The Frequency Number (Nf = f/F) representing the number of injection pulses per main flow pulse.

2) The Dilution Number (Nc = $\dot{v}/\dot{V}$); the volume fraction of the injection flow relative to the main flow.

3) The Pulse Number (Np = Vsmf/$\dot{V}$); the number of injection pulses per static mixer volume.

4) The Reynolds Number (Re = $4\rho\dot{V}/\pi\eta D$).

5) The Mixer Length/Diameter Ratio (Nld = 4 Vsm/πD^3).

6) The Resolution Number (Nr = Vcc fr/$\dot{V}$, where fr is the desired sampling rate).

Therefore, $\Delta = \psi$ (Nf, Nc, Np, Re, Nld, Nr, Ns, W, w, ϕ)

The same static mixer, model fluid, conductivity cell and injection pulse shape were used throughout, therefore Vsm, D, ρ, η, Vcc, Ns, Nld and w remained constant. A constant dilution of Nc = 0.01 was chosen to match that of the milk system. Nr represents the ratio of the sampling frequency to the volume of changeover frequency within the conductivity cell, which must be less than one to give the required sampling resolution. A sampling frequency fr, of 10 Hz was chosen to resolve events occuring every tenth of a pulse for the normal main flow frequency of 1Hz, giving Nr < 1 for these experiments.

3. EXPERIMENT

The effects of Nf, Np, W and ϕ on Δ were investigated for pulsatile flow at four Reynolds Numbers (1040, 1640, 2230, 2530) and the effects of Np and Re on Δ were investigated for steady flow.

A schematic diagram of the apparatus is shown in Fig. 1. The Macleod positive displacement diaphragm pump [see Knight (Ref. 2)], whose frequency (20-150 min^{-1}), stroke volume (0-100 ml) and systolic-diastolic duration ratio (1:2.5 – 2:1) can be varied independently, pumped water at $25^{\circ}C$ through a static mixer having six helical elements each 22.5 mm long and a 15.7 mm internal diameter, equal to that of the main flow pipe. A 5M aqueous NaCl solution was delivered along the main flow axis, via a 1 mm bore nozzle, by a diaphragm metering pump with variable frequency (15-100 min^{-1}) and stroke volume (0-0.6 ml). Two carbon electrodes (4 mm long x 7 mm wide) lying diametrically opposite each other, flush with the pipe wall, formed the conductivity cell with an Ultra-Violet Oscillograph recording the temporal variation of the conductivity and the pump phases.

4. RESULTS

The influence of Np is dominant when Np < 1 in all instances. (Fig. 2).

4.1 Harmonic frequencies : Nf = 0.25, 0.5 and 1.0; W = 1.0. (Fig. 2a)

    a) ϕ is important (best mixing with ϕ = 0), though this importance decreases as Np increases.

    b) For Nf = 0.5 and 1.0 satisfactory mixing is obtained when Np > 2.0, irrespective of ϕ.

    c) For Nf = 0.25 the effect of ϕ is less though Δ remains high at large Np values.

4.2 Non-harmonic frequencies : Nf = 0.75, 1.33 (Compared with Nf = 1); W = 1.0 (Fig. 2b).

    a) The effect of ϕ is small, since phase matching is impossible for all pulses.

    b) Δ remains relatively high, even at large Np values.

4.3 Pulse shape variations : W = 0.4, 1.0 and 2.0; Nf = 1 (Fig. 2c)

    a) There is no observable influence of the main fluid pulse shape on Δ.

4.4 Steady and pulsatile flows compared.

    a) In steady flow, increasing Re enhances the mixing quality.

    b) In pulsatile flow, changes in Re from 1040 - 2530 had no apparent effect on mixing quality.

    c) Pulsatile flow with Nf = 1 and ϕ = 0 gave more uniform mixing than steady flow at similar values of Re.

5. CONCLUSIONS

1) The pulse number has a dominating effect, especially when less than one where a large spread of residence times would be necessary to provide good mixing.

2) The effect of the frequency number relies both on its general magnitude and on its precise value, harmonics at higher frequency numbers giving better mixing than non-harmonic frequencies. The range and accuracy of the frequency numbers in this study was limited by the pumps' characteristics. As expected however, the best mixing was achieved when the pumps pulsed synchronously.

3) The phase angle had a greater effect at the harmonic values of the frequency number with this effect diminishing as the pulse number increased. For non-harmonic frequencies this independence of the phase angle is explicible since, for example with the frequency number set at 0.75, although the pumps may begin a cycle in phase, the second and third injections will be 120° and 240° out of phase and the effect on the degree of segregation of a small change in one of these angles will be compensated for by the corresponding change in the other two.

4) A high systolic-diastolic duration ratio should produce a large axial spread of
   the injection pulse, however in this study no effect of the main fluid pulse shape
   on the degree of segregation was observed.

5) The Reynolds number, although important in steady flow, has little influence in
   pulsatile flow, with pulsatile flow being capable of producing a higher quality of
   mixing at low Reynolds numbers.

6. REFERENCES

1. Lewis, J.M.O.: "A Blood Analogue for Thrombogenicity Assessment". PhD Thesis,
   University of Edinburgh, 1981.

2. Knight, C.J., Julian, D.G., Macleod, N., Taylor, D.E.M., Wade, D.: "An Artificial
   Heart with Independently Variable Pulse Parameters". Trans. Amer. Soc. Artif.
   Int. Organs, XVII, 1971, pp. 433-436.

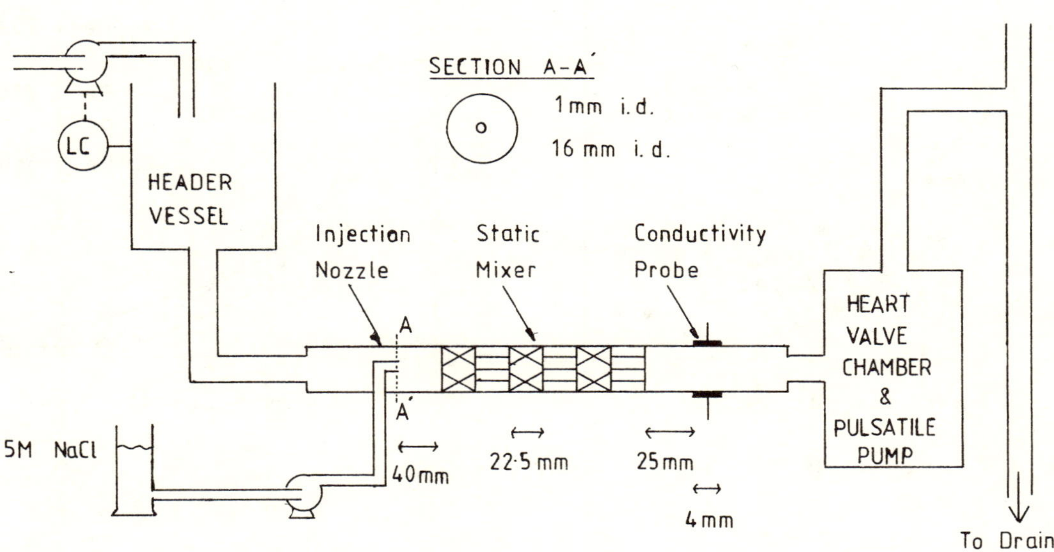

Figure 1   Schematic Diagram of Test Section

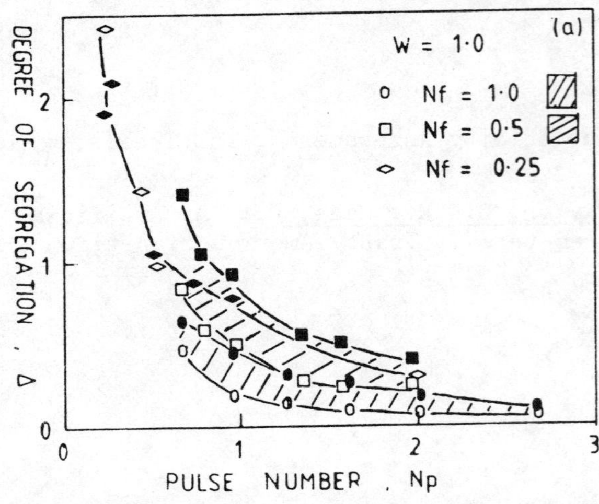

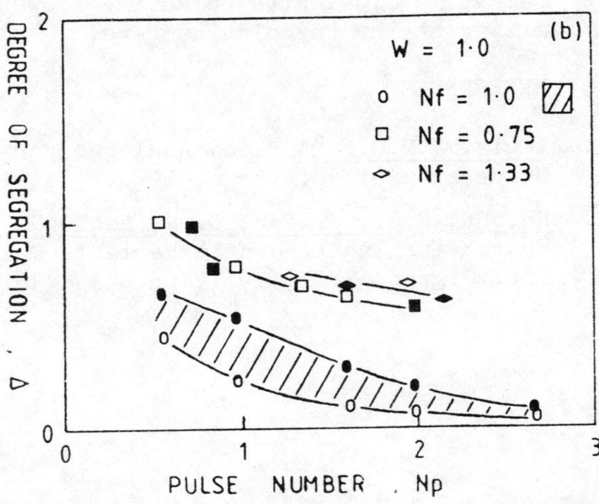

a) Effect of Harmonic Frequencies

b) Effect of Non-Harmonic Frequencies

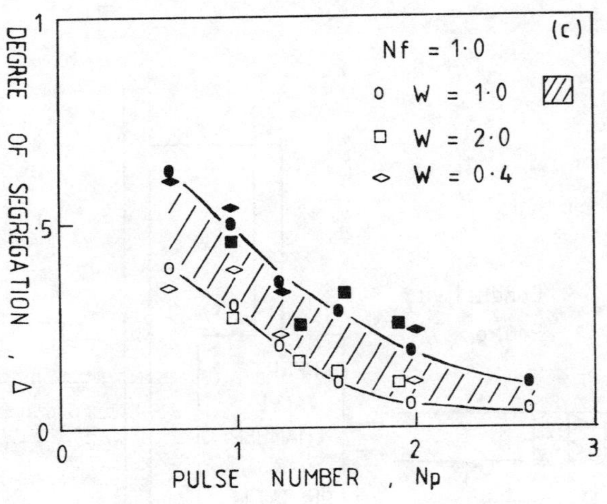

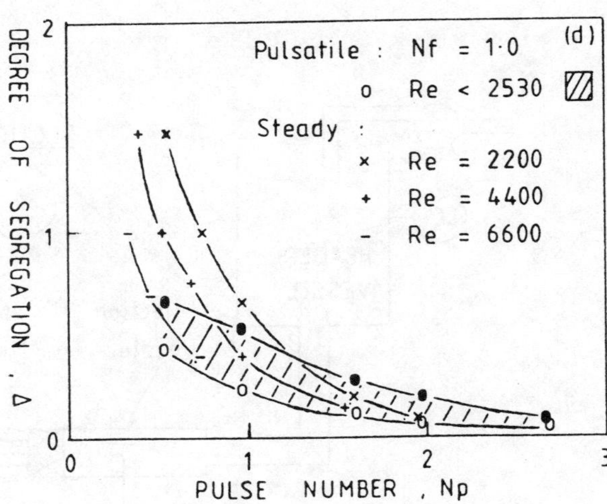

* The scale varies for Δ.
Open shapes : $\emptyset = 0°$
Closed shapes: $\emptyset = 180°$

c) Effect of Main Pulse Shape

d) Steady Versus Pulsatile Flow

<u>Figure 2</u>   Degree of Segregation Against Pulse Number for Various Operating
Conditions

FLOW-FIELDS IN A JET-MIXING VESSEL - MEASURING METHODS AND RESULTS

R.Fackler

TU-MÜNCHEN
Lehrstuhl für Mechanik Weihenstephan
Germany

Summary

The flow in a cylindrical vessel, created by a tangential inlet and an eccentric outflow, was studied experimentally. The ratio of the cylinder's height H to its diameter D was varied between $1<H/D<2$, the inlet-height h between $0,2<h/H<0,8$. The first results were the dependence between mixing time and geometry. To get these results investigations were made with air for measuring the flow-configuration in the rotating fluid and to determine the critical Reynolds number of the flow in the vessel for scale up reason.
In the second stage the vessel was filled with water which was recycled from the outflow to the inlet by a pump. Near the nozzle an indicator (NaCl) was injected as a Dirac-impulse -the system response was measured by the change of the conductivity . From the system response the mixing time was analysed and it was possible to derive the above mentioned relationship between the inlet-height, the level and the mixing time.
For better understanding the mixing-mechanism, the velocity-field was then detected by an optical measuring system. A small sphere of the same density as the fluid was used as a tracer photographed by two cameras. The optical axis of one camera was identic with the axis of the cylindric vessel, the other was perpendicular to the first one. The exposures were made by a stroboscope. The pictures were interpreted by a computer-program and from that the 3-dimensional velocity-field was obtained. With these results it is possible to explain the very complex mixing-mechanism in a vessel with tangential inlet and axial outflow.
As a further method the correlation analysis was used. Together with the above mentioned "sphere-experiments" interesting phenomena in the inlet-jet could be observed and measured.

Held at Wurzburg, 10-12 June, 1985.

Organised by DVCV· Deutsche Vereinigung für Chemie- und Verfahrenstechnik
(German Association of Chemical and Process Engineering).

Organisation: GVC·VDI-Gesellschaft Verfahrenstechnik und Chemieingenieurwesen. 

<u>NOMENCLATURE</u>

$\vec{a}$	Radiusvector	s	Probe-distance
d	Nozzle-diameter	S	Wave length of the periodical oszillation
d'	Diameter of the isotropic flow-core	S_1, S_2	Visual rays
D	Vessel-diameter	t	Parameter
D^+	characteristic length ($D^+ = V^{1/3}$)	t_n	Mixing-time
		T	Residence-time
f_o	Stroboscope-frequence	v^+	Dimensionless velocity
H/D	Slenderness-ratio	$v(r)$	Velocity
h/D	Inlet-height	V	Volume of the vessel
$\vec{P_s}$	Vector of intersecting point	ϱ_{fl}	Fluid-density
$\vec{r}$	Directional-vector	ϱ_s	Density of the sphere
$\hat{r}$	Perpendicular-vector	ϱ_{max}	Cross-correlations maxima

1. INTRODUCTION

Experiments have been made to investigate the fluid-motion in a cylindrical vessel, generated by a tangential inlet and axial outlet. Measurements have been made of the mixing-time, determined as a function of the inlet height h/H and the slenderness-ratio H/D (Kittner et al.1983), and the velocity-field.

This note presents in addition a technique useful for the quantitative measurement of small fluid velocities. The technique uses a sphere with the same density as the fluid ($\varrho_{fl}=\varrho_s$). It permits visualisation and measurements of three-dimensional flow-fields.

The results about the mixing-time investigations have already been published in Kittner et al. (1983), and shall be cited only as far as it is necessary for understanding the following text.

Some results of the work of Kittner et al. (1983) are shown in Fig. 1. It is obvious that the mixing-time is dependent on the slenderness-ratio H/D and the inlet-height h/H. It was shown that the energy of the inlet-flux is the reason for the quasi-stationary main-stream. The velocity-distribution v(r), measured with the CTA, shows that the rotating-stream is essentially dependent on the inlet-height. For fluid-inlet near the bottom, respectively in the middle of the vessel, there are presented the velocity-profiles for various measurement-heights presented (Fig. 2 and 3).

There is an interrelationship between the form of the main-stream in the vessel and the mixing-mechanism. Therefore it will be supposed that the mixing-time is dependent on the inlet-height.

For a fixed slenderness-ratio H/D and nozzle-diameter d, but variable inlet-height the mixing-time t_n was determined. The Reynolds-number Re for all experiments was greater than the critical Re-number which was obtained by former experiments. The results of mixing experiments are shown in Fig. 1 for H/D=1,0; 1,5; 2,0 and d/D^+=0,024. The dimensionless mixing-time t_n is outlined over the inlet height h/H.

Fig. 1 displays that for fluid-inlet h/H=0,5 the mixing-time is more than twice as big as for fluid-inlet near the bottom, respectively near the cover. Similar results are obtained for different slenderness-ratio H/D, certainly the influence of the inlet-height is decreasing with smaller H/D.

The mixing-time results were obtained from black-box experiments (conductimetry), so that no interpretation of the mixing-mechanism is possible. Therefore new experiments were made.

2. EXPERIMENTS
2.1 Experimental Equipment

For a better understanding of these results, the velocity-field was detected by an optical measuring system. A small sphere of the same density as the fluid was used as a tracer (it could be measured that

there was no slip between fluid and sphere), photographed by two cameras
(Fig. 4). The lighting was made by two stroboscopes, with the strobos-
cope-frequency f_o. The optical axis of one camera was identic with the
axis of the cylindrical vessel, the other was perpendicular to the first
one. On the two negatives of the cameras a series of points of the
sphere's flow-line were obtained. The points of two negatives belonging
together were scanned two by two. The position components of the sphere
are entered into the computer.

2.2 Computing

To find the value of the original point P(x,y,z) in the vessel, the two
vectors of the visual ray were computed from equation 1 and 2, Fig. 5.

$$S_1 \equiv \vec{a}_1 + \vec{r}_1 \cdot t_1 \quad (1) \qquad\qquad S_2 \equiv \vec{a}_2 + \vec{r}_2 \cdot t_2 \quad (2)$$

Afterwards the two visual rays were brought to intersecting point and
the co-ordinates of the point P(x,y,z) were calculated by equation 3.

$$\vec{P}_S = \vec{a}_1 + \frac{(\vec{a}_2 - \vec{a}_1) \cdot \hat{r}_2}{\vec{r}_1 \cdot \hat{r}_2} \cdot \vec{r}_1 \qquad\qquad (3)$$

From the distance of two points, and the stroboscope-frequence f_o, the
velocity-vector v was subsequently determined. With these results it is
possible to explain the very complex mixing-mechanism described above.

3. RESULTS
3.1 Flow-Field

Velocity components can also be computed at quite a large number of
positions in the field. Interpolation and smoothing then provide velocity components at other positions. From these data, there are several
instructive displays that can be generated. The schematic flow-field
obtained from these measurements is shown in Fig. 6. There are eight
flow-regions distinguished. Region 1 is a ring-flow which transports
fluid in the very stable regions 3 and 5. From there fluid goes in the
overflow region respectively bottom-layer and the central-region 6. This
flow region 6 is nearly a homogeneous, isotropic flow-field with a small
dominate upwards velocity. From the bottom-layer fluid flows screw-
shaped upwards (region 4) until the inlet height, where the streamlines
are turning round. The fluid-motion in region 2 is similar to that of
region 4, with one difference: there is a downward motion. In the inlet
height the fluid-streams of region 2 and 4 border on one another, without any permeability. A second type of display that can be generated
from the data is a flow-line, the way of the sphere. Figure 7 presents
an example of the computer-constructed flow-line in the lower half of
the vessel. In this type of display, different line width is printed
throughout the field denote different directions of motion (upwards or
downwards).

A third type of display (Fig. 8) prints tails and heads of arrows

indicating the magnitude and direction of the fluid velocity at each
point in the field, where the sphere has been measured. The lining
region in the middle indicates the homogeneous, isotropic flow-field
with the small dominating axial velocity.

3.2 Mixing-Mechanism

The above mentioned partition of the flow field is the explanation of
the mixing-behaviour, shown in Fig. 1. Each of the zone's 1 or 2 is
mixed thoroughly and very quickly. But the interchange of fluid extends
the mixing-time, because the interchange of fluid between zone 1 and
zone 2 (Fig. 6) takes only place about the slender central region 6. For
extreme h/H thereforee the mixing-time is shorter than for fluid-inlet
h/H 0,5.

4. CORRELATION-ANALYSIS - EXPERIMENTS AND RESULTS

With the described optical measuring method in the inlet a periodically
oszillation was observed which was measured and analysed by the correla-
tion-method. Into the same vessel now two hot-wire anemometers were
placed in the height of the fluid inlet near the wall (now there was
measured in air). Two experiments were made to find the periodically
oszillations.

In the first stage, probe 1 was fixed and probe 2 was carried on an arc
of a circle, in the same height as probe 1. The distance between the two
probes was than changed step by step. From the signals of the two probes
the cross-correlations functions were computed and outlined above the
probe-distance s (Fig. 9).

In the second stage probe 1 was also fixed but now probe 2 was carried
on a screw-shaped line on the jet border of the inlet flux, which was
known from former experiments.

The graphical illustration is the same as for experiment 1, Fig. 9. The
two graphs are plotted in the same figure. It is shown that the pronounced nature of the structures of the inlet flow is reproduced in the
developement of cross-correlations maxima. The oszillation is much in
evidence for both experiments disclosed by the wavelength S, which is
independent of the penetration depth of the inlet jet.

5. CONCLUDING REMARKS

It has been shown that the sphere technique is a useful tool for velocity measurement, without interfering the flow-field. According to the
experiments discussed above, this technique has proved to provide reliable estimates of the small axial-velocities in the central flow core of
the rotating flow and to recognize the mixing mechanism.

6. REFERENCE
1.   Kittner R., R. Fackler, V.Denk: "Ein Beitrag zur Strahlmischung mit
     äußerer Umwälzung". Brauwissenschaft August 1983 pp 316-324

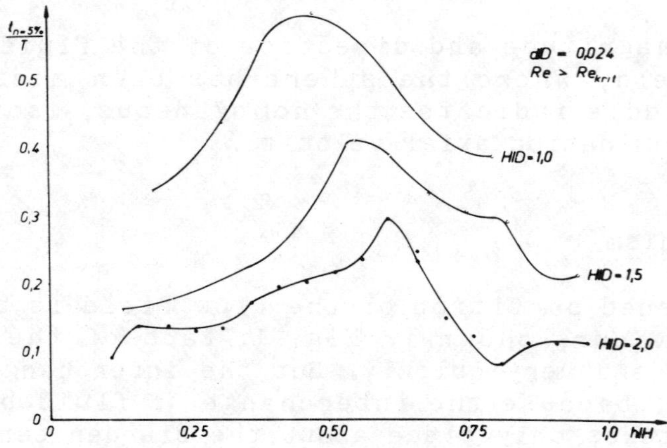

Fig. 1 Mixing time $t(hIH)$ for several slenderness-ratio HID

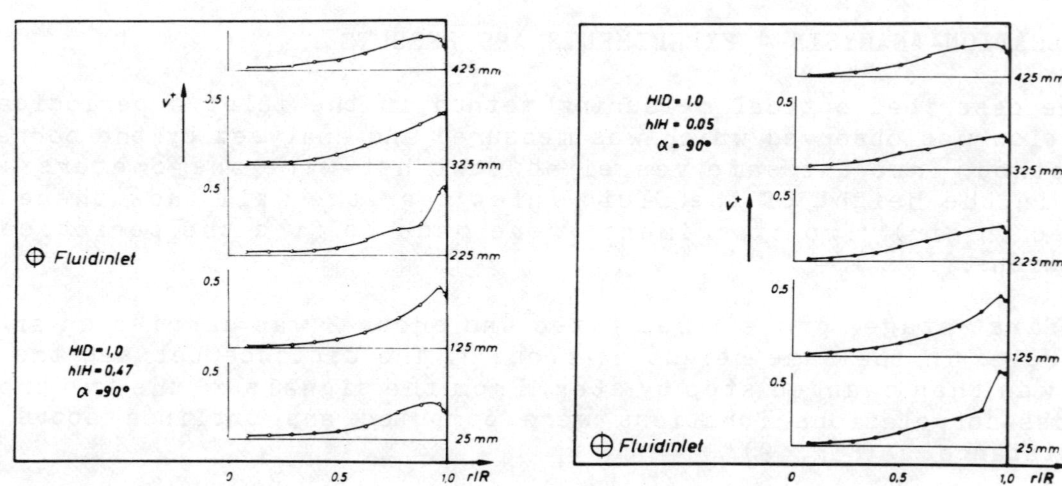

Fig. 2 and 3 Velocity profiles inlet-height hIH = 0,47 and hIH = 0,05

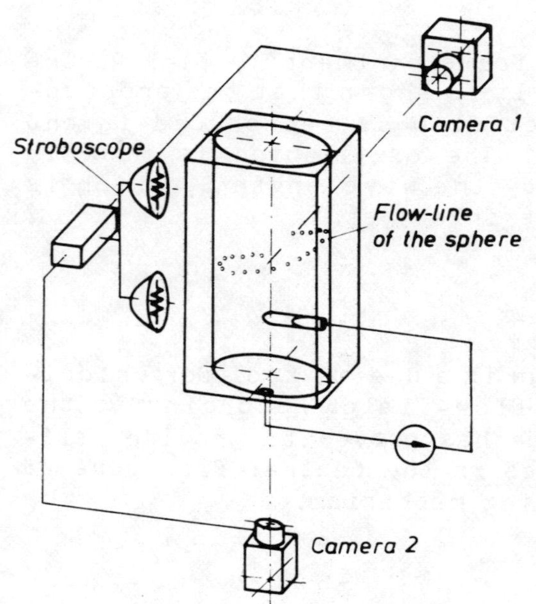

Fig. 4 Schematic of the experimental set-up

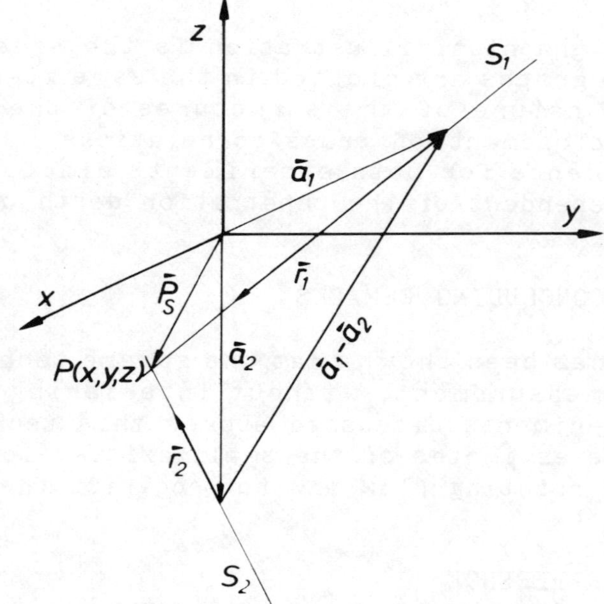

Fig. 5 Vektor-representation
of the intersecting-point

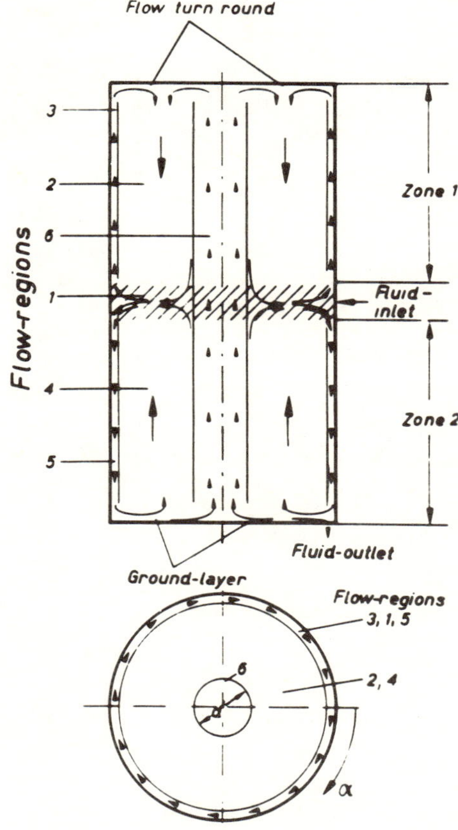

Fig. 6 Schematic of the flow-regions

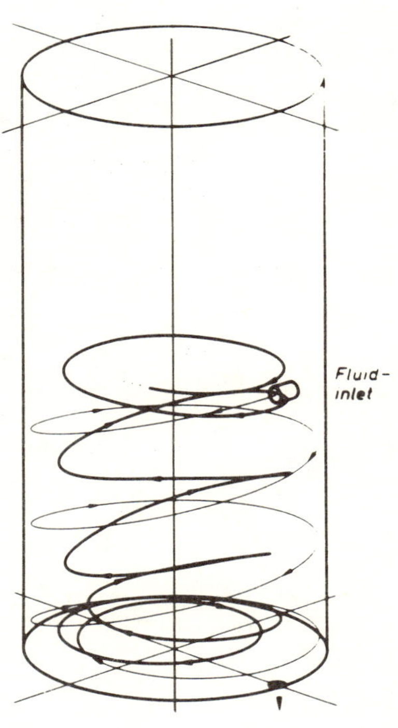

Fig. 7 Computer-constructed flow-line of the sphere

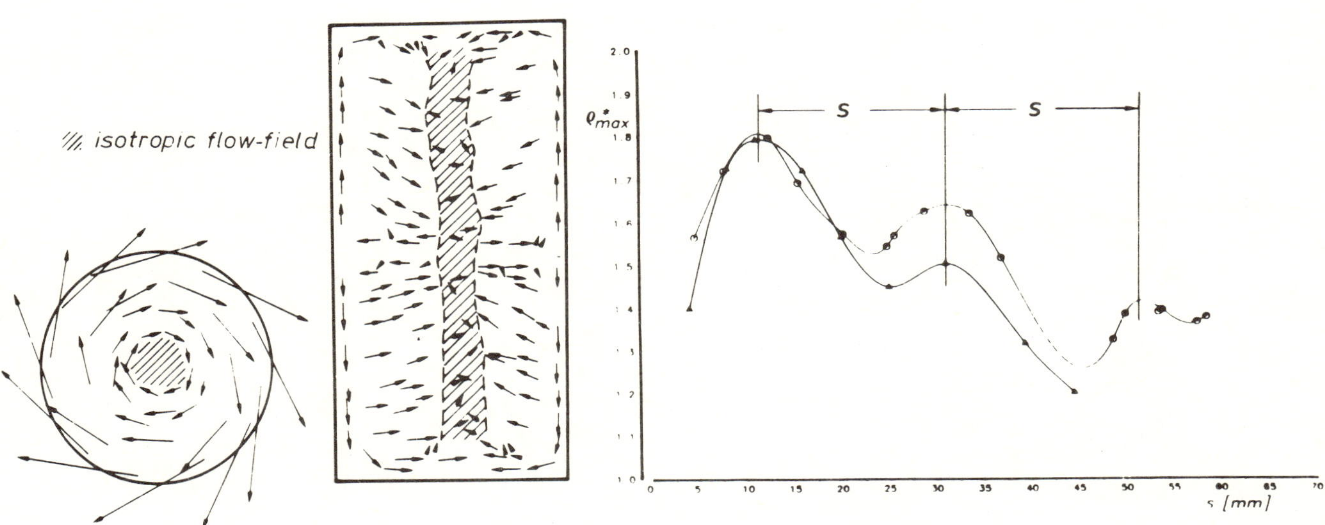

Fig. 8 Velocity-field in a vessel – H/D=2, h/H=0,5

Fig. 9 Maxima of the cross-correlations as funktion of the probe-distance s

Improvement of Homogeneous Tubular Reactors Operated at Low Reynolds Number

E. Flaschel, K.T. Nguyen, A. Renken

Institute of Chemical Engineering
Swiss Federal Institute of Technology
CH-1015 Lausanne, Switzerland

Summary

Motionless mixers are used as internals to improve the behaviour of tubular reactors for homogeneous liquid phase reactions in the range of low Reynolds number. In this range small gradients in temperature or density normally lead to large stochastic backmixing of the fluid. Such flow perturbations can effectively be suppressed in addition to a flattening of the flow profile if regularly assembled internals are used. For the motionless mixers applied, types SMX and SMV of Sulzer, CH, the ratios of the effective fluid velocity to the axial dispersion coefficient are found to be ca. 80 and 70 m^{-1}, respectively, in the range of Reynolds numbers from 0.005 to 100 for tubular reactors with a diameter of 40 mm.

Held at Wurzburg, 10-12 June, 1985.

Organised by DVCV· Deutsche Vereinigung für Chemie- und Verfahrenstechnik
(German Association of Chemical and Process Engineering).

Organisation: GVC·VDI-Gesellschaft Verfahrenstechnik und Chemieingenieurwesen.

<u>Nomenclature</u>

Bo	–	Bodenstein number: $u \cdot L/D_{ax}$
d_h	mm	hydraulic diameter of the reactor (including packing)
D_{ax}	$mm^2 \cdot s^{-1}$	axial dispersion coefficient
d_R	mm	reactor diameter
E	s^{-1}	residence time distribution (RTD) function
g	s^{-1}	RTD weight function
L	mm	reactor length
Pe	–	Peclet number: $u \cdot d_h/D_{ax}$
Re	–	Reynolds number: $u \cdot d_R/\nu$
t	s	time
$\bar{t}$	s	mean residence time
u	$mm \cdot s^{-1}$	effective linear fluid velocity: $L/\bar{t}$
ν	$mm^2 \cdot s^{-1}$	kinematic viscosity
τ	s	time

1. Introduction

The performance of tubular reactors for liquid phase reactions is impaired by the occurrence of unfavourable flow profiles and, even more, by convective fluctuations due
to temperature or density gradients. The latter effect is important at low fluid velocities and leads to large stochastic backmixing which has to be suppressed in order to
achieve a constant axial fluid dispersion as low as possible.

Different means have been used to circumvent these problems, which all have in common
that an enforced radial exchange of fluid elements is superimposed to the longitudinal
fluid flow. Cascading and compartmentation of reactors with and without stirring are
used as well as tower packings of diverse shapes. A comparison of these means on the
behaviour of tubular reactors with low fluid velocity may be found in Langensiepen
(Ref. 1).

Since some normal tower packings like rings and saddles may lead to the creation of
dead-zones in the reactor and give rise to high pressure drops for viscous media,
regularly arranged rigid packings with high voidage would be preferred. Such characteristics are provided by some types of motionless mixers. There is only few information
available on the influence of such internals on the dispersion in tubular reactors
(see: Giger et al., Ref. 2; Streiff and Jancic, Ref. 3; Flaschel et al. Ref. 4; Nguyen
et al., Ref. 5). For that reason a detailed study has been undertaken to characterize
two different types of motionless mixers with particular emphasis on the range of low
Reynolds numbers.

2. Experimental

The reactor consists of jacketed tubes with an inner diameter of 40 mm. It is divided
into three sections: The bottom part of 600 mm serves for distributing the tracer injected near its entrance, a central part of 1500 mm representing the reactor to be
characterized and a top part of again 600 mm for avoiding any flow disturbance and
feedback by the outflow. Probes for tracer detection are mounted into the interconnec-
tions of the tubes. A fiber optic photodetection system (Ref. 6) or conductimetric
probes are used to monitor the tracer response before and after the central reactor
section. The three reactor sections are entirely filled with motionless mixers either
of the type SMX or SMV (Sulzer, CH) characterized by hydraulic diameters of 11 and 9
mm, respectively.

The main flow of liquid is doted by means of an HPLC-valve and a constantly operating
metering pump with a pulse of tracer solution. For photometric detection an aqueous
solution of Patent Blue V is used. For conductimetric detection either a KCl or a HCl
solution is used as tracer. Both tracer signals monitored at the entrance and at the
outlet of the central reactor section are taken for the calculation of the axial dispersion coefficient of the tracer by means of the momentum method.

The mean residence time ($\bar{t}$) and the Bodenstein number (Bo) obtained are used
as starting parameters for an evaluation according to the convolution method. This
method is applied because it is less sensitive towards problems of proper signal limit
settings than the momentum method. The dispersion model for large extents of dispersion and open undisturbed boundaries (Levenspiel and Smith, Ref. 6) is used as weight
function:

$$g(t) = \frac{1}{2\bar{t}} \left(\frac{\bar{t}\,Bo}{\pi t} \right)^{1/2} \exp\left(-\frac{\bar{t}\,Bo}{4t} \left(1 - \frac{t}{\bar{t}} \right)^2 \right) \qquad (1)$$

A theoretical outlet signal is calculated by the convolution of the normalized entrance signal with the weight function:

$$E(t) = \int_0^\infty E_{in}(t - \tau)\, g(\tau)\, d\tau \qquad (2)$$

The sum of squares of the deviations of the theoretical outlet signal E(t) from the
normalized outlet signal measured is taken as objective function for the estimation of
the parameters ($\bar{t}$ and Bo) according to the algorithm of Marquardt (Ref. 7).

The viscosity of the aqueous solutions is modified by means of the temperature and by
addition of PEG 20'000 or Thylose H 4'000 (both: Hoechst, D). If PEG is used together

with HCl tracers, a calibration of the conductivity is necessary. The conditions for experiments undertaken with both internals SMX and SMV are gathered in Tab. 1 and Tab. 2. The experiments with KCl tracers have only been evaluated according to the momentum method.

<table>
<tr><td colspan="4">

Tab. 1: Experimental conditions
        for SMX-equipped reactors

</td><td colspan="4">

Tab. 2: Experimental conditions
        for SMV-equipped reactors

</td></tr>
</table>

Fluid		ν $mm^2 \cdot s^{-1}$	Tracer	Fluid		ν $mm^2 \cdot s^{-1}$	Tracer
50%	PEG	3000	KCl	50%	PEG	2500	HCl
40%	PEG	820	KCl	40%	PEG	950	HCl
1%	Thylose	150	KCl	30%	PEG	120	HCl
0.75%	Thylose	50	KCl	3%	PEG	2.18	HCl
7%	PEG	5.58	HCl	water		0.80	HCl
3%	PEG	2.55	HCl				
Water		1.00	Dye**				
Water		0.80	HCl				
Water		0.55	HCl				

** Patent Blue V

3. Results

The dispersion characteristics in tubular reactors equipped with motionless mixers SMX is given in Fig. 1. The Bodenstein number is not given, but the ratio of the effective linear fluid velocity and the axial dispersion coefficient, because it represents a length-independent quantity contrary to the Bodenstein number. A pronounced change of the behaviour is observed at Re = 100. The dispersion in terms of u/D_{ax} is almost constant up to this Reynolds number, beyond this value the ratio increases with a slope of $d(u/D_{ax})/d(lgRe) = 250$ m^{-1}. In the quasi-laminar flow range the mean ratio u/D_{ax} is 79.6 as shown in Fig. 2. From this the axial dispersion coefficient should depend linearly on the effective fluid velocity as shown in Fig. 3.

The corresponding dependencies for reactors equipped with the motionless mixer SMV are given in Fig. 4, Fig. 5 and Fig. 6. A pronounced change from a quasi-laminar to a quasi-turbulent region is not found in this case and the behaviour seems to be somewhat different from that for the internals SMX. When all the values of u/D_{ax} are gathered as in Fig. 5, a mean value of 68.3 is calculated. This leads to a linear relation of the axial dispersion from the effective linear fluid velocity with an accuracy of about ±20% as shown in Fig. 6.

4. Conclusion

The results show that liquid phase tubular reactors equipped with motionless mixers of the types SMX and SMV compare well with packed bed reactors with respect to the dispersion behaviour of the liquid. Data for packed bed reactors give Peclet numbers in the range of 0.02 to 0.2 in the quasi-laminar flow range (Ref. 1). When the hydraulic diameter of the motionless mixers is taken for the calculation of the corresponding Peclet numbers, the mixer SMX is characterized by a value of 0.88 in the quasi-laminar flow range and the mixer SMV by a mean value of 0.62.

<u>Literature</u>

1. Langensiepen, H.-W.: "RTD in homogeneous flow reactors for a maximum mean residence time of 3 hours". (Verweilzeitverteilung von einphasig durchströmten Reaktoren mit maximal drei Stunden mittlerer Verweilzeit). Chem.-Ing.-Tech., <u>52</u>, 2, 1980, pp. 176-177 (MS 771/80).(In German)
2. Giger, G.K., Habegger, E. and Richarz, W.: "RTD and pressure drop of viscous liquids in motionless mixers". (Untersuchungen der Verweilzeitverteilung und des Druckabfalles einer viskosen Flüssigkeit in einem statischen Mischer). Chimia, <u>32</u>, 9, 1978, pp. 334-339. (In German)
3. Streiff, F.A. and Jancic, S.J.: "Use of Static Mixer Packing in Countercurrent Extraction Columns". Ger. Chem. Eng. <u>7</u>, 1984, pp. 178-183
4. Flaschel, E., Raetz, E. and Renken, A.: "Development of a Tubular Recycle Membrane Reactor for Continuous Operation with Soluble Enzymes". In: Enzyme Technology, Lafferty, R.M. (ed.), Berlin, D, Springer Verlag, 1983, pp. 285-295
5. Nguyen, K.T., Flaschel, E. and Renken, A.: The Thermal Bulk Polymerisation of Styrene in a Tubular Reactor". In: Polymer Reaction Engineering, Reichert, K.H. and Geiseler, W. (eds.), Munich, D, Hanser Publ. 1983, pp. 175-205
6. Levenspiel, O. and Smith, W.K.: "Notes on the Diffusion-Type Model for the Longitudinal Mixing of Fluids in Flow". Chem. Eng. Sci., <u>6</u>, 1957, pp. 227-233
7. Marquardt, D.W.: "An algorithm for least-squares estimation of nonlinear parameters". J. Soc. Indust. Appl. Math., <u>11</u>, 2, 1963, pp. 431-441

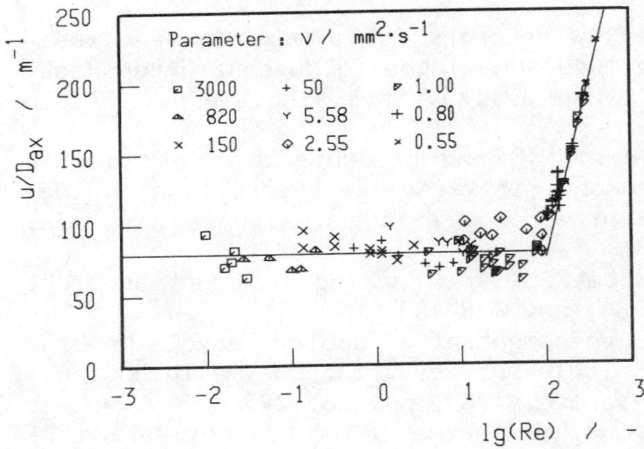

Fig. 1: Dispersion characteristics of tubular reactors equipped with the motionless mixer SMX

Fig. 2: Dispersion characteristics at low Reynolds numbers – motionless mixer SMX

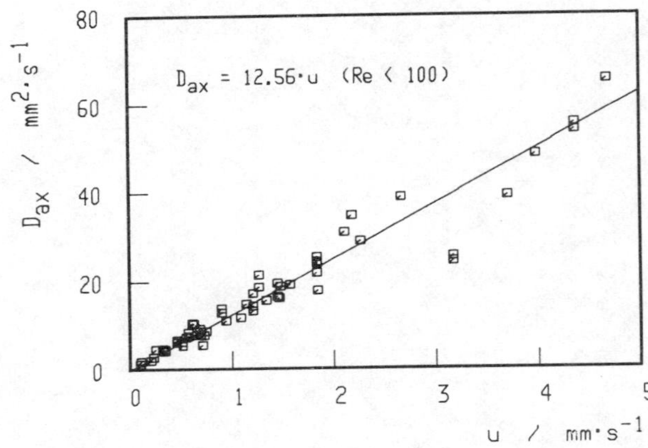

Fig. 3: Axial dispersion at low Reynolds numbers in the presence of the motionless mixer SMX

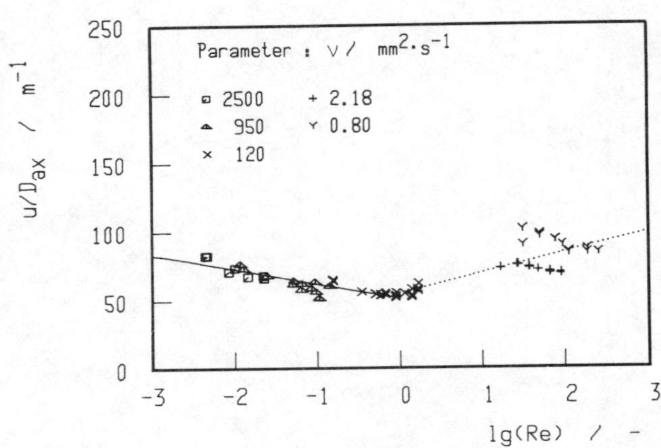

Fig. 4: Dispersion characteristics of tubular reactors equipped with the motionless mixer SMV

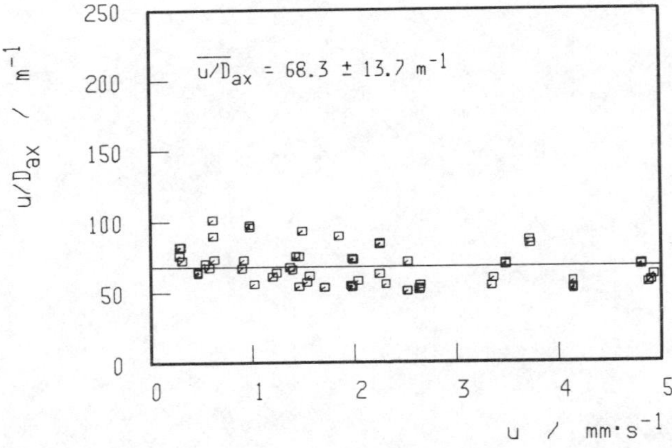

Fig. 5: Dispersion characteristics in the presence of the motionless mixer SMV

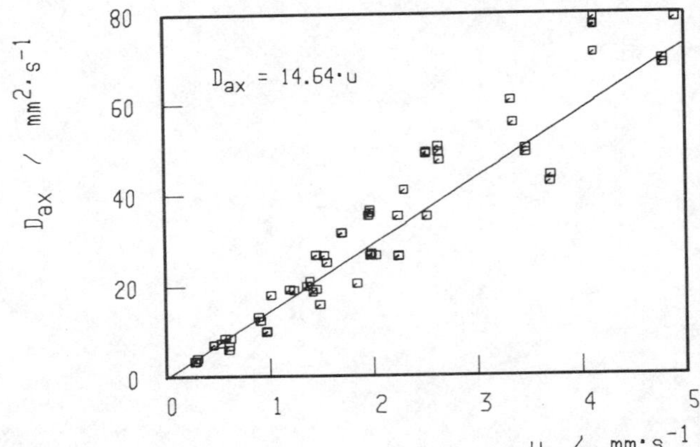

Fig. 6: Axial dispersion in the presence of the motionless mixer SMV

PROBLEMS WITH MEASURING MIXING TIMES

Julius W. Hiby

Institut für Verfahrenstechnik, Technische Hochschule, Aachen
Federal Republic Germany

Summary

A few problems are discussed related to measurement and evaluation of mixing times.

When making use of the colour change of a chemical indicator, there are two alternative methods. The reagent added to the bulk can cause monotonous decolouring, or it can produce the colour which then finally disappears. With the latter method one cannot detect dead zones.

Intensity of segregation and coefficient of variation are compared as to their usefulness as a scale of homogeneity.

Although an instantaneous reaction alters the concentration profiles, one can prove that mixing times measured with a chemical indicator apply also to non-reactive mixing.

Held at Wurzburg, 10-12 June, 1985.

Organised by DVCV· Deutsche Vereinigung für Chemie- und Verfahrenstechnik
(German Association of Chemical and Process Engineering).

Organisation: GVC·VDI-Gesellschaft Verfahrenstechnik und Chemieingenieurwesen. 

©BHRA, The Fluid Engineering Centre, Cranfield, Bedford MK43 0AJ, England.

NOMENCLATURE

A	mixing component (smaller input volume, $\bar{a} \leq 0.5$)
a	local volume fraction of A in mixture
$\bar{a}$	mean value of a in total liquid
a_e	volume fraction a, associated with equivalence of reagents
B	mixing component
c	molar concentration
d, L	tube diameter and length
f(a)	frequency distribution of a in the total volume
S	intensity of Segregation
t, t_m	time, mixing time
y	coordinate in direction of diffusive flow
σ	standard deviation

Index α: initial

1. TWO ALTERNATIVE DECOLOURING METHODS WITH A CHEMICAL INDICATOR

The two liquids to be mixed are A and B. The local volume fraction of A in the mixing field shall be called "a"; as the liquid A is assumed to be the one with the smaller initial volume, the average value in the total mixture is $\bar{a} \leq 0.5$.

There are two alternative methods for measuring the mixing time by decolouring a chemical indicator in the course of a neutralization or redox reaction. In both cases the decolouring reagent is in stoichiometric excess, but the two methods differ in the initial conditions.

With the first method which may be called "monotonous decolouring" the decolouring reagent is originally in the liquid A (smaller volume), Fig. 1a. During the mixing process the liquid B, initially coloured, is gradually decoloured. With the alternative "colouring/decolouring method", liquid B is initially decoloured, Fig. 1b; when liquid A which contains the colouring reagent is added, it causes a temporary and partial colouring of the mixture which finally disappears.

For a nearer insight one must look at the frequency distribution f(a) in the mixing volume, fig. 2. In a sufficiently advanced state of homogenization and in the absence of dead zones, f(a) is a symmetrical curve which slims with time and converges towards the final value $\bar{a}$. A certain value a_e on the abscissa pertains to those volume elements which contain the two reagents in equivalent amounts; this value a_e depends on the choice of the initial normalities of the reagents in A and B and therefore can be predetermined (Ref. 1,2).
With the monotonous method, final decolouring can be achieved only if $a_e < \bar{a}$ (Fig. 2a). Only volume elements with $a < a_e$ are coloured because they contain less than the equivalent amount of the decolouring reagent; the shaded area corresponds to their fraction in the total volume. At the moment when the last colour fades away, one knows the maximum deviation from homogeneity: $\bar{a} - a_{min} = \bar{a} - a_e$. This maximum deviation is approximately equal to 3σ (Ref. 3).

With the colouring/decolouring method, on the other hand, liquid A carries the colouring reagent, and one must choose $a_e > \bar{a}$, Fig. 2b. At the moment of decolouring one measures $a_{max} - \bar{a} = a_e - \bar{a}$.

In the case of a symmetrical distribution f(a), both methods lead to the same results. But if dead zones are present, the characteristic time for mass exchange between bulk and dead zone being larger than the mixing time in the bulk, the situation is different. After the liquid A is added, it first spreads in the well-agitated bulk before it penetrates into a dead zone. Therefore the dead zones correspond to a left "tail" of the distribution curve, broken line in Fig. 2a. Obviously the prolongation of mixing times by dead zones can be detected and measured only by the monotonous decolouring method.

2. INTENSITY OF SEGREGATION S AND COEFFICIENT OF VARIATION $\sigma/\bar{a}$

The status of a mixing field is usually described either by the intensity of segregation S:

$$S = \frac{\sigma^2}{\sigma_\alpha^2} = \frac{\sigma^2}{\bar{a}(1-\bar{a})} \tag{1}$$

or by the coefficient of variation $\sigma/\bar{a}$.

The intensity of segregation has the advantage that by definition it varies from 1 towards 0 during the mixing process. Another interesting relation was pointed out by Toor (Ref. 4): When stoichiometrically equivalent partners of an instantaneous reaction are mixed, $1 - S$ is equal to the fractional conversion.

On the other hand, there are also arguments in favour of the coefficient of variation. The two quantities are interrelated (Fig. 3):

$$\sqrt{S} = \frac{\sigma}{\bar{a}} \sqrt{\frac{\bar{a}}{1 - \bar{a}}} \tag{2}$$

During a mixing process, S and $\sigma/\bar{a}$ both vary as a function of time (or of the length of the mixing path, respectively), and in addition they depend on the Reynolds number and on $\bar{a}$, the average volume fraction of the component A in the total volume. This will be illustrated by examples from literature for turbulent mixing in straight tubes, figures 4a and b.

If $\bar{a} = 0.5$, $\sqrt{S}$ and $\sigma/\bar{a}$ are identical according to Equation (2) (straight lines 1 in Figures 4a and b). But in the case $\bar{a} \leq 0.01$ the two diagrams differ strongly. The plot for $\sqrt{S} = f(L/d)$, fig. 4a, gives the impression that with small values of $\bar{a}$ (lines 2 and 3, $\bar{a} \leq 0.01$) mixing proceeds quicker. But the corresponding lines 2 and 3 of the plot for $\sigma/\bar{a} = f(L/d)$, Fig. 4b, show that when a small relative volume flow of $A(\bar{a}<<1)$ is to be dispersed to a good relative homogeneity, the approach to a given value of $\sigma/\bar{a}$ requires a larger length of mixing conduit.

At small tube lengths, large values of $\sigma/\bar{a} > 1$ are observed, indicating a strong asymmetry of the distribution function f(a). With curve 3 this is exaggerated by the fact that the authors (Ref. 5) measured σ as local standard deviation along the centreline of the tube, while with line 2 (Ref. 6) the values of σ were averaged across the tube cross-section.

These effects do not show up in Fig. 4a; they are hidden by the formal constraint of an initial value S = 1. Therefore, the comparison of the two diagrams shows that a plot of $\sigma/\bar{a}$ gives more information on the homogenization process than a plot of S.

3. MIXING TIMES FOR NON-REACTIVE AND FOR REACTIVE COMPONENTS

When mixing times are measured by the decolouration method, the question arises whether the results are relevant also for mixing of non-reactive substances. One might suspect that the large concentration gradients caused by the instantaneous reaction could accelerate diffusive mass exchange between adjacent lamellae of the components to be mixed.

The profiles of molar concentration c of non-reactive components A and B along the coordinate axis y normal to the plane interface of the two layers will be plotted. The coordinates y = 0 and 1 are located at the planes of symmetry of layers A and B respectively. Actually the thickness of such layers in turbulent flow is continously reduced by streching, but for the purpose of mere comparison between non-reactive and reactive mixing the thickness of the layers may be regarded as constant.

In Fig. 6 the profiles c(y) of the interdiffusing components at an arbitrary time t_1 are plotted (curves A_{dif} and B_{dif}). As in this case the layers of A and B are of equal thickness, and as equal initial concentrations and equal diffusivities are assumed, the profiles are symmetrical.

The problem whether the mixing time is shortened by an instantaneous reaction reduces to the following question. Two cases are compared: 1.) The hypothetical case of diffusion transport up to a time t_1, then at all points of the mixing field instantaneous reaction, 2.) the realistic case of combined diffusion and instantaneous reaction. Do both cases result in the same final concentration profiles c = f(y) for A and B?

The answer is affirmative; it can be verified in a simple manner. Conceive first the diffusive transport of inert components, leading to the concentration profiles A_{dif} and B_{dif} at time t_1 (Fig. 6). If the components would now react instantaneously, the ultimate profiles A_{ult} and B_{ult} result as the difference between the former profiles.

Now the actual process of simultaneous transport and reaction which is assumed to be a simple combined reaction $A + B \rightarrow AB$; A and B could be atoms, AB a molecule. It will be assumed that all diffusion coefficients are equal; for the components used with decolouration experiments this can be accepted as a first approximation.

As to diffusion transport, one must realize that in an isothermal mixture without convection every particle (molecule, atom, ion) has at any place and any moment an isotropic probability of random motion, independent of whether it maintains its individuality or is integrated into a larger molecule. Consequently, at time t_1 the profile of the sum of concentrations of A and AB, and correspondingly the profile of B and AB, are identical to the profiles for non-reactive components (Fig. 6). Because of the instantaneous reaction, A and B cannot exist at the same coordinate. In the left, A-rich half of the plot, therefore, the curve for A results as the difference between the curves for A + AB and for B + AB; correspondingly the curve for B results in the right half of the plot. This proves that the ultimate profiles for A and B in Figures 6 and 7 are identical.

The simple argumentation obviously holds for any ratio of thickness of the two layers and also for any ratio of initial concentrations of the two components. The conclusion is therefore that mixing times measured with a chemical indicator are valid also for inert mixing components. Of course, differences between the relevant diffusion coefficients will have an effect.

4. REFERENCES

1. Hiby, J.W.: "Definition and measurement of the quality of mixing in liquid mixtures". (Definition und Messung der Mischgüte in flüssigen Gemischen). Chem. Ing. Tech. 51, 1979, pp. 704-709. (In German).

2. Hiby, J.W.: "Definition and measurement of the degree of mixing in liquid mixtures". Intern. Chem. Engng. 21, 1981, pp. 197-204.

3. Hartung, K.H. and Hiby, J.W.: "Promotion of turbulent mixing in tubes". (Beschleunigung der turbulenten Mischung in Rohren). Chem. Ing. Tech. 44, 1972, pp. 1051-1056.

4. Toor, H.L.: "Mass transfer in dilute turbulent and non-turbulent systems with rapid irreversible reactions and equal diffusivities". A.I.Ch.E. Journal 8, 1962, pp. 70-78

5. Lee, J. and Brodkey, R.S.: "Turbulent motion and mixing in a pipe". A.I.Ch.E. Journal 10, 1964, pp. 187-193

6. Hartung, K.H.: "Investigation of technical mixing processes by use of pH-indicators". (Untersuchung technischer Mischvorgänge mittels pH-Indikatoren). Dr.-Ing.-Thesis, Technische Hochschule Aachen, 1975.

7. Hiby, J.W.: "Turbulent mixing in pipes". (Mischung in turbulenter Rohrströmung). Verfahrenstechnik 4, 1970, pp. 538-543. (In German).

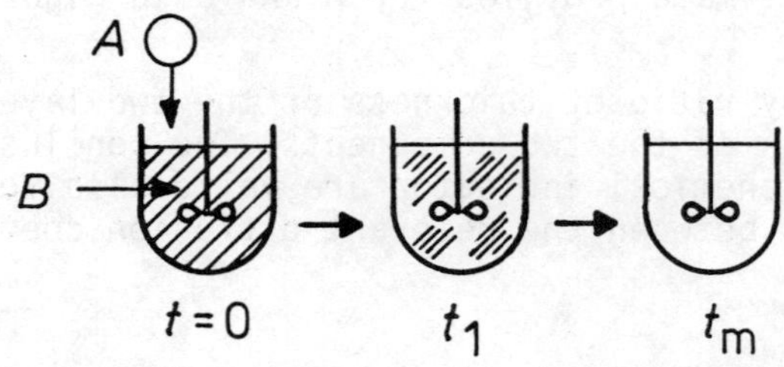

Fig. 1a. Monotonous decolouring method.

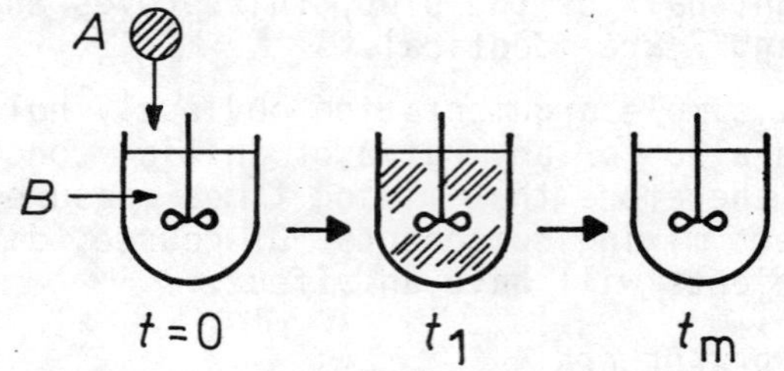

Fig. 1b. Colouring/decolouring method.

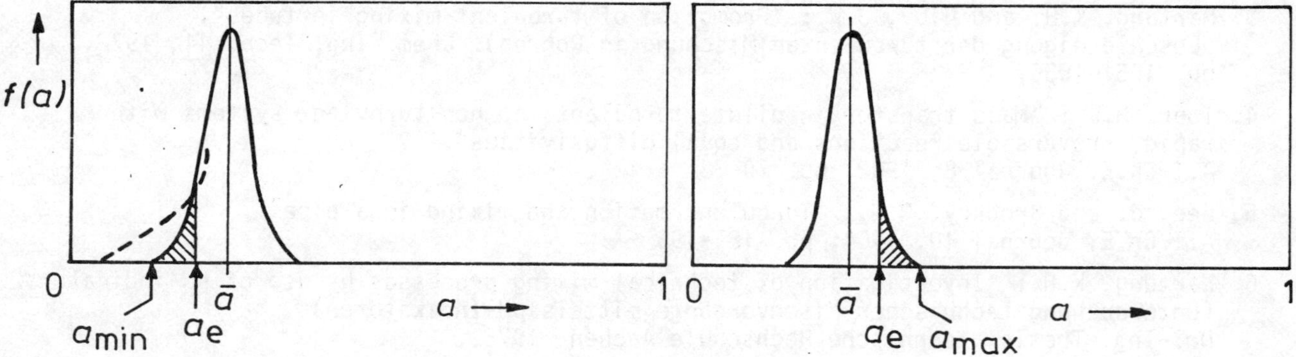

Fig. 2a. Frequency distribution with monotonous decolouring.

Fig. 2b. The same with colouring/decolouring method.

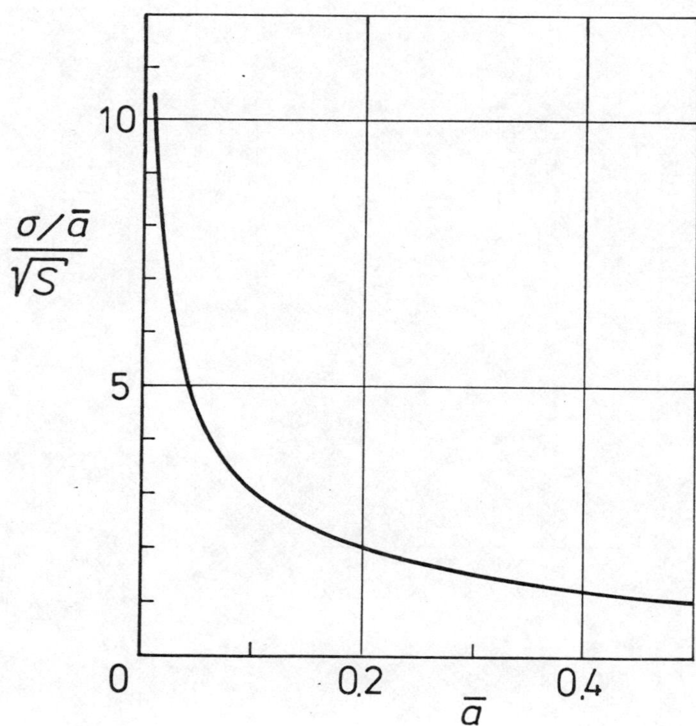

Fig. 3. Relation between $\sigma/\bar{a}$ and $\sqrt{S}$ as a function of $\bar{a}$.

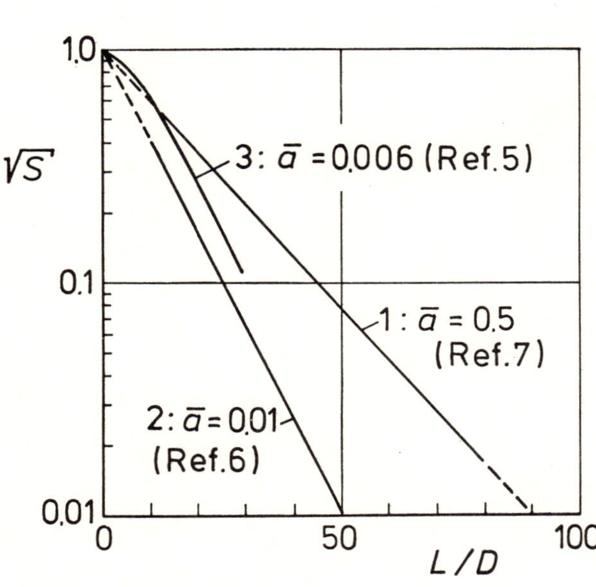

Fig. 4. √S̄ as a function of relative tube length.

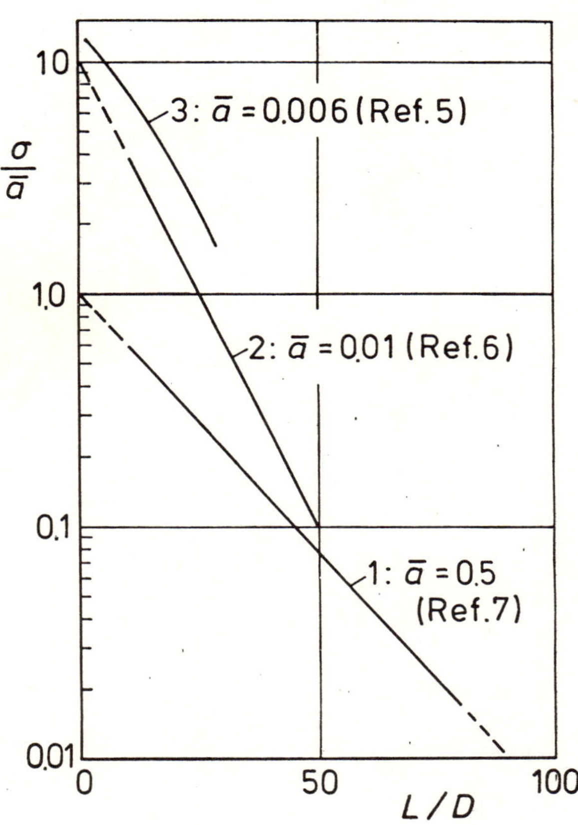

Fig. 5. Results as fig. 4, expressed as σ/ā.

ā = 0.5: Initially A filled half of tube cross-section: 

ā ≦ 0.01: Liquid A entered through axial capillary:

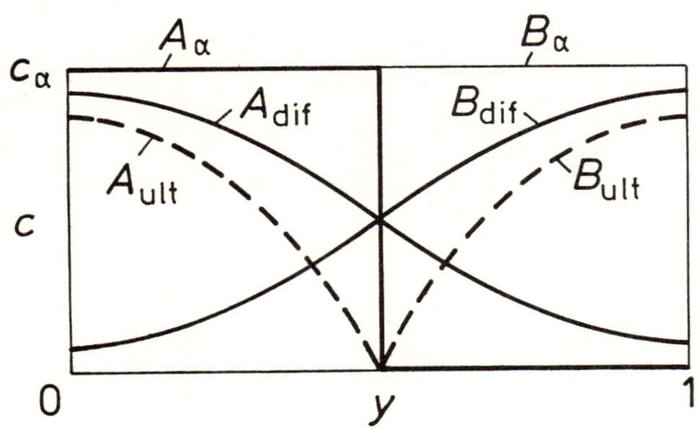

Fig. 6. Concentration profiles for consecutive diffusion and instantaneous reaction.

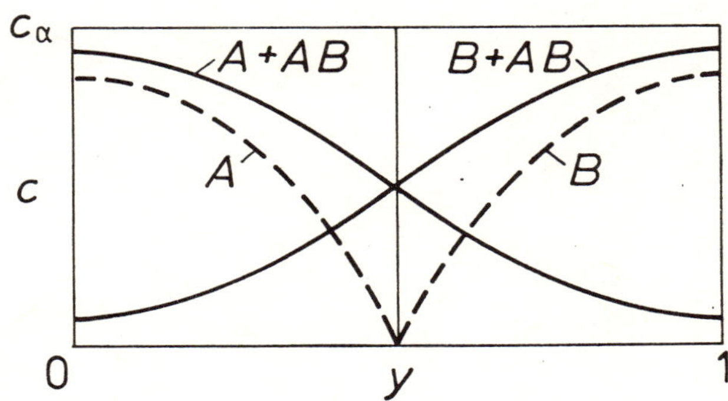

Fig. 7. Profiles for simultaneous diffusion and instantaneous reaction.

On-line parameter estimation of dispersion coefficients with a Fast-Fourier-Analyzer

Dipl.-Ing. P.Krizan, o.Prof. Dr.-Ing. E.Blaß
Lehrstuhl A für Verfahrenstechnik der TU München, Arcisstr. 21, D-8000 München, FRG

SUMMARY

The prior aim in chemical engineering is to design multiphase mass transfer reactors based on informations found in small scale laboratory plants. The determination of the necessary column height is still subject to considerable uncertainty. This fact is primarily attributable to fluiddynamic effects within the contacting phases, generally named dispersion. Dispersion leads to an irreversible reduction of the concentration driving force which is, in turn, important for the efficiency of mass transfer contactors. Several models have been developed to describe the influence of fluiddynamics on mass transfer operations. A traditional and widely used description is in terms of the one-dimensional axial dispersion model. The parameters are determinded by comparing experimental data of transient response experiments with model predictions and picking them so that the data and predictions match or fit in some sense. We know quick but inaccurate short-cut estimation methods and exact but time consuming methods only realizable in connection with large scale computers. The best method up to date to determine dispersion coefficients from tracer response measurements for an arbitrary input signal is a time domain analysis. An analytical solution of the axial dispersion model is fitted with discrete experimental data using a nonlinear regression procedure based on the convolution integral. A simpler and even more accurate way of the parameter estimation is possible based on a frequency response analysis performed on-line with a Fast-Fourier-Analyzer. The residence time density function can be evaluated from arbitrary input/output signals from the measured frequency response function.

Held at Wurzburg, 10-12 June, 1985.
Organised by DVCV· Deutsche Vereinigung für Chemie- und Verfahrenstechnik
(German Association of Chemical and Process Engineering).

Organisation: GVC·VDI-Gesellschaft Verfahrenstechnik und Chemieingenieurwesen. 

NOMENCLATURE

Bo	Bodensteinnumber $(\dot{v}_1 L/(\varepsilon_1 D_{ax,1})$	–	c	concentration	mol/m^3
C	norm. conc. in frequency domain	$1/s^2$	d_K	column diameter	m
D_{ax}	axial dispersion coefficient	m^2/s	H_K	column height	m
h	coordinate of height	m	j	$(-1)^{1/2}$	–
n	molar flux	$mol/(m^2 s)$	S	spectral density function	$1/Hz$
t	time	s	t_R	mean residence time	s
$\dot{v}$	superficial velocity	m/s	z	dim.less coordinate of height	–
θ	dim.less time t/t_R	–	μ_1	1. moment of distribution fct.	s
ω	circular frequency	rad/s	Ω	dim.less frequency $\omega \cdot t_R$	–

<u>1. INTRODUCTION</u>

In the field of reaction engineering two ideal dynamic reactor behaviours are known: the perfect mixing and the plug or piston flow. The former means thorough and complete mixing. The composition of the volume is throughout uniform; material leaving the system is identical to material at any point within the system. A perfect mixer always has an exponential distribution of residence times and is a first order lag process element. In the second case, the phases move like a piston through the column. All fluidelements have the same residence time. No diffusion and no mixing with fluidelements entering at different times will take place. After a time delay all elements entering at the same moment will appear togehter in the exit. The piston flow is a dead time process element. However, the real behaviour of mass transfer reactors lies somewhere between these two ideal behaviours and the residence times of the fluid follow an arbitrary distribution function.

<u>2. THE DISPERSION MODEL</u>

The common treatment of the influence of residence time distribution on mass transfer operations is based on the dispersion model. The effective flux of a solute in flow direction is assumed to be the superposition of a purely convective transport and a normal distributed stochastic transport. The later one is mathematically expressed in analogy to the Fick's law of diffusion by

$$\dot{n}^D = -D_{ax} * (dc/dh) \tag{1}$$

D_{ax} is named dispersion coefficient and is a global parameter to describe all physical effects on dispersion as molecular, eddy and Taylor diffusion (dispersion due to radial velocity profiles), bypassing, channeling, entrainment, gross fluid circulations etc. Therefore the extent of dispersion is strongly dependent on column geometry and operating conditions. The unsteady state balance of inert injected tracerparticles across an infinitesimal column element yields together with equation (1) to the following dimensionless partial differential equation:

$$\partial c/\partial\theta = 1/Bo \cdot \partial^2 c/\partial z^2 - \partial c/\partial z \tag{2}$$

Solutions for (2) for boundary conditions corresponding to the physical configuration have been described in the literature (Ref. 1)

<u>3. TRANSIENT RESPONSE FUNCTION FOR OPEN BOUNDARY CONDITIONS</u>

Only for the special case when a delta function of tracer particles is injected uniformily across the inlet area to an open system and the response at the outlet is registered as area-averaged concentration an analytical solution is possible (Ref. 2)

$$c(t,z=1) = 1/2 \cdot (Bo \cdot t_R/(\pi t^3))^{1/2} \cdot \exp(-Bo(t_R-t)^2/(4t \cdot t_R)) \tag{3}$$

In such an open system particles may move counter to the predominant flow so that they leave the system at the inlet or enter the system at the outlet. The random nature of dispersion effects ensures that such reverse flow must exist at microscopic and at macroscopic level of scrutiny. Since eq.(3) describes the duration between the first entry of all particles and their final exit, the mean μ_1 is not the residence time since time spent on temporary excursions outside the system boundaries are contributed. t_{max} is the mean of all forward movement. The relations

$$\mu_1 = t_R \cdot (1 + 2/Bo) \tag{4a}$$
$$t_{max} = t_R \cdot (1 + 2/Bo)^{-1} \tag{4b}$$

must be considered. The condition of a delta function injection of tracer particles cannot be realized experimentally. Therefore arbitrary input functions are used. The inlet and outlet concentrations of conserved particles are related by convolution

$$c_{out}(t) = \int_0^\infty c_{in}(t'-t)h(t')dt' \tag{5}$$

If $c_{in}(t)$ and $c_{out}(t)$ are area-averaged concentrations then $h(t)$ is equal to eq.(3) and is called the weighting function (impuls response). The time domain parameter estimation method for D_{ax} ist based on eq.(5) together with eq.(3).

4. FREQUENCY DOMAIN RELATIONS

In the frequency domain the nonlinear convolution gets a simple multiplication

$$C_{out}(j\omega) = C_{in}(j\omega) \cdot H(j\omega) \tag{6a}$$

If we rearange eq. (6a) and multiplicate with the complex conjugated input signal $C^*(j\omega)$ we get a more handy form, because no division with complex numbers is necessary

$$H(j\omega) = C_{out}(j\omega) \; C^*_{in}(j\omega) / ((C_{in}(j\omega) \; C^*_{in}(j\omega)) = S_{io}(j\omega)/S_{ii}(j\omega) \tag{6b}$$

The frequency response function and the weighting function are related via the Plancherel theorem.

$$h(t) \quad FFT \; \underset{IFT}{\overrightarrow{}} \quad H(j\omega) \tag{7}$$

The method for parameter estimation is based on eq.(6b) and relation (7) and following steps must be performed:

1) time signals; sample frequency and sample time must be se- $c_{in}(t)$; $c_{out}(t)$
   selected in relation to the expected Bo-number
2) Fast-Fourier-Transformation of time signals $C_{in}(j\omega)$; $C_{out}(j\omega)$
3) Autospectral densitiy function of input signal $S_{ii}(j\omega)$
4) cross-spectral densitiy function $S_{io}(j\omega)$
5) Frequency response function $H(j\omega)=S_{io}(j\omega)/S_{ii}(j\omega)$
6) inverse Fourier transform of $H(j\omega)$ $h(t)$
7) fitting measured $h(t)$ with eq.(3) Bo, t_R
8) real residence time t_R from eq.(4a) and (4b) t_R

In order to select an adequate sample frequency, we have evaluated the center frequency and the minimum and maximum frequency for which the gain factor changes $-3dB$/decade. The result is given in table 1

Bo	center freq.	max. freq.	min. freq.
100	1.33	4.0	0.1
50	0.9	4.0	0.1
1	0.4	10.0	0.01
0.5	0.6	20.0	0.01
0.1	2.0	90.0	0.01
0.01	20.0	600.0	0.002

<u>table 1</u>: normalized frequencies as a function of Bo-number

Hence the bias error is to be limited to $<= 0.1$ the required record length T is (Ref. 3)

$$T = (bandwith \cdot 0.01)^{-1} \tag{8}$$

In fig. 1 the measured $c_{in}(t)$ and $c_{out}(t)$ for an bubble column ($d_K=0.1m$, $H_K=2.86m$; $\dot{v}_D=0.04$ m/s; $\dot{v}_C=0.04$ m/s) are shown. In fig. 2 the measured weighting function and the best model fit according to eq.(3) are compaired. The parameter may also be evaluated from the Bode plot (fig.3a and 3b) or the Nyquist plot (fig.4), but the curve fitting in time domain with the weighting function is mathematically more suitable. The compairison with results using the convolution of eq.(5) was within +/- 3%. The on-line fitting via the FFT takes about 1 minute on a desktop computer when 2048 data points were used. The off-line fitting in time domain via the convolution takes about 5 minutes on a Cyber 175 when using only 600 data points.

6. REFERENCES

Ref. 1   J.Sawinsky, J. Hunek, L. Podmaniczky, "A diffusional model for longitudinal mixing", Int. Chem. Engng. 20(1980)4, pp.681-687
Ref. 2   E.B. Nauman, B.A. Buffham, "mixing in continous flow systems", London, John Wiley & Sons (1983), pp. 94-109
Ref. 3   J.S. Bendat, A.G. Piersol, "Random Data: analysis and measurement procedures", London, John Wiley & Sons (1971), pp. 210/211

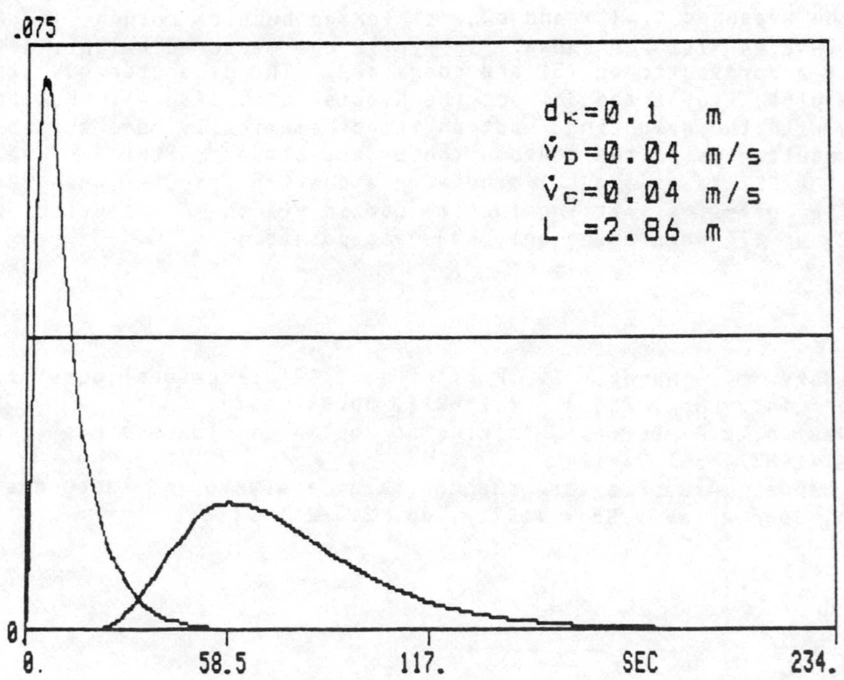

Fig. 1: Input/output concentration of a
NaCl-tracer in a bubble column

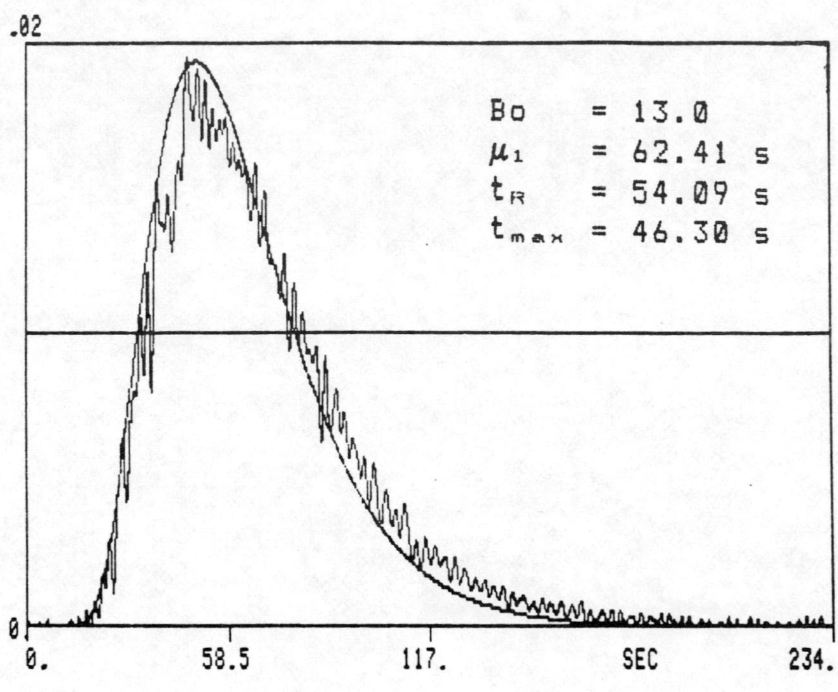

Fig. 2: measured weighting function and
best model fit

568

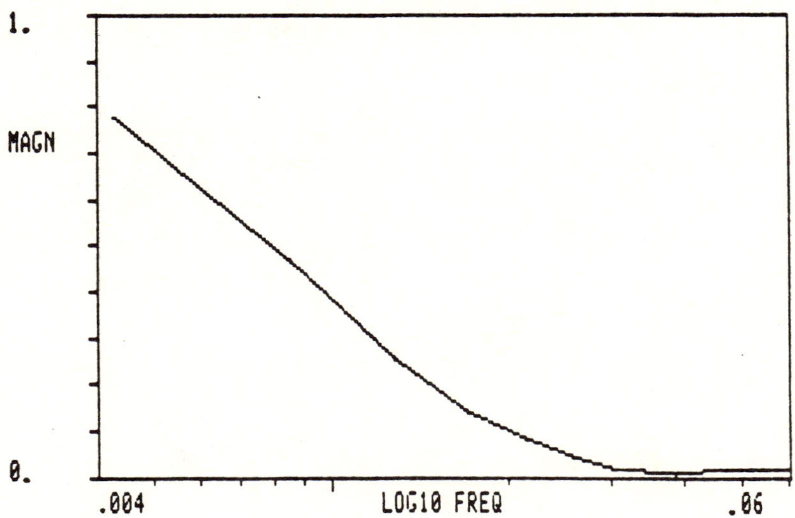

Fig. 3a: Bode plot, gain factor

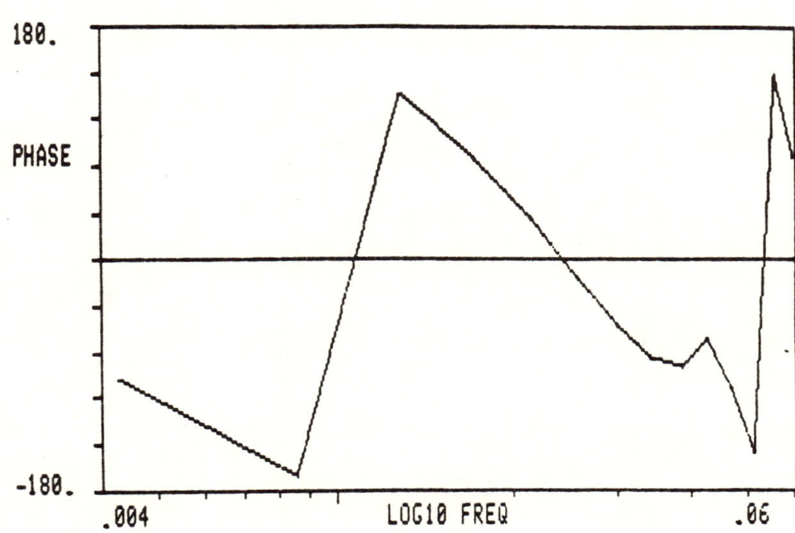

Fig. 3b: Bode plot, phase angle

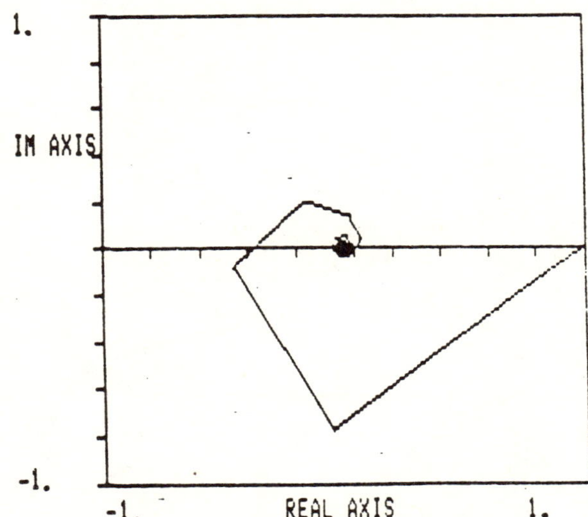

Fig. 4: Nyquist plot of Frequency
response function

HOMOGENIZATION EFFICIENCY OF MOTIONLESS MIXERS

V.Novák, V.Jandourek[*], F.Rieger

Czech Technical University Prague, Czechoslovakia

[*]Research Institute of Synthetic Rubber, Kralupy, Czechoslovakia

Summary

This paper presents structual details and performance charakteristics of a new type static mixer designed at the Department of Chemical and Food Equipment Design, Czech Technical University, Prague. The influence of geometry on the pressure drop and homogenization length has been studied.

Experimental results are presented in the form of dimensionless dependence of pressure drop coefficient λ_{SM} upon the Reynolds number Re and the dependence of relative mixing length L_{SM}/D upon Re. Dimensionless criterion λ_{SM} Re $(L_{SM}/D)^2$ which expresses specific energy consumption, was calculated from experimental data and homogenization efficiency of various types and geometries of static mixers was compared.

Held at Wurzburg, 10-12 June, 1985.

Organised by DVCV· Deutsche Vereinigung für Chemie- und Verfahrenstechnik
(German Association of Chemical and Process Engineering).

Organisation: GVC·VDI-Gesellschaft Verfahrenstechnik und Chemieingenieurwesen. GVC
©BHRA, The Fluid Engineering Centre, Cranfield, Bedford MK43 0AJ, England.

<u>NOMENCLATURE</u>

A = constant in eqn.(3)

C_1 = constant in eqn.(5)

C_2 = constant in eqn.(6)

D = pipe diameter

L = total length of static mixer

L_{SM} = mixing length

u = mean liquid velocity in the pipe

Re = Reynolds number, $Re = uD\varrho/\eta$

$\dot{V}_1, \dot{V}_2$ = flow rate of mixed components

λ_{SM} = pressure drop coefficient

Δp = pressure drop

π_c = criterion defined by eqn. (7)

η = viscosity

ϱ = density

1. INTRODUCTION

In connection with the development of continuous processing of highly viscous materials motionless mixers are often used for mixing operations. The number of commercially available static mixers has incresed rapidly in the past years. Their description and main characteristics (pressure drop and mixing length) can be found in literature e.g. (Ref. 1, 2). The new type of static mixer designated as Koax was developed for mixing of viscous materials (Ref. 3). This paper is devoted to the investigation of the effect of geometry upon the pressure drop and homogenization length of this new design. The comparison of the homogenization efficiency of all static mixers investigated was made on the basis of dimensionless specific energy requirements

2. EXPERIMENTAL

The types of motionless mixers investigated and their main geometrical parameters are shown in Fig. 1. Each element of the Koax mixer consists of a pair of oppositely oriented conical segments protruding mutually through slits on the lateral areas and connected in the plane of symetry. The basic geometry designated Koax is characterized by the ratio of segments surface S_s to slits area S_o, $S_s/S_o = 5$ and the cone angle $\alpha = 60^o$. The geometries designated Koax I resp. Koax III have the ratio of $S_s/S_o = 1$ and the angle $\alpha = 60^o$ resp. 90°.

In all experiments presented here the static mixers were placed in the glass tube of inner diameter $D = 16.5$ mm. The layout of experimental equipment is shown in Fig. 2. The pressure drop was measured by the pressure gauge.

The decolouration method was used in the mixing length measurements. The one liquid component was coloured by iodine and the second component contained sodium thiosulfate as decolourizing reagent in 5 % excess to equivalent amount. Starch as chemical indicator was used. Mixing length L_{SM} was defined as a distance at which the last traces of the colour disappeared. Aqueous solutions of corn syrup, glycerol and distilled water were used in the experiments. The flow rate ratio of mixed components was $\dot{V}_1/\dot{V}_2$ and viscosity $\eta_1 = \eta_2$.

3. RESULTS

The pressure drop of motionless mixers can be calculated using the modified Darcy-Weissbach equation:

$$\Delta p = \lambda_{SM} \; \frac{L}{D} \frac{u^2}{2} \varrho \tag{1}$$

where pressure drop coefficient λ_{SM} for given type and geometry of static mixer is a function of Reynolds number.

The graphical form of the relationship

$$\lambda_{SM} = f(Re) \tag{2}$$

for mixers investigated is shown in Fig. 3 wherefrom it can be seen that in the creeping flow regime (low values of Reynolds number) the logarithmic plot λ_{SM} vs Re is a straight line with the slope -1 and the relationship (2) takes the form:

$$\lambda_{SM} = A \, Re^{-1} \tag{3}$$

The values of constant A in eqn. (3) for investigated types of static mixers are summarized in Table I.

From Fig. 3 it follows that the lowest pressure drop exhibits the Kenics mixer and the highest pressure drop has the geometry Koax. Among the Koax geometries which were investigated, the the lowest pressure drop exhibits the Koax I geometry.

The results of homogenization measurements

The results of homogenization measurements were processed in the form of dimensionless dependence of relative mixing length L_{SM}/D on Reynolds number

$$L_{SM}/D = f(Re) \tag{4}$$

The graphical form of relationship (4) for mixers investigated is presented in Fig. 4. From this figure it can be seen that in the creeping flow regime and also in the fully turbulent regime the value of relative mixing length is constant.

In the creeping flow regime we can write

$$L_{SM}/D = C_1 \tag{5}$$

and in a fully turbulent regime

$$L_{SM}/D = C_2 \tag{6}$$

The values of C_1 and C_2 constants are also summarized in Table I. From the values of C_1 it follows that the shortest relative mixing length in the creeping flow regime exhibits the Koax mixer and the longest mixing length L_{SM}/D has the geometry Koax I. From Fig. 4 it is obvious that in transition region the value of the relative mixing length L_{SM}/D decreases with increasing Reynolds number for all types of mixers investigated with exception of Kenics. The last mentioned type of static mixer exhibits an increase of mixing length in the region of Reynolds numbers $3 < Re < 50$ which is in agreement with results reported by Hartung and Hiby (Ref. 4) and Šír (Ref. 5).

From the values of constant C_2 it can be seen that in the turbulent region the shortest mixing length exhibits the geometry Koax III and the longest mixing length has the Sulzer SMX mixer.

Homogenization efficiency

For comparison of energy required for homogenization by different type of static mixers the following dimensionless criterion was used (Ref. 2).

$$\pi_c = \lambda_{SM} \, Re(L_{SM}/D)^2 \tag{7}$$

The dependence of this criterion on Reynolds number is shown for investigated static mixers in Fig. 5. From this Fig. it can be seen that in agreement with eqs. (3) and (5), the criterion π_c is constant in the creeping flow regime.
The values of the criterion π_c in the creeping flow regime are for investigated mixers summarized in Table I.
From values listed in this table and also from Fig. 5 it follows that in the creeping flow regime the best efficiency (the lowest value of π_c) exhibits the Kenics mixer followed by Sulzer SMX and Koax whereas in the fully turbulent region the best efficiency has the geometry Koax I.

Even if the Kenics mixer has the best efficiency in the creeping flow regime it has in comparison with Koax and Sulzer SMX mixers, a disadvantage of long mixing length and low efficiency in transition region.

REFERENCES

1. Pahl,M.H., Mushelknautz,E.: "Static mixers and their applications". International Chemical Engineering $\underline{22}$, 2, Apr. 1982, pp. 197-205.

2. Streiff,F.A.: "Adapted motionless mixer design". In: Proc. 3rd Europ. Conference on Mixing (York, U.K.Apr. 4-6, 1979) Cranfield U.K., BHRA Fluid Engineering, 1979, Paper C2, 18 pp.

3. ČSSR pat. No 22 281

4. Hartung,K.H., Hiby,J.W.: "Continuous laminar mixing in Kenics-tube" (Kontinuierliche Laminarmischung im Kenics-Rohr), VDI - - Berichte 1975, No 232 pp. 383-389. (In German)

5. Šír,J.: " Pressure drop and homogenization effects of static mixer Kenics" (Tlaková ztráta a homogenizační účinky statických směšovačů Kenics) Thesis, VŠCHT Pardubice 1980. (In Czech).

TABLE I   Experimental results

Mixer type	A	C_1	$\Pi_c \times 10^{-6}$	Re<	C_2	Re>
Kenics	548.0 ± 7.8 [*]	33.7	0.622	35	8.85	3200
Koax	1674.0 ± 61	26.4	1.16	60	6.35	470
Koax I	839.5 ± 14.8	46.1	1.78	40	4.44	1900
Koax III	1690 ± 28	28.3	1.35	35	4.17	1000
Sulzer SMX	1306.0 ± 30	29.7	1.15	10	16	300

[*] 98 % confidence limits

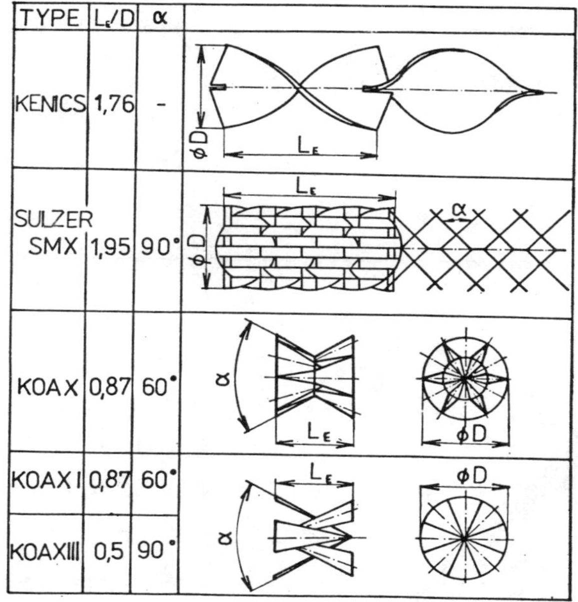

Fig. 1   Main geometrical parameters of investigated static mixers

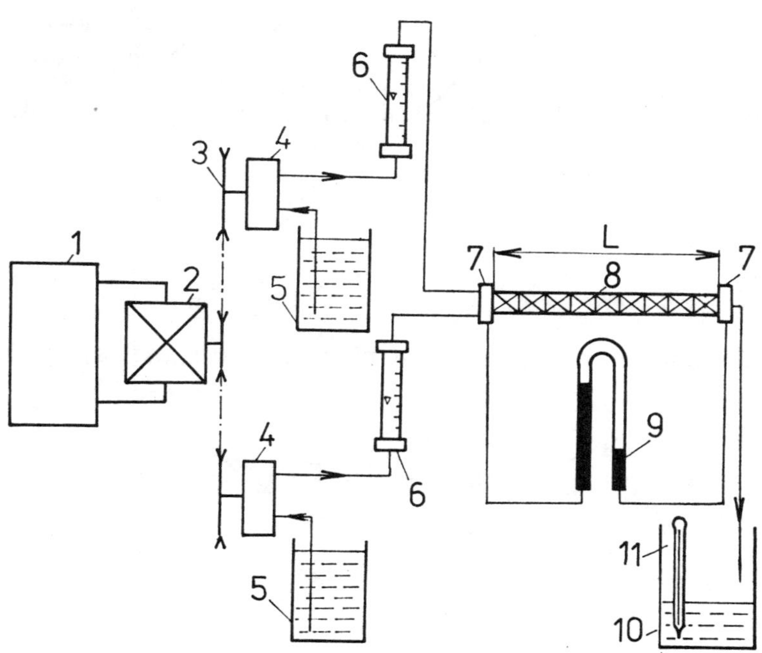

Fig. 2   Layout of experimental equipment

1 - hydraulic power unit, 2 - hydromotor, 3 - chain gearing
4 - gear pumps, 5 - storage vessels, 6 - flow meters, 7 - flanges
8 - static mixer, 9 - pressure gauge, 10 - collecting tank 11 - thermometer

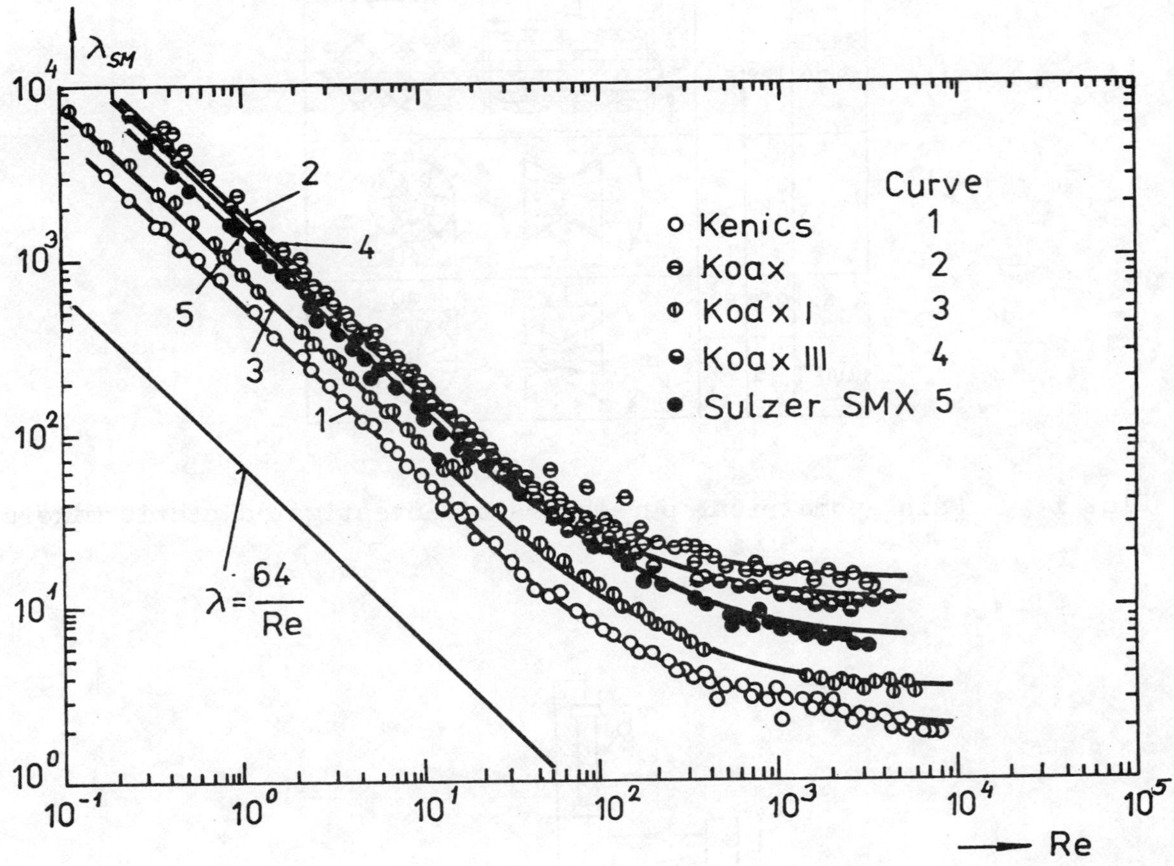

Fig. 3   Pressure drop characteristics for investigated mixers

578

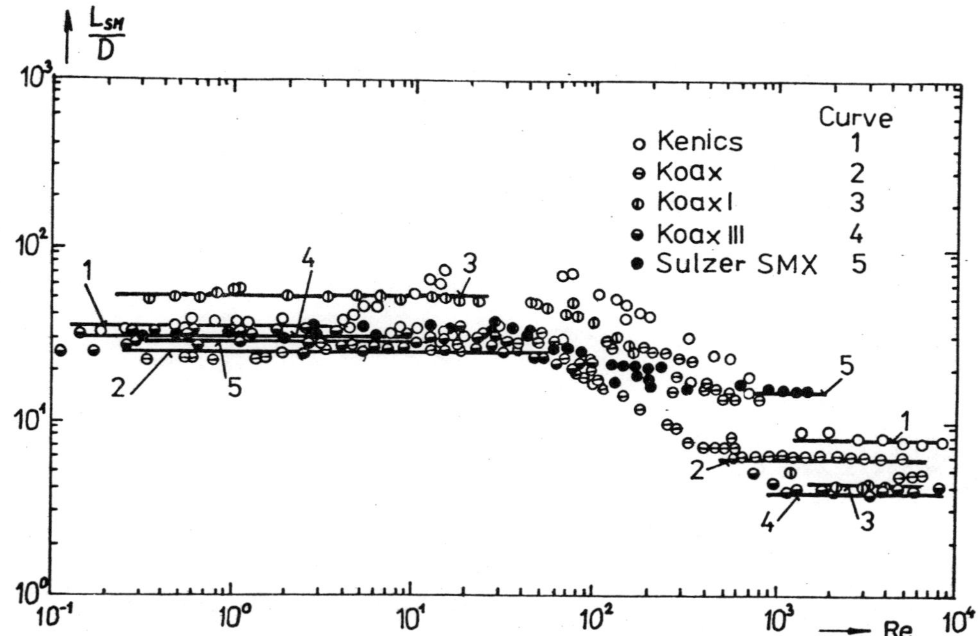

Fig. 4   Dependence of dimensionless mixing length versus Reynolds number

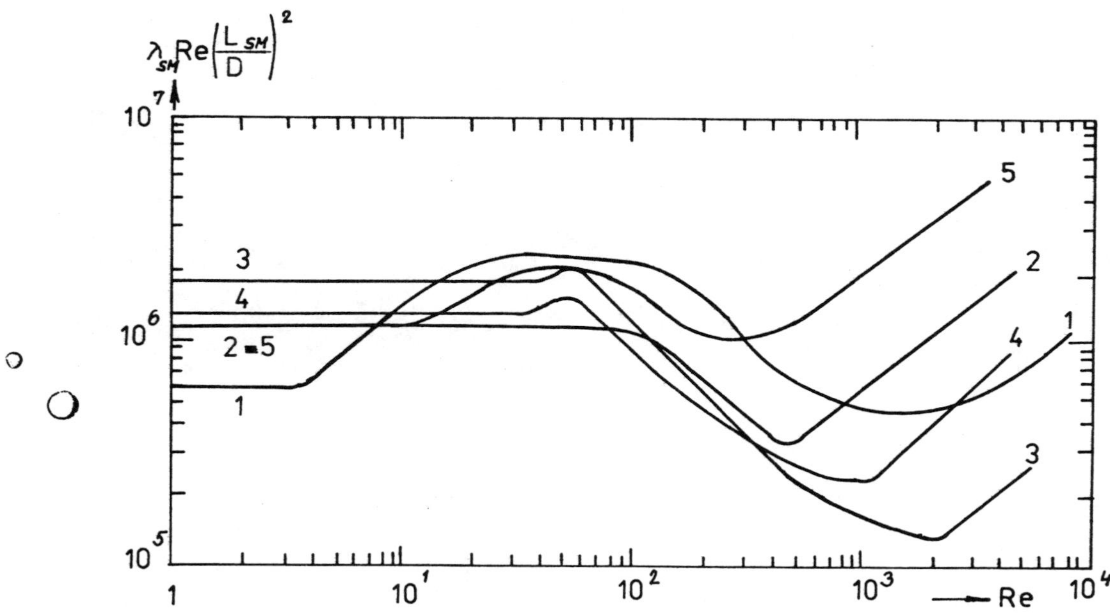

Fig. 5   Comparison of investigated static mixers with respect to energy consumption

1 – Kenics, 2 – Koax, 3 – Koax I, 4 – Koax III, 5 – Sulzer SMX

POSITIVE AND NEGATIVE SHEAR VISCOSITY
OF CLAY SUSPENSIONS INVOLVED BY MIXING

Z. Sobotka

Mathematical Institute
Žitná 25, 110 00 Praha 1,
Czechoslovakia

Summary

The orientated structure of suspensions as well as of other noo-Newtonian liquids, involved by rotational mixing, gives rise to the mechanical asymmetry in the shear flow. By analogy with the positive and negative shear of structured solids, the author has introduced the notion of positive and negative shear viscosities for the positive and negative flow of liquids with orientated structure. In the positive flow, with higher values of shear viscosity, the structured liquids resist movement more than in negative flow which is characterized by lower viscosities. Since the structure of liquids can be affected or destroyed by external or internal factors, such as chemical reactions, the differences between positive and negative viscosities are not constant but they vary with time.

The author has confirmed the existence of positive and negative viscosities by the results of tests of suspensions of clay as well as of those of clay and cement in water. These suspensions were first stirred in various ways in a rotational stirred tank before being tested at 12 angular velocities in the rotational viscometer Rheotest. If the moving cylinder of the viscometer rotated in the opposite sense to that of the agitator, the suspensions exhibited higher viscosities than when the cylinder and agitator were rotated in the same direction. The differences between the positive and negative shear viscosities are due to the orientated structure of suspensions, which has been formed in the course of stirring. They increase with the density of suspensions. The difference between positive and negative shear viscosities beyond the yield point of a clay suspension with the clay-water ratio 1.75 was about 90%. The viscosities of suspensions of clay and cement in water were affected by the setting of cement.

Held at Wurzburg, 10-12 June, 1985.

Organised by DVCV· Deutsche Vereinigung für Chemie- und Verfahrenstechnik
(German Association of Chemical and Process Engineering).

Organisation: GVC·VDI-Gesellschaft Verfahrenstechnik und Chemieingenieurwesen. **GVC**

NOMENCLATURE

b_{ijkl} = dimensionless fourth-rank tensor of anisotropy

$\dot{e}_{kl}$ = flow tensor or strain-rate tensor

$\mathring{\dot{\beta}}_{ij}$ = transformed flow tensor

$\left.\begin{array}{c} \mathring{\dot{\beta}}_{I} \\ \mathring{\dot{\beta}}_{II} \\ \mathring{\dot{\beta}}_{III} \end{array}\right\}$ = principal values of the flow tensor

σ_{ij} = stress tensor

$\left.\begin{array}{c} \phi_{0} \\ \hat{\phi}_{1} \\ \Phi_{2} \end{array}\right\}$ = scalar functions of invariants

1. INTRODUCTION

By analogy with the different moduli of the positive and negative shear of structured solids, having different mechanical properties in tension and compression, the author introduces the notion of positive and negative viscosities for the positive and negative flow of liquids with orientated structure. These phenomena are related to the structural viscosity of non-Newtonian liquids and to the orietation of dispersed particles. They are also affected by the orientated roughness of the walls.

It can be shown that the rotational mixing gives rise to the orientation of suspended clay particles in water, causing mechanical asymmetry in the shear flow. The degree of the orientation of particles depends on the time of mixing as well as on the time and various effects after mixing. Then the differencies between positive and negative viscosities are not constant but they can vary with time since the structure of liquids may be affected or destroyed by external or internal factors such as chemical reactions.

2. ANALYTICAL RELATIONS

The orientated structure of liquids involves the anisotropy. For the analysis of anisotropic liquids, Sobotka (Ref. 1) introduces the transformed flow tensor or strain-rate tensor which has the same principal axes as the stress tensor:

$$\dot{\beta}_{ij} = b_{ijkl}\, \mathring{e}_{kl}. \tag{1}$$

Then the stress tensor σ_{ij} can be expressed by an isotropic tensor function of the transformed flow tensor:

$$\sigma_{ij} = f_{ij}(\dot{\beta}_{kl}). \tag{2}$$

Expanding this function, according to the rules of tensor algebra, we obtain

$$\sigma_{ij} = \Phi_0\, \delta_{ij} + \Phi_1\, \dot{\beta}_{ij} + \Phi_2\, \dot{\beta}_{i\lambda}\, \dot{\beta}_{\lambda j}. \tag{3}$$

The scalar function Φ_0 of invariants corresponds to the volume effects, Φ_1 represents the shear-viscosity function and Φ_2 the cross-viscosity function. Substituting the relations between the principal values of the stress and transformed flow tensors

$$\sigma_I = f(\dot{\beta}_I), \qquad \sigma_{II} = f(\dot{\beta}_{II}), \qquad \sigma_{III} = f(\dot{\beta}_{III}), \tag{4}$$

which are based on the coincidence of the principal axes of both the tensors, into Eq. (3) and writing this equation for all principal stresses, we obtain the set of three algebraic equations:

$$f(\dot{\beta}_I) = \Phi_0 + \Phi_1\, \dot{\beta}_I + \Phi_2\, \dot{\beta}_I^2, \tag{5}$$

$$f(\dot{\beta}_{II}) = \Phi_0 + \Phi_1\, \dot{\beta}_{II} + \Phi_2\, \dot{\beta}_{II}^2, \tag{6}$$

$$f(\dot{\beta}_{III}) = \Phi_0 + \Phi_1\, \dot{\beta}_{III} + \Phi_2\, \dot{\beta}_{III}^2, \tag{7}$$

for determining the scalar functions Φ_0, Φ_1 and Φ_2. Substituting them into Eq. (3), we obtain, after rearranging the terms:

$$\sigma_{ij} = \frac{1}{(\dot{\beta}_I - \dot{\beta}_{II})(\dot{\beta}_{II} - \dot{\beta}_{III})(\dot{\beta}_{III} - \dot{\beta}_I)} \{[\dot{\beta}_{II}\, \dot{\beta}_{III}(\dot{\beta}_{III} - \dot{\beta}_{II})f(\dot{\beta}_I) +$$

$$+ \dot{\beta}_I\, \dot{\beta}_{III}(\dot{\beta}_I - \dot{\beta}_{III})f(\dot{\beta}_{II}) + \dot{\beta}_I\, \dot{\beta}_{II}(\dot{\beta}_{II} - \dot{\beta}_I)f(\dot{\beta}_{III})]\, \delta_{ij} +$$

$$+ [(\dot{\beta}_{II}^2 - \dot{\beta}_{III}^2)f(\dot{\beta}_I) + (\dot{\beta}_{III}^2 - \dot{\beta}_I^2)f(\dot{\beta}_{II}) + (\dot{\beta}_I^2 - \dot{\beta}_{II}^2)f(\dot{\beta}_{III})]\dot{\beta}_{ij} +$$

$$+ \left[(\mathring{\dot\beta}_{III} - \mathring{\dot\beta}_{II})f(\mathring{\dot\beta}_{I}) + (\mathring{\dot\beta}_{I} - \mathring{\dot\beta}_{III})f(\mathring{\dot\beta}_{II}) + (\mathring{\dot\beta}_{II} - \mathring{\dot\beta}_{I})f(\mathring{\dot\beta}_{III}\right]\mathring{\dot\beta}_{i\lambda}\mathring{\dot\beta}_{\lambda j}\Big\}. \qquad (8)$$

This relation yields the shear stress in the plane 12:

$$\mathring{\sigma}_{12} = \frac{1}{(\mathring{\dot\beta}_{I} - \mathring{\dot\beta}_{II})(\mathring{\dot\beta}_{II} - \mathring{\dot\beta}_{III})(\mathring{\dot\beta}_{III} - \mathring{\dot\beta}_{I})}\Big\{\left[(\mathring{\dot\beta}_{II}^{2} - \mathring{\dot\beta}_{III}^{2})\,f(\mathring{\dot\beta}_{I}) +\right.$$

$$+ (\mathring{\dot\beta}_{III}^{2} - \mathring{\dot\beta}_{I}^{2})\,f(\mathring{\dot\beta}_{II}) + (\mathring{\dot\beta}_{I}^{2} - \mathring{\dot\beta}_{II}^{2})\,f(\mathring{\dot\beta}_{III})\left]\mathring{\dot\beta}_{12} + \right[(\mathring{\dot\beta}_{III} - \mathring{\dot\beta}_{III})\,f(\mathring{\dot\beta}_{I}) +$$

$$+ (\mathring{\dot\beta}_{I} - \mathring{\dot\beta}_{III})\,f(\mathring{\dot\beta}_{II}) + (\mathring{\dot\beta}_{II} - \mathring{\dot\beta}_{I})\,f(\mathring{\dot\beta}_{III})\left]\left[\mathring{\dot\beta}_{12}(\mathring{\dot\beta}_{11} + \mathring{\dot\beta}_{22}) + \mathring{\dot\beta}_{13}\mathring{\dot\beta}_{32}\right]\Big\}.$$
$$(9)$$

The asymmetry of the function $f(\mathring{\dot\beta}_{K})$ with respect to the origin, characterized by the inequality $f(\mathring{\dot\beta}_{K}) \neq -\,f(-\mathring{\dot\beta}_{K})$, involves the differencies between positive and negative shear stresses.

3. EXPERIMENTAL INVESTIGATION

The author has confirmed the existence of differences between positive and negative viscosities by tests of various suspensions of clay as well as of those of clay and cement in water. These suspensions were first stirred in the rotational stirred tank at 1700 r. p. m. during 5 minutes and then at 600 r. p. m. during consecutive three and half hours before being tested at 12 angular velocities in the rotational viscometer Rheotest. If the moving cylinder of the viscometer rotated in the opposite sense to that of the agitator, the suspensions exhibited higher viscosities then those which were stirred and tested in the same sense of rotation. An illustrative example of the dependence of the positive and negative shear stresses on shear rates of a suspension of 17.5 g clay in 10 g water, tested in the rotational viscometer, is represented by the curves P and N in Fig. 1. These flow curves have slightly pronounced yield points.

4. REFERENCE

1.	Sobotka, Z.: "Rheolgy of Materials and Engineering Structures". Amsterdam, Oxford, New York, Tokyo, Elsevier, 1984, pp. 184-190.

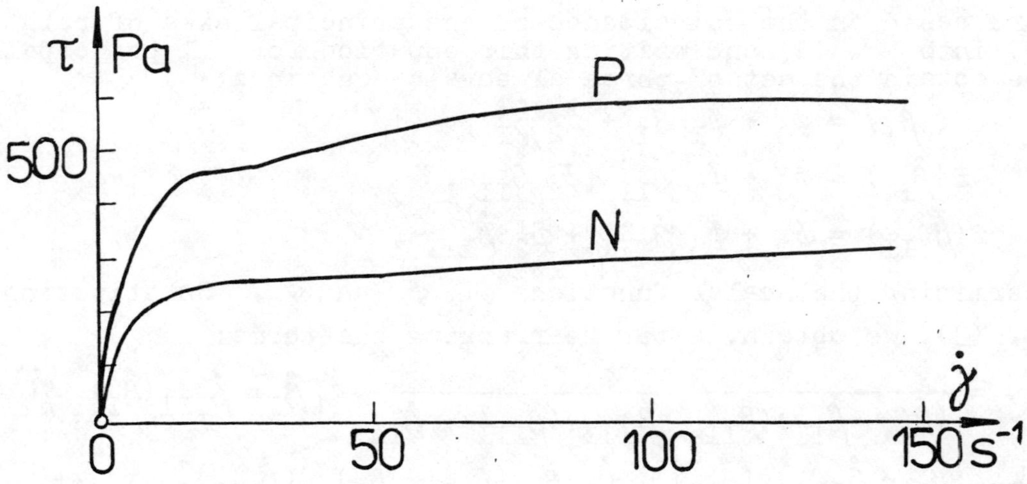

Fig. 1. Flow curves of the positive and negative shear of a suspension of clay in water

THE MECHANICAL AERATION PROCEEDING OF WASTEWATERS

Octavian Floarea, Gheorghița Jinescu and Paul Vasilescu

Polytechn. Inst., Chem. Enang. Dept., București, Romania

In this paper it was studied the aeration proceding of the wastewaters, containing phenol or detergent, using a mechanical aerator of original construction. The experimental results have permited to establish mathematical relations of polinominal expresion to reveal the oxigenation capacity depending on the impurificator concentration.

Held at Wurzburg, 10-12 June, 1985.

Organised by DVCV· Deutsche Vereinigung für Chemie- und Verfahrenstechnik (German Association of Chemical and Process Engineering).

Organisation: GVC·VDI-Gesellschaft Verfahrenstechnik und Chemieingenieurwesen. 

©BHRA, The Fluid Engineering Centre, Cranfield, Bedford MK43 0AJ, England.

NOTATIONS:

$\dot{m}$ - mass flowrate of the oxygen transfered in water, kg/h;

K_L - total mass transfer coefficient of oxygen in water, h^{-1};

t - time, h;

c - concentration of oxygen transfered in water, kg/m^3;

c_o - initial concentration of the oxygen in water, kg/m^3;

c_{ww} - concentration of oxygen transfered in the wastewater, kg/m^3;

c_{eq} - saturation concentration of the oxygen in water, kg/m^3;

$c_{eq,ww}$ - saturation concentration of the oxygen in the wastewater;

  , - correction factors defined by the (5) and (6) equations.

  - temperature correction factor;

V - water volume, m^3;

p - pressure, torr.

The high development of the wastewaters purification technologi-
es in the last years has led to the designing and achievement of high
efficiency mechanical devices for (waste) water aeration – an important phase in the technological processeses of wastewaters biological or bio-chemical treatment.

The achievement of the **surface** quick mechanical aerators –
in various alternatives – permanently **improved** – allowed increased
performances in the wastewaters oxygenation process be obtained.

The oxygenation capacity in **such** aerators is about 3 times
higher than that of the slow mecanical aerators.

This paper presents the aeration process for the **wastewaters**
containing phenol or detergent in a o.5 x o.5 m basin using a surface axial aerator, modified, improved alternative of the BSK (G.D.R.)
type. The characteristics of the aeration device to be used: o.1 m
rotor size, o.37 kw motor powre, 1000 rpm motor speed.

The testing of the achieved aeration device in standard conditions required determination of the variation of the oxygen concentration in water – in time and in various points of the basin.

The determination of the concentration **has** been both chemically
by winkler tests and by means of oxygen analyser ( ) achieved.

From the experimental data obtained for the variation – in
time – of the oxygen concentration in water – it is found ont that
the aeration process is represented by gas – liquid mass transfer
process in transitory condition expressed by the equation:

$$\frac{dc}{dt} = K_L(c_{eq} - c) \tag{1}$$

separating the variables and integrating the following is obtained:

$$\frac{c_{eq} - c}{c_{eq} - c_o} = \exp(-K_L t) \tag{2}$$

By this equation, the value of the K_L total mass transfer coefi-
cient has been determined.

The oxygenation capacity of the aerator has been calculated
by the equation:

$$m = K_L \, c_{eq} \, V \qquad [kg/h] \qquad (3)$$

The oxygenation capacity of the aerator depends on the depth
the device is sunk in liquid; the experimental determinations showed
that - for the aerator used - at h = 55 mm, a high oxygenation capacity is obtained. By all the experiments achieved - there have been
calculated - the total mass transfer coefficient, the actual efficiency and the time necessary for a recirculation of the liquid,
ensuing the following values: 88.14 h^{-1}, 3.76 kg O_2/kwh and respectively 14.4 sec.

As the achieved aerator is mini - dimensioned, is has been
put on a bigger scale - to comply with the 2 m rotor size, 30 kw
motor power and 1000 rpm speed, an actual efficiency of 2.25 kg O_2/
kwh is obtained, higher than that of the aerators manufactured by
the firms: BSK(GDR), ABC - SEM(France), GYROX - LURGI (RRG), VIDUS(H),
and quite similar to that manufactured by DORR OLIVER(USA), 2.3 kg O_2/
kwh - mixed surface mechanical aerator with air blastis.

The aerator - tested in standard conditions - has been used
for the oxigenation of the wastewaters containing phenol, in the
(1.1 - 3.3) g/m^3 concentration range and detergent, in the (0.05 -
0.2) g/m^3 concentration range.

The variation - in time and space - of the oxygen concentration in water- has been watched finding out a perfect mixing of the
air bubbles in the liquid mass, practically the concentrations of
the samples taken in various points of the basin being equal.

From the experimental determinations carried out, there en-
sured the values of the total gas - liquid mass transfer coefficients
and the saturation concentrations of the wastewater for various concentrations and type of impurifying substance.

The relation $\overline{/1/}$ has been used for the calculation of the homogenization coencentrations:

$$\dot{m} = \alpha \, K_{w,w} (\beta \, c_{eq,ww} - c_o) V \cdot \theta \cdot \frac{p}{760} \qquad (4)$$

where the α and β correction coefficients are given by the relations:

$$\alpha = \frac{K_{L,ww}}{K_L} \qquad (5)$$

$$\beta = \frac{c_{eq,ww}}{c_{eq}} \qquad (6)$$

It has been experimentally found out the variation of the , K_{ww} parameters with the concentration of the impurifying substance in water is linear and can be given by the equation of a straight line:

$$P_i = a_i + b_i c_{ww} \qquad (7)$$

where i index is α , β , K.

The variation of the impurified water saturation concentration is given by the equation:

$$P_c = c_{eq} = n_o \, c_{ww}^n \qquad n < 1 \qquad (8)$$

Havig the a_i, b_i, n_o and m coefficients determined for a type of wastewater and in the conditions of using the proposed aerator, the equation for the oxygenation capacity calculation becomes an equation depending only on the concentration of the impurifying substance, namely on the general form:

$$m = (C_1 - C_2 c_{ww}^n + C_3 c_{ww} - C_4 c_{ww}^{1+n} + C_5 c_{ww}^2 - C_6 c_{ww}^{2+n} + C_7 c_{ww}^3 - C_8 c_{ww}^{3+n}) \cdot$$
$$\cdot V \cdot \theta \cdot \frac{p}{760} \qquad (9)$$

where the C coeficients are calculated by the following relations:

$$C_1 = a_\alpha a_K (c_{eq}^2 - c_o) \qquad (10a)$$

$$C_2 = a_\alpha a_K n_o c_{eq} \qquad (10b)$$

$$C_3 = a_\alpha a_K b_\beta c_{eq} + (a_\alpha b_K + a_K b_\alpha)(c_{eq}^2 - c_o) \qquad (10c)$$

$$C_4 = n_o(a_\alpha a_K b_\beta + c_{eq}(a_\alpha b_K + a_K b_\alpha)) \qquad (10d)$$

$$C_5 = b_\beta c_{eq}(a_\alpha b_K + a_K b_\alpha) + b_\alpha b_K((c^*)^2 - c_e) \tag{10e}$$

$$C_6 = n_o(b_\alpha b_K c_{eq} + b_\beta(a_\alpha b_K + a_K b_\alpha)) \tag{10f}$$

$$C_7 = b_\alpha b_K b_\beta c_{eq} \tag{10g}$$

$$C_8 = b_\alpha b_K b_\beta n_o \tag{10h}$$

The walues of the constants from the relations (10) for the wat waters impurifying with the phenol are:

$a = 0.536$; $a_K = 49.1$; $a = 1$; $b = -1.95 \times 10^{-3}$; $b_K = -0.172$; $b = -1.31 \times 10^{-3}$; $n_o = 0.049$; $n = 0.63$

Knowing thath $c_{eq} = 9.18$ $g O_2/ m^3$ and that the water used hed $c_o = 0.22$ $g O_2/ m^3$, the ecuation (9) becomes known.

The obtained relation can be used both for the desisning of the wastewater treatment basinns when the necessary production and inpurifying substances content are known and for operation, for the quick determination of existing basin production capacities when the impurififying substance contend in the wastewater subjected to oxygenation is modified.

REFERENCES:

[1] M.K.Stenstrom and R.Gary Gilbert; Water Research, vol.15,pp. 643-54, 1981.

AN INDUSTRIAL MIXING LABORATORY

W. N. Ford, A. W. Etchells III, and D. G. R. Short
E. I. du Pont de Nemours & Co., Inc.
Engineering Department
Wilmington, DE 19898   USA

Summary

The industrial mixing laboratory is a useful facility for seeking answers to mixing questions which cannot be answered by currently available correlations and theory. The main purpose of the laboratory is quick answers to pressing problems. Once a solution is found, the work is stopped. Thus, very little of the data is such that high-quality correlations can be developed and published. This paper will describe one such facility to show what classes of problems are studied and what techniques are used. The need for improved correlations and techniques will be discussed.

Held at Wurzburg, 10-12 June, 1985.

Organised by DVCV· Deutsche Vereinigung für Chemie- und Verfahrenstechnik
(German Association of Chemical and Process Engineering).

Organisation: GVC·VDI-Gesellschaft Verfahrenstechnik und Chemieingenieurwesen. 

©BHRA, The Fluid Engineering Centre, Cranfield, Bedford MK43 0AJ, England.

<u>**THE CHEMICAL INDUSTRY MIXING PERSPECTIVE**</u>

In the end, someone must specify the mixer to do the industrial mixing job. This person must be willing to share the credit for the successes and take all of the blame for the failures. In some cases, the person works for the vendor of the mixing equipment. In some cases, the end user specifies the mixer. In other cases, the person is a third-party consultant. Our case is that of working directly for the end users of the mixing equipment as consultants. In all cases, an important decision-making tool is the ready availability of a mixing laboratory.

Our perspective is that of internal mixing consultants in the Engineering Department of the Du Pont Company, a large multifacility corporation. The paper we gave at the Fourth European Conference on Mixing presented our views of the present state of mixing knowledge. Two years is too short a time to expect much to change or for any of the problems to have been solved. But much genuine progress has been made.

The fact that there are still unsolved problems in mixing does not stop researchers from inventing new chemicals and processes which have difficult mixing demands. We feel that a laboratory for in-house problem-solving is a necessary item for a mixing consultant to perform his job.

In our position as consultants, we work on a wide variety of problems ranging from polymer additive blending to wastewater treatment to very small-scale biotechnology operations. Companies like Du Pont have traditionally made their products in equipment which is in the 0.2 to 60 cubic meter size. The largest vessel on which we have actively consulted in recent years has been 40,000 m^3 while a recent project in the biological sciences has involved mixing problems on the 100 ml scale. It is our belief that despite the variety of contradictory data in the literature, the fundamental mixing concepts for all these sizes remain the same.

The vast majority of mixing problems still occur in these traditional sizes of vessels. It is toward this scale of processing that our laboratory facilities are geared. Each new product our scientists develop seems to have mixing requirements which are nontraditional. No matter what has been studied before, someone will invent a new process or product which requires an even more severe test of mixing knowledge. This is where the in-house mixing laboratory has its major function.

Another important aspect of industrial mixing is that while the mixing step can play the controlling role in the manufacture of a product, the chemical industry has always demanded that the mixer be practically invisible. The demand on the industrial mixing specialist is to design it once and never fix it. If it does need fixing, it usually is a crisis which cannot tolerate a long investigation. The demands in the future will be to maintain the same high standards with nonoptimum vessel geometries and at lower investment and operating cost.

The majority of problems we study in the laboratory are troubleshooting and rating problems. Thus, the modeling problem is accurate scaledown of the process so that the key mixing steps are preserved.[1] An important feature of the mixing apparatus is flexibility and the ability to handle a wide variety of scales.

The investment and operating cost of mixing equipment is much misunderstood by those outside the industry.[2] We do not willingly waste energy. Our greatest costs in most mixing operations are the yield losses of high-value chemical feedstock and intermediates. It is our experience that if better mixing will achieve a 1% yield improvement, then the mixer can be over-designed by 10,000% before the extra energy cost negates the chemical saving. For all but a few designs with non-Newtonian fluids, the design safety factors used today are 10-15%.[3]

THE NEED FOR THE INDUSTRIAL LABORATORY

A previous paragraph touched on the justification for the in-house laboratory. It can be argued that with all the equipment available in the vendor
and university laboratories, the process industry has no need for its own
in-house laboratory. Our experience is that the outside laboratories are
not capable of meeting the needs of the process industry. Besides having to
meet the professional standards given in the introduction, the answers to
mixing problems must be obtained quickly and safely. The typical facility
outside the process industry is usually not capable of safely handling the
chemicals nor maintaining the proper security of the information.

Some specific examples of types of problems which required industrial
laboratory effort in order to solve the mixing problems are as follows:

o One method to recover synthetic fiber waste is to convert it to a paper.
  The paper mills have a wide variety of mixers in their existing pulping
  facilities but little fundamental understanding of the key variables of
  the mixing requirements in the stock chests. Extensive in-house work
  had to be done to determine the best designs for the stock chest mixers.

o Paint mixing has been done for decades, yet little of the key knowledge
  needed for the design of a paint blender appears in the mixing literature. This includes the effects of variable volume in a tank, drawdown
  of added ingredients, and addition of materials with significantly
  different viscosities from the tank material.

o We found that the best mixer for our TiO_2 slurry customer tanks is a
  four-bladed, radial turbine mounted 5 to 10 cm off the bottom, and a
  four-bladed, pitched-bladed turbine mount at the tank's midpoint. It
  took extensive trial-and-error testing to evolve the current design.

o We do not have many three-phase systems. However, a few of our systems
  reported great difficulty with solids suspension even though all conventional knowledge concluded everything was properly designed. Tests
  showed that the problem was poor solids suspension while the vessels
  were being gassed. The solution was to develop an overdesign factor for
  the three-phase systems.

CURRENT FACILITIES

The Du Pont Company has two tank-type mixing laboratories. One facility is
based on a 0.9m diameter by 1.8m high vessel made of clear acrylic. The
bottom head is dimensionally consistent with the ASME Code for the heads of
pressure vessels. The motor is a 4 kw, variable-speed unit. The impellers
are the industry standard four-bladed, open turbines. This unit pioneered
the use of FM telemetry in mixing for transmitting torque and bending moment
measurements to the recording instruments.[4]

The second facility has a 1.5 kw, variable-speed drive and a variety of
impellers. Several tanks are available ranging in size from 15 to 660 cm in
diameter. Thus, typical scaledowns are 10:1 based on diameter. The fluids
used normally are water or some other model fluid of the process. If we do
not feel that a model fluid will be satisfactory, then the process material
is used. The main instrumentation is a torque meter, rpm meter, people, and
television cameras and recorders.

In addition to these two facilities, there are a variety of piping loops
which can be used for testing motionless mixer applications. Other facilities include a variety of twin screw processors which are primarily used for
polymer processing studies.

Our mixing laboratories are not research laboratories. The emphasis is on
accurate answers at low cost in a minimum amount of time to pressing problems. Rarely is there the opportunity to gather enough data to make generalized correlations. Once sufficient experiments have been done to define
and solve the problem, the equipment is converted for use with the next
problem.

TYPE OF TESTS

The main tests are process result tests. Most of these are blending studies
and solids suspension. The tests consist of building a scale model of the
equipment, using a fluid which models the important aspects of the process
and observing how well the mixer works. Our eyes are our main measuring
tool.

Blending Tests

It is our experience that the measurement of blending in an agitated tank is
the most difficult of all mixing measurements.[5] This is confirmed by a
review of the literature where everyone wants to define their personal
measurement of blending using their proprietary impeller. A major failure
of mixing research is the lack of a well-defined standard for characterizing
blending in a stirred tank. This has been one of the first problems which
the BHRA/FMP researchers had to address.

The most common problem is developing a design which will ensure homogeneity
of the product. The last can of paint filled from the blender must be
exactly the same as the first can. Thus, the typical blending test is to
measure the length of time for a color change to occur throughout the
vessel. The point blending measurement techniques only ensure good mixing
for turbulent flow in Newtonian fluids.

Other uses for blending-type studies is the determination of flow patterns
and stagnation in a vessel. Typical problems involve blending operations
with large volume and large viscosity differences in the fluids. An interesting generalization of some of our studies is that there is not a major
difference between mixing methods for batch blending, but there can be a
substantial difference for continuous blending - an interesting research
project for a laboratory better equipped to handle such a problem. Another
interesting area is semibatch mixing which occurs during the filling and
emptying of a vessel. The amount of reported work in this area is miniscule.

Slurry Tests

Slurry handling needs to be modeled for much the same reason as the blending
studies. While everyone will agree that the Zwietering's correlation[6] gives
a good prediction of off-bottom suspension, many of the high-value products
of the chemical industry require homogeneous suspension. In addition, the
results are physical property dependent. The best mixer for one type of
slurry is not necessarily the best for another. There are no good generalized correlations for predicting the behavior of homogeneous slurries in
stirred tanks, hence we must test most cases in the laboratory.

The need for further research is clearly in the area of producing homogeneous suspensions in batch and continuous mixers. The tendency is to use
multiple-impeller systems, but this is based almost entirely on empiricism.
The thought that "It seemed to work before so let's use it again" motivates
much of the mixer specification and laboratory testing for making homogeneous suspensions.

This brings up another point. Many of our tests are devoted to rating
traditional multiple-impeller designs for new applications. The main
incentive for using the multiple impeller designs is to maintain a
homogeneous slurry.

Other Fluids

The gas-liquid and liquid-liquid systems are rarely modeled in our laboratories. When the mixing consultants get involved, it is usually due to the
non-Newtonian nature of the fluid. One study which was mentioned earlier on
comparative performance of gas-liquid contacting devices is a rather rare
example of a three-phase study. It is currently being used for rating a
variety of sparger and impeller combinations in gas-dispersion applications.
The experimental technique is the computerized version of the method
developed by Chapman, et al.[7]

The one case which is nearly always modeled is mixing involving non-
Newtonian fluids. Many of our more profitable products involve the handling
of non-Newtonian material. Most of the time, the rheology is that of a
Bingham plastic or shear-thinning behavior. While the Metzner-Otto[8] study
was an outstanding piece of research for the prediction of power, there is
still a lot to be done. Particular areas which we often study are elimination of stagnant areas in the tank and the effect of vessel internals on the
mixing.

Vessel Internals

A common item in the majority of our testing is achieving homogeneity in
the vessel. Very often this means special internals or multiple impellers.
We have previously estimated that 80% of the vessels have two or more
impellers.[9] Because there is little information in either the open literature or in vendors' private reports on the behavior of multiple impellers,
we find ourselves all too often testing these configurations.

TYPICAL PROCEDURES

Obviously, a motor, a shaft, a variety of impeller types and sizes, and a
range of vessel sizes are needed. Our experience is that these are not
costly items since they typically only have a limited service. Thus,
substantial construction beyond the minimum need for safe operation is
usually not important. The critical item is the variety of vessel sizes and
shapes. The mixer designer rarely has the privilege of specifying the
vessel geometry. In addition, the influence of vessel size is the most
understudied variable in the literature. The vast majority of published
research thinks that vessel geometry effects are properly accounted for by a
D/T ratio. Failure to carry out mixing research in a variety of vessel
sizes usually indicates a failure to understand the important mixing
mechanisms.

The most important measurements which are made are visual observation of the
mixer operation, shaft speed, and shaft torque. For the vast majority of
mixing problems, the human eye is the best measuring instrument. The chief
aid is television so that later playback of the events can be done if more
precise timing and recording of events is needed. The standard "home"-style
video equipment has proven adequate for nearly all studies. If there is a
need to record the fluid motions induced by the impeller, then the Dove
prism[10] or rotating turntable[11] is superior to high-speed photography. We
have not found a need for either device to date.

Torque is important if the purpose of the experiment is to write a mechanical specification on the prototype mixer. The shaft torque and speed
determine the motor size which is used as the basis for all mechanical
calculations.

The most common impeller we use is the pitched-bladed turbine. While this is the most widely used impeller, it has only been in the last few years that much research has been done on the behavior of the units. The pitched-bladed turbine is widely used because it does not have a real weakness in the different features which are used to compare mixers. It is easy to fabricate. And its energy consumption for a given task is not that much worse than the best of the hydrofoils. Thus, if the precise impeller design is uncertain, the pitched-bladed turbine will usually do the job.

DATA CORRELATION AND MIXER DESIGN

We at Du Pont feel that the best mixers are designed, not scaled up. Scale-up implies that there is a dimensionless group or rule of thumb which ensures equivalent performance between the various sizes. Geometric similarity is rigidly followed.

We define design as incorporating into the production vessel those features which are truly important for proper operation. The majority of our new mixers are actually designed without the use of the laboratory. When our literature knowledge and experience indicate that the design risk is too high, the laboratory is used to confirm the design.

For a new design, the laboratory vessels will be made geometrically similar to proposed full-scale design. The typical experiments are to determine which impeller geometry and shaft speed give the desired result. Different vessel sizes are used to differentiate between size versus size ratio effects. The actual process material is often too toxic or corrosive to be used so model fluids are used. Depending on the situation, a test of the proposed system may be run using the actual fluids to confirm the results.

Once a geometry and speed are found which give acceptable results, the mixer torque for a range of operating conditions is determined. With this information, the proper geometry and equipment size for the full-scale mixer can be specified.

The other problem is rating of existing designs for new service. The sequence is the same with a series of experiments in geometric models of the full-scale vessel. It is very important that more than one size vessel be used to ensure that the proper geometry is determined. It is usually necessary to modify the vessel internals to meet the demands of the new service.

FUTURE LABORATORY FEATURES

While everything can be improved, we feel that the Du Pont facilities are quite good for the problem-solving task. Anything which makes the laboratory personnel more productive will be incorporated. Recently, we have found on-line computers to be a useful addition to the facilities. Improving our torque and visual measuring techniques will be an ongoing task.

The design of the mixer for an agitated tank requires knowledge of five items. These are (1) the shaft speed and mixer geometry needed to get the required homogeneity of the entire vessel, (2) the torque required to turn the shaft, (3) the hydraulic forces on the shaft, (4) the state of fluid motion around the impeller, and (5) the mixing at the molecular scale; i.e., micromixing.

Two areas we do not study on a regular basis are Items 3 and 4. We rely mostly on the equipment vendors for proper estimates of the hydraulic forces. Since the vendor has to guarantee the mechanical integrity of the equipment, their judgment is a valuable item in the proper specification of the mixer.

Item 4 would require a Dove prism or rotating turntable apparatus so that
the detailed motions of the fluid relative to the impeller can be studied.
Fortunately, the excellent research work at Delft and Birmingham has not
made it necessary for Du Pont to invest in this type of equipment.

A major improvement to everyone's laboratory efforts will be better con-
ceptual models for correlating data. Our own laboratory data make every
air-water dispersion correlation in the literature suspect for design of
large-scale equipment.[12] We continue to have to study such simple things
like blending small amounts of low-viscosity material into high-viscosity
liquids. We continue to observe that CSTR's do not give the same result for
similar conditions to batch vessels, and we continue to have to study
multiple-impeller systems.

CONCLUDING REMARKS

The intent of this paper is to show the need for a laboratory in the process
industry to solve mixing problems. As long as new products are invented
which require mixing technology outside the realm of current knowledge,
there will always be a need for the laboratory. To be able to predict the
mixing needs of the future is the same as predicting the future.

A second objective is to inform our research colleagues of major areas which
need further investigation. In our paper in the previous European Mixing
Conference, we voiced some frustration that little academic research was
being done with the types of mixers which are most widely used in the
process industries. In this paper, the underlying message is that a lot of
industrial work is done because there is still only limited fundamental
knowledge of producing homogeneous fluids and slurries in agitated tanks.

Remember that the purpose of a mixing tank is to produce a homogeneous
material at the discharge or the mixer. The concepts of minimum speed to
achieve some internal condition such as off-bottom suspension of solids or
pumping rate are energy-saving concepts, not necessarily process-result
concepts. The need to guarantee the process result at an acceptable raw
materials and energy cost is the reason for the industrial laboratory. It
is also the reason for our interest and wishes of good luck in your own
mixing research.

REFERENCES

1) Garrison, C. M., Chem. Eng. _68_ (Feb. 7, 1983)

2) Short, D. G. R., and Etchells, A. W., 4th European Conference on Mixing,
   Noordwijkerhout, Netherlands, 1982

3) Gates, L. E., et al., Chem. Eng., Dec. 8, 1975, Jan. 5, Feb. 2, Apr. 26,
   May 24, July 19, Aug. 30, Sept. 27, Oct. 18, Nov. 8, Dec. 6, 1976

4) Short, D. G. R., Eng. Foundation Conf. on Mixing, Henniker, New
   Hampshire, 1979

5) Notable exceptions are:
   Ogawa, K., and Ito, S., J. Chem. Eng. Japan (8), 148, 1975
   Khang, S. J., and Levenspiel, O., Chem. Eng. Sci. (31), 569 (1976)

6) Zwietering, T. N., Chem. Eng. Sci. (8), 244 (1958)
   Nienow, A. W., Chem. Eng. Sci. (23), 1453 (1968)

7) Chapman, C. M., Ph.D. Thesis, University College, London, 1981

8) Metzner, A. B., and Otto, R. E., AIChE Journal (3), 3 (1957)

9) See Reference 2

10) Nienow, A. W., Private Communication (1983)

11) van't Riet, K., Ph.D. Thesis, Delft University, 1975

12) Etchells, A. W., Eng. Foundation Conference on Mixing, Henniker, New
    Hampshire, 1983

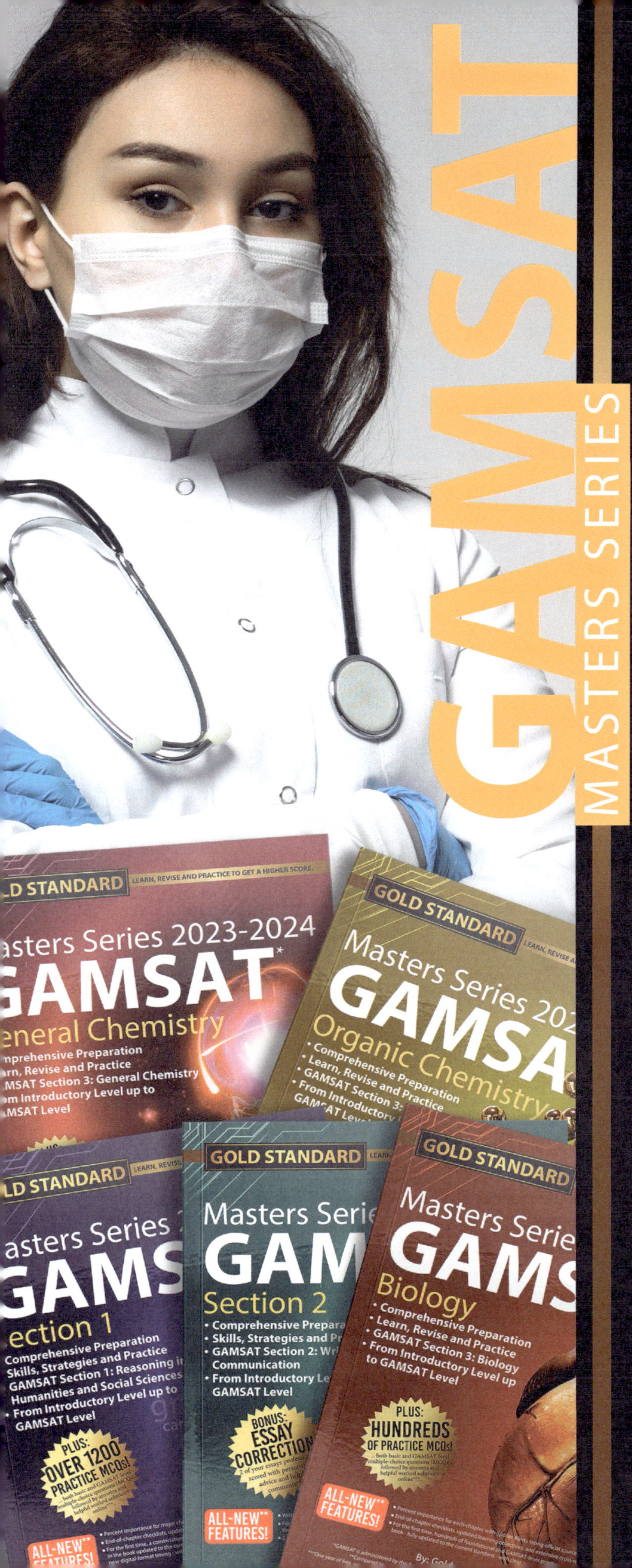

GOLD STANDARD

01 Editor and Author

Brett Ferdinand BSc MD-CM

02 Contributors

Lisa Ferdinand BA MA
Sean Pierre BSc MD
Kristin Finkenzeller BSc MD
Ibrahima Diouf BSc MSc PhD
Charles Haccoun BSc MD-CM
Timothy Ruger BA MA
Jeanne Tan Te

03 Illustrators

Harvie W. Gallatiera BS CompE
Gilbert Rafanan BSc

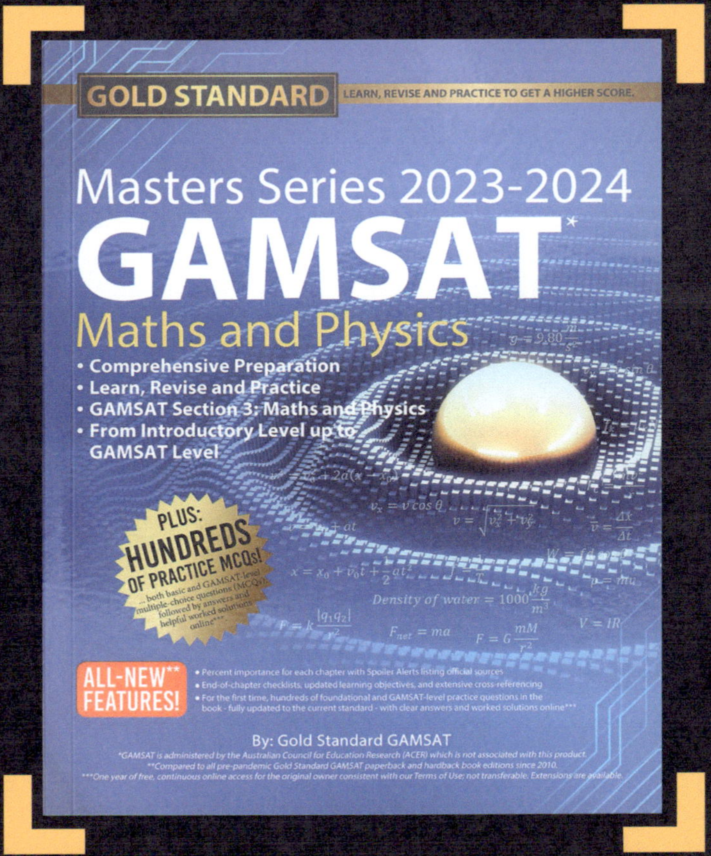

Free Online Access*

Answers and detailed worked solutions for hundreds of end-of-chapter practice questions, as well as the full-length practice test GS-Free which has the new digital GAMSAT format and cross-references to this Masters Series book.

*One year of continuous access for the original owner of this textbook upon online registration at
www.gamsat-prep.com/gamsat-maths-physics
If you purchased this textbook or the eBook directly from www.gamsat-prep.com, then your online access is automated.

Gold Standard GAMSAT Product Contact Information

Distribution in Australia, NZ, Asia

Woodslane Pty Ltd 10 Apollo Street
Warriewood NSW 2102 Australia
ABN: 76 003 677 549
learn@gamsat-prep.com

Distribution in Europe

Central Books 99 Wallis
Road LONDON,
E9 5LN, United Kingdom
orders@centralbooks.com

Distribution in North America

RuveneCo Publishing
334 Cornelia Street #559
Plattsburgh, New York 12901, USA
buy@gamsatbooks.com

RuveneCo Inc. is neither associated nor affiliated with the Australian Council for Educational Research (ACER) who has developed and administers the Graduate Medical School Admissions Test (GAMSAT). Printed in Australia.